GREEN CHEMISTRY AND ENGINEERING

GREEN CHEMISTRY AND ENGINEERING

A Practical Design Approach

CONCEPCIÓN JIMÉNEZ-GONZÁLEZ
DAVID J. C. CONSTABLE

WILEY

A JOHN WILEY & SONS, INC., PUBLICATION

Published by John Wiley & Sons, Inc., Hoboken, New Jersey.
Published simultaneously in Canada.

Library of Congress Cataloging-in-Publication Data:

Jiménez-González, Concepción Conchita.
Green chemistry and engineering : a practical design approach / Concepción Conchita Jiménez-González, David J. C. Constable.
p. cm.
Includes index.
ISBN 978-0-470-17087-8 (cloth)
1. Environmental chemistry–Industrial applications. 2. Sustainable engineering. I. Constable, David J. C., 1958- II. Title.
TP155.2.E58J56 2010
660–dc22

2010003431

10 9 8 7 6 5

CONTENTS

PREFACE

In the last decade, interest in and understanding of green chemistry and green engineering have increased steadily beyond academia and into the business world. Industries within different sectors of the economy have made concerted efforts to embed these concepts in their operations. Given our experience with green chemistry and green engineering in the pharmaceutical industry, we were initially approached by the publishers to edit a book on green chemistry in the pharmaceutical industry. This was a worthy proposal, but we felt that we had a greater opportunity and worthier endeavor to produce a book that would attempt to fully integrate green chemistry and green engineering into the academic curricula and that at the same time could serve as a practical reference to chemists and engineers in the workplace.

Green chemistry and green engineering are still relatively new areas that have not been completely ingrained in traditional chemistry and engineering curricula, but classes and even majors in these topics are becoming increasingly common. However, most classes in green chemistry are taught from an environmental chemistry perspective or a synthetic organic chemistry perspective, with neither approach addressing issues of manufacturing or manufacturability of products. Green engineering classes, on the other hand, tend to emphasize issues related to manufacturing, but do not treat reaction and process chemistry sufficiently, so these disciplines still seem to be disconnected. This lack of integration between chemistry, engineering, and other key disciplines has been one of the main challenges that we have had within the industrial workplace and in previous academic experiences.

As a consequence of these experiences, we decided to write this book to bridge the great divide between bench chemistry, process design, engineering, environment, health, safety, and life cycle considerations. We felt that a systems-oriented and integrated approach was needed to evolve green chemistry and green engineering as disciplines in the broader context of sustainability. To achieve this, we have organized the book in five main sections.

- Part I. Green Chemistry and Green Engineering in the Movement Toward Sustainability. Chapters 1 to 4 set the broader context of sustainability, highlighting the key role that green chemistry and green engineering have in moving society toward the adoption of more sustainable practices in providing key items of commerce.
- Part II. The Beginning: Designing Greener, Safer Chemical Synthesis. Chapters 5 to 8 address the key components of chemistry that will contribute to the achievement of more sustainable chemical reactions and reaction pathways. They also provide an approach to materials selection that promotes the overall greenness of a chemical synthesis without diminishing the efficiency of the chemistry or associated chemical process.
- Part III. From the Flask to the Plant: Designing Greener, Safer, More Sustainable Manufacturing Processes. Chapters 9 to 15 provide those key engineering concepts that support the design of greener, more sustainable chemical processes.
- Part IV. Expanding the Boundaries: Looking Beyond Our Processes. Chapters 16 to 20 introduce a life cycle thinking perspective by providing background and context for placing a particular chemical process in the broader chemical enterprise, including its impacts from raw materials extraction to recycle/reuse or end-of-life considerations.
- Part V. What Lies Ahead: Beyond the Chemical Processing Technology of Today or Delivering Tomorrow's Products More Sustainably. Finally, Chapters 21 to 25 provide some indication of trends in chemical processing that may lead us toward more sustainable practices.

To help provide a practical approach, we have included examples and exercises that will help the student or practitioner to understand these concepts as applied to the industrial setting and to use the material in direct and indirect applications. The exercises are intended to make the book suitable for both self-study or as a textbook, and most exercises are derived from our professional experiences.

The book is an outgrowth of our experience in applied and fundamental research, consulting, teaching, and corporate work on the areas of green chemistry, green engineering, and sustainability. It is intended primarily for graduate and senior-level courses in chemistry and chemical engineering, although we believe that chemists and engineers working in manufacturing, research, and development, especially in the fine-chemical and pharmaceutical areas, will find the book to be a useful reference for process design and reengineering. Our aim is to provide a balance between academic needs and practical industrial applications of an integrated approach to green chemistry and green engineering in the context of sustainability.

Acknowledgments

We thank all our colleagues who have contributed directly or indirectly to our journey toward sustainability, and whose ideas and collaborations throughout the years have contributed to our own experience in the areas of green chemistry and green engineering. We also express our gratitude to GlaxoSmithKline, in general, and to James R. Hagan, in particular, for their support and encouragement.

We also give special thanks to Rafiqul Gani and Ana Carvalho from the Computer Aided Process-Product Engineering Center, Department of Chemical and Biochemical Engineering at the Technical University of Denmark, for their comments, reviews, and contributions to Chapter 12; to Mariana Pierobon and BASF for their helpful comments and for allowing us to use one of BASF's eco-efficiency assessments as an example in the life cycle chapters; to Sara Conradt for allowing us to use a sample of her masters thesis as an example of LCA outputs; and to Tom Roper and John Hayler at GSK for their feedback on green chemistry throughout the years. Finally, we want to thank *Chemical Engineering* magazine, the American Chemical Society, Springer Science and Business Media, Elsevier, John Wiley & Sons, the Royal Society of Chemistry, and Wiley-VCH for permission to reproduce some printed material.

CONCEPCIÓN JIMÉNEZ-GONZÁLEZ
DAVID J. C. CONSTABLE

December 2009

PART I

GREEN CHEMISTRY AND GREEN ENGINEERING IN THE MOVEMENT TOWARD SUSTAINABILITY

1

GREEN CHEMISTRY AND ENGINEERING IN THE CONTEXT OF SUSTAINABILITY

What This Chapter Is About Green chemistry and green engineering need to be seen as an integral part of the wider context of sustainability. In this chapter we explore green chemistry and green engineering as tools to drive sustainability from a triple-bottom-line perspective with influences on the social and economic aspects of sustainability.

Learning Objectives At the end of this chapter, the student will be able to:

- Understand the need for the development of greener chemistries and chemical processes.
- Identify sustainability principles and associate standard chemical processes with the three areas of sustainability: social, economic, and environmental.
- Identify green chemistry and green engineering as part of the tools used to drive sustainability through innovation.
- Understand the need for an integrated approach to green chemistry and engineering.

1.1 WHY GREEN CHEMISTRY?

$$A + B \rightarrow C \tag{1.1}$$

Reactant A plus reactant B gives product C. No by-products, no waste, at ambient temperature, no need for separation. Is it really that easy?

Green Chemistry and Engineering: A Practical Design Approach, By Concepción Jiménez-González and David J. C. Constable

If industrial chemical reactions were that straightforward, chemists and engineers would have significantly more time on their hands and significantly less excitement and fewer long hours at work. Chemists know that this hypothetical reaction is not the case in real life, as they have less-than-perfect chemical conversions, competing reactions to avoid, hazardous materials to manage, impurities in raw materials, and the final product to reduce. Engineers know that in addition to conquering chemistry, there are by-products to separate, waste to treat, energy transfer to optimize, solvent to purify and recover, and hazardous reaction conditions to control. At the end of this first reality check, we see that our initial reaction is a much more complicated network of inputs and outputs, something that looks more like Figure 1.1.

Green chemistry and green engineering are, in a very simplified way, the tools and principles that we use to ensure that our processes and chemical reactions are more efficient, safer, cleaner, and produce less waste by design. In other words, green chemistry and green engineering assist us in first thinking about and then designing synthetic routes and processes that are more similar to the hypothetical reaction depicted in equation (1.1) than to the more accurate reflection of current reality shown in, Figure 1.1.

What are the drivers in the search for greener chemistries and processes? Engineers and scientists have in their capable hands the possibility of transforming the world by modifying the materials and the processes that we use every day to manufacture the products we buy and the way we conduct business. However, innovation and progress need to be set in the context of their implications beyond the laboratory or the manufacturing plant. With the ability to effect change comes the responsibility to ensure that the new materials, processes, and designs have a minimum (or positive) overall environmental impact. In addition, common sense suggests that there is a strong business case for green chemistry and engineering: linked primarily to higher efficiencies, better utilization of resources, use of less hazardous chemicals, lower waste treatment costs, and fewer accidents.

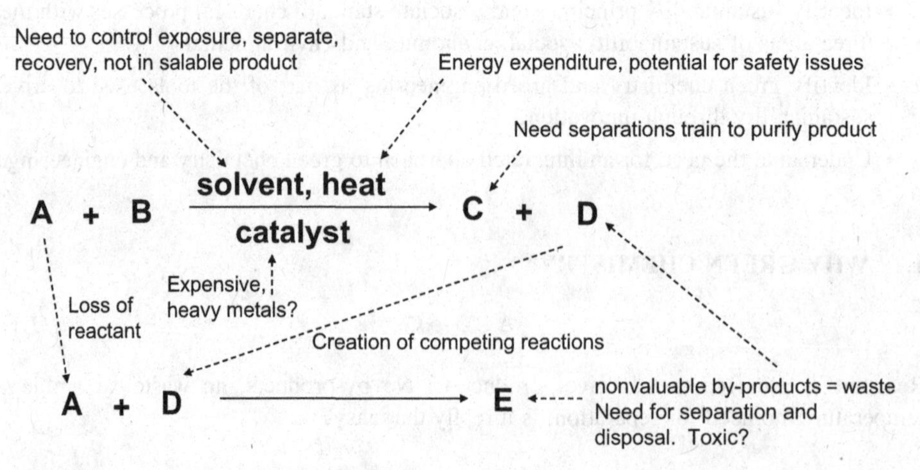

FIGURE 1.1 Simplified vision of some of the challenges and realities of designing a chemical synthesis and process.

Example 1.1 Potassium hydroxide is manufactured by electrolysis of aqueous potassium chloride brine,[1] as illustrated by the following net reaction:

$$2KCl + 2H_2O \rightarrow 2KOH + Cl_2 + H_2$$

How is this simple inorganic reaction different from the more complex challenges of the real world? Identify some of the green chemistry/green engineering challenges.

Solution The electrolysis reaction can be carried out in diaphragm, membrane, or mercury cell processes. The complexity of the reactions depend on the process that is used. Let's explore the mercury cell process, which has, historically, been the most commonly used method to produce chlorine.[1,2] In this case, potassium chloride is converted to a mercury amalgam in a mercury cell evolving chlorine gas. The depleted brine is recycled to dissolve the input KCl. The mercury amalgam passes from the mercury cell to the denuder. In the denuder, fresh water is added for the reaction and as a solvent for the KOH. Hydrogen gas is evolved from the reaction and mercury is recycled to the electrolysis cell:

$$\text{Mercury cell}: \underset{\substack{\text{potassium}\\ \text{chloride}}}{KCl} + \underset{\text{mercury}}{Hg} \rightarrow \underset{\substack{\text{potassium}\\ \text{mercury}\\ \text{amalgam}}}{K-Hg} + \underset{\text{chlorine}}{0.5Cl_2}$$

$$\text{Denuder}: \underset{\substack{\text{potassium}\\ \text{mercury}\\ \text{amalgam}}}{K-Hg} + \underset{\text{water}}{H_2O} \rightarrow \underset{\substack{\text{potassium}\\ \text{hydroxide}}}{KOH} + \underset{\text{hydrogen}}{0.5H_2} + \underset{\text{mercury}}{Hg}$$

Our simple net reaction has become a bit more complex, but it does not end there. We've not talked about a key input— energy. Electricity is required to drive the reaction forward; it represents the major part of the energy requirement for these types of reactions, and there is a need to optimize it. As a matter of fact, as of 2006 the chlor-alkali sector was the largest user of electricity in the chemical industry.[2]

But energy is not the only thing that we need to worry about. In addition to energy inputs, there is a need to eliminate impurities. To do that, the brine can be treated with potassium carbonate[3] to precipitate magnesium and heavy metals, and barium carbonate is often used to precipitate sulfates.[4] Also, hydrochloric acid needs to be added, as an acidic pH is required to drive the reaction to produce the desired chlorine gas, which can then be recovered from the solution, as shown in the following equilibrium reaction:

$$H^+ + OCl^- + HCl \rightleftharpoons H_2O + Cl_2$$

Besides using a large quantity of electricity, we have to worry about potential emissions from the reaction. Mercury is present in the reaction cell and the purged brine. Mercury emissions from the cell and the brine have long been a target for significant reduction. The purged brine is typically treated with sodium hydrosulfide to precipitate mercury sulfide, and the mercury-containing solid wastes need to be sent for mercury recovery. Other emission concerns include management of the environmental, health, and safety (EHS) challenges related to the gases in the reactions. Both the chlorine and hydrogen gas streams must be processed further. Chlorine is cooled and scrubbed with sulfuric acid to remove water, followed by compression and refrigeration. The hydrogen gas is cooled to remove water, impurities, and mercury, followed by further cooling or treatment with activated carbon for more complete mercury removal.[5] In addition, hydrogen is often burned as fuel at chlor-alkali plants.

The membrane process was introduced in the 1970s and it is more energy efficient and more environmentally sustainable, which is making it the technology of choice. However, a typical mercury-based plant can contain up to 100 cells and has an economic life span of 40 to 60 years. A long phase-out is required to convert an existing mercury plant. For example, as of 2005, 48% of the European chlor-alkali capacity was mercury cell–based.[2]

Additional Point to Ponder Chemistries and processes described in most textbooks normally don't give you all the information you need to consider the mass and energy inputs and outputs associated with a given reaction. In reality you won't always have the data you need and will have to use estimations to generate data, run experiments, perhaps use "nearest neighbor" approaches and/or make assumptions based on your experience. Sometimes, you will just have to use "simple" common sense.

1.2 GREEN CHEMISTRY, GREEN ENGINEERING, AND SUSTAINABILITY

The modern understanding of sustainability began with the United Nations World Commission on Environment and Development's report *Our Common Future*,[6] also known as the *Brundtland Report*. The Brundtland Commission described *sustainable development* as "development that meets the needs of the present without compromising the ability of future generations to meet their own needs." What does this actually mean? This definition doesn't give us many clues or supply much practical guidance as to how to implement sustainable development or move toward more sustainable activities, but it does provide us with a powerful aspiration. It has been up to society collectively and up to us as individuals to develop guidance and tools that will help us to design systems and processes that have the potential to achieve the type of development described in the definition.

The first thing to remember is that sustainability or sustainable development is a complex concept with which many people are still attempting to come to terms. In 1998, John Elkington, one of the early innovators of sustainable development, coined the phrase *triple bottom line*.[7] Elkington did this in an attempt to make sustainable development more understandable and palatable to business people, to encourage them to see it as a logical extension of the traditional business focus on economic performance. By using this term, Elkington was trying to highlight the need to consider the intricate nterrelationships among environmental, social, and economic aspects of human society and the world. In a way, sustainability can be seen as a very delicate balancing act among these three factors, and not always with a strong one-to-one relationship. Table 1.1 provides a summary of several approaches to sustainable development principles. It should be noted that the Carnoules statement includes an organizational principle framework, in addition to the overarching social aspects widely recognized to be an integral part of sustainability. This organizational principle is useful when relating the operational aspects of sustainability within the sphere of controls defined by company culture and policy.

When talking about sustainability, one cannot focus on only a single aspect, as this necessarily limits and biases one's view. For a system to be sustainable, there is the need to balance, insofar as possible, social, economic, and environmental aspects, ideally having each area "in the black," that is, with no single aspect optimized to the detriment of the others.

TABLE 1.1 Summary of Several Approaches to Sustainable Development Principles

Alcoa[8]	International Chamber of Commerce[9]	Chemical Associations[10]	Carnoules Statement[11]	Hanover Principles[12]	Natural Step[13]	UN Global Compact[14]
Supporting the growth of customer businesses. Standing among the industrial companies in the first quintile of return on capital among S&P Industrials Index companies. Elimination of all injuries and work-related illnesses and the elimination of waste. Integration of EHS with manufacturing. Products designed for the environment. EHS as a core value. An incident-free workplace (an incident is any unpredicted event with the capacity to harm human health, the environmental, or physical property).	*Corporate priority:* To recognize environmental management as among the highest corporate priorities and as a key determinant to sustainable development; to establish policies, programs, and practices for conducting operations in an environmentally sound manner. *Integrated management:* To integrate these policies, programs, and practices fully into each business as an essential element of management in all its functions. *Process of improvement:* To continue to improve corporate policies, programs, and environmental performance, taking into account technical developments, scientific understanding, consumer needs, and community expectations, with legal regulations as a starting point; and to apply the same environmental criteria internationally.	**Responsible Care** *Policy:* We will have a health, safety, and Environmental (HS&E) policy that will reflect our commitment and be an integral part of our overall business policy. *Employee involvement*: We recognize that the involvement and commitment of our employees and associates will be essential to the achievement of our objectives. We will adopt communication and training programs aimed at achieving that involvement and commitment. *Experience sharing:* In addition to ensuring that our activities meet the relevant statutory obligations, we will share experience with our industry colleagues and seek to learn from and incorporate best practice into our own activities. *Legislators and regulators:* We will seek to work in cooperation with legislators and regulators.	**Environmental Principles** Protect ecosystems' functions and evolution. Enhance (genetic, species, and ecosystem) biodiversity. Reduce anthropogenic resource throughput and degradation of land and sea. *Minimize the burden for the environment:* Improve resource productivity (mass, energy, land). *Minimize the impacts on health and environment:* minimize the outputs of known (eco)toxics. *Minimize damage for the economy:* reduce costs related to environmental degradation (damage costs, compliance costs, administrative costs, avoidance costs, etc.). **Social Principles** Social cohesion and social security.	Insist on rights of humanity and nature to coexist in a healthy, supportive, diverse, and sustainable condition. Recognize interdependence. The elements of human design interact with and depend on the natural world, with broad and diverse implications at every scale. Expand design considerations to recognize even distant effects. Respect relationships between spirit and matter. Consider all aspects of human settlement, including community, dwelling, industry, and trade in terms of existing and evolving connections between spiritual and material consciousness.	*System condition 1:* Substances from the Earth's crust must not increase in nature systematically. In a sustainable society, natural resources should not be extracted at a faster pace than their re-deposit into the ground. *System condition 2:* Substances produced by society must not increase in nature systematically. In a sustainable society, man-made substances should not be produced at a faster pace than they can be naturally degraded or re-deposited into the ground. *System condition 3:* The physical basis for the productivity and diversity of nature must not be diminished systematically.	To support and respect the protection of internationally proclaimed human rights. To avoid complicity in human rights abuses. To uphold freedom of association and the effective recognition of the right to collective bargaining. To eliminate all forms of forced and compulsory labor. To effectively abolish child labor. To eliminate discrimination with respect to employment and occupation. To support a precautionary approach to environmental challenges. To promote greater environmental responsibility.

(continued)

TABLE 1.1 (*Continued*)

Alcoa[8]	International Chamber of Commerce[9]	Chemical Associations[10]	Carnoules Statement[11]	Hanover Principles[12]	Natural Step[13]	UN Global Compact[14]
Increased transparency and closer collaboration in community-based EHS initiatives.	*Employee education:* To educate, train, and motivate employees to conduct their activities in an environmentally responsible manner. *Prior assessment:* To assess environmental impacts before starting a new activity or project and before decommissioning a facility or leaving a site. *Products and services:* To develop and provide products or services that have no undue environmental impact and are safe in their intended use, that are efficient in their consumption of energy and natural resources, and that can be recycled, reused, or disposed of safely. *Customer advice:* To advise and, where relevant, educate customers, distributors, and the public in the safe use, transportation, storage, and disposal of products provided; and to apply similar considerations to the provision of services.	*Process safety:* We will assess and manage the risks associated with our processes. *Product stewardship:* We will assess the risks associated with our products and seek to ensure that these risks are properly managed throughout the supply chain through stewardship programs involving our customers, suppliers, and distributors. *Resource conservation:* We will work to conserve resources and reduce waste in all our activities. *Stakeholder engagement:* We will monitor our HS&E performance and report progress to stakeholders; we will listen to the appropriate communities and engage them in dialogue about our activities and our products.	Access to education. Identity and self-realization. Security. Equitable access to food, drinking water, and natural resources. Healthy and secure shelter. Readjusted demand for resource consumption, and the environmental impact of household consumption. Secure environmental quality for the health of human beings. **Economic Principles** Sufficient supply and goods and services Efficient wealth creation Economic system's evolution and competitiveness Enhance the distributional justice (equity principle) Efforts (paid and unpaid) should be devoted fairly to generate sustainable incomes. Provide opportunities for paid labor to all willing and able to work. Increase knowledge intensity. Refocus innovation and adapt its speed to societal demands.	Accept responsibility for the consequences of design decisions on human well-being, the health of natural systems, and their right to coexist. Create safe objects of long-term value. Do not burden future generations with requirements for maintenance or vigilant administration of potential danger due to the careless creation of products, processes, or standards. Eliminate the concept of waste. Evaluate and optimize the full life cycle of products and processes to approach the state of natural systems in which there is no waste. Rely on natural energy flows. Human designs should, like the living world, derive their creative forces from perpetual solar income. Incorporate this energy efficiently and safely for responsible use.	In a sustainable society, nature's productivity should not be diminished in either quality or quantity, nor should more be harvested than can be recreated. System condition 4: We must be fair and efficient in meeting basic human needs. In a sustainable society, basic human needs must be met with the most resource-efficient methods possible, including the just distribution of resources.	To encourage the development and diffusion of environmentally friendly technologies.

Facilities and operations: To develop, design, and operate facilities and conduct activities taking into consideration the efficient use of energy and materials, the sustainable use of renewable resources, the minimization of adverse environmental impact and waste generation, and the safe and responsible disposal of residual wastes.

Research: To conduct or support research on the environmental impacts of raw materials, products, processes, emissions, and wastes associated with the enterprise and on the means of minimizing such adverse impacts.

Precautionary approach: To modify the manufacture, marketing, or use of products or services or the conduct of activities, consistent with scientific and technical understanding, to prevent serious or irreversible environmental degradation.

Management systems: We will maintain documented management systems which are consistent with the principles of responsible care and which will be subject to a formal verification procedure.

Past, present, and future: Our responsible care management systems will address the impact of both current and past activities.

Social Principles

Ethical trade: to ensure that all business, wherever companies trade, is conducted to the highest global ethical standards.

Public understanding: to play their part in helping people understand and appreciate relevant science and technology.

Part of the community: to play an active role in their communities by interacting with schools, local government, and other bodies.

Organizational Principles

Ensure structural change to reflect the need for societal development.

Improve societal interchange, communication, and intercultural learning.

Protect cultural diversity

Achieve distributional fairness and justice, equity and sufficiency.

Develop anticipatory capacities for the democratic process.

Understand the limitation of design. No human creation lasts forever and design does not solve all problems. Those who create and plan should practice humility in the face of nature. Treat nature as a model and mentor, not as an inconvenience to be evaded or controlled.

Seek constant improvement by the sharing of knowledge. Encourage direct and open communication among colleagues, patrons, manufacturers, and users to link long-term sustainable considerations with ethical responsibility, and reestablish the integral relationship between natural processes and human activity.

(*continued*)

TABLE 1.1 (*Continued*)

Alcoa[8]	International Chamber of Commerce[9]	Chemical Associations[10]	Carnoules Statement[11]	Hanover Principles[12]	Natural Step[13]	UN Global Compact[14]
	Contractors and suppliers: To promote the adoption of these principles by contractors acting on behalf of the enterprise, encouraging and, where appropriate, requiring improvements in their practices to make them consistent with those of the enterprise; and to encourage the wider adoption of these principles by suppliers.	*Employability:* to ensure that all employees have access to training and development opportunities to enable them to fulfill their role in the organization and to keep them up to date with the labor market.				
	Emergency preparedness: To develop and maintain, where significant hazards exist, emergency preparedness plans in conjunction with the emergency services, relevant authorities, and the local community, recognizing potential transboundary impacts.	*Equality of treatment and opportunity:* to ensure that all employees are free from discrimination and have the opportunity to develop their careers and themselves, subject only to business needs and personal ability.				
	Transfer of technology: To contribute to the transfer of environmentally sound technology and management methods throughout the industrial and public sectors.	*Participation:* to ensure that all employees have access to the information needed for them to do their job, be consulted about matters that affect them, and have the opportunity to participate, to the appropriate level, in the management of their company.				
		Balance between work and life: to provide all employees with the opportunity to balance the requirements of their work and their life outside work so as to enhance work effectiveness and personal well-being.				

Contributing to the common effort: To contribute to the development of public policy and to business, governmental and intergovernmental programs, and educational initiatives that will enhance environmental awareness and protection.

Openness to concerns: To foster openness and dialogue with employees and the public, anticipating and responding to their concerns about the potential hazards and impacts of operations, products, wastes, or services, including those of transboundary or global significance.

Compliance and reporting: To measure environmental performance; to conduct regular environmental audits and assessments of compliance with company requirements, legal requirements, and these principles; and periodically to provide appropriate information to the board of directors, shareholders, employees, the authorities and the public.

Economic Principles

Sustainable profitability: generating profits to satisfy shareholders' expectations and to invest in the future through R&D, capital expenditure, and employee development.

Competitiveness: achieving long-term competitiveness through the spread of international best practice, in a climate of fair competition

Innovation: continuing to research, develop, and market innovative products that help improve economic well-being and quality of life.

Wealth generation: generating wealth, thereby sustaining employment, improving the UK's trade balance, and contributing to government revenue to fund public expenditure.

Economic growth: continuing their key role in supporting sustained UK economic growth throughout the entire manufacturing supply chain.

Resource efficiency: making the most efficient use of resources, whether they be land, water, raw materials, or energy.

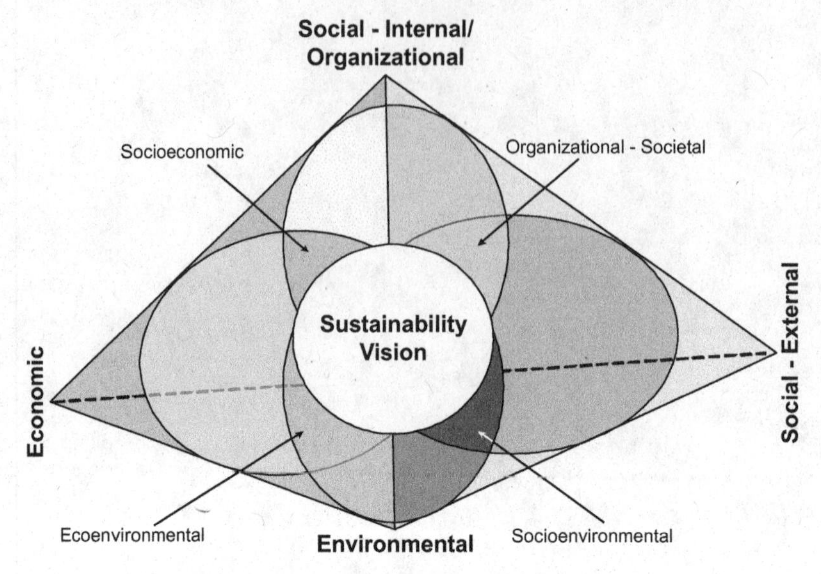

FIGURE 1.2 Spheres of action of sustainability.

One of the most puzzling, challenging, and exciting characteristics in the study of sustainability is the inherent complexity of the concept. There are synergies, trade-offs, a variety of shared values of what constitutes a sustainable practice, and so on. Figure 1.2 displays those interrelations graphically.

Green chemistry and green engineering represent some of the many concepts, tools, and disciplines that come into play in helping to move society toward more sustainable practices. They do this by focusing scientists and engineers on how to design more environmentally friendly, more efficient, and inherently safer chemistries and manufacturing processes. However, some might suggest that when talking about green chemistry and green engineering in the context of sustainable development, we can honestly say simply that the primary focus area is what has come to be known as *environmental sustainability*. Is this really true? Whereas green chemistry and green engineering may be seen as being related primarily to the environmental aspects of sustainability, they also have strong ties to the eco-environmental (or eco-efficiency) sub-area of sustainability by virtue of the fact that they include resource conservation and efficiency. By the same token, green chemistry and green engineering are related to the social aspects of sustainability because they promote the design of manufacturing processes that are inherently safer, thereby ensuring that workers and residential neighborhoods close to manufacturing sites are protected.

Example 1.2 Explain how reaction (1.1) relates to the three aspects of sustainability.

Solution Several of the issues related to green chemistry and green engineering were highlighted in the solution to Example 1.1. Table 1.2 provides examples of how they relate to the three aspects of sustainability.

TABLE 1.2 Issues Related to Sustainability

Environmental	Social	Economic
Mercury emissions from a cell and in the purged brine	Worker safety issues related to chlorine and hydrogen management	Jobs and wealth created by a potassium chloride plant
Energy consumption	Safety and well-being of communities adjacent to manufacturing plant	Economic resources needed to operate the plant in a safe and efficient manner
Water consumption		
Emissions released during energy production	Potential for process accidents, incidents, and lost-time injuries	Investment that will be necessary to replace mercury cells for an alternative technology
Fugitive chlorine emissions		
Waste management of carbonate precipitates	Issues related to safely transporting chlorine	Supply chain implications for other products that utilize KCl or chlorine
Environmental impacts resulting from mercury mining	Working conditions in mercury mines to extract the metal	

Additional Point to Ponder Most textbook examples and problems have only one correct answer, although many examples have several possible answers. In real-world manufacturing processes, it is common to have difficulties in defining what the true problem is—and when this is defined, several "not-quite-optimal" answers may be found. When this happens, a decision must be made that accounts for or balances all the important factors and, hopefully, leads to the optimal or "best" answer.

1.3 UNTIL DEATH DO US PART: A MARRIAGE OF DISCIPLINES

What does it mean to have an integrated perspective between green chemistry and green engineering? Just imagine the following not-so-hypothetical scenario. A chemist works at a large company and after years of hard work discovers a novel synthesis to produce a valuable material. At this point, hundreds of engineering questions are formulated and need to be addressed, such as:

- What is the best design for the reactor? Which material?
- Does the reaction need to be heated? Cooled? How fast are heating and cooling transferred?
- What types of separation processes are needed?
- How could the desired purity be achieved?
- How fast is the reaction? Is there a risk of an exothermic runaway?
- What can possibly go wrong? How can we prepare for problems?

- Are there inherent hazards in the materials?
- Are there any incompatibilities with materials?
- How much waste is produced? How toxic is it? Can it be avoided?
- Where should the reactants be procured? Is it more efficient to make them or to buy them?
- How much would this process cost?
- What types of preparations and skills would future operators need?

Imagine how difficult it would be to answer these and other questions if the chemist doesn't work closely with a chemical engineer. How efficient would the final process be? To truly understand the impacts of this novel chemistry in the real-world manufacturing environment, the chemist will need to involve engineers beginning at the earliest stages of development.

Similarly, a chemical engineer working on transforming a laboratory synthesis into a scalable, effective production process will need to collaborate closely with a chemist to understand how the chemical synthesis might be changed. A myriad of chemically related questions must be answered to design and scale-up a good manufacturing process:

- What function is the solvent performing in the reaction?
- Are there alternative reaction pathways that can be used to:
  Avoid uncontrollable exotherms?
  Substitute reactant A for B to avoid safety issues?
  Eliminate hazardous reagents?
- If we recirculate part or all of the reaction mother liquors, how much of material X can be tolerated by the reaction system before we are not able to do this?
- Are there any reactivity issues by introducing solvent Y as a mass separating agent?
- What are the potential side reactions?
- Are there any alternative catalytic methods that we might be able to use?

The decisions that are made in the design of synthetic chemistry pathways affect and either enable or restrict the engineering opportunities, and vice versa. Chemists and chemical engineers should operate in an integrated fashion if the goal is to design an efficient process, in the widest sense of the term and in the context of green chemistry and engineering.

Hopefully, we have made a good case for integrating green chemistry and green engineering, but our effort to integrate disciplines is not over. Carrying on with our original scenario, the chemist and engineer have successfully identified a chemical they want to make and the synthetic route or pathway to be used to make it, and have some idea of the critical process parameters that they need to focus on if they are to optimize the process from a green chemistry and green engineering perspective. So, is anything missing? What about knowledge of how the various reactants, reagents, catalysts, solvents, by-products, and so on, used in the process affect living organisms and the environment? One might be tempted to ask who really cares about such things, since most of the materials may be consumed in the process and the product we are making is a valuable material that others need or want.

These questions are not merely rhetorical; the answers are very important for current and future generations. Human beings have and continue to affect the world in very significant ways, and it is critical that all chemists and engineers understand how material choices, process designs, energy use, and so on, affect the world. Chemists and engineers need to design and choose synthetic strategies that minimize the potential for causing short-, medium-, and long-term harm not only to humans, but to other environmental organisms as well. To do this correctly, they need to collaborate with toxicologists and environmental, health, and safety professionals to discuss and develop appropriate options for syntheses. In short, a host of disciplines are required to bring a product to market appropriately and successfully and to ensure that this is done in a sustainable fashion. It is no longer acceptable practice for chemists to isolate themselves in a laboratory and design reactions that are chemically interesting but, because it is expedient to do so, utilize reagents, reactants, and solvents that are inherently hazardous.

PROBLEMS

1.1 How do green chemistry and green engineering differ from chemistry and engineering?

1.2 Examples 1.1 and 1.2 refer to the environmental, health, and safety challenges related to mercury, chlorine, and hydrogen. What are those challenges?

1.3 The primary route for making copper iodide is by reacting potassium iodide with copper sulfate:

$$2CuSO_4 + 4KI + 2Na_2S_2O_3 \rightarrow 2CuI + 2K_2SO_4 + 2NaI + Na_2S_4O_6$$

Identify potential green chemistry and green engineering challenges of the reaction.

1.4 From a sustainability framework, identify environmental, social, and economic impacts derived from the chemistry shown in Problem 1.3.

1.5 Using reaction system of example (1.1), provide some examples of how the chemistry can affect decisions made in engineering.

1.6 What are some potential barriers for an effective, close collaboration between a chemist and an engineer when designing a novel process. Provide some ideas on how to circumvent these obstacles.

REFERENCES

1. Schultz, H., Günter Bauer, G., Schachl, E., Hagedorn, F. Schmittinger, P. Potassium compounds. In *Ullmann's Encyclopedia of Industrial Chemistry*. Wiley-VCH, New York, 2000.
2. *Chlorine Industry Report: 2005–2006*. Euro Chlor, Brussels, Belgium, 2006.
3. McKetta, J. Potash, caustic. In *Kirk–Othmer Encyclopedia of Chemical Technology*, 2nd ed. Wiley, New York, 1970.
4. U.S. Environmental Protection Agency. *Profile of the Inorganic Chemical Industry*. EPA Office of Compliance Sector Notebook Project. EPA 310-R-95-004. U.S. EPA,Washington, DC, 1995.

5. European Commission, 2001. Integrated Pollution Prevention and Control (IPPC). Reference document on best available techniques in the chlor-alkali manufacturing industry. BREF 12.2001. ftp://ftp.jrc.es/pub/eippcb/doc/cak_bref_1201.pdf.
6. World Commission on Environment and, Development. *Our Common Future*. Oxford University Press, Oxford, UK, 1987, p.43.
7. Elkington, J. *Cannibals with Forks: The Triple Bottom Line of 21st Century Business*. New Society Publishers, Gabriola Island, New Brunswick, Canada, 1998, p.416.
8. Alcoa. 2020 Framework.http://www.alcoa.com/global/en/about_alcoa/sustainability/2020_Framework.asp.
9. International Chamber of Commerce. The Business Charter for Sustainable Development: 16 Principles.http://www.iccwbo.org/policy/environment/id1309/index.html.
10. International Council of Chemical Associations. Responsible Care Web site.http://www.responsiblecare.org/page.asp?p=6341&l=1, accessed Sept. 27, 2009.
11. Bartz, P., et al. Pignans Set of Indicators Statement: Carnoules Statement on Objectives and Indicators for Sustainable Development. Governance for Sustainable Development, Carnoules/Pignans, Provence, France, May 1–4, 2003.
12. McDonough and Partners. *The Hanover Principles*. McDonough and Partners, Charlottesville, VA, 1992.
13. The Natural Step Web site.http://www.naturalstep.org/, accessed Sept. 27, 2009.
14. United Nations Global Compact.http://www.unglobalcompact.org/, accessed Sept. 27, 2009.

2

GREEN CHEMISTRY AND GREEN ENGINEERING PRINCIPLES

What This Chapter Is About Following several decades of increased awareness of the human impact on the environment, there was a need to spur chemists and engineers on toward a deeper consideration of how they might facilitate pollution prevention beyond "end-of-pipe" solutions. True pollution reduction at its source would require a revised approach that emphasized new chemistries and technologies. There have been a number of notable attempts to define what it means for scientists and engineers to be "green," and in this chapter we outline the major contributions to the discussion.

Learning Objectives At the end of this chapter, the student will be able to:

- Identify the principles of green chemistry and green engineering.
- Understand the interrelationships between the principles of green chemistry and green engineering.
- Contrast the differences between some the principles postulated by Anastas and Warner, Anastas and Zimmerman, Winterton, and the San Destin declaration.
- Critique and analyze chemical reactions as related to the principles of green chemistry.

2.1 GREEN CHEMISTRY PRINCIPLES

What is a principle, and why do we develop principles? *Merriam-Websters, Collegiate Dictionary* defines a principle as "1a: a comprehensive and fundamental law, doctrine, or

Green Chemistry and Engineering: A Practical Design Approach, By Concepción Jiménez-González and David J. C. Constable

assumption; b(1): a rule or code of conduct; (2): habitual devotion to right principles <a man of principle>; c: the laws or facts of nature underlying the working of an artificial device."[1] Now that we know what a principle is, why would someone want to have principles for green chemistry and/or green engineering? To answer that question, it may be valuable to begin by providing just a bit of historical context. As the story goes, John Warner, formerly on the staff of the research and development department at the Polaroid Corporation, was working on novel chemistries related to dyes used in photographic films. John is not your usual chemist and was aware of many environmental regulations that might stand in the way of getting a new product to market (see Figure 2.1). In addition to being a great chemist, John is a very creative person, so he began to wonder how he might design novel molecules and chemical synthetic processes to make them in a way that would avoid creating and/or using toxic and/or regulated materials along the way. With this simple thought in mind, he contacted Paul Anastas, formerly a division head in the Office of Pollution Prevention and Toxics at the U.S. Environmental Protection Agency, to discuss what would now seem to be obvious to many, but at that time, was quite revolutionary: What might the average synthetic chemist do to make molecules that do not harm the environment or people? Thus began a continuing dialogue and fruitful collaboration between John and Paul that resulted in the publication of the Twelve Principles of Green Chemistry, first published in 1998.[2] Let's look at these principles for a moment and think about some of the broader issues and implications that they present. We should also ask ourselves whether or not they promote movement toward more sustainable behaviors and actions.

THE TWELVE PRINCIPLES OF GREEN CHEMISTRY

1. It is better to prevent waste than to treat or clean up waste after it is formed.
2. Synthetic methods should be designed to maximize the incorporation into the final product of all materials used in the process.
3. Wherever practicable, synthetic methodologies should be designed to use and generate substances that possess little or no toxicity to human health and the environment.
4. Chemical products should be designed to preserve efficacy of function while reducing toxicity.
5. The use of auxiliary substances (e.g., solvents, separation agents) should be made unnecessary whenever possible and innocuous when used.
6. Energy requirements should be recognized for their environmental and economic impacts and should be minimized. Synthetic methods should be conducted at ambient temperature and pressure.
7. A raw material feedstock should be renewable rather than depleting whenever technically and economically practical.
8. Unnecessary derivatization (blocking group, protection–deprotection, temporary modification of physical/chemical processes) should be avoided whenever possible.
9. Catalytic reagents (as selective as possible) are superior to stoichiometric reagents.
10. Chemical products should be designed so that at the end of their function they do not persist in the environment and break down into innocuous degradation products.

(*continued on page 20*)

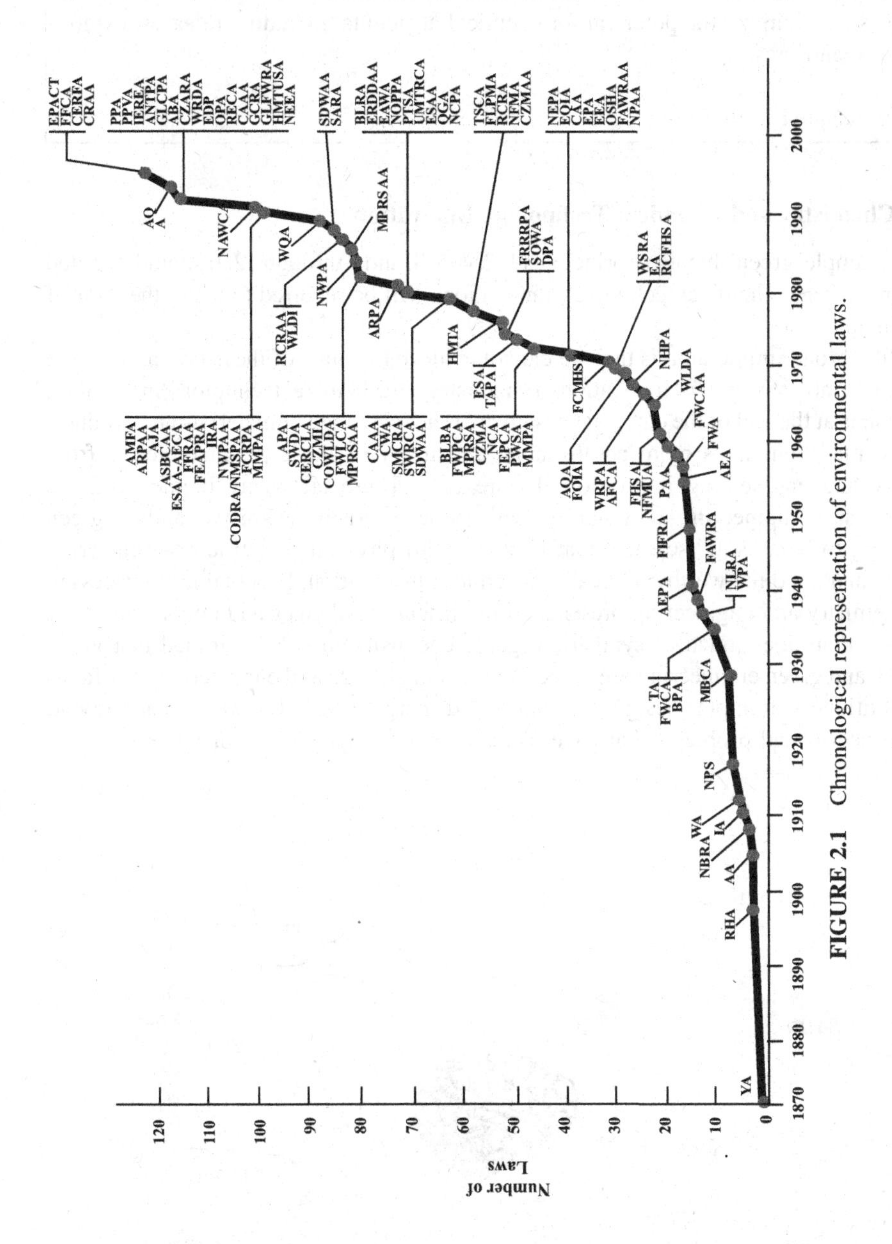

FIGURE 2.1 Chronological representation of environmental laws.

11. Analytical methodologies need to be further developed to allow for real-time in-process monitoring and control prior to the formation of hazardous substances.
12. Substances and the form of a substance used in a chemical process should be chosen so as to minimize the potential for chemical accidents, including releases, explosions, and fires.

Source: Adapted from ref. 2.

2.1.1 Chemistry and Chemical Technology Innovation

See, for example, green chemistry principles 1, 2, 4, 5, 8, and 9 through 12. It should be noted that chemistry and chemical technology innovations will be required to foster the aims of each principle.

The first broad implication is that we cannot achieve the aims of these principles if we are not constantly striving for innovation in chemistry and chemical technology innovation. Innovation is at the end of the day what has given us a wide range of materials and products that have made our lives more comfortable, and many of those products derive from chemistry and engineering innovation (Figure 2.2). There are many in the synthetic chemistry and engineering community who believe green chemistry and/or green engineering to be a "soft" science: that is, not a hard physical scientific or engineering discipline and not quite worthy of "real" academic consideration. In actual fact, success in green chemistry and engineering presents more difficult challenges and opportunities for innovation than does much of synthetic organic chemistry. It can be argued that green chemistry and green engineering are indeed "smart" chemistry and engineering, insofar as their practitioners attempt to design more mass- and energy-efficient processes and to avoid design concerns and problems that have plagued chemical processes for decades.

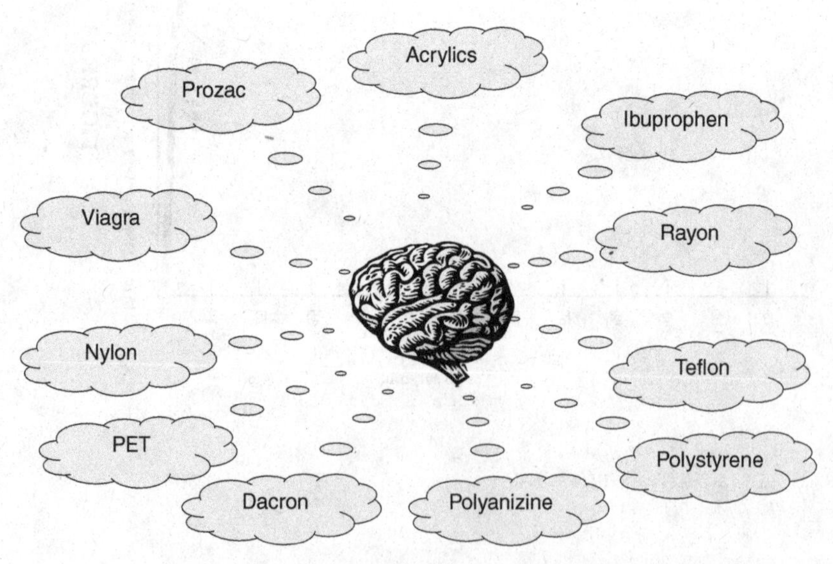

FIGURE 2.2 Some of the many products in use.

Think about it for a moment. How difficult is it to design a synthesis with a very reactive molecule such as an azide or acetylene, where the reaction is overwhelmingly thermodynamically and kinetically favorable, using whatever solvent you want, in as dilute a solution as you like, when you don't care how you'll separate the product from the reaction mixture? This is approximately equivalent to being proud of hitting the ocean when you throw a stone into it.

In contrast to the above, think for a moment about the green chemistry design challenge. You are being asked to make a complex chemical such as a drug or an advanced liquid crystal with the following design constraints: Use as little extra chemical material as possible, with as little energy as possible, using compounds that are nontoxic and safe to handle, that are either biodegradable or recoverable and reusable; and extract the desired product without resorting to a large amount of solvent or energy, all the while causing no long-term impacts to people and or to the environment as you do it.

Green chemistry and green technology therefore require the best and the brightest to rethink and challenge existing paradigms and push the limits of our knowledge. This require people who understand and embrace different academic disciplines within chemistry, engineering, mathematics, and interrelated sciences (e.g., toxicology, biology, biochemistry). Imagine, for example, a synthetic organic chemist who understands thermodynamics and kinetics (largely the domain of the physical chemist), but who knows enough biochemistry to use enzymatic transformations while making use of process analytical technologies to develop reaction and process understanding and control. But this is not enough. Chemists and engineers also need to understand enough about other disciplines, such as biology, toxicology, engineering, and geology, that they are able to use chemistry more knowledgeably and design products, processes, separation technologies, and manufacturing plants based on greener, safer principles. Above all, it requires intellectual flexibility to provide continual innovation and change on a rapid scale.

2.1.2 Mass and Energy Efficiency

See, for example, green chemistry principles 1, 2, 5, 6, 8, and 9.

Only fairly recently has society become more aware of its impact on the global environment. Although it is true that different societies have become more or less aware of local or regional impacts on the environment (it is, after all, somewhat difficult to ignore a burning river, deforestation in parts of the northeastern United States, a large explosion, etc.), society is only beginning to become aware that human beings are engaged in earth systems engineering on a grand scale.[3] This has been spurred on, perhaps, by publication of a report by the UN International Panel on Climate Change, which has amassed sufficient and conclusive evidence for the impact on the climate of the increase in carbon dioxide (and other greenhouse gas) concentrations in Earth's atmosphere.[4]

If one thinks on a global scale and begins to ask where materials come from to make the products that society uses on a daily basis, it is not difficult to see evidence of our insatiable need for materials of commerce, such as plastics, electronics, clothing, food, and housing. Producing these materials requires increasingly complex global supply chains to meet the needs and wants of developed and developing nations. Increases in the costs of a variety of key materials, including fossil fuels for energy and petrochemical feedstocks, are a reflection of the demands being placed on supplying chemicals that are increasingly more difficult to find and transform into the desired materials. In addition, the production of materials is intimately related to emissions and resource depletion; in general, the more materials are needed to produce a good, the more resources that will be needed along the supply chain and the more emissions to the air, water, and land that will need to be controlled.

It could be argued that these trends are pushing society toward increasing material and energy efficiencies in relation to the material and energy utilized for every product produced. The consequence of low material and energy efficiencies, is, of course, the production of waste. Roger Sheldon pointed out the relative waste of different industrial sectors and coined the phrase *E-factor*.[5] The *E*-factor is related to the mass intensity (MI) as follows:

$$E_{\text{factor}} = \text{MI} - 1 \tag{2.1}$$

where

$$\frac{\text{kg waste}}{\text{kg product}} = \frac{\text{kg input}}{\text{kg product}} - \frac{\text{kg product}}{\text{kg product}}$$

As can be seen from Table 2.1, the farther one is from raw material extraction, as is true in the pharmaceutical sector, the greater the waste that is produced. The challenge for green chemistry and green engineering is to decrease significantly the material intensities observed in all industrial sectors: for example, decrease the mass intensity by at least an order of magnitude, if not more. It is interesting to note that in many cases, material and energy intensity are very highly correlated. If one thinks about this for a moment, it makes a certain intuitive sense that if I decrease the volume of material I am handling, I should use less energy to produce, use, reuse, and hopefully, dispose of it.

Historically, in response to regulations, industry has focused on waste (*E*-factor) and its elimination, as opposed to preventing waste generation through innovations in chemistry and chemical technology (principle 1). As the U.S. Congress opined in 1986, "the major obstacles to increased waste reduction are institutional and behavioral rather than technical." Although this is perhaps understandable, in many respects it is unfortunate because a focus on end-of-pipe solutions is generally costly and only increases the overall mass and energy intensity associated with the production of any product. Looking at mass and energy efficiency instead, we can shift our mindset from a treatment, end-of-pipe viewpoint, to a efficiency-increasing, revenue-generating solution.

2.1.3 Toxicity and Persistence

See, for example, green chemistry principles 3, 4, and 10.

Although a decrease in the amount of energy and materials used for our products is critical, it is important to understand that the nature of the materials we use is also critically

TABLE 2.1 Mass Intensity of Various Sectors of the Chemical Industry

Industry	Mass Intensity (kg total/kg product, excluding H_2O)
Oil refining	1.1
Bulk chemicals	1–5
Fine chemicals	5–50
Pharmaceuticals	25–100

Source: ref. 5.

important. Once again, there is a certain intuitive sense in this, as we might ask why society would want to use a chemical that might render us sterile or incapacitated while consigning us at some point in the future to a slow painful death by cancer or some other chronic illness (e.g., emphysema, heart disease). Most of us would, of course, say that this is probably not a good thing, yet this is exactly what is done, and generally done safely, on a daily basis. We use a large number of materials that are extremely toxic and difficult to handle because they happen to be extremely useful to us chemically. However, one might ask if this is a practice that we wish to continue if we can devise a way to avoid these inherently hazardous chemicals.

Legislation such as the Regulation, Evaluation, and Authorization of Chemicals Act (REACH)[6] approved by the European Commission suggests that at least some societies are interested in obtaining a better understanding of the environmental, health, and safety (EHS) hazards associated with existing and new chemicals. It may be surprising to many readers that for a large number of chemicals, despite a long history of use in a range of industries, an understanding of the EHS hazards associated with many compounds is not sufficient. The long-term objective of REACH is to obtain that EHS understanding, and once this better understanding is obtained, it is likely that certain chemicals will be banned if the risk associated with their use is deemed to be too great.

In addition to legislative restriction, to operate processes safely with chemicals that are highly toxic, the appropriate controls should be in place, and the more toxic a material is, the cost to design, set, validate, and maintain the appropriate controls normally increases. Thus, eliminating, substituting, or reducing the amounts of toxic chemicals is also tied to economic engineering and the economic bottom line of processes.

Green chemistry principles 3, 4, and 10 anticipated chemicals legislation and challenge chemists to design molecules and their basic building blocks in such a way that toxicity is eliminated or reduced sufficiently to eliminate high risk. These principles are arguably among the most difficult for chemists to address, for two reasons. First, synthetic chemists generally lack any understanding of toxicity, and for the most part, the relationship of molecular structure to toxicity is not well known for many chemicals represented by the myriad of potential combinations of the usual elements (i.e., C, H, O, N, S, Cl). A second thorny issue is that the efficacy of a molecule, as in pesticides, herbicides, and drug substances, among others, is related directly to their ability to exert a toxic effect on a target organism. Indeed, it is a tall order just to find and then make a compound of interest that works as intended without adding additional design constraints related to reducing potential toxicity!

Finally, after discovering an efficacious molecule of interest with no or minimal associated toxicity hazards, it must be designed either for reuse or for biodegradation. Implicit in any consideration of biodegradation is an aspect of risk management that is often poorly understood: *chemical fate*. Chemical fate concerns itself with where a molecule ends up once it is released to the environment, either in air, water, or on land, and will have a different degradation pathway depending on where it is distributed to, as is shown in Figure 2.3 for a household detergent. If the compound is chemically degradable or biodegradable, the degradation by-products must themselves be nontoxic. All of this emphasis on fate and toxicity should drive anyone who wants to introduce a new chemical, to EHS hazard testing on a very large scale unless the science to model fate and environmental effects, explosivity, flammability, and so on, in silico improves dramatically. For the time being, however, EHS hazard testing is generally the only way that we can adequately assess potential risks, and this will necessarily increase the cost and

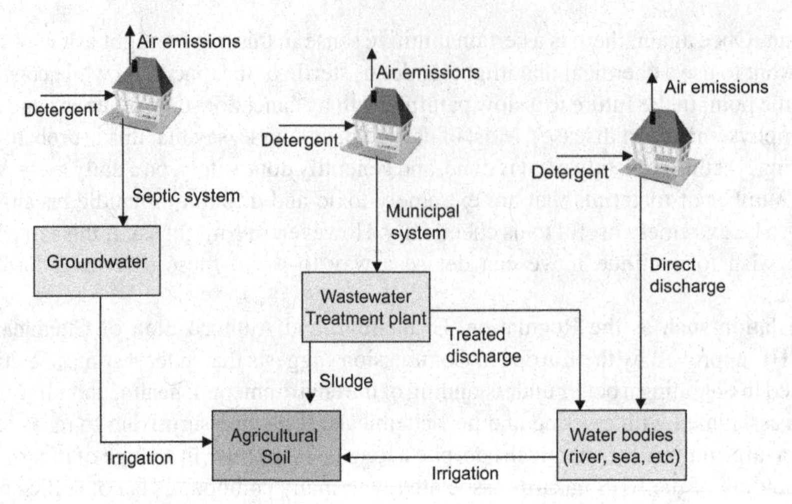

FIGURE 2.3 Fate and effects of a common household detergent.

complexity of bringing new products to society. Fate and effects are covered in more detail in Chapter 3.

2.1.4 Renewability of Feedstocks

See, for example, green chemistry principles 7 and 10.

One of the most exciting areas to think about is how to change the way we make and use the items we need and want in such a way that all Earth's species can continue to live at as good or at a better standard of living. Although very exciting for some to think about, this is still largely simply a nice thought. In actual fact, we are living in ways that are not sustainable. Human beings are depleting raw materials at an alarming rate or are having to expend more energy and to inflict greater environmental damage to obtain many of the key minerals, raw materials, and energy that we require to maintain a Western standard of living. Stated differently, that is a high standard of living for only about one-fourth of Earth's population. What about the rest of the human world and of all species that live in what are arguably less than ideal conditions?

In a sense, principle 7 draws a line in the sand and asks chemists and engineers to find, develop, and provide the materials and energy that we need and want in ways that reverse current trends. This is a very tall order. Think about the petrochemical industry for a moment; it did not start out being a highly efficient industry, but has developed and evolved over the course of more than 100 years. Chemists have to replicate for a biologically derived supply chain what took 100 years (not counting a few hundred millions of years to form) to optimize using a completely different type of feedstock. And they must do this in a shorter time frame and without major environmental damage if we are to preserve Earth's biodiversity and ability to maintain large human and nonhuman populations.

Some surprise might be registered by seeing that principle 10 is included here, and one may think that it does not belong in a category about renewability, so some explanation is warranted. This principle is, after all, about biodegradability or persistence. Some have argued, most notably Bill McDonough and Michael Braumgartner in their book *Cradle to*

Cradle,[7] that society could do with having some materials considered to be technical nutrients. By *technical nutrients* they mean materials that are used for a certain period of time, then after a given service life can be collected and returned to their original state and formed into new products. An example they give is replacing paper with a polymeric substance that can be reused repeatedly without a considerable amount of energy or loss.

The point here is that persistence is sometimes a useful characteristic if it is possible to have a closed-loop recycling system. The problem is, of course, that it is very likely that there will never be a completely closed loop, so some of this material will end up in the environment. In that case, such materials would have to be either biodegradable or completely nontoxic to all organisms. In either case, there is still a need to develop materials that are renewable and which do not cause environmental degradation. A lot of work needs to be done.

Example 2.1 Dimethyl carbonate can be produced by the following reaction[8]:

$$2CH_3OH + COCl_2 \rightarrow 2NaOH + CH_3OCOOCH_3 + 2NaCl + H_2O$$

Describe which of the green chemistry principles postulated by Anastas and Warner you could apply to this reaction to improve its greenness given the information provided.

Solution

Principle 1. It is better to prevent waste than to treat it or clean up after it is formed. In the reaction above, looking at the stoichiometry, there will be an aqueous waste stream with sodium hydroxide and sodium chloride in significant concentrations. The sodium hydroxide being formed is corrosive and will need to be neutralized and treated. Is there a way to produce the desired carbonate while avoiding the generation of this waste stream? How about separating and purifying the final product and the related waste? Is there a way to obtain a final product that is close to being pure?

Principle 2. Synthetic methods should be designed to maximize the incorporation into the final product of all materials used in the process. This is the concept of *atom economy*. In reactions with 100% atom economy, all the materials added to the chemistry are incorporated into the final product. Can we design an addition reaction that can produce the carbonate with no by-products?

Principle 3. Wherever practicable, synthetic methodologies should be designed to use and generate substances that possess little or no toxicity to human health and the environment. This reaction requires phosgene, a highly acute toxicant. Can we devise a different synthetic pathway that avoids the use of phosgene and doesn't replace it with another toxic material?

Principle 9. Catalytic reagents (as selective as possible) are superior to stoichiometric reagents. This reaction is stoichiometric. Is there a way that this chemical can be produced by catalytic means?

Principle 12. Substances and the form of a substance used in a chemical process should be chosen so as to minimize the potential for chemical accidents, including releases, explosions, and fires.

Additional Points to Ponder Although no information was provided on yields and conversion, mass efficiency is another factor to consider in improving the greenness of this reaction. Also, what about separation processes needed to purify the product which are not included in the information? What about energy requirements to run the reaction and to purify and separate the product?

2.2 TWELVE MORE GREEN CHEMISTRY PRINCIPLES

Since the publication of the Twelve Principles of Green Chemistry, there have been a variety of publications, symposia, and conferences dedicated to increasing our understanding of green chemistry and how it might be advanced. Reflecting on many of these presentations and publications, in 2001 Neil Winterton published 12 More Principles of Green Chemistry.[9] Critics of these additional principles have argued that it was not necessary to add to the Anastas and Warner list, as the principles outlined by Winterton can, in their minds, be subsumed within the original 12 principles. Supporters have argued that while the original 12 principles are useful, the Winterton list represents a practical, pragmatic, and industry-driven expansion of great value because the principles are not well understood or appreciated by the academic synthetic chemistry community. They are also an excellent bridge to chemical engineers, as they highlight the tight relationship between green chemistry and green engineering and how some of the most elementary principles that chemical engineers have learned and applied routinely for generations are also fundamental in designing safer, greener chemical processes (e.g., performing full mass balances, quantifying utilities, measuring losses, investigating heat and mass transfer limitations).

TWELVE MORE PRINCIPLES OF GREEN CHEMISTRY

1. Identify and quantify by-products.
2. Report conversions, selectivities, and productivities.
3. Establish full mass balances for a process.
4. Measure catalyst and solvent losses in aqueous effluent.
5. Investigate basic thermochemistry.
6. Anticipate heat and mass transfer limitations.
7. Consult a chemical or process engineer.
8. Consider the effect of the overall process on the choice of chemistry.
9. Help develop and apply sustainability measures.
10. Quantify and minimize the use of utilities.
11. Recognize where safety and waste minimization are incompatible.
12. Monitor, report, and minimize laboratory waste emitted.

Source: Adapted from ref. 9.

Example 2.2 How do the Anastas and Warner green chemistry principles relate to Winterton's additional green chemistry principles?

Solution To illustrate the point of how the Winterton principles might be subsumed in the Anastas and Warner principles, the accompanying box contains a combination of the two sets. These lists are complementary to one another and are useful for focusing on some important practical aspects of the principles as they apply to industry.

THE TWELVE PRINCIPLES OF GREEN CHEMISTRY (COMBINED)

1. It is better to prevent waste than to treat or clean up waste after it is formed.
   a. Consider the effect of the overall process on the choice of chemistry.
   b. Recognize where safety and waste minimization are incompatible.
   c. Monitor, report, and minimize laboratory waste emitted.
   d. Consult a chemical or process engineer.
2. Synthetic methods should be designed to maximize the incorporation into the final product of all materials used in the process.
   a. Identify and quantify by-products.
   b. Report conversions, selectivities, and productivities.
   c. Establish full mass balances for a process.
   d. Measure catalyst and solvent losses in aqueous effluent.
   e. Consult a chemical or process engineer.
   f. Help develop and apply sustainability measures.
3. Wherever practicable, synthetic methodologies should be designed to use and generate substances that possess little or no toxicity to human health and the environment.
   a. Help develop and apply sustainability measures.
   b. Consult a chemical or process engineer.
4. Chemical products should be designed to preserve efficacy of function while reducing toxicity.
5. The use of auxiliary substances (e.g., solvents, separation agents) should be made unnecessary whenever possible and, innocuous when used.
6. Energy requirements should be recognized for their environmental and economic impacts and should be minimized. Synthetic methods should be conducted at ambient temperature and pressure.
   a. Investigate basic thermochemistry.
   b. Quantify and minimize use of utilities.
   c. Consult a chemical or process engineer.
7. A raw material feedstock should be renewable rather than depleting whenever technically and economically practical.
   a. Help develop and apply sustainability measures.
   b. Consult a chemical or process engineer.

8. Unnecessary derivatization (blocking group, protection/deprotection, temporary modification of physical/chemical processes) should be avoided whenever possible.
9. Catalytic reagents (as selective as possible) are superior to stoichiometric reagents.
10. Chemical products should be designed so that at the end of their function they do not persist in the environment and break down into innocuous degradation products.
    a. Help develop and apply sustainability measures.
11. Analytical methodologies need to be developed further to allow for real-time in-process monitoring and control prior to the formation of hazardous substances.
    a. Consult a chemical or engineer.
12. Substances and the form of a substance used in a chemical process should be chosen so as to minimize the potential for chemical accidents, including releases, explosions, and fires.
    a. Investigate basic thermochemistry.
    b. Recognize where safety and waste minimization are incompatible.
    c. Consult a chemical or process engineer.

2.3 TWELVE PRINCIPLES OF GREEN ENGINEERING

In an attempt to engage the engineering community more broadly, Paul Anastas teamed up with Julie Zimmerman and published the Twelve Principles of Green Engineering, shown in the accompanying box. As is readily apparent, there are some principles, as shown in Table 2.2, that are in part related to the 12 principles of green chemistry. It is interesting to see these repetitions between the principles of green chemistry and the principles of green engineering. One very apparent shortcoming of these two lists is that in a way they seem to have been published as if they were independent, but in reality the principles should not be considered separately. When designing products and processes, the chemistry should be designed with the real-life process in mind, which is beginning to be known as *design for manufacturability*. Green engineering should be able to feed the chemistry back to designers and provide ideas of what is feasible and the trade-offs between safety and toxicity. In addition, the 12 principles of green engineering take into account several additional concepts that are worth a moment's consideration.

THE TWELVE PRINCIPLES OF GREEN ENGINEERING

1. Designers need to strive to ensure that all material and energy inputs and outputs are as inherently nonhazardous as possible.
2. It is better to prevent waste than to treat or clean up waste after it is formed.
3. Separation and purification operations should be designed to minimize energy consumption and materials use.
4. Products, processes, and systems should be designed to maximize mass, energy, space, and time efficiency.
5. Products, processes, and systems should be "output pulled" rather than "input pushed" through the use of energy and materials.

6. Embedded entropy and complexity must be viewed as an investment when making design choices on recycle, reuse, or beneficial disposition.
7. Targeted durability, not immortality, should be a design goal.
8. Design for unnecessary capacity or capability (e.g., "one size fits all") solutions should be considered a design flaw.
9. Material diversity in multicomponent products should be minimized to promote disassembly and value retention.
10. Design of products, processes, and systems must include integration and interconnectivity with available energy and materials flows.
11. Products, processes, and systems should be designed for performance in a commercial "afterlife."
12. Material and energy inputs should be renewable rather than depleting.

Source: Adapted from ref. 10.

2.3.1 Thermodynamics: Limits and Potential for Innovation

See, for example, green engineering principle 6.

Although principle 6 is seen primarily as an engineering principle, it is unfortunate that it is not also a part of the green chemistry principles. It is quite clear that a good understanding of thermodynamics is required even of synthetic chemists if they are to be successful in green chemistry. An example would be designing a reaction to take advantage of phase differences for separations (gravity separation) rather than on distillations (energy intensive), or favoring a bias toward homogeneous reactions when heterogeneous reactions could work with better or different reactor design, or order of addition. Moreover, it is also clear that except for a few instances (e.g., Heusemann[11]), very few people in green chemistry and engineering consider thermodynamic limits to sustainable chemistry and engineering. There are limits to what is possible under existing practice and state of the art.

Principle 6 is very useful in drawing attention to the tendency on the part of some people to assume that recycling/reuse is a preferred option in most cases. In actual fact, this is not always the case, at least with current chemistries and technologies. This principle reminds us to broaden our boundaries and look at the entire system or life cycle of a product to ensure that proposed processes and products achieve an appropriate consideration of thermodynamic opportunities and limits. In some instances, recycling to a certain point in the supply chain might be more effective than recycling the raw materials. From a chemist's perspective, recent reconsideration of standard protection–deprotection schemes for making naturally occurring marine products resulted in the exploitation of cascade reactions to take advantage of thermodynamically favored reaction sequences.[12]

TABLE 2.2 Broad Themes in Green Chemistry and Green Engineering Principles

Principles	Chemistry and Technology Innovation	Mass and Energy Efficiency	Toxicity and Persistence	Renewability of Feedstocks
Green chemistry	2, 4, 5, 8, 9–12	1, 2, 5, 6, 8, 9	3, 4, 10	7, 10
Green engineering	3, 11	2–5, 10	1, 7	12

TABLE 2.3 Mass and Energy Intensity of an iMac

Year	Product Mass (kg/computer)	Packaging Mass (kg/computer)	Idle Power Consumption (W)
1998	18.3	3.7	93
2002	10.5	3.6	70
2006	7.0	1.9	55

Source: ref. 14.

2.3.2 Complexity

See, for example, green engineering principles 6 and 9.

The degree of complexity embedded in some products and the processes used to make them is nothing less than astounding. There are few industries that embody this better than the semiconductor industry, where Moore's law[13] has driven innovation ever closer to the fundamental limits of physics and the chemicals used (Si, Ge, In, etc.). Very large scale integration has worked to reduce the size and number of parts of modern electronics while increasing their capability dramatically. In this instance, complexity is generally considered to be a good thing in that electronics are doing more, with less embedded mass and energy to make them and use them. For example, Table 2.3 shows the mass intensity, packaging mass intensity, and idle energy consumption of an iMac.

In addition, in response to product-take-back legislation, many leaders in the industry are beginning to think about how they might simplify the overall design of a product (e.g., a photocopying machine) so that it might be disassembled easily and many of the parts either reused or easily recycled. In this instance, assembly, disassembly, and end-of-life considerations must be accounted for in the up-front design of a product if the overall complexity of the product is to be reduced.

2.3.3 Use, Reuse, and End-of-Life Considerations

See, for example, general engineering principles 8, 9, and 11.

No industry better embodies a living example of principle 8 than the pharmaceutical industry. In this industry there is a considerable degree of structural and chemical complexity in the molecules that ultimately become products. However, although it is true that there is tremendous complexity in discovering and delivering drug candidates to the market, it is also true that most of the chemistries used to make these molecules, and the processes employed to synthesize and then formulate them into products, would be recognizable to anyone living 100 to 150 or more years ago.

Because of the phenomenal attrition rate of most drug candidates (i.e., 1 in 10,000[15] candidates makes it to market), there is a tendency for manufacturing processes to be designed and implemented in a multipurpose batch chemical operation. Invariably volume estimates for drugs are very poor and either under- or overestimated, leading to "making do" with existing equipment until additional capacity can be brought online. Then, because of short patent lives following compound registration, there is little appetite for optimizing processes that will be lost to manufacturers of generic drugs. The consequence of this is that the pharmaceutical industry has some of the worse mass and energy efficiencies of any industry.

Example 2.3 How can an electronic product can be designed for ease of disassembly?

Solution Applying life cycle techniques to electronics design can help engineers create features that enable the recovery of materials for reuse or recycling. Disassembly features allow for the quick sorting and removal of components and materials for servicing. For example, according to Sun Microelectronics,[16] some of the strategies that Sun incorporates in product design to enable ease of disassembly, reuse, and recycle are:

- Product upgrades are planned intentionally to prevent the premature retirement of materials.
- Many components, such as boards, memory, and disk drives, can be added or replaced by the latest technology improvements.
- Once recovered, these components can be refurbished and sold as re-marketed equipment, or can be disassembled to separate valuable components for reuse elsewhere.
- Instead of using permanent methods such as ultrasonic welding or spray coatings to unite components, engineers can design shields with the minimum number of heatstakes (bonding points), or they can snap-fit materials so that metal shields and plastic housings are easy to separate and recycle.
- Embedded ISO 11469 identification codes for plastic type on plastic parts increase the chances of reuse and make it easier to sort materials that are in demand.
- Thin-walled plastic design conserves the amount of material needed while maintaining strength requirements and yields extra environmental benefits by reducing the amount of fuel needed to transport new, lighter products.
- Nonpainted plastics make recycling and recovery easy.

Other computer manufacturers (e.g., Apple, Hewlett-Packard, Dell) have similar schemes and strategies that include end-of-life considerations.

Additional Points to Ponder Computers are complex machines and some substances might be released during the recycling process. For example, nickel–cadmium batteries, used previously for backup power, can release cadmium at the end of the useful life of the battery and as a result were phased out. When this is the case, substitution for these types of substances should be investigated.

2.4 THE SAN DESTIN DECLARATION: PRINCIPLES OF GREEN ENGINEERING

In the spring of 2003, a very heterogeneous mix of chemists and engineers from industry, government and academia met in San Destin, Florida to discuss principles of green engineering. This group was intending to appeal to a slightly larger engineering audience beyond that generally associated with the chemical industry, in addition to potentially broadening the scope of previous work to incorporate principles of sustainability. The output is shown in the accompanying box.

THE SAN DESTIN DECLARATION: PRINCIPLES OF GREEN ENGINEERING

1. Engineer processes and products holistically, use systems analysis, and integrate environmental impact assessment tools.
2. Conserve and improve natural ecosystems while protecting human health and well-being.
3. Use life cycle thinking in all engineering activities.
4. Ensure that all material and energy inputs and outputs are as inherently safe and benign as possible.
5. Minimize depletion of natural resources.
6. Strive to prevent waste.
7. Develop and apply engineering solutions while being cognizant of local geography, aspirations, and cultures.
8. Create engineering solutions beyond current or dominant technologies; improve, innovate and invent (technologies) to achieve sustainability.
9. Actively engage communities and stakeholders in development of engineering solutions.

Source: Adapted from ref. 17.

As with the previous green chemistry and green engineering principles, several of the San Destin principles are similar to previous approaches; for example, principles 6 and 8, which fits nicely with the theme of chemical and chemical technology innovation. There was also an element of pragmatism, or perhaps pessimism, as in principle 5, which asserts that natural resource depletion should be minimized, the implicit assumption being that we will never attain a situation where society will not deplete natural resources and achieve a cradle-to-cradle sustainable society.

Apart from these three principles, which were discussed previously, there are several concepts that are brought out in this declaration that are worth a moment's consideration.

2.4.1 Systems and Life Cycle Thinking

See, for example, principles 1 and 3.

In general, most human beings are reductionist thinkers; that is, we cut things down into small bits that are easily grasped or understood so that we are not overwhelmed by the considerable complexity that attends most things in our world. However, in our attempts to reduce complexity, we are sometimes guilty of not seeing the bigger picture, or optimizing one small corner of our universe to the detriment of a broader part or another aspect of the system. This is where systems and life cycle thinking come into play. Certainly, biologists and environmentalists are more attuned than most chemists and engineers to systems thinking, given their attempts to understand entire ecosystems: for example, the interplay of microorganisms, plants, invertebrates, vertebrates (animals), and humans across space and time. It is this ability to look at the big picture and discern the key interactions, responses, and impacts that is so important in green chemistry and green engineering.

One of the disciplines that helps us to better understand some of the interactions and impacts from chemicals and the processes used to make them is life cycle inventory and assessment (LCI/A). Although LCI/A is covered in more detail in Chapter 16, it is worth just a moment's explanation. LCI/A is a rigorous analytical methodology developed to evaluate the environmental impacts of a product or activity, starting with the product or functional unit (e.g., a car, a single dose of a drug, a can of paint, a service) and works back through all the unit operations and materials that are used to make the product, all the way back to raw material extraction. A person does this by performing an input/output inventory or an accounting of all the mass and energy used for each unit operation or production process. It also includes a look at the product in use and its ultimate disposition (i.e., what happens to it when it no longer performs the function originally intended.) Such an exercise generally forces one to look across entire systems because most products are not simple extractions of raw materials followed by immediate use with no emissions. LCI/A broadens our perspective as we try to understand the material and energy flows and their impacts.

The tie-in of these principles to green chemistry and green engineering is hopefully apparent. If one considers or optimizes only one reaction or one part of a process, it is easy to miss the rest of the process, or perhaps the use of a very nasty chemical in another part of the supply chain. By looking across the entire supply chain, one can optimize material use so that only the best materials, the appropriate chemistries, and the best processes are selected.

2.4.2 Community or Societal Engagement

See, for example, principles 7 and 9.

As mentioned in Chapter 1, a key component of sustainability is the inclusion of societal issues or concerns, in addition to environmental and economic issues, in any decision-making process. Principles 7 and 9 are important contributions to green engineering and the San Destin declaration because they bring to any discussion an explicit consideration, from an engineering (and/or chemistry) perspective, of societal, community, or cultural issues regarding what it means to be green. Scientists are, in general, very good at pursuing science for the sake of science, at challenging the scientific status quo, or at pushing the limits of current scientific knowledge. However, scientists are generally not as well equipped to grapple with the implications of their science in terms of impacts to communities, societies, or to the broader environment. However, the capability of assessing the implications of science and technology must increasingly be embedded into scientific education as a matter of routine.

These principles are therefore key principles, insofar as they do highlight those issues and challenge scientists and engineers to practice their disciplines not only in the context of the communities they live and work in, but in the broader context of regional, national, and global communities. For example, the platinum group catalyst that may enable an interesting and potentially novel asymmetric hydrogenation probably comes at the cost of enormous impacts in others part of the world, where people, their culture, and the environment are in some instances affected significantly by the mining of that metal. At some point we need to ask if that is justifiable and if the local benefit of greening a particular process is balanced by the impact of mining. Local societies affected by mining may in fact agree that the benefits of mining outweigh the impact, but they should be a part of that decision. Similarly, chemists and engineers should be aware of the broader impacts of their science and engineering and ask themselves if those impacts could in some way be avoided, mitigated, and/or minimized.

Example 2.4 According to the San Destin declaration of green engineering, which option is better as a fuel for the operation of vehicles: conventional gasoline or corn-derived bioethanol?

Solution It depends on which impacts one is considering. On the one hand, bioethanol represents a good renewable alternative to fossil fuel use, shows a considerable better profile for global warming and ozone depletion, but in general fares worse regarding smog formation, acidification and eutrophication, and ecotoxicity.[18,19] The key here is to analyze the effects of using corn-derived ethanol from a life cycle viewpoint, accounting for the general impacts.

Additional Points to Ponder Consider using corn as a crop to produce fuel, vs. other crops, such as sugarcane, from which the substrate can be extracted significantly more efficiently. The other point to consider is the societal trade-offs of using corn or sugarcane for fuel vs. using them for food.

2.5 SIMPLIFYING THE PRINCIPLES

In this chapter we have studied various sets of principles for green chemistry and green engineering that have been proposed to guide scientists and engineers in developing greener processes. These sets of principles are in general accepted by most people and have proven very powerful in disseminating the intent and guidelines of green chemistry and green engineering. However, these principles can be simplified.[20]

At the end of the day, applying green chemistry and green engineering principles to your work as an engineer or process chemist means that when designing novel chemical routes, selecting reactors or separations, designing chemical processes, building plants, and so on, one should strive to:

- Maximize resource efficiency
- Eliminate and minimize hazards and pollution
- Design systems holistically, using life cycle thinking

But are these three principles enough? Do they cover all the previous advice that we have given? The principles we have studied can be mapped to these three overarching principles, as can be seen in Table 2.4. At the end of the day, the various strategies should drive us to design better, cheaper, faster, and cleaner processes from the beginning. In the remaining chapters, we discuss how many of these strategies can be utilized to design greener processes.

TABLE 2.4 Simplifying Green Chemistry and Engineering Principles

Principles	Maximize Resource Efficiency	Eliminate and Minimize Hazards and Pollution	Design Systems Holistically and Using Life Cycle Thinking
Green chemistry: Anastas and Warner	Synthetic methods should be designed to maximize the incorporation of all materials used in the process into the final product. The use of auxiliary substances (e.g., solvents, separation agents) should be made unnecessary whenever possible and innocuous when used. Energy requirements should be recognized for their environmental and economic impacts and should be minimized. Synthetic methods should be conducted at ambient temperature and pressure. Unnecessary derivatization (blocking group, protection–deprotection, temporary modification of physical/chemical processes) should be avoided whenever possible. Catalytic reagents (as selective as possible) are superior to stoichiometric reagents.	It is better to prevent waste than to treat or clean up waste after it is formed. Wherever practicable, synthetic methodologies should be designed to use and generate substances that possess little or no toxicity to human health and the environment. Chemical products should be designed to preserve efficacy of function while reducing toxicity. Chemical products should be designed so that at the end of their function they do not persist in the environment and break down into innocuous degradation products. Analytical methodologies need to be developed further to allow for real-time in-process monitoring and control prior to the formation of hazardous substances. Substances and the form of a substance used in a chemical process should be chosen so as to minimize the potential for chemical accidents, including releases, explosions, and fires.	A raw material feedstock should be renewable rather than depleting whenever technically and economically practical.

(*continued*)

TABLE 2.4 (*Continued*)

Principles	Maximize Resource Efficiency	Eliminate and Minimize Hazards and Pollution	Design Systems Holistically and Using Life Cycle Thinking
Green chemistry: Winterton	Identify and quantify by-products. Report conversions, selectivities, and productivities. Establish full mass balances for a process. Anticipate heat and mass transfer limitations. Quantify and minimize the use of utilities.	Measure catalyst and solvent losses in aqueous effluent. Investigate basic thermochemistry. Recognize where safety and waste minimization are incompatible. Monitor, report, and minimize laboratory waste emitted.	Consult a chemical or process engineer. Consider the effect of the overall process on the choice of chemistry. Help develop and apply sustainability measures. Recognize where safety and waste minimization are incompatible.
Green engineering: Anastas and Zimmerman	Separation and purification operations should be designed to minimize energy consumption and materials use. Products, processes, and systems should be designed to maximize mass, energy, space, and time efficiency. Products, processes, and systems should be "output pulled" rather than "input pushed" through the use of energy and materials. Design of products, processes, and systems must include integration and interconnectivity with available energy and materials flows.	Designers need to strive to ensure that all material and energy inputs and outputs are as inherently nonhazardous as possible. It is better to prevent waste than to treat or clean up waste after it is formed.	Embedded entropy and complexity must be viewed as an investment when making design choices on recycle, reuse, or beneficial disposition. Targeted durability, not immortality, should be a design goal. Design for unnecessary capacity or capability (e.g., "one size fits all") solutions should be considered a design flaw. Material diversity in multicomponent products should be minimized to promote disassembly and value retention.

			Products, processes, and systems should be designed for performance in a commercial "afterlife." Material and energy inputs should be renewable rather than depleting.
Green engineering: San Destin	Minimize depletion of natural resources. Strive to prevent waste. Conserve and improve natural ecosystems while protecting human health and well-being.	Ensure that all material and energy inputs and outputs are as inherently safe and benign as possible. Strive to prevent waste. Conserve and improve natural ecosystems while protecting human health and well-being.	Engineer processes and products holistically, use systems analysis, and integrate environmental impact assessment tools. Use life cycle thinking in all engineering activities. Develop and apply engineering solutions while being cognizant of local geography, aspirations, and cultures. Create engineering solutions beyond current or dominant technologies; improve, innovate and invent (technologies) to achieve sustainability. Actively engage communities and stakeholders in development of engineering solutions.

PROBLEMS

2.1 Explain why it was important for several people to create green chemistry and green engineering principles.

2.2 Why is innovation so important for advancing green chemistry and green engineering?

2.3 Explain how you might be able succeed at green chemistry and green engineering without getting multiple degrees in different disciplines?

2.4 How would you assess the mass efficiency of a process?

2.5 Why is chemical persistence such a problem?

2.6 Find an example of a renewable feedstock and critique it using the 12 principles of green chemistry and/or the 12 principles of green engineering.

2.7 Why is it important to use a life cycle approach when evaluating processes? What are the *trade-offs*, and how can one deal with them?

2.8 Dimethyl carbonate can be produced by the catalytic oxidative carbonylation of methanol:

$$2CH_3OH + CO + 0.5O_2 \rightarrow CH_3OCOOCH_3 + H_2O$$

According to the green chemistry principles, which one is a better option, this reaction or the reaction of Example 2.1? Why?

2.9 A traditional way to separate compounds in an azeotropic mixture is to use azeotropic or extractive distillation, in which a mass separating agent is added to a first distillation column to break the azeotrope, followed by a separator and a second distillation column. An alternative is to use a separation train consisting of a distillation column followed by vapor permeation (see Figure P2.9). Based on the principles of green chemistry and green engineering, which system is greener, and why?

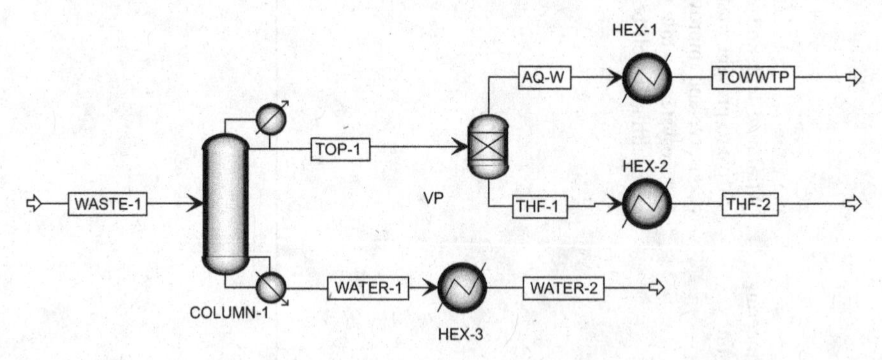

FIGURE P2.9 Atmospheric distillation followed by vapor permeation.

REFERENCES

1. *Merriam-Webster's Collegiate Dictionary*, 11th ed. Merriam-Webster, springfield, MA, July 2003.
2. Anastas, P., Warner J. *Green Chemistry: Theory and Practice*. Oxford University Press, New York, Dec. 1998.
3. Allenby, B. R. Grand challenges and research needs. In *Sustainability in the Chemical Industry*. A Workshop Report. National Research Council, National Academy of Sciences. National Academies Press, Washington, DC, 2005.
4. Intergovernmental Panel on Climate Change. Climate Change 2001: Synthesis Report; Summary for Policymakers. http://www.ipcc.ch/pub/un/syreng/spm.pdf.
5. Sheldon, R. A. Consider the environmental quotient. *Chemtech*, 1994, Vol 24(3), 38–46.
6. European Commission. Regulation EC 1907/2006.
7. McDonough, W., Braumgartner, M. *Cradle to Cradle: Remaking the Way We Make Things*. North Point Press, New York, Apr. 2002.
8. Matlack, A. S. *Introduction to Green Chemistry*. Marcel Dekker, New York, 2001.
9. Winterton, N. Twelve more green chemistry principles? *Green Chem.*, 2001, 3, G73–G75.
10. Anastas, P. T., Zimmerman, J. B. Design through the twelve principles of green engineering. *Environ. Sci. Technol.*, 2003, 37(5), 94A–101A.
11. Huesemann, M. H. The limits of technological solutions to sustainable development. *Clean Technol. Environ. Policy*, 2002, 5(1), 21–34.
12. Baran, P. S., Maimone, T. J., Richter, J. M. Total synthesis of marine natural products without using protecting groups. *Nature*, 2007, 446, 404–408.
13. Schaller, R. R. Moore's law: past, present and future. *IEEE Spectrum*, 1997, 34(6), 52–59.
14. Apple Computers. Apple and the Environment. http://www.apple.com/environment/design/.
15. Pisano, G. P. *The Development Factory: Unlocking the Potential of Process Innovation*. Harvard Business School Press, Boston, 1997.
16. Sun Microsystems company information. Environment Health and Safety. http://www.sun.com/aboutsun/ehs/ehs-design.html.
17. Abraham, M.A., Nguyen, N. Green engineering: defining the principles—results from the San Destin Conference, 2003, 22(4), 233–236.
18. Kim, S., Dale, B. E. Life cycle assessment of various cropping systems utilized for producing biofuels: bioethanol and biodiesel. *Biomass Bioenergy*, 2005, 29(6), 426–439.
19. Von Blottnitz, A. H., Curran, M. A. A review of assessments conducted on bio-ethanol as a transportation fuel from a net energy, greenhouse gas, and environmental life cycle perspective. *J. Cleaner Prod.* 2007, 15, 607–619.
20. Beckman, E. J. Using principles of sustainability to design "leap-frog" products. Keynote presentation at the 11th Annual Green Chemistry and Engineering Conference, June 26–29, 2007.

3

STARTING WITH THE BASICS: INTEGRATING ENVIRONMENT, HEALTH, AND SAFETY

What This Chapter Is About In Example 1.1 we identified some high-level green chemistry and green engineering challenges related to the production of potassium hydroxide. Those challenges are related to the effects of materials and processes on the environment and the health and safety of workers, on property such as the reactors, and on other unit operations in the manufacturing facility and on the community. In Chapter 2 we discussed some well-established green chemistry and green engineering principles that encourage us to avoid and minimize the use of hazardous materials. However, how do we know what makes a material hazardous? How do we determine which hazards are more serious than others? How do we identify which materials or operations might pose a risk to the environment, the workers, or the community? In this chapter we present a high-level overview of the environmental, health, and safety (EHS) issues generally considered to be of importance and discuss the need to take an integrated view of these issues when considering what is more sustainable in the realm of green chemistry and green technology. However, before we get into the details of EHS hazards, we need to understand a bit more about the important distinction between hazard and risk.

Learning Objectives At the end of this chapter, the student will be able to:

- Identify and understand the most common environmental issues of importance.
- Identify and understand the most common health issues of importance.
- Identify and understand the most common safety issues of importance.
- Understand the importance of an integrated approach to environment, health, and safety aspects.

Green Chemistry and Engineering: A Practical Design Approach, By Concepción Jiménez-González and David J. C. Constable

3.1 ENVIRONMENTAL ISSUES OF IMPORTANCE

3.1.1 Climate Change

The climate on Earth has changed through natural geological and biological processes, over time, including cooling and warming periods. In the last century, however, average world temperatures have been rising constantly, and there is general scientific agreement that the warming trend has been accelerating in recent years. For example, the period between 1996 and 2005 contains the warmest years ever recorded, with the exception of 1996 and 2000,[1] and the average temperature for the continental United States from January through June 2006 was the warmest first half of any year since records began in 1895.[2] The concern expressed is not only for changing temperatures but also for the rate of climate change to which entire ecosystems have been accustomed. Although the extent of potential future impacts is difficult to predict, some effects of climate change that have been observed already include sea-level rise, shrinking glaciers, changes in the range and distribution of plants and animals, trees blooming earlier, lengthening of growing seasons, ice on rivers and lakes freezing later and breaking up earlier, and thawing of permafrost.[3]

Scientists agree that the main cause of this rise in temperature and the resulting climate change is an enhanced greenhouse effect. When solar energy reaches the Earth, a part of this energy warms Earth's surface and the rest is emitted back to the atmosphere, where gases such as water vapor, carbon dioxide, methane, and nitrogen oxides absorb some of this heat and reflect it back to Earth, adding to the warming of Earth's surface. This effect is very similar to what happens in a greenhouse, where glass panels act like some gases in the atmosphere; hence this phenomenon has been called the *greenhouse effect* and the gases involved are known as *greenhouse gases*.[4] The greenhouse effect is a natural phenomenon essential for life on Earth as we know it, as it keeps the Earth warm and habitable. However, steadily increasing atmospheric concentrations of greenhouse gases traps more heat, causing global temperatures to rise.

Scientists across a range of disciplines are in broad agreement that the recent increase in carbon dioxide and other greenhouse gases is related directly to human activity.[5] Water vapor is the most abundant greenhouse gas, but it is not produced directly by human activities. Greenhouse gases related directly to human activity have reached new highs, with globally averaged mixing ratios of carbon dioxide (CO_2), nitrous oxide (N_2O), and methane (CH_4) higher than those in preindustrial times by 35.4%, 18.2%, and 154.7%, respectively, although methane concentration growth rates have slowed during the past decade. Carbon dioxide is the most prevalent anthropogenic greenhouse gas and is responsible for about 62% of the radiative forcing on Earth by long-lived greenhouse gases and about 90% of the increase in radiative forcing during the past decade.[6] Estimation of the potential impacts of emissions on greenhouse gas effects is normally reported in terms of *global warming potential* (GWP). The GWP is a simplified index based on the radiative properties of a gas relative to carbon dioxide over a chosen time span and is expressed in units of mass of carbon dioxide equivalents (e.g., kg CO_2-equivalent). GWP has been defined formally as the ratio of the time-integrated radiative forcing from the instantaneous release of 1 kg of a trace substance relative to that of 1 kg of a reference gas.[7] Values of GWP selected over different time spans are reported in Table 3.1.

TABLE 3.1 Direct Global Warming Potentials of Selected Gases (Mass Basis) Relative to Carbon Dioxide

		Global Warming Potential, Time Horizon		
Chemical	Formula	20 Years	100 Years	500 Years
Carbon dioxide	CO_2	1	1	1
Methane	CH_4	62	23	7
Nitrous oxide	N_2O	275	296	156
	Chlorofluorocarbons			
CFC-11	CCl_3F	6,300	4,600	1,600
CFC-12	CCl_2F_2	10,200	10,600	5,200
CFC-13	$CClF_3$	10,000	14,000	16,300
CFC-113	CCl_2FCClF_2	6,100	6,000	2,700
CFC-114	$CClF_2CClF_2$	7,500	9,800	8,700
CFC-115	CF_3CClF_2	4,900	7,200	9,900
	Hydrochlorofluorocarbons			
HCFC-21	$CHCl_2F$	700	210	65
HCFC-22	$CHClF_2$	4,800	1,700	540
HCFC-123	CF_3CHCl_2	390	120	36
HCFC-124	CF_3CHClF	2,000	620	190
HCFC-141b	CH_3CCl_2F	2,100	700	220
HCFC-142b	CH_3CClF_2	5,200	2,400	740
HCFC-225ca	$CF_3CF_2CHCl_2$	590	180	55
HCFC-225cb	$CClF_2CF_2CHClF$	2,000	620	190
	Hydrofluorocarbons			
HFC-23	CHF_3	9,400	12,000	10,000
HFC-32	CH_2F_2	1,800	550	170
HFC-41	CH_3F	330	97	30
HFC-125	CHF_2CF_3	5,900	3,400	1,100
HFC-134	CHF_2CHF_2	3,200	1,100	330
HFC-134a	CH_2FCF_3	3,300	1,300	400
HFC-143	CHF_2CH_2F	1,100	330	100
HFC-143a	CF_3CH_3	5,500	4,300	1,600
HFC-152	CH_2FCH_2F	140	43	13
HFC-152a	CH_3CHF_2	410	120	37
HFC-161	CH_3CH_2F	40	12	4
HFC-227ea	CF_3CHFCF_3	5,600	3,500	1,100
HFC-236cb	$CH_2FCF_2CF_3$	3,300	1,300	390
HFC-236ea	CHF_2CHFCF_3	3,600	1,200	390
HFC-236fa	$CF_3CH_2CF_3$	7,500	9,400	7,100
HFC-245ca	$CH_2FCF_2CHF_2$	2,100	640	200
HFC-245fa	$CHF_2CH_2CF_3$	3,000	950	300
HFC-365mfc	$CF_3CH_2CF_2CH_3$	2,600	890	280
HFC-43-10mee	$CF_3CHFCHFCF_2CF_3$	3,700	1,500	470
	Chlorocarbons			
1,1,1-Trichloroethane	CH_3CCl_3	450	140	42
Chloroform	CCl_4	2,700	1,800	580

(*continued*)

TABLE 3.1 *(Continued)*

		Global Warming Potential, Time Horizon		
Chemical	Formula	20 Years	100 Years	500 Years
Trichloromethane	$CHCl_3$	100	30	9
Methyl chloride	CH_3Cl	55	16	5
Methylene chloride	CH_2Cl_2	35	10	3
	Bromocarbons			
Methyl bromide	CH_3Br	16	5	1
Methylene bromide	CH_2Br_2	5	1	<<1
Bromodifluoromethane	$CHBrF_2$	1,500	470	150
Halon-1211	$CBrClF_2$	3,600	1,300	390
Halon-1301	$CBrF_3$	7,900	6,900	2,700
	Fully Fluorinated Species			
Sulfur hexafluoride	SF_6	15,100	22,200	32,400
Carbon tetrafluoride	CF_4	3,900	5,700	8,900

Source: ref. 8.

3.1.2 Stratospheric Ozone Depletion

The *stratosphere* is an atmospheric layer that lies at an altitude between approximately 10 and 50 km and is characterized by an inverted (increasing) temperature profile. About 90% of atmospheric ozone resides in this layer, and it is being produced and destroyed continuously as part of a natural cycle that when in balance keeps the overall amount of ozone stable. Stratospheric ozone filters out ultraviolet (UVB) radiation, a form of radiation that has been linked to skin cancer, cataracts, damage to materials such as plastics, and harm to certain crops and marine organisms. It is obviously in our best interests to preserve the ozone layer to prevent human, property, and ecological damage.

Ozone is produced in the stratosphere by the action of the sun's UV light on oxygen. Nitrogen oxides naturally present in the atmosphere are responsible for most of the natural destruction of ozone by catalytic cycles involving free radicals, as shown in the following reactions[9]:

$$NO + O_3 \rightarrow NO_2 + O_2 \tag{3.1}$$

$$NO_2 + O \rightarrow NO + O_2 \tag{3.2}$$

$$\text{Net: } O + O_3 \rightarrow 2O_2 \tag{3.3}$$

The balance of the natural ozone cycle is upset by the presence of other free radicals, such as chlorine or bromine, which can catalytically destroy ozone faster than natural processes can replenish it. Chlorofluorocarbons and other ozone-depleting substances are extremely stable and are not removed by the natural processes that destroy pollutants. However, after reaching the stratosphere they are decomposed by the very energetic short-wavelength UV light, where they rapidly release chlorine or bromine atoms. These atoms then participate in catalytic reactions that decompose ozone.

The thinning of the ozone layer had been postulated since the early 1970s,[10] but significant depletion of the ozone layer was confirmed over Antarctica and is popularly known as the "Antarctic ozone hole." This is a seasonal phenomenon occurring only during the Antarctic spring. In addition to the global ozone depletion that occurs homogeneously in the upper and lower atmosphere, this localized thinning of the ozone layer is a very interesting phenomenon caused in part by the very low temperatures during the Antarctic winter that contribute to heterogeneous chemistry on ice surfaces and the combination of meteorological effects that are specific to Antarctica from August to November.

Similar to GWP, stratospheric *ozone depletion potential* (ODP) is reported as the ratio of the ozone depletion potential of a chemical relative to that of another chemical: in this case, CFC-11. Table 3.2 shows values of ODP for selected ozone-depleting substances.

TABLE 3.2 Ozone Depletion Potential of Selected Gases, Expressed in CFC-11 Equivalents

Compound	Lifetime (years)	Ozone Depletion Potential, Montreal Protocol
Trichlorofluoromethane	45	1
Dichlorodifluoromethane	100	1
1,1,2-Trichlorotrifluoroethane	85	0.8
Dichlorotetrafluoroethane	300	1
Monochloropentafluoroethane	1700	0.6
Bromochlorodifluoromethane	16	3
Bromotrifluoromethane	65	10
Dibromotetrafluoroethane	20	6
Chlorotrifluoromethane	640	1
Pentachlorofluoroethane	—	1
Tetrachlorodifluoroethane	—	1
Heptachlorofluoropropane	—	1
Hexachlorodifluoropropane	—	1
Pentachlorotrifluoropropane	—	1
Tetrachlorotetrafluoropropane	—	1
Trichloropentafluoropropane	—	1
Dichlorohexafluoropropane	—	1
Chloroheptafluoropropane	—	1
Carbon tetrachloride	26	1.1
1,1,1-Trichloroethane	5	0.1
Methyl bromide (CH_3Br)	0.7	0.6
Chlorobromomethane	0.37	0.12
$CHFBr_2$	—	1
CHF_2Br	—	0.74
CH_2FBr	—	0.73
Dichlorofluoromethane	1.7	0.04
Monochlorodifluoromethane	12	0.055
Monochlorofluoromethane	—	0.02
Tetrachlorofluoroethane	—	0.01–0.04
Trichlorodifluoroethane	—	0.02–0.08
Dichlorotrifluoroethane	1.3	0.02–0.06
Monochlorotetrafluoroethane	5.8	0.02–0.04
Trichlorofluoroethane	—	0.007–0.05

(*continued*)

TABLE 3.2 (*Continued*)

Compound	Lifetime (years)	Ozone Depletion Potential, Montreal Protocol
Dichlorodifluoroethane	—	0.008–0.05
Monochlorotrifluoroethane	—	0.02–0.06
Dichlorofluoroethane	9.3	0.11
Monochlorodifluoroethane	17.9	0.065
Hexachlorofluoropropane	—	0.015–0.07
Pentachlorodifluoropropane	—	0.01–0.09
Tetrachlorotrifluoropropane	—	0.01–0.08
Trichlorotetrafluoropropane	—	0.01–0.09
Dichloropentafluoropropane	1.9	0.025
Dichloropentafluoropropane	5.8	0.033
Monochlorohexafluoropropane	—	0.02–0.1
Pentachlorofluoropropane	—	0.05–0.09
Tetrachlorodifluoropropane	—	0.008–0.1
Trichlorotrifluoropropane	—	0.007–0.23
Dichlorotetrafluoropropane	—	0.01–0.28
Monochloropentafluoropropane	—	0.03–0.52
Tetrachlorofluoropropane	—	0.004–0.09
Trichlorodifluoropropane	—	0.005–0.13
Dichlorotrifluoropropane	—	0.007–0.12
Monochlorotetrafluoropropane	—	0.009–0.14

Source: ref. 11.

As a reaction to the ozone layer depletions, the global treaty from the Vienna Convention for the Protection of the Ozone Layer was signed, followed by the Montreal Protocol on Substances That Deplete the Ozone Layer, which went into force in 1989. The Montreal protocol currently has the participation of over 180 countries, which agreed on a timetable for phasing out the production and consumption of major halocarbons.[11]

3.1.3 Photochemical Ozone Formation

In the stratosphere, ozone acts as a protective filter from UVB rays. At ground level, the atmospheric layer known as the *troposphere*, ozone is a harmful pollutant. *Volatile organic compounds* (VOCs) participate in producing tropospheric ozone through photochemical reactions with nitrogen oxides, as shown in the following reactions[12, 13]

$$RO_2 + NO \rightarrow RO + NO_2 \tag{3.4}$$

$$NO_2 + h\nu \rightarrow NO + O \tag{3.5}$$

$$O + O_2 + M \rightarrow O_3 + M \tag{3.6}$$

$$\text{Net}: \quad RO_2 + O_2 \rightarrow RO + O_3 \tag{3.7}$$

where R = H, CH_3, or other organic peroxy radical. These reactions produce smog, which is exacerbated during the summer months, thanks to the combination of strong sunlight and heat, and they show that the effect of NO in atmospheric chemistry is twofold. Above 25 km, where oxygen radical concentrations are high, ozone destruction [reactions (3.1) and (3.2)] dominates over ozone production [reactions (3.4), (3.5), and (3.6)]. The latter set of reactions can also take place with carbon monoxide or methane, but these reactions will also require sufficient concentrations of NO and NO_x.

Nitrogen oxides and VOCs are released to the atmosphere mainly through fuel combustion processes of many types, from such sources as vehicles, utilities (electric, steam, etc.), and a host of industrial processes.[14] Tropospheric ozone is responsible for a series of short-term vertebrate effects, such as irritation of the respiratory tract, lung function reduction, increases in asthma symptoms, and inflammation of the lungs' lining. Releases of VOCs to air are generally compared on the basis of their potential to create ozone relative to ethylene and expressed as photochemical ozone creation potential (POCP). Table 3.3 shows POCP values for selected substances.[15]

TABLE 3.3 Photochemical Ozone Creation Potential Values for Selected Compounds

Compound	POCP 100 kg ethylene equivalents
Alkanes	
Methane	3.4
Ethane	14
Propane	41.1
n-Butane	59.9
i-Butane	42.6
n-Pentane	62.4
i-Pentane	59.8
n-Hexane	64.8
2-Methylpentane	77.8
3-Methylpentane	66.1
2,2-Dimethylbutane	32.1
2,3-Dimethylbutane	94.3
n-Heptane	77
2-Methylhexane	71.9
3-Methylhexane	73
n-Octane	68.2
2-Methylheptane	69.4
n-Nonane	69.3
2-Methyloctane	70.6
n-Decane	68
2-Methylnonane	65.7
n-Undecane	61.6
n-Dodecane	57.7
Cyclohexane	59.5
Methylcyclohexane	73.2
Alkenes	
Ethylene	100
Propylene	107.9
1-Butene	113.2
2-Butene	99.3
2-Pentene	95.3

(*continued*)

TABLE 3.3 (*Continued*)

Compound	POCP 100 kg ethylene equivalents
1-Pentene	104.1
Butylene	70.3
Alkynes	
Acetylene	28
Aromatics	
Benzene	33.4
Toluene	77.1
o-Xylene	83.1
m-Xylene	108.8
p-Xylene	94.8
Ethylbenzene	80.8
n-Propylbenzene	71.3
Isopropylbenzene	74.4
1,2,3-Trimethylbenzene	124.5
1,2,4-Trimethylbenzene	132.4
1,3,5-Trimethylbenzene	129.9
Aldehydes	
Formaldehyde	55.4
Acetaldehyde	65
Propionaldehyde	75.5
Butyrlaldehyde	77
Isobutylaldehyde	85.5
Ketones	
Acetone	18.2
Methyl ethyl ketone	51.1
Methyl-*i*-butyl ketone	84.3
Cyclohexanone	52.9
Alcohols	
Methyl alcohol	20.5
Ethyl alcohol	44.6
Isopropanol	21.6
n-Butanol	62.8
Isobutanol	59.1
s-Butanol	46.8
t-Butanol	19.1
Esters	
Methyl acetate	4.6
Ethyl acetate	32.8
n-Propyl acetate	48.1
Isopropyl acetate	29.1
n-Butyl acetate	51.1
s-Butyl acetate	45.2

TABLE 3.3 (*Continued*)

Compound	POCP 100 kg ethylene equivalents
Organic acids	
Formic acid	0.3
Acetic acid	15.6
Propionic acid	3.5
Ethers	
Butyl glycol	62.9
Propylene glycol methyl ether	51.8
Dimethyl ether	26.3
Methyl *tert*-butyl ether	26.8
Halocarbons	
Methyl chloride	3.5
Methylene chloride	3.1
Methylchloroform	0.2
Other	
Nitric oxide	−42.7
Nitrogen dioxide	2.8
Sulfur dioxide	4.8
Carbon monoxide	2.7

3.1.4 Acidification

When emissions such as sulfur and nitrogen oxides combine with water, oxygen, and other chemicals in the atmosphere, they form acidic compounds: nitric and sulfuric acids. These acids are subsequently deposited onto land or water either incorporated into dust and smoke as *dry deposition* or mixed with rain, snow, fog, or mist in *wet deposition*. The combined phenomenon of acid deposition is commonly known as *acid rain*.

There are multiple effects from acid deposition, including acidification of lakes and rivers, rendering them unfit for plant or animal life, accelerated decay and corrosion of buildings and property (e.g., damage to automotive paint), and indoor-quality issues. There is also damage to agricultural crops and forests, as the increased soil acidity can lead to the displacement of calcium ions and inhibit growth of plants, or plants can simply be defoliated in extreme cases of acid deposition.

Although the main contributors of acidification are sulfur and nitrogen oxides, certain emissions that are precursors of some of the most typical inorganic acids can be assigned an *acidification potential*, an index of the capacity of releasing protons per unit mass relative to that of sulfur dioxide (Table 3.4). The acidification potential can also be applied to water deposition.[16]

3.1.5 Eutrophication

Eutrophication is a natural response of plants to increased nutrients, typically nitrogen and phosphorus compounds. This nutrient enrichment can occur on land or water, and it normally

TABLE 3.4 Acidification Potentials of Selected Compounds

Compound	Acidification Potential (Sulfur Dioxide Equivalents)
Ammonia	1.88
Chloromethane (methyl chloride)	0.634
Dichloromethane (methylene chloride)	0.744
Hydrogen chloride	0.88
Hydrogen fluoride	1.6
Hydrogen sulfide	1.88
Nitric acid	0.508
Nitrogen dioxide	0.7
Sulfur dioxide	1
Sulfuric acid	0.653
Tetrachlorocarbon (tetrachloromethane)	0.83
Trichloroethane (1,1,1-trichloroethane)	0.72
Trichloroethene	0.72
Trichloromethane (chloroform)	0.803

TABLE 3.5 Eutrophication Potentials of Selected Compounds

Species	Aquatic Eutrophication Potential (Phosphate Equivalents)	Species	Terrestrial Eutrophication Potential (Phosphate Equivalents)
Phosphate (PO_4^-)	1	Phosphate (PO_4^-)	1
Ammonia (NH_3)	0.33	Ammonia (NH_3)	0.33
Ammonia ion (NH_4^+)	0.33	Nitrogen monoxide (NO)	0.2
Nitrate (NO_3^{-2})	0.42	Nitrous oxides (NO_x)	0.13
Chemical oxygen demand (COD)	0.022		

Source: ref. 16.

promotes excessive plant growth and decay. In aquatic environments, excessive plant growth can manifest itself among lower plant forms such as algae, or higher forms such as aquatic plants. In either case, the excess growth and subsequent decay of these plants results in a reduction of dissolved oxygen in the water, which in turn usually causes other organisms (e.g., fish) to die.

Eutrophication is a natural phenomenon that is generally associated with aging water bodies such as ponds, lakes marshes, and estuaries, but additional human activity from agriculture, industry, or residences can accelerate this phenomenon significantly. This is particularly true in water bodies such as lakes and estuaries, as the slow flow allows the nitrogen and phosphorus nutrients from runoff to accumulate at a higher rate over time (Table 3.5).

3.1.6 Organic Matter in Water

Organic matter in water normally undergoes oxidation processes by either biological or chemical processes. The presence of organic matter in water from sewage and other sources is of importance because natural degradation processes create high oxygen demand in the receiving bodies, potentially harming aquatic ecosystems. This demand for oxygen is an indirect measure of the organic waste concentrations in water, and typical indicators for this are the biological oxygen demand (BOD), chemical oxygen demand (COD) and total organic carbon (TOC).

BOD determines the amount of oxygen needed to oxidize the organic matter present in water by biological means, and *COD* determines the amount of oxygen required to oxidize organic matter by chemical means. Since the biological process is time dependent, BOD normally is reported at a specific time from the start of the bioassay, normally 5 days (BOD_5). *TOC* is the amount of carbon that is organically bound.

In addition to organic matter, there is a long list of potential water and air emissions from anthropogenic sources that can cause adverse effects to the environment or to human health and therefore require monitoring and treatment. Among these emissions we can mention heavy metals such as zinc, lead, mercury, chromium, and others; organic compounds such as benzene, toluene, and xylene; gases such as carbon monoxide; and many more substances. Normally, these pollutants are subject to local and other air and water quality regulations.

3.1.7 Ecotoxicology

For the purposes of inherent hazard, fate, and effects, and a consideration of appropriate metrics, it is helpful to have a good understanding of a compound's chemical properties, as this will fundamentally drive the overall environmental risk. Fate and effects are very important components of environmental risk that are worth a moment's further consideration. The environmental fate of a chemical is concerned with where a compound will go (air, water, soil) once it is released. Once in the environment, one needs to know what potential effects a chemical may have on organisms, including humans. In general, insufficient attention is paid to the chemical mechanisms in the environment that affect chemical fate or distribution, and most attention is given to the potential effects that a chemical will have on a variety of organisms. Just because a compound is inherently hazardous to plants or animals does not mean that once released it will necessarily present a great risk to the environment. It is important to evaluate carefully distribution and degradation mechanisms that will directly affect the potential for exposure to any given chemical. The reader is referred elsewhere for a more complete treatment of this important topic.[17–19]

From an environmental hazard assessment perspective, most chemicals in recent years have been categorized according to their potential for persistence, bioaccumulation, and toxicity. *Persistence* is associated with whether or not a chemical will be resistant to chemical (e.g., hydrolysis, photolysis) or biological (e.g., biodegradation, metabolism) degradation or breakdown. Various tests are used to determine a chemical's environmental depletion mechanism. *Bioaccumulation* is the tendency for chemicals to become increasingly concentrated, usually in fat, as one moves up the food chain from microorganisms to large fish, birds, or mammals. The water/octanol partition coefficient, log K_{ow} or D_{ow} (corrected for pH and ionizability of a compound in water), is generally used to estimate the tendency for bioaccumulation. *Toxicity* is generally the most contentious area of concern and there are a wide variety of toxicity tests that might be used to assess toxicity. Generally, one conducts multiple acute toxicity tests (the endpoint is lethality) at three levels of the food chain: for examples, algae, aquatic flea (e.g., *Daphnia magna*), and fish [e.g., *Pimephelas* (fathead minnow), *Onchorynchus* (trout)]. There are variations on the type of test depending on the fate of the compound (e.g., whether it will be released to a freshwater or marine environment, or applied to land as for a pesticide) and the type of application for the chemical. Recently, there has been increasing concern about chronic exposures to chemicals, so there has been a movement toward requiring toxicity testing that assesses chronic exposures and different endpoints (e.g., fecundity, endocrine disruption) The reader is referred elsewhere for an extensive treatment of ecotoxicity testing.[34]

3.1.8 Hazardous and Nonhazardous Waste

A waste is *hazardous* if it has properties that make it dangerous or capable of having an adverse effect on human health or the environment. This type of waste normally requires special considerations for handling, recycling, transport, storage, treatment, and disposal. *Nonhazardous* waste normally requires either recycling, treatment, or disposal systems that demand resources (i.e., landfill, energy).

Example 3.1 A manufacturing process produces the potential pollutants shown in Table 3.6 prior to any treatment processes. Estimate the photochemical ozone creation potential, global warming potential, acidification potential, eutrophication potential, and total organic carbon for this manufacturing process.

Solution Calculation basis: 1000 kg of final product. The calculations require you to multiply the value of the impact potentials times the mass of the emission and then add the contributions for each of the impacts. For example, to estimate the global warming potential related to the methane emissions:

$$\begin{aligned} \text{Mass} \times \text{GWP} &= (1.89\ \text{kg methane})(62\ \text{kg CO}_2\ \text{equivalents/kg methane}) \\ &= 117.43\ \text{kg CO}_2\ \text{equivalents} \end{aligned}$$

The factors for each impact are shown in a calculating table, Table 3.7 taken mainly from Tables 3.1,3.3,3.4, and 3.5. NO_x indirect effects in global warming potential are difficult to calculate, as they are dependent on the place where they are emitted. For example, the emissions of NO_x from aircraft are characterized by far greater GWPs than those of surface

TABLE 3.6 Potential Pollutants

Pretreatment	Total (kg/1000 kg product)
Air emissions	
CH_4	1.89
CO	0.94
CO_2	1143.52
NO_x	4.10
SO_x	4.79
HCl	1.37
Methyl chloride	74.52
Methylene chloride	14.35
Water emissions	
COD	0.46
Methylene chloride	21.56
Trichloromethane	2.82

TABLE 3.7 Calculating Table

	Total (kg per 1000 kg)	Impact Potentials [per kg of material]				Calculations [Impacts per 1000 kg of product]			
		POCP	GWP (20 yr)	ACID	EUTR	POCP (kg ethylene equivs)	GHG (kg CO_2 equivs)	ACID (kg SO_2 equivs)	EUTR (kg phosphate equivs)
Air emissions									
CH_4	1.89	0.034	62.0	—	—	0.064	117.43		
CO	0.94	—	—	—	—	—	—		
CO_2	1143.52	—	1.0	—	—	—	1143.52		
NO_x	4.10	—	40	0.70	0.130	—	163.80	2.87	0.53
SO_x	4.79	—	—	1.00	—	—	—	4.79	
HCl	1.37	—	—	0.88	—	—	—	1.21	
Methyl chloride	74.52	0.035	55	0.634	—	2.608	4098.82	47.25	
Methylene chloride	14.35	0.031	35	0.744	—	0.445	502.32	10.68	
Water emissions		—	—						
COD	0.46	—	—		0.022	—	—	—	0.01
Methylene chloride	21.56	—	—	0.634		—	—	13.67	
Trichloromethane	2.82	—	—	0.803		—	—	2.26	
Total (per 1000 kg of product)		—	—		3.118	6,025.1	82.72	0.54	

sources, due primarily to the longer lifetime of the NO_x emitted at higher altitudes,[7] but for the purposes of this example, we are using a GWP of 40, as reported by the IPCC in 1997.

The TOC can be estimated using a relationship of the molecular weights of each compound going to water relative to the mass of organic carbon in the molecule:

$$\text{TOC} = 21.56\ \text{kg methylene chloride}\ (\text{MeCl}_2)\frac{12\ \text{kg C}}{84.93\ \text{kg MeCl}_2}$$

$$+\,2.82\ \text{kg Trichloromethane}\ (\text{TCM})\frac{12\ \text{kg C}}{119.38\ \text{kg TCM}}$$

$$= 3.05 + 0.28 = 3.33\ \text{kg Total Organic Carbon}$$

Additional Point to Ponder In addition to the environmental impacts calculated here, why should we be concerned about emissions of methylene chloride, methyl chloride, and trichloromethane arising from this process?

3.2 HEALTH ISSUES OF IMPORTANCE

3.2.1 Toxicology

Toxicology is defined as the study of the toxic properties of agents or substances. Modern toxicology includes both qualitative and quantitative evaluation of substances or agents. The qualitative evaluation is concerned with the types of adverse effects that can potentially be caused by a chemical or agent. The quantitative evaluation estimates the amount of a chemical or agent that can cause such an adverse effect. One of the most commonly cited maxims is from Philippus T. A. B. von Hoenheim (also known as Paracelcus), who stated that "all substances are poisons; there is none which is not a poison. The right dose differentiates a poison and a remedy" or as it is better known, "the dose makes the poison." Quantitative toxicological evaluations are the basis for the concepts of dose–response or exposure, integral parts of risk assessment.

There are several ways in which an agent or chemical can cause effects in relation to the periods of time to which a person is exposed to a chemical. The same chemical can have immediate toxic effects, or acute effects, and if the exposure continues during a longer period of time, chronic or subchronic effects can be developed. The terms *acute, subchronic*, and *chronic* refer not only to the time frame of the exposure, but to the type of toxicological tests that are needed to characterize the hazards and risks of different chemicals, including specific tests for the potential of a chemical to cause death (e.g., LD_{50}), eye irritation, skin irritation, sensitization, and allergic reactions. Data from acute studies such as those on eye and dermal toxicity are often used to characterize and communicate health hazards in product labels. Data from subchronic and chronic studies are often used to establish reference doses or regulatory standards, such as permissible exposure limits (PELs).[20]

The objective of an acute study is to characterize short-term effects due to relatively high exposures such as might result from accidental spillage of a chemical. To determine the acute toxicity of a given chemical, toxicity tests are normally performed with single or repeated doses within a 24-hour period. Many people are familiar with the term *lethal dose 50* (LD_{50}), which is the dose of a substance that can be expected to cause the death of 50% of the population exposed for a determined time. For example, LD_{50}'s are used to rank different degrees of acute toxicity, as in the U. S. Department of Transportation's standards,[21] where hazardous materials are classified for transport purposes. In Division 6.1, poisonous materials are defined as:

a liquid with an LD_{50} oral not more than 500 mg/kg,
or a solid with an LD_{50} oral not more than 200 mg/kg,
or a compound with a LD_{50} dermal not more than 1000 mg/kg,
or a dust/mist with a LC_{50} or not more than 10 mg/L.

Subchronic toxicity is evaluated over a longer period, generally up to 90 days, under conditions of repeated exposure, which would be similar to the repeated ingestion of a medicine for a limited time to cure a disease. Chronic tests are carried out over longer periods of time, generally under relatively low-dose conditions, and would be appropriate for studying the effects of prolonged daily exposure to a toxic chemical that had achieved a steady-state equilibrium concentration in the water supply.

The effects of chemical or material (e.g., mineral, soot) exposure will vary depending on the way the exposure occurs or through which route the person comes into contact with a toxic material: by dermal contact, inhalation, or oral ingestion. These means of being exposed to a chemical are called *routes of exposure* or *exposure routes*. There are testing protocols that are targeted to identify the effects of chemicals by means of various routes of exposure. For example, skin absorption is a common route of exposure in some work conditions, as in the case of working with chemicals in a manufacturing setting. Inhalation is generally the most significant route of workplace exposure, as chemicals can volatilize or workers may be handling materials that consist of fine dust or where dust is produced as part of a manufacturing process. In some instances, toxicological data for the specific route of exposure are not available, and in these cases, if enough data are available and the route-to-route extrapolation is judged to be appropriate, it might be possible to utilize information from other routes of exposure.

Carcinogenicity *Cancer* refers to a group of diseases involving abnormal, uncontrolled tissue or cell growth resulting from a series of defects in the genes that control cell growth, division, and differentiation. Cancer can develop through several different, complex, and generally not-well-understood mechanisms, and carcinogenic agents can exert their effects in a number of ways.

Chemical carcinogens may be either *genotoxic* (those causing permanent change in DNA) or nongenotoxic. Genetic defects leading to cancer may occur because a carcinogenic agent damages DNA directly. Alternatively, a carcinogenic agent may have indirect effects that increase the likelihood or accelerate the onset of cancer without interacting directly with DNA. For example, an agent might interfere with DNA repair mechanisms, increasing the likelihood that cell division will give rise to cells with damaged DNA. A carcinogenic agent might also promote an increase in cell division rates and thereby increase the potential for genetic errors to be introduced as cells replicate their DNA in preparation for division.

Genotoxic agents are believed not to have a threshold amount (i.e., in theory a single molecule of a genotoxic agent could alter DNA), while nongenotoxic carcinogens are thought to have a threshold quantity. Chemicals can work as both genotoxic and nongenotoxic agents, but sometimes it is difficult to differentiate between the two categories. Genetic toxicology as a discipline plays the important role of estimating the potency of a given carcinogenic agent through a determination of the dose–response curve slope for tumor induction.

A wide variety of chemical compounds can cause cancer. Some examples of classes of compounds with known carcinogenic effects include:

- *Polycyclic aromatic hydrocarbons (PAHs)*. Many PAHs, including benzo[*a*]pyrene, can be produced by incomplete combustion of organic matter. A common source of human exposure to these chemicals is tobacco smoke.
- *Aromatic amines and azo dyes*. These compounds are found primarily in occupational settings, such as aniline dye manufacturing. Examples are 2-naphthylamine and 2-acetylaminofluorene.
- *N-Nitroso compounds*. Many of these compounds produce tumors in a wide variety of organs. Humans can be exposed to these compounds in the environment and in occupational settings. Examples include hydrazines, nitrosamines, and N-nitrosodimethylamine.
- *Halogenated aliphatic hydrocarbons*. These include carbon tetrachloride, vinyl chloride, and ethylene dibromide. Human exposure occurs mainly in occupational settings.
- *Other carcinogenic compounds*. The list above is not exhaustive, of course, as carcinogenic properties are associated with materials that include urethane and related compounds, metals and metalloids (e.g., beryllium, hexavalent chromium, nickel, asbestos), miscellaneous organics (e.g., thiourea, thioacetamide), and natural compounds (e.g., aflatoxin B1, safrole, isatidine).

Developmental Toxicity *Teratology* is concerned with the study of the causes, mechanisms, and manifestations of adverse outcomes of pregnancy. The major manifestations of developmental toxicology include death, malformations, altered growth, or functional deficits of offspring. The causes of some birth defects have been associated with genetic transmission, chromosomal aberrations, and environmental factors (e.g., ionizing radiation, infections, chemicals). Although drugs and chemicals account for no more than about 6% of congenital defects, there are about 18 proven teratogenic substances [e.g., alcohol, thalidomide, lead, polychlorinated biphenyls (PCBs), Coumadin] and about 1000 chemicals that have some measure of developmental toxicity in animals.[20]

The determination of whether or not a chemical agent is a human teratogen is a difficult task. Studies in animals help to provide useful models for understanding the mechanisms of action for potential human developmental toxicants, but these studies are not able to help predict potential human effects since interspecies extrapolations cannot be performed with enough certainty. As a result, most developmental toxicants have been discovered through astute observations by physicians and biomedical scientists, followed by epidemiological studies that analyze trends and elucidate whether a causal relationship can be established.

3.2.2 Occupational Exposure Limits

Eliminating workplace exposures to hazardous agents is the best way to reduce the human health risks from exposure to these substances. However, this is still an aspiration in most cases, and it has been necessary to set acceptable ranges and limits of exposure within which a worker's health is not jeopardized during routine work with materials used in a process. The

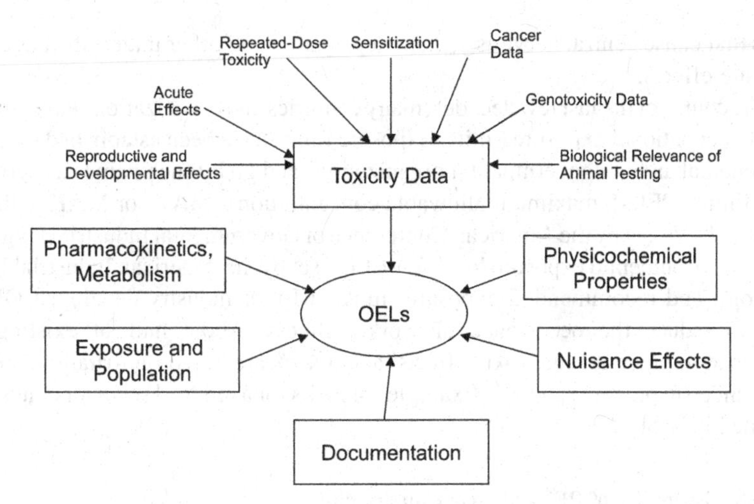

FIGURE 3.1 Factors in setting occupational exposure limits.

concept of exposure limits has been used for some time, but has become an increasingly important consideration over the past several decades. There are two main categories of worker exposure limits that may be established: limits based on measurement of the concentration of a chemical in air; and limits based on measurements of a chemical or its metabolites in a worker's blood or urine.

Occupational exposure limits (OELs) based on workplace air concentrations are more common, due mainly to the relative simplicity in collecting and analyzing air samples compared to invasive measurements of chemicals or their metabolites in blood or urine. Although it is a basic tenant of toxicology that all substances are toxic, OELs are developed under the premise that there is a level of exposure that would cause no harmful effect (i.e., there is a threshold).[22] These OELs are generally established with the objective of protecting everyone in the workforce, but given the fact that some people are more susceptible than others to specific chemicals or other hazardous agents, some workers could experience adverse effects even when airborne concentrations in the workplace are below the occupational limits.

OELs are typically based on a combination of the inherent toxicological hazard of a chemical and a series of safety factors that account for variations or uncertainties in setting limits (Figure 3.1). These safety factors include such considerations as inter- and intraspecies variability in test results, extrapolations from subchronic to chronic tests, the nature and severity of the effect, and the adequacy and quality of the information. To set an OEL, all the available physicochemical, toxicological, and clinical data are typically reviewed carefully by a cross-functional team composed of occupational toxicologists, occupational hygienists, medical doctors, and others to ensure the appropriateness of a proposed limit.

An OEL is usually expressed in terms of a time-weighted average (TWA), although a few limits are short-term or ceiling limits. *A time-weighted average* is defined as the average exposure over a specified period of time, usually a nominal 8 hours, and assumes a healthy workforce exposed to the chemical 5 days a week during a 40-hour workweek. A *short-term exposure limit* (STEL) is the maximum average concentration to which workers can be exposed for a short period (usually, 10 or 15 minutes), and a *ceiling limit* is one that must never be exceeded. Ceiling limits are applied to irritants, sensitizers, materials that cause central nervous system depression, and other materials that can have severe acute effects.

Over the course of the last few decades, many countries and organizations have developed a series of occupational exposure limits. Exposure limits have been established by a variety of governmental and nongovernmental organizations and go by such names as permissible exposure limits (PELs), maximum allowable concentrations (MACs or MAKs), threshold limit values (TLVs, set by the American Conference of Governmental Industrial Hygienists), workplace environmental exposure levels (WEELs, set by the American Industrial Hygiene Association), and recommended exposure limits (RELs). Industry-developed OELs are intended to reduce the occurrence of worker illness. About half of existing OELs are developed to prevent acute toxic effects, about 40% are based on irritation, and about 2% are intended to prevent cancer.[22] Examples of PELs for a reduced set of air contaminants are provided in Table 3.8.

TABLE 3.8 Examples of PELs for Air Contaminants

Substance	CAS No	ppm[a,b]	mg/m[a]	Skin Designation
Acetaldehyde	75-07-0	200	360	
Acetic acid	64-19-7	10	25	
Acetic anhydride	108-24-7	5	20	
Acetone	67-64-1	1000	2400	
Acetonitrile	75-05-8	40	70	
Acrylamide	79-06-1	—	0.3	X
Allyl alcohol	107-18-6	2	5	X
Allyl chloride	107-05-1	1	3	
Ammonia	7664-41-7	50	35	
Benzyl chloride	100-44-7	1	5	
Bromine	7726-95-6	0.1	0.7	
n-Butyl acetate	123-86-4	150	710	
sec-Butyl acetate	105-46-4	200	950	
n-Butyl alcohol	71-36-3	100	300	
sec-Butyl alcohol	78-92-2	150	450	
tert-Butyl alcohol	75-65-0	100	300	
Carbon dioxide	124-38-9	5000	9000	
Carbon monoxide	630-08-0	50	55	
Chlorine	7782-50-5	1	3	
Chloroform	67-66-3	50	240	
Cyclohexanol	108-93-0	50	200	
Cyclohexanone	108-94-1	50	200	
o-Dichlorobenzene	95-50-1	50	300	
1,1-Dichloroethane	75-34-3	100	400	
Diisopropylamine	108-18-9	5	20	X
Dimethyl acetamide	127-19-5	10	35	X
Dimethylamine	124-40-3	10	18	
Dimethylformamide	68-12-2	10	30	X
1,1-Dimethylhydrazine	57-14-7	0.5	1	X
Ethanolamine	141-43-5	3	6	
Ethyl acetate	141-78-6	400	1400	
Ethyl acrylate	140-88-5	25	100	X
Ethyl alcohol (ethanol)	64-17-5	1000	1900	

TABLE 3.8 (*Continued*)

Substance	CAS No	ppm[a,b]	mg/m[a]	Skin Designation
Ethyl benzene	100-41-4	100	435	
Ethyl bromide	74-96-4	200	890	
Ethyl ether	60-29-7	400	1200	
Fluorine	7782-41-4	0.1	0.2	
Formic acid	64-18-6	5	9	
Heptane (*n*-heptane)	142-82-5	500	2000	
2-Hexanone	591-78-6	100	410	
Methyl isobutyl ketone	108-10-1	100	410	
Hydrazine	302-01-2	1	1.3	X
Hydrogen bromide	10035-10-6	3	10	
Hydrogen chloride	7647-01-0	(C)5	(C)7	
Hydrogen cyanide	74-90-8	10	11	X
Isobutyl alcohol	78-83-1	100	300	
Isopropyl acetate	108-21-4	250	950	
Isopropyl alcohol	67-63-0	400	980	
Isopropyl ether	108-20-3	500	2100	
Methyl acetate	79-20-9	200	610	
Methyl alcohol	67-56-1	200	260	
Methylamine	74-89-5	10	12	
Methyl bromide	74-83-9	(C)20	(C)80	X
Methyl chloride	74-87-3	—	2	
Methylene chloride	75-09-2	—	2	
Methyl hydrazine	60-34-4	(C)0.2	(C)0.35	X
Methyl isocyanate	624-83-9	0.02	0.05	X
Methyl mercaptan	74-93-1	(C)10	(C)20	
Methyl methacrylate	80-62-6	100	410	
Nicotine	54-11-5	—	0.5	X
Nitric acid	7697-37-2	2	5	
Nitric oxide	10102-43-9	25	30	
Nitrogen dioxide	10102-44-0	(C)5	(C)9	
Phenol	108-95-2	5	19	X
Phenylhydrazine	100-63-0	5	22	X
Phosgene	75-44-5	0.1	0.4	
Phosphoric acid	7664-38-2	—	1	
n-Propyl acetate	109-60-4	200	840	
n-Propyl alcohol	71-23-8	200	500	
Pyridine	110-86-1	5	15	
Sodium hydroxide	1310-73-2	—	2	
Styrene	100-42-5	—	2	
Sulfur dioxide	7446-09-5	5	13	
Sulfuric acid	7664-93-9	—	1	
Sulfuryl fluoride	2699-79-8	5	20	
Tetrachloronaphthalene	1335-88-2	—	2	X
Warfarin	81-81-2	—	0.1	
Xylenes (*o*-, *m*-, *p*-isomers)	1330-20-7	100	435	
Zirconium compounds (as Zr)	7440-67-7	—	5	

Source: ref. 23.

[a]The PELs are 8-h TWAs unless otherwise noted; a (C) designation denotes a ceiling limit.

[b]ppm = parts of vapor or gas per million parts of contaminated air by volume at 25 °C and 760 torr.

Compliance with an OEL is considered to be achieved if the workplace exposure level is not exceeded over an 8-hour time-weighted period and does not include allowances for the use of personal protective equipment (e.g., respirators). Exposure limits should be used as an important element in the design of manufacturing options for a chemical or manufacturing plant in such a way that the potential for exposure to chemicals is contained or controlled by engineering means. If engineering controls are not sufficient to achieve the exposure limits, protective equipment may need to be employed. However, the use of personal protective equipment as the primary means of controlling exposures is not acceptable and certainly is not aligned with the philosophy of greener chemistries and processes.

Example 3.2 A pharmaceutical process uses the chemicals in Table 3.9 as solvents and raw materials. What are the health hazards of these materials?

Solution By consulting the material safety data sheets of the solvents and reactants in Table 3.9, we have the summary of health hazards and occupational exposure limits presented in Table 3.10.

From Table 3.10 we can see that:

- Lithium hydroxide causes severe burns and is harmful by inhalation and if swallowed. 2,6-Lutidine is harmful by inhalation, in contact with skin, and if swallowed.
- Ethyl chloroformate is very toxic by inhalation and causes burns.
- Toluene is harmful by prolonged exposure through inhalation, and it is a possible risk of harm to a fetus.
- Acetic acid may cause severe burns.
- Heptane may cause lung damage if swallowed, and its vapors may cause drowsiness and dizziness.
- All solvents are irritating to either skin or eyes.

Additional Points to Ponder Judging from the OELs, which material in the process is of the greatest concern from a health viewpoint? According to the hazard list, toluene has a possible risk of harm to a fetus. To what type of toxicity does this refer?

TABLE 3.9 Solvents and Reactants

CAS No.	Name
	Solvents
64-19-7	Acetic acid
64-17-5	Ethanol
142-82-5	Heptane
1634-04-4	Methyl *tert*-butyl ether (MTBE)
108-88-3	Toluene
75-65-0	*t*-Butanol
	Reagents and Reactants
572-09-8	Acetylbromoglucose
1310-66-3	Lithium hydroxide monohydrate
108-48-5	2,6-Lutidine
541-41-3	Ethyl chloroformate
110-86-1	Pyridine

TABLE 3.10 Health Hazards

CAS No.	Name	OEL	Principal Hazards
		Solvents Used	
64-19-7	Acetic acid	10 ppm; TWA; 15 ppm STEL	R10: Flammable R35: Causes severe burns
64-17-5	Ethanol	1000 ppm	R11: Highly flammable R36: Irritating to eyes
142-82-5	Heptane	400 ppm; 500 ppm STEL	R11: Highly flammable R38: Irritating to skin R50/53: Very toxic to aquatic organisms, may cause long-term adverse effects in aquatic environment R65: Harmful: may cause lung damage if swallowed R67: Vapours may cause drowsiness and dizziness
1634-04-4	MTBE	50 ppm	R11: Highly flammable R19: May form explosive peroxides R38: Irritating to skin
108-88-3	Toluene	50 ppm; skin	R11: Highly flammable R38: Irritating to skin R48/20: Harmful; danger of serious damage to health by prolonged exposure through inhalation R63: Possible risk of harm to a fetus R65: Harmful: may cause lung damage if swallowed R67: Vapors may cause drowsiness and dizziness
75-65-0	*t*-Butanol	100 ppm	R11: Highly flammable R20: Harmful by inhalation R36: Irritating to eyes
		Reagents and Reactants	
572-09-8	Acetylbromoglucose		
1310-66-3	Lithium hydroxide monohydrate		R20/22: Harmful by inhalation and if swallowed R35: Causes severe burns
108-48-5	2,6-Lutidine		R10: Flammable R20/21/22: Harmful by inhalation, in contact with skin, and if swallowed
541-41-3	Ethyl chloroformate		R11: Highly flammable R22: Harmful if swallowed R26: Very toxic by inhalation R34: Causes burns
110-86-1	Pyridine	1 ppm	R52: Harmful to aquatic organisms R36: Irritating to eyes R11: Highly flammable R20/21/22: Harmful by inhalation, in contact with skin, and if swallowed

3.3 SAFETY ISSUES OF IMPORTANCE

Material properties and specific process conditions can cause adverse consequences in the workplace, such as falls, slips, uncontrolled releases of hazardous materials (e.g., leaks, spills), explosions, fires, and a series of accidents and near misses. Some of the most common workplace hazards include chemical agents (such as dioxins and heavy metals), biological agents (such as bacteria and viruses), physical agents (such as electricity, vibration, noise, and radiation), and physical hazards (such as slips, trips, and falls).

3.3.1 Basic Safety

A large number of accidents arise from working conditions or activities that are common to most places of work, such as handling objects and materials, working with mechanical equipment or hand tools, falls, and slips. Preventing these types of accidents is normally described as a basic safety measure. Most of these safeguards are related to good housekeeping, workplace order, appropriate procedures, and the like. Table 3.11 shows some considerations for basic safety in the workplace.

TABLE 3.11 Basic Safety Considerations for Preventing Accidents in the Workplace

Handling Materials and Objects	Operating Mechanical Equipment	Using Hand Tools	Prevention of Trips, Falls, and Slips
Safety and work procedures training	Guarded machinery and operating equipment	Appropriate tool maintenance	Smooth, not slippery floors
Maintenance of equipment used for lifting, handling, and transportation	Follow lockout/tag out procedures	Providing the right tools for the job	Safe portable equipment
Proper illumination and working conditions	Maintenance of equipment	Safety and work procedures training	Safe means of reaching or serving overhead equipment
Good housekeeping	Safety and work procedures training	Proper illumination and working conditions	Proper illumination and working conditions
Safety integrated in standard procedures	Appropriate hazard communication	Appropriate hazard communication	Safety and work procedures training
Appropriate hazard communication	Safety integrated into standard procedures	Good housekeeping	Good housekeeping
			Appropriate hazard communication

3.3.2 Process Safety

Catastrophic consequences of accidents in industrial processes have been reported for many years and continue to occur. Some of the better known major disasters have included unexpected releases of toxic, reactive, or flammable liquids and gases in processes involving highly hazardous chemicals. Fires and explosions of uncontrolled processes are other examples of events that pose a significant threat to workers, environment, and property. In addition to the harm to people and the environment, accidents are costly to the economy, and substantial savings can be made through prevention. The importance of preventing accidents has brought international and national attention to the development of methodologies, processes, and regulations to eliminate or minimize the potential for such events.

To identify safety hazards, especially as they relate to a chemical process, a *process hazard analysis* can be conducted.[24] This is a thorough, orderly, systematic approach to identifying, evaluating, and controlling the hazards of processes. In other words, it is a careful review of what could go wrong in a process and what types of controls must be implemented to prevent accidents. There are several methodologies for hazard analysis,[25] such as:

1. *Checklists.* A checklist analysis is a list of items to verify the status of a system. The checklists have a varied degree of detail and are very often used to verify compliance with standards or practices. The output of this analysis is mainly only as good as the checklist it is based on, and the use of checklists is frequently combined with other hazard analysis methods.
2. *What-if.* A what-if analysis is an approach whereby a group of experienced people who are very familiar with a system brainstorm about its potential undesired effects. This technique is not structured by nature, but it is used widely throughout all phases of an analysis to identify hazards, hazardous situations, and undesired consequences.
3. *What-if/checklists.* This technique combines creative thinking with the methodical focus of a prepared checklist. A review team examines a process methodically, generating a list of "what-if" questions regarding the hazards and safety of the operation. After preparing an initial list after the process review, the team goes through a prepared checklist systematically to stimulate additional questions and provides the answers to all the questions. From these answers, a list of recommendations is developed that specifies the need for additional actions or controls.
4. *Hazard and operability study (HAZOP).* This is a structured methodology to investigate each element of a process systematically for all the potential deviations that can cause hazards and operability issues. The effect(s) that deviations from design conditions cause are examined for each key process parameter (e.g., flow, volume, pressure, temperature) using a list of key words such as *more of*, *less of*, *none of*, and *part of* to describe potential deviations. The deviations are noted and the causes of potential failure and existing controls are identified. Finally, an assessment is made weighing the consequences, causes, and protection requirements. Figure 3.2 shows the nodes evaluated in a HAZOP for a hydrogenation system and an example consisting of a hydrogenation reactor, a nitrogen delivery system, a hydrogen delivery system, venting, cooling water, air, and a catch tank. Figure 3.3 shows the output of the analysis for one deviation (high temperature) in the hydrogenation reactor.

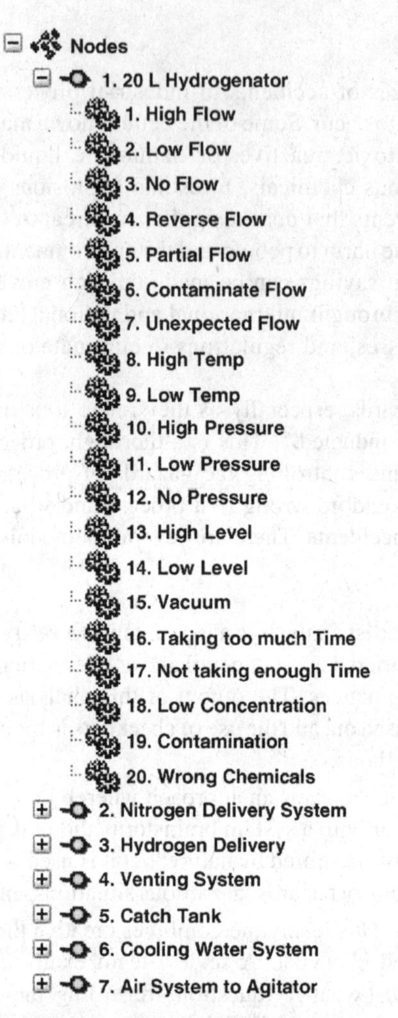

FIGURE 3.2 Nodes analyzed in a HAZOP of a pilot plant hydrogenation system. Boldface type indicates deviations that are applicable to the hydrogenation reactor node.

	A	B	C	D	E	F	G
104							
105	Item/Substep: 1.8. High Temp						
106	Guide Word	More Of		Parameter:	Temperature		Des
107	Comment						
108	Causes	Consequences	Impact			Safeguards	
109			Pers	Env	Proc		Actions
110						1.1. Operator monitoring	
111		1.1. Overheat batch, decomposing the batch. Possibility of becoming				down the heating element at I9	
112	1. Failure of temperature control module of the reactor	highly exothermic and generating non-condesable gases.	Trivially Low	Trivially Low	Medium	with heater element (I11 & I12)	1.1. Challenge interlocks during commissioning
113	2. Failure of the cooling coil during exothermic reaction	2.1. Overheat batch, decomposing the batch. Possibility of becoming highly exothermic and generating non-condesable gases.	Trivially Low	Trivially Low	Medium	2.1. Operator monitoring temperature in the reactor and stoping Hydrogen supply.	2.1. Evaluate process and conduct risk assessment on proposed chemistry
114						3.1. Inert Vessel with Nitrogen	3.1. Include in batch emergency procedure an A,B,C,D portable extinguisher
115	3. Fire in the catalyst	3.1. Lost of batch and potential fire in reactor suite.	Medium	Low	Medium	3.2. Use of wet catalyst	3.2. Add lid for charging and fire control.
116		4.1. Lost batch	Medium	Medium	Medium	handling catalyst and solvents	
117	4. External fire in the reactor suite	4.2. Peripheral damage to reactor and other equipment in suite	Medium	Medium	Medium	4.2. Sprinkler system to mitigate fire	4.1. Provide portable fire extinguisher compatible with catalyst.

FIGURE 3.3 Screenshot of a HAZOP study showing the outcome of one deviation (high temperature) in the hydrogenation reactor node of Figure 3.2.

5. *Failure mode and effects analysis (FMEA).* This is a methodical study of component failures. A FMEA normally starts with a diagram of the process, showing all the components that could fail. Typical examples are controllers, valves, and pumps. These components are listed in a table and analyzed individually to document potential causes of failure (e.g., leak, close, off), the consequence(s) of the failure (e.g., fire, explosion, spillage), the severity of the consequence(s) (high, moderate, low), the probability of the failure, the detection modes, and the mitigation measures. The last step is to perform an analysis of each component or multiple-component failure in the table and to develop a list of recommendations to prevent potential failures.
6. *Fault tree analysis.* A fault tree analysis is a quantitative assessment of all the undesirable outcomes, such as a toxic gas release or explosion, that could result from a specific initiating event. It begins with a graphic representation (using logic symbols) of all possible sequences of events that could result in an incident. The resulting diagram looks like a tree with many branches, each branch listing the sequential events (failures) for different independent paths to the top event. Probabilities (using failure rate data) are assigned to each event and then used to calculate the probability of occurrence of the undesired event. This technique is particularly useful in evaluating the effect of alternative actions on reducing the probability of occurrence of the undesired event.
7. *Event tree analysis.* An event tree shows graphically the potential results of an accident in response to a specific event or failure. An event tree analysis describes the accident sequence as the series of errors leading to an accident. Event tree analysis is useful to identify various adverse events that can occur in a complex process.

Other equivalent and complementary techniques can be used, such as computer modeling to estimate the severity of potential explosions, relative rankings, safety reviews, cause–consequence analysis, and human reliability analysis. As hazard analysis can be rather complex, with sophisticated details, the process hazard analysis methodology selected must be appropriate to the complexity of the process and must identify, evaluate, and control the hazards involved in the process. Several factors influence the decision of the type of technique, the most important being the underlying motivation for the assessment (i.e., the purpose of the assessment and what problem is being addressed). Other considerations include the information and resources available and the preferences of the analyst.

3.3.3 Inherent Safety

In Section 3.3.2 we discussed how hazards can be identified and mitigated within an industrial process. Traditionally, much emphasis has been placed on the design and implementation of additional systems to ensure the safety of a process rather than focusing on eliminating or minimizing hazardous materials and conditions. This "bolt-on" approach works, but it begs the question of whether or not there is a way to prevent the occurrence of hazardous conditions by design.

The safety of a process can be achieved by inherent (internal) and external means. Inherent safety focuses on the intrinsic properties of a process and attempts to "design out" hazards rather than trying to control hazards through the application of external protective

TABLE 3.12 Inherent Safety Principles

Principle	Description
1. Intensification	Reduce or eliminate hazardous materials or hazardous conditions.
2. Substitution	Replace hazardous materials or conditions with safer ones.
3. Attenuation or moderation	Use a hazardous material under less hazardous conditions.
4. Simplification	Design plants as simply as possible so they provide fewer opportunities for error and less equipment that can fail.
5. Limitation of effects	Change the design or the reaction conditions to reduce the severity of adverse effects.
6. Avoiding knock-off effects	Design to avoid domino effects if adverse events do occur.
7. Preventing incorrect assembly	Design equipment so that improper construction or reassembly is impossible or very difficult.
8. Making status clear	Design and use equipment that indicates clearly if it is on, off, open, shut, or assembled correctly or incorrectly.
9. Error tolerance	Design and use more robust equipment that endures poor operation or poor conditions without failing.
10. Ease of control	When possible, control by physical principles instead of add-on systems.
11. Computer control	Design simple, user-friendly, thoroughly tested software by people who understand the process.
12. Instructions	Write clear and short instructions.
13. Life cycle thinking	Consider construction and demolition phases as well as the operation phase.
14. Passive safety	If hazards cannot be avoided and there is a need to add safety measures, use passive protective measures instead of active or procedural measures.

Source: refs. 26 and 27.

systems. Trevor Kletz has postulated some basic principles of inherent safety, as presented in Table 3.12.

Inherently safer processes rely on chemistry and physics (properties of materials, quantity of hazardous materials) instead of control systems (interlocks, alarms, procedures) to protect workers, property, and environment. It would be inappropriate to talk about an inherently safe process, as an absolute definition of *safe* is difficult to achieve in this context since risk cannot be reduced to zero. However, we can talk about a process or chemical being inherently safer than other(s). Water can be an extremely hazardous chemical under certain conditions (e.g., floods), but in the context of a chemical process, water is an inherently safer solvent than other chemicals.

Inherent safety principles applied during the design stage of a process provide a larger payoff than if applied when the process has been operating for some time in industry, as it is more difficult and costlier to retrofit inherent safety concepts (i.e., elimination of a hazardous material). In the earliest phases of design, the engineer has greater freedom to consider changes in basic process chemistry or technology that can result in an inherently safer plant. Although these principles seem to be primarily common sense, they haven't been put into practice in the industrial world as widely as have engineering or procedural measures.[28] This is probably related to the fact that designing processes that are inherently safer requires a fundamental change in the way that chemists and engineers approach process development, and also requires a very close collaboration among disciplines.

A series of indexes have been developed to evaluate how inherently safer a process or plant is, including the Prototype Index for Inherent Safety(PIIS),[29] the Inherent Safety Index (ISI),[30, 31] and the i-Safe Index.[32] In the context of designing greener chemistries and processes, the principles of intensification, substitution, attenuation, simplification, and limitation of effects align particularly well with the philosophy of designing out hazards instead of having to add corrective measures to the original design.

Example 3.3 You have just been hired as the lead engineer who will be responsible for the design of the main reactor used in the manufacture of a new chemical. You are attending the implementation team kickoff meeting for this chemical process when the chemistry team leader tells you that they have not yet finalized work on the reaction. He mentions that the starting material for this chemical is very hazardous (toxic), that *N*-methylpyrrolidone is used as the solvent, and that the main production reaction is very exothermic. He also mentions that when working at the lab scale (about 50 mL), the reaction goes to completion very rapidly, but when trying the reaction in the pilot plant, there are problems with new impurities forming and the reaction takes considerably longer to go to completion. He is confident that the chemistry team can get this situation under control and he wants to start discussing the manufacturing equipment. All the experiments and pilot plant work have been conducted in low-volume batch equipment, so they are thinking that the process can use an existing 3000-gal batch reactor (about 11 m^3) that was left in the plant after a previous process moved. This existing batch reactor could be used or adapted for the main reaction, and the chemistry team wants to know your opinion about starting to scale-up to this existing piece of equipment. You became concerned about safety issues when you heard "exothermic" and "3000 gal" mentioned in the meeting and would like to make your first reactor design as inherently safe as possible. What would be your suggestion to the chemistry team?

Solution At this point you have limited knowledge of the reaction, the reaction kinetics, and the reaction thermodynamics. It all depends on the reaction and reaction conditions, but for the limited information above you could suggest exploring options that would incorporate some of the principles of inherently safer design:

Option 1: Intensification. The team could explore options of intensification, such as continuous stirred tank reactors (CSTRs), plug flow reactors (PFRs), tubular reactors, spinning disk reactors, and static mixer reactors, among others. A potential runaway reaction in about 3000 gal (11,355 liters!) of material can have significantly larger adverse consequences than if the reaction were carried out in a considerably smaller reaction system. It has been reported that the same amount of material can be manufactured in either a 3000-gallon batch reactor or a 100-gal (about 480 L) CSTR. For example, nitroglycerine was manufactured in batch reactors containing more than 1 ton of material, but newer CSTRs reduced the inventory considerably.[33]

Option 2: Attenuation and ease of control. There is a good chance that the creation of impurities that the chemistry team is observing during scale-up are related to poor mass and energy transfer, and that an alternative reactor configuration or type would probably help. Mixing in a CSTR is generally better, and continuous reactors (e.g., CSTRs, PFRs, tubular reactors) tend to have greater heat transfer surface per unit of reactor volume, thereby improving temperature control and reducing the risk of runaway reactions.

Option 3: Substitution. *N*-Methylpyrrolidone has been identified as having the hazard of potentially causing harm to a fetus. Without knowing the specifics of the reaction, the suggestion would be to explore different possibilities to use inherently safer materials: Can a different solvent be used? Does the reaction progress at all in an aqueous system? Can the product be manufactured using a different chemistry altogether?

Nonetheless, these are just starting suggestions for applying inherently safer design principles. It would appear that the chemistry team and you as a newly hired engineer still have a lot of work to do. A comprehensive understanding of the reaction kinetics, reaction mechanism, chemistry, and process conditions, among others, is imperative to ensure a truly efficient, inherently safer design.

Additional Point to Ponder Are there any potential trade-offs when striving for an inherently safer design? For example, would more severe process conditions in a smaller-scale reactor produce a greater risk than less severe process conditions in a larger reactor?

3.4 HAZARD AND RISK

A *hazard* is an inherent physical, chemical, or biochemical characteristic or property of a chemical or material that has the potential of causing harm to or adverse effects in people, property, or the environment. *Risk*, on the other hand, is the likelihood or probability that harm or an adverse effect will occur. Risk is also usually expanded to include some indication of the consequences, degree, or severity of adverse effects if they do occur: In other words, how often could the adverse effect occur, and what would be the magnitude of the damage? For example, one could characterize a human exposure risk in terms of the probability that a chemical exposure in the workplace over a period of years might result in a cancer rate of one case in 100,000 people exposed. One common way to express risk, which shows that risk is a function of both hazard and exposure, is

$$\text{risk} = f(\text{hazard, exposure})$$

Risk assessment is the characterization of the potential adverse effects resulting from human and ecological exposures to environmental hazards. Risk assessments have the purpose of providing information to risk managers, policymakers, and regulators so that the best decisions can be taken to bring the risk down to an acceptable level. The risk assessment process is complex, but it can be divided into four major steps[34]:

1. *Hazard identification* determines if a particular chemical or agent is or not causally related to a particular adverse effect.
2. *Dose–response assessment* determines the relationship between the magnitude of exposure and the probability of occurrence of the adverse effects in question.
3. *Exposure assessment* measures the intensity, frequency, or duration of a human or ecological exposure to agents that are currently in the environment, or may be present in the future.
4. *Risk characterization* estimates the incidence of an adverse effect under the conditions of exposure described in the exposure assessment. It includes the description of the risk and the uncertainties in the preceding steps.

Risk assessments can contain all or some of these steps and can include a process that may be qualitative or quantitative in nature to estimate the magnitude of risk. Quantitative risk assessment provides a degree of objectivity and a facility for ranking risks and priorities, although it is not always feasible to arrive at a numerical value of risk and does involve some degree of subjectivity, as assessments rely to a certain extent on past events and/or experience.

The term *risk assessment* has been used to describe the likelihood of a vast array of adverse effects: explosions, equipment failure, workplace injuries, natural catastrophes (e.g., tornadoes), diseases, death due to natural causes (e.g., diabetes), lifestyles (e.g., smoking), personal activities (e.g., skydiving) and many others; it is not used only to describe the probability of adverse response to chemicals.[19]

Example 3.4 Testing emissions from an incinerator indicates potentially high levels of dioxins.

(a) Why is this a concern (i.e., what is the hazard)?
(b) Explain how you would estimate the risk related to these emissions.

Solution

(a) One of the main hazard concerns is cancer in adults. Several studies suggest that workplace exposure to high levels of dioxin carries an increased risk of cancer. Also, based on animal studies, there is some concern of reproductive or developmental hazards with chronic exposure to low levels of dioxin. Effects of exposure to a large amount of dioxin include chloracne (a severe skin disease with acnelike lesions), skin rashes, skin discoloration, excessive body hair, and possibly, mild liver damage.[35]

(b) The risk would depend on the hazard we are evaluating and the magnitude of the exposure. For example, if we are interested in evaluating the risk of cancer, the general equation for estimating lifetime excess cancer risk is

$$\text{risk} = 1 - \exp\left(-q^* \cdot \text{LADD}\right) \tag{3.4}$$

where

$$\text{LADD (lifetime average daily dose)} = \text{exposure media concentration} \times \text{contact rate} \times \text{contact fraction} \times \frac{\text{exposure duration}}{\text{body weight} \times \text{lifetime}}$$

and q^* for the toxic equivalents (TEQs) of dioxin is 0.156 ng/kg $\cdot$ day^{-1}.

Additional Points to Ponder What does q^* represent? Exactly what is a TEQ?

We can also use quantitative risk assessment when we are interested in estimating possible adverse effects of human actions on the nonhuman environment, known as *environmental risk assessment*. Environmental risk assessments can be either predictive or retrospective. *Predictive risk assessments* estimate risks from proposed future

TABLE 3.13 Phases of Environmental Predictive Risk Assessments

Hazard Definition	Risk Quantification	Risk Management
Source description (e.g., pollutant fate properties)	Exposure assessment	Course of action decision
	Effects assessment	Mitigation solutions
Environment description (e.g., fate and transport modeling)	Risk characterization	Risk vs. benefit
Endpoint definition (e.g., organisms potentially affected)		

actions (e.g., marketing a new chemical, operating a new process), whereas *retrospective risk assessments* estimate the risks posed by actions that had already occurred that might have ongoing consequences (e.g., spills, existing effluents). Predictive risk assessments involve three major phases: hazard definition, risk quantification, and risk management (Table 3.13).

3.5 INTEGRATED PERSPECTIVE ON ENVIRONMENT, HEALTH, AND SAFETY

In previous sections of the chapter we covered high-level environmental, health, and safety issues and aspects. However, to truly design safer, greener processes, we need to resist the temptation of viewing these issues in a compartmentalized fashion. By changing a part of the process with the aim of reducing an environmental impact, we can adversely affect the overall safety of the process, or vice versa. For example, by attempting to use a nonvolatile solvent to avoid overpressurization in a runaway reaction, we could be introducing a more toxic material. Another example is the use of chlorofluorocarbons as refrigerants; although inherently safer than ammonia because they are less toxic and explosive, but they have a strong environmental impact as ozone-depleting agents, and some are greenhouse gases.

There are also trade-offs among various environmental, health, or safety burdens associated with product production and/or use. For example, the use of hydrofluoroalkanes (HFAs) as replacements for CFC propellants in metered-dose inhalers used in the treatment of asthma has reduced the ozone-depleting impact of these medicinal products but has posed an interesting problem in that HFAs are potent global warming gases. As is frequently the case, moving to one solution by eliminating a significant impact in one environmental category can lead to a different but potentially no less concerning impact in another environmental category.

In assessing the greenness or sustainability of a process or chemistry, and to reach the best decision regarding the option or options that offer the highest overall benefit or the lowest level or overall risk, it is imperative to have an integrated view of all the potential hazards and risks of a certain activity.

PROBLEMS

3.1 Estimate the acidification potential of acetic acid, hydrogen bromide, hydrogen cyanide, hypochlorous acid, and sulfuric acid.

3.2 Estimate the eutrophication potential of pyridine.

3.3 Figure P3.3 represents a conventional distillation system. Develop a series of recommendations to apply inherent safety principles in the design of this unit operation.

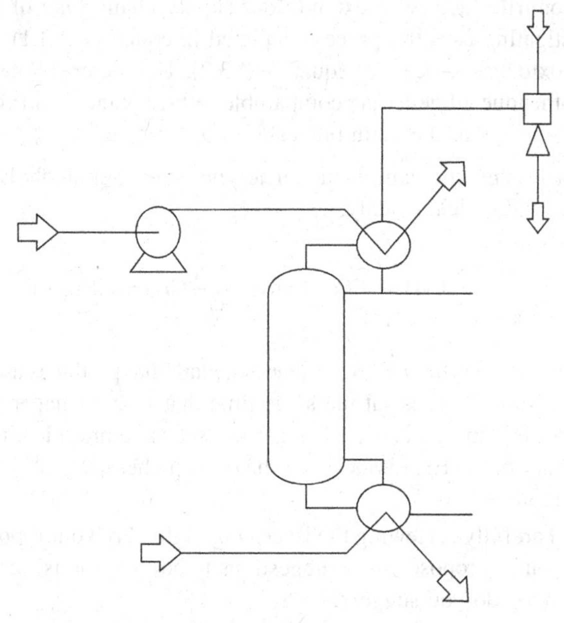

FIGURE P3.3

3.4 Provide three examples where substitution of hazardous materials has resulted in inherently safer designs or chemistries. Why are they safer alternatives?

3.5 You are in the solvent selection stage of designing a chemical reaction. Dimethylformamide, dichloromethane, methyl ethyl ketone, and toluene seem to work well. You have looked at the properties of the solvents and have found the EHS information needed:

(a) Using the OEL of the solvent as a surrogate of health hazard and the vapor pressure as a surrogate for exposure potential, which solvent would you recommend from a health perspective?

(b) Using the flash point as a surrogate of flammability hazard, which solvent would you recommend from a safety perspective?

(c) Are there additional factors that would need to be considered when selecting among these solvents?

3.6 Acrylonitrile can be manufactured by reacting acetylene with hydrogen cyanide:

$$C_2H_2 + HCN \rightarrow CH_2 = CHCN \qquad (P3.1)$$

The ammoxidation process uses propylene and ammonia:

$$CH_2 = CH - CH_3 + NH_3 + 1.5O_2 \rightarrow CH_2 = CHCN + 3H_2O \qquad (P3.2)$$

(a) Identify and compare the environmental, health, and safety hazards of each process.

(b) You inherited a handsome amount of money from a rich distant relative, the only condition being that you must invest the money in the options that cause the least impact on the environment, health, and safety. At this point, you need to buy an acrylonitrile factory to extend your supply chain. One of the plants you are investigating uses the process depicted in equation (P3.1); the other uses the ammoxidation process of equation (P3.2). The factors related to cost, location, and other due diligence are comparable. Which plant would you buy based on the EHS impacts of the main process?

3.7 The *Reppe process* for manufacture of acrylic esters uses alcohols, acetylene, carbon monoxide, and a nickel catalyst:

$$C_2H_2 + CO + ROH \xrightarrow[HCl]{Ni(CO)_4} CH_2 = CHCO_2R$$

(a) You are hired as the EHS manager of a plant that produces an isopropyl acrylate by the Reppe process at the same time that your manager is hired. The plant director is also new and asks you as a first task to provide him with a high-level summary of the EHS issues with the current chemistry. What do you include in your report?

(b) After carefully reviewing the issues highlighted in your report, the site director asks you to propose some suggestions to address the issues and minimize the risk. What do you suggest?

3.8 A manufacturing process produces the potential pollutants shown in Table P3.8 prior to any treatment processes. Estimate the photochemical ozone creation potential, global warming potential, acidification potential, eutrophication potential, and total organic carbon for this manufacturing process.

TABLE P3.8 Potential Pollutants (kg/h)

CAS No.	Name	Gas Flow	Liquid Flow	Solid Flow
106-89-8	Epichlorohydrin	10.3	19.0	0
1310-73-2	Sodium hydroxide	0	0.768	0
56-81-5	Glycerin	0	5.02	0
67-66-3	Chloroform	0.238	2.40	0
7647-14-5	Sodium chloride	0	7.50	0

3.9 Identify other environmental, health, and safety issues related to the compounds being emitted in Problem 3.8.

3.10 Identify the environmental and safety issues of the materials of Example 3.2.

3.11 Develop recommendations to apply inherent safety principles for Example 3.2 based solely on the materials used.

3.12 A production process uses the materials listed in Table P3.12.

TABLE P3.12 Process Materials

CAS No.	Name
64-19-7	Acetic acid
67-64-1	Acetone
141-78-6	Ethyl acetate
142-82-5	Heptane
109-66-0	Pentane
7664-41-7	Anhydrous ammonia
471-34-1	Calcium carbonate
56-81-5	Glycerol
60-33-3	Linoleic acid
1310-73-2	Sodium hydroxide
61-90-5	*S*-Leucine
7664-93-9	Sulfuric acid

(a) Identify the environmental, health, and safety hazards of the materials using data available through material safety data sheets.

(b) With the green chemistry principles in mind, what would you propose to improve the process?

3.13 Lorber et al. performed a screening risk assessment of the indirect impacts from dioxin emissions from a waste incinerator in Ohio.[36] They estimated the lifetime average daily doses listed in Table P3.13.

(a) Explain why they chose the exposure pathways presented in Table P3.13.

(b) Using the risk equation in Example 3.4, estimate the excess risk for each exposure pathway.

(c) What do the numbers you calculated mean?

TABLE P3.13 Lifetime Average Daily Doses

Exposure Pathway	Lifetime Average Daily Dose, Excess Cancer Risk (ng/kg·day)
Soil dermal contact	6×8^{-8}
Vegetable ingestion	1×10^{-5}
Inhalation	6×10^{-6}
Beef ingestion	1×10^{-3}
Milk ingestion	5×10^{-4}

3.14 Equation (3.4) was an estimate of cancer risk for a set of compounds (dioxins). Investigate how risk is evaluated for noncancer toxicants.

REFERENCES

1. World Meteorological Organization. WMO Statement on the Status of the Global Climate in 2005. http://www.wmo.int/pages/index_en.html.

2. National Climatic Data Center. Climate of 2006—June in Historical Perspective http://www.ncdc.noaa.gov/oa/climate/research/2006/jun/jun06.html.
3. U. S. Environmental Protection Agency. Climate Change. 2006, http://www.epa.gov/climatechange/basicinfo.html#science.
4. Pew Center on Global Climate Change. Climate Change 101: Understanding and Responding to Global Climate Change. 2006. http://www.pewclimate.org/docUploads/Climate101%2DFULL%5F121406%5F065519%2Epdf.
5. Intergovernmental Panel on Climate Change. Climate Change 2001: Synthesis Report; Summary for Policymakers. 2001. http://www.ipcc.ch/pub/un/syreng/spm.pdf.
6. World Meteorological Organization. The state of greenhouse gases in the atmosphere using global observations through 2005. *Green House Gas Bull.*, Nov. 2006, 2(1) http://www.wmo.ch/web/arep/gaw/ghg/ghg-bulletin-en-11-06.pdf.
7. Houghton, J. T., Callander, B. A., Varney, S. K., Eds. *IPCC Climate Change 1990*. Cambridge University Press, New York,1990.
8. Intergovernmental Panel on Climate Change. *IPCC Climate Change 2001: The Scientific Basis*. Cambridge University Press, New York, 2001. Available at http://www.grida.no/climate/ipcc_tar/wg1/248.htm.
9. Crutzen, P. J. The influence of nitrogen oxides on atmosphere ozone content. *Q. J. R. Meteorol. Soc.*, 1970, 96, 320–325.
10. Molina, M. J., Rowland, F. S. Stratospheric sink for chlorofluoromethanes: chlorine catalysed destruction of ozone. *Nature*, 1974, 249, 810–814.
11. United Nations Environment Programme, Ozone Secretariat. *The Montreal Protocol on Substances That Deplete the Ozone Layer*. UNEP, Nairobi, Kenya, 2000. Available at http://hq.unep.org/ozone/Montreal-Protocol/Montreal-Protocol2000.shtml.
12. Fishman, J., Ramanathan, V., Crutzen, P. J., Liu, S. C. Tropospheric ozone and climate. *Nature*, (1979), 282(5741), 818–820.
13. Fishman, J., Solomon, S., Crutzen, P. J. Observational and theoretical evidence in support of a significant in situ photochemical source of tropospheric ozone. *Tellus*, 1979, 31, 432–446.
14. U. S. Environmental Protection Agency. *Good Up High, Bad Nearby*. EPA 451-K-03-001. U.S. EPA, Washington, DC. 2003. Available at http://www.epa.gov/ozone/ods.html.
15. European Chemical Industry Council. Responsible Care: Health, Safety and Environmental Reporting Guidelines. 1998. http://www.cefic.be/activities/hse/rc/guide/01.htm.
16. De Leeuw, R., Fam, A. M. (1993) *Assessment of the Atmospheric Hazards and Risks of New Chemicals: Procedures to Estimate "Hazard Potentials."* RIVM Rapport 679102017. National Institute of Public Health and Environmental Protection, Amsterdam, The Netherlands, 1993, S.1324ff, 22 p, in English. *Chemosphere*, 1993, 27(8), 1313–1328.
17. MacKay, D., Shiu, W. Y., Ma, K., Lee, S. C. *Handbook of Physical–Chemical Properties and Environmental Fate for Organic Chemicals*, 2nd ed. CRC Press, Boca Raton, FL, 2006.
18. Cronin, M., Livingstone, D., Eds. *Predicting Chemical Toxicity and Fate*.CRC Press, Boca Raton, FL, 2004.
19. Organisation for Economic Co-operation and Development. *OECD Guidelines for the Testing of Chemicals*, 13th addendum, OECD Publishing, Paris, 2001.
20. Fan, A. M., Chang, L. W., Eds. *Toxicology and Risk Assessment: Principles, Methods and Applications*. Marcel Dekker, New York, 1996.
21. U.S. Department of Transportation, Pipeline and Hazardous Materials Safety Administration. International Standards. http://www.phmsa.dot.gov/hazmat/regs/international, accessed Sept. 27, 2009.
22. Paustenback, D. J., Ed. *Human and Ecological Risk Assessment: Theory and Practice*. Wiley, Hoboken, NJ, 2002.

23. U. S. *Occupational Safety and Health Administration. Occupational Safety and Health Standards, Air Contaminants.* 29 CFR 1910.1000. OSHA, Washington, DC, April 6, 2006.

24. U. S. Occupational Safety and Health Administration. *Process Safety Management*. OSHA 3121. OSHA, Washington, DC, 2000 (reprint).

25. American Institute of Chemical Engineers, Center for Chemical Process Safety. *Guidelines for Hazard Evaluation Procedures,* 2nd ed. AIChE, New York, 1992.

26. Kletz, T. A. Inherently safer plants. *Plant Oper. Prog.*, 1985, 4 (3), 164–167.

27. Kletz, T. A. *Process Plants: A Handbook of Inherently Safer Design.*CRC Press, Boca Raton, FL, 1998.

28. Amyotte, P. R., Khan, F. I., Dastidar, A. G. Solids handling: reduce dust explosions the inherently safer way. *Chem. Eng. Prog.*, Oct. 2003, pp. 36–43.

29. Edwards, D. W., Lawrence, D. Assessing the inherent safety of chemical process routes: Is there a relation between plant costs and inherent safety? *Trans. IChemE, B*, 1993, 71, 252–258.

30. Heikkilä, A.-M., Hurme, M., Järveläinen, M. Safety considerations in process synthesis. *Comput. Chem. Eng.*, 1996, 20, S115–S120.

31. Heikkilä, A.-M. Inherent Safety in Process Plant Design. D.Tech. dissertation, VTT Publications 384. Technical Research Centre of Finland, Espoo, Finland, 1999. http://www.inf.vtt.fi/pdf/publications/1999/P384.pdf.

32. Palaniappan, C., Srinivasan, R., Tan, R. Selection of inherently safer process routes: a case study. *Chem. Eng. Process.*, 2004, 43, 647–653.

33. American Institute of Chemical Engineers, Center for Chemical Process Safety. *Guidelines for Engineering Design for Process Safety*. AIChE, New York, 1993.

34. National Research Council. *Risk Assessment in the Federal Government: Managing the Process*. National Academies Press, Washington, DC, 1983.

35. U.S. Environmental Protection Agency. Questions and Answers About Dioxins. Jan. 2003; updated Oct. 2003, Oct. 2004, July 2006. http://www.cfsan.fda.gov/~lrd/dioxinqa.html.

36. Lorber, M., Cleverly, D., Schaum, J. A screening level risk assessment of the indirect impacts from the Columbus waste to energy facility in Columbus, Ohio. In *Solid Waste Management: Thermal Treatment and Waste-to-Energy Technologies*, VIP-53. National Center for Environmental Assessment, U.S. EPA, Washington, DC,1996, pp. 262–278.

4

HOW DO WE KNOW IT'S GREEN? A METRICS PRIMER

What This Chapter Is About This chapter is about metrics for green chemistry. This is an important thing to think about because many scientists and engineers make many claims about how their chemistry or technology is "greener" or "cleaner" than are existing chemistries or technologies. By looking at a given approach from a systems-wide, life cycle-based perspective, you will be more likely to derive a fair and accurate measure of a new chemistry, process, or product's comparative "greenness."

Learning Objectives At the end of this chapter, the student will be able to:

- Understand metrics that are commonly used in measuring what is "green."
- Understand which metrics are best for chemistry, processes, and products.
- Make meaningful comparisons between different chemistry or process options.

4.1 GENERAL CONSIDERATIONS ABOUT GREEN CHEMISTRY AND ENGINEERING METRICS

Much has been written about the characteristics of metrics, or what constitutes a good metric.[1–4] It is generally agreed that metrics must be clearly defined, simple, measurable, objective rather than subjective, and must ultimately drive the desired behavior.[5] A considerable amount has been written about assessing the economic or commercial viability of chemical processes.[6,7] Some things have also been written about how to assess chemistries,

Green Chemistry and Engineering: A Practical Design Approach, By Concepción Jiménez-González and David J. C. Constable

processes, or products comparatively from a green chemistry or engineering perspective,[8–15] and there have been several reviews of various metric systems.[16–19] All of these have value in helping to answer the general question of which metrics are best, but it can ultimately be confusing to decide which ones should be used for any given comparison, situation, or context. It is quite clear to anyone in the green chemistry and green engineering field that there are immense difficulties with most metric discussions and with the application of metrics, especially in finding the few key metrics that have the greatest impact or potential to "green" any given process.

In general, our difficulty is perhaps best seen as a boundary problem, that is, where does one draw the boundary for a consideration of how green a process is? In the case of a given chemistry, is the boundary drawn to include only reactants, or does it include everything in the flask, or does it include all the materials required to make what is in the flask? In the case of a product, are we concerned with the final step to produce the product, or with all the steps that led to the final step? At this point, it is our opinion that a guiding principle of green metrics must be the comprehensive application of a life cycle approach if one is to evaluate appropriately the greenness of chemistry, processes, or products. You would be well advised to consult Chapter 16 and have a look at the literature to develop a greater understanding of life cycle inventory and assessment methodologies.[20–23]

Another practical consideration for green metrics is the tendency for many not to take a systems, systems-wide, or holistic view of a chemical synthesis and the associated processes. For both undergraduate- and graduate-level training in chemistry, there is almost no practical training in the industrial processes that are based on the chemistries that are studied in the laboratory. In the case of engineering students and/or those early in their careers, when reviewing a process flow and instrumentation diagram, it is probably the case that you do not really know what a "typical" chemical or petrochemical plant might look like. It would take some time in a plant to understand that processes are generally, but not always, a closely linked set of unit operations carried out across as small a space as possible and in as short a time as possible. Changes in one particular unit operation, say a reactor, would probably bring a cascade of effects in another unit operation, such as the separation train. It would be a mistake to have a set of metrics that do not take into consideration the closely knit network of cause-and-effect relationships that comprise synthetic chemical processes.

A systems-wide view will also require you to collect more than one type of metric. A univariate approach is popular because it is simple, but this simplicity often leads to erroneous conclusions. The world of green chemistry and engineering is a complex one and requires a multivariate view of a system. There is an underlying complexity in most chemical operations and it is hard work to find just a few key metrics to describe a given system or operation. When evaluating the greenness of a chemistry, process, or product, it is not uncommon to encounter trade-offs among metrics. For example, a change in a critical process parameter that yields an improved environmental profile can have an adverse effect on the overall safety of the process. It can also be the case that a change to reduce one environmental impact can increase the impact in another area. A systems approach, with the right set of carefully chosen metrics, should allow you to evaluate trade-offs and make appropriate judgments about the greenness of a chemistry, process, or product.

This leads nicely into another general principle about metrics—one size does not fit all. Metrics must be adapted for their context and evaluated continuously as to their utility, applicability, appropriateness, and so on. Metrics should also be tested and validated to ensure that they are driving the desired behaviors and objectives successfully. They must also be easily understood and accepted by those who have an interest in what is being evaluated.

Another general principle for good green chemistry and engineering metrics is that they should promote strategic analysis and continuous improvement. If the metrics are being collected but not evaluated on a regular basis, and decisions based on the metrics results are not made, there is no point in collecting them. This may seem to be an obvious point, but metrics are not always questioned, assessed, evaluated, and evolved routinely to help make strategic decisions or make them more useful.

4.2 CHEMISTRY METRICS

4.2.1 Selected Metrics Used in the Past

Over the past five to 10 years, a number of metrics have been proposed as chemists have become more aware of green chemistry and the need to change the art and science of chemical syntheses. However, it is useful to pause and take a moment to consider how chemists generally evaluate the success of their reactions in terms of efficiency (i.e., how much of the chemicals they start out with ends up in the molecule they are trying to make). The most obvious and prevalent measure of chemical efficiency for a chemist for many years has been yield. For any given reaction;

$$A + B \rightarrow C$$

$$\begin{aligned}\text{theoretical yield} &= (\text{[A]moles of limiting reagent}) \times (\text{stoichiometric ratio: [C/A]} \\ &\quad \text{desired product/limiting reagent}) \times (\text{FW of desired product[C]})\end{aligned}$$

$$\text{percentage yield} = \frac{\text{actual yield}}{\text{theoretical yield}} \times 100$$

Example 4.1 Calculating Yield 0.90 g of *n*-pentanol is combined with 1.55 g of NaF, and 4 g of sulfuric acid is added. What is the theoretical yield? If the actual yield is 0.80 g of pentyl fluoride, what is the percentage yield?

Solution

$$\text{n-pentanol (OH)} + NaF + H_2SO_4 \longrightarrow \text{pentyl fluoride (F)} + NaHSO_4 + H_2O$$

n-pentanol	NaF	H_2SO_4	pentyl fluoride
0.9 g	1.55 g	4 g	0.92 g (theoretical yield)
0.010 mol	0.037 mol	0.049 mol	0.010 mol (theoretical yield)

The *n*-pentanol is the limiting reagent, so

$$\begin{aligned}\text{theoretical yield} &= (0.9\text{g pent.})(\text{mol pent.}/90\text{ g pent.})(\text{mol pentyl F/mol pent.}) \\ &\quad \times (92\text{ g pentyl F/mol pentyl F}) \\ &= 0.92\text{g pentyl F}\end{aligned}$$

$$\begin{aligned}\text{percentage yield} &= \frac{\text{actual yield}}{\text{theoretical yield}} \times 100\\ &= \frac{0.8\text{g}}{0.92\text{g}} \times 100\\ &= 87\%\end{aligned}$$

As is readily apparent, yield does not take into account other materials that participate in the reaction but which do not end up in the final molecule. In the case of NaF, there is an almost fourfold stoichiometric excess, and in the case of the sulfuric acid, there is nearly a fivefold excess in the reaction mixture. Now in this reaction, the amounts of NaF and H_2SO_4 are not optimized, so there is an opportunity to reduce the amount of the reagents that are used. However, there is still going to be residual starting material, product, and $NaHSO_4$ that will end up as waste. Several chemists have made an effort to account for these other materials.

Additional Points To Ponder Do you think that a metric such as yield will drive businesses toward sustainable practices? Is yield a good metric for green chemists to use?

Effective Mass Yield In 1999, Hudlicky et al.[24] proposed a metric known as *effective mass yield*. This was defined as "the percentage of the mass of desired product relative to the mass of all nonbenign materials used in its synthesis." Or, stated mathematically,

$$\text{effective mass yield} = \frac{\text{mass of products}}{\text{mass of nonbenign reagents}} \times 100$$

This metric is an attempt to define yield in terms of that proportion of the final mass (i.e., the mass of the product that is made from nontoxic materials). It should be noted that the introduction of reagent and reactant toxicity was an extremely important consideration in determining what is "green," and it is something that is absent from traditional yield measures. Although Hudlicky et al. define *benign* (i.e., "those by-products, reagents or solvents that have no known environmental risk associated with them: for example, water, low-concentration saline, dilute ethanol, autoclaved cell mass, etc.") the explanation suffers from a lack of definitional clarity.

Defining *nonbenign* is difficult in practice, especially when you are working with complex reagents and reactants that have limited environmental or occupational toxicity information. Until human toxicity and ecotoxicity information or believable estimates are routinely available for the wide diversity of chemicals used, it would be difficult to use this metric for most synthetic chemical operations. In addition, depending on the situation, saline, ethanol, and autoclaved cell mass all have environmental impacts of one kind or another that would have to be evaluated and addressed.

E-Factor A second and earlier metric, the *E*-factor, was proposed by Roger Sheldon[25,26] in the early 1990s and is defined as follows:

$$E\text{-factor} = \frac{\text{total waste(kg)}}{\text{kg product}}$$

It is uncertain whether this metric includes or excludes water, but it could be used to describe either case. This metric is relatively simple and easy to determine and has done a good job at drawing attention to the quantity of waste that is produced for a given quantity of product. Because Sheldon had worked in industry and had consulted with different industrial sectors once in academia, he was able to produce a comparison of the relative wastefulness of the various parts of the chemical processing industries, including industries as diverse as petrochemicals, specialities, and pharmaceuticals. This metric has been used by industry and has spurred efforts to reduce waste.

The metric may, however, be subject to a lack of clarity, depending on how waste is defined; it is once again a question of where one draws the boundaries in any given comparison. For example, is waste that passes over the fence the only waste considered? Is waste that is produced as a result of emissions treatment (e.g., acid gas scrubbing, pH adjustment in wastewater treatment plants) included? Is waste that is produced as a result of energy use (e.g., heating or cooling reactions, abatement technology) included? Finally, is waste solvent that is passed on to a waste handler to be burned in a cement kiln included? From an operational perspective, these types of questions complicate use of this metric.

It is also generally true that drawing attention to waste does not always drive chemists to think about what might be done to avoid producing wastes. Instead, the tendency on the part of many chemists is to focus on "good science," novelty (for patent protection), and precedent (using what they know works). By focusing solely on these areas, they ignore the waste generated in a chemical reaction or process and leave it to others to focus on waste treatment at a later date.

Atom Economy When developing chemical processes, chemists often focus on maximizing selectivity and yield. We reviewed briefly above the concept of chemical yields, so let's spend a moment reviewing a bit about how chemists think about selectivity. This is an important digression since much of what is not green about chemistry has to do with a lack of chemical selectivity in the collection of synthetic tools used by chemists. Following are some important definitions of chemical selectivity.

- *Chemoselectivity:* a chemoselective reagent that reacts with one functional group (e.g., a halide, R–X) but not another (e.g., a carbonyl group, R–C=O).
- *Enantioselectivity:* a chemical reaction in which an inactive substrate (a molecule of interest) is converted selectively to one of two enantiomers. Enantiomers are *isomers* (compounds with the same numbers and types of atoms but possessing different structures, properties, etc.) that differ only in the left and right handedness of their orientations. Enantiomers rotate polarized light in equal but opposite directions and react at different rates with other chiral compounds.
- *Stereoselectivity:* any reaction in which only one of a set of stereoisomers (isomers whose relative spatial positions of atoms or functional groups differ) is formed exclusively or predominantly.
- *Regioselectivity:* when a reaction can potentially give rise to two or more structural isomers (e.g., R–O–C=N or R–N=C=O) but actually produces only one.

As you can see, there are several important concepts around selectivity that inform the discussion and debate around what is green. Clearly, the more generally chemoselective a reaction is, the better. In addition, when working with chiral molecules (i.e., those that have stereochemical centers), chemical selectivity becomes even more important. In these instances, we not only have to worry about reacting with a particular type of bond or functional group but have to do it in such a way that the only bond of interest that forms preserves or creates the desired isomer. In practice, there is still a considerable amount of chiral chemistry that suffers from a lack of selectivity (stereo-, regio-, and enantioseleclectivity) and in many cases more than 50% of the starting material ends up as waste. More about this in subsequent chapters.

Example 4.2 Calculating Reaction Selectivity—Chemoselectivity

OH, CH_3 + OMe —catalyst→ OH + CH_3OH

A new heterogeneous alkylation catalyst (a zeolyte) is discovered and a chemist desires to test it on a simple system. She combines *p*-cresol (15.5 g, 0.141 mol) with methyl *tert*-butyl ether (MTBE, 12.4 g, 0.141 mol) and the zeolite catalyst (2 wt%) and heats them at 125°C for 2 h. After cooling, the products were identified by gas chromatograpy. The main product was 2-*tert*-butyl-*p*-cresol (2-*t*-BpC) (10.5 g), with 7.95 g of *p*-cresol remaining unreacted. What is the selectivity of this particular catalyst?

Solution

$$\text{theoretical yield} = 15.5\text{g } p\text{-cresol} \times \frac{166\text{g/mol 2-}t\text{-BpC}}{110\text{g/mol } p\text{-cresol}} = 23.4\text{g 2-}t\text{-BpC}$$

$$\%\,\text{yield} = 100 \times \frac{10.5\text{g actual}}{23.4\text{g theoretical}} = 44.9\%$$

$$\%\,\text{selectivity} = 100 \times \frac{\text{yield of desired product}}{\text{fraction of substrate converted}}$$

$$= 100 \times \frac{12}{23.4[(15.5-6.95)/15.5]}$$

$$= \frac{12}{12.9} = 93\%$$

Additional Point to Ponder Do you think atom economy is useful as a stand-alone metric? Why or why not?

Based on Example 4.2, you might be prone to conclude that the reaction isn't so bad if your measure of "bad" was based on selectivity. The zeolyte catalyst in this case is considered to be fairly selective in transforming *p*-cresol into 2-*t*-butyl-*p*-cresol since no other by-products are formed. You could imagine a system where perhaps the *p*-cresol is recycled back into a reactor and the yield is increased, but in this particular batch process, the yield is a bit disappointing. What else might you say about how good a reaction this is?

In recent years, another metric that some chemists have been considering is *atom economy*,[27] a term introduced by Barry Trost in an attempt to prompt synthetic organic chemists to pursue greener chemistry. At its simplest level, atom economy is a calculation of how much of the reactant remains in the final product. The final product applies equally to a single chemical transformation, a series of chemical transformations in a single stage of a multistage synthetic route, or to the entire route to a final product. The method of calculating the atom economy is kept deliberately simple by making certain key assumptions: It ignores reaction yield and the usual molar excesses of reactants, and it does not account for solvents and reagents. Before discussing the pros and cons of atom economy, further clarification will be useful.

4.2.2 Key Assumptions About Atom Economy

Reactants A *reactant* is a substance some part of which is incorporated into a reaction product, although not necessarily into the final product. The process of calculating atom economy may be simplified by considering only key reactants. For example, catalysts used in stoichiometric quantities, or the acid or base used for hydrolysis, are considered to be reactants. These examples are in contrast to common inorganic reagents, even when used in stoichiometric quantities (e.g., potassium carbonate in a Williamson ether formation), which have been ignored. Inorganic reagents and/or other materials are not included in the calculations as long as at least two other reacting substances are identified.

Reactants also include those materials incorporated in a reaction intermediate. Even if no part of a reactant is present in the final product itself (e.g., in the case of addition and removal of a protecting group), it was part of an intermediate and is therefore included in the calculation. An example would be an acid chloride formed in situ during an N-acylation reaction. The chlorinating agent would be included in the calculation even if the chlorination were not a distinct step in the process. Similarly, in a Vilsmeier formylation, since the reacting species is a chemical complex formed from the reaction between dimethylformamide and oxaloyl chloride, both would be included in the calculation.

Stoichiometry *Reagent stoichiometry* (i.e., using an excess of either or both reactants to maximize reaction yield or selectivity) is not considered in the calculation of atom economy. *Reaction stoichiometry*, on the other hand, has been taken into account. Thus, when two molecules of one substance combine with a single molecule of another to form a new molecule (either a reaction or process intermediate), the relevant ratio would be used.

Working with Chiral Selectivity (i.e., Regio-, Stereo-, or Enantioselectivity) In calculating atom economy for syntheses that employ a resolution step (i.e., a step to obtain the desired isomer), the reaction stoichiometry needs to be adjusted to account for the fact that some portion of the mass will be discarded as the unwanted enantiomer. This includes those cases where the resolving agent is in a 1 : 1 or 2 : 1 ratio with respect to the desired enantiomer, or 1 : 2 as in the case where the desired enantiomer is difunctional.

How Atom Economy Is Calculated For a generic reaction,

$$A + B \rightarrow C$$

$$\text{atom economy} = \frac{\text{m.w. of product C}}{\text{m.w. of A} + \text{m.w. of B}} \times 100$$

where m.w. represents the molecular weight. The calculation considers only the reactants used and ignores the intermediates that are made in one stage and consumed in the next. *Because of this, it is not possible to multiply the atom economy of each stage to give an overall process atom economy*. Process atom economy must be calculated as follows: For a generic linear synthetic process,

$$\text{(1)}\quad A+B \rightarrow C$$
$$\text{(2)}\quad C+D \rightarrow E$$
$$\text{(3)}\quad E+F \rightarrow G$$

$$\text{atom economy} = \frac{\text{m.w. of product G}}{\text{m.w. of A} + \text{m.w. of B} + \text{m.w. of D} + \text{m.w. of F}} \times 100$$

Processes with two or more separate branches are treated analogously by taking into account all of the reactants but none of the intermediates. Thus, for the branched synthetic process

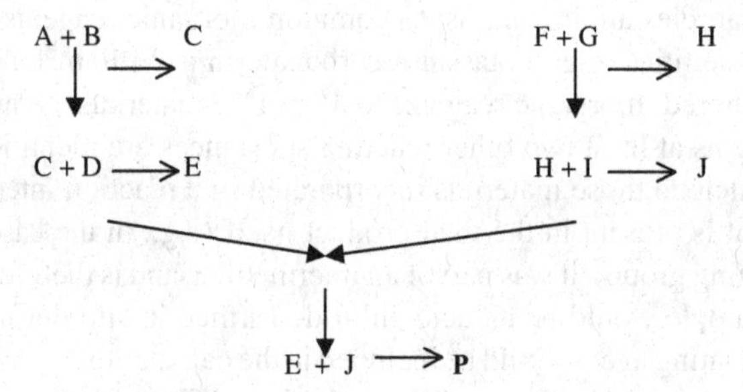

where C, E, H, and J are intermediates and E and H are coupled in the final step, atom economy is calculated as follows:

$$\text{atom economy (\%)} = \frac{\text{FW } P}{\Sigma(\text{FW A, B, D, F, G, I})} \times 100$$

Example 4.3 Calculate atom economy for Example 4.1.

Solution You will recall the reaction in Example 4.1:

OH + NaF + H_2SO_4 → F + $NaHSO_4$ + H_2O

FW: 90 42 82 92 103 18

$$\text{atom economy} = \frac{92}{90+42} \times 100$$

$$= 70\%$$

Additional Point to Ponder Do you think atom economy is useful as a stand-alone metric? Why or why not?

Because atom economy makes the assumptions that it does and includes only those materials that are reactants and products while ignoring stoichiometry and yield, another metric has been proposed that includes atom economy stoichiometry (the actual molar masses that are used in the reaction) and yield. This measure is known as *reaction mass efficiency*.

4.2.3 Reaction Mass Efficiency

When calculating reaction mass efficiency, atom economy (AE), yield, and the stoichiometry of reactants are included. RME is the percentage of the mass of the reactants that remain in the product. There are two ways to calculate RME. For a generic reaction A + B → C,

$$\text{reaction mass efficiency} = \frac{\text{m.w. of product C}}{\text{m.w. of A} + (\text{m.w. of B} \times \text{molar ratio B/A})} \times \text{yield}$$

or, more simply,

$$\text{reaction mass efficiency} = \frac{\text{mass of product C}}{\text{mass of A} + \text{mass of B}} \times 100$$

This particular measure is thought to be one step closer to determining a bit more accurately the efficiency of a given chemical reaction, since it includes stoichiometry and yield. Why would we want to do this? Clearly, some reactions require significant molar excesses to drive the reaction to completion. In addition, because of the problem of poor chemical selectivity in all its forms, chemists often end up with yields that are considerably less than 100%, especially when seeking to make chiral molecules.

So if we wanted to be a bit more accurate in accounting for all the materials it takes to make a particular molecule, what might be missing from the equations above? As is readily apparent in the examples above, there are reagents and, frequently, solvents that promote the reaction and/or provide a means for heat and mass transfer. The *E*-factor is a good measure perhaps, but it does suffer from focusing on waste and not on the product (i.e., it looks at the end of the process, not at the beginning). For this reason, a different metric that is almost the same as the *E*-factor has been proposed. It is known as *mass intensity*.

4.2.4 Mass Intensity and Mass Productivity (Efficiency)

Mass intensity has been discussed elsewhere[4] and is defined as follows:

$$\text{mass intensity(MI)} = \frac{\text{total mass in reaction vessel(kg)}}{\text{mass of product(kg)}}$$

It may also be useful to compare MI with the *E*-factor:

$$E\text{-factor} = \text{MI} - 1$$

Mass intensity takes into account the yield, stoichiometry, solvent, and reagent(s) used in a reaction mixture, and expresses this on a mass basis rather than as a percentage. In the ideal situation, MI would approach 1. Total mass includes everything that is put into a reaction vessel *with the exception of water* (i.e., reactants, reagents, solvents, catalysts, etc.). Total mass includes all mass used in acid, base, salt, and organic solvent washes, in organic solvents used for extractions, in crystallizations, or for solvent switching.

Water has initially been excluded from the equation above since in many processes it skews mass data. Including water in mass metrics can be a somewhat contentious issue in many instances. Water by itself generally does not constitute a significant environmental impact. However, in the case of highly purified water there are generally significant life cycle impacts related to the chemicals and equipment used to purify the water. This is especially true for such industries as the semiconductor industry, pharmaceuticals, and some parts of the fine-chemical industry. There is also the problem of mixed aqueous–organic reaction mixtures, where separations and subsequent disposal of wastes can lead to increased use of solvent, additional unit operations, and waste treatment operations, and much more energy will certainly be consumed. Finally, in many parts of the world, competition for potable (drinking) water is becoming more of an issue and will continue to be a greater issue in the future. Consequently, metrics for water use are being included more frequently.

By expressing mass intensity as its reciprocal and making it a percentage, it is in a form similar to effective mass yield and atom economy. This metric is generally known as *mass productivity* or *mass efficiency*:

$$\text{mass productivity} = \frac{1}{\text{MI}} \times 100 = \frac{\text{mass of product}}{\text{total mass in reaction vessel}} \times 100$$

Example 4.4 A chemist reacts benzyl alcohol (10.81 g, 0.10 mol, FW 108.1) with *p*-toluene sulfonyl chloride (21.9 g, 0.115 mol, FW 190.65) in toluene (500 g) and triethylamine (15 g) to give the sulfonate ester (FW 262.29) isolated in 90% yield (0.09 mol, 23.6 g). Calculate atom economy (AE), reaction mass efficiency (RME), mass intensity (MI), and mass productivity (MP).

Solution

$$\text{atom economy} = \frac{262.29}{108.1 + 190.65} \times 100 = 87.8\%$$

$$\text{reaction mass efficiency} = \frac{23.6}{10.81 + 21.9} \times 100 = 70.9\%$$

$$\text{mass intensity} = \frac{10.81 + 21.9 + 500 + 15}{23.6} = 23.2\text{g/g} = 23.2\text{kg/kg}$$

$$\text{mass productivity} = \frac{1}{\text{Mass intensity}} \times 100 = 4.3\%$$

Additional Points to Ponder What key elements of chemistry and process do reaction mass efficiency combine? Does this make it a useful metric for chemists, process chemists, or chemical engineers? How does a pursuit of reaction mass efficient reactions influence energy intensity?

As is evident from this example, the atom economy is 88%, due to the formation of HCl as a by-product, but it is, nevertheless, pretty good as reactions go. The reaction mass efficiency takes into account the 90% yield and the need for a 15% molar excess of *p*-toluene sulfonyl chloride, so it is only 70.9% efficient—a closer reflection of the reality of the reaction. The

TABLE 4.1 Comparison of Metrics for Different Chemistries

	Stoichiometry of B mole (%)	Yield (%)	Atom Economy (%)	Reaction Mass Efficiency (%)	Mass Intensity Excluding Water (kg/kg)	Mass Productivity (%)
Acid salt	135	83	100	83	16.0	6.3
Base salt	273	90	100	80	20.4	4.9
Hydrogenation	192	89	84	74	18.6	5.4
Sulfonation	142	89	89	69	16.3	6.1
Decarboxylation	131	85	77	68	19.9	5.0
Esterification	247	90	91	67	11.4	8.8
Knoevenagel	179	91	89	66	6.1	16.4
Cyanation	122	88	77	65	13.1	7.6
Bromination	214	90	84	63	13.9	7.2
N-acylation	257	86	86	62	18.8	5.3
S-alkylation	231	85	84	61	10.0	10.0
C-alkylation	151	79	88	61	14.0	7.1
N-alkylation	120	87	73	60	19.5	5.1
O-arylation	223	84	85	58	11.5	8.7
Epoxidation	142	78	83	58	17.0	5.9
Borohydride	211	88	75	58	17.8	5.6
Iodination	223	96	89	56	6.5	15.4
Cyclization	157	79	77	56	21.0	4.8
Amination	430	82	87	54	11.2	8.9
Lithal	231	79	76	52	21.5	4.7
Base hydrolysis	878[a]	88	81	52	26.3	3.8
C-acylation	375	86	81	51	15.1	6.6
Acid hydrolysis	478	92	76	50	10.7	9.3
Chlorination	314	86	74	46	10.5	9.5
Elimination	279	81	72	45	33.8	3.0
Grignard	180	71	76	42	30.0	3.3
Resolution	139	36	99	31	40.1	2.5
N-dealkylation	2650[a]	92	64	27	10.1	9.9

[a] Inflated by use of solvent as reactant.

mass intensity is on the order of 23 kg/kg, which is about what one would expect for the fine-chemical industry (*E*-factor, 22), but in terms of mass efficiency of the entire reaction mixture, it is only about 4% efficient. That means that for every kg of product, we throw out 96% of what we brought in. Is this really sustainable?

Table 4.1 contains averaged data for 28 different types of commonly used chemistries in industry. For each type of chemistry, the stoichiometry, yield, atom economy, reaction mass efficiency, mass intensity, and mass productivity are shown. Detailed analysis of these data has revealed a number of things:

1. These data show that most reactions are run at significant stoichiometric excesses, and this is not accounted for in atom economy.
2. Another observation is that reaction yield, a metric used almost universally by synthetic chemists, does not account for poor reaction mass efficiencies and a correspondingly significant waste of resource (mass or energy). Although this may

TABLE 4.2 Comparison of Three Different Chemistries with Similar Mass Intensities

Chemistry	Stoichiometry of B mole (%)	Yield (%)	Atom Economy (%)	Reaction Mass Efficiency (%)	Mass Intensity Excluding Water (kg/kg)
S-alkylation	230.9	85.3	83.8	61.3	10.0
Chlorination	313.6	86.0	73.6	45.8	10.5
Acid hydrolysis	478.4	92.4	75.6	50.0	10.7

be an obvious statement, it should be noted that wasted resource may be expensive from both a direct materials cost and a more comprehensive life cycle costing perspective.

3. Data for mass intensity, yield, atom economy, and stoichiometry do not correlate with each other in any meaningful way. These appear to be discretely different types of metrics, and following one metric in isolation of the others may not drive the best behavior for greening reactions. As an example, Table 4.2 shows three different chemistries with similar mass intensities that have generally different and conflicting trends in the values for the other metrics.
4. Because reaction mass efficiency accounts for all mass used in a reaction (excluding water) and includes yield, stoichiometry, and atom economy, the combined metric is probably the most helpful metric for chemists to focus attention on how far from green current processes are being operated.
5. Mass productivity may be a useful metric for businesses since it highlights resource utilization. This may be illustrated by comparing the average atom economy with the average mass productivity for 38 drug-manufacturing processes having an average of seven stages.

If you asked most synthetic organic chemists about the data in Table 4.3, they would probably agree that an average atom economy of 43% for a seven-stage synthesis of a complex drug would not appear to be unreasonable. Remember that this means the average atom economy for each individual stage would be somewhere in the mid-80% range, and that would represent the current state of the art for complex syntheses. However, before patting ourselves on the back, we should think about the fact that the average mass productivity for these synthetic processes is only 1.5%. This means that 98.5% of the total mass used to make your average drug is being wasted. Even if the atom economy for individual steps of the process were raised above 95%, this may not necessarily increase the overall average mass intensity of the process to a significant extent. Since a majority of the mass in a given process is not accounted for by atom economy, it may be argued that atom economy may not be the most robust measure or the best measure of sustainability for industrial use.

TABLE 4.3 Comparing Atom Economy and Mass Productivity for 38 Processes

	Overall Process Average (%)	Range (%)
Atom economy	43	21–86
Mass Productivity	1.5	0.1–7.7

4.3 PROCESS METRICS

It is useful to remember that chemistry on an industrial scale does not take place in the round-bottomed flasks of the laboratory. As the amount of material we try to make increases, there is an inevitable drive for increased efficiency; we simply cannot afford the high cost of buying and disposing of large quantities of materials. At the industrial level it could therefore be argued that process considerations take precedence over chemistry. This is something of a false dichotomy because chemistry will invariably be inextricably linked to the process, but it is useful to think about them separately for just a moment and make a few distinctions between them.

For the sake of simplicity, Table 4.4 contains a small collection of general categories of interest or concern that should be considered for metrics applied to any route or process development. As what is hopefully readily apparent from the table, the list contains primarily categories that would be found in many texts on process metrics. It should be noted that each category and each example could be further dissected for additional attributes or areas of concern in ever-increasing levels of detail. Table 4.4 was not intended to be comprehensive but illustrative and high level. It should also be noted that life cycle considerations affect each category and aspect contained in the table.

Figure 4.1 is presented as an attempt to show the interrelationships between the process metrics categories shown in Table 4.4. Material (chemicals, solvents, reagents, etc.) choices will directly affect the choice of unit operations, and the combination of materials and equipment will directly affect the operability of the process. Materials, equipment, and operability all have associated environmental, health, and safety (EHS) impacts and opportunities and ultimately affect product and product quality. The remainder of this chapter is an extended discourse on the categories in Table 4.4 and Figure 4.1 and suggested approaches to green metrics development.

4.3.1 Materials

Physical Form and Properties Physical form and properties are directly linked to and affect other aspects of a process: reactor type, type of mixing, throughput (i.e., how quickly

TABLE 4.4 General Areas of Interest for Process Metrics

Materials	Equipment	Operability	EHS Risk
Physical form and properties (i.e., gas, liquid, solid)	Unit operation type	Throughput/cycle time	Occupational exposure
Mass (i.e., total, solvent, reactant, process, etc.)	Number of unit operations	Robustness	Environmental air, water, land
Inherent hazard (e.g., toxicity, stability, reactivity)	Size (volume)	Energy (i.e., total, heating, cooling, recovery, treatment, etc.)	Safety/process safety
Cost	Scalability	Ease of cleaning and maintenance	Cost
Renewability	Controllability	Cost	
Recyclability	Cost		
Quality: purity/impurity profile			

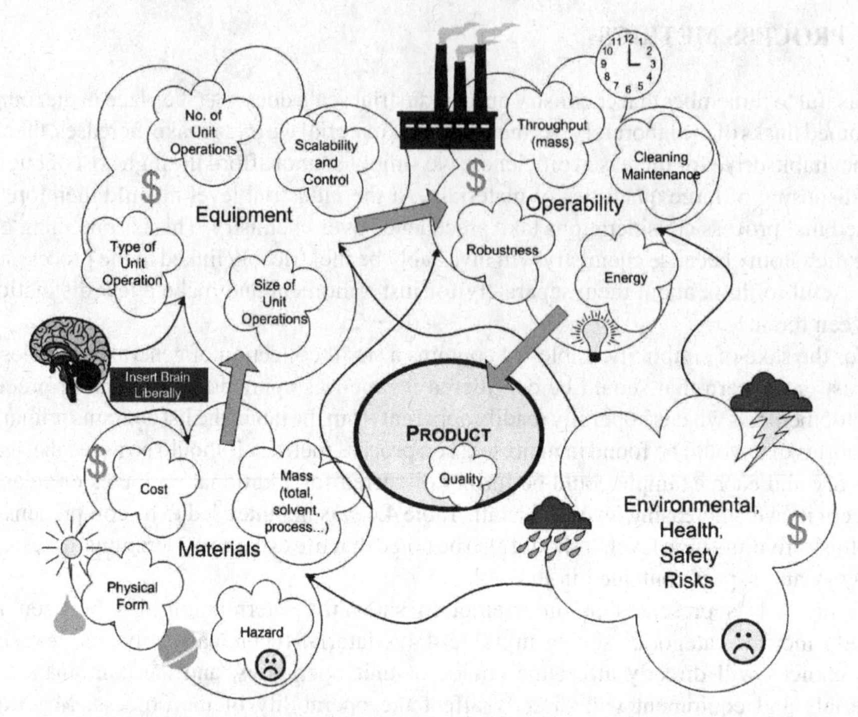

FIGURE 4.1 Interrelationships between process metrics categories.

a chemical will dissolve in a solvent or precipitate out), the ease of separation of two liquids, and so on. They also affect the energy required for heating or cooling, recovery if applicable, cleaning, and wastes.

In terms of metrics for this aspect, one could profitably divide this into three different approaches, depending on the question that metrics are trying to answer:

- Summing the number of materials possessing a given attribute
- Summing the mass of materials possessing a given attribute
- A combination of these to find a general score for material complexity

In the case of summing the numbers of materials, one might consider the number of materials of different phase, the number of different solvents containing azeotropes or close boiling points, and so on. One could also sum on the basis of mass. Both number and mass could be proxies for the degree of complexity associated with the process. Larger numbers and masses of different materials are likely to necessitate more equipment or energy use, slow throughput, and increase general materials management. Another option is devising a combined metric to provide an index of complexity. Complexity generally drives up cost and increases waste in one part of the life cycle or another.

Mass This has been discussed in detail above, but some additional metrics for further consideration are:

$$\text{solvent intensity} = \frac{\text{total solvent input, excluding water}}{\text{total mass input}}$$

$$\text{waste intensity} = \frac{\text{total waste produced}}{\text{total mass input}}$$

$$\text{specific compound } i \text{ released} = \frac{\text{amount of compound } i \text{ released as an emission}}{\text{total mass input}}$$

Inherent Hazard This aspect of materials is a major driver in assessing the overall ESH risk associated with a process. It is also a key driver in equipment design and material of construction, chemical storage and handling, throughput, and process robustness. For example, highly reactive materials (e.g., vinyl chloride monomer, aziridines, acetylenes, hydrogen fluoride) are used routinely and safely, but they do require considerable care during manufacture, storage, transport, and use to maintain thermodynamic and kinetic control without causing any adverse events.

As with the case of mass, there are several approaches to metrics for this aspect. One can simply sum numbers and/or mass of chemicals possessing hazards in different areas: process safety, occupational exposure, or environmental hazard. Typically, most companies will use a banding approach for materials, which allows a quick identification of the hazard category and usually marries hazard with a suggested control approach (e.g., layers of protection, pressure relief valves). One is then rapidly able to identify issues and potential opportunities for elimination, substitution, or control.

An example of how this might look for a group of materials is shown in Figure 4.2. Materials are listed according to type, and then a hazard ranking is applied. This hazard ranking is generally based on an assessment of a variety of potential hazards associated with a given material. It may be a composite ranking based on multiple types of hazards, or the ranking can be made for each hazard category individually. In the example shown in Figure 4.2, two different potential means of scoring are illustrated: either a simple weighted average (hazard ranking times mass of material used) and a score based on weighted averages for each category of materials (hazard ranking times mass of solvent, reagent, or process chemical used) and taking a geometric mean of the scores for each category. As can be seen, a category score may help to focus efforts on areas of the process most in need of attention and accounting for the perceived priorities at a given time.

A note of caution is appropriate at this juncture, as many confuse the inherent hazard of materials with risk. This has been covered in great detail in Section 3.4, so this serves only as a reminder that it is critically important to distinguish between hazard and risk. Risk is defined as being a function of the inherent hazard of a material and the potential, or likelihood, for exposure:

$$\text{risk} = f(\text{inherent hazard of a material, the potential or likelihood for exposure})$$

It is often expanded to include a severity rating at given probabilities. This is useful to bear in mind from a green perspective because one can use highly hazardous materials but design a process that has less impact and lower risk. For example, one might consider in situ generation of highly hazardous materials (e.g., phosgene) in small volumes that are consumed immediately.[28]

Composite score as weighted average		Score by type of material
7	Solvent	7
	Reactant	3
	Process Chemical	8
	Composite Score (Geometric Mean)	6

Material	Ranking	Mass - kg/kg API	Material class
Acetone	8	15.7	solvent
Acetonitrile	6	37.9	solvent
DMF	2	0.5	solvent
Heptane	9	25.8	solvent
Hexane	4	3.1	solvent
N-Propanol	5	7.1	solvent
TBME	6	3.2	solvent
Reactant 1	4	0.9	Reactant
Reactant 2	1	0.8	Reactant
Reactant 3	4	0.8	Reactant
Reactant 4	4	0.9	Reactant
Reactant 5	1	0.9	Reactant
Reactant 6	4	1.9	Reactant
5% Pd/C	4	0	Process chemical
Acetic anhydride	7	0.7	Process chemical
Activated Charcoal	10	0.1	Process chemical
Hexyl Lithium	7	1.3	Process chemical
Potassium Carbonate	10	2.3	Process chemical
Potassium Hydrogen Sulfate	4	1.1	Process chemical
Sodium Hydroxide	10	0.6	Process chemical

High hazard material - selection of lower hazard material recommended; if substitution is not feasible, perform health risk assessment and adopt exposure control strategy to reduce health risks

Hazard material; perform health risk assessment and adopt exposure control strategy to reduce health risks

Relatively low hazard material - perform health risk assessment and adopt exposure control strategy to manage health risks

FIGURE 4.2 Generic example of hazard scoring for process materials.

Renewability It is fair to say that very few chemicals currently available in routine commercial operations are derived from renewable sources; or if from renewable sources, they are often associated with considerable life cycle impacts or exhibit trade-offs of environmental impact, such as reduction of global warming potential at the same time that the eutrophication potentials increase. A contemporary example of this debate is found in the production of corn-based ethanol. Considerable controversy surrounds the sustainability of this particular use for corn, and this controversy is not without merit.

The take-away message from the debate is twofold. First, assessments of renewability have to be done from a life cycle perspective. Second, there is considerable difficulty in comparing highly developed chemical processing routes based on nonrenewable sources with processes that use potentially renewable materials with processing approaches that are not as fully developed as is a process that has been optimized over the course of many years. This is the current dilemma facing many biologically based fermentation or biotransformation processes compared to chemical synthetic processes, which is not significantly different from the dilemma facing new vs. established technologies.

The state of the art of new technologies is evolving continuously, and any comparisons one might wish to make are not always able to be carried out on an "apples to apples" basis. It is also not uncommon for materials and energy use in a biological or potentially renewable process to be several times as large as a chemical synthesis given the differences in process intensity, separability of the product from the biological matrix, and the substrate conversion efficiency of the organisms.

Recyclability Process recycle includes in-process and post-processing recycle. In general, *in-process recycle* is the preferred approach if at all possible, but it is clearly more common in the petrochemical and commodity chemicals context. For batch chemical operations commonly found in the fine-chemical, pharmaceutical, and agrochemicals businesses, in-process recycle is generally very difficult to carry out. The promise in those industries, however, is to reduce the scale of batch operations, make them continuous, and number-up to achieve the required volumes. If this can be done successfully, it is more likely to be able to incorporate in-process recycle.

Out-of-process or *post-process recycle* is commonly employed when possible and is used extensively on- and off-site. It may be important to distinguish between the two options, as each has a different set of issues. In general, off-site recycle is likely to have a greater number of potential impacts related to the transport and management of the waste to be recycled on and off the site.

Approaches to metrics for recyclability are similar to approaches for other aspects. One can sum the masses of potentially recyclable chemicals, solvents, water, and so on. Recycling can, however, be potentially problematic, depending on the type of reaction mixture, the scale of the process, and throughput. For example, aqueous–organic reaction mixtures containing solvents with similar boiling points or solvents that form azeotropic mixtures may, from a balanced perspective, be less desirable to recycle than simply obtaining virgin solvents. Clearly, one needs to balance costs and impacts across the life cycle to evaluate recycling opportunities appropriately. In such instances, however, recyclability metrics should drive chemists and engineers to utilize different solvent systems that enable the chemistry while ensuring facile recovery and reuse.

Whether the recycling occurs in- or out-of-process, amounts of recycled materials will necessarily affect the mass metrics, as the net amount of materials used is bound to be reduced by the amount recycled. From a life cycle viewpoint, it is important to account not only for the reduced impacts, but also for the environmental and resource usage impacts that result from recycling materials (e.g., distillation energy for solvent recovery). Nonetheless, in most cases the benefits of avoiding manufacturing impacts tend to dwarf the energy and resources utilized for recycling the materials.

Quality and Purity Purity is certainly a key driver for most chemical synthetic processes. The synthetic chemist is obviously looking to use reactions that result in very high yields with as few impurities in the isolated intermediate or product as possible. In the ideal situation, a purity as close as possible to 100% is sought, but this is often a very difficult target, especially for complex target molecules and the processes used to make them. In general, as the complexity of a chemical synthesis increases, the greater the occurrence of difficult separations and the need for one or more recrystallizations of the final product and/or intermediates to achieve the desired purity. Recrystallizations are essential in some cases to achieve the desired purity, crystal structure (including polymorphs), or particle size characteristics, but they obviously lead to an increase in spent solvent. While spent solvent can hopefully be recycled, it usually comes at a cost of energy and the additional capital invested in storage and distillation infrastructure.

With the above in mind, one might track purity in the usual manner of stating the purity or impurity profile of the final product or output of any given process. From a green perspective, however, one might be more interested in the number of isolations and/or recrystallizations required for an overall process to achieve the desired purity. The other point to consider is that *quality* has been defined as the capacity of satisfying the customer's requirements. In some instances, a higher purity will go beyond the customer's needs, so additional purification steps to increase purity can in some cases add to the mass and energy intensity without necessarily adding value for a customer, and therefore are wasteful steps. Alternatively, one might look at combinations such as net waste per isolation, or the quantity of energy per isolation as a means of distinguishing between processes. However, whereas these are more insightful metrics to derive and use, they are more difficult and time-consuming metrics to develop.

4.3.2 Equipment and Operability Intertwined

At the beginning of this chapter, several points were made about general metric principles that are particularly applicable within the context of equipment and operability. It is worth to revisit them for a moment to say that good process metrics for these categories are especially dependent on an understanding of the overall process and the optimization of a process should be done from a multivariate perspective. Metrics in these categories should be seen as being considerably dependent on each other and on the materials and chemicals used in the process.

It is also generally not the case that industry is building new plants to accommodate each new process or product. This means that within any industrial sector of the chemical enterprise, the type of unit operations in use is generally fairly small and fixed. Within a multipurpose chemical plant commonly found in the batch chemical industry, it is common practice for process designers to "make do" with what is available on a given site to avoid capital expenditure, plant shutdown for modifications, and so on. In terms of metrics, the approach recommended is generally to try to understand at a high level what the complexity is that arises from a given process. There is therefore a great opportunity to take relatively standardized and simple approaches to the evaluation of processes.

Type and Number of Unit Operation These two metrics provide an indication of the complexity of a process. There are a number of different unit operations required in any but the simplest processes, and there are a number of options for achieving a desired end. For example, in isolating a solid from a liquid reaction mixture, there are different strategies that one might employ, including gravity settling and decanting, solvent switching, filtration, centrifugation, and variations on or combinations of all of the above. In the batch-processing context, apart from the complexity of timing campaigns and overall throughput, each of the types of unit operations above comes with a different set of impacts. These impacts affect primarily the overall mass and energy efficiency of a process and the potential for various EHS impacts (e.g., occupational exposure risk, process safety risk, and/or environmental risk). For example, process simplicity is one of the principles of inherent safety, and thus the more complex a process is, the farther it is from being inherently safer.

Size of Unit Operations In general, the size of a unit operation can make a difference in the efficiency of mixing, the reaction rate, by-product formation, control of exo- or endotherms, separations, and so on. From a process engineering standpoint, one is interested in obtaining and maintaining a high degree of control over mass and energy transfer in any unit operation, but this may be difficult to achieve unless a chemical plant is designed to be fit for purpose. As this is often not the case, there is an inherent inefficiency associated with many processes.

Scalability *Process scalability* is an indication of how well a process would handle the rigors of moving to different production sizes as the process develops. Process scalability is widely used in the software industry to describe the ability to produce mass customization of products. *Scalability* in chemical processes implies sufficient process understanding and control to ensure that when a process moves to different and/or larger or smaller equipment, there is some confidence that product quality and yield will not be affected adversely. In the continuous processing context of a large chemical plant, scalability is achieved routinely, or else a product would never make it to market, although scaling-up a new technology is by no means a trivial endeavor.

In a multipurpose chemical plant context, facile transitions from R&D laboratory or bench scale to pilot scale to commercial size are frequently not easy to achieve. Although the reasons for this are many, the consequences of this lack of scalability can be a reduction in product quality and yield, increased by-product formation, longer cycle times, and in some cases an inability to reproduce key product properties such as color, size, or crystal structure.

From a green perspective, decreased yields, by-product formation, and inability to reproduce key product properties will invariably increase waste and require greater materials and energy use. Longer cycle times invariably will lead to increased energy use and, in some cases, increased material use.

To measure the scalability of a process, one needs to define the critical quality attributes and to understand the chemistry and processes involved in order to find the limits of acceptability of these critical attributes and thus the limits of the scalability of a process. For a chemical process to be functional at a large scale, it should also be operationally simple, safe, and straightforward.

Controllability In petrochemical and bulk commodity chemical manufacture, real-time process control has been a fact of life for many years. There is considerable understanding of processes, and control of process parameters is usually maintained within tight specifications to ensure statistical process control to within *six sigma* (6σ), the occurrence of one defect in 1 million. This has been enabled through the use of real-time analytical capability that works with programmable logic circuits to make small changes to various process inputs and physical parameters as required. In batch chemical operations, this level of real-time process control is rarely achieved, although there are increasing attempts in recent years to achieve greater statistical process control, the industry is generally only able to operate at about 3σ, or occasionally, 4σ, or one defect in 1000 to 10,000.

From a green perspective, processes that are not under tight control are obviously going to produce a greater quantity of waste, consume more materials and energy per unit of finished product, and lead to reduced throughput and cycle time. In some cases, not holding the process in control will lead to a failure to meet product specifications with the follow-on need to either reprocess the off-spec product or discard the product entirely. Either way, through the additional waste production or the materials consumed, an out-of-control process is a problem. Metrics for this might include the number of excursions from statistical process control, but one very useful metric for controllability is *process capability*, or more accurately, *process capability indices*. Process capability compares the output of an in-control process to the specification limits by using capability indices. The comparison is made by forming the ratio of the spread between the process specifications (the specification "width") to the spread of the process values. In a 6σ environment, this is measured by 6 standard deviation units for the process (the process "width"). A process under control is one where almost all the measurements fall inside the specification limits. The general formula for the *process capability index* is

$$\text{process capability} = C_p = \frac{\text{USL} - \text{LSL}}{6\sigma}$$

where USL is the upper specification limit, LSL the lower specification limit, and σ the standard deviation of the process. Clearly, if $C_p > 1.0$, the process specifications cover almost all the process measurements. Indirect proxies of controllability could be the amount of waste produced per kilogram of product, or the amount of materials and/or energy consumed per kilogram of product caused by excursions outside the control zone; for example, a rejected batch will become waste, and additional mass and energy will be required to replace (or rework) the batch rejected.

Robustness Simply speaking, process *robustness* is characterized by the extent to which process excursions adversely affect product quality and yield. A process that is not greatly affected by variations in process temperatures, mixing, minor variations in rates of addition, or other factors would be considered robust. Good process understanding through appropriate statistical design of experiment to test process inputs and parameters is the key to understanding process robustness.

The main difference between robustness and controllability is that a capable or controlled process will stay within the desired parameters. A robust process can exhibit excursions outside the control parameters, but the critical attributes will not be affected significantly by the excursion beyond specifications. When assessing the robustness of a process, several factors that can affect it adversely include nonselective or side reactions that might produce adverse effects and impurities, physical and chemical stability of the materials involved, and complexity of the separation train of the processes.

Throughput and Cycle Time *Throughput* is, in simple terms, the average salable production output in a given time unit. *Cycle time* is the average time between the release and completion of a job: in other words, the rate at which products are manufactured. Some of the key parameters that affect throughput in a chemical plant are the chemical conversion, yield, capacity and availability of existing equipment, process time, cycle time, number of chemical steps, number of unit operations, plant layout, warehouse processes, raw material availability, process bottlenecks, and labor availability.

In general, the goal of any well-run in-plant manufacturing operation is to ensure that all unit operations are utilized at capacity for as much time as possible before cleaning and/or maintenance is required. The design goal for any process is to ensure that reactions proceed to completion as rapidly as possible without losing control of the reaction and ensuring yields as close to theoretical as possible. Once again, for continuous processing, high throughputs must be maintained to achieve profitability and economic viability, so throughput is generally at capacity or nearly so.

In the multipurpose batch chemical plant context, continuous utilization of multiple unit operations in-plant is extremely difficult and rarely achieved. In addition, for many batch chemical operations, it is not uncommon for there to be long reaction times, extended periods of time at reflux, extended filtration times, drying times, and so on. Each of these process parts, steps, or stages lengthens the cycle time and is likely to lead to increased consumption of materials, energy, and labor. For example, a survey of processes by AstraZeneca established a clear relationship between the number of steps in a synthesis and the throughput; when the number of steps decreased linearly, the throughput increased exponentially.

There can be different approaches to tracking throughput, such as either volume of product or material manufactured per unit of time, number of products per unit of time, or mass per unit of time. In the campaign context of a batch plant, it is important to note that throughput and cycle time should include time required for cleaning, transportation, and maintenance, as these can dramatically affect the overall times and the utilization of any part of a plant. For example, a simple model developed from the AstraZeneca survey calculates the time for a manufacturing campaign as

$$\text{time for manufacturing} = T_m = \frac{N_B}{P} + \text{misc.}$$

where N_B is the number of batches, P the productivity, and "misc." accounts for any time needed for additional activity, such as plant cleaning and maintenance.

Cleaning and Maintenance In a batch chemical plant, because individual unit operations are utilized for multiple products, many pieces of kit may be subjected to long clean-out periods using large solvent volumes and/or aqueous detergents, or both. If possible, clean-in-place protocols are preferred to break down and rebuild. These cleaning materials are often not considered as part of a process, so their use is not optimized in the same manner as are other process-related materials and solvents. Frequency of cleaning, length of cleaning, volumes of solvent, water, detergent, energy use, and other factors are all important parameters that affect the real mass intensity of a process, as well as cycle time and throughput, among others. In general, a combination of volume per unit of time and/or energy would be most useful. However, one should also consider these materials in terms of their intrinsic hazards, just as for any process reagent, solvent, or reactant.

Energy In a large petrochemical and chemical manufacturing complex, energy reduction through process integration and modification are constantly pursued given the enormous amounts of energy consumed by many processes. Energy efficiency is comparatively much greater than what is found in the batch chemical environment. For many batch chemical operations it is not unusual for both heating and cooling to be required for any given step or stage of the synthesis. Although this can be avoided through closer attention to the combination of chemistry with reactor type and configuration, it is generally not achieved routinely, for a variety of reasons. However, it is generally true that the existence of a large installed base of reactors with their supporting unit operations prevents the implementation of newer technologies. Existing in-ground capital that has been paid for many times over is difficult to stop using unless the gains in efficiency or the reduction in costs are overwhelming.

Energy metrics are usually very similar to those for mass; that is, one can slice the total energy used per kilogram of product to highlight key materials use, or forms of energy such as those used to heat or to cool. As mentioned before, it is also important to account for life cycle energy requirements, which would include not only processing energy but also the energy required to produce raw materials, recycle of materials, and waste treatment. Some potential metrics might include:

$$\text{energy intensity} = \frac{\text{total process energy (MJ)}}{\text{kg of final product}}$$

$$\text{life cycle energy} = \frac{\text{life cycle energy requirements (MJ)}}{\text{kg of final product}}$$

$$= \frac{\Sigma(\text{process, material manufacturing, recovery, treatment})}{\text{kg of final product}}$$

$$\text{waste treatment energy} = \frac{\text{waste treatment energy requirements (MJ)}}{\text{kg of final product}}$$

$$\text{solvent recovery energy} = \frac{\text{solvent recovery energy requirements (MJ)}}{\text{kg of final product}}$$

or the requirements can be expressed as a fraction of total energy input, such as

$$\text{solvent energy ratio} = \frac{\text{total energy for solvent use and recovery}}{\text{total energy input}}$$

$$\text{waste energy ratio} = \frac{\text{total waste produced}}{\text{total energy input}}$$

Example 4.5 *n*-Butanol is obtained from hydroformylation reaction of propylene by the following reaction:

$$CH_3CH = CH_2 + CO + 2H_2 \rightarrow CH_3CH_2CH_2CH_2OH$$

Isobutanol is also produced in the process by means of

$$CH_3CH = CH_2 + CO + 2H_2 \rightarrow CH_3CH_3CHCH_2OH$$

The process uses 609 kg of propylene, 399 kg/h of CO, and 58 kg/h of H_2 to produce 1000 kg/h of 1000 kg/h of 99% pure *n*-butanol and 50 kg/h of 85% isobutanol. The process also requires 749 MJ/h of electricity and 2270 MJ/h of steam. What is the energy intensity of this process if the only useful product is the *n*-butanol?

Solution Since in this case the only useful product is *n*-butanol, we have

$$\text{energy intensity} = \frac{\text{total process energy}}{\text{mass of product}} = \frac{(749 + 2270)\text{MJ/h}}{1000\,\text{kg/h}} = 3.2\text{MJ/kg}$$

Additional Points to Ponder How would this result change if isobutanol were also a usable product? If we wanted to evaluate the life cycle energy, which other energy requirements will need to be added for this process?

4.3.3 EHS Hazards and Risk

Chapter 3 provided a good overview of EHS hazards and risks, and you may want to refer back to it while reading this section of the chapter. In general, there is a tendency on the part of many to focus on one portion of the EHS risk universe to the exclusion of other areas. For example, as extremely potent or toxic compounds are handled, there may be greater concern about occupational hygiene exposures while less consideration is given to potential process safety risks such as dust explosions. In another situation, there may not be enough of an appreciation of potential environmental risks that might arise while disposing of cleaning wastes. Consequently, it is extremely important to look at EHS risks together as an integrated problem and not as isolated disciplines.

A second pitfall is to try to list all of the regulatory constraints or lists on which a particular chemical has the misfortune to be found. In this day and age, most chemicals are on one list or another, or if not, one is lured into a false sense of security because the particular chemical of interest appears to be relatively free of potential issues. However, this usually has more to do with the fact that insufficient testing has been done on the material to determine its hazard, or it is not made in sufficient volumes to exceed a regulatory bright line. Recent legislation such as REACH should in time overcome this lack of EHS information, but this will take some time, and for chemicals not manufactured in large quantities, a lack of credible EHS hazard information is likely to continue.

A third pitfall is to confuse hazard and risk. There are those that suggest that a very hazardous material cannot or should not be used under any circumstance. Although application of the precautionary approach is perhaps desirable, there are situations where materials of higher hazard may be used at a lower level of risk than a larger quantity of material of lesser hazard. Once again, it is a question of taking a more holistic view and carefully weighing alternatives. The classic example of this is in the Boots ibuprofen process, where hydrogen fluoride and carbon monoxide were used to reduce the number of synthetic steps and increase the mass efficiency while dramatically decreasing waste.[29] In many countries it is now common practice in industry to group or band chemicals into certain ranges of hazard and to apply a certain control technology or set of technologies with that band. Following the completion of appropriate risk assessment, the task of determining appropriate control technology is therefore much more straightforward. Banding is generally done in each of the EHS risk areas (e.g., occupational hygiene, process safety, and environmental). This generally simplifies the job of the bench scientist or engineer who is not an expert in assessing EHS hazards and risks but must still ensure that the process they are developing has a minimal amount of associated EHS risk.

A final note of caution for developing process metrics and their relationship to toxicology concerns is the idea of acute and chronic effects. In general, most attention is directed toward substances that possess acute hazards. For both human and environmental hazards, chronic exposures to chemicals that cause ill health in humans or other species are often difficult to predict, and in many cases, conclusively determine cause and effect. Apart from avoiding the use of chemicals known to cause ill health from chronic exposures, there are several strategies that one may employ to avoid potential problems. The obvious choice and the one most often expressed by environmentalists is to apply the precautionary principle uniformly (i.e., until the hazard potential is known, avoid using the materials in question). If the precautionary principle cannot be followed, one may assess chemicals by analogy (i.e., if the chemical is structurally similar to a chemical of known chronic hazards, you may want to avoid it). A more sophisticated approach[30,31] is to use quantitative structure–activity relationships (QSARs) to screen out chemicals, but this not only requires an understanding of the compound of interest but also requires a higher degree of understanding of human or ecotoxicity to make good use of the predictive capability of any type of modeling.

Occupational Exposure Hazard and Risk The easiest means for assessing occupational exposure hazards associated with materials used in a process is through the use of permissible or occupational exposure limits (OELs or PELs), as described in Section 3.2.2. One can assess the potential for occupational exposure risk through close attention to the materials and the unit operations employed in a process, by performing an exposure assessment coupled to a given hazard to estimate the occupational exposure risk (e.g., Dow's exposure index[32]). A variety of approaches are possible here, from simply summing the number of materials in a given hazard band through more sophisticated approaches that take into account additional toxicological concerns such as the potential for carcinogenicity, mutagenicity, and reproductive effects. It is also possible simply to sum the mass of materials in a given band or to do a high-level assessment of potential risk based on the mass used, its physical form, the type of unit operations in the process, or the potential for accidental release into the workspace.

Process Safety Hazards Risk There are a variety of process safety risks that one needs to assess with chemical processes. In general, these risks will lead to an evaluation of the potential for the process to have precipitous changes in temperature and/or pressure that lead to secondary events such as detonations, explosions, overpressurizations, and fires. The most cost-effective way of avoiding these sorts of risks is through the adoption of inherent safety principles.[33] Inherent safety principles are very similar, and complementary, to pollution prevention principles, where one attempts to use a hierarchy of approaches to avoid and/or reduce the risk of an adverse event (see Section 3.3.3 and Chapter 14 for more details on inherent safety).

For processes under development, the most cost-effective means of avoiding potential risk is to eliminate those materials that are inherently unsafe (i.e., materials whose physical or physicochemical properties lead to them being highly reactive or unstable). This is somewhat difficult to achieve, for several reasons. First, without a full battery of tests to determine, for example, flammability, upper/lower explosivity limits and their variation with scale, minimum ignition temperatures, and so on, it is almost impossible to tell how a particular chemical will behave in a given process. Second, chemical instability may make a compound attractive to use because its inherent reactivity ensures a reaction proceeds to completion at a rapid enough rate to be useful (i.e., the reaction is kinetically and thermodynamically favored).

The approach to developing metrics for process safety is analogous to those that might be used to assess occupational exposure risk. One can also cite several indices that have been developed as metrics for estimating and ranking the safety of a given process or chemical reaction, such as the Dow fire and explosion index,[34] the Stoessel index[35] for hazard assessment and classification of chemical reactions, the inherent safety index, and the prototype index for inherent safety, among others.[36,37]

Environmental Hazards and Risk There are a number of strategies that one might employ to assess the collective environmental risk associated with a process. Whatever strategy that is employed, however, there are a number of general areas that should be considered as part of the overall assessment. First, one should consider the inherent hazard, fate, and effects of materials used in the process. Second, one needs to consider the potential for release from the process and its unit operations. Third, one should consider issues related to the transportation, storage, and disposal options related to the materials used in the process. Finally, one should consider the life cycle impacts of producing those materials.

In terms of environmental metrics to assess processes, it is hopefully clear that a considerable testing burden exists to assess potential environmental hazards that lead to a credible risk assessment. At a first pass, one would typically screen compounds from an environmental hazard perspective to assess their tendency for persistence, bioaccumulation, and toxicity. Depending on the final application of the compound, one might avoid commercial production of a particular compound, or one might devise processes that would use the compound but control the environmental risk to acceptable levels. For this, it is important to perform a process-specific risk assessment, as the impact of a given chemical, or set of chemicals, will be affected by the inherent hazard, environmental fate, and the specific characteristics of the process, available treatment, and volumes, among others.

4.4 COST IMPLICATIONS AND GREEN CHEMISTRY METRICS

It would be a mistake to leave a discussion of green chemistry metrics at the point of only considering chemistry and mass implications without due consideration of cost. Clearly, wasted resources usually have significant cost implications, both in terms of the chemistry metrics and in terms of process metrics.

For the moment, there is a need to point out a few observations that are important for bringing a green perspective to metrics and cost. First, the current cost of materials is generally not an accurate reflection of the true or overall cost required to deliver that material. Apart from the actual purchase cost that a business pays a supplier for a chemical, there are costs that are usually not considered. Happily, there has been considerable development of approaches to total cost assessment[38] or environmental accounting.[39] Total cost accounting is covered in Chapter 20, but it is strongly encouraged to seek out more information about how this is done.

4.5 A FINAL WORD ON GREEN METRICS

As can be seen from the preceding discussion, developing green metrics for chemical processes requires a holistic, systems point of view across a range of disciplines. Metrics are also generally context dependent; that is, one type or one set of metrics does not fit all situations. Instead, different organizations or companies will have to undertake some very hard work to identify, assess, and implement metrics that are most applicable to their needs. The good news is that there are a large number of metrics that have already been identified, and many of these will meet the needs of most organizations or companies. Any one person is unlikely to possess sufficient knowledge in all areas of interest to identify key metrics, so it should be common practice for green metrics to be developed drawing on the resources of cross-disciplinary teams. In addition, to truly drive the design of greener, safer processes, there is a need to resist the temptation of addressing metrics in a compartmentalized manner, as many of these metrics are interrelated. Finally, one should apply the 80/20 rule liberally; that is, don't strive for the perfect set of metrics that covers all situations if a few metrics meet your needs most of the time.

PROBLEMS

4.1 Why is, or isn't, mass productivity (mass efficiency) a useful metric for businesses? Explain your answer.

4.2 Do you think that there is a correlation between any of the metrics discussed in this chapter and the molecular complexity of a particular molecule?

4.3 Do you think that it would be possible to create a model for molecular complexity that could be correlated with reaction mass efficiency?

4.4 Could realistic targets be set for reaction mass efficiency based on this complexity model?

4.5 Estimate the reaction mass efficiency, mass intensity, mass productivity, and *E*-factor for Example 4.1.

4.6 Estimate the reaction mass efficiency, mass productivity, *E*-factor, and atom economy for Example 4.2.

4.7 Calculate selectivity toward *n*-butanol for Example 4.5.

4.8 Calculate energy intensity for Example 4.5 assuming that isobutanol is also a salable material.

4.9 Calculate reaction mass efficiency, mass productivity, mass intensity, and yield for Example 4.5, assuming that:

(a) Both *n*-butanol and isobutanol are useful, salable products.
(b) Only *n*-butanol is a usable, salable product.

4.10 It has been estimated that the process for Example 4.5 has the emissions shown in Table P4.10. Estimate the intensity of emissions (kg/kg) of each of the chemicals emitted.

TABLE P4.10 Emissions

Gas Emissions	Amount (kg/h)
n-Butanol	9.97
Propylene	11.9
Isobutanol	0.5
Hydrogen	0.302
Carbon monoxide	2.00

4.11 Dimethyl malonate is produced by the following chemistry:

$$\underset{\text{monochloracetic acid}}{Cl-CH_2COOH} + NaOH \rightarrow \underset{\text{sodium monochloroacetate}}{Cl-CH_2COONa} + H_2O$$

$$\underset{\text{sodium monochloroacetate}}{Cl-CH_2COONa} + NaCN \rightarrow \underset{\text{sodium cyanoacetate}}{NC-CH_2COONa} + NaCl$$

$$\underset{\text{sodium cyanoacetate}}{2(NC-CH_2COONa)} + 4CH_3OH + 5H_2SO_4 \rightarrow \underset{\text{dimethyl malonate}}{2C_5H_8O_4} + Na_2SO_4 + 2NH_3 + 4H_2SO_4$$

Monochloroacetic acid in water is mixed with cracked ice. Sodium hydroxide is added until the solution is made alkaline. Subsequently, sodium chloroacetate is formed. Sodium cyanide in water is added carefully to form a solution of sodium cyanoacetate. This solution is evaporated under reduced pressure to form a crude sodium cyanoacetate cake. The cake is hydrolyzed and esterified in the presence of methanol and sulfuric acid. Three extractions are performed with toluene. The dried product is distilled, at first under atmospheric pressure, and finally under reduced pressure, to remove any remaining toluene. To produce 1000 kg/h of 97% pure monomethyl malonate, 3.5 MJ/h of electricity, 101 MJ/h of steam, and 57 MJ/h of refrigeration are needed, and it is required to dissipate 3165 MJ/h of heat using cooling water.

(a) Estimate the energy intensity of the process.

(b) How is the required energy produced?

4.12 Estimate the yield, mass intensity, mass productivity, atom economy, reaction mass efficiency, and waste intensity (*E*-factor) for the dimethyl malonate process of Problem 4.11. Assume that the amounts of materials listed in Table P4.12 are required to produce 1000 kg/h of monomethyl malonate.

TABLE P4.12 Amounts of Chemicals

Chemical	Amount (Kg/hr)	Comments
Toluene	55	
Chloroacetic acid	715	
Sodium sulfate	40	
Methanol	470	
Sodium cyanide	372	
Sodium hydroxide	688	50% purity (344 kg/h water)
Sulfuric acid	1835	98% purity (37 kg/h water)
Water	3254	

REFERENCES

1. Bennett, M., James, P., Eds. *Sustainable Measures*. Greenleaf Publishing, Lebanon, TN, 1999.
2. Committee on Industrial Environmental Performance Metrics. National Academy of Engineering, National Research Council. *Industrial Environmental Performance Metrics: Challenges and Opportunities*. NCR, Washington, DC, 1999.
3. *Corporate Environmental Performance 2000*, Vol. 1, *Strategic Analysis*. Haymarket Business Publications, London, UK, 1999.
4. Curzons, A. D., Constable, D. J. C., Mortimer, D. N., Cunningham, V. L. So you think your process is green, how do you know? Using principles of sustainability to determine what is green—a corporate perspective. *Green Chem.*, 2001, 3, 1–6.
5. Constable, D. J. C., Curzons, A. D., Freitas dos Santos, L. M., Geen, G. R., Hannah, R. E., Hayler, J. D., Kitteringham, J., McGuire, M. A., Richardson, J. E., Smith, P., Webb, R. L., Yu, M. Green chemistry measures for process research and development. *Green Chem.*, 2001, 3, 7–9.
6. Douglas, J. *Conceptual Design of Chemical Processes*. McGraw-Hill, New York, 1988.
7. Peters, M. S., Timmerhaus, K. D., West, R. E. *Plant Design and Economics for Chemical Engineers*, 5th ed., McGraw-Hill, New York, 2002.

8. Center for Waste Reduction Technologies. Collaborative Projects – Focus Area: Sustainable Development, American Institute of Chemical Engineers (AIChE), New York, NY, 2000.
9. Azapagic, A., Howard A., Parfitt, A., Tallis, B., Duff, C., Hadfield, C., Pritchard, C., Gillett, J., Hackitt, J., Seaman, M., Darton, R., Rathbone, R., Clift, R., Watson, S., Elliot S. The Sustainability Metrics. Institution of Chemical Engineers, 30 pp., Rugby, UK, 2003, http://cms.icheme.org/.
10. Constable, D. J. C., Curzons, A. D., Cunningham, V. L. Metrics to green chemistry: Which are the best? *Green Chem.*, 2002, 4, 521–527.
11. Jiménez-González, C., Curzons, A. D., Constable, D. J. C., Overcash, M. R., Cunningham, V. L. How do you select the "greenest" technology? Development of guidance for the pharmaceutical industry. *Clean Products Process.*, 2001, 3, 35–41
12. Jiménez-González, C., Constable, D. J. C., Curzons, A. D., Cunningham, V. L. Developing GSK's green technology guidance: methodology for case-scenario comparison of technologies. *Clean Technol. Environ. Policy*, 2002, 4, 44–53.
13. Jiménez-González, C., Curzons, A. D., Constable, D. J. C., Cunningham, V. L. Expanding GSK's solvent selection guide: application of life cycle assessment to enhance solvent selections. *Clean Technol. Environ. Policy*, 2005, 7, 42–50.
14. Jiménez-González, C., Curzons, A. D., Constable, D. J. C., Cunningham, V. L. Cradle-to-gate life cycle inventory and assessment of pharmaceutical compounds: a case-study. *Int. J. Life Cycle Assess.*, 2004, 9(2), 114–121.
15. Butters, M., Catterick, D., Craig, A., Curzons, A. D., Dale, D., Gillmore, A., Green, S. P., Marziano, I., Sherlock, J. P., White, W. Critical assessment of pharmaceutical processes: a rationale for changing the synthetic route. *Chem. Rev.*, 2006, 106(7), 3002–3027.
16. Marteel, A. E., Davies, J. A., Olson, W. W., Abraham, M. A., Green chemistry and engineering: drivers, metrics, and reduction to practice. *Annu. Rev. Environ. Resour.*, 2003, 28, 401–428.
17. Wrisberg, N. D., Haes, H. A. U., Bilitewski, B., Bringezu, S., Bro-Rasmussen, F., Clift, R., Eder, P., Ekins, P., Frischknecht, R., Triebswetter, U. In *Analytical Tools for Environmental Design and Management in a Systems Perspective*, Wrisberg, N., de Haes, H. A. U., Triebswetter, U., Eder, P., Clift, R., Eds. Kluwer Academic, Dordrecht, The Netherlands, 2002, pp. 45–73.
18. Lapkin, A. In *Renewables-Based Technology: Sustainability Assessment*, Dewulf J., von Langenhove, I. H., Eds. Wiley, Hoboken, NJ, 2006, pp. 39–53.
19. Brunner, N., Starkl, M. Decision aid systems for evaluating sustainability: a critical survey. *Environ. Impact Assess. Rev.*, 2004, 24, 441–469.
20. de Haes, H. U., Ed. *Life Cycle Assessment: Striving Towards Best Practice*. SETAC Press, Brussels, Belgium, 2002.
21. Barnthouse, L., Fava, J., Humphreys, K., Hunt, R., Laibson, L., Noesen, S., Norris, G., Owens, J., Todd, J., Vigon, B., Weitz, K., Young, J., Eds. *Life-Cycle Impact Assessment: The State-of-the-Art*. SETAC Press, Brussels, Belgium, 1998.
22. Jiménez-González, C., Kim, S., Overcash, M. R. Methodology for developing gate-to-gate life cycle inventory information. *Int. J. Life Cycle Assess.*, 2000, 5(3), 153–159.
23. Jiménez-González, C., Overcash, M. R. Energy sub-modules applied in life-cycle inventory of processes. *Clean Products Process.*, 2000, 2(1), 57–66.
24. Hudlicky, T., Frey, D. A., Koroniak, L., Claeboe, C. D., Brammer, L. E. Toward a "reagent-free" synthesis—tandem enzymatic and electrochemical methods for increased effective mass yield (EMY). *Green Chem.* 1999;1(2), 57–59.
25. Sheldon, R. A. Organic synthesis; past, present and future. *Chem. Ind. (London)*, 1992, 903–906.
26. Sheldon, R. A. Catalysis and pollution prevention, *Chem. Ind. (London)*, 1997, 12–15.

27. Trost, B. M. The atom economy–a search for synthetic efficiency. *Science*, 1991, 254(5037), 1471–1477.

28. Ajmera, S. K., Losey, M. W., Jensen, K. F., Schmidt, M. A. Microfabricated packed-bed reactor for phosgene synthesis. *AIChE J.*, 2001, 47, 1639–1647.

29. Cann, M. C., Connelly, M. E. *Real World Cases in Green Chemistry*. American Chemical Society, Washington, DC, 2000. Available at: http://www.chemistry.org/portal/resources/ACS/ACSContent/education/greenchem/case.pdf.

30. Anastas, N. D., Warner, J. C. The incorporation of hazard reduction as a chemical design criterion in green chemistry. *Chem. Health Saf.* 12(2), Mar–Apr 2005, 9–13.

31. Anastas, N. D., Incentives for using green chemistry and the presentation of an approach for green chemical design. In *Green Chemistry Metrics*, Lapkin, A., Constable, D. J. C., Eds. Blackwell, London, 2008, pp. 27–40.

32. American Institute of Chemical Engineers. *Dow's Chemical Exposure Index Guide*. AIChE, New York, 1998.

33. See, e.g., Hurme, M., Rahman, M., Implementing inherent safety throughout process lifecycle. *J. Loss Prev. Process Ind.*, 2005, 18(4–6), 238–244.

34. American Institute of Chemical Engineers. *Dow's Fire and Explosion Index Hazard Classification Guide*, 7th ed., AIChE, New York, 1994.

35. Stoessel, F. What is your thermal risk? *Chem. Eng. Prog.*, 1993, 89(10), 68–75.

36. Edwards, D. W., Lawrence, D. Assessing the inherent safety of chemical process routes: Is there a relation between plant costs and inherent safety? *Trans. IChemE B*, 1993, 71, 252–258.

37. Heikkilä, A.-M., Hurme, M., Järveläinen, M. Safety considerations in process synthesis. *Comput. Chem. Eng.*, 1996, 20, S115–S120.

38. Center for Waste Reduction Technologies. 1999. Total Cost Assessment Methodology. American Institute of Chemical Engineers, New York, NY. http://www.aiche.org/IFS/Products/TotalCostAssessmentMethodology.aspx.

39. U.S. Environmental Protection Agency. *An Introduction to Environmental Accounting as a Business Management Tool: Key Concepts and Terms*. EPA 742-R-95-001. U.S. EPA, Washington, DC, June 1995. Available at http://www.epa.gov/oppt/library/pubs/archive/acct-archive/resources.htm.

PART II

THE BEGINNING: DESIGNING GREENER, SAFER CHEMICAL SYNTHESES

5

ROUTE AND CHEMISTRY SELECTION

What This Chapter Is About This chapter is about chemical synthesis and how to incorporate green chemistry principles into route design and selection. A synthetic organic chemist is faced with many challenges and choices in constructing a molecule; therefore, an efficient route design is imperative.

Learning Objectives At the end of this chapter, the student will be able to:

- Understand the synthetic chemistry process for chemical design.
- Understand some basic terminology about chemical synthesis.
- Understand a few basics about how decisions in chemistry and route selection may affect the greenness of a product.
- Understand what is meant by a chemical route, and differentiate it from a chemical process.
- Identify opportunities for the application of green chemistry principles into route selection and design.

5.1 THE CHALLENGE OF SYNTHETIC CHEMISTRY

Ask any synthetic organic chemist and you are likely to be told that the art and science of synthetic chemistry requires an encyclopedic knowledge of many different types of chemistries, named reactions, and approaches to making molecules. At the same time that

Green Chemistry and Engineering: A Practical Design Approach, By Concepción Jiménez-González and David J. C. Constable

this is a skill that can be learned, it demands a considerable amount of practice and experience before one becomes proficient at chemical synthesis. Once proficient, the synthetic organic chemist must stay continually abreast of a plethora of new reactions, constantly scanning the literature for new ideas and approaches to making molecules in new and hopefully more efficient ways.

Many synthetic organic chemists will also tell you that they already practice green chemistry, at least in part, because green chemistry is just "good" synthetic chemistry. After all, they say, who would knowingly start out to design a molecule that is the next thalidomide (an antinausea drug prescribed in the 1950s that caused multiple birth defects), or a process for a pesticide that produces dioxin as an added "bonus"? Although you can understand why the chemist would take this position, it should really be seen for what it is: a fallacious argument. In actual fact, green chemistry adds an additional level or two of difficulty to synthetic organic chemist's jobs because now they must not only design a molecule, they must do so in such a way that it has as few adverse impacts as possible, and do this using materials that are derived from sustainable sources. This is no small feat!

In this chapter we do not attempt to teach advanced organic synthesis or abstract graduate-level courses about organic synthesis but, instead, lay out some general green chemistry principles that practicing synthetic organic chemists might want to consider as they design a route and optimize synthetic processes. We provide a rough outline of the steps that a synthetic organic chemist goes through, so that those who are not trained as synthetic organic chemists might see ways in which they can support such chemists, or perhaps think of using technology to remove some of the barriers the chemists face.

5.2 MAKING MOLECULES

So how does a synthetic organic chemist go about making a molecule? Typically, the chemist will be given a target molecule on which to work. For example, in the pharmaceutical industry, such target molecules are usually discovered by medicinal chemists who are exploring particular ways to treat a disease, kill or weaken an invasive organism (a virus or bacteria), or effect some change in a metabolic process. In agrochemicals, the target molecule may affect a particular stage of development of a pest, might be a selective inhibitor of a key enzyme or protein, or may prevent a pest from reproducing. In the electronics industry, a molecule might be designed as a photoresistor, as a liquid crystal for displays, or for a desirable optoelectronic property.

The point is that the target molecule may be out of the control of many synthetic organic chemists; that is, the chemists are given a target molecule and are typically asked to design an efficient chemical route to obtain it. This limits chemists in what they are able to do, since they cannot change a functional group or add another part of the molecule to change it, since its function has already been determined. Those types of changes, changes that reduce the hazard of a molecule, actually have to be thought about as the molecule is being designed for a particular function. Recently, there have been several good attempts to think about green design rules and how chemists might apply them.[1,2]

So once the chemist receives a target molecule, what general steps might they take to design a robust and hopefully green synthesis? In general, many chemists will perform a retrosynthetic analysis of the target molecule. A *retrosynthetic analysis* is simply a process of working backward from the target molecule to progressively simpler molecules. This is done by systematic dissection of the molecule by means of strategic disconnections (i.e., places where the molecule can be taken apart). The chemist also makes use of what are known as *functional group interconversions*. These are operations that correspond to well-known named reactions or reactions they have used in the past, or reactions for which they can find a precedent in the literature. The idea is to obtain as your building block as simple a starting material as you can buy, usually at as low a cost as possible. Note here that the feasibility of the synthesis and cost are typically the principal drivers, not whether or not the synthesis is green. The synthetic plan that the chemist follows is simply the reverse of the retrosynthetic analysis that he or she has performed.

In the following paragraphs, some general steps the chemist would take in making a molecule are presented. Those who have an interest in learning more are referred to selected books and journals in the reference section.[3–15]

5.2.1 Study the Molecule

One alternative for the chemist is to make a model or look at a three-dimensional representation on a computer screen and assess the overall size and shape, conformation of rings, and so on, looking for symmetry elements (or regions containing pseudosymmetry). In some cases, it can be a simple molecule such as cyclohexyl bromide:

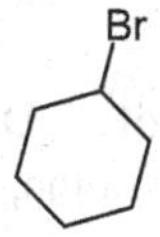

In other cases, it will be slightly more complicated:

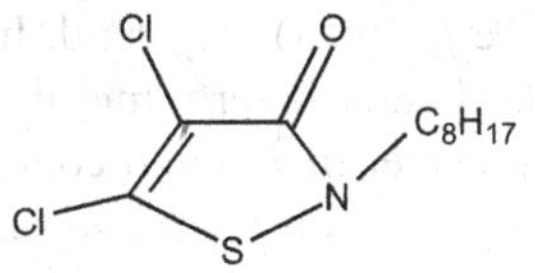

4,5-dichloro-2-*n*-octyl-4-isothiazolin-3-one
(Sea Nine 211: antifoulant)

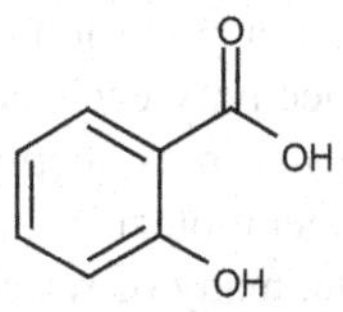

salicylic acid

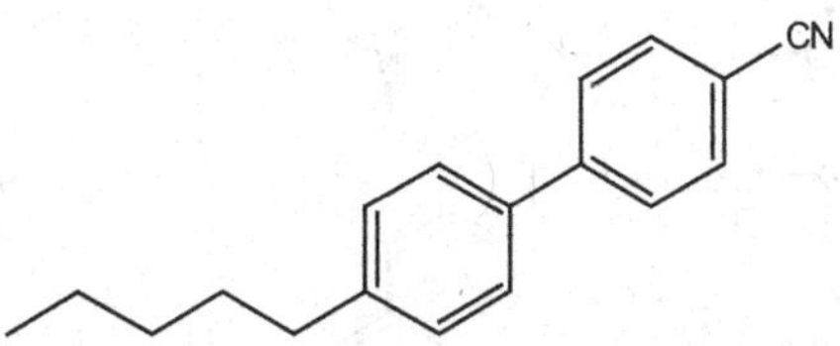

5 CB (liquid crystal)

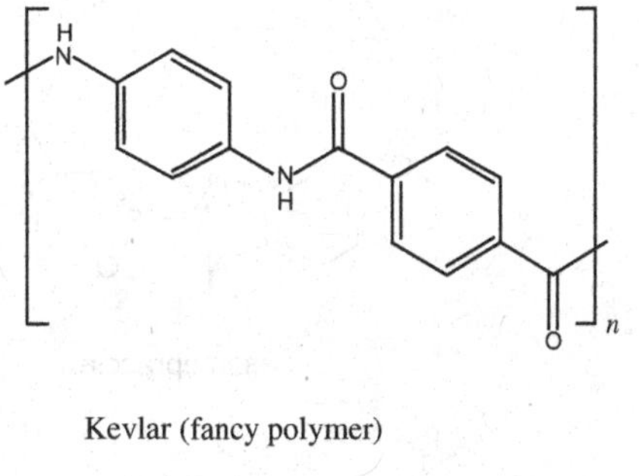

Kevlar (fancy polymer)

In other cases, the molecules may be very complicated indeed!

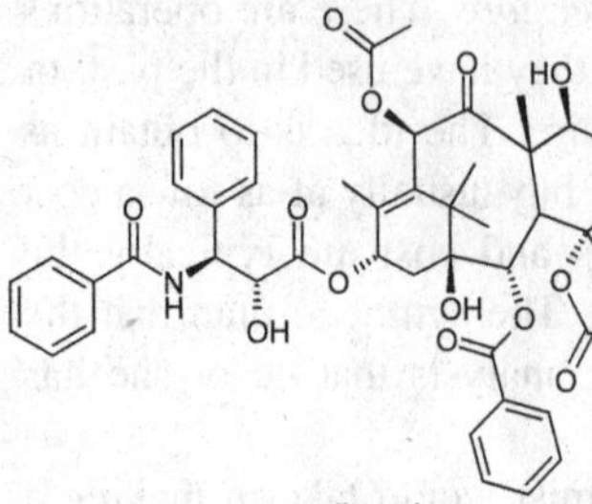

paclitaxel 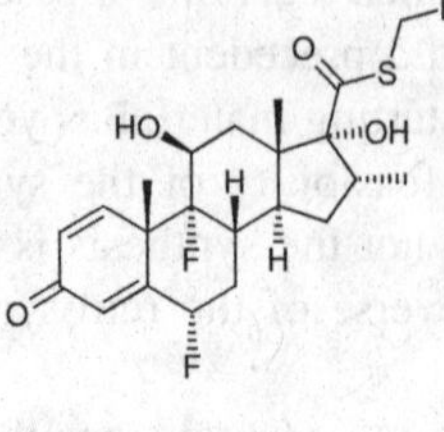 fluticasone 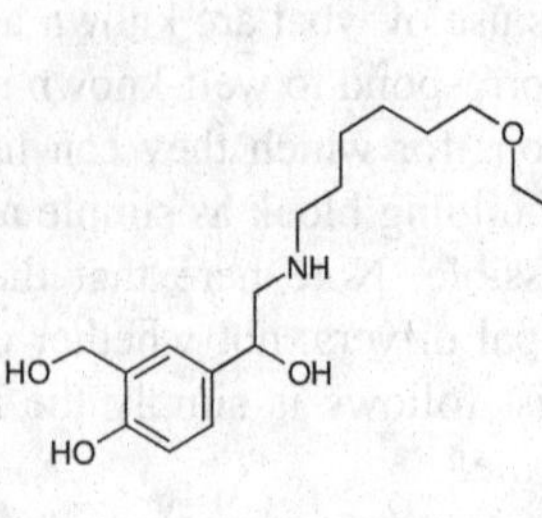 salmeterol

5.2.2 Assess Bond Connectivity

The chemist will try to understand the topology of the molecule: those parts of the molecule that remain invariant under certain chemical transformations. It is particularly important to understand the different types of atoms and how they are connected to one another, especially in polycyclic molecules [e.g., molecules containing multiple ring systems, such as alkaloids or steroids (see the structure for fluticasone above)]. The chemist will also assess the degree of branching and the different types of bonds present in the target molecule.

Example 5.1: Topology and Connectivity Describe features of interest for the two molecules below, esomeprazole and sertraline, in terms of their topology and bond connectivity.

Solution In terms of topology, each molecule has fused ring systems that give the molecule a particular geometric configuration. If you made a model of each molecule to see what the geometry is like, one portion of each molecule would be geometrically rigid. In terms of bonds and connectivity, each molecule has multiple double bonds. In sertraline, there are two chiral centers, and in esomeprazole, there is one chiral center that would complicate the synthesis of either molecule. In esomeprazole, there are oxygen, nitrogen, and sulfur atoms in the molecule, many of which are likely to be reactive.

Another interesting characteristic of the molecules is the stereochemistry. Some part of both molecules is planar (e.g., the benzene-like rings), which means that they can be set neatly atop a table. Other parts of the molecules are arranged in a tridimensional space.

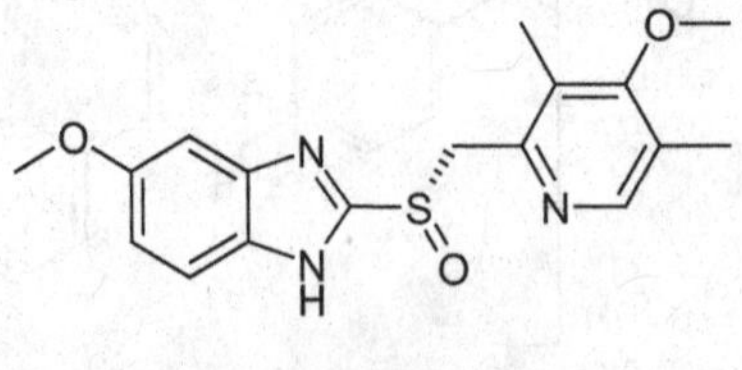

esomeprazole

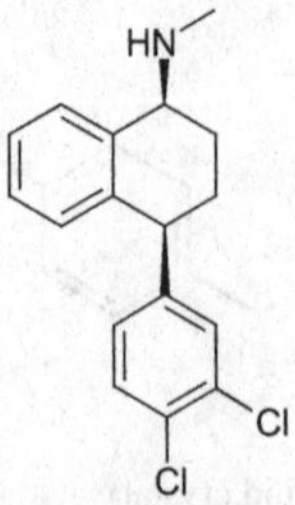

sertraline

Additional Points to Ponder How are the molecules oriented? What type of green chemistry implications can the molecular topology represent?

Chemists will also look at molecules and assess which bonds might become strategic points of disconnection. Points of disconnection are the reverse of how an actual chemical reaction would occur (i.e., how you would put the molecule together). For example, bonds to heteroatoms (C–O/N/S/P) are comparatively easy to form and often provide the desired strategic points of disconnection. Carbonyl groups (C=O) are very useful and are used extensively as points of disconnection. Generally, the chemist would start from the middle of the molecule; in the case of esomeprazole, the sulfonyl group would be a likely point of disconnection.

The chemist would also look to disconnect ring systems from carbon chains. In the ideal situation, as many bonds as possible would be formed for any given set of reaction conditions. The intent would be to add reactants and reagents sequentially or at the same time and in such a way that there would be no need to change solvents or reactors or perform isolations of any kind.

Example 5.2: Multicomponent Synthesis Reaction A chemist desires to make the following moderately complex molecule, 5-aminooxazole.

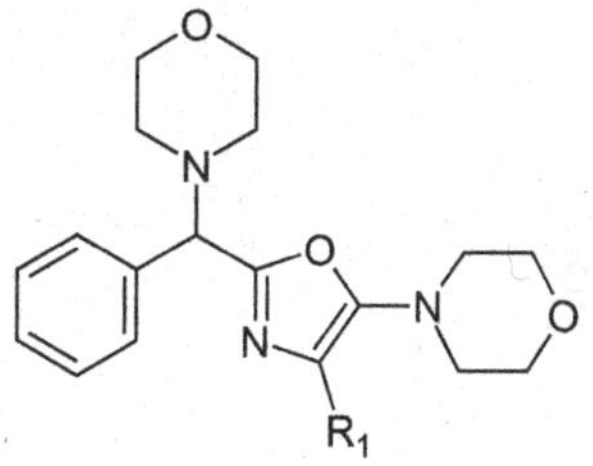

How might she do this?

Solution

Step 1. Describe several strategic points of disconnection and how the chemist might take the molecule apart.

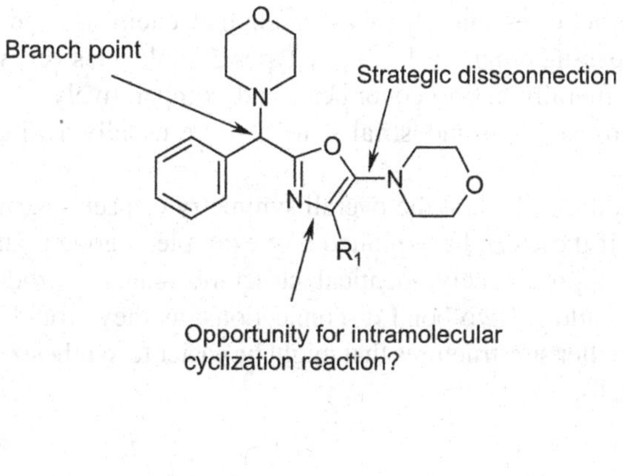

Step 2. Find some reasonably simple starting materials that might be used to make the molecule.

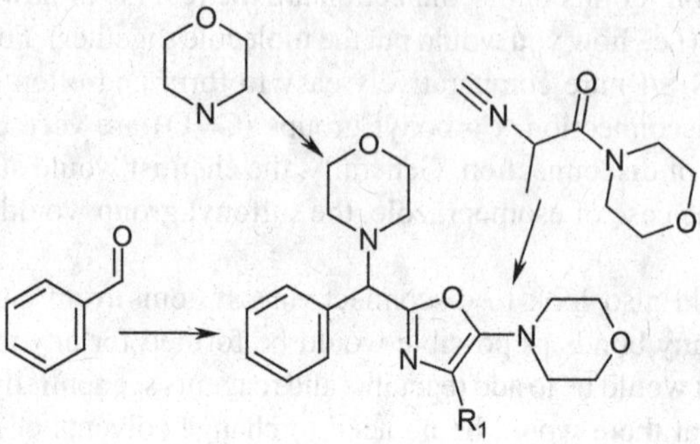

Step 3. Try the synthesis[16]:

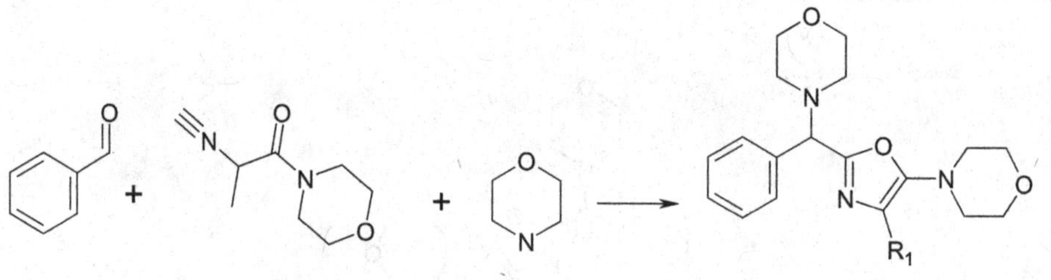

Additional Point to Ponder From a green chemistry point of view, what are the advantages and disadvantages of this synthesis?

In addition to domino reactions and other synthetic strategies, multicomponent reactions such as the above have been developed in recent years to yield very elegant total syntheses of natural products. There is an extensive specialist literature on these strategies, and the reader is referred elsewhere for further study.

Double (alkenes) and triple bonds (alkynes) can also be strategic sites for disconnection, although alkynes are generally very energetic materials with a degree of inherent hazard due to the pi bonds in the structure, and therefore represent a considerable process safety risk. Academic synthetic organic chemists, medicinal chemists, and those involved in high-throughput screening programs like these types of molecules very much because they are very reactive, generally atom economical, and comparatively "clean," but they are generally not safe to use on an industrial scale and are usually avoided in an industrial setting.

The chemist would also look at the overall symmetry or pseudosymmetry elements in the molecule to see if these can be exploited. For example, a good point of disconnection is one where two approximately identical structures can be produced. Even better in some instances would be two-bond disconnections, as they would generally simplify the molecule into smaller substructures that might be easier to synthesize or may already be available commercially.

Example 5.3: Synthesis of a Bicyclopentyl Ketone You work in a chemical company and are tasked by your manager to synthesize the following molecule:

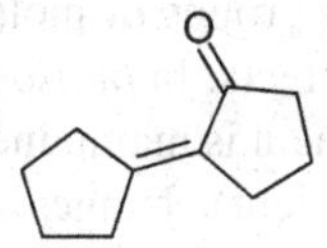

What steps might you suggest to synthesize this compound?

Solution

Step 1. Look at the symmetry of the compound. There are two cyclopentyl rings that could be disconnected at the double bond that joins the rings to form two approximately equivalent fragments.

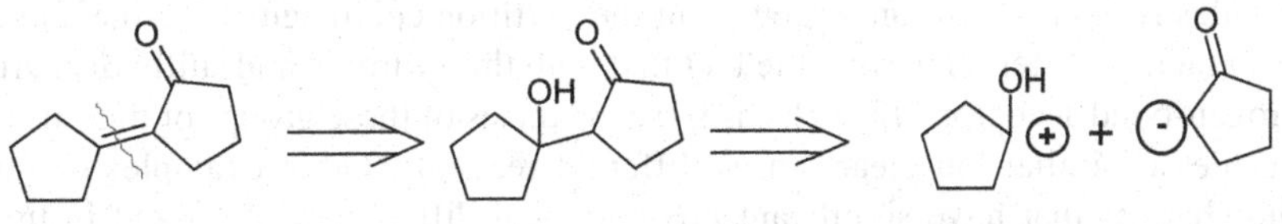

Step 2. Design a synthesis that makes use of commonly available starting materials to make the target molecule.

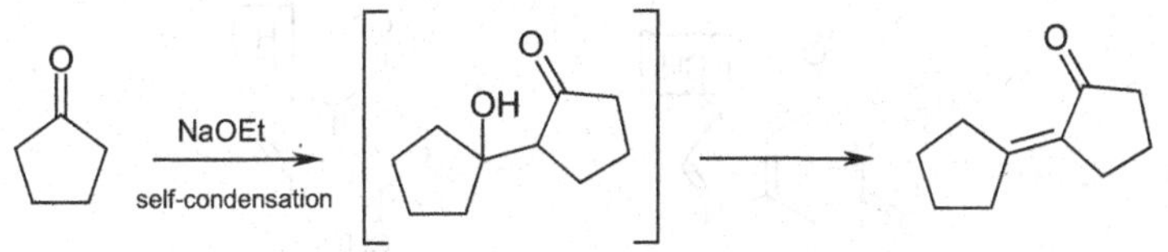

NaOEt

self-condensation

Additional Point to Ponder From a green chemistry perspective, why is it a good idea to use a self-condensation for this synthesis?

5.2.3 Identify Functional Groups and Their Relative Disposition

The chemist will typically recognize key structural motifs (recognize different patterns, arrangements of atoms, or substructures). Below is a series of compounds with a very similar ring structure but which differ in the pattern of heteroatoms and substituents, known as *congeners*:

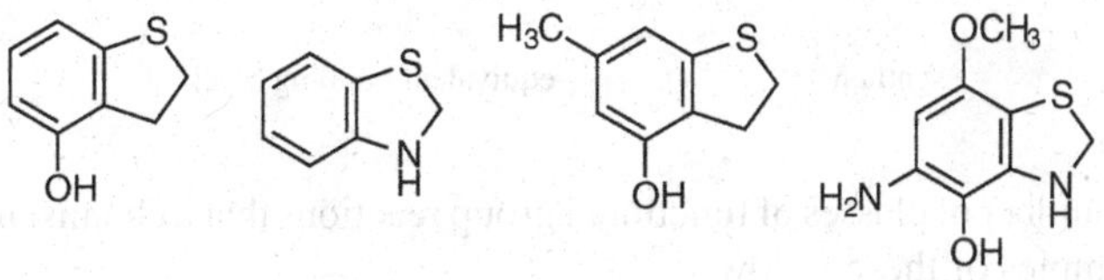

Although these molecules, have similar structures, they vary in their physicochemical properties and in the ease of their synthesis.

Another example of the importance of functional group analysis in designing molecules is illustrated by *bioisosteres*, subunits or groups of molecules that possess physicochemical properties having similar biological effects. In *bioisosteric replacement* the interaction of the functional group with the environment is maintained (i.e., hydrogen-bond acceptor) but the physical property (partition coefficient) changes.

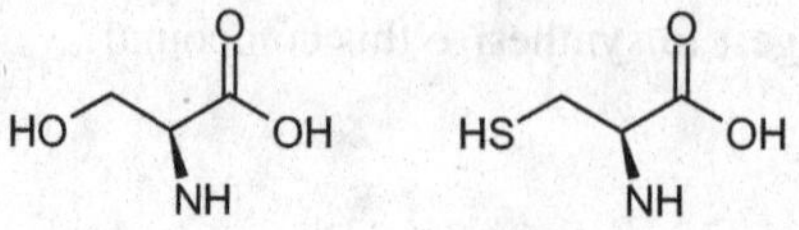

serine cysteine

In the case of the two simple amino acids above, the only difference between these molecules is in one functional group, but the partition coefficient of serine and cysteine will, in fact, be different despite the fact that both the hydroxyl and sulfhydryl groups are hydrogen-bond acceptors. In each case, the synthesis of these groups or their use in other molecules is similar but clearly uses different reagents. Other examples of functional group changes that have significant physiological differences are found in the simple example below. These molecules are structurally very similar, but there is a slightly different synthetic strategy for making each. They also have very different physiological effects!

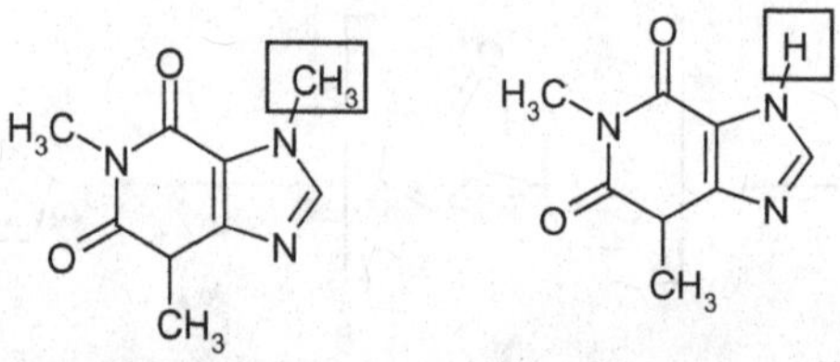

caffeine theophyline

In looking at these molecules, the chemist would try to identify key structural building blocks that could be put together using familiar classes of reactions (e.g., Claisen, aldol, Mannich, Diels–Alder, Heck, Suzuki). These structural building blocks, known as *retrons* or *synthons* are idealized fragments in the form of a cation or anion. They represent the points where the chemist will make strategic bond connections or disconnections.

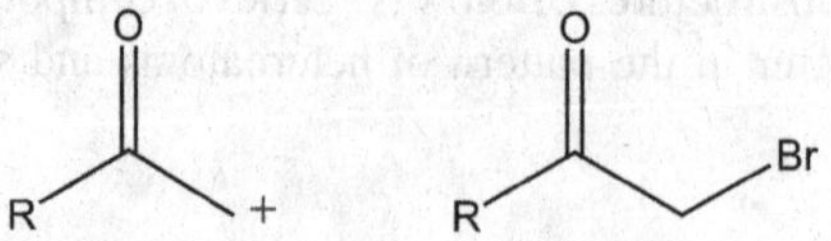

synthon equivalent building block

There are a number of classes of functional group reactions that a chemist uses, and a few illustrative examples of these follow.

Functional group interchange (FGI):

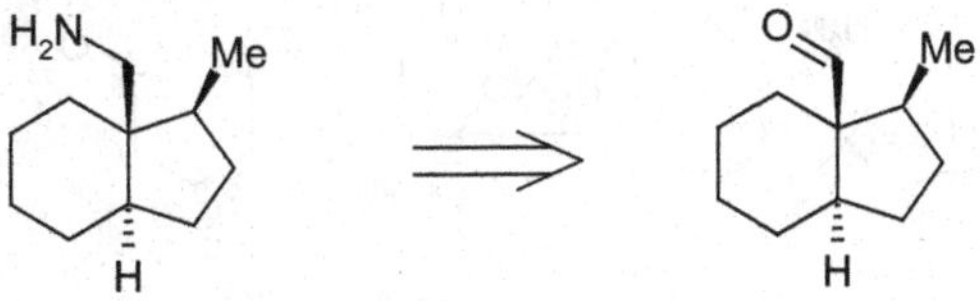

Functional group removal (FGR):

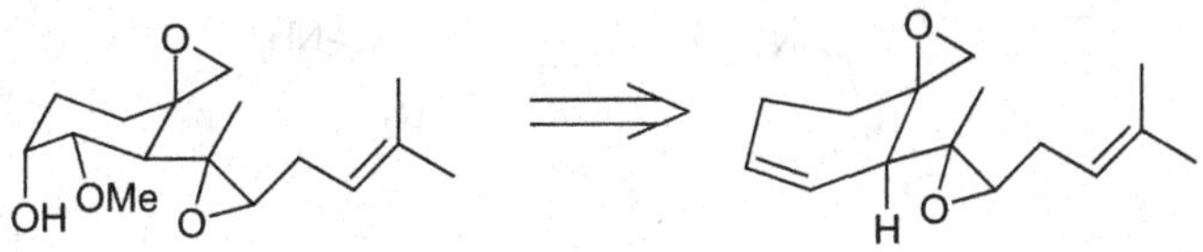

Functional group addition (FGA):

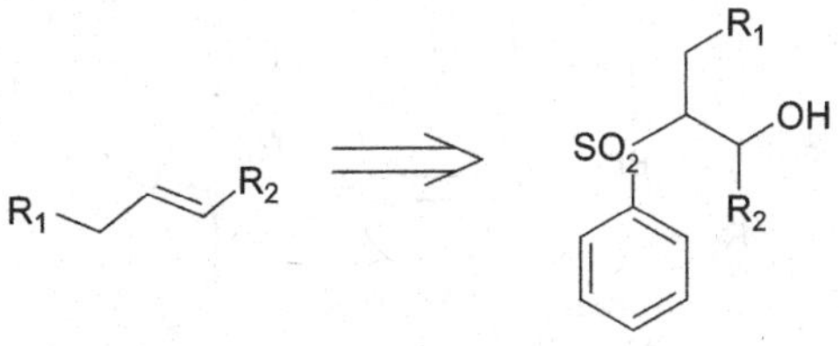

Single functional group transforms:

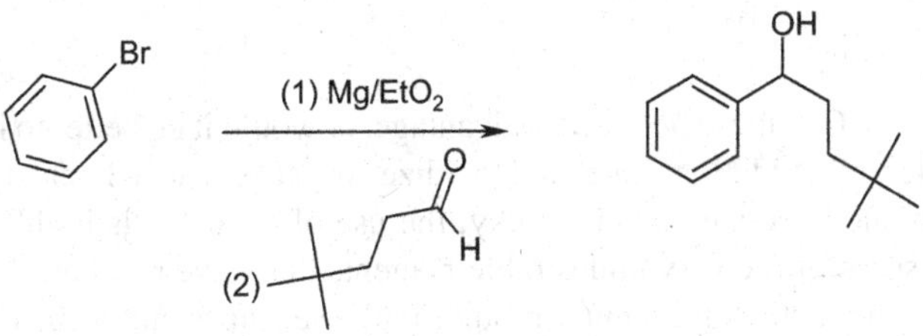

Multiple functional group transforms:

xylocaine

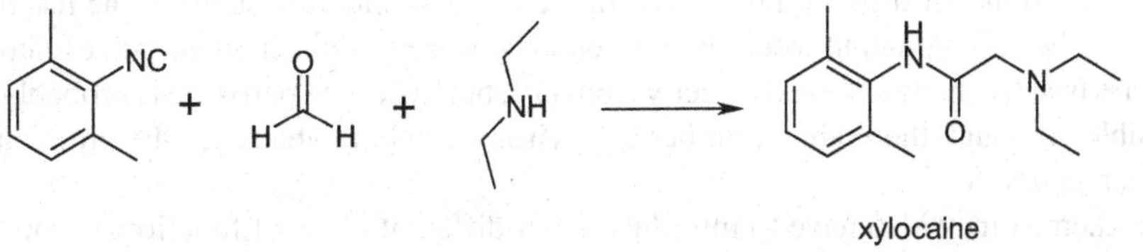

Rearrangement transforms:

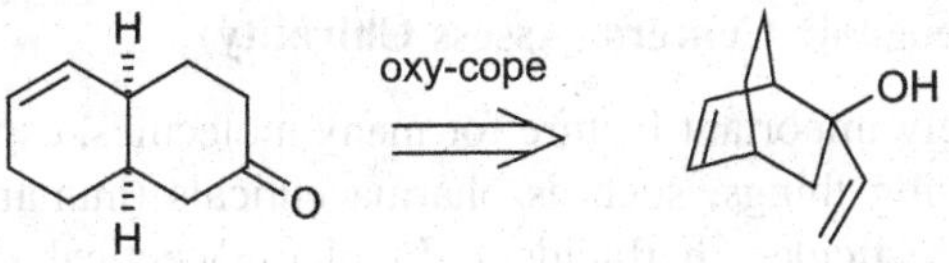

oxy-cope

Ring transforms:

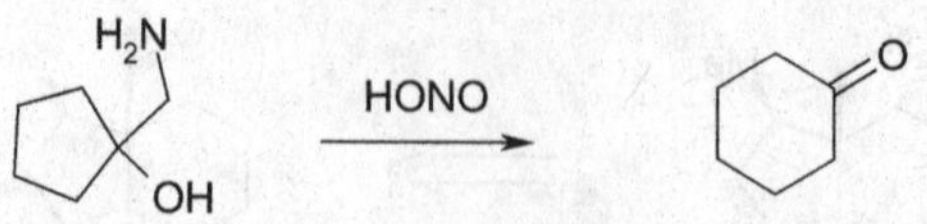

H_2N HONO OH O

The chemist would also want to analyze the oxidation states of functional groups contained in the molecule. For example:

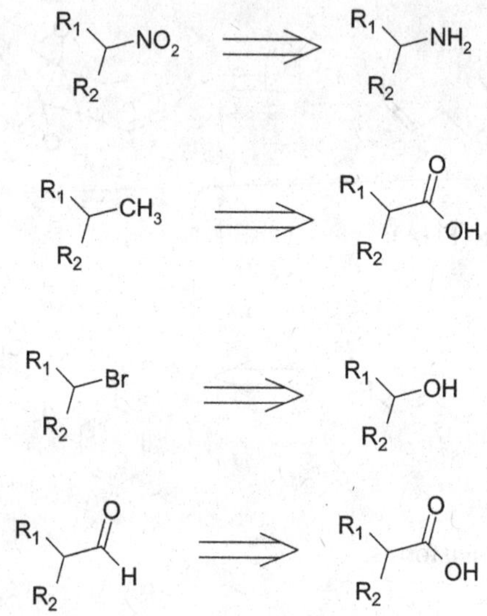

R_1 NO_2 R_2 ⟹ R_1 NH_2 R_2

R_1 CH_3 R_2 ⟹ R_1 O OH R_2

R_1 Br R_2 ⟹ R_1 OH R_2

R_1 O H R_2 ⟹ R_1 O OH R_2

A key question is: Can these be used to advantage, or would it be better to introduce these groups in a different oxidation state and oxidize or reduce at a later stage? Selective oxidations and reductions can be very tricky; the use of catalysts is highly desirable and may get around some of the very undesirable reagents that have been used historically. In both cases, these chemistries frequently require multiple stages and historically have very difficult waste streams [e.g., $LiAl(OH)_4$, MnO_4^-, $Cr_2O_7^{2-}$] that would require treatment and careful disposal. In general, the chemist will want to try to design the synthetic route so that key functional groups are introduced in the desired oxidation state. If one has to do a reduction, the best reductant would be elemental hydrogen; if one has to oxidize, elemental oxygen is best (perhaps obtained directly from air), but hydrogen peroxide is probably more accessible, although there are a number of environmental drawbacks to the production of hydrogen peroxide.

The chemist may also have to introduce other different types of functional groups in a masked or protected form, although we would want to avoid that if at all possible. More will be said about protection and deprotection a little later.

5.2.4 Look for Stereogenic Centers (Assess Chirality)

Chirality is an extremely important feature for many molecules, especially those that are destined for use in living things, such as pharmaceuticals (human and/or animal) and agrochemicals (e.g., pesticides, herbicides). In pharmaceutical compounds, different enantiomers can produce dramatic changes in the therapeutic properties of a molecule, with one enantiomer not being therapeutically active or being opposite in effect to the other enantiomer.

Chemists must recognize if there are multiple chiral centers and assess if and how they are related. As a general rule, it would be most desirable to introduce these chiral centers through stereo- or regioselective multicomponent reactions.

The choice of the strategy to introduce chirality has effects from the green chemistry viewpoint, as a racemic chemistry might represent increased waste production during resolution steps if the nondesired enantiomer cannot be re-racemized and recycled.

The degree of chirality in any given molecule will determine the approaches available for installing these centers. For example, chemists might be able to rely on the shape of the substrate to dictate the stereochemical outcome of the reaction (substrate control, as in the case of an intramolecular conversion, where there is significant steric hindrance that forces a particular configuration), or they may need to rely on external sources (reagent control, as in the case of classical resolving agents, or the use of asymmetric catalysts, or more recently, ionic liquids).

5.2.5 Assess the Overall Stability of the Molecule

Chemical stability is a key characteristic that chemists repeatedly exploit throughout a synthesis. For example, chemists will identify labile groups and think about the best times to incorporate these. Usually, if there is a labile group, it is best to introduce that group at later stages of the synthesis so that it does not need to be protected. If a portion of the molecule does need to be protected, the final stages of a synthesis often involve removal of the protecting group, so the stability of the product or the penultimate intermediate to the deprotection conditions is a key element of the synthetic design. The chemist also needs to assess whether or not the molecule is prone to rearrangement or decomposition under acidic or basic conditions, an especially key characteristic for hydrolysis reactions which generally precede salt formation.

5.3 USING DIFFERENT CHEMISTRIES

Although it is true that synthetic organic chemists are challenged continuously by the very new and different target molecules they are asked to make, it is also true that they will generally design their synthesis using reactions and chemistries with which they are familiar and know are robust (i.e., they know that the chemistries are likely to work at an industrial scale). By way of example, in the case of forming C–C bonds, a number of possible reactions could be used, such as a Knoevenagel condensation, a Grignard reaction, cyanation, C-acylation, and C-alkylation (electrophilic or nucleophilic). Clearly, selection of the preferred way of making a C–C bond will depend on such factors as the reactivity of other functional groups present, the stability of the substrate and product under the reaction conditions, requirements for onward processing, and others.

Table 5.1 gives you an idea of some classes of reactions that chemists use. This particular short list was taken from a review of 299 different reactions used to make about 38 different drug products. It is interesting and perhaps significant to note that about 28 different chemistries accounted for 227 of those reactions, or 76% of the total number of reactions studied. This suggests that it may be possible to focus on a limited number of reaction classes and to develop new greener chemistries for each.

TABLE 5.1 Categorizations of Chemistries

Purpose	Reaction Category	Reaction Name/Type
Forming a new carbon–oxygen bond	O-Alkylation	Ether synthesis
Forming a new carbon–carbon bond	C-Alkylation	Alkylation of an aromatic
	Addition to C=O	Knoevenagel
Forming a new carbon–nitrogen bond	N-Acylation	Amidation
	N-Alkylation	Of heterocycle
	N-Alkylation	Of amine
Forming a new carbon–sulfur bond	S-Alkylation	Thio ether synthesis
Reduction	Catalytic hydrogenation/ hydrogenolysis	
	Metal hydride	Lithal
Cyclization	Heterocycle synthesis	By miscellaneous ring closure
Elimination	C=C formation	
Hydrolysis	Acid catalyzed	
	Base catalyzed	
Halogenation	Halogenation of an alcohol	
Salt formation	Acid or base	
Neutralization		
Resolution	Using diastereoisomers	Acid or amine

You may recall that Table 4.1 showed a selection of key metrics for the types of chemistries shown in Table 5.1. By expanding on these metrics, it is possible to rank chemistries according to their preference of use. Using the data set described above, this was done for each of the reactions. Figure 5.1 is a chart of targeted bond types to be formed and the types of chemistries employed to produce these bonds, and ranks these according to the efficiency of the chemistry, their mass efficiency, the use of energy, and solvent use. One could, of course, argue that the particular metrics and categories used were not the best to use, but at least the output depicted in Figure 5.1 represents a good first step in taking a multivariate view of chemistries.

Example 5.4: Ether Synthesis A chemist desires to protect an alcohol by making an ether as one step in a drug synthesis and uses an O-alkylation reaction to combine an alcohol from a preceding step with benzyl bromide. The complete reaction looks as follows:

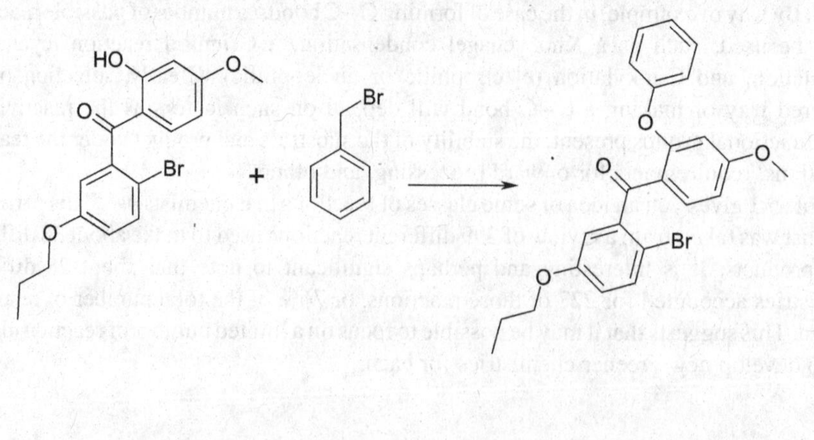

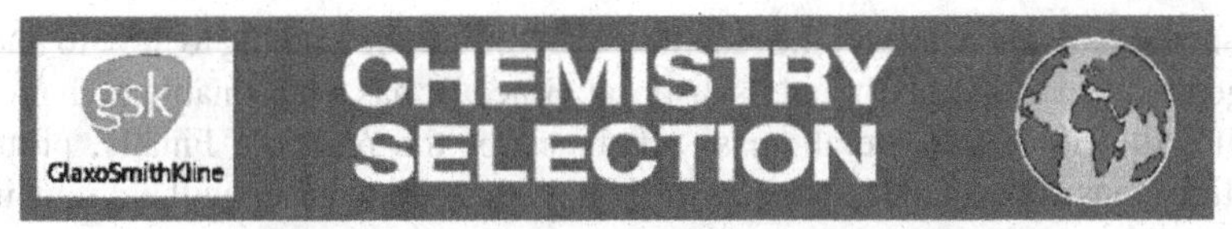

A guide for the integration of environmental, health and safety factors into the selection of chemistries

Bonf Type	Chemistry	Chemistry	Mass	Energy	Solvent
C-C	Knoevenagel	14	10	9	10
	Cyanation	8	7	8	6
	C-acylation	4	6	9	6
	C-alkylation	4	4	3	8
	Grignard	1	1	2	4
C-H	Hydrogenation	7	5	4	8
	Borohydride	6	5	4	7
	Lithal	5	3	1	4
C - Halogen	Chlorination	7	8	9	8
	Bromination	9	7	6	8
C-N	N-alkylation	8	7	5	9
	Amination	5	8	5	8
	N-acylation	5	5	8	7
C-O	O-alkylation	6	8	10	8
	Epoxidation	4	5	7	6
C-S	S-alkylation	7	9	9	8
General	Esterification	6	8	9	9
	Base salt	9	4	8	8
	Hydrolysis (Acid)	8	8	3	7
	Acid salt	7	6	6	8
	Sulfonation	8	6	4	7
	Elimination	5	7	6	6
	Cyclisation	4	4	3	6
	Hydrolysis (Base)	7	1	2	3

Footnotes:
1. Chemistry evaluates yield and reaction mass efficiency
2. Mass is based on the mass of materials (excluding water) needed to make product
3. Energy is based on the energy needed
4. Solvent is based on recoverability and the number of different solvents required

EXPLANATION

Composite scores over a range of 1 - 10 are given in 4 key areas.
The higher the score, the better the chemistry.

Major issues identified.

Issues identified.

No major issues.

This chart was derived from an evaluation of approximately 250 examples of chemical reactions used in the most optimised processes developed within R&D and manufacturing over the last 10 years.

For each chemistry there is:
- a summary sheet that details key issues and provides selected information about the chemistry, e.g., on the typical solvents used
- detailed process descriptions for the best examples of each chemistry ("best in class")

The summary sheets and additional information may be found in the Green Chemistry Guide Intranet Site at:
http://kopsalt01/eps/Gcg/default.htm or http://greenchemistry.gsk.com

Corporate Environment, Health and Safety, Environmental Health & Safety Product Stewardship
CHEMISTRY GUIDELINES: Version 1, 1st June 2002
For GSK internal use and guidance only. For further information or comments Email environment.matters@gsk.com

FIGURE 5.1 Example of chemistry types ranked by their relative greenness.

Using the GSK green chemistry chart, what conclusions could be drawn about this synthesis? What might be some concerns with this synthesis?

Solution Looking at the chart, the chemist sees that for an O-alkylation reaction, the scores are generally quite good. As this is a comparatively simple reaction, the chemist is confident that she could make this into a high-yielding mass-efficient reaction. However, because of the molecular weight, there is some concern about the best solvent, so she chooses a dipolar aprotic solvent to test the reaction. She starts with dimethylformamide but notes that these types of solvents carry an R61 phrase, harm to the unborn child, so she will have to see if she can find another solvent to use for the reaction. In addition, she is going to have to find a way to manage the formation of HBr and its disposal. Since Br has a reasonably high molecular weight, it will reduce the overall mass efficiency of the reaction. Finally, protection–deprotection steps are best to avoid since the benzyl protecting group will not remain in the final molecule.

Additional Points to Ponder Is there another reason to avoid protection–deprotection steps? What strategy can be followed to avoid the need to protect the alcohol?

5.4 ROUTE STRATEGY

The concept of the ideal synthesis is not a new one, and it is one for which most synthetic organic chemists strive. A good definition for the ideal synthesis was put forward by Paul Wender as "the target molecule is made from readily available starting materials in one simple, safe, environmentally acceptable and resource effective operation that proceeds quickly and in quantitative yield."[17] Wender has also extensively promoted the related idea of *step economy*, which he defines as follows: "Step economy is the drive to reduce the number of synthetic steps (and hence the associated waste such as solvents used in the reaction, solvents used in workup and product isolation, solvents used in product purification, silica gel or related substances used in chromatographic purification, contaminated aqueous waste generated from cleaning equipment and glassware, etc.) in the synthesis of any given target molecule."[18] Step economy has been closely tied to the idea of using only those steps that greatly increase the overall complexity of the molecule.

So how might an ideal synthetic route look? For the sake of simplicity, we examine two general approaches that the synthetic chemist may take in designing a complex synthesis. In the first instance, the chemist can string together a series of reactions in a sequential fashion as follows:

$$A + B \rightarrow AB + C \rightarrow ABC + D \rightarrow ABCD + E \rightarrow ABCDE + F \rightarrow ABCDEF$$

This would be known as a linear route. In the second instance, the chemist could create a series of smaller fragments or parts of the molecules in simple, high-efficiency chemical reactions:

$$A + B \rightarrow AB$$
$$C + D \rightarrow CD$$
$$E + F \rightarrow EF$$

Then combine these initial molecular fragments in two more reaction steps,

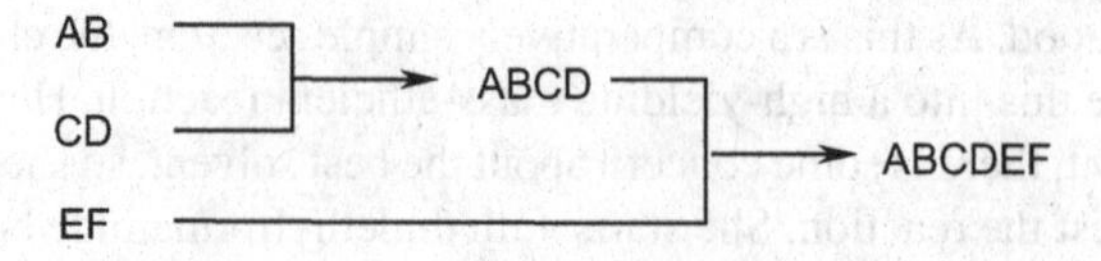

to yield the overall synthesis and the desired target molecule:

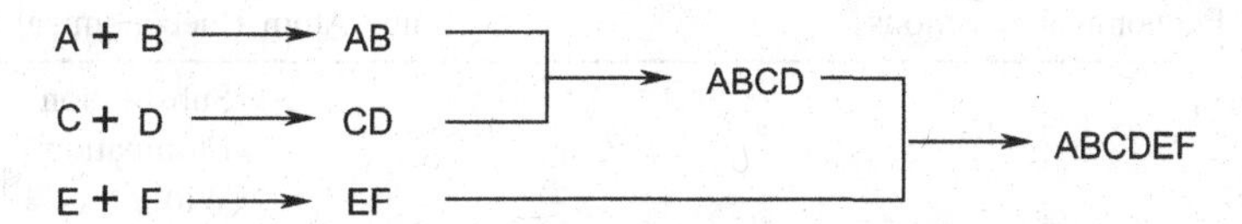

Now, which approach is best? Well, conventional wisdom, based on a mathematically driven yield perspective, would undoubtedly say that the convergent synthesis would appear to be the best bet. Let's see why.

First, we take the linear route, with each step having an average 80% yield, which is not a bad first approximation for what happens in the real world:

$$\mathrm{A+B} \rightarrow \underset{80\%}{\mathrm{AB}} + \mathrm{C} \rightarrow \underset{64\%}{\mathrm{ABC}} + \mathrm{D} \rightarrow \underset{51\%}{\mathrm{ABCD}} + \mathrm{E} \rightarrow \underset{40\%}{\mathrm{ABCDE}} + \mathrm{F} \rightarrow \underset{32\%}{\mathrm{ABCDEF}}$$

approx. overall yield

and we see that to obtain the desired molecule, we have a yield of about 32%.

Now let's take a look at a convergent route and once again assume that each step has an average 80% yield:

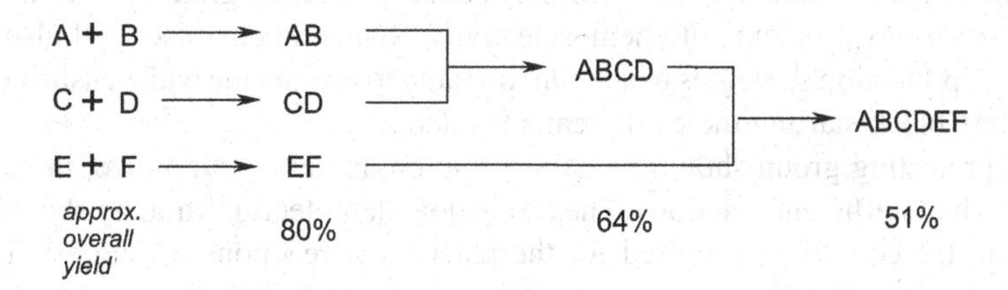

We see that on the basis of yield, the chemist would undoubtedly choose the convergent synthesis; there are fewer steps, and the overall yield of the final product is 51%. But is this really an accurate measure of how good the synthesis is? As discussed in Chapter 4 and more rigorously elsewhere, yield is probably not the best measure for determining whether or not a particular route is green. In addition, we may want to consider a few other points about linear vs. convergent route designs.

1. In actual fact, all routes are essentially linear. Just because we introduce a molecular fragment through a coupling reaction for the sake of reducing the number of steps does not necessarily mean that we reduce the complexity of the synthesis or automatically have a more mass-efficient reaction. For example, in the idealized convergent synthesis above, making AB, CD, and EF could use very atom-uneconomical reactions and mass-intensive reactions, but they could still have relatively high yields.
2. A linear route that has a sequential series of high-yield atom-economical, mass-efficient reactions can have greater overall yields than a convergent synthesis of mass-inefficient, low-atom-economical reactions which also possess lower yields. In Table 5.2 one could imagine a convergent synthesis that employs reactions in the right-hand column while a linear synthesis employs reactions in the left-hand column.

TABLE 5.2 Examples of Atom-Economical and Atom-Uneconomical Reactions

Some Atom-Economical Reactions	Some Atom-Uneconomical Reactions
Rearrangement	Substitution
Addition	Elimination
Diels–Alder	Wittig
Other concerted reactions	Grignard

3. A highly convergent route using complex starting materials may be masking mass-inefficient, low-atom-economical and low-yielding processes to produce the starting materials you are working with. This is a case of paying attention to where you draw the boundaries of your evaluation of the overall greenness of your reaction.

5.5 PROTECTION–DEPROTECTION

A *protecting group* is a reagent that is used by chemists to keep a specific atom or functional group from reacting. They may not want the functional group to react because it may interfere with another reaction of interest, or they may need to preserve a particular functionality in the substrate molecule. Simply stated, protecting groups allow the chemist to overcome various problems of chemoselectivity. Synthetic chemists will also employ protection–deprotection strategies to add functionality to a molecule while ensuring that the basic building block/starting molecule remains intact.

A good protecting group should be easy to put on and easy to remove, in both high-yielding and mass-efficient reactions. The protection–deprotection strategy should also be unreactive to the conditions required for the reaction or reactions of interest. Table 5.3 contains a summary of commonly protected functional groups and the types of groups they are protected as.

Example 5.5: Protection–Deprotection of a Ketone You work in a chemical company and have been asked by your manager to produce the primary alcohol from a ketoester as shown in the following reaction, using $LiAlH_4$ as a reducing agent:

R1 (O) (O) OH —?→ R1 (O) OH

How would you achieve that?

TABLE 5.3 Commonly Used Protecting Strategies

Functional Group to Protect	Protected as:
Alcohols (carbohydrates, sugars, R–OH)	*t*-Butyl ethers
Amines (first and second degree, e.g., $R{-}NH_2$, R–NH–R)	Acetamides
Aldehydes	Acetals
Ketones	Ketals
Carboxylic acids	*t*-Butyl esters

Solution $LiAlH_4$ will reduce the ester, but since $LiAlH_4$ is a very good reducing reagent that would reduce the ketone as well, it would be an unwanted reaction. We can avoid this problem if we first "change" the ketone to a different functional group. Conceptually, this is like being able to put a cover over the ketone while we do the reduction, then remove the cover. In reality, the "molecular cover" is a protecting group.

Step 1. In this example, you could first protect the ketone as an acetal (which is an ether that doesn't react with $LiAlH_4$):

HO OH
R1 OEt → TsOH/toluene → R1 OEt

Step 2. Then you could reduce the ester:

LiAlH4/THF
R1 OEt → acidic workup → R1 OH

Step 3. Finally, you will have to remove the acetal to obtain the desired primary alcohol:

H_3O^+
R1 OH → heat → R1 OH

Additional Point to Ponder What alternative to this strategy would you suggest to your manager to avoid protection–deprotection steps?

In this simple example it should be readily apparent that protection–deprotection has features that render it less than desirable from a green chemistry perspective. Let's list a few of these:

1. Protection–deprotection introduces at least three additional steps to the synthetic route. This works against the desire for step economy and the desire to find reactions that add complexity to the molecule. Any complexity that is added in these reactions is taken away.
2. Additional solvents and reagents are required for each step that can add rather significantly to the mass intensity of the reaction.
3. The molecule that the chemist uses to protect the functional group of interest is not retained in the final product. In some instances, protecting groups can represent extremely large amounts of mass.

To illustrate this last point, we can use the case of amino protecting groups, for which three commonly used protecting reagents are

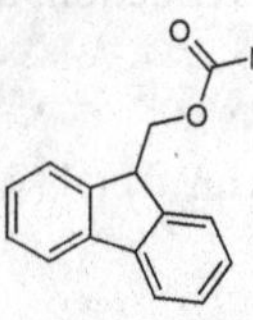

9-fluorenylmethyl carbamate
(Fmoc-NRR′)

FW = 279.4

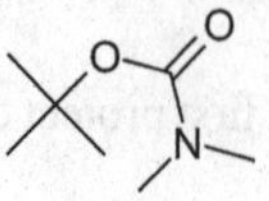

t-butyl carbamate
(Boc-NRR′)

FW = 145.2

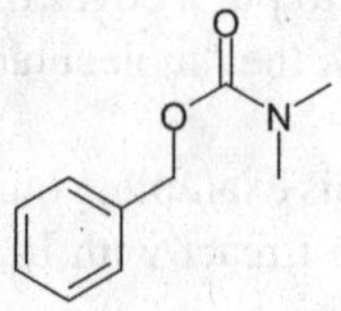

benzyl carbamate
(Z-NRR′, Cbz-NRR′)

FW = 179.2

As you can see from these molecules and their formula weights (FWs), they represent reasonably significant fragments to end up in a waste stream. Protection–deprotection steps are sometimes unavoidable, but as a general rule, one should do everything possible to avoid having to use this approach in a synthetic strategy.

5.6 GOING FROM A ROUTE TO A PROCESS

Once the synthetic chemist is able to put together a route that appears to work, the next job is to see how well the chemistry actually works, and turn it into a workable process. Let's take the first step of a drug synthesis as an example and see what a process chemist might start with:

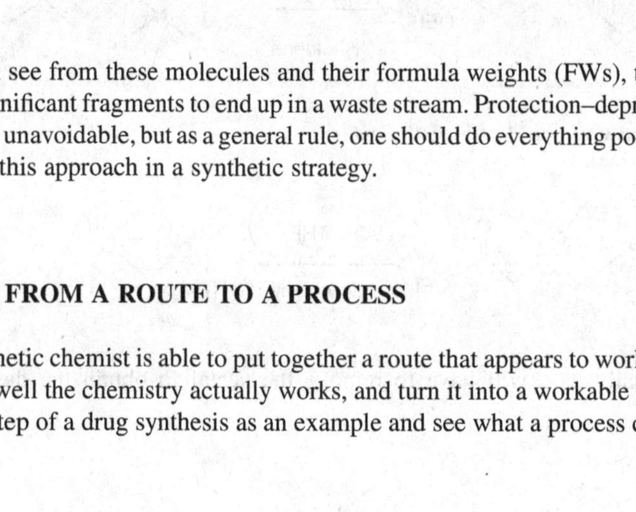

Br, Me, F
(1) Mg/THF
(2) $(t\text{-BuCO})_2O$
$t\text{-BuO}_2C$, Me, F

What is missing from this simple chemical route schematic? Do you know, for example, how much of each material is required, whether or not heating or cooling is involved, and if any separations are required? If you included all this information, what might this look like?

Figure 5.2 is a process flow diagram of this single step. As, hopefully, is readily apparent, a great many steps are actually required to carry out what is summarized in the schematic above. In Chapter 9 we go into greater detail about how to take all the various components of a process flow diagram and turn it into a mass and energy balance (i.e., account for everything that goes into and out of a process). For our purposes at this point, it is sufficient to understand that more than just the basic chemistry is depicted in most chemical reaction flowsheets involved in making the actual material.

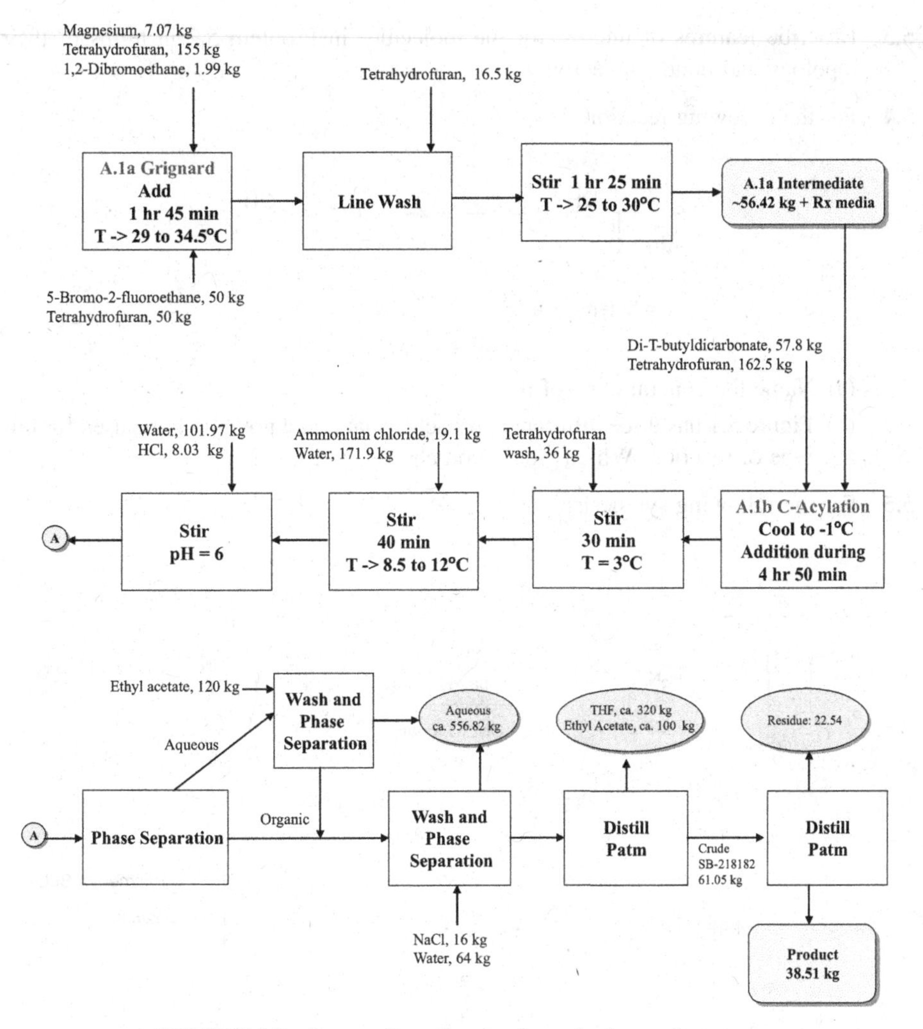

FIGURE 5.2 Process flow diagram for a single reaction step.

PROBLEMS

5.1 In your own words, describe what a retrosynthetic analysis is and how this can be used to design a greener synthesis.

5.2 Look at the following molecules and identify key functional groups, points of disconnection, symmetry elements, and key bonds.

(a) (b) (c)

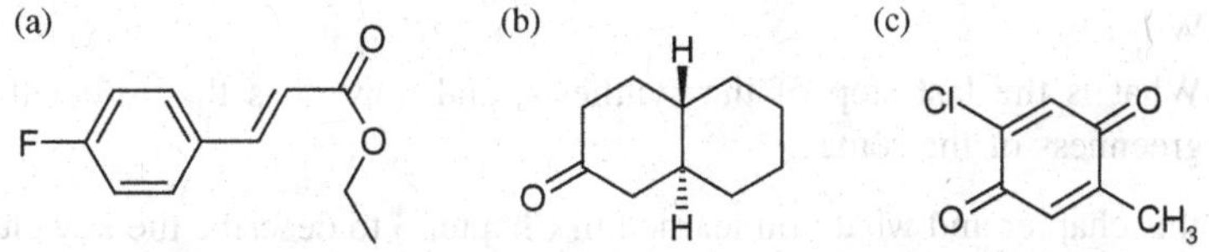

5.3 Describe features of interest for the molecules in Problem 5.2 in terms of their topology and bond connectivity.

5.4 For the following reaction:

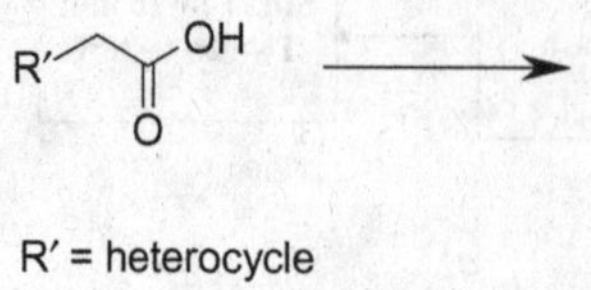

R′ = heterocycle

(a) Name the general class of reaction.

(b) Figure 5.1 has three different approaches that could potentially be used for this type of reaction. Which would you choose, and why?

5.5 For the following synthesis:

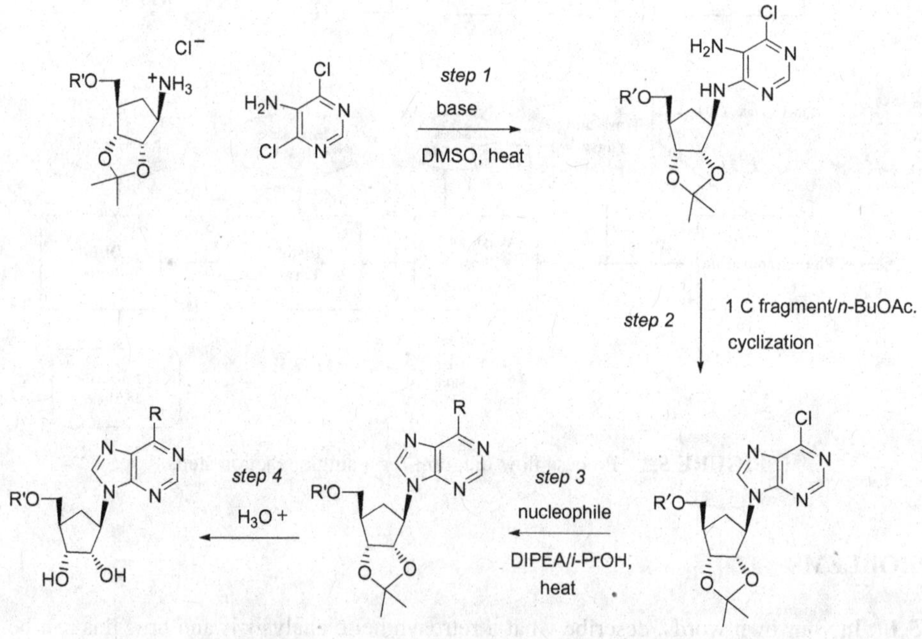

step 1
base
DMSO, heat

step 2
1 C fragment/*n*-BuOAc.
cyclization

step 3
nucleophile
DIPEA/*i*-PrOH,
heat

step 4
H_3O^+

(a) Draw an idealized synthesis route for this synthesis.

(b) Draw a retrosynthetic analysis from the final product.

(c) The chiral ammonium salt starting material is already a complex molecule. Why?

(d) What is the last step of the synthesis, and how does this affect the overall greenness of the route?

5.6 Use this chapter and what you learned in Chapter 3 to describe the key elements of an ideal synthesis.

5.7 Using the Craig synthesis of nicotine, we have

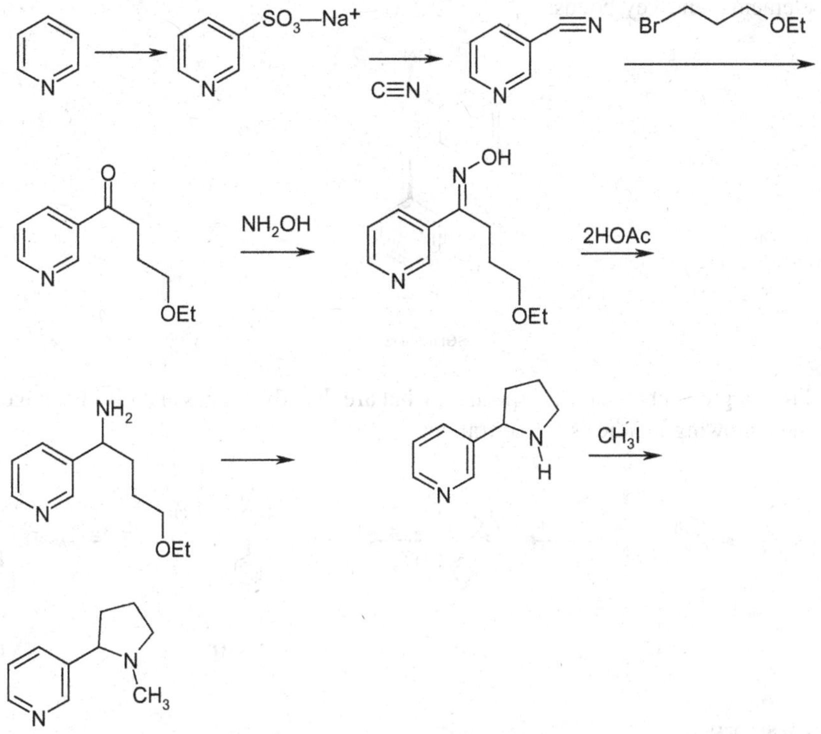

SO$_3$—Na$^+$ C≡N ≡N Br OEt NH$_2$OH N OH OEt 2HOAc NH$_2$ OEt N H CH_3I N CH_3

(a) Is this a linear or a convergent synthesis?
(b) If every step had a yield of 75%, what would the overall yield be?
(c) Draw a retrosynthetic representation of this synthesis.
(d) Which functional groups have a change in oxidation state?
(e) Compute the overall atom economy for this synthesis. Is this good? Why or why not?
(f) What are the key functional groups exploited in this synthesis?

5.8 Based on the information presented in Figure 5.2, estimate the mass efficiency, yield, and reaction mass efficiency for the production of the product.

5.9 Using the GSK green chemistry guide in Figure 5.1, what conclusions could be drawn about the reaction of Example 5.3? What would be any green chemistry concerns?

5.10 For esomeprazole, identify key functional groups, points of disconnection, symmetry elements, and key bonds.

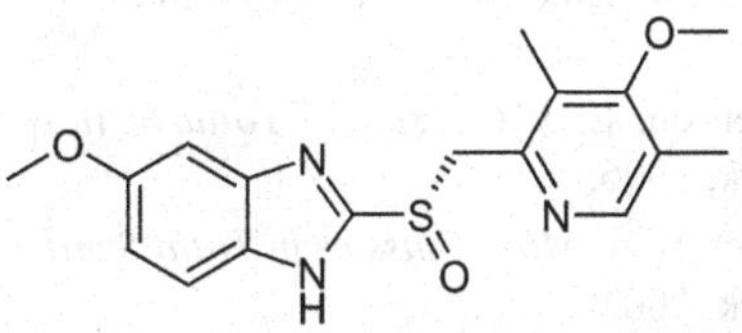

esomeprazole

5.11 For sertraline, identify key functional groups, points of disconnection, symmetry elements, and key bonds.

sertraline

5.12 From a green chemistry perspective, what are the advantages and disadvantages for the following synthesis for sertraline:

V —MMA/TiCl$_4$, THF→ VI (non-isolation) —1. H$_2$/Pd/THF 2. HCl→ VII —D(-) mandelic acid, EtOH→ XXIII —EtOAc/HCl→ VIII

REFERENCES

1. Anastas, N. D., Warner, J.C. The incorporation of hazard reduction as a chemical design criterion in green chemistry. *Chem. Health Saf.*, 12(2), Mar.–Apr. 2005, 9–13.
2. Anastas, N. D. Incentives for using green chemistry and the presentation of an approach for green chemical design. In *Green Chemistry Metrics*, Lapkin, A., Constable, D. J. C., Eds. Blackwell, London, 2008, pp. 27–40.
3. Warren, S. *Designing Organic Syntheses: A Programmed Introduction to the Synthon Approach*. Wiley, New York, 1977.
4. Warren, S. *Organic Synthesis: The Disconnection Approach*. Wiley, New York, 1982.
5. Warren, S. *Workbook for Organic Synthesis: The Disconnection Approach*. Wiley, New York, 1982.
6. Corey, E. J., Cheng, X. M. *The Logic of Chemical Synthesis*. Wiley, New York, 1989.
7. Carey, F. A., Sundberg, R. J. *Advanced Organic Chemistry*, Part B, 4th ed. Springer-Verlag, New York, 2007.
8. Smith, M. B., March, J. *Advanced Organic Chemistry*, 5th ed. Wiley, New York, 2001.
9. Anand, N., Bitra, J. S., Randanathan, S. *Art in Organic Synthesis*, 2nd ed. Wiley, New York, 1988.
10. Nicolaou, K. C., Sorenson, E. J. *Classics in Total Synthesis: Targets, Strategies, Methods*. Wiley-VCH, New York, 1996.
11. Nicolaou, K. C., Snyder, S. A. *More Classics in Total Synthesis: Targets, Strategies, Methods*. Wiley-VCH, New York, 2003.
12. Mundy, B. P., Ellerd, M. G., Favaloro, F. G. *Name Reactions and Reagents in Organic Synthesis*, 2nd ed. Wiley, Hoboken, NJ, 2005.

13. Wuts, P. G. M., Greene, T. W. *Greene's Protective Groups in Organic Synthesis*, 4th ed. Wiley, Hoboken, NJ, 2006.
14. Ho, T.-L. *Tactics of Organic Synthesis*, Wiley, New York, 1994.
15. Nicolaou, K. C. The art and science of total synthesis at the dawn of the twenty-first century. *Angew. Chem. Int. Ed.*, 2000, 39, 44–122.
16. Sun, X., Janvier, P., Zhao, G., Bienayme, H., Zhu, J. Three-component synthesis of 5-aminooxazole, *Org. Lett.*, 2001, 3, 877–880; *J. Am. Chem. Soc.*, 2002, 124, 2560–2567.
17. Wender, P. A., Handy, S. T., Wright, D. L. Towards the ideal synthesis. *Chem. Ind. (London)*, 1997, 765–769.
18. Wender, P. A., Miller, B. L. In *Organic Synthesis: Theory and Applications*, Vol. 2, Hudlicky, T., Ed. JAI, Greenwich, CT, 1993, pp. 27–66.

6

MATERIAL SELECTION: SOLVENTS, CATALYSTS, AND REAGENTS

What This Chapter Is About In this chapter we address one of the key components of chemistry that contributes to the achievement of more sustainable chemical reactions and reaction pathways: material selection. We also provide an approach to selection of solvents, reagents, and catalysts that promotes the overall greenness of a chemical synthesis without diminishing the efficiency of the chemistry or associated chemical process.

Learning Objectives At the end of this chapter, the student will be able to:

- Understand the role that solvents, reagents, catalysts, and other ancillary materials play in green chemistry and process design.
- Understand and apply the hierarchy of eliminating, replacing, minimizing, and mitigating the environmental, health, and safety hazards related to solvents and other ancillary materials.
- Understand and apply various material selection strategies.
- Contrast, analyze, and evaluate potential material alternatives based on green chemistry and green engineering principles and metrics.

6.1 SOLVENTS AND SOLVENT SELECTION STRATEGIES

Solvents have been used in all type of industries throughout history for processing, manufacturing, and formulation of all type of products (Figure 6.1). In the chemical processing industries, solvents are used in all types of unit operations, including all types of separations (gas–liquid, liquid–liquid, solid–liquid), situations where the solvent acts as

Green Chemistry and Engineering: A Practical Design Approach, By Concepción Jiménez-González and David J. C. Constable

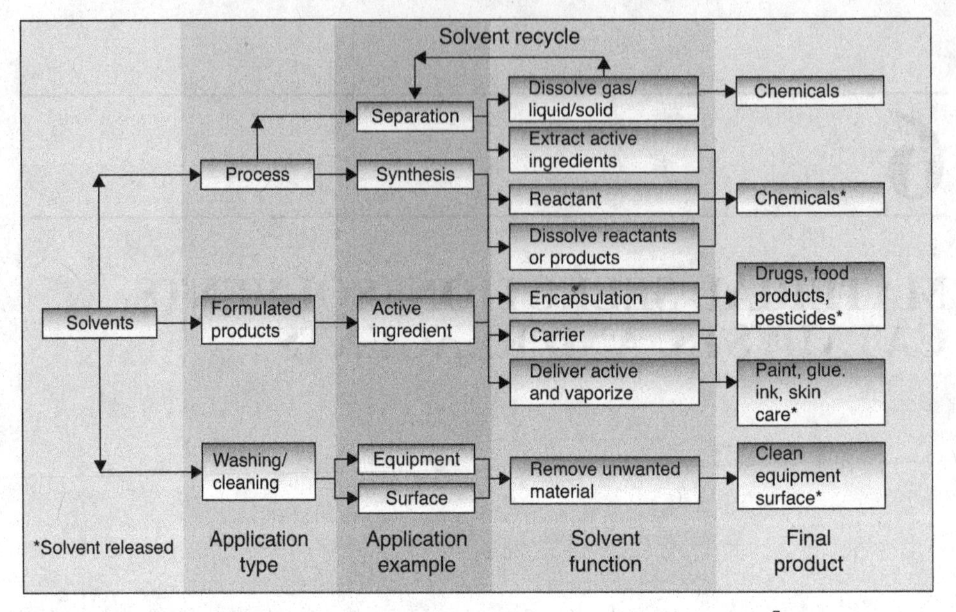

FIGURE 6.1 Some solvent uses and examples of applications. (From Gani et al.[7], with permission of *Chemical Engineering*. Copyright © 2006 Access Intelligence, New York.)

a reaction medium or as a carrier or reactant, in equipment cleaning, and in many more processes. Solvents are also widely used as part of product formulations, such as in paints, textiles, rubber, adhesives, and thousands of other products found at work and home. In general, solvents are used for many applications in industry, academia, and domestic situations.

While solvents have a very wide range of applications, many commonly used solvents pose serious environmental, health, and safety (EHS) hazards, including human and ecotoxicity hazards, process safety hazards, and waste management issues. In addition, solvents are subject to increasing and continuous regulatory scrutiny.[1,2] The issues cited above, among others, highlight the need to avoid the use of organic solvents as much as possible. As covered in Chapter 2, the use of ancillary materials not integrated in the product should be avoided, and when organic solvents and other ancillary materials must be employed, their use must be minimized and optimized to enhance reactions with minimal environmental and operational concerns, as shown graphically in Figure 6.2.

In other words, chemists and engineers designing a process are faced with the challenge of avoiding the use of organic solvents where possible, and when elimination is not feasible, solvent selection plays an important role in ensuring that the EHS impacts are minimized. Once a solvent is selected, green chemistry and green engineering principles direct efforts to optimize and minimize solvent use, and recycle solvents as much as possible in order to reduce the EHS impact and move toward more sustainable practices.

6.1.1 Environmental, Health, and Safety Hazards of Organic Solvents

In this section the major EHS concern is related to the use of organic solvents. These environmental, health, and safety hazards and potential impacts will also have repercussions in the way that companies deal with waste management.

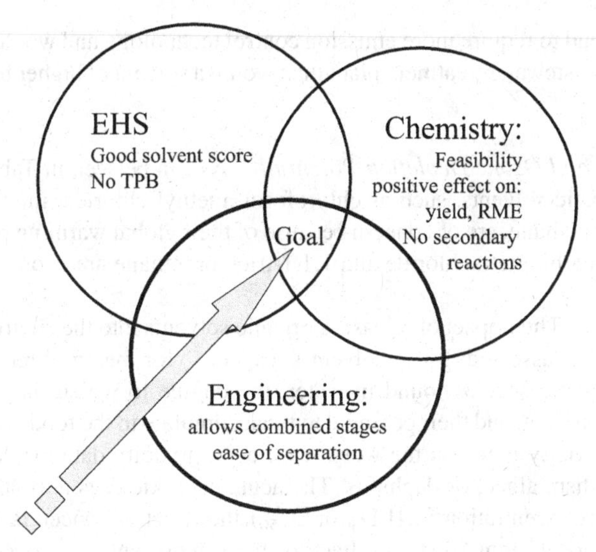

FIGURE 6.2 The goal is to select a solvent that promotes chemical reactivity and efficient downstream processing with minimal EHS impacts. TPB, toxic, persistent, bioaccumulative; RME, reaction mass efficiency. (Reprinted with permission from Gani et al.[11] Copyright © 2005 Elsevier.)

Environmental Issues

Photochemical Ozone Creation Potential (POCP) Most organic solvents are volatile compounds that present concerns regarding their potential to produce photochemical ozone. As discussed in Chapter 3, volatile organic compounds (VOCs) are capable of producing tropospheric ozone by reacting with nitrogen oxides, contributing to smog formation. POCPs of selected common solvents are shown in Table 6.1, although more comprehensive lists are available elsewhere.

To prevent or reduce VOCs, emissions control technology needs to be used, which adds to the capital and operating costs of a manufacturing plant. It is important to determine the amount of control technology that will be needed based on the intrinsic physicochemical properties of the solvent. For example, a solvent with a low boiling point and high vapor

TABLE 6.1 POCPs of Selected Organic Solvents

Organic Solvent	POCP (kg of ethylene equivalent/100 kg of solvent)
Acetone	18.2
Ethanol	44.6
Ethyl acetate	32.8
Isopropanol	21.6
Methyl ethyl ketone	51.1
Methyl isobutyl ketone	84.3
Toluene	77.1
o-Xylene	83.1
m-Xylene	108.8
p-Xylene	94.8

Source: ref. 3

pressure would tend to require more emission control technology and would be more likely to volatilize in a wastewater treatment plant than would a solvent of higher boiling point and vapor pressure.

Climate Change and Ozone Depletion Potential As can be seen in Tables 3.1 and 3.2, a handful of organic solvents, such as chloroform, methyl chloride, methylene chloride, and 1,1,1-trichloroethane, are of concern because of their global warming potentials, while solvents such as carbon tetrachloride and 1,1,1-trichloroethane are ozone depleters.

Impact on Water The impact of releasing organic solvents into the environment through water is generally assessed by a solvent's capacity for being directly ecotoxic to representative aquatic species found in water, the degree to which they may persist in the aquatic environment, and their potential to bioaccumulate in the food chain. One way of prioritizing ecotoxicity is to use the worst–case acute toxicity data available for several species, such as, fish, algae, or daphnids. The acute ecotoxicity is normally expressed in terms of the lethal concentration 50, (LD_{50} or LC_{50}), the effective concentration 50 (EC_{50}) or the inhibition concentration 50 (IC_{50}). Each of these represents the concentration of the

TABLE 6.2 Acute Toxicity and Log K_{ow} Data for Selected Solvents

	Acute Toxicity			
Solvent	LC_{50} Fish (mg/L)	EC_{50} Daphnia (mg/L)	IC_{50} Algae (mg/L)	Log K_{OW}
Acetic acid	75	47	90	−0.17
Acetone	5,500	10,000	7,500	−0.24
Acetonitrile	1,850	—	520	−0.34
1-Butanol	2,291	1,862	76	0.88
Butyl acetate	100	44	320	1.82
Cyclohexane	34	3.8; 400	3.8	3.44
Dimethylformamide	7,143	12,350	890	−1.01
Ethanol/IMS	13,000	9,190	1,450	−0.31
Ethyl acetate	422	556	556	0.73
Heptane	4	1.5	—	4.66
Hexane	4	3.8	1,079	4.11
2-Propanol	5,000	9,974	997	0.05
Isopropyl acetate	—	1,256	1,410	1.02
Methanol	15,300	10,132	—	−0.77
Methyl acetate	250–350	700–1,000	—	0.18
Dichloromethane	220	224	548	1.25
Methyl ethyl ketone	1,690	8,870	10	0.29
Methyl isobutyl ketone	460	4,273	—	1.31
Petroleum spirits	21	0.4–2	—	4.66
Propionic acid	188	50	—	0.33
Propyl acetate	—	510	536	1.23
Pyridine	650	940	28	0.7
t-Butyl methyl ether	887	542	184	0.94
Tetrahydrofuran	10	—	225	0.46
Toluene	13–72	19.6	12.5	2.73
p-Xylene	22	3.2	—	3.15

solvent (or other chemical) at which 50% of the test subjects die, or fail to reproduce, or whose growth is inhibited, depending on the test and the aquatic species being tested. The persistence of a material is the degree to which a material will remain in the environment in its original form without modification. Persistence is a function of the potential for a compound to be degraded through biological mechanisms (biotic degradation) or degraded through chemical or physical chemical phenomena (abiotic degradation) such as hydrolysis or photolysis. The search for a degradation pathway for chemicals is commonly known as identifying a depletion mechanism for the compound of interest and is expressed in terms of the degradation half-life of a substance in the environment. The potential for a chemical to bioaccumulate is represented by the octanol : water partition coefficient (log P or log K_{ow}), a measure of the affinity of a material to the organic vs. aqueous phases. Table 6.2 contains acute toxicity and log K_{ow} data for selected solvents.

There are several key messages to take away from a review of this table. First, it is apparent that not all solvents have a complete data set, and this is fairly typical. In these instances, one is faced with estimating potential effects through nearest-neighbor approaches, through quantitative structure–activity relationships, or with actually performing the desired test to obtain the data. Second, there is interspecies variation in the magnitude of the toxic effect that a solvent will have. This may potentially complicate any environmental risk assessment, but generally, the risk assessment will be based on the most sensitive species of the three trophic levels [i.e., the different parts of the food chain: the plant (algae), the organism that eats the plant, and a higher species that preys on the organism that eats the plant]. Finally, within a class of solvents, you will see variations in the magnitude of the ecotoxic effects.

Life Cycle Impacts In addition to the environmental challenges involved in managing and handling solvents, considerations over the entire life cycle of the solvent need to be taken into account. These include the emissions and resources involved in the manufacture of the solvent, transportation, recycling, and treatment:

$$\begin{aligned}&\text{total environmental footprint of solvent use}\\&= \text{manufacturing impacts} + \text{solvent use impacts} + \text{solvent disposal impacts}\end{aligned}$$

In some instances, the emissions and footprint for the production and transportation of solvents is significantly larger than the emissions related to solvent use.[4] For example, Figure 6.3 is a *chemical tree* for the manufacture of ethyl ether that shows graphically the materials that need to be produced or extracted throughout the supply chain of this material. To estimate the true environmental impact, the total amount of emissions and resources consumed during the production of the solvent need to be estimated and accounted for across all the chemicals shown in the chemical tree.[5] In addition to manufacturing-related emissions, the emissions to treat, recycle, or dispose of a solvent need to be considered as part of the overall environmental footprint of a solvent. When possible, a reaction should be chosen so that it may be carried out in a solvent which ensures that separation and recycle of a solvent are possible. A more detailed discussion in life cycle assessment is presented in Chapter 16.

Health Issues Because a series of health hazards are posed by solvents, specific controls need to be in place to minimize worker and community exposure to solvents used in processes and other solvent-handling operations. Health hazards resulting from solvent

Ether	Ethanol			
Ether 1,000	Ethanol 1,325	Corn 2,144	K fertilizer 14.8	Brine 12.2
				Sylvanite rock 61.8
			N fertilizer 39.7	Air (untreated) 83.7
				Natural gas (unprocessed) 14.4
				Petroleum product 4.36
				Phosphate rock (in ground) 14.3
				Water (untreated) 39.0
			P fertilizer 18.0	Air (untreated) 56.1
				Natural gas (unprocessed) 0.936
				Petroleum product 8.41
				Phosphate rock (in ground) 28.1
				Water (untreated) 16.1
		Water for reaction 138	Water (untreated) 138	

FIGURE 6.3 Chemical tree of ethyl ether. Each block in the tree represents a manufacturing process, and the numbers denote mass of material (in kilograms) to produce 1000 kg of the solvent. The tree is read from left (final product) to right (cradle material).

exposure range from irritancy effects to skin, eyes, and the respiratory track, to solvents that are sensitizers, carcinogens, mutagens, or reproductive toxins. For example, some solvents currently classified as carcinogens include benzene and 1,2-dichloroethane, while solvents such as dimethylformamide, *N*-methylpyrrolidone, and dimethylacetamide are classified as toxic to reproduction.

Health hazards are taken into consideration when setting occupational exposure limits for solvents, but an additional factor that plays a significant role is the release potential of the solvent. The release potential is considered to be a function of a solvent's volatility, the temperature, and latent heat effects, all of which increase the potential for respiratory exposure when using solvents.

Safety Issues One source of concern with the use of organic solvents is flammability. Ethers such as ethyl ether, methyl ether, and hydrocarbon solvents such as hexane and cyclohexane are extremely flammable. The key solvent property for determining flammability is the *flash point* defined as the minimum temperature at which its vapors can form an ignitable mixture with air near the surface of a liquid. The lower the flash point, the easier it is to ignite the material. For example, hexane has a flash point of −22°C, and it is considerably more flammable than ethylene glycol (the main component of antifreeze), whose flash point is 111°C.

Another key solvent property to consider as part of any safety risk assessment and management of processes, solvent-handling operations, or storage of solvents is the

TABLE 6.3 Lower and Upper Explosion Limits of Selected Solvents

Solvent	LEL (%)	UEL (%)
Acetone	3%	13%
Benzene	1.2%	7.8%
Diethyl ether	1.9%	36%
Ethanol	3%	19%
Ethylbenzene	1.0%	7.1%
Heptane	1.05%	6.7%
Hexane	1.1%	7.5%
Toluene	1.2%	7.1%
Xylene	1.0%	7.0%

autoignition temperature defined as the lowest temperature at which a substance will ignite spontaneously without the presence of an external source of ignition such as a flame or spark. For example, the autoignition temperature of heptane is 225°C, which is considerably lower than for acetic acid, whose autoignition temperature is 426°C.

Another safety concern related to solvent use, storage, and handling is the potential for reaching a range of concentrations in air in which the solvent vapors can ignite and explode. The extremes of this range of concentrations are known as the *lower explosive limit* (LEL) and the *upper explosive limit* (UEL). At concentrations in air below the LEL, there is not enough solvent to propagate an explosion; at concentrations above the UEL, the solvent has displaced so much air that there is not enough oxygen to begin a reaction. Table 6.3 shows the LEL and UEL of several solvents.

In addition to fire and explosion properties, some solvents present chemical reactivity and material compatibility concerns. Examples include solvents that might form explosive peroxides (e.g., 1,2-dimethoxyethane, diisopropyl ether) or might react violently with certain materials (e.g., formamide reacts exothermically with oxidizers, dimethyl carbonate reacts violently with potassium *t*-butoxide), or solvents that might corrode or react with the materials in containers, reactor surfaces, or overheads/piping.

Example 6.1 The reaction below is a peptide coupling between compounds A (a carboxylic acid) and B (an amino ester salt) with diisopropyl carbodiimide (DIC) as a coupling agent and *N*-hydroxybenzotriazole as a catalyst. This liquid-phase reaction runs in a solvent mixture of 1 : 1 dimethylformamide (DMF) and methylene chloride and because of limited solubility of reactant A, the reaction runs over several hours.

R–COOH + HCl·NH_2–R′–C(=O)–OCH_2CH_3 + diisopropyl carbodiimide

A (solid) B (solid) diisopropyl carbodiimide

(1) 1:1 DMF/CH_2Cl_2
(2) 0.2 equiv. *N*-hydroxybenzotriazole
(3) ambient temp., 6 h

→ R–C(=O)–NH–R′–C(=O)–OCH_2CH_3 + diisopropyl urea + HCl

C (dissolved) diisopropyl urea (dissolved)

This reaction has been reported as an example of solvent selection and replacement.[6] Why would solvent replacement be considered in this case?

Solution In this example, both dimethylformamide and dichloromethane (DCM) present health and environmental concerns. Dimethylformamide is a reproductive toxicant (may cause harm to a fetus), and methylene chloride has limited evidence of carcinogenic effects. On the environmental side, DCM is volatile and will result in fugitive air emissions, and DMF will probably be discharged as an aqueous waste possessing high biological oxygen demand and a significant nitrogen loading for the wastewater treatment plant to nitrify (convert to nitrates by way of ammonia). Incomplete nitrification will lead to discharge of ammonia, which is toxic and can lead to overfertilization of a receiving body of water and eventual eutrophication.

Independent of the health and environmental issues related to DMF and DCM, there would seem to be a strong case for finding a substitute for this solvent system, as the limited solubility of reactant A is causing the reaction to proceed over 6 h, thereby significantly limiting the potential throughput of the reaction.

Additional Points to Ponder Are there any problems with the reagents and by-products of the reaction?

6.1.2 Solvent Selection

The appropriate selection of solvents from the process viewpoint depends to a large extent on the application intended: more specifically, on what needs to be dissolved and under what conditions. In the chemical process industries, solvents are used to enhance the reaction rate, and serve as reaction media, as carriers of reagents, as mass separating agents to perform extractions or to break azeotropes, as cleaning agents, and as part of product formulations (as in paints, adhesives, etc). Therefore, as can be seen in Figure 6.1, the solvent selection problem is defined in terms of the application.[7] The type of application then drives the solvent properties to be considered for solvent selection. For example, if we want to use a solvent to dissolve a polar compound we will probably be looking for polar solvents; if we are looking for solvents to facilitate separation through physical means, we might want to look for solvents that form biphasic systems with the other components of the mixture, or for a solvent with good separability to be recovered by distillation and eventually recycled.

In a general sense, the solvent selection process consists of four main iterative steps that are closely aligned with the scientific method, as shown in Figure 6.4.

1. *Problem formulation.* The initial step is concerned primarily with problem identification. Do we really need a solvent? Why do we need a solvent? What properties do we want the solvent to have? One example of a problem definition might be: "We need a solvent in the main reaction that performs as well as the current solvent, dichloromethane, without the unacceptable environmental, health, and safety properties."

2. *Selection of search method and constraints.* In this step the search criteria are defined in terms of the measurable properties that a solvent needs to exhibit to fulfill the task defined in the problem formulation. The importance of solvent properties is illustrated in Figure 6.5. The objective is to start with a wide range of potential solvents that might be suitable and then use the solvent properties to help reduce the size of the search space. Examples of search criteria include the following: The solvent must dissolve the reactants and/or the

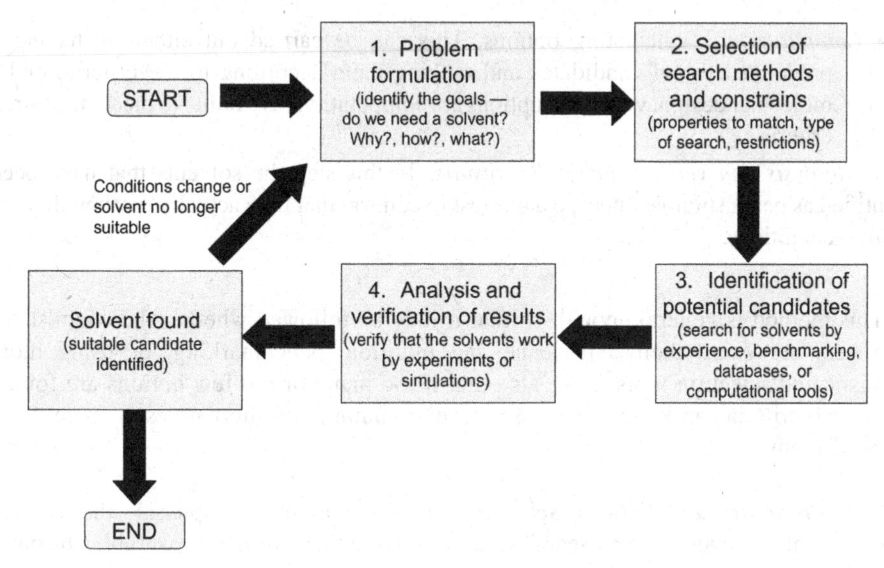

FIGURE 6.4 General iterative solvent selection process.

product, the solvent must have a boiling point between 70 and 120°C, the solvent may not be carcinogenic, and the solvent should not be miscible with water, so that it will form two phases for ease of separation.

3. *Identification of potential candidates.* In this step the search is performed using the search criteria chosen and may be performed in a number of ways:

- *Experience*: identifying options based on intuition and previous work.
- *Benchmarking*: identifying solvents based on those ones used in similar systems.
- *Databases*: searching databases for options that fulfill the required criteria.

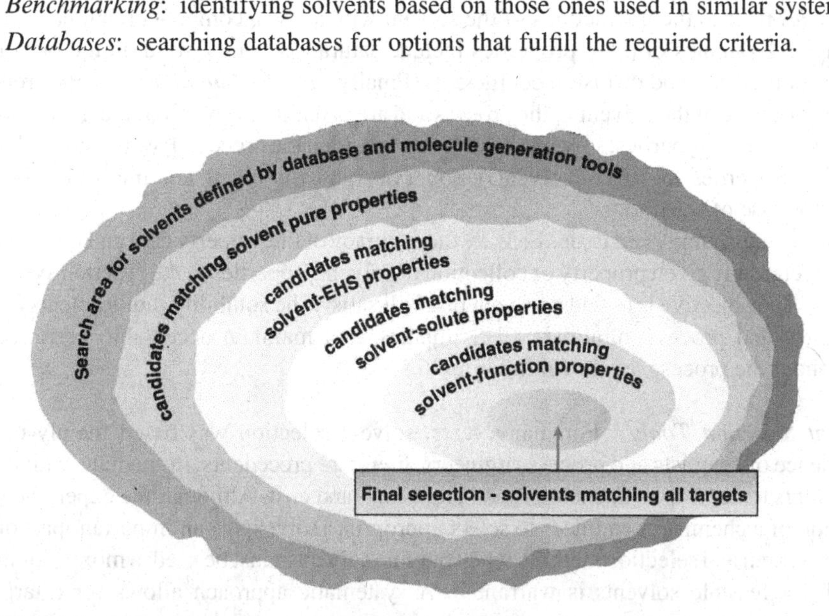

FIGURE 6.5 Solvent properties and their role in solvent selection. (From Gani et al.[7], with permission of *Chemical Engineering*. Copyright © 2006 Access Intelligence, New York.)

- *Computational*: calculating options. This can be carried out either by having a predefined list of candidates and ranking them according to the criteria; or by molecular design, where the options are computationally "built" to meet the desired criteria.

4. *Analysis and verification of the results.* In this step the solvents that have been identified as being suitable undergo a use test to confirm that they actually work within the desired conditions.

This method is general enough so that it may be followed whether the chemists or engineers are using their experience and intuition, benchmarking, or using more sophisticated selection tools. Note also that if too many or too few options are found, the search criteria can be modified to reduce the number of alternatives or to generate more of them.

Solvent Properties and Solvent Selection It is important to emphasize that solvent properties play a major role as search criteria because they are the measurable attributes that would allow for a targeted selection.

Solvent properties can be further described as pure solvent properties, as solvent–solute properties, and as solvent–function properties. Pure solvent properties are the intrinsic properties of the solvent, whether they are primary solvent properties or temperature- and pressure-dependent properties. For example, boiling point, flash point, and standard enthalpy of vaporization at 298 K are examples of primary solvent properties (one value per chemical), whereas enthalpy of vaporization, vapor pressure, and viscosity are pure solvent properties that are dependent on temperature, pressure, or both.

Most environmental, health, and safety attributes of solvents are a function of *pure solvent properties*, such as odor threshold, toxicity (LD_{50}), and explosivity limits. *Solvent–solute properties* refer to the interactions of the solvent with a given component that needs to be dissolved. Examples of these properties include saturation points, solid–liquid or liquid–liquid equilibrium, and diffusion coefficients. Finally, *solvent–function properties* refer to the performance of the solvent in the process and are in reality the result of the interaction of the pure-solvent properties, solvent–solute properties and the process. Examples of solvent–function properties are loss of solvent, amount of solute dissolved, and immiscibility with water for ease of separation.

The two most important issues here are the accuracy of the property data and how artfully we can relate any given property or collection of solvent properties to the reaction system of interest. The objective is to find a solvent that will satisfy the solubility requirements for the chemistry and process, minimize EHS impacts, and maintain acceptable performance throughout the process, as shown in Figure 6.5.

Solvent Selection Tools For many years, solvent selection was based mainly on the experience of chemists and process engineers, literature precedents, immediate availability of solvents to development chemists and engineers, and cost. Although the experience and intuition of a chemist or engineer to select appropriate solvents is an important part of the solvent search and selection method, when organic solvents must be used, a more systematic search for feasible solvents is warranted. A systematic approach allows for a targeted selection of solvents that can perform the required task efficiently while minimizing environmental, health, and safety impacts and optimizing solvent separability and recovery. Over the course of many years, methods and tools have been developed in a systematic attempt to identify and select appropriate solvents. Table 6.4 shows several of these tools and databases and provides a description of their main features.

The solvent selection tools that appear in Table 6.4 fall within the classification of either databases, benchmarking tools, or computational tools. Next we provide more detail and examples as to how these tools can be used when selecting a solvent.

Databases Using databases for solvent selection can be either through a direct search or through the use of some sort of methodology to make differences between solvents more readily apparent. For example, when using a direct search of a database to select solvents, one can look for solvents that meet the requirements of the desired application. Let's assume that we need to search for a water-immiscible solvent that has a boiling point between 60 and 90°C. A direct database selection will, in a first pass, probably screen solvents that comply with the boiling-point parameter (pure solvent property) and then look for those solvents that are not miscible with water (solvent–solute properties).

TABLE 6.4 Several Solvent Selection Tools and Databases

	Main Features and Comments
	Solvent Selection Tools
GSK-solvent selection guides	Methodology for selecting solvents for organic (green) synthesis. A visual, Web-based solvent selection tool highlighting EHS issues to be managed, solvent compatibility, solvent separability, and physical properties.
ProCAMD (ProPred, PDS, SoluCalc, CAPEC-Database); SMSwin	Hybrid computer-aided technique for solvent selection and design; covers a very wide range of applications, methods, and tools [ProPred is a tool for pure component property estimation; PDS is a tool for design and analysis of distillation operations; SoluCalc is a tool for model creation and analysis of solid solubility; CAPEC-database; SMSwin is specialized software for property estimation and solvent classification].
	http://www.capec.kt.dtu.dk/Software/ICAS-and-its-Tools/ and http://www.capec.kt.dtu.dk/documents/software/SMSWIN.htm
NRTL-SAC and eNRTL-SAC	Activity coefficient method based on segment contributions. Predictive based on a small set of solubility data. The electrolyte method treats salts. Useful for crystallization solvent selection and extends to LLE and VLE.
	ASPEN Properties, Aspen Plus: http://www.aspentech.com
ISSDS and SAGE	ISSDS (integrated solvent substitution data system) facilitates access to solvent alternative information from multiple data systems through a single, easy-to-use command structure.
	SAGE (solvent alternative guide) is a comprehensive guide designed to provide pollution prevention information on solvent and process alternatives for parts cleaning and degreasing. SAGE does not recommend any ozone-depleting chemicals.
	U.S.EPA: http://es.epa.gov/ssds/ssds.html

(*continued*)

TABLE 6.4 (*Continued*)

	Main Features and Comments
COSMO-RS; COSMO-SAC	Ab initio quantum mechanical methods using a continuum solvation approach. COSMO-SAC is implemented in Aspen Properties. The constituent molecular sigma profiles must be generated first in a quantum mechanics package (e.g., Gaussian '98). COSMO-RS is implemented in COSMO*therm*, available at http://www.cosmologic.de and http://www.aspentech
	Solvent Database
CambridgeSoft ChemFinder	Searchable data and hyperlink index for thousands of compounds; the ideal starting point for Internet data mining. http://chemfinder.cambridgesoft.com/
CRC Handbook of Chemistry and Physics	Library Network Database: http://www.hbcpnetbase.com/
DECHEMA Chemistry Data Series	A 15-volume data collection; see volume 15 for solubility data for large complex chemicals. http://www.dechema.de/CDS.html
DETHERM, CHEMSAFE	Comprehensive collection of thermophysical and mixture properties data, including Dortmund DDB and ELDAR DDB; data related to flammability and explosivity. http://i-systems.dechema.de/detherm/mixture.php? and http://i-systems.dechema.de/chemsafe/mixture.php
SSDS; HSSDS	SSDS (solvent substitution data systems) contains a collection of solvent-related data (solvent-pure properties, solvent-EHS properties, etc.). http://es.epa.gov/ssds/ssds.html
Solvents Database	A solvent database in one comprehensive source, containing data vital in many solvent applications; see also the *Handbook of Solvents* from the same source. http://www.chemtec.org/cd/ct_23.html
DIPPR	Critically evaluated thermophysical data. http://www.aiche.org/TechnicalSocieties/DIPPR/About/Mission.aspx
Knovel Science and Engineering Resources	Library Network Database (Polymers—Property Database, Handbook of Thermodynamic and Physical Properties of Chemical Compounds, Solvent Properties Database, etc.)
SOLVDB	SOLVDB (solvents database) contains data on the physical/chemical properties, environmental fate, health and safety, and regulations (137 data fields) of chemicals. http://solvdb.ncms.org/index.html
TAPP	TAPP (thermochemical and physical properties database) may be accessed at http://www.chempute.com/tapp.htm
The NIST Webbook	An excellent source of physical and chemical data. http://webbook.nist.gov
CAPEC Database	Contains pure as well as mixture properties data, including a solvent–solute database. http://www.capec.kt.dtu.dk/Software/ICAS-and-its-Tools/

Principal Components Analysis In addition to a direct search of the database, principal components analysis (PCA) can be used to identify those solvent types that give the best results for the process in question. PCA is of considerable use in many areas of chemistry, as it allows one to extract chemical information from large data sets containing multiple variables or factors that seemingly are unrelated. PCA transforms the data to a new coordinate system so that the data that account for the greatest variance lie on the first coordinate (called the *first principal component*), the data that describe the second greatest variance lie on the second coordinate, and so on. This is defined mathematically as an orthogonal linear transformation that results in a new and reduced factor space. PCA reduces the number of factors or dimensions in a data set by retaining those aspects that contribute the most to its variance, retaining the lower-order (principal) components and ignoring the higher-order components. For example,[8] PCA can be used in solvent selection by constructing a database of known solvents that contains solvent descriptors of interest, such as boiling point, melting point, dielectric constant, solubility parameter, density, water solubility, and dipole moment. PCA is used to reduce these solvent descriptors to several new components that represent several of the descriptors in a single component, or all of the descriptors in just a few components. A simple plot of these new components in factor space allows one readily to identify solvent groupings and those solvents that possess similar properties. Therefore, if a solvent is already known to be a good fit for a given application but more alternatives or a replacement are needed, the results of the PCA can be used to identify other possibilities. Figure 6.6 shows an example of the use of PCA for solvent selection that integrates environment, health, and safety considerations into the data set. In this figure the variables of the solvent data set

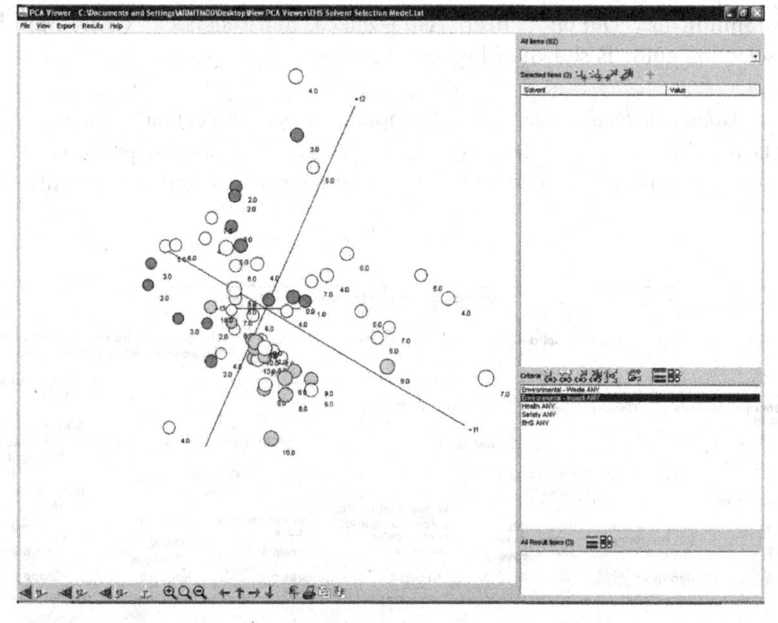

FIGURE 6.6 PCA used for solvent selection. Each dot represents a solvent in tri-dimensional solvent space. Solvents that are located near each other have similar characteristics given the statistical assessment.

have been reduced to three principal components t_1, t_2 and t_3 that form a three-dimensional solvent space. Each solvent is represented by a colored dot, and solvents that are located near each other in the solvent space should exhibit similar characteristics. A color code is used to differentiate between solvents that have been ranked according to their environmental, health, and safety profile. The numbers denote the specific score for the EHS profile of a solvent. Thus, if a solvent does not exhibit a good profile for a given EHS characteristic, the user can select another solvent with similar solvent properties but without the EHS concerns associated with the first solvent. Other examples can be found in the literature, where a two-dimensional solvent space has been produced for solvent selection using PCA.[9]

Benchmarking Tools There are a variety of tools that classify solvents according to their performance in defined categories, which are in turn based on important and well-known physical properties of a solvent, such as hydrogen bonding, polarity, and Hildebrand solubility parameter. In most cases, these tools quickly generate a list of potential solvents whose performance will need to be corroborated in the context of the chemical reaction or process for which the solvent is being searched.

For example, a protocol derived by Britest Ltd. (http://www.britest.co.uk) seeks to use mechanistic principles to guide solvent selection, as shown in Figure 6.7. The objective is to follow the arrows using the problem definition and a search criterion to reach the endpoint, the identification of the candidate solvent's group type. Solvents contained in a solvents database that correspond to the desired group type are evaluated in the test system and a final selection is made.

Another example of this type of tool includes solvent selection guides such as those developed by various companies to help their scientists and engineers to select the proper solvent for a specific use. Examples of these solvent selection guides include those developed by GlaxoSmithKline, DuPont, Pfizer, AstraZeneca, and others.[5,10] GlaxoSmithKline's solvent selection guide is shown in Figure 6.8.

Computer-Aided Molecular Design Computer-aided molecular design (CAMD) represents a hybrid computer technique that integrates databases, property-estimation models, and solvent-search procedures into a computer-aided tool for solvent selection,

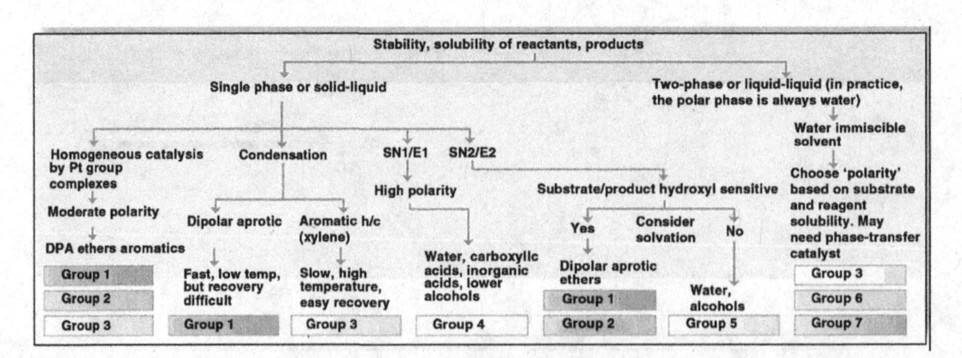

FIGURE 6.7 Mechanism-based solvent selection procedure (Britest Ltd., http://www.britest.co.uk). (From Gani et al.[7], with permission of *Chemical Engineering*. Copyright © 2006 Access Intelligence, New York.)

SSG class	Solvent	Environmental - Waste	Environmental - Impact	Health	Safety	LCA ranking
Alcohols	Ethylene glycol	4	9	8	9	9
	1-Butanol	5	8	8	8	5
	Diethylene glycol butyl ether	5	7	10	9	7
	Isoamyl alcohol	7	7	7	8	6
	2-Ethylhexanol	9	6	8	7	6
	2-Butanol	4	7	7	7	6
	1-propanol	3	7	5	8	7
	Ethanol	3	8	10	7	9
	2-propanol	3	9	9	7	5
	t-butanol	3	10	7	7	8
	Methanol	3	10	5	8	9
Esters	t-butyl acetate	7	10	7	7	7
	Butyl Acetate	7	8	9	8	5
	n-Propyl acetate	6	7	8	7	5
	Isopropyl acetate	5	8	8	7	6
	Ethyl Acetate	4	8	8	4	6
	Methyl acetate	2	10	7	5	7
	Dimethyl carbonate	3	7	8	7	8
Aromatics	p-Xylene	8	2	7	5	7
	Toluene	7	3	6	4	7
	Fluorobenzene	4	2	4	5	1
Ketones	Methyl isobutyl ketone	7	6	6	7	2
	Acetone	2	9	8	5	3
	Methyl ethyl ketone	3	6	8	5	3
Polar Aprotics	N-Methyl-2-pyrrolidone	4	6	8	9	3
	dimethyl acetamide	4	7	2	10	3
	Dimethyl formamide	4	6	2	8	6
	Dimethylpropylene urea	4	7	5	9	4
	Dimethylsulphoxide	4	4	8	3	6
	Formamide	3	7	2	10	8
	Acetonitrile	2	6	6	8	4
Acids	Propionic acid	5	8	4	9	7
	Acetic acid	3	8	4	8	8
Alkanes	Cyclohexane	5	6	8	2	7
	Methyl cyclohexane	7	5	8	2	7
	Heptane	6	3	9	1	7
	2-Methylpentane	5	3	5	1	7
	Hexane	5	2	4	1	7
	Petroleum ether	4	1	2	1	7
Chlorinated	Dichloromethane	2	5	3	10	7
Ethers	Methyl tertbutyl ether	4	4	6	2	8
	1-2-dimethoxyethane	3	5	3	2	7
	Tetrahydrofuran	2	6	7	2	5
	Bis(2-methoxyethyl) ether	6	5	1	3	6
	Diisopropyl ether	5	2	9	1	9
Basics	Triethylamine	4	5	2	4	7
	Pyridine	3	4	3	6	2

FIGURE 6.8 GlaxoSmithKline's solvent selection guide (http://www.gsk.com). (Reprinted with permission from Jimenez-Gonzalez et al.[5] Copyright © 2005 Springer Science and Business Media.)

design, and evaluation. In this type of tool, the molecule of a potential solvent is designed from scratch using molecular design and group contribution theory to match the desired set of properties. The main steps and rationale for CAMD techniques can be described as follows:

1. Define the type of problem that needs to be solved. For example: "We are looking for solvents with properties similar to those of chloroform." Define the set of constraints that will translate the problem constraints into computer numerical constraints.
2. Define the building blocks that will allow us to narrow the search for specific types of compounds, such as alcohols, ketones, aldehydes, or esters. For aliphatic alcohols the building blocks needed would be CH_3, CH_2, CH, and OH.
3. Design the molecular structure that fulfills the problem constraints and that utilizes the building blocks defined in steps 1 and 2.

Once the target properties have been identified and their goal values have been specified, some of the questions and challenges of the CAMD methods are:

- How should we generate and represent molecular structures?
- What level of molecular structural information will be used? This is especially important because of the number of isomers that can be generated in the search, so it is important to define the level of detail needed for the problem at hand.
- How will the target properties be obtained (calculated and/or measured)?

These questions are important because an efficient CAMD system should have a good set of integrated tools:

- To represent the molecular structures
- To generate the molecules based on a set of rules
- To analyze the stability of the molecules generated
- To estimate the properties of the molecules generated
- To integrate known experimental values of properties of molecules identified
- To solve the molecular design problem in an integrated fashion

For example, a method to generate solvent alternatives for organic reactions using CAMD methods has been developed by Gani et al.[11,12] This method was designed to select appropriate solvents for the promotion of a class of organic reactions and at the same time present an acceptable environmental, health and, safety profile. The method combines knowledge from industrial practice and physical insights with computer-aided property estimation tools for the selection and design of solvents. The method employs estimates of thermodynamic properties to generate a knowledge base of reaction-, solvent-, and environment-related properties that directly or indirectly influence the rate and/or conversion of a given reaction. Solvents are selected using a rules-based procedure in which the estimated reaction-solvent properties and the solvent–environmental properties guide the decision-making process. The objective is to produce a short list of potential solvents for a given reaction and rank them according to a scoring system that accounts for their performance and their greenness. This methodology uses a step-by-step algorithm consisting of a sequence of five steps through which the user is guided towards the solution, as illustrated in Figure 6.9. As can be seen, the search is performed using a series of CAMD tools to predict pure component properties (i.e, melting point, boiling point, etc.), verify reactivity through a solvent–reaction database, predict the equilibrium composition, predict whether or not there is a phase split, predict reactant–product solubility, and predict whether there are others parameters of potential interest.

Example 6.2 Find a replacement for dichloromethane, which is used as a solvent in the following general reaction:

$$—N{=}R—OH + H_3C—\overset{\overset{O}{\|}}{S}—O—Cl \longrightarrow —N{=}R—O—\overset{\overset{O}{\|}}{S}—O—Cl$$

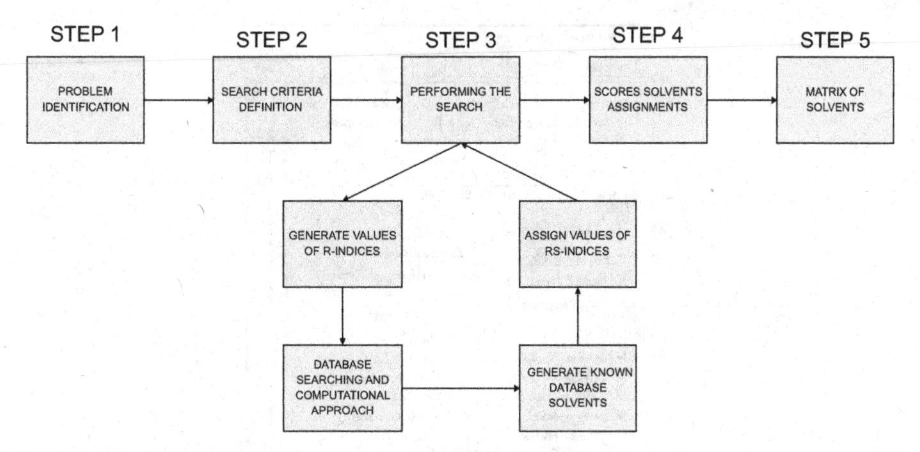

FIGURE 6.9 Methodology to select green solvents for organic reactions using computer-aided molecular design to perform the search.

Solution In this case, the method developed by Gani et al. is used to find potential replacements using dichloromethane (DCM) properties as a benchmark. The replacement solvent should match the performance and properties of DCM but have more favorable EHS properties.

Search criteria definition. Based on the property values for DCM, a series of search criteria were defined through variables called *R-indices*. These are used to formulate the solvent selection problem as follows:

R0 = 1	Solvent addition is needed to improve the miscibility of the reactants.
R1 = 1	The reaction is taking place in the liquid phase.
R2 = 278 K	The reaction temperature is 278 K.
R3 = 1	The solvent needs to dissolve the reactants.
R4 = 0	The solvent does not need to dissolve the products.
R5 = 1	The solvent needs to have a phase split with water.
R6 = 1	The solubility parameters for DCM are known.
R7 = 1	The solvent must be neutral toward the compounds in the reaction system.
R8 = 1	The solvent must not associate or dissociate.
R9 = 1	Favorable solvent EHS properties are desirable.

Following the indices above, the following solvent target properties are specified: melting point < 250 K, boiling point < 350 K, $16 <$ solubility parameter < 19, and $0.5 < \log K_{OW} < 4$. In addition, the solvent should be partially miscible with water to promote phase split and be neutral with respect to reaction chemicals.

Performing the search. CAMD techniques can be used to identify solvents that match the solvent target properties and the results can be verified using an appropriate

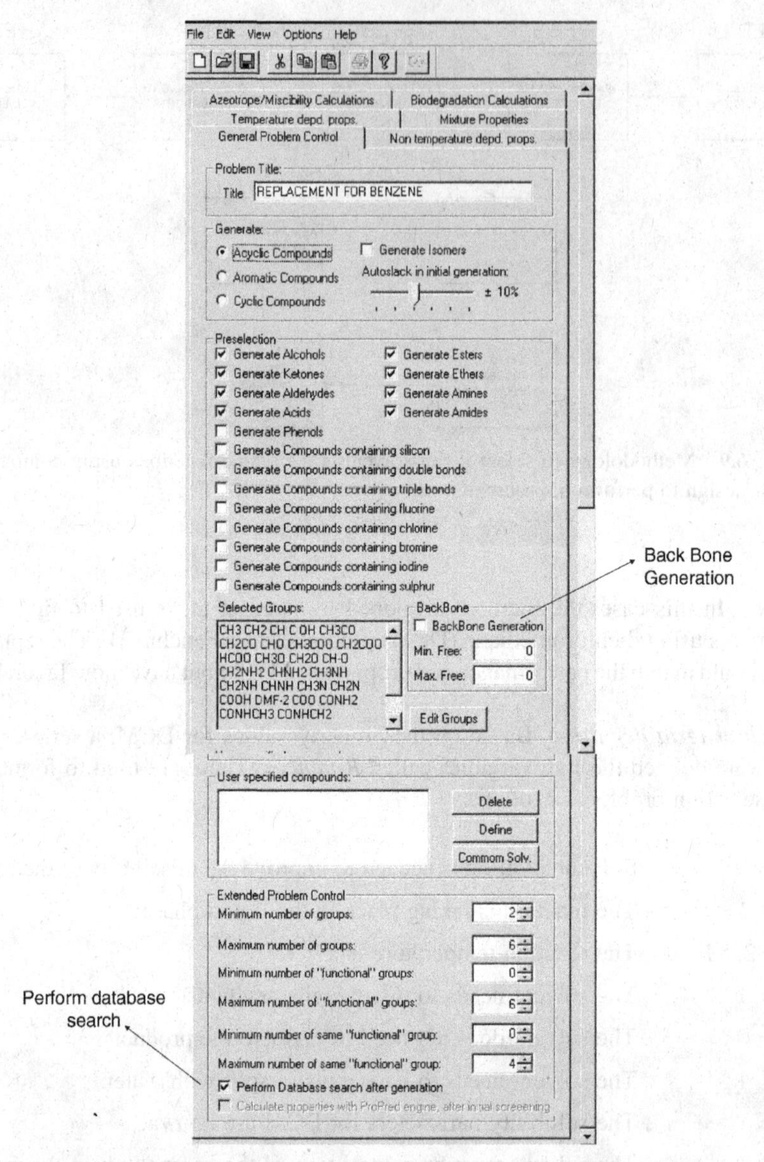

FIGURE 6.10 Screen shot of a CAMD search using ProCAMD software.

database. In this case, the CAMD software developed by Gani et al. was used to perform the search.[13,14] A screen shot of the search is shown in Figure 6.10, and it can be seen how the building blocks were selected.

Scoring the solvents. Solvents identified were scored based on how well the restrictions set in the R-indices are fulfilled. The scores are given in Table 6.5, but as the reactivity hasn't been evaluated, the score for restriction 7 is not provided. Based on the scores, all the solvents listed would be good candidates to replace dichloromethane. To corroborate this, further experimental work will be needed.

TABLE 6.5 Solvent Scores

Solvent	S_2	S_3	S_5	S_6	S_7	S_8	S_{91}	S_{92}	Action	Score
Ethyl acetate	10	10	10	8	N/A	10	6	8	Study	62
Isopropyl acetate	10	10	10	6	N/A	10	8	6	Study	60
Ethyl propionate	10	10	10	8	N/A	10	8	6	Study	62
Toluene	10	10	10	6	N/A	10	8	8	Study	62
Methylcyclopentane	10	10	10	6	N/A	10	8	8	Study	62

Additional Points to Ponder What are the pros and cons of the five solvents identified using this CAMD method? Which solvents are likely to be reactive in the given system?

6.1.3 Solvent-Less Reactions

Section 6.1.2 was dedicated to describing how you might go about selecting an organic solvent if a solvent needs to be used. However, the ideal target is to run production processes without utilizing organic solvents. Actually, many industrial processes are carried out without the use of organic solvents.

Gas-Phase Chemistry Methanol is made of synthesis gas (CO and H_2) in a catalytic reaction and the oxidation of methanol to formaldehyde is normally carried out in the gas phase (with an excess of methanol to keep the reaction mixture outside the explosivity limits)[15]:

$$CO + 2H_2 \longrightarrow CH_3OH + 0.5O_2 \longrightarrow CH_2 = O + H_2O \qquad \text{(in gas phase)}$$

Solid-State Chemistry Microwave-accelerated organic syntheses involve the exposure of neat reactants to microwave (MW) irradiation with supported reagents or catalysts (primarily of mineral origin, normally recyclable) such as $Fe(NO_3)_3$–clay (clayfen), $Cu(NO_3)_2$–clay (claycop), NH_4NO_3–clay (clayan), NH_2OH–clay, $PhI(OAc)_2$–alumina, $NaIO_4$–silica, CrO_3–alumina, MnO_2–silica, and $NaBH_4$–clay. This area has shown promising results in the laboratory in reactions such as condensation, cyclization, rearrangement, and oxidation and reduction reactions.[16] For example, some heterocyclic hydrazones have been synthesized in the laboratory using microwaves without the need for catalysts, solid support, or solvent.[17]

$$R_1NHNH_2 + R_2R_3C{=}O \xrightarrow{MW} R_2R_3C{=}NNHR_1$$

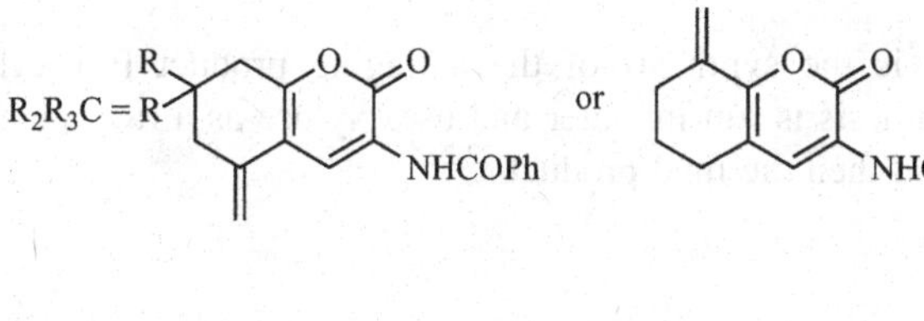

$R_2R_3C = R$ ... NHCOPh or ... NHCOPh

Another example of the use of microwave-assisted chemistry in industrial applications is the synthesis of ceramides for antiaging products by L'Oreal.[18] In this case, the acylation is achieved with microwave technology and without the need for a solvent. Additionally, this chemistry avoids acid activation and eliminates one step from the original synthesis, thereby providing easier separation and workup.

$$\text{R-CHOH-CH(NH}_2\text{)-CH}_2\text{OH} \xrightarrow[\text{MW., 15min (without solvent), 70–80\% yield}]{\text{R}_1\text{-COOH}} \text{R-CHOH-CH(NH-C(=O)-R}_1\text{)-CH}_2\text{OH} \quad \text{R}_1\text{=C}_{11} \longrightarrow \text{C}_{17}$$

In addition to the use of microwaves, solid-state chemistry can be performed with the help of ultrasound, grinding and heat. For example, solvent-free bromination of diquinoline has been performed using grinding and *N*-bromosuccinimide (NBS) in a regio and stereo-selective manner, avoiding the use of carbon tetrachloride solvent.[19]

NBS, solid-state grinding

1R = H → 2R = H 85%
3R = Br → 4R = Br 88%

Use of Technology A method to purify bisphenol A by fractional melt crystallization in a falling film dynamic crystallizer can be used instead of extractions with methylene chloride or toluene to remove impurities. The falling film crystallizer can produce highly pure bisphenol A without contamination by solvents. This process is also energy efficient, as it avoids the drying and distillation steps.[20]

Nonorganic Solvents The use of nonorganic solvents for synthesis, most notably water and supercritical CO_2, has received considerable attention recently. For example, the formation of diazonium salts from the corresponding primary aromatic amines and their subsequent reaction with potassium iodide to give aryl iodides has been performed in a CO_2–H_2O solvent system. Isolation of the final product was achieved by venting CO_2 without the utilization of organic solvents.[21]

$$\text{CH}_3\text{O-C}_6\text{H}_4\text{-NH}_2 \xrightarrow[\text{H}_2\text{O/CO}_2]{\text{NaNO}_2\text{, room Temp.}} \text{CH}_3\text{O-C}_6\text{H}_4\text{-N}^+\text{:NaHCO}_3^- \xrightarrow[\text{room Temp.}]{\text{KI}} \text{CH}_3\text{O-C}_6\text{H}_4\text{-I} + \text{N}_2$$

Another example is the synthesis of the antiaging product Pro-Xylane developed by L'Oreal. This synthesis is run in water and uses xylose as a raw material to produce first a C-glycoside and then the final product.[22,23]

xylose

$NaHCO_3$, H_2O, 90°C

A or B, H_2O

A = $NaBH_4$
B = catalytic reduction (industrial process)

Example 6.3 Around 1900, the German chemist Ludwig Roselius found a number of compounds that dissolved the natural caffeine in coffee beans without completely changing its taste. These chemicals included chloroform, ethyl acetate, benzene, and methylene chloride. If you could go back in time and talk to Roselius, which of these solvents would you recommend that he to use in the decaffeination process?

Solution None of the above.

Benzene and chloroform were known to be toxic at the time the decaffeination process was devised, so you would certainly not recommend those and would have no problem convincing Roselius.

Because you have traveled back in time from the present, you will also know that for about 70 years methylene chloride was the solvent of choice for decaffeination of coffee. However, when some toxicity testing suggested that dichloromethane is a suspected human carcinogen the need to migrate to different methods became apparent. So you wouldn't recommend using methylene chloride, even though it is still used in some specialty decafs (provided that the methylene chloride residues are less than 10 ppm), because you know that most of the major coffee players have abandoned that solvent.

Ethyl acetate is also still used as a decaffeination solvent, and you know that coffee decaffeinated with ethyl acetate can even carry the "all natural" label, as ethyl acetate is found naturally in fruits, and doesn't pose the toxicity issues posed by benzene, chloroform, or methylene chloride. However, since you are taking a green chemistry/green engineering class, you know that one of the principles of green chemistry and green engineering is to avoid the use of auxiliary materials, so you venture to look into alternatives that avoid the use of organic solvents, and find two:

- *Swiss water process.* This process is an interesting example of exploiting a simple and wellknown chemical principle to retain the function of a product while modifying its properties in a benign fashion. The process mixes green bean extract, which is effectively caffeine-free, with the coffee beans and uses the difference in concentration to drive the caffeine in the coffee bean (higher concentration) into the extract (lower concentration). Since the extract contains essential oils and other valuable components of the bean, most of the desirable components of the coffee remain in the beans. This method retains more of the flavor compounds than do the chemical methods, while removing the caffeine.

- *Supercritical CO_2 process.* This method uses supercritical carbon dioxide at about 250 atm and 200°F. Under these conditions carbon dioxide behaves like a liquid in terms of its density but has the viscosity of a gas. The caffeine is removed from the bean by the carbon dioxide and carried to an absorption column using water. The caffeine-rich solution is sold to soft-drink manufacturers, and the carbon dioxide is recirculated to the process.

Additional Points to Ponder. Would a genetically modified low-caffeine coffee plant[24] be more sustainable than the Swiss water or supercritical carbon dioxide method? What are the pros and cons?

6.2 CATALYSTS AND CATALYST SELECTION STRATEGIES

As discussed in Chapter 2, catalytic reagents (as selective as possible) are superior to stoichiometric reagents. You will no doubt recall from general chemistry that catalysts are substances that increase the rate of a reaction without being used up in the process. You will also recall that before a particular chemical reaction can take place, energy, called *activation energy*, is required to break and form bonds. It represents the thermodynamic energetic barrier that must be surpassed for a reaction to proceed. There are several ways in which a catalyst can lower the activation energy:

- By forming bonds with one or more of the reactants and so reducing the energy needed by the reactant molecules to complete the reaction
- By bringing the reactants together and holding them in a way that makes reaction more likely

Catalysts cannot make energetically unfavorable reactions possible because they have no effect on the chemical equilibrium of a reaction, as the rate of the forward and reverse reactions are affected equally. The net free energy change of a reaction is the same whether or not a catalyst is used; the catalyst just makes it easier to activate, as shown in Figure 6.11.

From the green chemistry/green engineering prospective, catalysts have a series of advantages related to the enhanced efficiency and use of resources, as shown in Table 6.6.

For example, the BHC Company (now BASF) developed a synthetic process to manufacture ibuprofen, a well-known nonsteroidal anti-inflammatory painkiller marketed under brand names such as Advil and Motrin. The new technology commercialized since 1992, involves only three catalytic steps with approximately 80% atom utilization, replacing technology with six stoichiometric steps and less than 40% atom utilization.

This process provides a great example of an elegant catalytic solution to eliminate large quantities of solvents and waste associated with the traditional stoichiometric use of auxiliary chemicals. Large volumes of aqueous waste (salts) normally associated with such

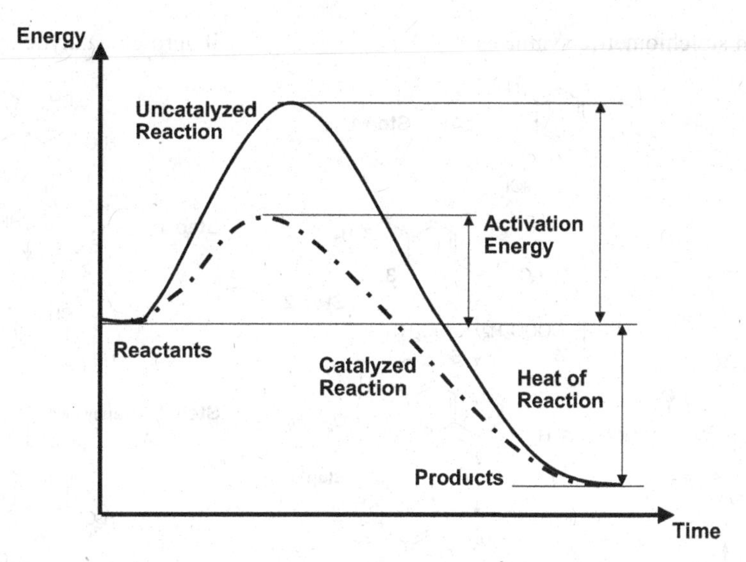

FIGURE 6.11 Catalyst effect on activation energy of a reaction. Note that the enthalpy of reaction (energy of products minus energy or reactants) remains unchanged.

manufacturing are virtually eliminated. The anhydrous hydrogen fluoride catalyst/solvent is recovered and recycled with greater than 99.9% efficiency, and no other solvent is needed in the process, simplifying product recovery. For its innovation, BHC was the recipient of the Kirkpatrick Achievement Award for "outstanding advances in chemical engineering technology" in 1993 and the EPA's Green Chemistry Challenge Award for Greener Synthetic Pathways in 1997.[25,26]

TABLE 6.6 Potential Advantages and Disadvantages of Catalysts from a Green Standpoint

Potential Advantages	Potential Disadvantages
Productivity	Productivity
Increased yield	If the turnover number or turnover frequency is too low, the productivity will decrease
Reduced cycle time	
Increased reactor volume efficiency	
Increased selectivity	
Reduced variability	
Environment	Environment
Elimination or reduction of solvent	Catalysts containing heavy metals have a significant impact on the environment if released and not recycled
Reduced by-products	
Increased resource efficiency	
Safety	
Control of exotherms	
Can reduce the use of hazardous materials	

Ibuprofen stoichiometric synthesis:

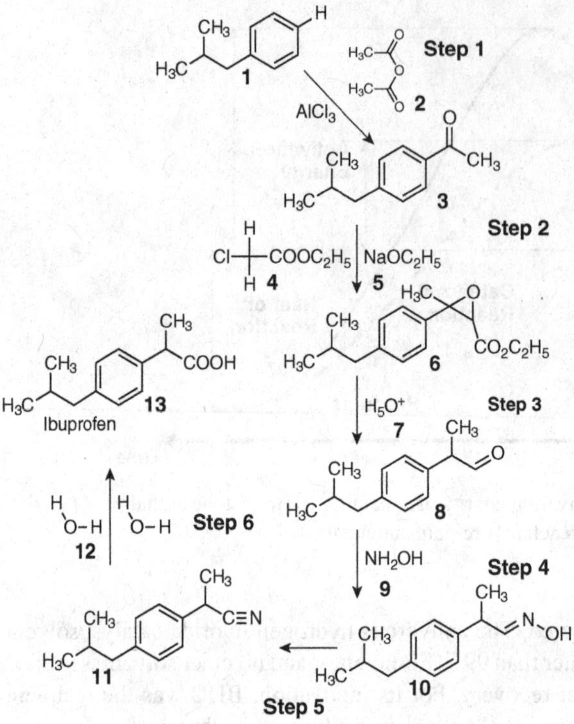

Ibuprofen catalytic synthesis:

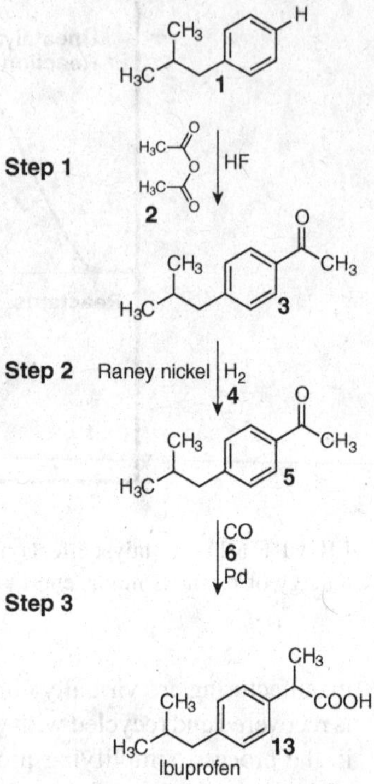

6.2.1 Selecting Catalyst–Catalyst Efficiency

When selecting catalysts it is important to compare how well the catalyst will perform for the desired reaction relative to the existing chemistry and associated process. The more efficient a catalyst is, the less waste that will be generated in the reaction and the more efficient the use of resources will be. There are several ways to measure catalytic efficiency that are commonly used to compare and select catalysts. In this section we cover several of them.

Catalyst Activity *Catalytic activity* is a measure of the catalyzed reaction rate per unit mass of catalyst. The SI-derived unit for measuring catalytic activity is the *katal*, which is the amount of catalyst that converts 1 mol of reactant or substrate per second under specified conditions (e.g., the reaction rate).[27,28] For example, trypsin is a natural enzyme found in the digestive system that helps to break down proteins (a serine protease). In recent times it has also been used in numerous biotechnological processes. One katal of trypsin is that amount of trypsin that breaks down 1 mol of peptide (protein) bonds per second under specified conditions. By mode of action, catalysts can be classified as chemical or enzymatic. A biochemical equivalent is the *enzyme unit* (U), defined as that amount of the enzyme that catalyzes the conversion of 1 μmol of substrate per minute at a given temperature, pH, and substrate concentration. Although the General Conference on Weights and Measures has recommended that the katal replace the enzyme unit (U), enzyme units are still used in practice more commonly than the katal, especially in biochemistry. Since 1 katal is the amount of enzyme that converts 1 mol of substrate per second, 1 U = 1/60 μ katal = 16.67 katal.

Turnover Number and Frequency The degree of activity of a catalyst can also be described by the turn over number (TON) and the catalytic efficiency by the turnover frequency (TOF). *TON* is defined as the number of synthesized molecules per number of catalyst molecules used (or catalytic sites when talking about enzymes). A different common use of the term in chemically catalyzed reactions is to express the maximum number of times a catalyst may be used for a specific reaction before there is a decay in its activity. In this context, the turnover number for industrial applications is between 10^6 and 10^7.[29]

TOF is a mass-independent unit that accounts for the time of conversion, defined as the number of molecules converted by each catalytic site in a unit of time:

$$\mathrm{TOF} = \frac{n_s}{t n_{\mathrm{cat}}}$$

where n_s are the moles of converted starting material, n_{cat} are the moles of active sites, and t is the time of conversion (s).[30] For example, carbonic anhydrase has a turnover frequency of 400,000 s^{-1}, which means that each carbonic anhydrase molecule can produce up to 400,000 molecules of product (CO_2) per second.

When using TON or TOF to compare catalysts, it is important to provide the specifics of the reaction. For most industrial applications, the TOF is in the range 10^{-2} to 10^2, and for enzymatic processes it is from 10^3 to 10^7. Although biological catalysts generally have higher TOF values than those of chemical catalysts, there is still a need to account for the differences in molecular masses and stability of the catalyst. For example, the Mn–Salen epoxidation and sulfidation catalyst has a TOF of 3 h^{-1}, while the chloroperoxidase (enzymatic counterpart) has a TOF of 4500 h^{-1}. However, the molecular masses are 635 and 42,000 g/mol respectively.[30]

Catalyst Selectivity Another important measure of catalyst efficiency is *selectivity*, defined as the number of molecules synthesized per number of molecules converted:

$$\text{selectivity} = \sigma = \frac{n_p - n_{p0}}{n_{s0} - n_s} \cdot \frac{\nu_s}{\nu_p}$$

where n_p represents the moles of product at the end of the reaction, n_{p0} the moles of product at the beginning of the reaction, n_s the moles of starting material at the end of the reaction, n_{s0} the moles of starting material at the beginning of the reaction, and ν_p and ν_s are the stoichiometric factors for the product and starting material, respectively. For a catalyzed reaction, the reaction yield would be the product between selectivity and conversion. In other words, selectivity is an indication of the accuracy of conversion to the product desired, and it therefore needs to be as close to unity as possible, to avoid waste. For example, a selectivity of 99.99% means that the catalyst renders one undesired product in 10,000 conversions. This is the performance achieved by enzymes in living systems, although most enzymes do better than this. Few synthetic industrial catalysts achieve this degree of control over the chemistry they catalyze. For example, Table 6.7 shows the selectivities of several zeolite catalysts for the isopropylation of naphthalene.[31]

TABLE 6.7 Selectivities of Natural Zeolite Catalysts for the Isopropylation of Naphthalene

	Product Distribution of Diisopropyl Naphthalene (%)						
Catalyst	1,3-	1,4-	1,5-	1,6-	1,7-	2,6-	2,7-
HY	23.7	0.6	0.2	6.8	4.9	32.6	31.2
HL	39.9	7.9	6.7	15.3	16.3	6.7	7.2
HM	5.3	3.8	1.9	7.1	6.1	50.8	24.9

Catalyst Stability Chemical, stability, temperature, and in the case of a heterogeneous catalyst, mechanical stability are important factors in the selection of a catalyst. A way to express the chemical stability of a catalyst is via the deactivation rate and half-life. The *deactivation rate* measures the loss of a catalyst activity per unit of time (s^{-1}), and the half-life is the time (s) in which the catalyst activity is halved.

Example 6.4 For the production of methyl propanoate via the methoxycarbonylation of ethene:[32]

$$CH_3OH + CO + CH_2{=}CH_2 \longrightarrow H{-}(CH_2CH_2\overset{O}{\overset{\|}{C}})_n{-}OCH_3$$

Recently there have been several reports of results using palladium-complexed catalysts. One of these complexes was reported as being capable of converting ethene, CO, and MeOH to methyl propanoate at a rate of 50,000 mol of product per mole of catalyst per hour with a selectivity of 99.98% under relatively mild conditions (80°C and 10 atm combined pressure of ethene and CO). Another catalyst has now also been operated under steady-state conditions for extended periods, giving total turnover numbers for palladium in excess of 100,000, with similar selectivity.

Additional Points to Ponder The same source reports TOF data for a second catalyst as being on the order of 12,000 mol of product per mole of catalyst per hour when operated in a batch mode. Why does the continuous process appear significantly more efficient than the batch process?

6.2.2 Selection by Types of Catalysts

Catalysts can be classified by several criteria, including their state of aggregation, their structure, their area of application, and their composition. A simplified general classification is given in Figure 6.12.

Heterogeneous and Homogeneous Catalysts When classifying catalysts by their state of aggregation, they fall into two classes, known as heterogeneous or homogeneous. *Homogeneous catalysts* are part of reacting systems in which reactant and catalyst are in the same state, most commonly liquid. Examples of homogeneous catalysts include liquid-phase acid or basic catalysts and transition metal complexes in solution. The most

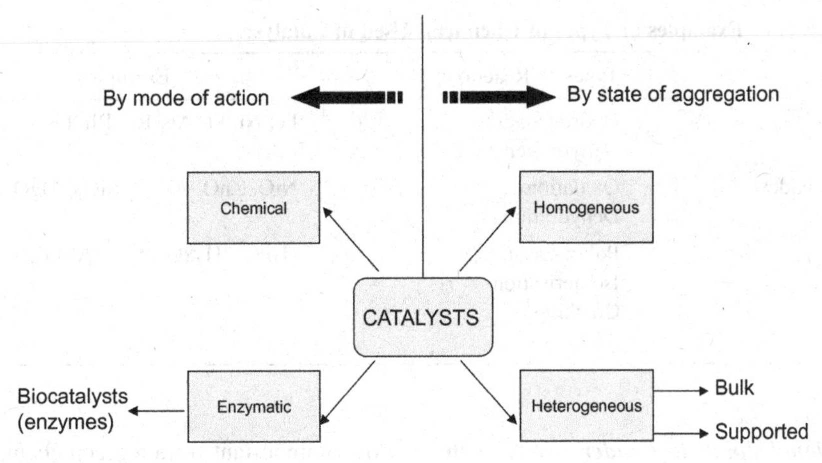

FIGURE 6.12 General classification of catalysts.

important industrial application of homogeneous catalysts is for the oxidation of hydrocarbons with oxygen or peroxides. Industrial processes using homogeneous catalysts include hydroformylation (with reported annual capacities of 2000 to 3000 tons), hydrocyanation (DuPont), acetic acid (Eastman Kodak), acetic acid anhydride (Eastman Kodak), Indenoxide (Merck, at an annual scale of about 600 kg), and the oxo synthesis, among others.

Heterogenous catalysts are generally solids (or supported on a solid framework or backbone), and the reactants are in either the liquid or gas phase. Examples of heterogeneous catalysts include metal oxides, transition metal catalysts, zeolites, or acid catalysts supported on a solid. Heterogeneous catalysts are by far the most important class of catalysts on the market. The market share of homogeneous catalysts is estimated to be only about 10 to 15%.

Example 6.5 Scientists Vladimir Dioumaev and Morris Bullock of Brookhaven National Laboratory in New York have developed a tungsten-based catalyst that facilitates a reaction and then easily separates out from the product and can be reused.[33] This homogeneous catalyst is initially dissolved in the ingredients and then begins to form oily clumps as the reaction progresses, to precipitate finally as a solid. They have focused on reactions that create alkoxysilanes-a common ingredient in ceramics and in organic compounds used in agriculture and pharmaceuticals:

$$R{-}C({=}O){-}R' + HSiEt_3 \xrightarrow{\text{catalyst}} R{-}CH(OSiEt_3){-}R'$$

In this reaction system they exploited the fact that a charged, or polar, catalyst would be soluble in the initial polar solvent but not in the nonpolar product. They created a catalyst that remained soluble and kept working until the very end of the reaction, when all of the ingredients had been used up. They used compounds that tend to form an oily mass in a nonpolar solution before separating out.

TABLE 6.8 Examples of Types of Chemicals Used in Catalysis

Catalyst	Types of Reactions	Examples
Metals	Hydrogenation Dehydrogenation	Fe, Ni, Pt, Ag, Ru, Rh, Os
Metal oxides	Oxidation Dehydration	NiO, ZnO, Al_2O_3, SiO_2, MgO
Acids	Polymerisation Isomerisation Cracking Alkylation	H_3PO_4, H_2SO_4, SiO_2/Al_2O_3, zeolites

Additional Points to Ponder Why is this discovery important from a green chemistry/green engineering perspective?

Chemical Catalysts There are many varieties of chemical catalysts, and they are used for a vast range of industrial applications. Some of the types of elements and chemicals commonly used in catalysts are given in Table 6.8.

Industrial processes that are run with catalysts include selective oxidations, alkane activation, stereo and regioselective synthesis, alkylation reactions, olefin polymerization, and others.

The most widely used catalysts are for acid-catalyzed reactions and are based largely on inexpensive Brønsted and Lewis acids, such as sulfuric acid, hydrogen fluoride, and aluminum chloride. These can be used for very diverse types of chemistries. A statistical survey looking at industrial applications of solid acid–base catalysts was completed in 1999 and found 127 industrial processes using either solid acid, solid base, or solid acid–base catalysts (Figure 6.13), with 81% of these processes using solid acid catalysts, 8% solid base catalysts, and 11% bifunctional solid acid-base catalysts. The survey also found that 180

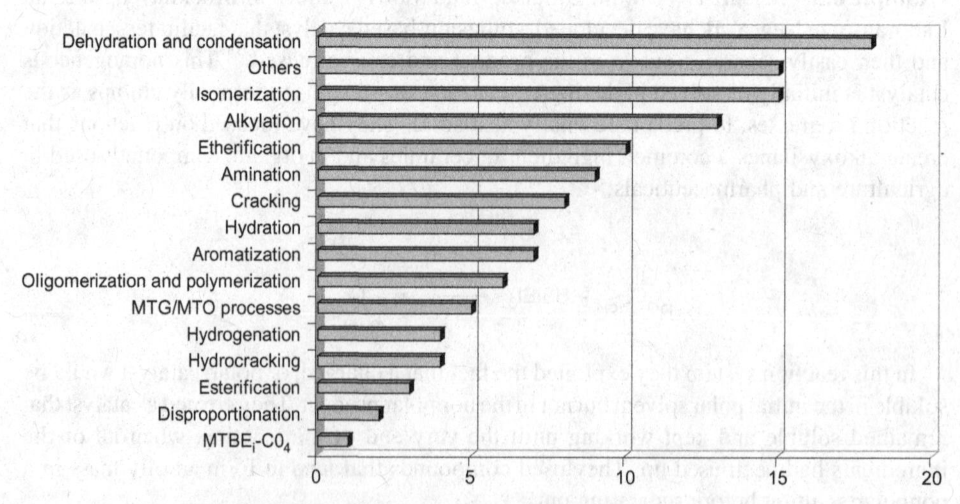

FIGURE 6.13 Industrial processes using acid–base catalysts.

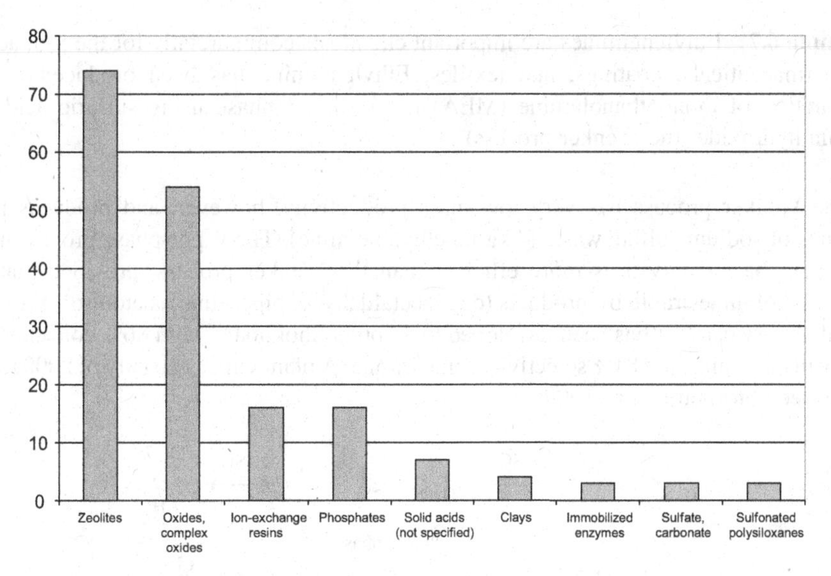

FIGURE 6.14 Type of catalysts used in industrial processes.

different types of catalysts were being used at an industrial level, with zeolites being the most used catalysts in industry, followed by oxides, ion-exchange resins and phosphates (Figure 6.14).[34]

Example 6.6 Cumene is produced from propylene and benzene. The total worldwide production capacity of cumene is about 6 million tons/year. The conventional processes use solid phosphoric acid (SPA) or aluminum trichloride as catalysts. SPA production is still heavily predominant.

H_2C CH_2

\+ H_2C— CH_2 catalyst →

Several companies have been involved in the development of new zeolite-based processes. The Mobil/Badger cumene process uses a novel zeolite catalyst developed by Mobil that offers higher yield and product purity than the traditional processes while eliminating problems with corrosion, catalyst handling, and disposal, such as significant amounts of acid waste. The pilot plant results show a 100% propylene conversion and nearly 100% selectivity in the alkylation reactor over a period of 5000 h of operation. Dow has also commercialized its process using a novel dealuminated mordenite with a pseudo-three-dimensional structure.[34]

Additional Points to Ponder Even after the process improvements with the catalytic process, are there any concerns with regard to the cumene route? Would it be possible to produce cumene from another source besides benzene?

Example 6.7 Ethyleneimines are important chemicals commercially for the production of pharmaceuticals, coatings, and textiles. Ethyleneimine has been produced by the dehydration of monoethanolamine (MEA) in the liquid phase using sulfuric acid and sodium hydroxide (the Wenker process).

The Wenker process has very low mass productivity, however, and produces large amounts of sodium sulfate waste (4 kg/kg ethyleneimine). The vapor-phase process using solid acid–base catalysts is more efficient than the Wenker process, provided that the formation of undesirable by-products (e.g., acetaldehyde, piperidine, acetonitrile) is minimized. A new catalyst has been developed by Nippon Shokubai[35] with 86% conversion of monoethanolamine and 81% selectivity for aziridine. A plant with a capacity of 2000 tones/yr has been onstream since 1990.

H_2N–CH_2CH_2–OH (MEA) $\xrightarrow{H_2SO_4}$ H_2N–CH_2CH_2–OSO_3H (liquid phase) $\xrightarrow{H_2SO_4}$ N–H aziridine (EI)

MEA → EI: solid catalyst (vapor phase)

Additional Points to Ponder How could the formation of by-products in the vapor-phase process be minimized or eliminated? What are the implications of the equipment and plant layout in the vapor-phase process?

Phase-Transfer Catalysts In a heterogeneous system phase-transfer catalysts (PTCs) facilitate the migration of an insoluble reactant from one phase into another phase by making the reactant soluble in both phases. These type of catalysts are often quaternary ammonium salts. The importance of PTCs to green chemistry and engineering is that these types of catalysts have the potential to dramatically increase reaction rates and thereby reduce the overall process energy requirements and energy-related emissions. They can also increase the specificity of a reaction system, reducing waste in general. PTCs are generally nontoxic, and they enable the use of less toxic solvents as substitutes for dipolar aprotic solvents such as dimethylformamide or facilitate the use of liquid reactants as solvents. In general, all these factors contribute to process efficiency and improve process design in terms of equipment size and process simplification. Some of the reactions and applications of PTCs are shown in Table 6.9.

There are many examples of PTC applications that have been commercialized. Following are a few examples of PTC applications.

Example 6.8 The synthesis of 2,6-diisopropylphenyl [(2,4,6-triisopropylphenyl)acetyl]-sulfamate, a pharmaceutical agent investigated in the treatment of high cholesterol and arteroesclerosis, initially required the production of 2,4,6-triisopropylbenzyl cyanide as an intermediate for the synthesis. The original synthesis included the use of dimethyl sulfoxide:[36]

$LiAlH_4$, Et_2O
98%

$SOCl_2$, toluene
95–98%

NaCN, DMSO
80–85%

As part of the development work, phase-transfer catalysis chemistry was used for the cyanation step, with tetra-*n*-butylammonium bromide (1.5% by weight) in a solution of toluene and water. Water had the added benefit of being a safer medium for both the handling and disposal of cyanide:

NaCN, H_2O,toluene
Bu_4NBr

The reaction was essentially quantitative, generating only trace amounts of new impurities. The reaction proceeded quickly and exothermically, but the exotherm was controlled through the rate of mixing instead of by stepwise cyanide addition. The reaction can now be run with high throughput and a reduction of sodium cyanide from 1.7 to 1.05 equivalents, decreasing the need to treat and handle cyanide-containing waste. The final product was sufficiently pure that the separate isolation, drying, and purification steps used were no longer necessary, enabling the toluene solution to be used directly in the hydrolysis step.

Additional Points to Ponder Why is the reduction of sodium cyanide important? What are the implications of not having to perform the isolation steps required in the original reaction?

Biocatalysts Enzymes or *biocatalysts* have increased in importance in recent years because they have extremely high selectivities and activities. In addition, they function under mild conditions: normally at room temperature, in aqueous media, and with a pH close to 7. Biocatalysts are generally more efficient than their chemical counterparts; for example, enzymes exhibit turnover numbers on the order of 100,000 s^{-1} in comparison to values of

TABLE 6.9 Applications of Phase-Transfer Catalysts

Commercial Applications of Phase-Transfer Catalysts	Reactions	Benefits
Polymers	O-Alkylation (Etherification)	Improved mass and heat transfer
Agricultural chemicals	N-Alkylation	Improved control of residence time and temperature distribution
Pharmaceuticals	C-Alkylation and chiral alkylation	Reduced solvent use
Monomers	S-Alkylation	Reduced work-ups
Additives	Dehydrohalogenation	Possibility of higher selectivity, yield, and quality
Flavors and fragances	Esterification	
Dyes	Displacement with cyanide hydroxide, hydrolysis	Linear scale-up (larger-scale vortex mixers will yield the same results given that the residence time and flow velocities used at lab level are matched)
Explosive	Fluoride thiocyanate, cyanate	Faster development of new processes or substances
Surfactants	Iodide sulfide, sulfite	Continuous operation of traditional batch processes
Petrochemicals	Azide nitrite, nitrate	
Commodity, specialty and fine chemicals	Other nucleophilic, aliphatic and aromatic substitution	
	Oxidation	
	Epoxidation and chiral epoxidation	
	Michael addition	
	Aldol condensation	
	Wittig	
	Darzen's condensation	
	Carbene reactions	
	Thiophosphorylation	
	Reduction	
	Carbonylation	
	HCl/HBr reactions	
	Transition metal cocatalysis	

about 1 s^{-1} for chemical catalysts. This increased efficiency has direct green chemistry/green technology implications given:

- The potential for using materials from renewable sources in the production of commodity and fine chemicals
- The potential reduction of waste and increased mass efficiency due to their unsurpassed enantio- and regioselectivity
- The elimination or reduction in the use of hazardous materials and reagents
- The possibility of running the reactions under mild conditions of temperature, pressure, and pH

- The use of aqueous systems instead of organic solvents, although some enzymes are also active in organic solvents, which are sometimes required to solubilize reactants
- The stereo- and regioselectivity of enzymes results in fewer by-products, which might simplify downstream processing and recovery
- The possibility of binding an enzyme, making it heterogeneous and therefore reusable

Biocatalysts come from a wide range of sources, including whole cells of animals, plants, and microbes, cell-free extracts, and enzymes. Table 6.10 shows some of the advantages and disadvantages of the use of whole cells vs. isolated enzymes.

Although biocatalyst use at a commercial scale is almost doubling with every decade, only a relatively small proportion of them are being used on a commercial scale. For example, from about 4000 known enzymes, only about 10% are commercially available and only about 1 % are used in commercial processes.

About 150 commercial processes are reported to use biocatalysts at a scale of 100 kg/yr or more, and about nine processes are reported to be run on a scale of 10,000 metric tons per year or more, and these are used primarily in the production of carbohydrates for the food industry. Non-food-related commercial bioprocesses on a large scale include the production of acrylamide and the production of 6-aminopenicillanic acid. The most important areas of application of biocatalysts include food, basic chemicals, pharmaceuticals, pulp and paper, medicine, energy production, and mining. In many cases, the combination of chemical and biocatalytic processes plays an important role. For example, production of the antibiotic cefalexin has been shortened from 10 steps to 6 steps using a chemoenzymatic procedure:[37]

TABLE 6.10 Some Advantages and Disadvantages of Biocatalysts

Type of Biocatalyst	Advantages	Disadvantages
Whole cells General	No cofactor recycling necessary	Low substrate concentration
		High biomass production
		Side reactions (low specificity)
		Instability at high temperatures and pH
Fermenting	High activities	Large biomass production
		More by-products
Non fermenting	Easier work up	Lower activities
Immobilized	Easier separation	Lower activities
	Easier purification	
Isolated enzymes General	Extremely high selectivity	Cofactor usually required and its recycling is necessary
	Easy to use	Availability for selected reactions only
	Simple work up	Instability at high temperatures and pH
	Higher substrate concentration	
Immobilized	High selectivity	Loss of activity during immobilization
	Easier separation	
	Easier purification	

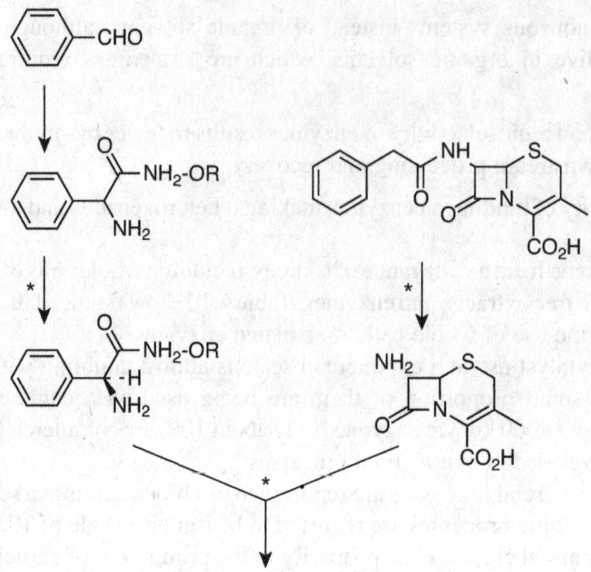

Cefalexin (six steps)
* Steps with biocatalysis

Not surprisingly, chirality plays an important role in biocatalysis, as biocatalysts are chiral by nature and exhibit an unsurpassed enantioselectivity. Therefore, most applications of biocatalysts involve enantioselective synthesis, either through enantiomeric pure precursors, kinetic resolutions, or asymmetric synthesis. In those cases where stereochemistry is not exploited, regioselectivity is an important characteristic of biocatalysts, and most examples of this type include regioselective oxidations, nitrile hydrolysis, and hydrations. Hydrolases are the enzymes used more commonly in nonchiral industrial biotransformations, with no particular class of enzyme dominating the applications. Some of the enzyme types used in industrial biotransformations include:

- Oxidizing cells
- Reducing cells
- Oxydoreductases, ketoreductases, etc.
- Isomerases
- Lyases
- Transferases
- Hydrolases
- Proteases
- Amidases
- Lipases
- Glycosidases
- Esterases
- Nitrilases

Biocatalysis is a highly interdisciplinary area. For a successful application of biocatalysis, there is a need for close interaction of practitioners in the areas of biology (enzymology, protein chemistry, molecular biology), chemistry (biochemistry and organic chemistry), and chemical engineering (transport phenomena, kinetics, reaction engineering, separations).

Example 6.9 2-Quinoxalinecarboxylic acid (QCA) is used in the synthesis of a variety of biologically active compounds. This compound was prepared from a di-N-oxide, via the chemical route shown below:[38]

1 → ($Na_2S_2O_4/H_2O$, EtOH) → 2 → (aq. HCl) → 3 → (35% H_2O_2, 10% NaOH) → 4

This route was demonstrated at the 25-kg scale (35% overall yield), but was deemed unsuitable for scale-up due to the mutagenic and thermal properties of the di-*N*-oxide. A biocatalytic synthesis was developed to run in aqueous media at about 30°C using whole cells of *Pseudomonas putida* grown on benzyl alcohol and had an impressive 86% yield. This whole-cell process is speculated to occur in the following three-enzyme process:

5 → (monooxygenase) → 6 → (benzyl alcohol dehydrogenase) → 3 → (benzaldehyde dehydrogenase) → 4

A comparison of the estimated material used per kilogram of QCA for both syntheses is provided in Table 6.11:

What advantages would the biocatalytic route have over the chemical route from a green chemistry/green engineering perspective? Quantify your answers as much as possible.

Solution In comparison with the chemical route, the biocatalytic route:

- Avoids hazardous di-*N*-oxide
- Uses four times less starting material (Table 6.12)
  Reduces organic solvent consumption (3.8 L/kg QCA) vs. the chemical process (196 L/kg QCA)
- Has a higher yield and higher mass productivity
- Operates under milder conditions

TABLE 6.11 Material Use

Chemical Process		*P. putida* Process	
DI-*N*-oxide	3.9 kg	2-Methylquinoxaline	0.97 kg
$Na_2S_2O_4$	5.7 kg	Benzyl alcohol	2.9 L
35% H_2O_2	6.5 L	*p*-Xylene	0.9 L
4 N HCl	13.6 L	4 N HCl	3.8 L
10% NaOH	11.7 L	10% NaOH	1.7 L
Chloroform	142 L	Inorganic salts	0.75kg
N,N-Dimethylacetamide	36 L	Trace elements	0.005 kg
Ethanol	18 L	H_2O	79 L

TABLE 6.12 Comparison of Mass Intensities and Efficiencies

	Chemical Route	Biocatalytic Route
Mass intensity (kg/kg) including water	301.9	90.0
Excluding water	275.5	6.2
Mass efficiency (%) including water	0.33	1.11
Excluding water	0.36	16.03

Additional Points to Ponder Are there any disadvantages to the biocatalytic route? Without additional data, what would you expect the energy requirements to be?

6.3 OTHER REAGENTS

After solvents, acid and bases are the most used auxiliary materials on a mass basis. We have covered solvents and catalysts in detail in this chapter. For the selection of other reagents and reactants, there are several aspects to consider, such as mass, downstream processing, their influence on the efficiency of a reaction, cost, renewability, recyclability, life cycle implications, and inherent safety.

PROBLEMS

6.1 In this chapter PCA was defined as an orthogonal linear transformation. Explain what this means and how this is done.

6.2 Several of the benchmarking tools presented in this chapter for solvent selection use a series of categories for ranking solvents. For example, GlaxoSmithKline's solvent selection guide shows solvents with bad scores in some categories but not others. How would you reconcile the trade-offs that this represents?

6.3 You are a chemical engineer working in the development of a new industrial process of a very durable plastic. You are having lunch with the lead chemist of that product and the chemist mentions that the best solvent for the process is tetrahydrofuran.

(a) What would you tell the chemist about the proposed solvent? Why?

(b) Over dessert, the chemist tells you that in addition to tetraydrofuran, they are considering pyridine and 2-methyltetrahydrofuran. What would be your opinion?

(c) What other recommendation would you give to your colleague?

6.4 Motorola Inc. is drastically reducing its emissions of perfluorocompounds (PFCs), solvents used in chip manufacturing as a stripper to clean polymer residues. The company has developed an alternative, "dry" process that dissolves the polymer residue with an ionized gas plasma of oxygen, hydrogen, nitrogen, and small amounts of nitrogen trifluoride (NF_3), followed by a final deionized-water rinse.[39]

(a) Why would Motorola be interested in reducing PFC emissions?

(b) Are there any issues that you foresee with use of this new process?

(c) How can you devise an even better process for cleaning chips?

6.5 The reaction

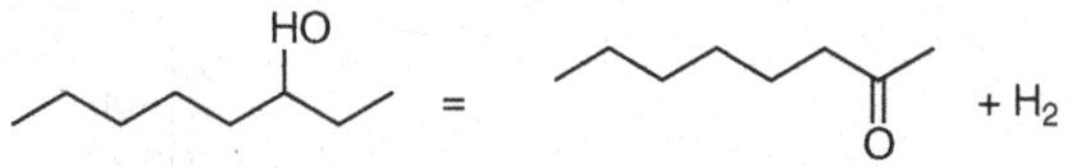

HO = O + H_2

is normally run in dichloromethane. Find a solvent that replaces dichloromethane while providing an inert reaction medium by dissolving the reactants and products and by keeping the reaction temperature low. The solvent needs to dissolve 3-octanol and 3-octanone, must be inert, and must be liquid at the reacting condition (temperature = 298 K). Also, it must have favorable EHS properties and must have a density close to that of water.

6.6 Provide alternatives for the solvent system of Example 6.1 using:

(a) Any solvent selection guide available

(b) Properties databases

(c) CAMD techniques

6.7 The product of Example 6.1 undergoes a saponification reaction that hydrolyzes the ethyl group in compound C with 2.5 N sodium hydroxide, with no isolation between the first reaction (Example 6.1) and the saponification reaction. The saponification reaction,

R–C(=O)–NH–R–C(=O)–OCH_2CH_3 + NaOH ⟶ R–C(=O)–NH–R–C(=O)–ONa + CH_3CH_2OH

C (dissolved) C (dissolved)

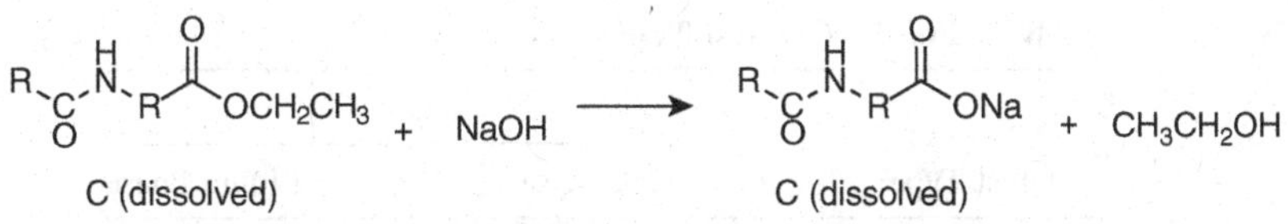

is followed by an extraction in methylene chloride, leaving the product in the aqueous phase. What other solvents would you propose for the saponification reaction and extraction?

6.8 2-Methyltetrahydrofuran, methyl cyclopentyl ether and *n*-alkyl tetrahydrofurfuryl, ethers (nATEs) are being advertised as greener solvents. Describe in detail why these solvents could be considered greener. Are there any instances in which they can be disadvantageous?

6.9 In Example 6.3 we describe the Swiss water process and supercritical carbon dioxide process for caffeine extraction in general terms.

(a) What are the disadvantages and advantages of these two extraction processes?

(b) Provide two examples of the use of surpercritical fluids for extractions.

(c) Provide two examples of the use of surpercritical fluids for reactions.

(d) What would be the difference between extractions using a supercritical fluid vs. liquid carbon dioxides?

(e) Provide two examples of the use of liquid carbon dioxide extractions.

6.10 Provide 10 examples from the literature of chemical reactions reported to proceed in the aqueous phase.

6.11 Find a feasible set of solvents that could be used in an enzymatic glycerolysis reaction which takes place in the presence of a catalyst (lipase enzyme).

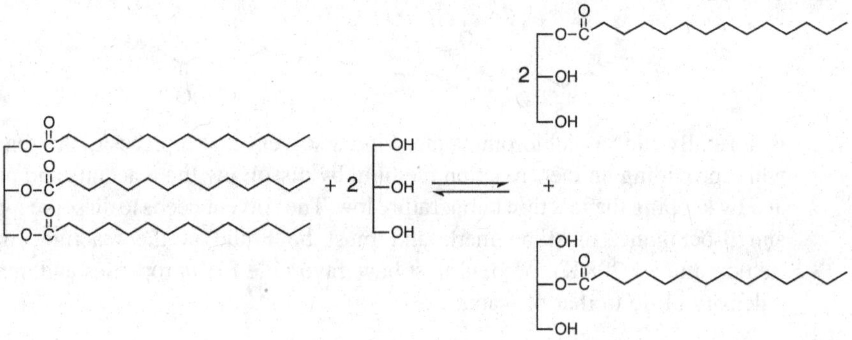

6.12 Table P6.12 shows turn over frequencies for the hydrogenation of cyclohexane at 25°C and 1 bar, using different supported catalysts.

(a) Based on the data, which phase would you recommend for the reaction? Why?

(b) Which catalyst would you recommend? Why?

(c) What other green chemistry or green engineering considerations need to be addressed with the system you are recommending?

TABLE P6.12 Catalyst TOFs

	TOF (s^{-1})	
Catalyst Type	Gas Phase	Liquid Phase
Ni	2.0	0.45
Rh	6.1	1.3
Pt	3.2	1.5
Pd	2.8	0.6

6.13 Describe the advantages and disadvantages of heterogeneous vs. homogeneous catalysts.

6.14 The principles of green chemistry state that catalytic reactions are preferable to stoichiometric reactions. Why is this? Describe three situations in which a catalyst is not green.

6.15 The route of caprolactam (a precursor of nylon 6) has evolved from the original chemical route:

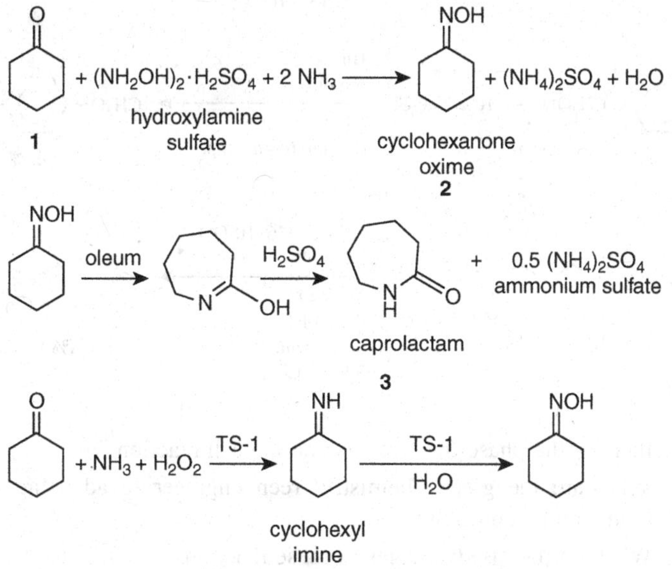

O
+ $(NH_2OH)_2 \cdot H_2SO_4 + 2\ NH_3$ → NOH + $(NH_4)_2SO_4 + H_2O$
hydroxylamine sulfate
1
cyclohexanone oxime
2

NOH
oleum
N
OH
H_2SO_4
N
H
O
+ 0.5 $(NH_4)_2SO_4$
ammonium sulfate
caprolactam
3

O
+ $NH_3 + H_2O_2$
TS-1
NH
TS-1
H_2O
NOH
cyclohexyl imine

to a one-step catalytic production:

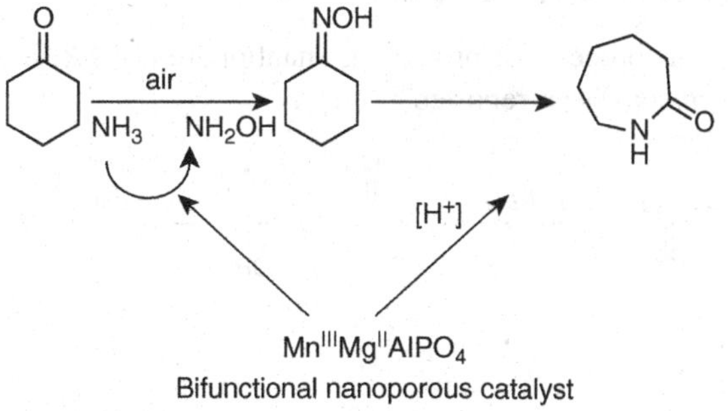

O
air
NH_3
NH_2OH
NOH
N
H
O
$[H^+]$
$Mn^{III}Mg^{II}AlPO_4$
Bifunctional nanoporous catalyst

What are the advantages of the catalytic route over the chemical route from a green chemistry/green engineering perspective.

6.16 Investigate the catalytic industrial process for production of the following chemicals:

(a) Linear alkyl benzenes

(b) Methylamine

(c) Cyclohexanol

(d) *t*-butylamine

Do these catalytic processes exhibit advantages from the green chemistry/green technology perspective? Are there disadvantages?

6.17 The following oxidations[40–42] are performed in phase-transfer catalysts:

Toluene (CH_3-benzene) + air(12–15 atm) → [$(C_{10}H_{21})_2N(CH_3)_2^+Br^-$ 0.15 mol%; $CoCl2{\cdot}6H_2O$, 0.2 mol%, no solvent, 3 h, 135–160°C] → benzoic acid (COOH), 99%

CH_3O–C_6H_4–CH_2OH + 10% NaOCl → [$Bu_4N\ HSO_4$, 5 mol%; ethyl acetate, 28 min, room temp.] → CH_3O–C_6H_4–C(=O)–H, 92%

Dodec-11-en-2-ol (OH) + 30% H_2O_2, 1.5 mol → [Aliquat 336 (HSO_4) 0.2 mol%; Na_2WO_4, 0.2 mol%, no solvent, 3 h, 90°C] → dodec-11-en-2-one (O), 97% + 0.4% 11,12-epoxide-2-one

(a) Identify the phase-transfer catalyst in each reaction.

(b) What are the green chemistry/green engineering advantages of these two oxidation reactions?

(c) What are the disadvantages of these reactions?

6.18 Identify the barriers to adopt biocatalyzed processes on a commercial scale. Provide some options to overcome these barriers.

6.19 A continuous process for producing enantiopure (2*R*,5*R*)-hexanediol using *Lactobacilus kefir* has been reported[43]:

1 (2,5-hexanedione) —*L. Kefir*→ **2** (5-hydroxyhexan-2-one) —*L. Kefir*→ **3** (2,5-hexanediol)

e.e. >99%
d.e. >99%

Conversion of (2,5)-hexanedione was nearly quantitative and the selectivity between product and intermediate was 78% for the product. Enantioselectivity and diastereoselectivity were >99% for the entire period. The productivity of *L. kefir* could be increased by factor of 30. (2*R*,5*R*)-Hexanediol was produced continuously over 5 days with a space–time yield of 64 gal/L per day. What are the advantages of using l kefir over the typical chemical routes to produce (2*R*,5*R*)-hexanediol? What are the advantages of running this process in continuous rather than batch mode?

6.20 When selecting a reaction for a chemical synthesis, list the aspects that are important from a green chemistry/green engineering perspective?

REFERENCES

1. European Commission. Council Directive 1999/13/EC. Mar. 11, 1999.
2. Toxic Chemical Release Reporting: Community Right-to-Know. 40 CFR Part 372.
3. European Chemical Industry, Council. Responsible Care: Health, Safety and Environmental Reporting Guidelines. 1998. http://www.cefic.be/activities/hse/rc/guide/01.htm.
4. Jiménez-González, C., Curzons, A. D., Constable, D. J. C., Cunningham, V. L. Cradle-to-gate life cycle inventory and assessment of pharmaceutical compounds: a case-study. *Int. J. Life Cycle Assess.*, 2004, 9(2), 114–121.
5. Jiménez-González, C., Curzons, A. D., Constable, D. J. C., Cunningham, V. L. Expanding GSK's solvent selection guide: application of life cycle assessment to enhance solvent selections. *J. Clean Technol. Environ. Policy*, 2005, 7, 42–50.
6. Vinson, J. CAMD for solvent selection in industry: II. In *Computer-aided molecular design: theory and practice*, Achenie, L. E. K., Gani, R., Venkatasubramanian, V., (Eds.), Elsevier, Amsterdam, The Netherlands, 2002.
7. Gani, R., Jiménez-González, C., ten Kate, A., Crafts, P. A., Jones, M., Powell, L., Atherton, J., Cordiner, J. A modern approach to solvent selection. *Chem. Eng.*, 2006, 113(3), 30–43.
8. Malinowski, E. *Factor analysis in chemistry*, 2nd ed. Wiley, New York, 1991.
9. Carlson, R., Lundstedt, T., Albano, C. Screening of suitable solvents in organic synthesis. *Acta Chemica. Scand.* 1985, B39(2), 79–91.
10. Curzons, A. D., Constable, D. J. C., Cunningham, V. L. Solvent selection guide: a guide to the integration of environmental, health and safety criteria into the selection of solvents. *Clean Products Process*, 1990, 1, 82–90.
11. Gani, G., Jiménez-González, C., Constable, D. J. C. Method for selection of solvents for promotion of organic reactions. *Comput. Chem. Eng.* 2005, 29(7), 1661–1676.
12. Gani, R., Arenas-Gómez, P., Folić, M., Jiménez-González, C., Constable, D. J. C. Solvents in organic synthesis: replacement and multi-step reaction systems. *Comput. Chem. Eng.* 2008, 32, 2420–2444.
13. Gani, R. *ICAS Documentations.* CAPEC Internal Report. Technical University of Denmark, Lyngby, Denmark, 2001.
14. Harper, P. M., Gani, R. A multi-step and multi-level approach for computer aided molecular design. *Comput. Chem. Eng.* 2000, 24, 677–683.
15. Matlack, A. *Introduction to Green Chemistry*. CRC Press, Boca Raton, FL, 2001.
16. Varma, R. S. Solvent-free organic syntheses using supported reagents and microwave irradiation. *Green Chem.*, 1999, 1, 43–55.
17. Jeselnik, M., Varma, R. S., Polanc, S., Kocevar M. Solid-state synthesis of heterocyclic hydrazones using microwaves under catalyst-free conditions. *Green Chem.*, 2002, 4, 35–38.
18. Semeria, D., Philippe, M., Patent EP0884305, to L'Oreal, 2001.
19. Rahman, A. N. M. M., Bishop, R., Tan, R., ShanGreen, N. *Green Chem.*, 2005, 7, 207–209.
20. Kissinger, G. M., Wynn, N. P. U. S. patent 5,475,152, to General Electric, Dec. 12, 1995.
21. Tundo, T., Loris, A., Selva, M. *Green Chem.* 2007, 9, 777–779.
22. Philippe, M., Semeria, D., WO/2002/051803 to L'Oreal, 2002.
23. Hersant, Y., Abou-Jneid, R., Canac, Y., Lubineau, A., Philippe, M., Semeria, D., Radisson, X., Scherrmann, M.-C. *Carbohydr. Res.* 2004, 339, 741–745.
24. Yarnell, A. Caffeine-free ethiopian coffee plants promise full-flavored cup of decaf. Chem. Eng. News, June 25, 2004.

25. Cann, M. C., Connelly, M. E. *Real World Cases in Green Chemistry*, American Chemical Society, Washington, DC, 2000.
26. EPA Presidential Green Chemistry Challenge Awards 1997. http://www.epa.gov/greenchemistry/pubs/pgcc/past.html.
27. Dybkær, R. Unit "katal" for catalytic activity (IUPAC Technical Report). *Pure Appl. Chem.*, 2001, 73(6), 927–931.
28. Dybkær, R. The tortuous road to the adoption of katal for the expression of catalytic activity by the General Conference on Weights and Measures. *Clin. Chem.*, 2002, 48, 586–590.
29. Hagen J. *Industrial Catalysis: A Practical Approach*. Wiley-VCH, New York, 1999.
30. Liese, A., Seelbach, K., Wandrey, C., Eds. *Industrial Biotransformations*. Wiley-VCH, New York, 2006.
31. Ramaswamy, A. V. Catalyst selectivity: shape selective catalysis over zeolites. *Bull. Catal. Soc. India*, 2003, 2, 140–156.
32. Clegg, W., Eastham, G. R., Elsegood, M. R. J., Tooze, R. P., Wang, X. L., Whiston, K. Highly active and selective catalysts for the production of methyl propanoate via the methoxycarbonylation of ethene. *Chem. Commun.*, 1999, 1877–1878.
33. Dioumaev, V. K., Bullock, R. M. A recyclable catalyst that precipitates at the end of the reaction. *Nature*, 2000, 424, 530–532.
34. Tanabe, K., Hölderic, W. F. Industrial application of solid acid-base catalysts. *Appl. Catal. A*, 1999, 181, 399–434.
35. Ueshima, M., Tsuneki, H. *Catal. Sci. Technol.* 1991, 1, 357.
36. Dozeman, G., Fiore, P., Puls, T., Walker, J. Chemical development of a pilot scale process for the ACAT inhibitor 2,6-diisopropylphenyl [(2,4,6-triisopropylphenyl)acetyl]sulfamate. *Org. Proc. Res. Dev.*, 1997, 1, 137.
37. Gotor, V. Biocatalysis applied to the preparation of pharmaceuticals. *Org. Process Res. Dev.*, 2002, 6, 420–426.
38. Wong, J. W., Watson, H. A., Bouressa, J. F., Burns, M. P., Cawley, J. J., Doro, A. E., Guzek, D. B., Hintz, M. A., McCormick, E. L., Scully, D. A., Siderewicz, J. M., Taylor, W. J,, Truesdell, S. J., Wax, R. G. Biocatalytic oxidation of 2-methylquinoxaline to 2-quinoxalinecarboxylic acid. *Org. Process Res. Dev.*, 2002, 6, 477–481.
39. Arensman, R,Clean equal $. *Green Electron.* Bus., May 1, 2003.
40. Lee, G., Freedman, H. U.S. Patent 4,079,075, to Dow Chemical, 1978.
41. Dakka, J., Zoran, A., Sasson, Y. US Patent 4965406, to Gadot Petrochemical Industriesl Inc., 1990.
42. Sato, K., Aoki, M., Takagi, J., Noyori, R. Organic solvent- and halide-free oxidation of alcohols with aqueous hydrogen peroxide. *J. Am. Chem. Soc.*, 1997, 119, 12386.
43. Haberland, J., Hummel, W., Daussmann, T., Liese, A. New continuous production process for enantiopure (2*R*, 5*R*)-hexanediol. *Org. Process Res. Dev.*, 2002, 6, 458–462.

7

REACTION CONDITIONS AND GREEN CHEMISTRY

What This Chapter Is About In this chapter we address what chemists and engineers can do about reaction conditions to promote greener reactions and chemistry. What is discussed in this chapter follows naturally out of the preceding two chapters and contains very practical advice about how to set up and carry out reactions in a way that leads to optimal reaction conditions. The chapter presupposes some understanding of statistical design of experiments.

Learning Objectives At the end of this chapter, the student will be able to understand the importance of:

- Paying attention to stoichiometry
- Environmental impacts from temperature control
- Impacts of solvent use
- Reaction time
- Order and rate of addition
- Mixing

in the context of designing greener chemistries and chemical processes.

Note: For those not familiar with common batch chemical operations, you may want to refer to the available literature and Appendix 7.1 for a general understanding of common practices found in batch chemical operations. You will want to become familiar with a number of the concepts and practices as a means of providing context for what you are reading. It should

Green Chemistry and Engineering: A Practical Design Approach, By Concepción Jiménez-González and David J. C. Constable

also help you in thinking about the examples in this chapter, and you may find it especially helpful when trying to answer the problems at the end of the chapter.

7.1 STOICHIOMETRY

Anyone who has studied basic chemistry is certain to understand the importance of balancing equations and paying attention to the stoichiometry of any given reaction. These are the types of things that are drilled into one's head as general chemistry is taught. It is also true that for many chemists, excess amounts of a given reactant or reagent are used to take advantage of LeChâtelier's principle and to ensure that a chemical reaction is pushed toward one direction and goes to completion. However, as we have seen from Chapter 4, if we use more than an equimolar amount of a reactant, in addition to noncatalytic amounts of reagents, our mass efficiency is likely to be affected significantly (i.e., it will go down and we will have to manage larger amounts of waste). A general principle that synthetic organic chemists would agree they strive for is that reactions need to use equimolar quantities, or as close to equimolar quantities of reactants as possible. This is just "good" chemistry, right? Another principle most chemists would agree that they follow is to employ small amounts of a catalyst, to avoid having to use large stoichiometric excesses of a catalyst or a reactant.

An example of the use of stoichiometric "catalysts" is found in the nitration of aromatic compounds, where it is common to employ large amounts of nitric and sulfuric acid with strong Lewis acids such as boron trifluoride. The Lewis acid is normally used in large stoichiometric excess and is destroyed in the aqueous quench with the production of large amounts of acidic waste. However, it was reported by Braddock[1] and shown for the general scheme in the reaction depicted below that by using catalytic quantities of lanthanide or group IV metal triflates [$M(COTf)_4$, where M = Hf or Zr] or metal triflides [$M(CTf)_3$, where M = Yb or Sc], one only needs to employ stoichiometric amounts of nitric acid and that the only by-product is water. The catalyst can be recycled and reused following evaporation of the aqueous medium (the nitroaromatic would be recovered from the organic solvent).

R–C_6H_5 —[10 mol% $Yb(OTf)_3$, 1 equiv. 69% HNO_3; $(CH_2Cl)_2$, reflux]→ R–C_6H_4–NO_2

nitration of an aromatic using metal triflate

Example 7.1 For the scheme shown above, Braddock reports the conversions listed in Table 7.1 for a range of aromatic substrates and reports data for recycling of the catalyst shown in Table 7.2.

Describe the positive green chemistry elements of this approach to avoiding the use of stoichiometric catalysts. Describe any elements of this approach that are not green.

Solution

Positive elements

This approach avoids the use of excessive volumes of acid, and the catalyst is able to be recycled. In this particular example, the recovery is not optimized. Conversion of the aromatic substrate is quite high.

TABLE 7.1 Nitration of Aromatics with Catalytic Quantities of Yb(OTf)$_3$[a]

Entry	Arene	Conversion (%)[b,c]	Product Distribution[c]		
			ortho	*meta*	*para*
1	Benzene	>95 (95)			
2	Toluene	>95 (95)	52	7	41
3	Biphenyl	89	38	trace	62
4	Bromobenzene	92	44	trace	56
5	Nitrobenzene	0			
6	*p*-Xylene	>95			
7	*p*-Dibromobenzene	8			
8	*m*-Xylene	>95	4-NO_2 : 85 2-NO_2 : 15		
9	Naphthalene	>95	1-NO_2 : 91 2-NO_2 : 9[d]		

Source: ref. 1.

[a]All reactions carried out on a 3-mmol scale with 10 mol% ytterbium(III) triflate and 1.0 equiv. of 69% nitric acid in refluxing 1,2-dichloroethane (5 mL) for 12 h.

[b]Isolated yields in parentheses.

[c]Determined by gas chromatography (GC) and/or ^{1}H nuclear magnetic resonance analysis.

[d]*Caution:* Nitronaphthalenes are potent human carcinogens.

TABLE 7.2 Recycled Ytterbium(III) Triflate for the Nitration of *m*-Xylene[a]

Run	Conversion (%)[b]	Mass of Catalyst (mg)[c]
1	89	190 (>100)
2	81	152 (82)
3	90	127 (68)
4	88	115 (62)

[a]All runs performed with 3 mmol of *m*-xylene, 10 mol% ytterbium triflate (run 1), and 1 equiv. of 69% nitric acid in refluxing 1,2-dichloroethane (5 mL) for 5 h.

[b]Determined by GC analysis. The isomeric ratio of 4- and 2-nitroxylene was unchanged throughout (85 : 15, respectively).

[c]Mass of ytterbium(III) triflate recovered from each run. The figures in parentheses indicate the percentage recovery that was not optimized.

Nongreen elements

- The solvent used in the reaction, dichloroethane, is not a solvent that is generally used outside academic labs, as most chlorinated solvents have multiple health effects and regulatory constraints.
- There is an excessive reflux time of 5 to 12 h; holding a reaction at reflux for long periods of time wastes a considerable amount of energy and is often unnecessary for the reaction to progress.
- The reaction is not particularly chemoselective for a number of substrates, and separation of isomers can be problematic at scale.

Additional Points to Ponder Look ahead to Chapter 16. What might be the drawback to using a group IV metal such as Yb or Sc? Where do these elements come from? Are they abundant, or is there a potential environmental impact in using these elements?

It should be noted that when using very expensive catalysts, or catalysts that are based on certain precious metals, it may be worth considering a yield reduction. Think about this for a moment. If the catalyst is expensive, it is likely to be difficult to make; either the synthesis required to make it is complicated or perhaps it is using a platinum group metal or another relatively rare metal. In either case, the environmental life cycle impacts associated with the catalyst are likely to be quite large (see Chapter 16). It is also necessary to explore the recovery and reuse of stoichiometric "catalysts", especially when molar excesses are required. One option is to use bound catalysts that may be regenerated. For example, many older industrial processes are catalyzed by $AlCl_3$ in stoichiometric amounts, creating large problems with waste as $Al(OH)_3$ and no easy method of recovering the $AlCl_3$. A number of Lewis solid acids (which are easier to recover and reuse), such as zeolites, clays, mixed metal oxides, inorganic–organic composite materials, functionalized polymers, and supported reagents, could be considered as alternatives.

One example of the use of zeolites as a solid acid catalyst was reported by Smith et al. for the selective *p*-bromination of phenyl acetate.[2] By using zeolites, the use of aluminum chloride or other Lewis acids is avoided and the zeolite removes the HBr formed by ion exchange, thus increasing the selectivity and preventing the formation of phenol:

$$C_6H_5\text{–OAc} + Br_2 + NaY \longrightarrow Br\text{–}C_6H_4\text{–OAc} + NaBr + HY$$

Such approaches are potentially useful if they avoid the large quantities of waste normally associated with the use of Lewis acids. However, such schemes need to be evaluated carefully to see if they are truly green.

7.2 DESIGN OF EXPERIMENTS

Traditionally, synthetic organic chemistry route development and the associated process optimization has been performed by changing one variable at a time; that is, the chemist will change the temperature, or pH, or the stoichiometry of the reagents or reactants one at a time while holding all other variables constant. However, with the advent of combinatorial chemistry,[3–6] there has been considerable interest in rapid and parallel methods of synthesis for developing large libraries of compounds of potential interest for pharmaceutically active materials, catalysts, agrochemicals, chemicals used in the electronics industry, and others. Such approaches have also been recognized for their potential in optimizing chemical processes.[7,8]

Rather than take a one-variable-at-a-time approach (i.e., change temperature, pressure, time, reagent addition, etc. one experiment at a time), it is possible to use statistics in such a way as to change multiple variables at a time using a reduced number of experiments. For example, by using a traditional two-level factorial design, it is possible to perform 2^n experiments, where n is the number of variables that one is interested in testing. In the simple situation where a chemist is interested in varying three factors—say, time, temperature, and pH—to see which combination of conditions delivers the most mass-efficient reaction, we can perform a simple design of 15 experiments, depicted in Figure 7.1(a) as the star points, middle, and vertices of a cube, to obtain an optimized process. In each experiment, multiple variables are changed, but this is done in a systematic way and it is possible to construct a response surface as shown in Figure 7.1(b).

It is actually possible that the results of the first set of experiments would lead us to design additional tests to further round out our knowledge of the reaction system. These additional

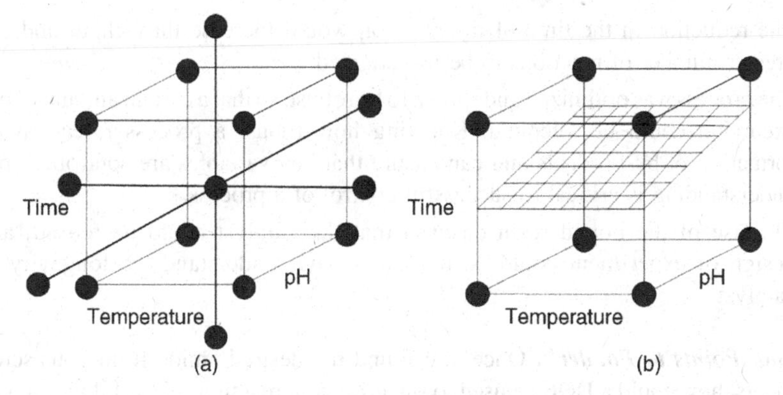

FIGURE 7.1 (a) Simple 2-level factorial design of experiment; (b) response surface for 2-level factorial design.

experiments could be depicted as either the vertices of the cube or at midpoints along the edges of the cube. In some cases these extra experiments may be needed to enhance our understanding of a particularly sensitive reaction system and ensure that we are not at a local maximum or minimum in a response surface.

Some chemists object to the rigor of a design of experiments (DOE) and often cite a number of objections to its use. However, a number of recent publications have answered these objections and have shown the utility of a DOE for optimizing synthetic approaches and processes.[9–11]

Example 7.2 The preparation of amides using a polymer-bound *N*-hydroxytriazole resin through an automated DOE has been described in the literature.[12] An example of a reaction of a bound *N*-hydroxytriazole resin for the synthesis of amides is shown below:

(1) RCO_2H coupling agent (2) wash → ; R_1NH_2 → R_1HN–C(=O)–R

In the original factorial design three of five parameters were found to be of importance: order of addition, solvent ratio, and the amount of diisopropyldicarbodiimide (DIC). In a second optimization, time was added to the solvent ratio, and the amount of DIC with the ranges shifted toward maximum loading and an additional variable of acid type was added to further investigate the robustness of the conditions. A third optimization run was used to validate the generality (robustness) of the conditions and test lot-to-lot variability. By using a DOE, the amount of carboxylic acid was reduced from 6 equiv. to 1.5 equiv. and the reaction time from 6 h to 2 h. These optimized conditions were then used to generate a 48-member structurally diverse library of amides with yields ranging from 54 to 93% and purities >87%. What are the green chemistry benefits in this approach?

Solution There are multiple benefits from a green chemistry approach:

1. The reduction of carboxylic acid from 6 to 1.5 equivalents would result in a significant increase in mass efficiency.

2. The reduction in the time of the reaction would increase throughput and allow a greater number of reactions to be investigated.
3. The process was optimized and shown to be robust, so that a certain amount of process excursion could be tolerated. Knowing how robust a process is can avoid the formation of by-products and can ensure that key variables are controlled; process understanding is critical to successful control of a process.
4. The use of the bound resin ensured that the catalyst could be reused, and the design of experiment could be used to further understand the longevity of the catalyst.

Additional Points to Ponder Once you found the desired amide from your screening experiments, how could a DOE be used to scale-up your reactions to the kilogram or larger scale? What additional process parameters might you be interested in investigating more closely?

When performing a DOE, there are a number of variables that one might want to investigate to ensure that a reaction is well understood. Key variables include:

Temperature	Mixing
Solvents	Catalyst reuse
Reaction time	Stoichiometry
Pressure	Order and rate of addition

Next we consider briefly some of these key variables to highlight a few issues and factors to consider in a green chemistry and green engineering context.

7.3 TEMPERATURE

Generally speaking, from a green chemistry perspective we will want to use reactions that can be run at nearly ambient temperatures and pressures. This will allow us to avoid using energy to heat or to cool our reaction mixtures. In the case of cooling, we especially want to avoid running reactions at extremely low temperatures (−20°C or lower), as holding temperatures this low requires a considerable amount of energy for refrigeration. In Chapter 18 we explore the environmental impacts of producing energy for either heating (e.g., steam) or cooling (e.g., cooling water, refrigeration).

The key is to balance temperature with reaction rates so that reaction kinetics are optimized while avoiding excessive exotherms, endotherms, gas evolution, rapid polymerizations, and other process conditions that can lead to one type of adverse event or another. For an exothermic reaction, you will want your design of experiments to investigate, at a minimum, the relationships between the reaction rate, order of addition, mixing, reaction temperature, and heat removal. You can estimate or quantify the heat of reaction (either endothermic or exothermic) in advance so that you are able to balance the heating or cooling requirements to a good first approximation. The maximum expected change in temperature of a reaction may easily be calculated by adding the process temperature plus the expected change in temperature. The maximum adiabatic temperature rise may be estimated by dividing the heat of reaction by the average heat capacity of the mixture.

As many reactions proceed, the rate of heat production or reduction may increase (or decrease, in the case of an endothermic reaction) exponentially, but the ability to remove the heat (e.g., cool the reaction mixture) or add it is considerably lower. In addition, an increase of 10°C in reaction temperature often represents a two- to threefold increase in the rate of reaction. This is a very rough rule of thumb for the Arrhenius relationship,

$$k = Ae^{-E_a/RT}$$

where k is the rate constant, denoted as the total frequency of collisions between reaction molecules A, times the fraction of collisions exp(−*Eâ*/*RT*) that have an energy beyond the reaction's activation energy (E_a) at a temperature of T. R is the universal gas constant. However, this rule of thumb might not apply to all the reactions, as the rate constant for some reactions would increase exponentially

Another useful rule of thumb is that the time needed for a 1°C drop in temperature by natural cooling from 80°C to 20°C is 230 min in a 25-m^3 reactor, 20 min in a 2.5-m^3 reactor, and 20 s in a 100-mL glass beaker.

Example 7.3 Estimate the temperature rise for a reaction with a standard heat of reaction of −50 kJ/mol and an average constant-pressure heat capacity of the reaction mixture of 250 J/mol·K. Assume that the entire reaction is carried in the liquid phase, there are no side reactions, no reactants are used in excess, and no heat is lost to the surroundings.

Solution Since the heat of reaction is negative, that means the reaction will liberate heat. A quick estimation of the temperature rise for the reaction with the data given could be as follows:

$$\Delta T_{\text{ad}} = \frac{\Delta H_r}{C_p} = \frac{50\ \text{kJ/mol}}{250\ \text{J/mol}\cdot\text{K}} \frac{1000\ \text{J}}{1\ \text{kJ}} = 200\ \text{K}$$

Now, an interesting point of this first approximation is that on the one hand, unless the heat is dissipated, the reaction will be taking place at a higher temperature instead of the standard temperature. Therefore, the heat of reaction will also change, the calculations above will need to be redone, and the process would need to be repeated through iterations until the heat of reaction and the final temperature do not change. When this is done in gaseous reactions, the heat capacity will also change. An experimental determination of the adiabatic temperature rise might be the best approach, but the means are not always available to do it. In addition, an experimental approach might not be recommended in all cases.

Additional Points to Ponder What is the relevance of assuming that the reaction happening in the liquid phase? What are examples of an experimental approach not being recommended for determination of adiabatic temperature rise?

One fruitful area of investigation for controlling reaction temperatures would be the use of micro and minireactors and alternative reactor designs. In one recent example, a loop reactor was used for a highly exothermic reaction. Six, 5-metric ton batch reactors originally used to produce an inorganic oxychloride from the oxide and chlorine, were replaced with one continuous loop reactor producing 500 m^3/h. The chemistry was changed to the oxidation of the chloride, which is rapid and exothermic, but the actual temperature rise is less than 20 K, since the product acts as a heat sink and a diluent in the loop.

7.4 SOLVENT USE

As discussed in Chapter 6 and elsewhere,[13–16] the use of solvents accounts for significant environmental life cycle impacts. For the most part, most synthetic organic chemists are not concerned about the solvents they choose to use apart from their potential reactivity in a given reaction of interest. In general, the environmental, safety, and health impacts of the solvents chosen are very often of little interest, despite the fact that solvents represent 80 to 90% of the mass in many batch chemical reactions commonly found in the fine chemicals and pharmaceutical industries. There is also a cavalier attitude about solvents; we use them, we know how to use them safely in the laboratory and at scale, so why should we worry about whether or not the solvent is obtained from renewable sources, is flammable, is explosive, or has any of a range of impacts? Chemists should, in fact, seek to minimize the quantity and number of solvents used. But how might they do this? There are actually a number of possible ways that this can be accomplished and some are outlined in Table 7.3.

As noted in Table 7.3, chemists should investigate the possibility of carrying out as much of their synthetic route without having to isolate intermediates. By combining steps and stages of the overall synthesis in a single reactor, they may be able to reduce the amount of solvent, the number of washes, and intermediate filtration and drying. Chemists might also consider running steps without solvent by investigating the potential for reactions in the solid state or by using microwave technology, or by using other processing approaches, such as grinding materials together. The latter approaches are often not considered in the fine-chemical industries, for reasons of impurity removal, process robustness/reproducibility, and so on.

As part of any good design of experiment, chemists will also want to investigate the boundaries of their reaction systems by running reactions at higher reactant and reagent concentrations. Reactants and reagents may not need to be dissolved to achieve optimal reactivity; in fact, they might exhibit an increased rate of reaction as heterogeneous mixtures under the optimal mixing regime. For endothermic reactions, chemists may want to investigate the use of more concentrated solutions run in smaller vessels with more precise temperature control (e.g., consider micro- or minireactors). As noted above, the use of DOE is critical in assessing a number of highly interactive variables to achieve optimal solvent use.

As noted in Table 7.3, another possible avenue to reducing solvent use is to investigate the reuse of mother liquors, especially in those situations where the mother liquors contain an excess of reagent, homogeneous catalyst, or reactant. As long as there is no significant by-product formation, impurity buildup, interference from product that remains in the mother liquors, or change in the reaction rate, it may be possible to reuse mother liquors for several batches. This is especially the case when there are stoichiometric catalysts or nonstoichiometric reagent uses for the reaction to proceed to completion. If successful, one might only need to replenish the reagent for the next batch.

The use of novel technologies may also provide a means of using less solvent. For example, one might consider a microwave synthesis that uses a limited amount of

TABLE 7.3 Strategies That Chemists Might Employ for Solvent Reduction

Strategy	Green Benefit
Reduce the number of solvents used in a reaction sequence	Avoids solvent mixtures and need for complex separations. Reduces amount of infrastructure required to handle, store, and manage solvents. Reduces energy used for solvent switching. Reduces the complexity of the synthesis (fewer solvents, fewer separation steps).
Look for opportunities for nonisolations	Increases throughput by reducing the number of steps. Reduces energy used for solvent switching. Avoids the need for filtration, drying, etc. Reduces the complexity of the synthesis.
Take advantage of reactant and product solubility	May permit simple gravity separations through phase splits. May permit product "drop" processes (i.e., product drops out of solution, avoiding the need for antisolvent to promote crystallization). May be able to run the reaction heterogeneously with reactants sparingly soluble and product going into solution for the next reaction. Solvent volumes could potentially be reduced dramatically if there is a significant difference in solubility. May be able to use reactants as solvents.
Run solid-phase reactions	Little or no solvent required for reaction to proceed.
Investigate biotransformations	May be able to run the reaction in water.
Investigate use of phase transfer catalysts	Phase-transfer catalysts may, in fact, allow a reduction of solvent use, as the catalyst will transfer reactants between immiscible phases to allow the reaction to occur; this is a variation on reactant and product solubility. General improvement of mass transfer.
Investigate potential for use of temperature or pressure	Pressure and temperature are critical for certain types of reactions, and it may be possible to use temperature to increase solubility and/or reactivity while decreasing the amount of solvent required. However, there is clearly a trade-off between solvent use and energy that needs to be considered, and such considerations require a higher level of process understanding and sophistication in design.
Investigate reuse of "mother" liquors (i.e., the reactant mixtures)	Reuse of mother liquors will obviously have the multiple benefit of solvent reduction and reactant, catalyst, and/or reagent reduction. Less material use will increase mass efficiency, especially in instances of large stoichiometric excesses.
Choose different solvents	By careful consideration of solvent selection, one might be able to select a solvent that not only permits nonisolation but increases the solubility of the reactants or products (thereby permitting process intensification), or decreases the miscibility with water or an antisolvent (thereby increasing the probability of recycle and reuse).

high-boiling solvent or no solvents. The reactions can be completed on a scale of several hundred grams in a few minutes, and larger-scale synthesis should be possible by using commercially available equipment such as the type used in the food industry. Some multistep syntheses, such as those for advanced intermediates used in lactam antibiotics, amino sugars, alkaloid, and some nitrations have been conducted using microwave technology.

One example of the use of microwave chemistry without the use of solvents is found in the synthesis of amino alcohols from epoxides:

$$\text{R-epoxide} \xrightarrow[\text{40–120 s}]{NH_4OAc,\ MW} \text{R–CH(OH)–CH}_2\text{–NH}_2 + \text{R–CH(NH}_2\text{)–CH}_2\text{–OH}$$

This reaction microwave-promoted synthesis of β-amino alcohols in dry media by Sabitha et. al.[17] takes less than 2 minutes, depending on the R-group, with reasonable yields (>70%) and good regioselectivity.

As you will see in Chapter 11, the use of continuous reactors, separators, and crystallizers offers another potential strategy for technology to reduce solvent use. For example, a mass transfer limited reaction that is highly exothermic may benefit by a novel microreactor that balances order and rate of addition with high shear mixing and heat transfer in such a way that the product formation, solvent use, and separability are optimized. This is clearly an area where chemists and engineers must work closely together and use design of experiments and scientific first principles to develop a high degree of process understanding.

Another example of a potential technology approach might be the use of an innovative reactor that uses pervaporation to remove water from toluene in a Knoevenagel reaction instead of using azeotropic distillation to do this. The pervaporation may, in fact, result in significant energy savings and may offer significant advantages for recycling and reusing the toluene. In other instances, a continuous process may facilitate the use of in-process recycle of solvent such that the net solvent use is drastically reduced. Once again, this will be an intricate balance of the reaction chosen, the processing technology that is available, considerable fundamental process understanding, and trade-offs between material and energy use. Making reactions and processes green is not an easy thing to do!

7.5 SOLVENTS AND ENERGY USE

As alluded to several times earlier, the use of solvents is often associated with significant energy use in chemistry, although this is generally not something that is apparent to most chemists while designing the process. Energy is required in many reactions for heating and cooling, it is used in solvent recovery, and it is used in solvent disposal. As a starting point, chemists should simply consider the boiling point of the solvent and how this can be used from a green engineering perspective. First, very volatile solvents (i.e., solvents whose boiling points are less than 80°C) should be distilled at atmospheric pressure whenever possible. However, vacuum distillation may be a better alternative for higher-boiling-point solvents, as this may not only be more energy efficient, but it may also be essential to ensure product stability or to prevent unnecessary side reactions that lead to impurities or unwanted by-products.

Whenever possible, solvents should be recovered and reused even if they have relatively high boiling points (e.g., dimethylformamide). It is extremely important to build this into the process as it is being scaled-up so that the recovered solvent may be use-tested in later campaigns. In addition, one should also consider separations technology other than distillation for recovering solvents. For example, one might consider flash evaporators, membrane process (e.g., pervaporation), thin-film evaporators, or the use of novel reactors that incorporate an in-process recycle loop employing any of the foregoing technology options.

Although it is true that, in general, the energy used to control reactions is minimal compared to the energy required to distill a solvent, one should still choose solvents from the perspective of the entire synthetic route, not just a single step. If heating a solvent to promote a reaction is necessary, it is extremely important to avoid extended periods of heating solvents at reflux. Design of experiments should be used to optimize reaction kinetics in such a way that solvents are maintained at just below their reflux temperature or lower, rather than at reflux. This reduces the energy required for heat of vaporization. The following example gives an idea of how important this principle is by showing the difference in holding a solvent at reflux vs. holding it at 5°C below reflux. For example, to distill toluene requires 1135 kJ/kg toluene distilled. A gentle reflux of toluene (a reflux rate of 20% per hour) for 8 h requires a theoretical total energy use of 680 kJ/kg toluene mixture, or about 60% of the energy required for distillation. At approximately 5°C below the reflux temperature, the energy will be reduced by at least a factor of 2.

Example 7.4 Holding a reaction at reflux,[18,19] calculate the amount of energy required to distill methanol for 1 h at a reflux rate of 20% per hour for a 450-L reactor, as shown in Figure 7.2. How much energy would be saved if the methanol were held at 5°C below the reflux temperature?

Key data:

- Solvent: methanol
- Molecular weight: 32 g/mol
- Boiling point: 65°C
- Specific gravity: 0.79
- Latent heat of evaporation: 35 kJ/mol
- Heat capacity: 3 J/g·°C
- Vessel is filled to 80% of its capacity
- 20% reflux in 1 h
- 85% overall energy efficiency
- Mean temperature difference (steam and solvent): 20°C
- *UA* (global heat transfer coefficient × the contact area) calculated experimentally = 251.9 W/K = 251.9 J/K·s (1 W = 1 J/s); heat transfer efficiency is already built into *UA*

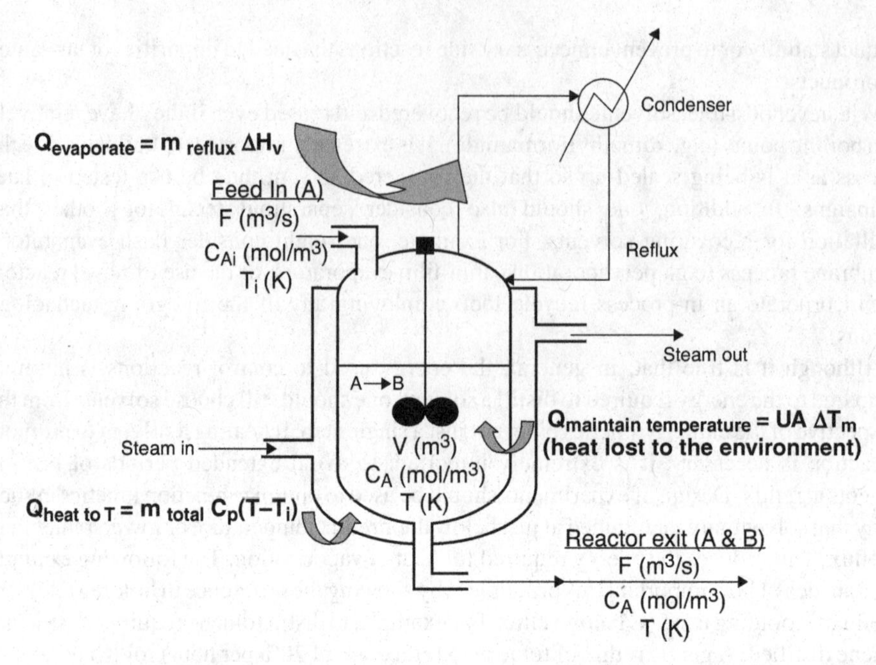

FIGURE 7.2 Figure for Example 7.4. Holding a reaction at reflux.

Solution In general, following the law of conservation of energy, for heating at reflux for 1 h (Table 7.4):

$$\begin{aligned}
\text{energy required} &= \text{heat to boiling point} + \text{heat to evaporate reflux} \\
&\quad + \text{heat to maintain temperature} + \text{energy lost} \\
&= Q\,\text{heat to}\,T_{\text{BP}} + Q\text{evaporate} \\
&\quad + Q\,\text{maintain temperature} + \text{energy lost} \\
\text{energy lost} &= (\text{energy required})(1 - \text{energy efficiency}) \\
\text{energy required} &= \frac{Q\text{heat to}\,T_{\text{BP}} + Q\,\text{evaporate} + Q\,\text{maintain temperature}}{\text{energy efficiency}} \\
&= \frac{mC_p\Delta T}{0.85} + \frac{m\Delta H_v}{0.85} + UA\Delta T \\
\text{heat to boiling point} &= \frac{(1000\,\text{g/kg})(3\,\text{J/g}\cdot{}^\circ\text{C})(65^\circ\text{C} - 20^\circ\text{C})/(1000\,\text{J/kJ})}{0.85} \\
&= 134\,\text{kJ/kg} \\
\text{energy to evaporate} &= \frac{(0.2)[(1000\,\text{g/kg})/(32\,\text{g/mol})](35\,\text{kJ/mol})}{0.85} = 259\,\text{kJ/kg} \\
\text{energy to maintain temperature} &= \frac{(UA\,\text{J/K}\cdot\text{s})(T_2 - T_1)}{\text{vol. (L) (specific gravity)}} \\
&= \frac{[(251.9\,\text{J/K}\cdot\text{s})(3600\,\text{s/h})]/(1000\,\text{J/kJ})(20^\circ\text{C})}{(0.8)(450\,\text{L})(0.79)} = 64\,\text{kJ/kg} \\
\text{total energy required} &= 134\,\text{kJ/kg} + 259\,\text{kJ/kg} + 64\,\text{kJ/kg} = 457\,\text{kJ/kg}
\end{aligned}$$

TABLE 7.4 Heating Requirements

	Heating a Vessel 5°C Below Boiling Point for 1 h	Heating a Vessel at Reflux for 1 h	Savings for Heating a Vessel 5°C Below Reflux
Energy required, (kJ/kg·h)	183	457	274
Energy required (MJ/vessel·h)	52	130	78
Savings in energy			60%

For heating 5°C below reflux for 1 h:

$$\begin{aligned}
\text{energy required} &= \text{heat to } 5^\circ\text{C below boiling point} \\
&\quad + \text{heat maintain temperature} + \text{energy lost} \\
&= Q \text{ heat to } T_{BP} + Q \text{ maintain temperature} + \text{energy lost} \\
\text{energy lost} &= (\text{energy required})(1 - \text{energy efficiency}) \\
\text{energy required} &= \frac{Q \text{ heat to } T_{BP} + Q \text{ maintain temperature}}{\text{energy efficiency}} \\
&= \frac{mC_p\Delta T}{0.85} + UA\Delta T \\
\text{heat to } 5^\circ\text{C below the BP} &= \frac{[(1000\,\text{g/kg})(3\,\text{J/g}\cdot{}^\circ\text{C})](60^\circ\text{C} - 20^\circ\text{C})/(1000\,\text{J/kJ})}{0.85} \\
&= 119\,\text{kJ/kg} \\
\text{energy to maintain temperature} &= \frac{(UA\,\text{J/K}\cdot\text{s})(T_2 - T_1)}{\text{vol. (L)(specific gravity)}} \\
&= \frac{[(251.9\,\text{J/K}\cdot\text{s})(3600\,\text{s/h})]/(1000\,\text{J/kJ})(20^\circ\text{C})}{(0.8)(450\,\text{L})(0.79)} = 64\,\text{kJ/kg} \\
\text{total energy required} &= 119\,\text{kJ/kg} + 64\,\text{kJ/kg} = 183\,\text{kJ/kg}
\end{aligned}$$

This represents a total energy savings of 60%. To estimate the energy per vessel, we have only to multiply by the mass contained in the vessel with the specifications of the problem.

Additional Points to Ponder For what additional unit operations would reaction or processing time be a critical parameter? What would the green chemistry or green engineering impacts be?

7.6 REACTION AND PROCESSING TIME

When in the laboratory performing exploratory chemical reactions, most chemists don't worry too much about how long a reaction takes; they generally are pursuing a reaction that works and produces the desired product at high yields with few, if any, by-products. This is not a bad thing in itself, as the chemist usually needs to determine, as quickly as possible, whether or not a reaction is viable and whether it can potentially be used to produce the desired amount of intermediate or final product.

So it is not uncommon for chemists developing processes to allow reactions to run for a period of time and to check periodically to see how the reaction is proceeding. In some instances, they may set up a reaction to run overnight so that when they arrive in the morning, they can proceed to isolate their product and set up for the next step of the reaction. This overnight period is then, in the worse-case scenario, written into processing instructions, and repeated at increasingly larger scales until it becomes a way of working in the manufacturing facility.

It is therefore critical that somewhere during subsequent campaigns to scale-up the chemistry, a good design of experiments is constructed to clearly determine the reaction kinetics and the exact endpoint of the reaction. But why is this important from a green chemistry and green engineering perspective? There are a number of reasons, but let's cover three obvious ones. First, the old adage that time is money is also true in this instance, but there are also major environmental impacts that can be monetized as well. Usually, reactions are not run at room temperature, so a reaction that is being run at an elevated or reduced temperature is consuming energy for temperature control and mixing. The longer the reaction is left to run, the more energy it will be consuming. Second, if you know the exact kinetics of the reaction and the endpoint of the reaction, you may be able to run the reaction in a different type of reactor and thereby actually increase the rate of the reaction or improve the overall operability of the reaction. Finally, the longer a reactor is occupied, the less throughput a facility can accommodate and the more utilities will be required. This will inevitably increase the baseload energy footprint of the facility. In the laboratory scale, this might not be a big impact, but as one scales up the size of the reactions, the impacts become more important.

7.7 ORDER AND RATE OF REAGENT ADDITION

Order and rate of addition can have profound effects on how a reaction proceeds and on the final product, yet they are often not considered immediately by many chemists. Chemists may, in fact, discard a perfectly good reaction because they neglect to test these factors adequately. By controlling the order and rate of addition, reactions may be improved radically. For example, in the fabrication of brightening agents for detergents, a 4% *p*-nitrotoluene sulfonic acid solution was heated slowly to 70°C in the presence of 4% NaOH solution and a manganese catalyst. Over 20 h there was an 80% conversion to 2,2′-dinitrostilbene 2,2′-disulfonic acid (DNS). Optimization of the reaction conditions and the catalyst, combined with a change to pure oxygen instead of air, reduced the reaction time, energy consumption, and waste considerably.

Order and rate of addition can also make reactions more controllable. For example, in controlling exothermic or endothermic reactions, one is able to control heat removal or addition in such a way that there is not excessive localized heating (as in the simple case of adding acid to water as opposed to water to acid) or cooling (as in the case of there being a rapid or uncontrolled precipitation or crystallization that leads to solvent or impurity occlusions). This is an especially attractive possibility for the use of alternative reactors and/or miniaturization of the reaction train, where it is possible to add reagent in a controlled fashion in a micro- or minireactor with high thermal mass to ensure the optimal reaction rate and temperature. Solvents or reagents might also be added at different temperatures such that careful addition of the reagent or reactant in heated or cooled solvents may also be used to control the reaction temperature.

Order and rate of addition may also be used to achieve optimum kinetic control. Basically, the order and rate of addition affect the concentration of the limiting reactant, so the effect on the reaction rate may be used to avoid the creation of unwanted impurities, as in the case of exothermic reactions or where a transient intermediate forms that is kinetically favored but not the thermodynamically favored form and reacts before the thermodynamically favored structure forms. The addition of reactants in reverse order can be also considered to avoid heat-of-reaction issues, impurity formation, and/or by-product formation. Order and rate of addition may also be used to control particle characteristics (size and morphology) and to facilitate separations. Order and rate of addition are also often used in both classical and dynamic kinetic chiral resolutions to ensure that optimal resolutions occur.

The order and rate of addition may also be used to good effect in the formation of multiphase reaction systems to control either the rate or the timing of phase formation. Whenever different phases are formed in a reaction mixture, there is the added concern of emulsion formation, so the order and rate of addition may be used to control and hopefully, to avoid emulsion formation.

Order and rate of addition should also be coupled with different mixing regimes to ensure the optimal reaction conditions. Speed, timing, and other mixing characteristics are keys to heat and mass transfer and should be coordinated with the addition order and rate. For example, in the nitration of an aromatic compound, an acid mixture was initially added to the reactor without stirring. When agitation started the mixture reacted violently, causing a fire.[20]

7.8 MIXING

Mixing is an extremely important issue in chemical processing, and its importance tends to be overlooked by most bench chemists. Mixing is also a major issue in scaling-up processes, and we have more to say about this in Chapter 10. In the meantime, let's think for a few moments about why mixing is so important. First, we need to overcome issues related to heat and mass transfer. Stated slightly differently, mixing allows us to avoid concentration gradients or temperature gradients in reactors of all sizes and shapes and ensures that the reactants are uniformly distributed and at a desired and uniform temperature. This is obviously important from a green chemistry standpoint since we need to make sure that all of the reactants are consumed and form product in equimolar amounts. In terms of heat transfer, we need to ensure that there are no thermal gradients that may lead to tarring, by-product formation, or in the worse case, localized hot spots and possibly thermal runaways, all of which will reduce mass efficiency and increase the process safety risk.

In terms of mass transfer, we need to ensure that reactants are dispersed in the reaction space to ensure that the reaction may proceed at the appropriate kinetic rate. Bourne[21] reports that in some instances, as in the case of a simple acid or base neutralization reaction, the kinetic rate (10^8 m^3/mol·s) means that the neutralization will take place on the order of nanoseconds, where as the mixing rate will rarely be faster than 1 ms in theory and on the order of 1 to 10 s operationally. In this case the mixing rate will always control the rate of the overall reaction. By comparison, the kinetic rate of a typical hydrolysis reaction, as in the case of the ester ethyl ethanoate ($CH_3COOC_2H_5$), is on the order of 10^{-4} m^3/mol·s, with a time scale on the order of minutes. In this case, the rate of mixing will almost certainly be fast enough to ensure that the reaction proceeds under kinetic control with no major issues related to mixing. Most reactions will be somewhere in between these two ranges, and therefore, reactions will proceed optimally when a balance between the kinetic rate and the mixing rate is achieved. Finally, for both heat and mass transfer, mixing will affect certain flow field phenomena, such as velocity distribution and turbulence. These phenomena are less critical for a batch-processing environment but become critically important in mini- and microreactor environments.

Example 7.5 Competitive Consecutive Reactions in a Single-Phase Reaction Mixture. Consider the following case (adapted from Bourne[21]), in which two reactants form an intermediate (I) that reacts further with one of the reactants to form the final product (P):

$$\mathrm{A+B \rightarrow I} \qquad \text{governed by rate constant } k_1$$

$$\mathrm{I+B \rightarrow P} \qquad \text{governed by rate constant } k_2$$

In a batch reactor where $n_{\mathrm{B,initial}}$ moles of B are added to $n_{\mathrm{A,initial}}$ of A such that $n_{\mathrm{B,initial}} < 2n_{\mathrm{A,initial}}$, the yields of the intermediate and final products can be shown to depend on the ratio of the rate constants (k_1/k_2) and the initial stoichiometric ratio $\mathrm{SR}_i = n_{\mathrm{B,initial}}/n_{\mathrm{A,initial}}$ when all B has been consumed and both reactions have gone to completion. In the special case where $k_1 = 2k_2$, the final composition of the reaction mixture is given by the following equations:

$$\frac{n_{\mathrm{A,final}}}{n_{\mathrm{A,initial}}} = (1-0.5\,\mathrm{SR}_i)^2$$

$$\frac{n\mathrm{I}}{n_{\mathrm{A,initial}}} = b-0.5b^2$$

$$\frac{n\mathrm{P}}{n_{\mathrm{A,initial}}} = (0.5\,b)^2$$

provided that $\mathrm{SR}_i < 2$.

Bourne[21] reports that for single-phase reactions, experiments on the selectivity of "mixing" show that mixing is more important when:

1. Reagents exhibit high reactivities, so that reactions are fast.
2. Concentrated reagent solutions are employed.
3. Stirring is weak (low stirrer speed, poor stirrer design).
4. Reagents are added to weakly agitated regions in a reaction vessel.
5. Viscosities are high.
6. Reagents are added at a rate faster than that at which they can mix completely.

When we have solids in the reactor (i.e., for heterogeneous catalytic reactions), we need to ensure that they, too, are uniformly distributed to ensure reproducible reactions, no by-product formation, and no localized hot spots in the reaction liquors. In the case of gas–liquid or liquid–liquid reactions, we need to ensure that the gases or liquids are mixed uniformly and the interfacial area is optimal for the reaction desired. Once again, ensuring optimum interfacial area will ensure that all the reactants are likely to be consumed to form product. This is especially important in multiphase liquid mixtures, where a lack of good mixing can reduce interfacial mixing rather dramatically and lead to reductions in yield. Phase changes that lead to increased viscosity in one phase or another can be especially problematic and must be compensated for to ensure that reactions go to completion. An additional issue in multiphase reaction mixtures is the formation of emulsions; unoptimized mixing can lead to the formation of emulsions and ultimately may lead to the undesirable effects of product loss, low yields, and increased energy.

Mixing can also have adverse effects on reactants or the final products. For liquid–solid mixtures, some of these adverse effects include breaking particles or crystals into smaller sizes in such a way that they can complicate and slow filtration and/or drying, both of which can lead to adverse effects on the product (e.g., solvent inclusion, formation of solvates), increases in the use of energy (e.g., longer drying cycles), or loss of product in filter aids. In other cases, poor mixing can lead to undesirable coagulation and/or agglomeration of particles. In the case of vigorously stirring gas–liquid mixtures, you may entrain liquid in a gas stream, or you may break up liquid droplets to an undesirable size.

To improve mass transfer, it is recommended to investigate the use of:

- Alternative solvents such as ionic liquids and supercritical fluids
- Alternative technology such as vortex or in-line mixers, and micro and minireactors
- Phase-transfer catalysts

APPENDIX 7.1 Common Practices in Batch Chemical Processing and Their Green Chemistry Impacts

Problem	Why This Is a Problem	Strategy to Solve Problem
Acid and base washes	Increases volume of wastewater requiring neutralization.	Investigate order and method of addition.
	Increases total dissolved solids in wastewater treatment plant and effluent discharges.	Optimize reactions to avoid the need for by-product, starting material, or impurity, removal.
	Products or intermediates are lost in the wash liquids.	If possible, recover acid or base.
	Risk to people and the environment associated with handling, storing, and using acids and bases.	
	Increases complexity and process setup.	
	Increases number of separations and unit operations.	
	Risk of losing desired products in the washing liquors	
pH adjustment	Increases volume of liquid waste requiring neutralization.	Investigate alternative chemistries to avoid pH extremes.
	Increases total dissolved solids in wastewater treatment plant and effluent discharges.	Investigate use of solid acid and base catalysts and use test regenerated and recovered catalysts.

(continued)

APPENDIX 7.1 (*Continued*)

Problem	Why This Is a Problem	Strategy to Solve Problem
pH adjustment (*continued*)	Increases the number of steps in a process and therefore affects throughput and capacity. Potential creation of sludge waste. Increases process complexity. Risk to people and the environment associated with handling, storing, and using acids and bases.	
Salt washes	Increases volume of wastewater requiring treatment. Increases total dissolved solids in wastewater treatment plant and effluent discharges. Products or intermediates are lost in the wash liquids. Increases process complexity. Increases processing time and capacity requirements.	Test whether or not salt wash is really necessary. If washes are necessary, consider using better washing technology. For example, centrifugal separators increase the separation efficiency of the phases and increase the throughput compared to traditional gravity separation.
Use of desiccants	Potential for this to become solid hazardous waste, depending on type of synthesis and process conditions. There is an embedded life cycle impact in the mining, transport, purification, and so on, of these materials. Disposal of wet desiccants will be increasingly difficult and costly unless they can be recycled and reused.	Consider alternative membrane technologies for water removal. Membrane technologies separate water with greater mass efficiency and fewer steps (just the membrane to separate water, versus the addition of desiccant and its separation). Pervaporation Reverse osmosis Direct osmosis
Chromatographic cleanups	Potential for this to become solid hazardous waste, depending on type of synthesis and process conditions. There is an embedded life cycle impact in the mining, transport, purification, and so on, of these materials Disposal of spent packing will be increasingly difficult and costly over time unless it can be recycled and reused. Chromatographic separations are extremely expensive and inefficient from a mass productivity perspective.	Attempt chiral syntheses. Investigate biotechnology solutions. If chromatographic separations need to be used: Maximize the recovery and reuse of the solvent. Consider the use of simulated moving-bed chromatographic separations. Consider the use of coupled columns. Optimize separations using traditional chromatographic separations best practices.

APPENDIX 7.1 (*Continued*)

Problem	Why This Is a Problem	Strategy to Solve Problem
Carbon cleanup	There is an embedded life cycle impact in the mining, transport, purification, and so on, of activated carbons Disposal of wet carbon will be increasingly difficult and costly unless it can be recycled and reused. Increased process complexity	Optimize chemistry to avoid by-product formation. Avoid extreme temperature. Use highest purity reagents necessary to avoid introduction of impurities. Optimize addition of reagents to control reaction kinetics. Ensure operational control over plant facilities and clean-in-place procedures.
Use of filter aids	There is an embedded life cycle impact in the mining, transport, purification, and so on, of activated carbons Disposal of wet carbon will be increasingly difficult and costly unless it can be recycled and reused.	Investigate use of continuous processing. Investigate nonisolation processes. Investigate use of alternative technology: Membranes with shear wave at water/membrane interface. Pulsed membranes.
Put and take distillations	Put and take distillations generally may:	Evaluate changes in chemistries so that a put and take distillation is not required.
	Require significant energy for heating, condensing, recycling, and recovering solvents Lead to contaminated and nonrecoverable solvent waste mixtures Lead to the formation of difficult emulsions Require azeotropic distillations leading to difficulties in recycling and recovering spent solvent Entail higher capital expense for additional unit operations and distillation equipment. There is also an embedded life cycle impact for any piece of equipment.	Evaluate chemistries that can be carried out in the same solvent across multiple steps. Evaluate the use of different process conditions: Nonisolation or combined steps to avoid having to change the solvent prior to crystallization. Solvents in which the reactants are soluble but the product is not. Reduce the volume of the solvent used to run the reaction. Consider employing slurry reactions to reduce the volume of solvent required (this requires good mixing).
	Leads to greater process complexity and reduces throughput.	Investigate combinations of phase-transfer catalyst(s) and solvent

(*continued*)

APPENDIX 7.1 (*Continued*)

Problem	Why This Is a Problem	Strategy to Solve Problem
Put and take distillations (*continued*)		Optimize solvent separability to ease the overall separation (e.g., azeotropes may enhance or adversely affect solvent replacement). Minimize the amount of mixed solvent use, since solvent mixtures increase the difficulty and cost of the separation and residual solvent may adversely affect the crystallization yield and quality. Distill as close to dryness as possible before carrying out the solvent change. This can only be done if there are no process safety risks or affects to the chemistry (i.e., by-product formation, impurity occlusion, etc.) in doing this. Avoid distilling alcohols and other hydrogen-bonding solvents (e.g., dipolar aprotics, water, etc.) since the energy required to distill these solvents is relatively high. Consider constant-level batch distillation to minimize the loss of replacement solvent. Significant reductions of solvent losses might be possible by using this variation Exploit solvent solubility differences in multiphase solvent systems: • Ionic liquids • Biphasic reaction solvents • Supercritical fluids Consider alternative technologies for separating solvents or solvent exchange: • Countercurrent liquid liquid extraction • Pervaporation • Separative reactors etc.

APPENDIX 7.1 (*Continued*)

Problem	Why This Is a Problem	Strategy to Solve Problem
		Physical means of separation: • Membranes • Filtration • Thin-film evaporators
Isolation steps/ crystallizations	Isolations are intimately connected with acid/base and or salt washes and the use of solvents for separations, washing, or setting up for the next reaction. See issues recorded above under acid and base washes, salt washes, and pH adjustment. Isolations generally lead to use of energy for heating, cooling, and/or drying. Decrease throughput and increase overall process complexity. Increased solvent use, especially in final stages when and if there is a recrystallization. Loss of product or intermediate. The more product or intermediate you lose, the greater the associated life cycle impacts.	Evaluate one-pot synthesis by combining multiple chemical steps. Consider route strategies using chemistries that can be carried out in the same solvent across multiple steps and stages. Avoid drying isolated wet cake; use wet cake directly in the next stage. Evaluate changes to chemistry or process conditions that minimize impurities and remove the need for isolation Increase the purity of reactants, solvents, or reagents. Optimize mixing. Optimize the reaction time (e.g., performing kinetics studies). This will also help to identify parallel or consecutive reactions to be avoided. Balance the reaction temperature and heat dissipation with the reaction rate and formation of the product desired. Consider the use of novel technologies: Centrifugal separators for stream washes coupled with phase separators can increase the separation efficiencies and throughput. Centrifugal separators can also increase the throughput when operating in a semicontinuous mode of operation. Evaluate the use of combined reaction/separation technology: for example, application of reactive columns.

(*continued*)

APPENDIX 7.1 *(Continued)*

Problem	Why This Is a Problem	Strategy to Solve Problem
Isolation steps/ crystallizations (*continued*)		Consider the use of continuous and semi-continuous processes Evaluate smaller-scale production equipment for conversion from batch to continuous manufacture.

PROBLEMS

7.1 A chemist reacts benzyl alcohol (10.81 g) with *p*-toluenesulfonyl chloride (21.9 g) in toluene (500 g) and triethylamine (15 g) to give the sulfonate ester (23.6 g).

(a) Determine the stoichiometry of this reaction.

(b) Is the stoichiometry optimized?

(c) Calculate the yield, reaction mass efficiency, and mass efficiency for this reaction.

(d) Is this a green reaction? Defend your answer.

7.2 The classic route to hydroquinone is

2 (aniline, NH_2) + 4 MnO_2 + 5 H_2SO_4 → 2 (p-benzoquinone) + $(NH_4)_2SO_4$ + 4 $MnSO_4$ + 4H_2O

(p-benzoquinone) + Fe + H_2O → (hydroquinone, OH / OH) + FeO

(a) Determine the stoichiometry of this reaction.

(b) Is the stoichiometry optimized?

(c) Calculate the yield, reaction mass efficiency, and mass efficiency for this reaction.

(d) Is this a green reaction? Defend your answer.

7.3 Vo et al.[22] report the development of a process to a chiral auxiliary (*S*)-4-(phenylmethyl)-2-oxazolidinone utilizing automated synthesis and a design of experiment:

(phenylalanine: NH_2, OH, O) → [$NaBH_4$; additive, solvent] → (NH_2, OH) → [triphosgene; base, solvent] → (oxazolidinone: N, O, O)

The discrete reactions can be written as follows:

$$2NaBH_4 + H_2SO_4 \longrightarrow 2BH_3 + Na_2SO_4 + 2H_2O$$

$$2NaBH_4 + I_2 \longrightarrow 2BH_3 + 2NaI + 2H_2$$

$$Cl_3C{-}O{-}C({=}O){-}O{-}CCl_3 \longrightarrow 3\ Cl{-}C({=}O){-}Cl$$

$$PhCH_2CH(NH_2)C({=}O)OH + BH_3 + 3MeOH \longrightarrow PhCH_2CH(NH_2)CH_2OH + B(OMe)_3 + H_2 + H_2O$$

$$PhCH_2CH(NH_2)C({=}O)OH + Cl{-}C({=}O){-}Cl + 2NaOH \longrightarrow \text{4-benzyl-oxazolidin-2-one} + 2NaCl + 2H_2O$$

(a) Write two balanced equations showing the stoichiometry for the overall reaction.

(b) This reaction was developed as a one-pot synthesis. What are some of the advantages of this based on the topics covered in this chapter?

(c) Table P7.3 contains the recipe for the one-pot synthesis.

  (1) Based on what you have learned in this chapter, what are the key variables in this synthesis? Do you think there is room for improvement from a green chemistry or engineering perspective? Describe any improvements or potential additional DOE studies that you might consider.

  (2) What would the waste stream composition be from this one-pot synthesis?

  (3) This synthesis was done on a gram scale. What green chemistry and green engineering issues might you encounter if you were to scale this process up to the 100- or 500-kg scale? Describe these issues and how you might overcome them.

TABLE P7.3 Recipe for One-Pot Synthesis

Step	Operation	Material	Settings
1	Manual charge	Phenylalanine	8.3 g, 1.0 equiv.
2	Manual charge	THF	50 mL, 6 mL/g of Phe
3	Start agitation		800 rpm
4	Set temperature		25°C
5	Manual charge	$NaBH_4$	3.4 g, 1.8 equiv.
6	Temperature-based feed	I_2 solution	20 mL, 1 mL/min, temp. ± 5°C, 0.9 equiv.
7	Wait		1 h
8	Manual charge	MeOH	10 mL, 5.0 equiv.
9	Manual charge	20% NaOH	50 mL, 5.0 equiv.
10	Set temperature	Distill organics	to 80°C
11	Set temperature		0 ± 5°C

(*continued*)

TABLE P7.3 (*Continued*)

Step	Operation	Material	Settings
12	Manual charge	DCM	50 mL, 6 mL/g Phe
13	Temperature-based feed	Triphosgene solution	20 mL, 2 mL/min, temp. $< 30°C$
14	Set temperature		25°C
15	Wait		1.5 h
16	Stop agitation		0 rpm
17	Manual separation	Remove aqueous	Dry (Na_2SO_4), return to reactor
18	Start agitation		150 rpm
19	Manual charge	Isopropyl Acetate	25 mL, 3 mL/g of Phe
20	Manual charge	Heptane	17 mL, 2 mL/g of Phe
21	Set temperature	Distill DCM	to 85°C
22	Set temperature		to 0°C, linear ramp over 2 h
23	Stop agitation		0 rpm
24	Manual filter		Remove product

7.4 In Braddock's description of work with group IV metal catalysts,[1] he describes the synthesis of a metal triflide catalyst:

$$Me_3Si\diagup\!\!\diagdown Li \xrightarrow[\text{pentanes, 0°C}]{Tf_2O} Tf\diagup\!\!\diagdown Tf \xrightarrow[Tf_2O]{\text{2-BuLi}} Tf_3C\text{-}CH_3 \text{ (Tf, Tf, Tf)} + LiOTf \xrightarrow{\text{aq. CsCl}}$$

$$CsCTf_3\text{ (ppt)} \xrightarrow{H_2SO_4} HCTf_3 \xrightarrow[H_2O,\text{ reflux}]{M_2O_3} M(CTf_3)_3 \quad M = Yb, Sc$$

(a) Describe the green elements of this synthesis.

(b) Describe the nongreen elements of this synthesis.

(c) What would you have to take into consideration in deciding whether or not to scale-up this reaction?

7.5 In the example of the use of sodium zeolites as a replacement for stoichiometric additions of Lewis acids, Smith et al.[2] used the following protocol:

> Phenyl acetate (10 mmol), the zeolite (3 g) and dichloromethane (30 ml) were mixed in a three-necked round-bottomed flask fitted with a calcium chloride guard tube and a magnetic stirrer and protected from light. The mixture was stirred at room temperature for 15 min, after which bromine (10 mmol) was added. After a suitable time (typically 2.5 h), the reaction mixture was filtered through a sintered funnel and the solid was rapidly washed with acetone. An aqueous sodium hydrogen sulfite solution (10%, 30 ml) was quickly added to the filtrate to remove bromine and hydrogen bromide. The organic layer was washed with distilled water (3330 ml), dried over anhydrous magnesium sulfate, and filtered. Dodecane (60 ml, 0.045 g) was added as an internal standard and the solution was subjected to quantitative GC analysis. The residual solid was stirred with acetone (15 ml) for 30 min in order to desorb any materials in the channels of the zeolite, then filtered. An aqueous sodium hydrogen sulfite solution (10%, 30 ml) was quickly added to the filtrate and the aqueous phase was then extracted with dichloromethane (3310 ml). The organic layer was treated and analyzed. . . .

The yields using this protocol are shown in Table P7.5A.

TABLE P7.5A Bromination of Phenyl Acetate in the Presence of Various Catalysts[a]

Catalyst	Phenyl Acetate (%)[c]	3[b] (%)[c]	p-Bromophenyl Acetate (%)[c]	Others[b] (%)[c]
None	100			
None [d]	67	—	7	26
HY	99	—	1	
NaY	32	—	68	
NaX	42	—	58	
Nab	96	—	3	
NaMord	97	—	2	
NaZSM-5	100	—	—	
$AlCl_3$	3	5	52	33

[a]Bromination of phenyl acetate **1,** (10 mmol) with bromine (10 mmol) over catalyst (3 g) in CH_2Cl_2 (30 mL) at room temperature for 2.5 h.
[b]Compound **3** is 2-bromophenyl acetate; others are phenol, 2-bromophenol, 4-bromophenol, 2,4-dibromophenol, and 2,4-dibromophenyl acetate.
[c]Absolute yields were determined by quantitative gas chromatography.
[d]20 mmol of **1** and 20 mmol of bromine were used in otherwise identical conditions.

(a) Describe the green aspects of this synthesis.
(b) Describe the nongreen aspects of this synthesis.
(c) How might this reaction be optimized and made greener?
(d) What would the waste streams contain, and what problems would they create for disposal?

Smith et al.[2] went on to try and optimize the reaction through the addition of bases to neutralize the acid using the following protocol:

Bromination of 1 in the presence of bases

The procedure was similar to the procedure for zeolite-catalyzed bromination of **1**, but with a base present instead of a zeolite and with the following components and conditions: phenyl acetate (50 mmol), bromine (50 mmol), base (50 mmol) and dichloromethane (10 mL), at 0°C for 24 h.

They obtained the results shown in Table P7.5B.

TABLE P7.5B Effect of Bases on the Selectivity of Bromination of 1[a]

Base	Amount of Base (mmol)	Time (h)	Phenyl Acetate (%)[c]	p-Bromophenyl Acetate (%)[c]	**2** (%)[c]	Others[d] (%)[c]
	0	1.5	12	5	76	7
$NaHCO_3$	38	24	21	3	76	
$NaHCO_3$	50	24	19	1	80	
NaOAc	50	24	26	1	73	
Na_2CO_3	25	24	22	1	77	

[a]Bromination of **1** (50 mmol) with bromine (50 mmol) in the presence of base and CH_2Cl_2 (10 mL) at 0°C.
[b,c]See corresponding footnotes to Table P7.5A. Compound **2** is 4-bromophenyl acetate.

TABLE P7.5C Additions to Reaction

Metal Acetate (mass/g)	Ac_2O (mmol)	Time (h)	Phenyl Acetate (%)[b]	*p*-Bromophenyl Acetate (%)[b]	**2** (%)[b]
$Cu(OAc)_2$ (3)	40	3	28	—	71
$Co(OAc)_2$ (5)	40	4	41	1	58
$Hg(OAc)_2$ (5)	40	5	2	1	96
$Zn(OAc)_2$ (5)	40	5	—	1	99
$Zn(OAc)_2$ (5)	—	5	—	1	99

[a]Reaction conditions: **1** (40 mmol), bromine (60 mmol), metal acetate and acetic anhydride at 0°C.
[b]See footnote *b* to Table P7.5A.

(e) How does the addition of the base change the greenness of the initial reaction?

(f) How might the reaction conditions be changed to improve the greenness of the process?

Smith et al.[2] went on to further optimize the reaction through the addition of metal acetates and acetic anhydride (Table P7.5C).

(g) What role does the acetic anhydride play?

(h) How does the addition of these metal acetates affect the overall greenness of the reaction?

(i) How might the reaction conditions be changed to improve the greenness of the process?

(j) In your opinion, is this work a good example of green chemistry? Explain and defend your answer.

REFERENCES

1. Braddock, C. Novel recyclable catalysts for atom economic aromatic nitration. *Green Chem.*, 2001, 3(2), G26–G32.
2. Smith, K., He, P., Taylor, A. Selective para-bromination of phenyl acetate under the control of zeolites, bases, acetic anhydride or metal acetates in the liquid phase. *Green Chem.*, 1999, 1(1), 35–38.
3. Corkan, L. A., Lindsey, J. S. Experiment manager software for an automated chemistry workstation, including a scheduler for parallel experimentation. *Chemometr. Intell. Lab Syst.*, 1992, 17, 47–74.
4. Plouvier, C.-J., Corkan, L. A., Lindsey, J. S. Experiment planner for strategic experimentation with an automated chemistry workstation. *Chemometr. Intell. Lab Syst.*, 1992, 17, 75–94.
5. Lindsey, J. S. Automated approaches toward reaction optimization. In *A Practical Guide to Combinatorial Chemistry*, Czarnik, A., Hobbs-Dewitt, S., Eds. American Chemical Society, Washington, DC, 1997, pp. 309–326.
6. Wagner, R. W., Li, F., Du, H., Lindsey, J. S. Investigation of cocatalysis conditions using an automated microscale multireactor workstation: synthesis of meso-tetramesitylporphyrin. *Org. Process Res. Dev.*, 1999, 3, 28–37.

7. Pilipauskas, D. R. Can the time from synthesis design to validated chemistry be shortened? *Med. Res. Rev.*, 1999, 5, 463–474.
8. Gooding, O. W. Process optimization using combinatorial design principles: parallel synthesis and design of experiment methods. *Curr. Opin. Chem. Biol.*, 2004, 8, 297–304.
9. Owen, M. R., Luscombe, C., Lai, L.-W., Godbert, S., Crookes, D. L., Embiata-Smith, D. Efficiency by design: optimisation in process research. *Org. Process. Res.*, 2001, 5, 308–323.
10. Lendrum, D., Owen, M. R., Godbert, S. DOE (design of experiments) in development chemistry: potential obstacles. *Org. Process. Res.*, 2001, 5, 324–327.
11. Parker, J. S., Bowden, S. A., Firkin, C. R., Mosley, J. D., Murray, P. M., Welham, M. J., Wisedale, R., Young, M. J., Moss, W. O. *Org. Process. Res.*, 2003, 7, 67–73.
12. Gooding, O. W., Vo, L., Bhattacharyya, S., Labadie, J. W. Use of statistical design of experiments (DoE) in the optimization of amide synthesis utilizing polystyrene-supported *N*-hydroxybenzotriazole resin. *J. Comb. Chem.*, 2002, 4, 576–583.
13. Jiménez-González, C., Curzons, A. D., Constable, D. J. C., Cunningham, V. L. Cradle-to-gate life cycle inventory and assessment of pharmaceutical compounds: a case-study. *Int. J. Life Cycle Assess.*, 2004, 9(2), 114–121.
14. Jiménez-González, C., Curzons, A. D., Constable, D. J. C., Cunningham, V. L. Expanding GSK's solvent selection guide: application of life cycle assessment to enhance solvent selections. *J. Clean Techno. Environ. Policy*, 2005, 7: 42–50.
15. Constable, D. J. C., Jiménez-González, C., Henderson, R. K. Perspective on solvent use in the pharmaceutical industry. *Org. Process Res. Dev.*, 2007, 11, 133–137.
16. Gentilcore, M. J. Reduce solvent usage in batch distillation. *Chem. Eng. Prog.*, 2002, 98(1), 56–69.
17. Sabitha, G., Reddy, B. V. S., Abraham, S., Yadav, J. S. Microwave promoted synthesis of beta-aminoalcohols in dry media. *Green Chem.*, 1999, 1(5), 251–252.
18. Perry, R. H., Green, D. W., Maloney, J. O. *Perry's Chemical Engineers' Handbook*, 7th Edn. McGraw-Hill, New York, 1997.
19. Woods, D. R. *Process Design and Engineering Practice*. Prentice Hall, Upper Saddle River, NJ, 1995.
20. Health and Safety Executive. *Designing and Operating Safe Chemical Reaction Processes*. HSE Books, Norwich, UK, 2000.
21. Bourne, J. R. Mixing and the selectivity of chemical reactions. *Org. Process Res. Dev.*, 2003, 7, 471–508.
22. Vo, L., Ciula, J., Gooding, O. W. Chemical development on the chiral auxiliary (*S*)-4-(phenylmethyl)-2-oxazolidinone utilizing automated synthesis and DoE. *Org. Process Res. Dev.*, 2003, 7, 514–520.

8

BIOPROCESSES

What This Chapter Is About In previous chapters we have largely restricted our discussions to traditional chemical processing. In this chapter we shift our attention to the use of fermentation, plants, and other biological systems to produce chemicals of interest. To most persons, the use of biological systems is considered a priori to be green; after all, most fermentation systems use only water, buffers, and trace minerals. In this chapter we discuss bioprocessing and potential issues and impacts associated with bioprocesses.

Learning Objectives At the end of this chapter, the student will be able to:

- Understand in general terms some biological processes used to make chemicals.
- Understand the potential issues and impacts of fermentations and the downstream processing associated with them.
- Identify and assess opportunities for improving the greenness of fermentation processes.
- Understand the trade-offs between traditional chemical manufacturing and fermentation from an environmental perspective.

8.1 HOW BIOTECHNOLOGY HAS BEEN USED

When we think about it for a moment, we realize that biotechnology has been used by people for a very long time. This includes the beer, glass of wine, or other alcoholic beverage that someone may have had in the not-so-distant past, and the piece of bread you had during the day. Or perhaps at one time you have been sick with a sinus or ear infection and have been on a

Green Chemistry and Engineering: A Practical Design Approach, By Concepción Jiménez-González and David J. C. Constable

TABLE 8.1 Use of Bioprocessing in Industry

Industry	Scale	Downstream Processing Complexity	Biocatalyst[a]	Types of Products
Basic chemicals	Very large	Low	MO/enzymes	Organic small molecules
Fine Chemicals	Medium	Medium	MO/enzymes	Organic small molecules
Detergents	Large	Low	MO	Enzymes
Personal care products	Small–medium	Medium–high	MO/enzymes/mammalian cells	Proteins and small molecules
Conventional pharma	Medium	Medium–high	MO/enzymes	Organic small molecules
Biopharma	Small	High	MO/mammalian cells	Proteins
Food/feed	Very large	Medium	MO/enzymes	Proteins and others
Metal mining	Very large	Low	MO	Metals and metal compounds
Waste treatment	Very large	Low	MO	Purified air, water, and soil

Source: ref. 1.

[a]MO, microorganisms.

course of antibiotics for a few days. All of these items are products of biological or fermentation processes of one kind or another. Admittedly, the technology used today is more advanced, but harnessing microorganisms to modify our foods and what we consume or use in commerce is not a recent phenomenon. What are recent phenomena, however, are the developments in molecular and synthetic biology over the past decade or so that have ushered in an explosive expansion in the number and complexity of applications of biotechnology in all areas of society. This is becoming so prevalent that it is difficult not to find examples of the intelligent use of biotechnology in many areas of society. Witness, for example, the development of bio-based ethanol or biodiesel for fuels, enzymes used in clothing detergents for stain removal, and the increasing number of very chemoselective chiral syntheses in the agrichemical and pharmaceutical industries. Even the basic feedstocks for existing polymers or new types of polymers are being developed from bio-based materials. Table 8.1 summarizes some processing industries that use bioprocessing technologies routinely, and Table 8.2 is a list of materials produced through bioprocesses and the volumes produced in 2005.

8.2 ARE BIOPROCESSES GREEN?

However, when we look at tables such as Tables 8.1 and 8.2, we must always ask the question of whether or not any chemical or other material that we may want to develop and use, either through chemical or biological means, is truly green. If you take a look at Figure 8.1, an overly simplified graphic of a bioprocess, you might be tempted to ask yourself, "What couldn't be green about that?"

Look at Table 8.3 and think for a moment about the 12 principles of green chemistry. Let's pause for a minute and see if there are any benefits to using biotechnology that are in

TABLE 8.2 Annual Production Volumes of Major Bio-derived Products, 2005

Product	Annual Production (tons)	Product	Annual Production (tons)
Bioethanol	26,000,000	L-Hydroxyphenylalanine	10,000
L-Glutamic acid (MSG)	1,000,000	6-Aminopenicillanic acid	7,000
Citric acid	1,000,000	Nicotinamide	3,000
L-Lysine	350,000	D-*p*-Hydroxyphenylglycine	3,000
Lactic acid	250,000	Vitamin F	1,000
Food-processing enzymes	100,000	7-Aminocephalosporinic acid	1,000
Vitamin C	80,000	Aspartame	600
Gluconic acid	50,000	L-Methionine	200
Antibiotics	35,000	Dextran	200
Feed enzymes	20,000	Vitamin B_{12}	12
Xanthan	30,000	Provitamin D_2	5
L-Threonine	10,000		

Source: ref. 2.

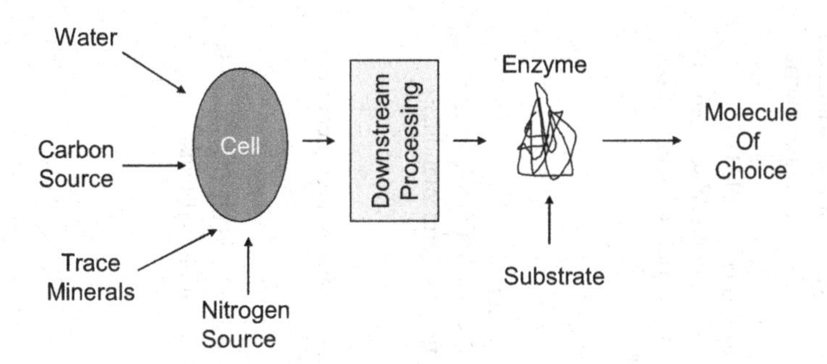

FIGURE 8.1 Simplified graphic representation of a bioprocess.

alignment with these principles. Then we can move on to a discussion of what bioprocessing actually entails.

8.3 WHAT IS INVOLVED IN BIOPROCESSING

Bioprocessing may be thought of as comprising discrete operations that usually start with some sort of fermentation to obtain various chemicals of interests, biocatalysts, proteins, enzymes, antibodies, and other products. Fermentation processes are, for the most part, very similar to batch chemical processes. Basically, you have materials that are mixed together for a period of time at a particular temperature and pH, and the material of interest is extracted, isolated, purified, and sent for additional reactions or for formulation into a final product.

TABLE 8.3 12 Principles of Green Chemistry and Potential Biotechnology Deliverables

Principle	Green Benefit of Biotechnology	Green Downside of Biotechnology
Prevent waste	Biological systems (enzymatic and/or whole cell) can lead to products of higher enantiomeric purity. Catalysts may be able to be recycled. Membrane–resin separation processes remove product and other materials of interest from waste and water.	Fermentation processes produce large quantities of biosolids that must be inactivated (heat or chemical treatment) and disposed of. Life cycle environmental issues associated with agrichemical feedstocks. Water used in fermentation processes is generally highly purified. Purification may require chemicals, membrane systems, resins, and so on, to achieve the desired purity, which increases the environmental impacts.
Design safer chemicals and products	Genetically modified organism strains are designed to be safe and not viable outside the controlled conditions of fermentation.	Time and resource required for genetic evolution, synthetic biology, or related methods to develop desired organism and expression of desired enzyme or product. There are related impacts for organism deactivation.
Design less-hazardous chemical syntheses	Majority of biotransformations are aqueous based and essentially benign.	Life cycle impacts of minerals used in fermentation processes; pH control often uses significant quantities of acids and bases (ammonia).
Use renewable feedstocks	Biological process carbon and energy source is derived from agricultural feedstocks.	Agriculture is material and energy intensive and has significant impacts on carbon, nitrogen, and phosphorus cycles. Competition for arable land for food and other items of commerce. Impacts associated with mining for minerals.
Use catalysts, not stoichiometric reagents	Biological enzymes, free, bound and whole cell are used as catalysts.	Bias toward homogeneous catalysis and inactivation of bound catalyst results in nonrecyclability of catalyst. Environmental life cycle impacts of catalyst production can be significant, especially if not recycled.

Avoid chemical derivatives	Identification of chemoselective transformation (e.g., asymmetric hydrogenation, esterification) is generally followed by strain improvement to deliver high yield and enantiomeric excess desired. Process thereby avoids the need for protection/deprotection steps.	
Maximize atom economy	Identification of chemoselective (region- and stereo-specific) transformation (e.g., esterases, lipases, nitrolases) is generally followed by strain improvement to deliver high yield, or enantiomeric excess desired, or in many cases, transformations that avoid multiple chemical synthetic steps. Use of genetics to eliminate pathways to by-products.	
Use safer solvents and reaction conditions	Make use of largely aqueous buffered systems to maintain neutral pH; membrane/resin processes used for concentration and downstream processing.	pH control is a major issue that may require large amounts of mass and energy resources. Downstream processing requires tremendous quantities of water and energy in addition to specialty resins and other materials. Downstream processing might require organic solvents for separations and purifications.
Increase energy efficiency	Enzyme reactions (coupling, hydrolysis, amidation, etc.) generally run at around 30 °C rather than at higher temperatures or pressures. Strain improvement can lead to reduction in waste as a result of process intensification (increased titer of product in fermentation broth).	Large volumes of water are used throughout fermentation processes. Heating and maintaining water at the desired temperature, even if only to 30 °C, can require large amounts of energy (see Chapters 9 and 18). Maintenance of sterility in large fermenters can require large volumes of steam. Requirement for oxygen necessitates either large volumes of air or pure oxygen. Delivery of either requires a large amount of energy.

(continued)

TABLE 8.3 (*Continued*)

Principle	Green Benefit of Biotechnology	Green Downside of Biotechnology
Design chemicals and products to degrade after use	Fermentation waste is generally easily degraded by biological waste treatment processes.	Wastewater treatment plants for biosolids have: Energy requirements: air needed for respiration and nitrification Nitrogen and phosphorus release (leads to eutrophication) pH control can require large amounts of Na_2CO_3 sludge or similar materials Biosolids with residual product, depending on the product, may require incineration of biosolids or some other process to ensure inactivation or destruction. Land application of biosolids can result in the introduction of residues that over time can cause salination or residual products (e.g., agrichemical-related antibiotics) that migrate into water supplies.
Analyze in real time to prevent pollution	Real-time analysis and automated process control ensures optimal quality, reactant and reagent use, yield, and reduced by-product formation.	
Minimize the potential for accidents	Replacing hazardous chemical use associated with chemical synthesis with more benign chemicals found in biological systems reduces needs for hazardous materials use, treatment, storage, and disposal.	

As with the chemical processing industries, you will find reactions being run mostly as batch and semibatch processes, although there are examples of some bioprocessing operations being carried out as a continuous process. Certainly, wastewater treatment plants are bioprocesses on a large scale, and these are run continuously, but we don't really extract chemicals of interest. There are also examples of solid-state fermentation processes in industry, but these are largely restricted to the food industry and are more common in Japan.

It would be useful to dissect bioprocessing in further detail, as it is a little different from what is encountered in traditional chemical processing industries. In general, it is common to break bioprocessing into two main parts, upstream and downstream processing.

8.3.1 Strain Development and Catalyst Selection

It is beyond our scope here to cover in detail the process of developing strains or selecting a catalyst, but it is extremely important to understand that the selection of a target organism for a biotransformation of interest is a key challenge in bioprocessing. In addition to in-house groups, there are a number of biotechnology companies that specialize in high-throughput screening of multiple libraries of microorganisms (e.g., fungi, bacteria, plant cells) or enzymes to obtain promising candidates for further development. Upon selection of appropriate target organisms or catalysts, organisms may be evolved through such techniques as gene shuffling, directed genetic evolution, or other molecular engineering techniques to enhance the desired catalyst production or conversion of substrate in a whole-cell transformation. There is also a component of developing the chosen organism in a manner that is robust for industrial use, such as making sure that it is relatively free of genetic drift or that it does not possess sensitivity to temperature fluctuations.

Once an organism has been selected, the process must be seated-up to the volume desired. Scaling-up a bioprocess has been aided tremendously through the recent development of metabolomic technologies. Understanding the metabolic profile of an organism through metabolomics can be used to good effect to ensure that reaction conditions are optimized to support cells in their growth phase and to express proteins and/or enzymes that enhance the bioreaction. Detailed discussion of this is beyond the scope of this book, but these technologies offer tremendous promise for the future.

8.3.2 Upstream Processing

Solution Preparation Depending on the type of process being employed, solution prep will vary with the feedstock. As most fermentation processes are sensitive to contamination from wild strains of microorganisms (e.g., viral, bacterial, fungi) of one kind or another, care must be taken to ensure proper sterility. For the most part, solutions are filtered through successively smaller filters to ensure that microorganisms are not introduced into the fermentation broths.

Sterilization Maintaining sterility is a major concern in many bioreactors, and a considerable amount of steam is used to clean and sterilize bioreactors before they are seeded. As the growth media, temperature, pH, and so on, provide an ideal environment for a host of microorganisms, exceptional care must be taken to ensure that contamination does not occur in the first place. In addition, some industrial genetically modified organisms (GMOs) are developed intentionally with resistance to an antibiotic so that when the antibiotic is added to the fermenter, potential competition from other organisms is avoided.

Seed Fermentation Industrial-scale fermenters range in size from hundreds of liters to those that are on the order of 100 to 200 m^3 or larger—practically large enough to drive a bus into. Typically, seed fermenters are employed to grow the organism of interest to a point where there is sufficient cell density and the organisms are in their growth phase. Seed fermenters also allow enough growth to take place to determine whether or not there has been any contamination of the fermentation broth.

Fermentation at Scale Fermentation at scale requires a considerable amount of mass and energy resource.

Raw Materials Raw materials choice varies with the nature of the organism, the product desired, and the type of fermentation. The principal carbon and nitrogen sources used in fermentation processes are shown in Tables 8.4 and 8.5. As can be seen from Table 8.4, the types of carbon substrates used as carbon sources are derived primarily from agricultural sources and are arguably renewable; however, renewability does not necessarily imply sustainable, nor does it come without significant impact. In addition, although many of these substrates may be considered to be of defined composition, there is generally greater variation in the feedstocks on a batch-to-batch basis and on a regional basis. As was said for carbon sources, the same sorts of caveats can generally be made for nitrogen sources. It would be wrong to conclude that bioprocessing is inherently greener than chemical synthetic processes, but they are certainly closer to renewable feedstocks.

Additives (Foam Control, pH, Minerals) There are a number of additives to fermentation processes that are required to maintain microorganisms in their growth phase. Control of pH is generally accomplished through the use of buffers with the direct addition of ammonia. Trace mineral composition is generally very similar, but the details of type and quantity are largely dependent on the type of organism. The evolution of foam is often associated with many fermentation processes, as microorganisms produce a variety of proteins and macromolecular substances that are excreted into the fermentation broths and become part of foams. The large amount of aeration and mixing required to maintain adequate mass and heat transfer in the bioreactor generally exacerbates the formation of foams and leads to the need for antifoaming agents.

Aeration Aeration needs to be maintained at a rate that supports an organism's respiration while maintaining optimal growth. There are two options for ensuring that sufficient oxygen is delivered to microorganisms: injection of air or direct injection of oxygen. In some situations the use of oxygen is warranted to support an intensified process, to reduce the mixing requirement, or to prevent the excessive formation of foams. The decision to use direct oxygen injection must be balanced between the increased energy use to obtain, store, and transport pure oxygen and the energy use from larger aeration rates using air, the increased mixing rates, and the potential for increased antifoaming agent use.

TABLE 8.4 Characteristics of Commonly Used Fermentation Carbon Source Substrates

Carbon Source	Composition	Defined Composition	Source	Green Issue
Glucose	$C_6H_{12}O_6$	Yes	Starch	Multiple environmental impacts from agricultural practices: arable land–competition for food
Starch	$(C_6H_{12}O_5)_x$	Yes	Corn/maize/ grain, potato, rice	*See* Glucose
Corn syrup	Different sugars, mainly glucose and dextrin	No	Hydrolyzed corn or potato starch	*See* Glucose
High-fructose corn syrup	Fructose, glucose, higher saccharides	Yes	Hydrolyzed corn starch	*See* Glucose
Molasses	Mainly carbohydrates	No	Sugar beet, sugar cane	
Cottonseed flour	Mainly carbohydrates	No	Cotton	
Corn-steep liquor	Lactic acids, sugar	No	By-product of corn wet milling process	
Soybean oil	Fat, fatty acids	Yes	Soybeans	
Palm oil	Fat, fatty acids	Yes	Oil palm tree	
Glycerol	$C_3H_8O_3$	Yes	Natural oils and fats	
Ethanol	C_2H_6O	Yes	Oil/gas or fermentative based on oil/ gas	
Methanol	CH_4O	Yes		

Source: Adapted from ref. 1.

Agitation As we have seen for chemical processes, agitation and mixing are also critically important for biological processes. Mixing rates that are too high can lead to shear forces that disrupt cell membranes, leading to cell death. Mixing rates that are too low will lead to inadequate mass transfer, and cell growth will not be maintained in the growth phase.

TABLE 8.5 Characteristics of Commonly Used Fermentation Nitrogen Source Substrates

Nitrogen Source	Composition	Defined Composition	Source	Green Issue
Cottonseed flour	Proteins	No	Cotton	Multiple environmental impacts from agricultural practices: arable land—competition for food
Soybean meal, flour, or grits	Mainly proteins	No	Soybeans	
Peanut meal	Mainly proteins	No	Peanuts	
Dried distillers' grains	Mainly proteins	No	Malt, hops, barley	
Whole yeast, yeast extract	Mainly proteins and peptides	No	Brewer's yeast	Positive use of "waste" from brewing
Yeast hydrolysates	Mainly proteins and peptides	No	Brewer's yeast	
Corn-steep liquor or its powders	Mainly proteins and peptides	No	By-product of corn wet milling process	Multiple environmental impacts from agricultural practices: arable land—competition for food
Corn gluten meal	Mainly proteins and peptides	No	Corn dry milling process	
Linseed meal	Mainly proteins	No	Linseeds	
Fish meal	Mainly proteins	No	Fish such as menhaden and related species	Fish protein used extensively for aquaculture, feed application, and so on; huge pressure on fish stocks and marine and avian food chain impacts
Urea	$H_2N-CO-NH_2$	Yes	Ammonia and carbon dioxide	High-energy process to derive ammonia
Ammonium sulfate	$(NH_4)_2SO_4$	Yes	Multiple nondiscretionary sources (by-products of multiple industrial processes)[5]	
Ammonia gas	NH_3	Yes	Natural gas (CH_4) and air (N_2)	

Heat Transfer For microorganisms to thrive they must be kept at a limited temperature range. Heat transfer becomes an issue as fermentation broth titers are increased, as many respiring organisms produce heat. Scale-up of processes requires very careful attention to the balance among heat transfer, aeration, and mixing. See Chapter 10 for additional details on scale-up aspects.

Cleaning in Place As with chemical processing, reactors must be cleaned carefully between batches to ensure no carryover of unwanted substances. Cleaning generally requires a large amount of water and energy, as usually, some cleaning and rinsing requires high-purity water.

8.3.3 Downstream Processing

Decanting and Biomass Removal At a certain point in time, the concentration of metabolic products, lysed cells, and other extracellular materials in the fermentation broth will reach a concentration that will begin to slow the growth of the organism. At this point the fermentation will be halted and the fermenter emptied. Depending on whether or not the product of interest is part of the extracellular mixture, part of the intracellular organelles or protoplasm, or bound as part of the membrane, different types of decantation will be used. Usually, decantation is primarily separation of the fermentation broth from the cells or fragments of cells contained in the broth. If the product of interest is in the broth, the cells will just be filtered or centrifuged out of the broth and the broth passed on for further downstream processing. Many times the simple act of decanting and filtering the broth will disrupt the cells.

Homogenization/Cell Disruption Where cells are not lysed and/or the material of interest is bound to part of the cell, the cells will be disrupted by a variety of means and homogenized. Disruption and homogenization will provide for more efficient extraction of the desired product. Disruption of the cells is accomplished by relatively simple procedures such as freezing/thawing, addition of detergents or other materials that change osmotic pressures in the cell rapidly and cause it to lyse, mechanical methods such as ball milling, or using pressure to force cells through a small orifice that causes the cells to lyse upon decompression on the other side of the orifice.

Concentration Usually, the material of interest is not present in the cells or in the spent broth at a very high concentration, so some concentration takes place before there is additional downstream processing. Concentration may also be used to remove some of the macromolecular compounds (e.g., lipids, proteins, cell membranes) that may interfere with further isolation and purification of the product desired. Concentration may also ultimately reduce the amount of extraction solvent or other materials required for further downstream processing. Concentration can be effected through relatively straightforward processing steps, such as evaporation of water or solvent, some type of filtration, or in some cases the material of interest can undergo crystallization or precipitation.

Phase Separation Depending on the nature of the product, the fermentation, and the processing steps to this point, phase separation may be required. Phase separation may be solid–liquid or liquid–liquid, depending on the need. Solid–liquid separations are usually effected through centrifugation, filtration, or through simple sedimentation, depending on the size of the solids to be separated, the throughput desired, and the viscosity of the fermentation broth. As fermentation broths are water-based, solvents for concentration and/or preliminary extraction will undoubtedly be chosen for their ability to solvate the desired product and their water immiscibility. Reactor-based liquid–liquid phase separations may be avoided if one is able to use countercurrent liquid–liquid extractors (CCLEs); otherwise, phase separations might have to take place in the reactor. Such phase separations can be problematic, depending on the nature of the extracellular material in the fermentation broth and the potential for formation of emulsions that are sometimes difficult to break.

Product Separation and Purification Isolation and purification of product from solutions containing biologically derived materials can be quite complex, depending on the nature of the product desired. Clearly, many desired products are chemically very similar to materials that are either starting materials (e.g., as in the case of a simple biotransformation) or the product desired (e.g., related metabolic materials). In other cases, the product desired will partition to other macromolecular materials, as in the case of a hydrophilic compound partitioning into lipid-based cell membranes. The exact strategy for isolation and purification needs to be considered carefully so that extractants, processing equipment, yields, and mass and energy efficiency can be optimized.

1. *Extraction.* We have discussed some extractions above. Extractions are in the simplest case undertaken as simple mixing and settling arrangements, or they may be slightly more complicated. It is not uncommon for several trains of CCLEs to be used to extract into an organic solvent, then to back-extract into water to either effect the desired purification or to prepare for a subsequent step, such as an enzymation or some other biocatalytic step. Extractions can be a source of considerable waste if solvents are not recovered in large amounts. Such a situation might arise if solvents are used to extract lysed cells, where a large amount of solvent may be retained in the biomass. Extractions may be further complicated by pH, the ionic strength of the fermentation broth, and temperature, so in any extraction optimization study these should be experimental variables. In addition, the use of organic solvents in extractions will invariably pose some health and safety concerns.
2. *Distillation, pervaporation, and wet film evaporators.* As more common building blocks for the chemical enterprise are derived from bioprocesses, distillation is likely to be used as extensively as it is currently in the petrochemical industry. Where boiling points between desired substances are sufficiently large (10 °C or greater) or where no azeotropes are formed, single column distillations will continue to be a great option for product separation and purification. They are also used extensively to recycle and reuse the solvents employed in bioprocessing for extractions. Where there are azeotropes or there is extensive water miscibility, pressure-swing distillations or pervaporation may be used to dewater certain solvents, or more sophisticated distillation schemes (e.g., multicolumn) may be employed. Finally, where boiling-point differences are very large or a lower purity of solvent is acceptable, wipe film evaporators may be used for separation.
3. *Electrodialysis.* For products of a certain molecular weight and ionic charge, electrodialysis may be a useful option for concentration and purification. Electrodialysis takes advantage of the fact that when a current is applied across a semipermeable, ion-selective membrane, ions in an aqueous stream will be selectively transported through the membrane into an ionic stream. In most cases, electrodialysis is used as a means of concentrating or preconcentrating organic acids or bases. The

usefulness of this particular strategy will be dependent on the flux of the desired ions (flux is usually measured in terms of $g/m^2 \cdot h$) across the membrane and the transport number, defined in terms of the ratio of the flux of desired ions to the flux of all ions across the membrane.

4. *Adsorption.* For some products, various adsorption options exist for extracting the desired material or removing interfering substances or color from a complex fermentation broth. Typically, adsorption steps are carried out using fixed-bed columns where the solution containing the substance of interest is passed across chemically modified resins that possess the desired properties for extracting the compound of interest. Fermentation broths, or perhaps partially processed solutions, are passed through the column and either the product, or the interferent, contaminant, or macromolecule, is retained. The column is washed to remove the undesirable components, and the compound of interest is removed from the column with a buffer solution of a particular pH, or with a solvent, or with a combination of extraction fluids. In some cases there may be several columns in series or in parallel, depending on the desired extraction or purity required. Perhaps the best known adsorption strategies take advantage of the cationic and anionic properties of some compounds to extract them using cationic and/or anionic exchange columns, but other approaches are found in the purification of vitamins and cyclodextrins. Performance of adsorption columns, as with most extraction methods, will be influenced by such factors as pH, temperature, and ionic strength.
5. *Chromatography.* Chromatography is related to adsorption, but is more akin to column distillation and the fractionation of compounds across theoretical distillation plates. Materials of interest are differentially retained as they pass through a packed column and are eluted over a discrete time period. The retention time on the column is a function of the packing material, the composition of the mobile phase, temperature, pH, and in some cases, ionic strength. Mobile-phase composition can either be held constant in what are known as *isocratic conditions* or it can change over time in what is known as a *gradient elution*. Gradient elutions are frequently used to either speed up or slow down the time a compound remains on the column so that interfering compounds are eluted from the column. Generally, there are five different types of chromatography that may be used, depending on the application. These include:
   a. Gel or size-exclusion chromatography, where the differences in molecular size are exploited to separate large from small molecules, or vice versa.
   b. Ion-exchange chromatography, where differences in ionic charge are used to effect separations. Ion exhange will be highly dependent on pH and ionic strength, so the buffers used for the mobile phase must be carefully matched with the appropriate packing materials to obtain the separation desired.
   c. Hydrophobic interaction chromatography, separations that are used principally for proteins and large macromolecules such as monoclonal antibodies. The relative hydrophobicity of any given protein is dependent on the amino acid content and distribution on the surface of the protein. By changing the ionic strength of the mobile phase, the protein will become more or less hydrophilic and therefore have greater or lesser affinity for the stationary phase or resin.
   d. Affinity chromatography, where separations are based on differences in stereochemistry between compounds and the stereoselectivity of the ligands bound to the packing materials. These are highly selective separations, and generally the packing materials are very expensive. Bound molecules are eluted by changes in pH, ionic strength, or buffer composition.

e. Reversed-phase chromatography, probably the best known chromatographic technique. Reversed-phase chromatography is based on portioning of solutes between the mobile phase and the stationary phase (usually, chemically modified silica-based). A wide range of ligands and mobile phases are used to tune separations and obtain the desired compound. Reversed-phase chromatography is used extensively in industry to obtain smaller quantities of materials (i.e., gram or low kilogram). Recent years have seen the introduction of large-scale reversed-phase separations in the form of multicolumn or simulated moving-bed chromatography that can deliver multi-kilogram quantities of materials.

8.3.4 Final Steps

Pharmaceutical products, especially those derived from mammalian cell cultures, usually require additional processing steps that are generally not required for most other products derived from bioprocesses. These additional processing steps provide evidence for inactivation of viral, prion, or other pathogenic organisms. While the processing steps used to obtain the desired product will radically reduce the occurrence of such organisms, additional micro- or ultrafiltration, ultraviolet inactivation, pasteurization (heat treatment), or chemical addition may be required to ensure freedom from pathogens and overall product integrity.

In addition to viral inactivation, it is sometimes necessary to perform additional steps to solubilize and refold proteins to their active forms. Although fermentation products from bacteria and fungi produce a protein that has the correct amino acid sequence and other secondary structures, the protein may contain certain inclusion bodies or water, and its overall three-dimensional structure is frequently incorrect. This requires that the protein be denatured (unfolded) to remove the inclusion bodies or water, and then refolded by careful removal of the denaturant.

For many materials of intermediate molecular weight (e.g., < 750 g/mol and > 150 g/mol) a further crystallization and sometimes recrystallization step is required to obtain the product in the desired purity. Crystallization for bio-derived chemicals is effectively no different from crystallization encountered in normal chemical processing.

Crystallization is carried out by physical and/or chemical processes such as simple volume reduction through distillation, changes in temperature, addition of an antisolvent, or crystal seeding, or a combination of these. Crystal form, the presence of one or more polymorphs, or another factor can be the source of considerable frustration and effort to obtain the desired form. Care must be taken in the design of crystallization processes to ensure robust results. In some cases, such as with proteins, enzymes, or biopharmaceutical products, stabilization is required through the addition of preservatives, buffers, or other stabilizing agents.

8.4 EXAMPLES OF PRODUCTS OBTAINED FROM BIOPROCESSING

A recent study on the application of biotechnology to industrial sustainability was conducted by the Organization on Economic Cooperation and Development.[3] The report found a number of cases where bioprocessing was replacing conventional chemical synthetic production processes or delivering new products. In the overwhelming majority of cases, an improved sustainability profile was not the driving force for changes in the processes or in the introduction of new products; rather, economics played a decisive role. Table 8.6 was

TABLE 8.6 Cost and Environmental Benefits Reported from Case Studies of the Application of Biotechnology to Industrial Sustainability

Product	Energy	Raw Materials	Air Emissions	Water Emissions	Operating Costs
Riboflavin (B_2)	Same	−75% (nonrenewables)	−50%	−66%	−50%
7-Aminocephalosporic acid	—	—	−90%	−33%	−90% (environment-related)
Cephalexin	Electricity +, steam −	—	−80%	−80%	Considerable reduction
Amino acids	Same	—	—	−43%	
Acrylamide	−80%	—	Down	Down	Down
Acrylic acid	—	Down	Down	Down	−54% (raw materials)
Enzyme-catalyzed synthesis of polyesters	—	—	Down	Down	Down
Vegetable oil degumming enzyme	−70%	—	—	−80%	−40%
enzymatic pulp bleaching	−30–40%	Down	—	Down	
On-site production of xylanase	—	−35% (Cl_2), −65% (ClO_2)	—	Down	
Gypsum-free zinc refinery	—	Down (recycle)	—	Down	

Source: ref. 3.

adapted from the report and contains a summary of some of the environmental and cost benefits identified for some of the case studies.

In the following paragraphs we discuss briefly some of the current industrial applications of bioprocesses.

8.4.1 Pharmaceuticals

The classic example of small-molecule pharmaceuticals produced through bioprocessing is found in antibiotics such as penicillins or cephalosporins. Of equal importance is the production of larger or more chemically complex molecules such as insulin or taxol, or the recent development and use of monoclonal antibody therapies. The success in producing these products has led to a dramatic increase in the life expectancy of many while contributing to a generally higher quality of life for those living with chronic diseases. Henderson et al.[4] have recently reported a comparison of a chemical synthetic route with a route that employs biocatalysis to a cephalosporin precursor known as 7-aminocephalosporic acid (7-ACA).

Example 8.1 Compare the chemical and biocatalytic routes to 7-ACA. Which route is greener, and why? Explain your answer.

Chemical route description. A four-step process is used to convert the potassium salt of cephalosporin C to 7-ACA, as shown in Figure 8.2(a). In the first step, a common protection strategy is used to convert the acid to an anhydride and the amine to an amide using chloracetyl chloride in the presence of the base dimethyl aniline. Next, phosphorous pentachloride is added to the mixed anhydride, which is held at −37 °C to form the imodyl chloride, which is followed sequentially by the addition of methanol to form the transient imodyl ether and then water to form 7-ACA. 7-ACA is precipitated by using ammonia to change the pH to the isoelectric point and the 7-ACA is recovered methanol wet and then dried under vacuum.

Biocatalytic route. A three-step process is used to convert the potassium salt of cephalosporin C to 7-ACA, as shown in Figure 8.2(b). A solution of cephalosporin C is stirred with the immobilized biocatalyst D-amino acid oxidase (DAO), while air is bubbled through the solution to supply the required oxygen. The by-product of the bioconversion, hydrogen peroxide, reacts spontaneously with the keto intermediate to give glutaryl 7-ACA. The reaction is carried out at a constant temperature (18 °C) and elevated pressure (5 bar) under controlled pH (starting at pH 7.3 and rising to 7.7 at completion) to ensure the desired conversion. Additional hydrogen peroxide may be added to promote greater conversion to glutaryl 7-ACA if desired. Upon completion, the solution containing glutaryl 7-ACA is separated from DAO, and immobilized glutaryl 7-ACA acyclase (GAC) is added at a pH of 8.4 and a temperature of 14 °C to obtain the desired 7-ACA. Dilution may be required to control the concentration, but upon completion of the reaction, the 7-ACA is separated from GAC and isolated. In both cases, the enzymes may be recovered and reused.

Solution Table 8.7 compares the chemical and biocatalytic processes from the work of Henderson et al.

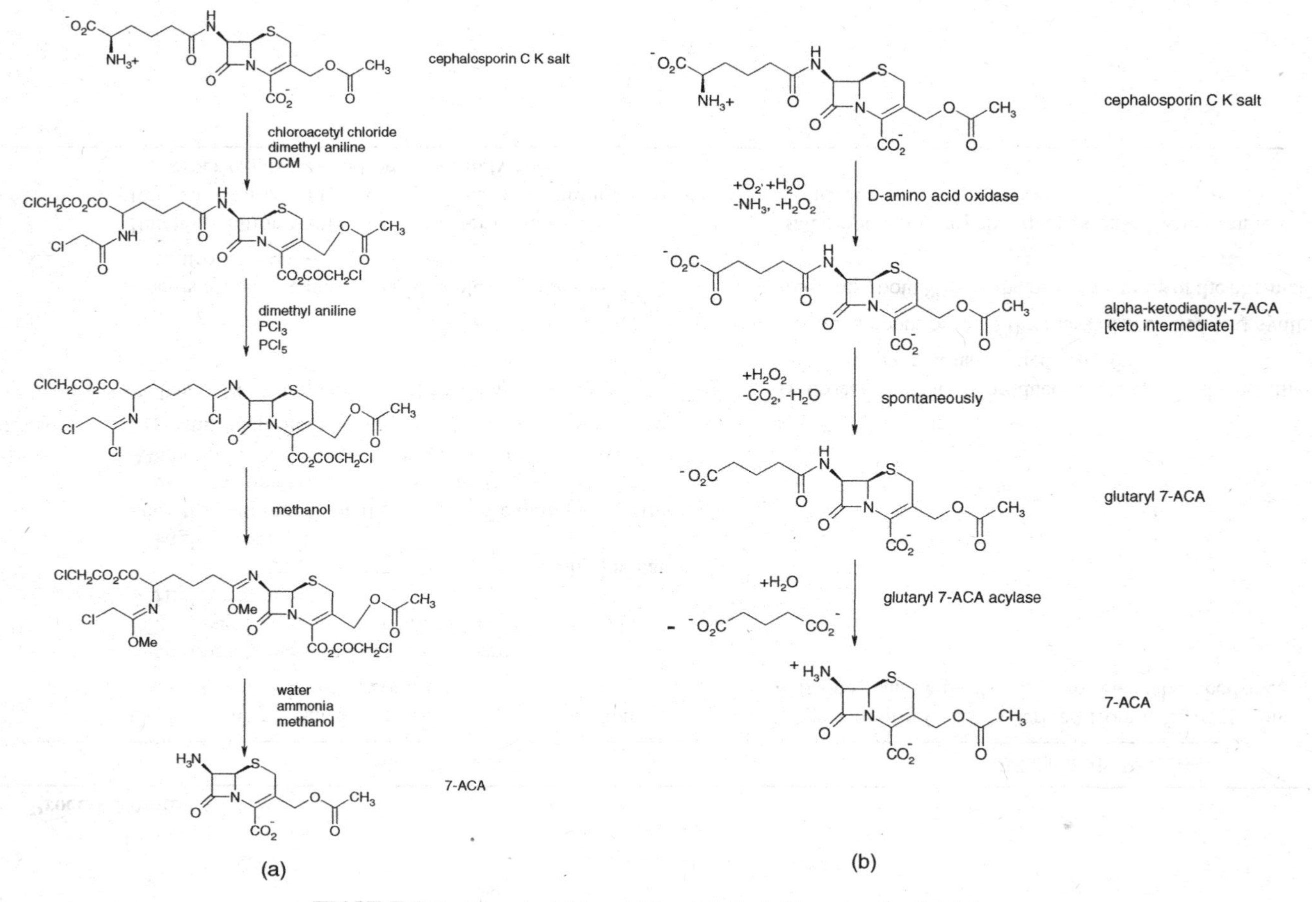

FIGURE 8.2 Chemical (a) and biocatalytical (b) routes for 7-ACA.

TABLE 8.7 Process Comparison

Attribute	Chemical Process	Biocatalytic Process
Renewability	Cephalosporin C salt is derived from a fermentation. Reagents and solvents are not renewable.	Cephalosporin C salt is derived from a fermentation. Enzymes can be produced from renewable feedstocks.
Toxics	Dichloromethane—suspect human carcinogen. PCl_3: Reacts violently with water, very toxic by inhalation. Highly reactive. PCl_5: Reacts violently with water, very toxic by inhalation. Highly reactive. Dimethyl aniline: limited evidence of carcinogenic effect, toxic by inhalation, ecotoxic effects—aquatic. Ammonia: corrosive, toxic by inhalation, explosive.	
Chemoselectivity	Protection/deprotection strategy required.	Chemoselective.
Process safety	PCl_3 and PCl_5: require special handling.	Hydrogen peroxide is produced and may be added, although there is a very low risk associated with this.
Mass efficiency		Requires about 50% of the mass of the chemical synthesis.
Energy	Chemical route requires considerable chilling (to $-37\,^{\circ}C$) to control exotherms.	Requires about 80% of the process energy of the chemical synthesis.
Complexity	Batch operation with greater number of steps. Reagent addition must be carried out with care to avoid worker exposures and process safety risks.	Batch operation largely aqueous based with simple mixing operations.

As can be seen in the table, the biocatalytic route exhibits a better environmental profile than the chemical route, so we could say that in this case the biocatalytic route is greener than the chemical route.

Additional Points to Ponder Given what you know to this point in the book, what can you say about the greenness of bioprocesses? What are the potential advantages? What are the potential drawbacks?

8.4.2 Biofuels

Given the recent volatility in petroleum availability and pricing, combined with the increased and steadily increasing concern about greenhouse gases leading to climate change, there has been considerable interest in the development of biofuels. Bioprocesses for the production of ethanol, butanol, biodiesel, and related products have been extensively reported in the literature, and bioderived ethanol is now a major business in the United States with an annual capacity exceeding 50 million gallons. In Brazil, ethanol derived from sugarcane is a very significant product. There is considerable controversy surrounding biofuel production and its impact on the environment and its societal impact, especially at the interface for its perceived competition with arable land and food production.[5,6] There are also major concerns about the use of marginal land and the continued influx of fertilizers, herbicides, and pesticides into the environment. This is discussed in more detail in Chapter 23.

8.4.3 Plastics

There has been considerable effort in recent years to develop polymers whose monomers are obtained from renewable resources. The following example discusses the production of poly(lactic acid) (PLA) from corn-derived glucose.[6]

Example 8.2 PLA is not a new polymer and was originally worked on by the pioneering DuPont polymer chemist Carothers in the 1930s.[7] Petrochemically derived lactic acid is generally undesirable because it is produced as a racemic mixture of the D- and L-isomers, which leads to an amorphous PLA. A simplified diagram of the petrochemical process for lactic acid is shown in Figure 8.3. However, by the 1980s, fermentation of corn-derived glucose was seen as a possible source of obtaining almost pure L-lactic acid and ultimately was first utilized in a joint venture between Dow and Cargill to provide high-molecular-weight PLAs. Cargill uses a solvent-free process that begins by taking lactic acid from its fermentation process followed by polymerization to a low-molecular-weight polymer. The

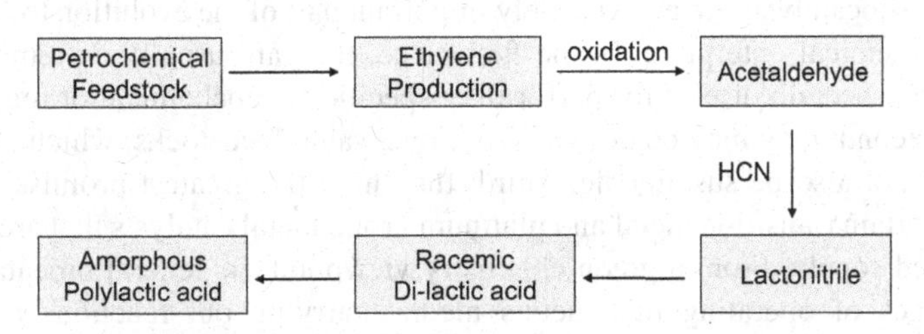

FIGURE 8.3 Petrochemical process for polylactic acid.

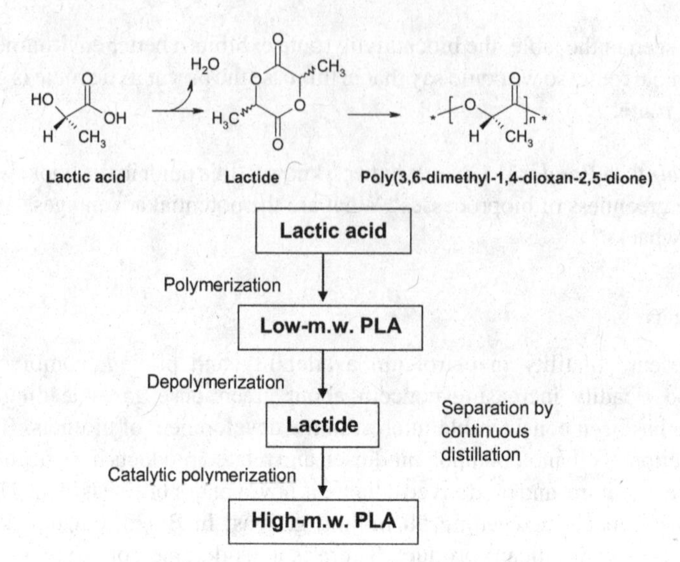

FIGURE 8.4 Polylactic acid production.

polymer undergoes a controlled depolymerization to produce a cyclic dimer known as the *lactide*, which is maintained as a liquid and distilled to increase its purity.[8] Catalytic ring opening of the lactide will produce a range of PLAs with controlled molecular weights. The entire process from the raw lactic acid to the production of the high-molecular-weight PLAs is continuous, with no need to separate the intermediate lactide. Figure 8.4 is a block diagram of PLA production.

Briefly describe the benefits of the PLA process developed by Cargill from a green chemistry and engineering perspective.

Solution The range of benefits of the Cargill PLA process are summarized in Table 8.8.

Additional Point to Ponder What other environmental impacts would the fermentation process for PLA have?

8.4.4 Biocatalysts

We learned in Chapter 6 that biocatalysts are beginning to be used extensively in a number of applications in chemical, pharmaceutical, and other parts of the chemical manufacturing enterprise. Table 8.9 is similar to Tables 8.3 and 8.4, but is specific to biocatalyst production, and Table 8.10 presents some advantages and disadvantages of biocatalysts. Biocatalysts are an extremely important part of the evolution toward a more sustainable chemical enterprise. In the first place, they are usually extremely chemoselective and especially useful for performing specific stereochemical or regiochemical syntheses. Second, they may be derived from renewable feedstocks, which at this point may or may not also be sustainable. Third, they hold the greatest promise for getting away from certain transition metal and platinum group metal catalysts that are both toxic and in limited supply. From a green chemistry viewpoint, in general biocatalysts have the advantages of operating in aqueous media, carrying out reactions at moderate temperatures, and having very good regio-, chemo-, and stereoselectivity. On the other

TABLE 8.8 Process Comparison

Attribute	Petrochemical Process	Cargill Process
Renewability	Fossil fuel (petroleum): No	Glucose (corn-based but could be derived from other plants, e.g., sugarcane)
Toxics	Ethylene oxide: human carcinogen Acetaldehyde: Potential human mutagen and carcinogen; possible sensitization; severe eye irritant HCN: acutely toxic	
Chemoselectivity	Racemic mixture: loss of unwanted isomer or production of inferior amorphous plastic	L-Lactic acid
Process safety	Ethylene: extremely flammable Ethylene oxide: explosive HCN: extremely flammable	
Energy	Range of 79–140 MJ/kg energy for a range of petrochemically derived plastics such as nylon 66, PET, PP, HDPE, and PE	First-generation PLA used about 57 MJ/kg energy
Complexity	Multiple unit operations of higher hazard, storage of toxics, etc.	Fewer unit operations.

hand, they have challenges on their operational stability and normally a very low volumetric productivity.

Example 8.3 Pfizer's Lyrica (Pregabalin) The initial chemical synthesis of Pfizer's Lyrica (pregabalin) is as follows:

CHO —3 steps→ NH$_2$ / CO$_2$H —(*S*)-mandelic acid→ NH$_2$ / CO$_2$H

29% overall
>99% enantiomeric
excess (e.e.)

The original chemical synthesis[1] began with isovaleraldehyde and diethyl malonate and in three steps arrived at the desired intermediate, (+)-3-(aminomethyl)-5-methylhexanoic acid. A classic resolution requiring stoichiometric addition of (*S*)-mandelic acid was used to obtain the desired (*S*)-3-(aminomethyl)-5-methylhexanoic acid, but the undesired enantiomer was difficult to recycle and over 75% of the input material became waste. The overall yield was low, as was the throughput, and the high cost to make this was undesirable.

Because of the undesirable aspects of the chemical synthesis, the development of a biotransformation route was undertaken.[2] Racemic 2-carbethoxy-3-cyano-5-methylhexanoic acid, ethyl ester (CNDE), an intermediate in the isovaleraldehyde process, was mixed with a lipase to obtain the desired isomer, the (3*S*)-3-cyano-5-methyl hexanoic

TABLE 8.9 Characteristics of Biocatalysts

Biocatalyst	Production Device	Raw Material	Time Scale	Purification	Examples[a]
Enzymes	Bioreactor	Pure substrates	Short	Simple	Cyclodextrin, acrylamide, L-dopa
Bacteria and yeasts	Bioreactor	Simple media	Short	Medium	Lysine, vitamin B_2, insulin
Fungi	Bioreactor	Simple media	Medium	Medium	Citric acid, antibiotics
Mammalian cells	Bioreactor	Complex media	Medium	Medium	Monoclonal antibodies, interferons
Plant cells	Bioreactor	Simple media	Medium	Medium	Taxol, shikonin, methyldigoxin
Transgenic plants	Bioreactor	Fertilizer, CO_2, various others	Long	Complex	Antibodies, antibody fragments, HSA, PHB
Transgenic animals	Whole plant	Various plants and animal materials	Long	Complex	α_1-Antitrypsin, HSA, lactoferrin
Extractive technology	Whole animal	Certain parts of plants, animals, and humans	Long	Complex	Plasma components, taxol

Source: ref. 1.

[a]HSA, human serum albumin; PHB, poly(3-hydroxybutyrate).

TABLE 8.10 Advantages and Disadvantages of Biocatalysts

Type of Biocatalyst		Advantages	Disadvantages
Whole cells	General	No need for cofactor recycling	Low substrate concentration High biomass production Low specificity Unstable at high temperatures and low pH
	Fermented	High activities	High biomass production More by-products
	Fermented	Easier to process	Low activities
	Immobilized	Ease of separation Ease of purification	Low activities
Isolated enzymes	General	Very high selectivities Ease of use Simple processing High substrate concentrations	Requires cofactor recycling Accesible for only few reactions Instable at high temperaturas and low pH
	Immobilized	High selectivity Ease of separation Ease of purification	Loss of activity during immobilization

acid, and in an additional step, taken to the desired (3*S*)-3-cyano-5-methylhexanoic acid, ethyl ester [(*S*)-CNE]. Reduction of the cyano group was accomplished with hydrogen in the presence of a spongy nickel catalyst to obtain the desired product, pregabalin:

recycling of *R*-1
NaOEt, 100%
EtO_2C CO_2Et
R-**1**, 85% ee
+
EtO_2C CO_2Et
racemic CNDE **1**
765 g/L of total volume
(3.25 kg/L of H_2O)
H_2O
lipolase
Step 1
^-O_2C CO_2Et
(*S*)-CNDE acid
>98 % ee
@ 45% conversion
H_2O
Step 2
CO_2Et
(*S*)-CNE
85–90%
H_2, Ni
H_2O
Step 3
NH_2
CO_2H
Pregabalin
90–95%
99% purity, >99.7% ee
overall 40–45% yields
after one recycling

This synthesis resulted in the elimination of 11 million gallons of solvent, better solvent profile use (see Figure 8.5), a reduction in CNDE use of more than 800 metric tons, and elimination of 1600 metric tons of (*S*)-madelic acid and 500 metric tons of nickel.

Additional Points to Ponder On the basis of Table 8.2 and Example 8.3, could you outline the disadvantages of the process described in Example 8.3 that are not mentioned? Do you think the process is green? How could you improve it?

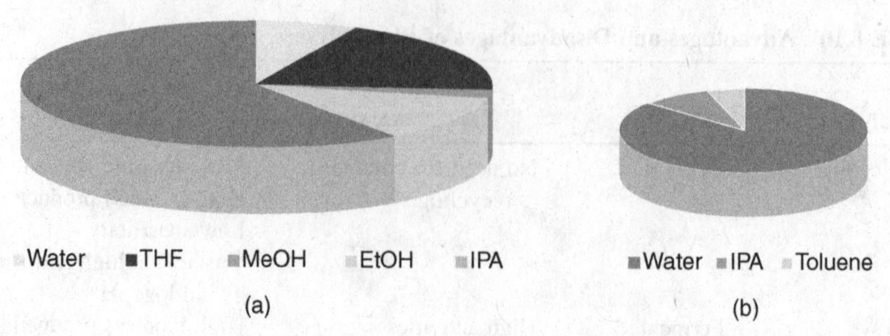

FIGURE 8.5 Solvent use of the chemical (a) and biocatalytic (b) routes to Pfizer's Lyrica (pregabalin).

PROBLEMS

8.1 Describe several examples of the beneficial use of bioprocessing not mentioned in Section 8.1.

8.2 Look at the applications in Table 8.1. Imagine that you are a policymaker and want to write legislation that would spur on particular applications of bioprocessing. Which applications would you choose to focus on? Defend your answer.

8.3 Look at the production volumes in Table 8.2. Which of these applications would you expect to grow? Would you expect any of these to decrease? What basic chemicals do you think might be useful targets for bioprocessing, and why?

8.4 Tables 8.4 and 8.5 list common carbon and nitrogen sources and cite a few general environmental issues associated with agriculture. Investigate and describe some of these potential impacts. How might these impacts be reduced through further improvements in bioprocessing?

8.5 Choose from the literature an application of an enzyme that is replacing a chemical synthesis. Evaluate its greenness. If you were a scientific director for a company, would you invest in developing the proposed replacement?

8.6 Adipic acid is a monomer produced in large quantities for the production of nylon 66 and polyurethane. We will look at three different routes to adipic acid.

Traditional chemical synthesis:

benzene —($Ni-Al_2O_3$, 370–800 psi)→ cyclohexane —($Co-O_2$, 120–140 psi)→ cyclohexanone + cyclohexanol —(HNO_3)→ adipic acid

Draths–Frost biotechnical synthesis:[11]

D-glucose —(*E. coli*)→ 3-dehydroshikimate —(*E. coli*)→ *cis-cis*-muconic acid —(Pt, H_2, 50 psi)→ adipic acid

Typical feed solution (in 1 L of water):

10 g Bacto tryptone	1 g NH_4Cl	Yield = 20.4 mmol
5 g Bacto yeast	10 g glucose (62 mmol)	% Yield = 33%
10.5 g NaCl	0.12 g $MgSO_4$	
6 g Na_2HPO_4	1 mg thiamine	
3 g KH_2PO_4		

Synthesis by Kazuhiko et al.[9]:

$$\text{cyclohexene} + 4H_2O_2 \xrightarrow{Na_2WO_4[CH_3(n\text{-}C_8H_{17})_3N]HSO_4} HO_2C(CH_2)_4CO_2H \text{ (adipic acid)} + 4H_2O$$

cyclohexene   adipic acid

A typical procedure for the oxidation and reuse of the water phase proceeds as follows. In the first run, a 1-L round-bottomed flask equipped with a magnetic stirring bar and a reflux condenser was charged with 4.1 g (12.2 mmol) of $Na_2WO_4{\cdot}2H_2O$, 5.67 g (12.2 mmol) of $[CH_3(n\text{-}C_8H_{17})_3N]HSO_4$ (used as a phase-transfer catalyst), and 607 g (5.355 mol) of aqueous 30% H_2O_2. The mixture was stirred vigorously at room temperature for 10 min and 100 g (1.217 mol) of cyclohexene was added. The biphasic mixture was heated successively at 75 °C for 30 min, at 80 °C for 30 min, at 85 °C for 30 min, and at 90 °C for 6.5 h, with stirring at 1000 rpm. The homogeneous solution was allowed to stand at 0 °C for 12 h, and the resulting white precipitate was separated by filtration and washed with 20 mL of cold water. The product was dried in a vacuum to produce 138 g (78% yield) of adipic acid as a white solid (with a melting point of 151.0 to 152.0 °C). A satisfactory elemental analysis was obtained without further purification. Concentration of the mother liquor produced 23 g of pure adipic acid; the yield determined by GC (OV-1 column, 0.25 mm × 50 m, GL Sciences, Tokyo) was 93%. The by-products identified were 1,2-cyclohexanediol (2% yield) and glutaric acid (4% yield). In the second run, a 2-l round-bottomed flask was charged with the water phase of the first run, which contained the W catalyst, 5.67 g (12.2 mmol) of $[CH_3(n\text{-}C_8H_{17})_3N]HSO_4$, and 552 g (4.868 mol) of aqueous 30% H_2O_2. After the mixture was stirred vigorously at room temperature for 10 min, 100 g (1.217 mol) of cyclohexene was added. This mixture was heated successively at 75 °C for 30 min, at 80 °C for 30 min, at 85 °C for 30 min, and at 90 °C for 46.5 h, with stirring at 1000 rpm; the homogeneous solution was allowed to stand at 0 °C for 12 h. The resulting white precipitate was separated, washed, and dried in a vacuum to produce 138 g (78% yield) of analytically pure adipic acid as a white solid.

(a) Which of the three syntheses would be the greenest? Explain your answer.

(b) Which synthesis is more mass efficient?

(c) Assess the recycle and reuse strategy in the synthesis by Kazuhiko et al. Do you think this would be commercially viable? Defend your answer.

(d) Which of these syntheses would have a greater environmental impact? Why?

8.7 A pharmaceutically important building block, pyruvic acid (2-oxopropanoic acid CAS No. 127-17-3), may be produced through a fermentation from glucose using *Escherichia coli.*[1,4] This is an important route, as traditionally the process used to produce pyruvic acid is via pyrolysis of tartaric acid and possesses a number of economic and environmental drawbacks. There are two alternatives for downstream processing: a solvent-based process, and recovery of product by electrodialysis. Tables P8.7A and P8.7B contain a mass balance and energy use for the downstream process.

TABLE P8.7A Mass Balance

	Input (kg/kg)		Output (kg/kg)	
Component	Extraction	Electrodialysis	Extraction	Electrodialysis
Acetic acid	0.09	0.09		
Ammonium	0.03	0.03		
Ammonium sulfate	0.19	0.19	0.18	0.18
Biomass	<0.01	<0.01	0.09	0.09
Carbon dioxide	—	—	0.05	0.05
Glucose	1.19	1.18	0.10	0.10
Hydrogen chloride	0.62	<0.01		
Solvent 1	0.31	—	−0.31	
Product loss	—	—	0.08	0.08
Other organic material	—	—	0.15	0.15
Oxygen	0.19	0.18		
Salts				
Inorganic	—	—	0.75	0.32
Mineral	0.14	0.14		
Sodium hydroxide	0.37	0.38		
Water	46.9	32.3	47.3	32.7
Mass Intensity	52.6	34.5	51.6	33.5
Without water	3.1	2.2	1.7	0.8

TABLE P8.7B Energy Use

Energy Use	Extraction	Electrodialysis
Process energy use (kWh/kg)	1.9	2.4
Steam (kg/kg)	50	15
Cooling (kg/kg)	265	50
Chilled water (m^3/kg)	2.7	1.5

(a) What would lead to the large differences in energy use between the two options?

(b) What would lead to a greater quantity of solvent and water being used for the extraction process?

(c) Based on these two tables, which downstream processing alternative is best, and why?

8.8 Citric acid is one of the few commodity chemicals produced by bioprocessing with a worldwide production on the order of 1.1 million tons. It has been produced for over 80 years, predominantly using the filamentous fungus *Aspergillus niger*.[1] Biwer et al. describe a process that uses starch as a feedstock instead of glucose, the starting material normally used.[10,11] Starch is a mixture of two different polymers of glycopyranose (amylose and amylopectin), whose only building block is glucose ($C_6H_{12}O_6$), linked by predominantly by α-1,4-glycosidic bonds. The basic process is depicted in Figure P8.8(a).

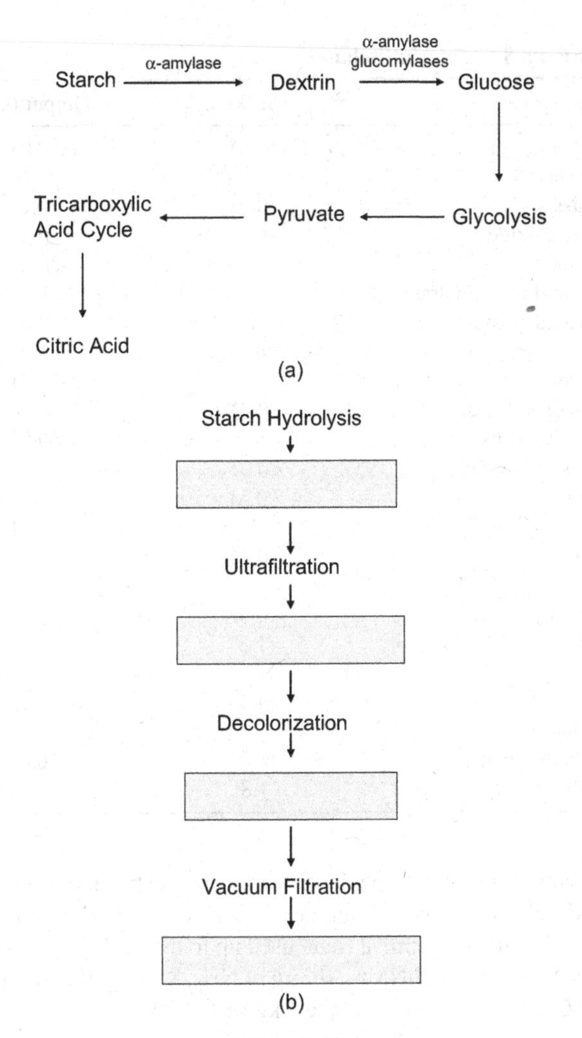

FIGURE P8.8

(a) What would be the advantage of using starch instead of glucose or molasses as a starting material?
(b) What other components should be added to the flow diagram to account more accurately for what is added and taken from the process?
(c) In the flow diagram in Figure P8.8(b), what might the missing unit operations be?
(d) How would you change the process to make it greener?
(e) The material balance for the citric acid process is as shown in Table P8.8. For every 100 kg of starch, the bioreaction yield is 84% (kg/kg) and the downstream processing yield is 94%. What is the overall yield?
(f) The authors report that 12,630 tons of citric acid monohydrate is produced over 330 days in 565 batches. What is the total CO_2 produced in a year from the bioreaction?What would be the composition of the aqueous waste?
(g) What would happen to the waste biomass? Outline some of the potential disposal options and the pros and cons of each.

TABLE P8.8 Material Balance

Component	Input (kg/kg)	Output (kg/kg)
Ammonium nitrate	<0.01	<0.01
α-Amylase	0.02	
Biomass	<0.01	0.16
Carbon dioxide	—	0.41
Chloride	—	<0.01
Citric acid monohydrate	—	1
Citric acid (loss)		0.07
Fats		0.01
Glucose		<0.01
Hydrogen chloride	<0.01	
Magnesium (dissolved)		<0.01
Magnesium sulfate	<0.01	
Oxygen	0.51	
Potassium (dissolved)		<0.01
KH_2PO_4	<0.01	
Sodium (dissolved)		<0.01
Sodium hydroxide	<0.01	
Starch	1.27	
Sulfate		<0.01
Water	14.98	15.12
Mass index		
Including water	16.8	16.8
Without water	1.8	1.65

8.9 The semisynthetic routes to penicillin and cephalosporin antibiotics via 6-aminopenicillanic acid are a classic example of enyzymatic biotransformations conducted at a large scale. In the chemical route used up to the mid-1980s, 0.6 kg of Me_3SiCl, 1.2 kg of PCl_5, 1.6 kg of $PhNMe_2$, 0.2 kg of NH_3, 8.41 kg of *n*-BuOH at −40 °C, and 8.41 kg of CH_2Cl_2 are used to make 1 kg of 6-APA:[12]

penicillin G

(1) Me_3SiCl

(2) PCl_5/CH_2Cl_2

$PhNMe_2$

(1) *n*-BuOH; −40°C

(2) H_2O; 0°C

6-APA

In the enzymatic route, water at 37 °C and 0.9 kg NH_3/kg 6-APA are used.

$$\text{penicillin G} \xrightarrow{\text{pen-acylase, } H_2O} \text{6-APA}$$

(a) Compare and contrast the relative greenness of the two routes.

(b) Assume a mass balance for the production of pen-acylase based on previous problems. How does this change the picture?

8.10 An early application of biotechnology to the production of a commodity chemical was for rayon production by Mitsubishi, where acrylonitrile was hydrated to form acrylamide using whole cells of *Rhodococcus rhodocrous* containing a nitrile hydratase.[13]

$$\text{acrylonitrile (CN)} + H_2O \xrightarrow[\textit{Rh. rhodocrous}]{\text{nitrile hydratase}} \text{acrylamide}$$

(a) Evaluate the relative greenness of the chemical process compared to the bioprocess (Figure P8.10).

(b) How might each process be improved?

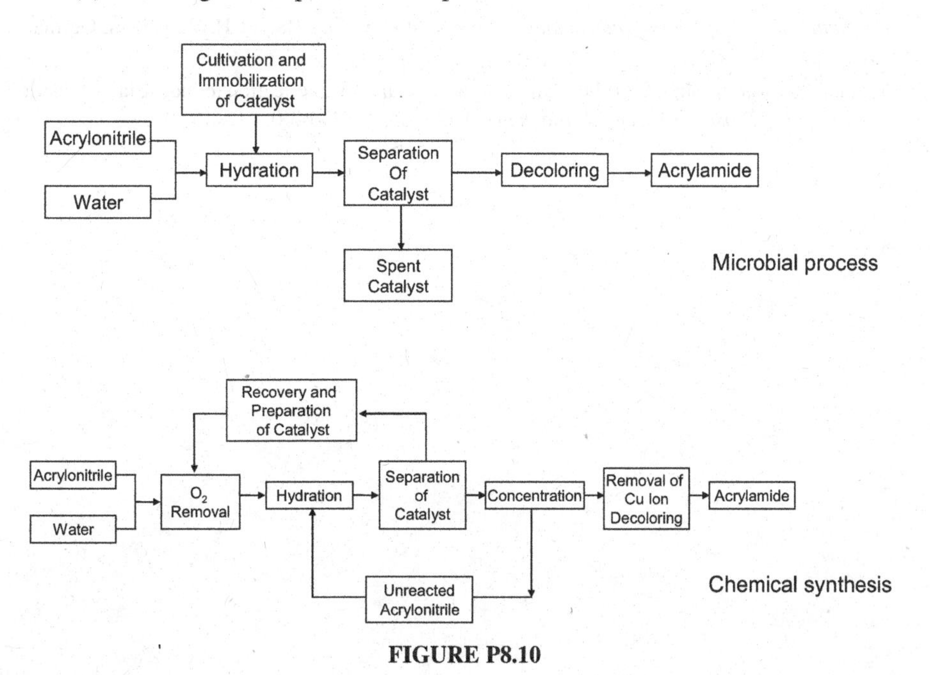

FIGURE P8.10

REFERENCES

1. Henzle, E., Biwer, A., Cooney, C. *Development of Sustainable Bioprocesses*. Wiley, Hoboken, NJ, 2006.
2. Gavrilescua, M., Chistib, Y. Biotechnology: a sustainable alternative for chemical industry. *Biotechnol. Adv.*, 2005, 23, 471–499.
3. Organisation for Economic Co-operation and Development. Sustainable Development: The Application of Biotechnology to Industrial Sustainability. 2001. http://www.oecd.org/sti/biotechnology.
4. Henderson, R. K., Jiménez-González, C., Preston, C., Constable, D. J. C., Woodley, J. M. EHS & LCA assessment for 7-ACA synthesis. A case study for comparing biocatalytic & chemical synthesis. *Ind. Biotechnol.*, 2008, 4, 180–192.
5. Fargione, J., Hill, J., Tilman, D., Polasky, S., Hawthorne, P. Land clearing and the biofuel carbon debt. *Science*, 2008, 319, 1235.
6. Searchinger, T., Heimlich, R., Houghton, R. A., Dong, F., Elobeid, A., Fabiosa, J., Tokgoz, S., Hayes, D., Yu, T. H. Land-use change in greenhouse gases through emissions from use of U.S. croplands for biofuels increases. *Science*, 2008, 319, 1238.
7. Holten, C. H. *Lactic Acid Properties and Chemistry of Lactic Acid and Derivatives*. Verlag Chemie, Weinheim, Germany, 1971.
8. Gruber, P. R., Kolstad, J. J., Ryan, C. M., Hall, E. S., Conn, R. S. E.Melt-stable amorphous lactide polymer film and process for manufacturing thereof. U.S. patent 5,484,881, 1996.
9. Kazuhiko, S., Masao, A., Ryoji, N. A "green" route to adipic acid: direct oxidation of cyclohexenes with 30 percent hydrogen peroxide. *Science*, 1998, 281, 1646.
10. Biwer, A. P., Zuber, P. T., Zelic, B., Gerharz, T., Bellman, K. J., Heinzle, E. Modeling and analysis of a new process for pyruvate production. *Ind. Eng. Chem. Res.*, 44, 3124–3133.
11. Biwer, A. Modelbildung, Simulation und oekologishe Berwertung in der Entwicklung biotechnologischer Prozesse. Ph.D. dissertation, Universitaet des Saarlandes, Saarbruecken, Germany, 2003.
12. *Ullmann's Encyclopedia of Industrial Chemistry*, 5th ed., Vol. B8. VCH, Weinheim, Germany, 1995, pp. 302–304.
13. Yamada, H., Kobayashi, M. (1996). Nitrile hydratase and its application to industrial production of acryamide. *Biosci. Biotechnol. Biochem.*, 1996, 60, 1391–1400.

PART III

FROM THE FLASK TO THE PLANT: DESIGNING GREENER, SAFER, MORE SUSTAINABLE MANUFACTURING PROCESSES

9

MASS AND ENERGY BALANCES

What This Chapter Is About In this chapter we cover mass and energy balances in the context of green chemistry and green engineering. Mass and energy balances are indispensable tools when estimating the greenness of chemical processes, as green chemistry and green engineering metrics cannot be calculated easily without a good understanding of the inputs and outputs of mass and energy for a given process. In this chapter we describe how to estimate mass and energy balances and how to apply them to identify and quantify inputs, by-products, emissions, and energy requirements for a given process, to evaluate its greenness and to identify opportunities for improvement.

Learning Objectives At the end of this chapter, the student will be able to:

- Understand the concepts of conservation of mass and energy in the context of designing more sustainable processes.
- Understand the importance of calculating mass and energy balances in the context of green engineering.
- Understand the relationship between mass balances, energy balances, and green chemistry and green engineering metrics.
- Understand the differences among batch, semicontinuous, and continuous processes and to understand the concept of steady-state and transient conditions.
- Given a process description for a single or multiple-unit operation process:
  - o Draw and understand process flow diagrams.

Green Chemistry and Engineering: A Practical Design Approach, By Concepción Jiménez-González and David J. C. Constable

  - o Understand and apply such concepts as recycle loop, conversion, yield, limiting reactant, percentage excess, and system boundary while setting mass and energy balance equations.
  - o Understand and apply the concepts of internal energy, enthalpy, heat capacity, sensible heat, and heat of reaction.
  - o Calculate mass and energy balances and apply them in the identification of improvement opportunities for the design of greener processes.
- Analyze and evaluate the results of mass and energy balances to develop potential process alternatives based on green chemistry and green engineering principles and metrics.

9.1 WHY WE NEED MASS BALANCES, ENERGY BALANCES, AND PROCESS FLOW DIAGRAMS

In Chapter 4 we explored metrics that helped to answer the question: How green is your process? In Chapters 5 through 8 we covered some of the strategies used to design more sustainable chemistries, such as material selection to promote the overall greenness of a chemical synthesis without diminishing the efficiency of the chemistry, route design, reaction conditions that would enable a greener chemistry, and other means of producing commodity (large volumes) chemicals such as fermentation and bioprocesses. But how do we know if these strategies are working? How do we measure and apply those metrics accurately beyond a single chemical step and in the context of an entire manufacturing production process?

Mass balances, energy balances, and process flow diagrams are indispensable tools to enable us to:

- Understand the:
  - o Process design
  - o Interrelationships among different parts of a process so we may address any issues arising from them
  - o Potential safety issues in a process so we may take action to minimize risks
- Estimate materials and energy flows, including:
  - o Resource consumption
  - o Energy requirements
  - o Raw materials
  - o Emissions and waste streams
- Plan for:
  - o Scale-up
  - o Procurement, inventories, and supply schemes
- Understand the need for material storage and transfers
- Estimate the relevant green metrics

The concept of mass and energy balances is pivotal in the understanding and practice of chemical and other types of engineering. Many good books devoted to these topics go into extensive detail on how to construct process flow diagrams, calculate mass and energy balances, and apply them to process design (the very good book by Felder and Rousseau[1] is highly recommended). This chapter is not intended to be a comprehensive review of mass and energy balances but, rather, a review of general concepts and principles to be used in the practical design of greener processes and analysis of current methods to identify improved opportunities from a sustainable engineering and chemistry viewpoint. One way to think of this is to imagine a very simplified green chemistry/green engineering step-by-step procedure to design, evaluate, and improve manufacturing processes:

1. Understand the desired output or function (i.e., what we really need).
2. Design a manufacturing process (this might or might not include chemical reactions).
3. Draw a process flow diagram.
4. Calculate the mass balance.
5. Calculate the energy balance.
6. Estimate life cycle impacts.
7. Estimate costs.
8. Identify green chemistry and green engineering issues.
9. Design and propose green chemistry and green engineering alternatives.
10. Return to step 1.

This is, of course, an oversimplified view, but it helps to illustrate just how pivotal flow diagrams and mass and energy balances are. Without steps 3 to 5, we cannot effectively and systematically apply green engineering and green chemistry concepts to process design.

9.2 TYPES OF PROCESSES

An important understanding to have prior to performing mass and energy balances is how processes are classified based on the type of operation that is being run. For example, processes can be run continuously, in semibatch, or batch mode. In a *batch process*, the materials are charged into a unit operation (i.e., a reactor, a filter press, a dryer, etc.), and after a period of time the product is removed. No material crosses the system or process boundary during that time. Routine examples of this type of operation may be found in the pharmaceutical industry, where chemical reactions to produce pharmaceutical intermediates on a small scale use reactors that are charged with materials, heated, and agitated for a period of time, after which the reaction mixture is withdrawn and sent to the next unit operation.

In a *continuous process*, the input materials, products, and all outputs flow continuously in and out of the process, crossing system boundaries. One example of this is a refinery process for cracking naphtha, a petroleum distillate. The crude distillate is fed continuously into a fractionation column, and the different cuts of the refined products and resulting waste are pumped continuously out of the column.

A *semibatch process* is a process that does not fall into either of these categories. One example of this type would be a fermentation process in which the input materials are charged continuously into a tank without removing any product for a long period of time.

Processes can also be run at steady state or in a transient state. If all the variables in a process (e.g., pressure, temperature, concentration) remain constant for the duration of the process, except for normal fluctuations around a mean value, the process is said to be in *steady state*. If any variable changes with time, the process is in a *transient state*. Continuous process are usually run as close to steady state as possible.

9.3 PROCESS FLOW DIAGRAMS

A process flow diagram (PFD) is a tool to help organize and visualize information about a manufacturing process. When a process is described, be it a chemical process or other type of manufacturing process, it is normal to find a paragraph that looks something like this:

> 3-Pentanone can be produced by catalytic ketonization of propionic acid over a thorium oxide or zirconium oxide catalyst at 350–380 °C. Hydroformylation of ethylene in the presence of cobalt carbonyl complexes at 100 °C is a further route. Propionic acid is preheated to 300 °C and then fed into a fixed bed reactor. The reactor is operated at 370 °C and 1 atm. Nitrogen is fed into the reactor as a blanket. The product mixture from the reactor is condensed and then pumped to a liquid/gas separator, where nitrogen and carbon dioxide are released to atmosphere. The product mixture is then pumped into a decanter where 3-pentanone is separated from the aqueous phase. Because 3-pentanone is slightly soluble in water, the purity of the product obtained from the decanter is about 91.6%. The product is further purified in a vacuum distillation column at 0.0482 atm and 25 °C.[2,3]

To efficiently determine something about a process such as mass efficiency, we need to know the type of unit operations (i.e., those unit operations that might need heating or cooling, the potential waste streams, etc.), and it is useful to organize this information in the form of a diagram. A *process flow diagram* is a diagram that uses boxes or specific symbols to represent the unit operations in a process (i.e., reactors, columns, etc.) and arrows to represent material flows. When a process flow diagram uses only boxes to represent the operations, it is normally called a *block diagram*.

Example 9.1 Draw a process flow diagram of the process described above for 3-pentanone.

Solution Figure 9.1(a) is a block diagram describing the unit operations, and Figure 9.1(b) is a process flow diagram that uses a set of specific symbols to denote the unit operations in the process and provides more detail on the streams.

Additional Points to Ponder Looking at the PFD for the 3-pentanone example above, why do you think these sorts of diagrams are relevant from a green engineering standpoint? Three streams are labeled as waste in the PFD. What types of materials could be in those streams?

Figure 9.2 shows common symbols used to denote unit operations, connectors, pumps, and so on which can be useful to construct and understand process flow diagrams. Most professional PFDs use these special symbols to denote the unit operations; however, simple block diagrams are very powerful, simple tools used to represent the process. There are numerous software applications that can be used to generate a PFD using more-or-less standard symbols, but the most important part of drawing a PFD is to gain an understanding of the process, the flow of the process and the resulting waste streams, and the potential improvements from a green engineering perspective.

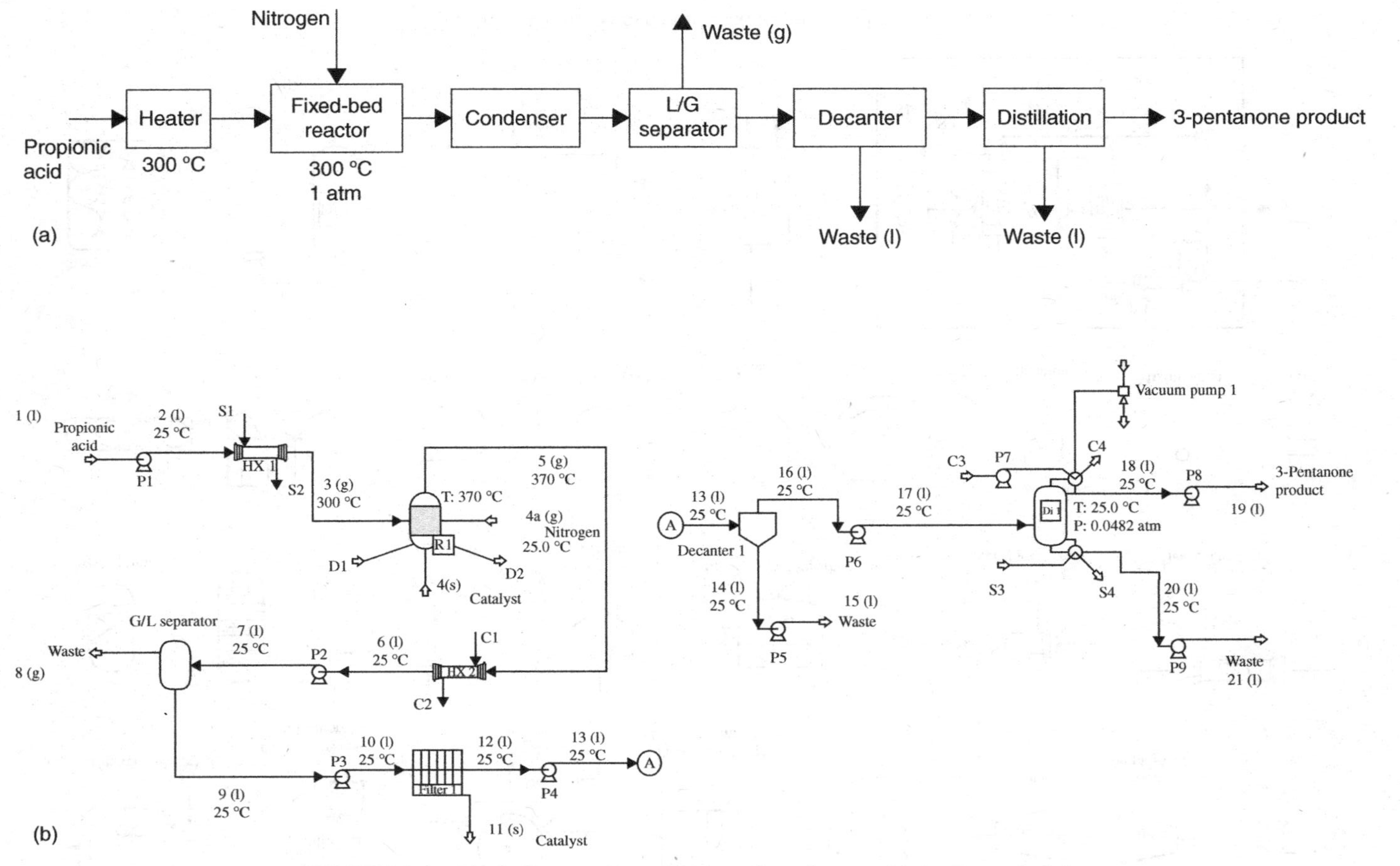

FIGURE 9.1 Block diagram (a) and process flow diagram (b) for Example 9.1.

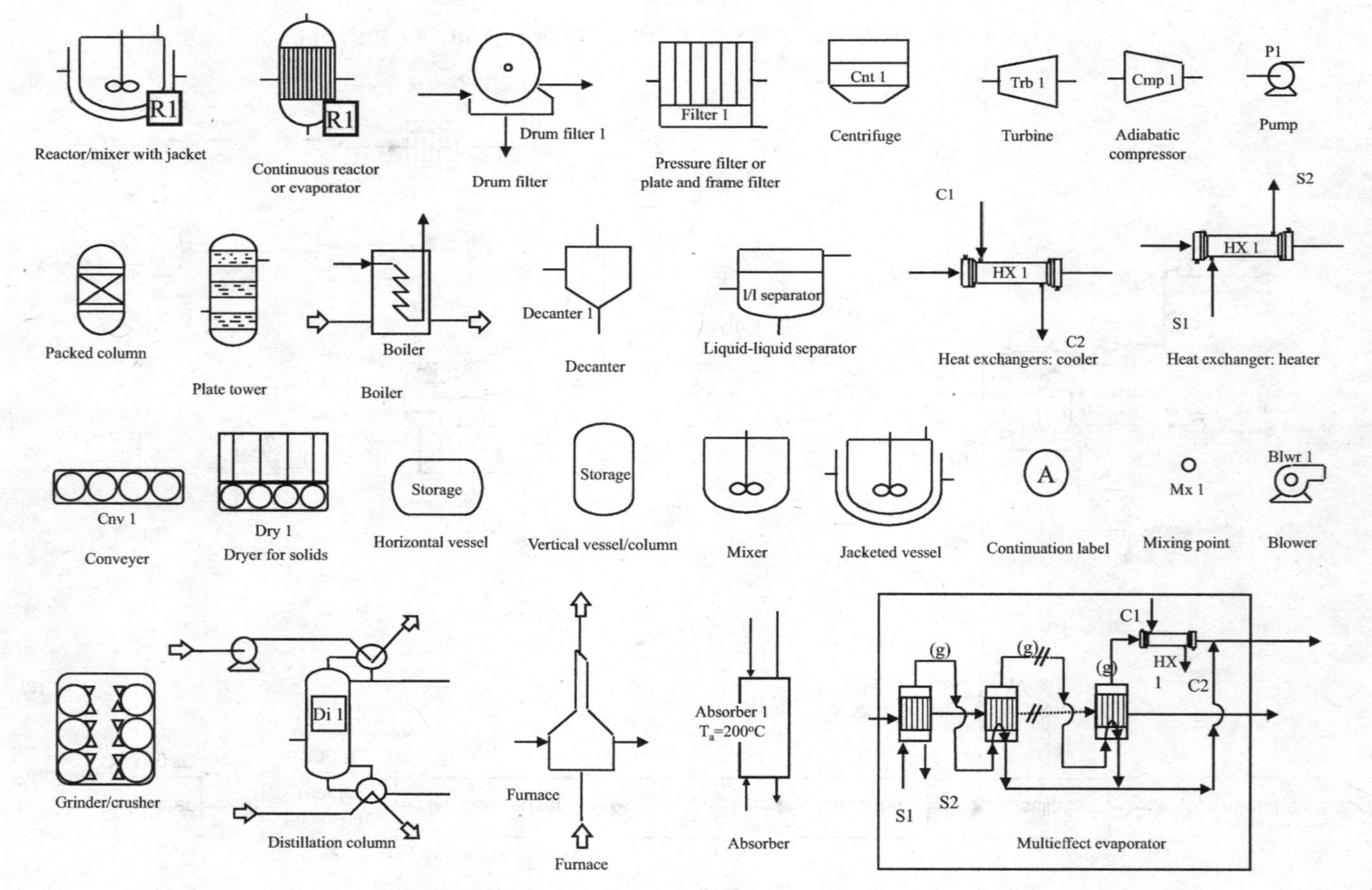

FIGURE 9.2 Common symbols used in process flow diagrams.

9.4 MASS BALANCES

The concept of *conservation of mass and energy* (or mass–energy equivalency), a well-known corollary from Einstein's theory of special relativity, expresses in short that the total amount of mass and that energy in the universe is constant, and that mass and energy can interconvert. Since mass–energy conversion is negligible in nonnuclear reactions, conservation of the atoms in nonnuclear chemistry for practical purposes means that there is mass conservation within the limits of our ability to measure it.

Mass and energy balances are powerful tools that help us understand manufacturing processes and identify opportunities for improvement from a green engineering viewpoint, as we can identify where we are losing efficiencies, how we can improve the use of resources, and where to improve our processes. To evaluate a manufacturing or chemical process from a green engineering perspective, and to calculate the green metrics of this process, the first thing to understand is where all the inputs and outputs of our processes are, and which add value and which can be eliminated or integrated. We can perform balances of any conserved quantity, such as total mass, mass of a particular substance, energy, and momentum. This can be done in a specific system such as an entire process, a given unit operation, a subpart of a process, or several processes linked together (such as an entire supply chain). To do this, we can use a general balance equation

$$\sum \text{inputs} + \sum \text{generation} - \sum \text{outputs} - \sum \text{consumption} = \text{accumulation} \qquad (9.1)$$

For example, generation and consumption can result from chemical reactions. For processes that are run in steady state, the equation is simplified, as accumulation can be set equal to zero.

There are two types of mass balances. *Differential mass balances* are applied to an instant of time and each term of the equation is a rate (mass/time, moles/time, people/year). Differential mass balances are widely used for continuous systems. *Integral mass balances* are normally carried out for an interval of time, and each term of the equation is an amount (mass, moles, people). Integral mass balances are used widely in batch and semibatch processes.

Example 9.2 A cleaning supplies manufacturer needs to produce 100 kg of a solution of sodium hypochlorite at 10% for a residential customer. He has two solutions in stock, one of 5% and one of 30%.

(a) To solve this problem, should you use a differential or an integral mass balance?
(b) How much of the two existing solutions would the manufacturer have to mix to produce the required 100 kg of 5% sodium hypochlorite solution?

Solution

(a) This is a mixing problem, probably carried out as a batch process, as the final product is expressed in kilograms. Therefore, we can use an integral mass balance, with the time interval between the time in which we input the materials and the time we extract the product. For semibatch and continuous processes the systematic procedure for writing an integral mass balance is to write a differential mass balance and then to integrate it between the time boundaries. In many cases (especially for

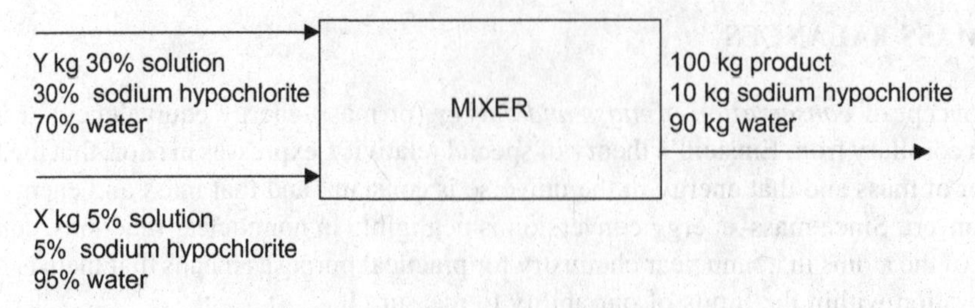

FIGURE 9.3 Block diagram for Example 9.2.

transient-state systems) the calculations can be rather complicated. However, for this batch process and some semibatch processes, the calculations can be straightforward.

(b) We can represent this process in a block flow diagram as in Figure 9.3.

Since there is no chemical reaction, the generation and consumption terms of the general equation disappear. We can assume that we remove all the product once it is mixed; therefore, equation (9.1) becomes

$$\sum \text{inputs} - \sum \text{outputs} = 0$$

If we perform a mass balance of sodium hypochlorite, we have

$$\sum \text{inputs} - \sum \text{outputs} = 0 \Rightarrow \text{NaClO}_{\text{in}} = \text{NaClO}_{\text{out}}$$

$$0.30Y\,\text{kg} + 0.05X\,\text{kg} = 10\,\text{kg}$$

which gives us one equation and two missing variables. Therefore, we need to do a global mass balance:

$$\sum \text{inputs} - \sum \text{outputs} = 0 \Rightarrow \text{total mass}_{\text{in}} = \text{total mass}_{\text{out}}$$

$$Y\,\text{kg} + X\,\text{kg} = 100\,\text{kg} \Rightarrow X = 100 - Y$$

$$0.30Y\,\text{kg} + 0.05(100 - Y)\text{kg} = 10\,\text{kg}$$

$$Y = \frac{10-5}{0.30-0.05} = \frac{5}{0.25} = 20\,\text{kg of 30\% NaClO solution}$$

$$X = 100 - Y = 100 - 20 = 80\,\text{kg of 30\% NaClO solution}$$

Closing the mass balance for sodium hypochlorite yields

$$0.30Y\,\text{kg} + 0.05X\,\text{kg} = 10\,\text{kg NaClO}$$

$$(0.30)(20)\,\text{kg} + (0.05)(80)\,\text{kg} = 10\,\text{kg NaClO}$$

$$6\,\text{kg} + 4\,\text{kg} = 10\,\text{kg NaClO}$$

$$10\,\text{kg NaClO} = 10\,\text{kg NaClO}$$

Additional Points to Ponder The calculations are done assuming perfect use of resources. What could go wrong that might make the manufacturer waste raw materials and need more? How about setup and cleanup waste?

Mass balances can be set for any defined system. A system can be defined at the molecular level, it can consist of a single step, a single unit operation, a few unit operations producing

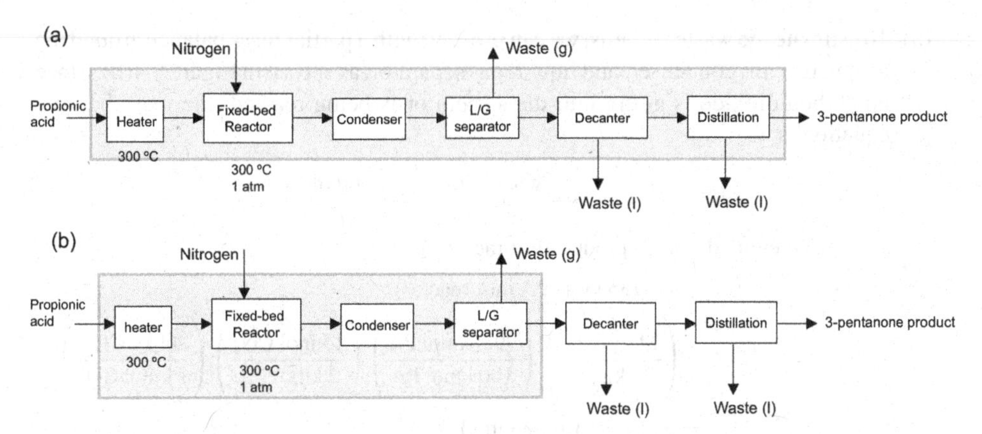

FIGURE 9.4 Two different system boundaries for the same 3-pentanone process.

intermediates, a full manufacturing process, an entire supply chain across countries; and so on. It all depends on what type of information is needed and what information is available at a certain time. It is the engineer who defines the system boundaries based on the objectives of the mass balance.

It is important to define the boundaries of the system since they define not only the system under study but also make it easy to see the inputs and outputs. Take the example of the 3-pentanone process flow diagram we discussed above. If the system is defined as the entire process and the system boundaries are drawn to illustrate that [Figure 9.4(a)], it is easy to see that there are two inputs: propionic acid and nitrogen, and four outputs: the 3-pentanone product, two liquid waste streams, and one gaseous waste stream. If for some reason the system needs to be defined as a subset of the process, such as in Figure 9.4(b), there will still be two inputs, the propionic acid stream and the nitrogen stream, and two outputs, one gaseous waste stream and one liquid process stream. Applying the concept of system boundaries is indispensable to estimating mass balances.

Example 9.3 The 3-pentanone process described in Example 9.1 has the following main reaction:

2 O OH → O + CO_2 + H_2O

Assuming no side reactions and a continuous process in steady state with a 99.5% conversion and 98.8% overall yield:

(c) How much carbon dioxide emission would be generated by the reaction?

(d) How much propionic acid do we need to purchase?

(e) How much propionic acid is wasted to produce 1000 kg/h of 3-pentanone with a purity of 99.6%?

Solution For the sake of simplicity, we can use 3P for 3-pentanone and PA for propionic acid.

(a) To estimate the waste streams, we can proceed with a partial mass balance around the heater, reactor, condenser, and liquid–gas separator, as shown in Figure 9.4(b). Since no carbon dioxide is going into the system or is being reacted, the mass balance equation is

$$\sum \text{generation} - \sum \text{outputs} = 0$$

$$\begin{aligned} CO_2 \text{ emitted} &= CO_2 \text{ produced in the reaction} \\ &\quad (1,735 \text{ kg/h PA into reactor}) \\ &\quad \left(\frac{1 \text{ kmol PA}}{74 \text{ kg PA}}\right)\left(\frac{99.5 \text{ kmol PA}}{100 \text{ kmol PA}}\right)\left(\frac{1 \text{ kmol } CO_2}{2 \text{ kmol PA}}\right)\left(\frac{44 \text{ kg } CO_2}{1 \text{ kmol } CO_2}\right) \\ &= 513 \text{ kg/h } CO_2 \text{ emitted} \end{aligned}$$

(b) For this example, we can start doing a mass balance around the entire process as it appears in Figure 9.4(a). To calculate how much propionic acid we need to purchase, we can assume 100% pure propionic acid for the moment and estimate that on the basis of the overall yield, since *yield* is defined as the ratio of moles of product out per mole of reactant. For that, we have

$$\text{yield} = \frac{\text{moles of product out}_{\text{out}}}{\text{moles of limiting reactant}_{\text{in}}} = \frac{\text{moles of 3P}_{\text{out}}}{\text{moles of PA}_{\text{in}}}$$

$$\begin{aligned} \text{PA required} &= 996 \text{ kg/h 3P in product} \left(\frac{1 \text{ kmol 3P}}{86 \text{ kg 3P}}\right)\left(\frac{2 \text{ kmol PA}}{1 \text{ kmol 3P}}\right)\left(\frac{1}{0.988}\right)\left(\frac{74 \text{ kg PA}}{1 \text{ kmol PA}}\right) \\ &= 1,735 \text{ kg/h propionic acid} \end{aligned}$$

(c) To calculate the amount of propionic acid wasted, we can perform a balance of propionic acid. Since this is a steady-state process there is no accumulation; since no propionic acid is being produced, there is no generation, so for propionic acid, Equation (9.1) becomes

$$\sum \text{inputs} - \sum \text{outputs} - \sum \text{consumption} = 0$$

$$\begin{aligned} \text{PA}_{\text{out}} &= \text{PA}_{\text{in}} - \text{PA}_{\text{reacted}} \\ &= 1735 \text{ kg/h PA} - (1735 \text{ kg/h PA})(0.995) \\ &= 1735 \text{ kg/h PA} - 1726.33 \text{ kg/h PA} = 8.67 \text{ kg/h PA} \end{aligned}$$

Additional Points to Ponder What are the other constituents of the waste streams? What is the probable fate of these streams? What is the global warming potential of the reaction based on the carbon dioxide produced? Is that the only source of greenhouse gases in the process?

These concepts are nothing new to chemical engineers and chemists, and right now there are many software applications that can facilitate the estimation and calculation of mass balances. However, in the context of green chemistry and green engineering, applying the concept of mass balances is indispensable to measuring the greenness of a process. As noted in Chapter 4, a series of metrics is used to measure how green a process or a chemistry route is. Mass-based metrics are very important, as they describe, for example, how efficient a process is in converting the raw materials into usable product and they also tell us how much of a specific hazardous material that process is emitting.

Example 9.4 Without accounting for the nitrogen needed for blanketing, what is the mass efficiency (mass productivity) of the 3-pentanone process described in Example 9.3?

Solution As seen in Chapter 4, *mass efficiency* is the ratio of the mass of all the products divided by the summation of the mass of all the inputs, expressed as a percentage. In other words, it is the mass percent of the inputs that a process is converting to sellable product. Since for this example we are not accounting for the nitrogen needed to blanket the reactor, we have

$$\text{mass efficiency} = \frac{\sum \text{mass of products}}{\sum \text{mass of inputs}} = \frac{1000\ \text{kg/h 3-pentanone product}}{1735\ \text{kg/h propionic acid}} = 57.5\%$$

That means that only about 58% of propionic acid is converted into useful product.

Additional Points to Ponder Why is mass efficiency so different from yield? How much lower would the mass efficiency be if we accounted for the nitrogen input? Could you devise changes that might make this process more efficient?

In the 3-pentanone example, we had a single reactant that condensed to form the main product and two by-products. However, this is a very simple process to illustrate mass balances. Most industrial processes are considerably more complicated and have several reactants and a myriad of side or competing reactions, and the process streams include purges and recycling loops. For example, as we saw in Chapter 1, it is very rare for the chemical reaction $A + B \rightarrow C$ to proceed with A and B fed in exact stoichiometric ratios without by-products and 100% conversion and yield. In the $A + B \rightarrow C$ example, A can be fed in less than its stoichiometric ratio; therefore, the reaction will stop once all A is consumed, leaving some B unreacted. A would then be a limiting reactant, which is a reactant that is present in less than its stoichiometric proportions compared to all the other reactants. We very frequently encounter the fact that the conversion for our main reaction is less than 100% and in addition to that, the $A + B \rightarrow C$ has competing reactions $2A \rightarrow D$ and $D + C \rightarrow E$. Therefore, we almost always end up with unreacted species and undesired by-products mixed with our valuable product.

In Chapters 5 through 8 we saw several strategies for using green chemistry concepts to maximize efficiencies of the main reaction and to minimize competing reactions and by-products; however, a perfect reaction might not be achieved. Despite the fact that a perfect reaction might elude us, we may, nonetheless, frequently design a way to separate the unreacted species and by-products from the reaction mixture and return them to the reactor to be turned into product. A stream carrying separated reactants back to the reactor is called a *recirculation* or *recycle loop*. It is especially useful when the conversion of a reactant is too low, as it is a way to increase the mass efficiency of a process with the same level of per-pass conversion. Observe that even though a large amount of material might be circulating within a system, this does not affect accumulation as long as the process is in steady state. That means that if 100 kg of material is entering the system each hour, an equal amount of material will leave the system each hour, even if there is 200 kg/h within the system. At the end of the day, there are several reasons to use recycle loops in a chemical process, and all of them converge toward more efficient process design:

- *Low per-pass conversion.* In this case the recycle loop will help to increase the mass efficiency for a low once-through conversion.
- *Catalyst recovery.* (especially in the case of a homogeneous catalyst). Catalysts could be a big component of the cost, and their production can result in large life cycle impacts, so recovery and recycling lower the overall impacts.
- *Circulation of working fluid.* (such as heating fluid or refrigerant). Instead of using once-through systems, this reduces the net consumption of resources.
- *Dilution.* When a reaction needs to be diluted (e.g., when reducing the concentration of reactants to control exotherms in a process), instead of using fresh water or solvent, the wastewater or spent solvent can be recovered and recycled.

Example 9.5 Hypochlorous acid is a highly reactive and relatively unstable compound in solution and in the gas phase. High concentrations of virtually chloride-free solutions of HOCl are obtained by reaction of atomized caustic with excess gaseous chlorine at temperatures above the dew point of the system. The reactor gases, containing Cl_2, Cl_2O, HOCl, and H_2O, are condensed to provide HOCl solutions of up to 60% concentration in high yield. The noncondensed gases are recycled to the reactor.[4]

Primary reaction:

$$\underset{\text{chlorine}}{Cl_2} + \underset{\text{sodium hydroxide}}{NaOH} \rightarrow \underset{\text{hypochlorous acid}}{HOCl} + \underset{\text{sodium chloride}}{NaCl} \qquad (1)$$

Side reaction:

$$\underset{\text{chlorine}}{3Cl_2} + \underset{\text{sodium hydroxide}}{6NaOH} \rightarrow \underset{\text{sodium chlorate}}{NaClO_3} + \underset{\text{Sodium chloride}}{5NaCl} + \underset{\text{Water}}{3H_2O} \qquad (2)$$

$$\underset{\text{chlorine}}{2Cl_2} + \underset{\text{sodium hydroxide}}{2NaOH} \rightarrow \underset{\text{dichlorine monoxide}}{Cl_2O} + \underset{\text{sodium chloride}}{2NaCl} + \underset{\text{water}}{H_2O} \qquad (3)$$

A sodium hydroxide solution of 40 wt% is fed continuously to an atomizer located in the upper portion of a cylindrical reactor. The atomizer sprays droplets of sodium hydroxide solution into a gaseous mixture of chlorine maintained at a temperature of 105°C. The molar ratio of chlorine to sodium hydroxide in the reaction mixture is maintained at 30 : 1. The product mixtures are fed to a cyclone, where sodium chloride is removed continuously. The vapor mixture passes through a refrigeration unit where the temperature is cooled to 5 °C. Hypochlorous acid and water are condensed after the refrigeration unit, and the vapor phase is separated from the liquid phase. The vapor phase has a composition of 20.7 vol% chlorine monoxide, 78.9 vol% chlorine, and 0.4 vol% water vapor. This stream is heated to 105 °C and then recycled to the reactor.[5] Assuming a steady-state process:

(a) Draw a process flow diagram following the description above.
(b) Assuming a yield of hypochlorous acid of 79.14% from sodium hydroxide, how much sodium hydroxide 40% aqueous solution is needed to produce 1000 kg/h of 40.2% hypochlorous acid solution?
(c) Assuming that all the sodium hydroxide is used in the reactor and that the 79.23% of sodium hydroxide is used in the main reaction (1), 20.04% in side reaction (2) and the rest in side reaction (3), how much sodium chlorate and sodium chloride are produced as by-products of the process?
(d) If sodium hydroxide is fed to the process with 9.7% molar excess with respect to the fresh chlorine, what mass of the recirculation stream is needed to maintain the 30 : 1 molar ratio as described above.

Solution

(a) A representation of the flow diagram is shown in Figure 9.5, including the figures from the overall mass balance.

(b) To estimate the amount of sodium hydroxide (NaOH) required to produce 1000 kg of 40.2% pure hypochlorous acid (HA), we have

$$\text{yield} = \frac{\text{moles of product out}_{\text{out}}}{\text{moles of reactant}_{\text{in}}} = \frac{\text{moles of HA}_{\text{out}}}{\text{moles of NaOH}_{\text{in}}}$$

$$\text{moles of NaOH}_{\text{in}} = \frac{\text{moles of HA}_{\text{out}}}{\text{yield}}$$

$$\begin{aligned}\text{NaOH}_{\text{in}} &= 402\,\text{kg/h HA}\left(\frac{1\,\text{mol HA}}{52.5\,\text{kg HA}}\right)\left(\frac{1\,\text{mol NaOH}}{1\,\text{mol HA}}\right)\left(\frac{1}{0.7914}\right)\left(\frac{40\,\text{kg NaOH}}{1\,\text{mol NaOH}}\right)\\ &= 387\,\text{kg/h NaOH}\left(\frac{100\,\text{kg solution}}{40\,\text{kg NaOH}}\right)\\ &= 967\,\text{kg/h of 40\% NaOH solution}\end{aligned}$$

which is equal to the amount of NaOH consumed. Since the system is in steady state, there is no generation of NaOH and we assume that no NaOH leaves the system:

$$\sum \text{inputs} - \sum \text{consumption} = 0$$

(c) Given that we are talking about a steady-state process, that no sodium chloride enters the processn or reacts, the general mass balance equation becomes

$$\sum \text{generation} - \sum \text{outputs} = 0$$

$$\begin{aligned}\text{NaCl}_{\text{out}} &= \text{NaCl}_{\text{produced}}\\ &= 387\,\text{kg/h NaOH}\left(\frac{1\,\text{mol NaOH}}{40\,\text{kg NaOH}}\right)\left[(0.7923)\left(\frac{1\,\text{mol NaCl}}{1\,\text{mol NaOH}}\right)\right.\\ &\quad \left. + (0.2004)\left(\frac{5\,\text{mol NaCl}}{6\,\text{mol NaOH}}\right) + (0.0073)\left(\frac{2\,\text{mol NaCl}}{2\,\text{mol NaOH}}\right)\right]\left(\frac{58.5\,\text{kg NaCl}}{1\,\text{mol NaCl}}\right)\\ &= 547\,\text{kg/h of NaCl}\end{aligned}$$

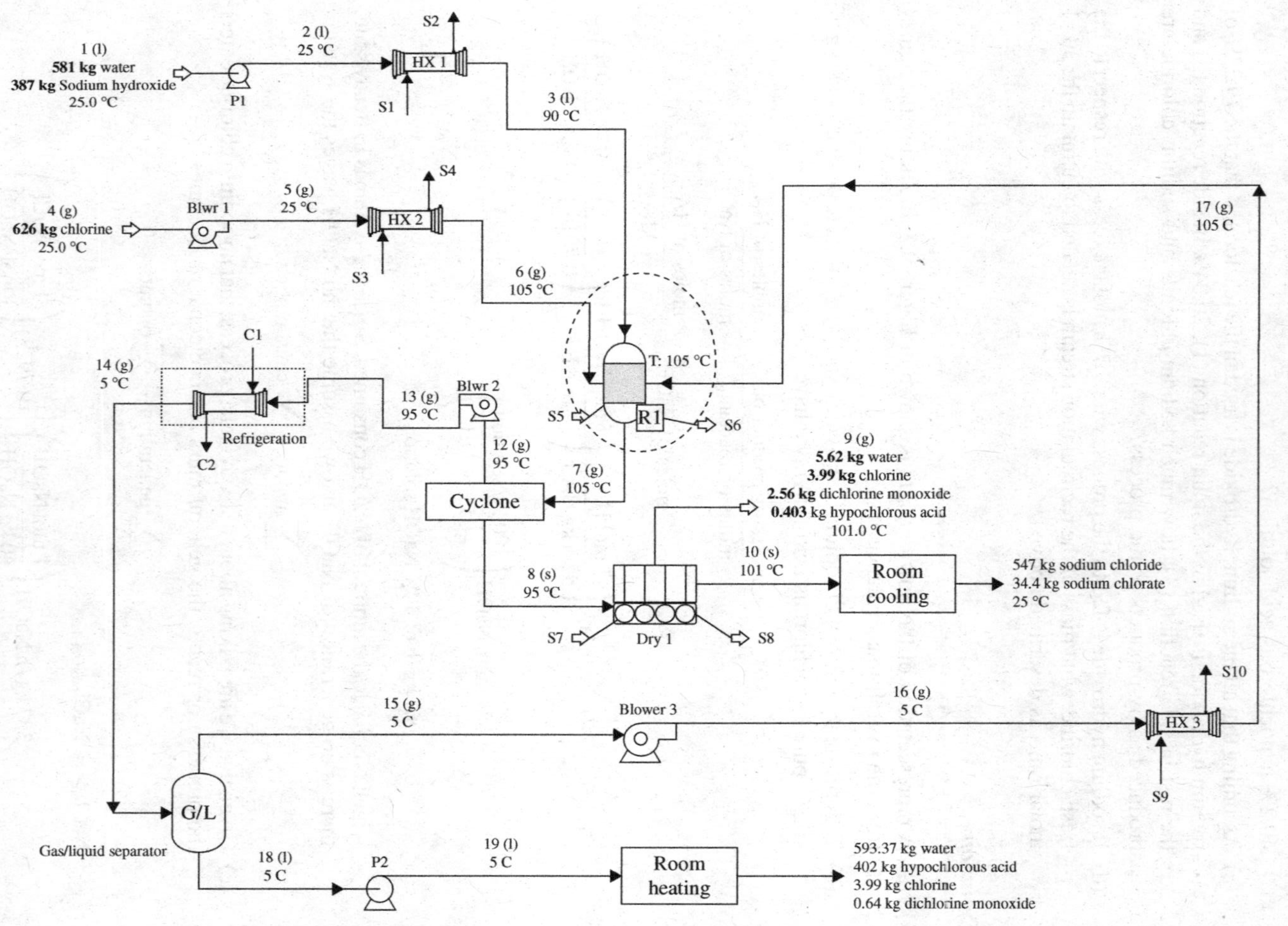

FIGURE 9.5 Process flow diagram for hypochlorous acid, Example 9.5.

(d) For this we need to perform a mass balance around the reactor. The boundaries of the system are marked by the dashed line in Figure 9.5. We can do a chlorine mass balance around the reactor, which gives us

$$\sum \text{inputs} - \sum \text{outputs} - \sum \text{consumption} = 0$$

$$\begin{aligned}
Cl_{2_{in}} - Cl_{2_{out}} &= Cl_{2_{consumed}} \\
Cl_{2_{in}} &= Cl_{2_{in}}|_{\text{raw material}} + Cl_{2_{in}}|_{\text{recirculation}} \\
Cl_{2_{in}}|_{\text{recirculation}} &= Cl_{2_{in}} - Cl_{2_{in}}|_{\text{raw material}} = 30\,\dot{m}_{NaOH} - \frac{\dot{m}_{NaOH}}{1.097} = 29.09\,\dot{m}_{NaOH}
\end{aligned}$$

$\dot{m}$ denotes molar flow (moles per unit time)

$$= 29.09\left[387\ \text{kg/h NaOH}\left(\frac{1\ \text{mol NaOH}}{40\ \text{kg NaOH}}\right)\right]$$

$$= 281.4\ \text{mol/h}\ Cl_2\left(\frac{71\ \text{kg}\ Cl_2}{1\ \text{mol}\ Cl_2}\right) = 19,981\ \text{kg/h}\ Cl_2$$

The recirculation stream has 78.9 vol% of chlorine, which for gases is equivalent to molar percent. We have

$$\begin{aligned}
\text{Recirculation} &= \frac{Cl_2|_{\text{recirculation}}}{0.789} = \frac{281.4\ \text{mol/h}\ Cl_2}{0.789} = 356.7\ \text{mol/h} \\
&= ClO|_{\text{recirculation}} + Cl_2|_{\text{recirculation}} + H_2O|_{\text{recirculation}} \\
&= 356.7\ \text{mol/h}\ (0.207)\left(\frac{86.9\ \text{kg}\ Cl_2O}{1\ \text{mol}\ Cl_2O}\right) + 19,981\ \text{kg/h}\ Cl_2 \\
&\quad + 356.7\ \text{mol/h}\ (0.004)\left(\frac{18\ \text{kg}\ H_2O}{1\ \text{mol}\ H_2O}\right) \\
&= 6\,416\ \text{kg/h}\ Cl_2O + 19,981\ \text{kg/h}\ Cl_2 + 26\ \text{kg/h}\ H_2O = 26,423\ \text{kg/h}
\end{aligned}$$

Additional Points to Ponder What would be the increase in mass efficiency for this process if a recycling loop were added? What are the waste streams of this process? What is the probable fate of the sodium chlorate/sodium chloride by-product stream? Is there a way to avoid this waste?

We need to keep in mind that the examples that are presented in textbooks tend to be perfect and have exact closure of the mass balance equation (9.1); however, industrial application of these concepts is less than perfect. In industry, some mass balances do not close perfectly, due to such things as measurement errors, human errors, impurities of the feed, incorrect assumptions of steady state, and approximations used during design, among others. However, the use of mass balance is an indispensable and very powerful tool when designing and analyzing production processes.

As hopefully is evident from a consideration of the additional questions in Examples 9.1 to 9.5, there is a direct application of process flow diagrams and mass balances into green engineering. Both process flow diagrams and mass balances help us to identify potential opportunities for improvement from the green engineering viewpoint. Some of the areas in which mass balances are applied directly in the practice of green engineering include:

- *Understanding the process and the flow of the various streams.* What types of separations are needed to purify the products desired? Are there streams that can be combined? Does it make sense to change the order of the separations?
- *Understanding the inputs and outputs of the process.* How much raw material is needed per unit of product (mass intensity, mass efficiency)? What are the waste streams? How much waste is being produced?
- *Identify the impacts on efficiency from the competing reactions and from the reactions desired.* How selective is this reaction? How does that translate in terms of tons of waste generated? Are any other materials being generated in the process?
- *Identify hazardous materials.* Are any toxic materials being produced? Is there a chance that any of them would get into the environment? How can that be avoided? Do we need to add any particular waste?
- *Identify the use of additional materials.* Are we using any unnecessary material in the process? Are we using any mass-separating agent that we can eliminate or change for a less toxic agent? Can we eliminate the use of solvents? If not, is it possible to recover them and recycle them into the process?

9.5 ENERGY BALANCES

Energy is not only expensive but its production invariably generates emissions and therefore environmental impacts. For example, the production of steam to heat industrial processes is achieved by combusting fossil fuels in boilers; in some cases the electricity required to power pumps and blowers requires the combustion of coal, or fuels, or nuclear reactions; refrigeration requires a large amount of electricity and generally requires the use of hazardous chemicals that may leak into the environment.

The first step in identifying energy optimization opportunities is to measure the energy requirements of a process. Energy balances can help us to:

- Identify and quantify:
  - o Minimum heating, cooling, and electricity requirements
  - o Potential-energy losses
  - o Opportunities for energy optimization
  - o Impact of energy use
- Compare and chose separation methods
- Compare different synthetic pathways
- Define plant layout

Similar to mass balances, the principle that governs energy balance is the *law of conservation of energy*, also known as the *first law of thermodynamics*. In general terms, the law states that the rate at which energy (kinetic + potential + internal) enters a system, plus the rate at which energy enters as heat, minus the rate at which it leaves the system and the rate at which it leaves the system as work, equals the rate of accumulation of energy in the system. This gives us the expected general relationship

$$\sum \text{inputs} - \sum \text{outputs} = \text{accumulation} \tag{9.2}$$

As we studied earlier, mass–energy conversion is negligible in nonnuclear reactions; therefore, for practical purposes this means that there is energy conservation within the limits of our ability to measure it in industrial processes.

9.5.1 Energy Balances in Closed Systems

In general, the total energy of a system has three components:

1. *Kinetic energy.* E_k, due to the translational motion relative to a frame of reference
2. *Potential energy.* E_p, due to the position of the system in a potential field such as gravitational or magnetic fields
3. *Internal energy.* U, all the energy of the system other than kinetic or potential, due to molecular, atomic, and subatomic motion, vibrations, and interactions.

In a closed system there is no mass transfer across the boundaries; however, energy can be transferred between a system and its surroundings by means of:

- *Heat.* Q, energy flow resulting from temperature differences. Heat is conventionally defined as positive when it is transferred to the system from the environment.
- *Work.* W, energy resulting from any driving force other than temperature difference (e.g., pressure, voltage, torque). The convention for the sign of work is arbitrary and not consistent; in this book, work is defined as positive when it is done by the system on the surroundings.

Therefore, for energy balances we cannot eliminate the input–output part of the equation, as energy can be transferred to the system as heat or work. Accumulation of energy equals the final value of the energy minus the initial value of the energy:

$$\text{accumulation} = \sum \text{inputs} - \sum \text{outputs}$$

$$\text{energy}_{\text{final}} - \text{energy}_{\text{initial}} = \text{energy}_{\text{in}} - \text{energy}_{\text{out}}$$

$$\text{energy}_{\text{in}} - \text{energy}_{\text{out}} = \text{energy transferred} = Q - W$$

$$\text{energy}_{\text{final}} = U_f + E_{kf} + E_{pf}$$

$$\text{energy}_{\text{initial}} = U_i + E_{ki} + E_{pi}$$

$$(U_f - U_i) + (E_{kf} - E_{ki}) + (E_{pf} - E_{pi}) = Q - W$$

$$\Delta U + \Delta E_k + \Delta E_p = Q - W \tag{9.3}$$

9.5.2 Energy Balances in Open Steady-State Systems

Flow Work and Enthalpy In an open system the flow of material net work performed by the system on its surroundings consists of the shaft work, $\dot{W}_s$ (the work the system performs on a moving part, such as a turbine or piston) and the flow work, $\dot{W}_{\text{fl}}$ (the difference between the work done by the fluid at the outlet, minus the fluid work at the inlet). Therefore, we have

$$\dot{W} = \dot{W}_s + \dot{W}_{\text{fl}} \tag{9.4}$$

where flow work for a fluid with pressure P_{out} and specific volume $\hat{V}_{\text{out}}$ at the outlet and pressure P_{in} and specific volume $\hat{V}_{\text{in}}$ at the inlet is defined as

$$\dot{W}_{\text{fl}} = \dot{W}_{\text{fl-out}} - \dot{W}_{\text{fl-in}} = P_{\text{out}}\hat{V}_{\text{out}} - P_{\text{in}}\hat{V}_{\text{in}} \tag{9.5}$$

A very important property for open systems is *enthalpy* (*H*), a thermodynamic function equivalent to the sum of the internal energy of the system plus the flow work of the system (as seen before, the product of its volume multiplied by the pressure exerted on it by its surroundings). Therefore, the specific enthalpy of a system is traditionally defined as

$$\hat{H} \equiv \hat{U} + P\hat{V} \tag{9.6}$$

Since it is not possible to know the absolute value of the specific internal energy or enthalpy of a material or system, we quantify the change of enthalpy, ΔH, or internal energy, ΔU, given a change of state (temperature, pressure, phase), related to a reference state of temperature, pressure, and phase, which is by convention assigned a value of zero enthalpy or zero internal energy. When performing energy balances, we can arbitrarily chose a reference state; in many ways the right reference state could simplify the calculations, as the internal energy or enthalpy at that state would be set to zero. Since both enthalpy and internal energy are state properties (i.e., properties related to a specific state of temperature, pressure, and phase), the changes of enthalpy or internal energy (ΔH or ΔU) are independent of the path that the system takes to get from state 1 to state 2. Thermodynamic tables with enthalpies of materials at different temperatures and pressures are widely available. *Perry's Chemical Engineer's Handbook*[8] and NIST[6] are very good references for this purpose.

Energy Balance Equation for an Open Steady-State System For an open steady-state system, the general energy balance equation can be written as

$$\sum \text{inputs} = \sum \text{outputs} \qquad \dot{Q} + \sum E_{\text{in}} = \dot{W} + \sum E_{\text{out}}$$

where Q is the heat entering the system and W is the total work performed by the system. Since E is the summation of $U + E_k + E_p$ and $\dot{W} = \dot{W}_s + \dot{W}_{\text{fl}} = \dot{W}_s + \left(P_{\text{out}}\hat{V}_{\text{out}} - P_{\text{in}}\hat{V}_{\text{in}}\right)$, we can derive the energy balance equation for an open steady-state system as

$$\Delta\dot{H} + \Delta\dot{E}_k + \Delta\dot{E}_p = \dot{Q} - \dot{W}_s \tag{9.7}$$

Example 9.6 Data for water at 1 atm between 300 and 400 K are provided in Table 9.1.

TABLE 9.1 Data for Example 9.6

Temperature (K)	Phase	Specific Internal Energy (kJ/mol)	Specific Enthalpy (kJ/mol)	Specific Volume (l/mol)
300	Liquid	2.0277	2.0295	0.018078
350	Liquid	5.7961	5.7980	0.018501
373.12	Liquid	7.5475	7.5494	0.018798
373.12	Gas	45.146	48.200	30.143
400	Gas	45.898	49.187	32.463

Source: ref. 6.

(a) 200 g of water is contained in a closed vessel. Assuming that no heat is lost to the container or the surroundings, how much heat is necessary to raise the temperature from 300 to 350 K?

(b) Steam at 400 K and 1 atm pressure flows through a horizontal condenser/cooler at a rate of 1 kg/h. Assuming perfect isolation and no significant changes in linear velocity, how much heat needs to be removed from the steam to leave the condenser as water at 350 K?

Solution

(c) This is a closed system, and the changes in potential and kinetic energy are negligible, as we are just heating a closed vessel with water. There are no moving parts, so no work is performed. The energy balance equation can be written as

$$\Delta U = Q$$

Since no heat is lost to the surroundings, the heat needed to heat the water is equal to the change in internal energy:

$$Q = \Delta U = m\left(\hat{U}_{\text{final}} - \hat{U}_{\text{initial}}\right) = m\left(\hat{U}_{350\text{ K}} - \hat{U}_{300\text{ K}}\right)$$

$$= 200\text{ g}\left(\frac{1\text{ mol}}{18\text{ g}}\right)(6.7961 - 2.0277)\text{ kJ/mol} = 41.8\text{ kJ}$$

(d) As this is an open system and a horizontal condenser, we can assume no significant changes in potential energy. Since the linear velocity does not change significantly, we can neglect changes in kinetic energy, and since there are no moving parts that can produce work, there is no shaft work. We therefore have the following energy balance equation:

$$\Delta\dot{H} = \dot{Q}$$

Since no heat is lost to the surroundings, the heat that needs to be removed to cool and condense the steam is equal to the change in enthalpy $|\times|$ the mass involved. This is an interesting problem, as the material in question changes phase and we must therefore account for the latent heat associated with that phase change. We can write the energy balance in the following two forms:

$$\dot{Q} = \Delta \dot{H} = \Delta \dot{H}^{\text{gas, 373.12 K}}_{\text{gas, 400 K}} + \Delta \dot{H}^{\text{liquid, 373.12 K}}_{\text{gas, 373.12 K}} + \Delta \dot{H}^{\text{liquid, 350 K}}_{\text{liquid, 373.12 K}}$$

$$= m(\hat{H}_{373.12\text{ K},g} - \hat{H}_{400\text{ K},g}) + m(\hat{H}_{373.12\text{ K},l} - \hat{H}_{373.12\text{ K},g}) + m(\hat{H}_{350\text{ K},l} - \hat{H}_{373.12\text{ K},l})$$

$$= 1000\text{ g}\left(\frac{1\text{ mol}}{18\text{ g}}\right)(48.2 - 49.187) + (7.5494 - 48.2) + (6.798 - 7.5494)\text{kJ/mol}$$

$$= 2,410.5\text{ kJ}$$

but this can also be done directly by reading from the thermodynamic information as above:

$$\dot{Q} = m\Delta \hat{H} = m(\hat{H}_{\text{out}} - \hat{H}_{\text{in}}) = m(\hat{H}_{350\text{K}} - \hat{H}_{400\text{K}}) = 1000\text{g}\left(\frac{1\text{ mol}}{18\text{g}}\right)(6.798 - 49.187)\text{kJ/mol}$$

$$= 2410.5\text{kJ}$$

Additional Points to Ponder In the real world we know that heat would be lost to the container and the surroundings. What are the consequences of losing heat to the surroundings? Can the heat that the steam loses be utilized in a productive way in a chemical process?

9.5.3 Sensible Heat and Heat Capacities

Sensible heat is the heat that needs to be applied or released to raise or lower the temperature. For cases in which kinetic and potential energy can be neglected and no work is produced, then the sensible heat is $Q = \Delta U$ for a closed system and $\dot{Q} = \Delta \dot{H}$ for an open system. For a closed system, imagine that we raise or lower the temperature in such a way that the volume remains constant.

As can be seen in Figure 9.6, as ΔT tends to zero, the slope of the curve $(\Delta \hat{U}/\Delta T)$ approaches the limit, which by definition is the heat capacity at constant volume, C_v, where C_v is also a function of the temperature:

$$C_v(T) = \left\{\lim_{\Delta T \to 0} \frac{\Delta \hat{U}}{\Delta T}\right\} = \left(\frac{\partial \hat{U}}{\partial T}\right)_V \tag{9.8}$$

$$d\hat{U} = C_v(T)dT \tag{9.9}$$

$$\Delta \hat{U} = \int_{T_1}^{T_2} C_v(T)dT \tag{9.10}$$

Formula (9.10) is valid at a constant volume. If the volume changes, it is still exact for ideal gases, it is a good approximation for liquids and solids, but it is not valid for nonideal gases (for which it is best to use tabulated thermodynamic data).

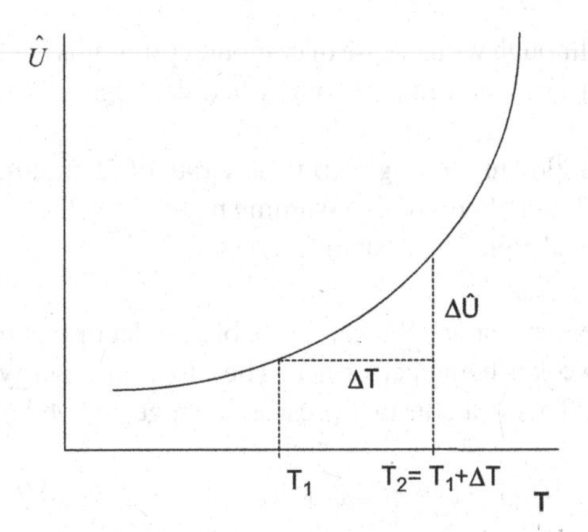

FIGURE 9.6 Changes in temperature at a constant volume.

Imagine that we now heat a material at constant pressure in an open system and want to consider the change of enthalpy. As with internal energy, enthalpy is dependent on temperature, and a curve similar to the one shown in Figure 9.6 can be constructed, having similar relationships for a heat capacity at constant pressure:

$$C_p(T) = \left(\lim_{\Delta T \to 0} \frac{\Delta \hat{H}}{\Delta T} \right) = \left(\frac{\partial \hat{H}}{\partial T} \right)_P \tag{9.11}$$

$$d\hat{H} = C_p(T)dT \tag{9.12}$$

$$\Delta \hat{H} = \int_{T_1}^{T_2} C_p(T)dT \tag{9.13}$$

Equation (9.13) is valid for processes at constant pressure in all cases. If the pressure changes, the equation is still exact for ideal gases. For solids and liquids we need to account for the changes of pressure if this is not constant. This may be expressed as

$$\Delta \hat{H} = \hat{V}\,\Delta P + \int_{T_1}^{T_2} C_p(T)dT \tag{9.14}$$

When both the pressure and temperature change in a system with nonideal gases, the best approach to calculating changes in enthalpy is to use thermodynamic tables. If these are not available, there are some relationships for variations of $\hat{H}$ with P that can be used in combination with equation (9.13). A good reference for those types of relationships is the book by Reid et al.[7]

To be able to integrate formulas (9.10) or (9.13), the heat capacity function is needed. They normally are given in polynomial form ($C_p = a + bT + cT^2 + dT^3$) and can be found in many references, such as Reid et al. or Perry's.[8] There are also simple relationships between C_p and C_v. For solids and liquids C_p is roughly equal to C_v. For ideal gases $C_p = C_v + R$, where R is the gas constant.

To simplify calculations, average values of heat capacities for a range of temperatures can be used as an approximation if the heat capacity remains relatively constant, but these average values cannot be used outside the temperature range for which they are calculated or measured. These average heat capacities are normally available in tabular form and are used throughout this book, although with the use of computer programs it is advisable to utilize the full equation whenever available, in particular when dealing with gases.

Example 9.7 Oxygen flowing through a pipe at a rate of 50 mol/min needs to be cooled from 150°C to 25°C at 1 atm of pressure. Assuming no losses to the environment, how much heat needs to be removed from the system?

Solution This is an open system. We can probably neglect potential and kinetic energy changes, and no shaft work is being performed. Therefore, the energy balance equation can be written as $\dot{Q} = \Delta\dot{H}$. This is a constant pressure change, so we have

$$\dot{Q} = \Delta\dot{H} = \dot{n}_{\text{oxygen, out}}\hat{H}_{\text{oxygen, out}} - \dot{n}_{\text{Oxygen, in}}\hat{H}_{\text{oxygen, in}} = \dot{n}_{\text{oxygen}}\Delta\hat{H} = \dot{n}_{\text{oxygen}}\int_{T_1}^{T_2} C_p(T)dT$$

The expression for the heat capacity at constant pressure is $C_p = a + bT + cT^2 + dT^3$ and from the literature[9] we find the coefficients for the equation to be: $C_p = 29.659 + 6.137261\,T - 1.186521T^2 + 0.095780T^3$ for the temperatures range 298 to 6000 K, C_p expressed in J/mol· K and T in K/1000.

$$\dot{Q} = \dot{n}_{\text{oxygen}}\Delta\hat{H} = 50\ \text{mol/min}\int_{423\ \text{K}}^{298\ \text{K}} (29.659 + 6.137261T - 1.186521T^2 + 0.095780T^3)dT$$

$$= 50\ \text{mol/min}\left[29.659T + 6.137261\frac{T^2}{2} - 1.186521\frac{T^3}{3} + 0.095780\frac{T^4}{4}\right]_{T=0.423}^{T=0.298}$$

$$= 50\ \text{mol/min}\left[(29.659)(0.298 - 0.423) + 6.137261\frac{(0.298^2 - 0.423^2)}{2}\right.$$

$$\left.-1.186521\frac{(0.298^3 - 0.423^3)}{3} + 0.095780\frac{(0.298^4 - 0.423^4)}{4}\right]$$

$$= 50\ \text{mol/min} - 3.7074 - 0.2766 + 0.0195 - 0.0006\ \text{kJ/mol}$$

$$= -198.25\ \text{kJ/min of heat that needs to be dissipated}$$

Additional Points to Ponder How might this stream be put to good use in a production process? Why are we interested in heating and cooling systems in a green engineering setting?

9.5.4 Latent Heat and Heats of Solution

Latent heat is the specific enthalpy change related to the phase change of a substance, which is an isobaric and isothermal process. The latent heat of a phase change varies considerably with the temperature of the phase change, but very little with the pressure. In Example 9.8 a phase change between gas and liquid occurs for water at 1 atm and 373.12 K (100°C). The difference between the enthalpies of the gas and the liquid phases of water is the specific *heat of vaporization* (or latent heat of vaporization) at 1 atm. If water is being condensed, the latent heat of vaporization, $\Delta\hat{H}_V$, would be negative, as the final enthalpy of the process (liquid) is less than the initial enthalpy (gas).

Another example of latent heat is the *heat of fusion*, $\Delta\hat{H}_m$, which is the enthalpy difference between the solid and liquid phases at a given temperature and pressure. Values of latent heats for most common materials are normally tabulated at their normal boiling or melting point, which are called *standard heats of fusion* or *vaporization* (ΔH°_{vap} or ΔH°_m). Again, there are several references on the Web and in the literature that provide experimental values for latent heats.

Example 9.8 A liquid stream of pure ethanol at approximately room temperature (25°C) and 1 atm flows into a vessel at a rate of 10 kg/min and leaves the vessel as a gas at 130°C and 1 atm.

(a) How much heat needs to be transferred to the stream of ethanol if we disregard potential and kinetic energy changes?
(b) If this heat is produced by condensing saturated steam at 100°C, what would the steam flow (kg/min) need to be, assuming no heat losses to the environment?

Solution

(a) Since there are no moving parts and we can disregard the kinetic and potential energy, we have $\dot{Q} = \Delta\dot{H} = n(\hat{H}_{out} - \hat{H}_{in})$. Ethanol's boiling point is about 78°C, so there will be a phase change at around that temperature. Since enthalpy is a state property, independent of the pathway that a system takes, we could use the pathway shown in Figure 9.7 to estimate the heat required. Therefore,

$$\dot{Q} = \Delta\dot{H} = \dot{n}(\hat{H}_{out} - \hat{H}_{in}) = \dot{n}(\Delta\hat{H}_{T1\,-\,Tb} + \Delta H^{\circ}_{vap} + \Delta\hat{H}_{Tb\,-\,T2})$$

$$= \dot{n}\left[\int_{T_1}^{T_b} C_p(T)dT + \Delta H^{\circ}_{vap} + \int_{T_b}^{T_2} C_p(T)dT\right]$$

The standard heat of vaporization of ethanol is about 42.2 kJ/mol. The heat capacity at constant pressure for the liquid ethanol is reported to be constant at about 112 J/mol·K. The heat capacity of the gas for the temperature range 0 to 1200°C is $C_p = 0.06134 + 15.72 \times 10^{-5}T - 8.749 \times 10^{-8}T^2 + 19.83 \times 10^{-12}T^3$, where T is in °C and C_p is in kJ/mol·°C.

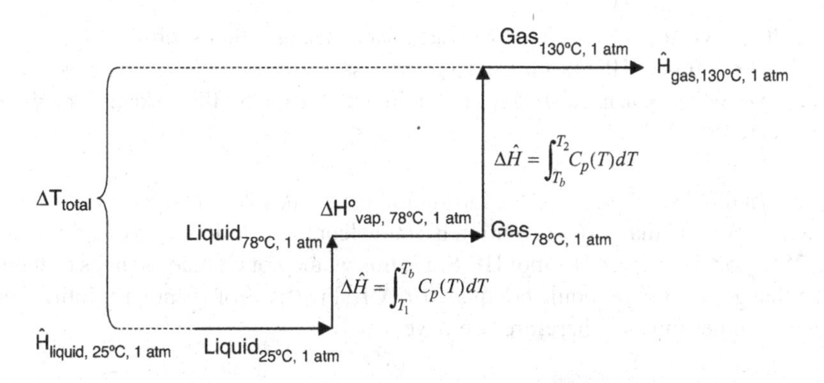

FIGURE 9.7 Changes in temperature at a constant volume Example 9.8.

$$\dot{Q} = \Delta\dot{H} = \dot{n}\left[C_{p_{\text{liquid}}}(T_b - T_1) + \Delta H^{\circ}_{\text{vap}} + \int_{T_b}^{T_2} (0.06134 + 15.72 \times 10^5\, T\, 8.749 \times 10^8\, T^2 + 19.83 \times 10^{12} T^3)dT\right]$$

$$= \frac{10,000\ \text{g/min}}{46\ \text{g/mol}}\left[(0.112\ \text{kJ/mol/K})(53.5\ \text{K}) + 42.2\ \text{kJ/mol} + \int_{78.5}^{130} (0.06134 + 15.72 \times 10^5\, T\, 8.749 \times 10^8\, T^2 + 19.83 \times 10^{12}\, T^3)dT\right]$$

$$= 217.4\ \text{mol/min}[6.992\ \text{kJ/mol} + 42.2\ \text{kJ/mol} + (6.16 + 0.844 - 0.05 + 0.0012)]$$

$$= 11,336\ \text{kJ/min} \approx 11.3\ \text{MJ/min}$$

(b) The saturation pressure of water at this temperature is 15.346 atm. The enthalpy of saturated vapor at 200°C is 2792 kJ/kg, and saturated water at the same temperature has an enthalpy of 852.3 kJ/kg (NIST[9]). $\dot{Q} = \Delta\dot{H} = \dot{m}(\hat{H}_{\text{out}} - \hat{H}_{\text{in}})$ since we are assuming no heat losses:

$$\dot{m} = (\hat{H}_{\text{out}} - \hat{H}_{\text{in}})\dot{Q} = \frac{(852.3 - 2\,792)\text{kJ/kg}}{11,336\ \text{kJ/min}} = 0.171\ \text{kg/min of steam}$$

Additional Points to Ponder If the heat transfer process was 80% efficient (thus, 20% of the heat is lost), how much steam do we need to produce? What types of emissions might be related to this small unit operation process?

The heat of solution, $\Delta\hat{H}_s(T, r)$, is the change in enthalpy in which 1 mol of solute is dissolved in r moles of solvent at a given temperature. Heat of mixing occurs when instead of a solute and a solvent, two fluids are mixed instead of dissolving a solute. Some tabulated data for heats of solution are expressed as the molar ratio (r) of the solution and temperature, but some of them are at infinite dilution (when r becomes really big) and room temperature. *Perry's Chemical Engineers' Handbook*[8] and the *CRC Handbook of Chemistry and Physics*[10] are good references for this information; one just needs to be careful about how the units are expressed and the dilution ratio r.

Example 9.9 Assume you need to calculate the change in enthalpy observed in preparing 5 L of a 10 M solution of HF. Assume that the process is isobaric and that changes in volume are negligible. Would you need any type of cooling or heating for this to keep the solution at room temperature (25°C)?

Solution In this case, we can get the information from the *CRCs Handbook of Chemistry and Physics*. We find that a 10 M solution is equivalent to a dilution ratio of 5.551 and that $-\Delta\hat{H}_s(25°\text{C}, 5.551)$ is 13.87 kJ/mol HF. Since this would not include changes of temperature, the change of enthalpy would be equal to n, $\Delta\hat{H}_s$. Five liters of 10 molar solution implies $10 \times 5 = 50$ moles in total; therefore, we have

$$\Delta H = n\,\Delta\hat{H} = 50\ \text{mol}(-13.87\ \text{kJ/mol}) = -694.5\ \text{kJ} \approx 0.7\ \text{MJ}$$

Since this is a closed system, then

$$Q = \Delta U = \Delta H - \Delta(PV)$$

Since changes in pressure and volume are negligible, we have

$$Q = \Delta U = \Delta H - \Delta(PV) = \Delta H \approx 0.7\,\text{MJ}$$

of heat will need to be removed to maintain the temperature of the solution at room temperature.

Additional Points to Ponder Do you see any EHS concerns related to preparing the HF solution described above? If you had to do this at a large scale, what would you suggest from a green engineering viewpoint?

9.5.5 Heats of Reaction

Reactors need to be cooled or heated to ensure that a reaction goes to completion, to minimize side reactions, and to minimize the potential process safety risk, among others. Therefore, the heat of reaction should be included in energy estimations. The heat or enthalpy of reaction, $\Delta\hat{H}_r(T, P)$, is the enthalpy change for a process in which the reactants are in stoichiometric quantities and the reaction progresses to completion at a given temperature and pressure. If the heat of reaction is negative, the reaction will be exothermic, and if it is positive, the reaction will be endothermic. At moderate values of pressure, the heat of reaction is nearly independent of pressure; however, it will depend heavily on the state of aggregation of the reactants and products (remember that enthalpy is a state property). The standard heat of reaction, ΔH_r°, is the enthalpy of reaction when both the reactants and products are at a specified reference temperature, usually 1 atm and 25°C. Standard heats of reactions can be estimated from the heat of formation ΔH_f° of the products and reactants and the absolute value of the stoichiometric coefficients:

$$\Delta\hat{H}_r^{\circ} = \sum_{\text{products}} v_i \Delta\hat{H}_{fi}^{\circ} - \sum_{\text{reactants}} v_i \Delta\hat{H}_{fi}^{\circ} \tag{9.15}$$

By convention, the standard heat of formation of elemental species (e.g., N_2) is zero. Since many reactions are carried out at conditions different than 25°C and 1 atm, the heat of reaction at the reaction temperature needs to be estimated. This can be done using a theoretical path (in dashed lines in Figure 9.8) in which the reactants are "brought" to 25°C from the inlet temperature, "reacted" at that temperature, and the products are then "taken to" the outlet temperature. Note that the enthalpy related to any phase change during this theoretical pathway needs to be accounted for.

For example, for the 3-pentanone process (Examples 9.1 to 9.4):

2 [propanoic acid: O, OH] → [3-pentanone: O] + CO_2 + H_2O

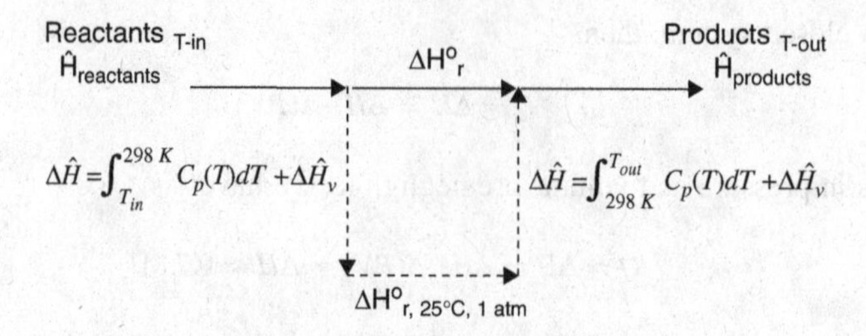

FIGURE 9.8 Heats of reaction.

If not readily available, the standard heat of reaction can be estimated as

$$\Delta\hat{H}_r^{\circ} = \sum_{\text{products}} \nu_i \Delta\hat{H}_{fi}^{\circ} - \sum_{\text{reactants}} \nu_i \Delta\hat{H}_{fi}^{\circ} = \left[(1)\Delta\hat{H}_{f-\mathrm{H_2O}}^{\circ} + (1)\Delta\hat{H}_{f-\mathrm{CO_2}}^{\circ} + (1)\Delta\hat{H}_{f-\text{3-pentanone}}^{\circ}\right] - \left[(2)\Delta\hat{H}_{f-\text{propionic acid}}^{\circ}\right]$$

For the 3-pentanone example, since propionic acid enters the reactor at 25°C, the temperature of the reaction is 300°C, and water exits the reactor as a vapor, the enthalpy of reaction at 300°C could be calculated as

$$\begin{aligned}\Delta\hat{H}_r^{300^\circ\mathrm{C}} &= \text{heat to "cool" reactants} + \text{standard heat of reaction} \\ &\quad + \text{heat to "heat" products} + \text{latent heats} \\ &= \Delta\hat{H}_r^{\circ} + \int_{25^\circ\mathrm{C}}^{300^\circ\mathrm{C}} C_p(T)dT\bigg|_{\text{3-pentanone}} + \int_{25^\circ\mathrm{C}}^{300^\circ\mathrm{C}} C_p(T)dT\bigg|_{\mathrm{CO_2}} \\ &\quad + \int_{25^\circ\mathrm{C}}^{300^\circ\mathrm{C}} C_p(T)dT\bigg|_{\mathrm{H_2O}} + \Delta H_{\text{vap}}^{\circ}\bigg|_{\mathrm{H_2O}}\end{aligned}$$

In some instances, the heat of formation and other thermodynamic properties cannot be found easily in tables or reference books. For example, to perform a full energy balance of the reaction that takes place in the production of 7-aminocephalosporanic acid (7-ACA),[11]

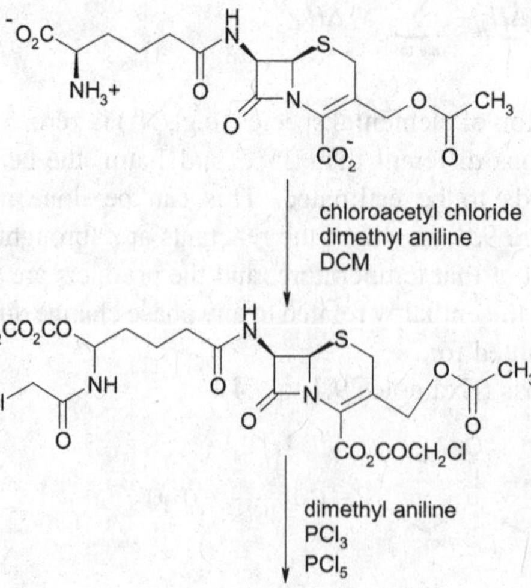

cephalosporin C K salt

chloroacetyl chloride
dimethyl aniline
DCM

dimethyl aniline
PCl_3
PCl_5

methanol

water
ammonia
methanol

7-ACA

molecular modeling was utilized to estimate heats of reactions, as no experimental data was available for the product or any of the organic intermediates. This type of physical and thermodynamic data is frequently not available for energy balances, and the materials need to be estimated so that energy requirements may be determined. Molecular modeling is often used to obtain these data. Most molecular modeling estimations are based on group contribution methods that are described in several reference books, journal articles and software packages.[12–14] The book by Reid et al.[7] is a very good reference for molecular modeling methods.

9.6 MEASURING GREENNESS OF A PROCESS THROUGH ENERGY AND MASS BALANCES

At this point, we have reviewed the general concepts of mass and energy balance and through the examples in this chapter have seen how mass and energy balances are the first step to be able to:

- Understand the process
- Identify green chemistry/green engineering issues and potential improvements
- Quantify resource needs
  - Raw materials
  - Utilities (heating, cooling, electricity)
  - Ancillary materials such as nitrogen, solvents, and catalysts
- Identify and quantify recycling loops needed
- Identify and quantify hazardous materials

  - o By-products
  - o Unreacted raw materials
  - o Ancillary materials
- Quantify process waste and emissions
  - o Liquid, solid, and gaseous waste streams
  - o Purges and potential leaks
  - o Potential fugitive emissions
- Estimate process efficiencies
  - o Energy efficiency
  - o Mass efficiency
- Identify potential energy optimization opportunities
- Quantify the impact of side and competing reactions
- Estimate emissions derived from energy and raw material production

In short, it wouldn't be possible to evaluate processes for green engineering purposes (or in general) without performing a mass and energy balance in order to understand the process, measure efficiencies, and identify potential opportunities for improvement. One example of the direct application of mass and energy balance is the comparison of two routes for the production of 7-aminocephalosporanic acid. 7-ACA had been done traditionally with the chemical route shown above (the block process diagram is shown in Figure 9.9).

A new enzymatic route was developed and there was a question about which route was greener. The first step in making that comparison was to perform a mass and energy balance for both processes so that the green metrics for each process could be contrasted and evaluated. A sample of the calculations for one of the steps in the chemical synthesis is shown in Table 9.2.

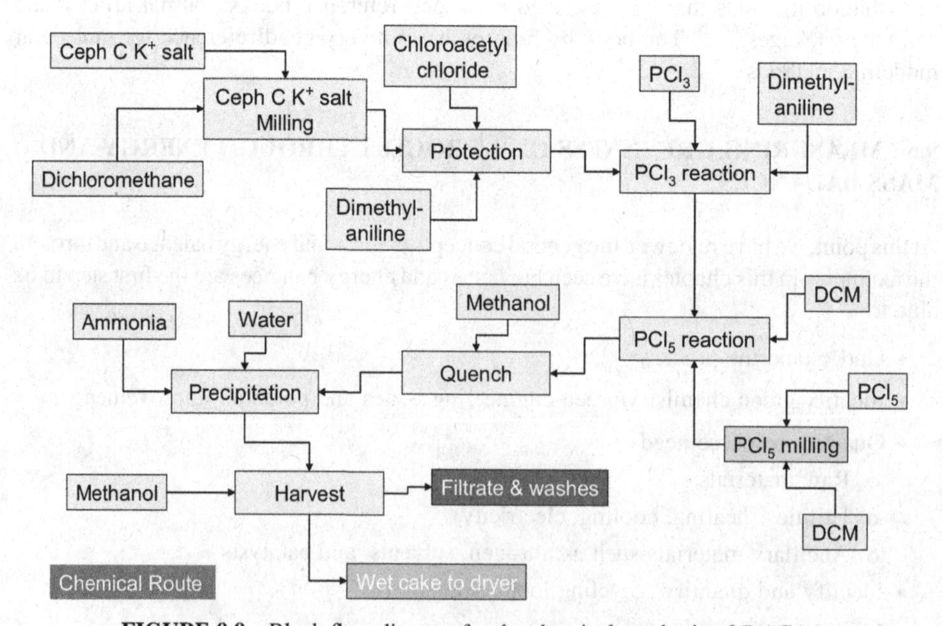

FIGURE 9.9 Block flow diagram for the chemical synthesis of 7-ACA.

TABLE 9.2 Sample Calculation

Description	Mass In (kg)	Mass Out (kg)	T_{input} (°C)	T_{sol} (°C)	$T_{initial}$ (°C)	T_{final} (°C)	Step Time (hr)	ΔH_{mixing} (kJ)	$\Delta H_{r\times c}$ (kJ)	$\Delta H_{heat/cool}$ (kJ)
Charge DCM	1064		20	20	20	20				
Charge cephalosporin C potassium	144	—	20	20	20	20	—	0		
Milling, 15–25°C; chilling if necessary, 15–60 min	—	—	20	20	20	20	0.625	2,621	−8,499	−8,499
Transfer, line washing with DCM	93	—	20	20	20	20				
Cool batch	—	—	—	—	20	3.5	—	0	—	−50,657
Charge chloroacetyl chloride	161	—	20	3.5	20	3.5	0.08	423	—	−3,442
Charge dimethylaniline	182	—	20	3.5	3.5	16	0.67	3,803	—	−1,287
Chill to −12°C and stir	—	—	—	—	16	−12	0.21	352	—	−58,173

TABLE 9.3 Mass and Energy Balance Metrics

Metric	Chemical Route	Enzymatic Route
Mass		
Yield (mol%)	75	67
Reaction mass efficiency (%)	14	46
Mass intensity [kg/kg 7-ACA (excluding water)]	81	44
Mass productivity [% (excluding water)]	1.2	2.3
Solvent intensity [kg/kg 7-ACA (excluding water)]	74	41
Organic phase waste (kg waste/kg 7-ACA)	80	43
E-factor including aqueous waste (kg waste/kg 7-ACA)	93	172
Energy		
Electricity (MJ/kg 7-ACA)	4.2	1.8
Cooling above 20°C (MJ/kg 7-ACA)	0.1	3.4
Cooling below 20°C (MJ/kg 7-ACA)	6.0	2.8
Heating, steam or hot air (MJ/kg 7-ACA)	1.3	1.3
Electricity for refrigeration (cooling below 20°C) (MJ/kg 7-ACA)	3.0	0.7
Total process energy Requirements (MJ/kg 7-ACA)	11.5	9.3

Once the mass and energy balances were completed, it was possible to estimate the green metrics for both the chemical and enzymatic routes, and to compare them. A sample of some of the metrics determined through the mass and energy balance is shown in Table 9.3.

Also, by performing complete mass and energy balances it is possible to identify EHS issues associated with the process, such as the amounts of hazardous materials that are either produced or utilized (e.g., dichloromethane, phosphorous chloride), and extreme operating conditions (very high or low temperatures and pressures). After the mass and energy balances are completed, we might be in a better position to answer the question of which process is greener. At this point we can say from the mass and energy balances that the enzymatic route is generally more resource efficient, requiring about half the mass of materials per kilogram of product and about 80% of the process energy. This is despite having a lower yield than that by the chemical route. In addition to that, we can say that the enzymatic route also avoids the use of hazardous materials and solvents (dichloromethane, phosphorous pentachloride, phosphorous trichloride, dimethyl aniline) required by the chemical route.

But can we really answer the question of which route is greener? In the 7ACA example, the enzymatic process uses considerable amounts of water and produces the associated aqueous waste, as can be seen by the aqueous waste produced. We can perhaps get an indication, but to answer the question fully, we will need to redraw the system boundaries to include the production of raw materials, the production of utilities (steam, refrigeration), and the treatment of waste, and that is done using a life cycle approach. That type of analysis is covered in (Chapters 16 through 20) in greater detail, but in a way, it can be explained as a mass and energy balance with expanded boundaries. After all, we have the ability to redraw the boundaries wherever they will render the appropriate answer to the right question. Perhaps the greater difficulty is in defining the right question.

PROBLEMS

9.1 It is said that batch processes, by their nature, are always in a transient state. Why? A continuous process is usually run as close to steady state as possible. Give an example of when this is not the case for a continuous process.

9.2 The commercial production of acrolein by heterogeneously catalyzed gas-phase condensation of acetaldehyde and formaldehyde was established by Degussa in 1942. Today, acrolein is produced on a large commercial scale by heterogeneously catalyzed gas-phase oxidation of propene.[15,16]

(a) Draw a block diagram of the process according to the following description: Propylene, air, and steam are compressed to 2 atm and then mixed at a molar ratio of 1 : 8 : 4. The gas mixture is fed to a multitubular fixed-bed reactor, which is operated at 350°C and 2 atm. The conversion rate of propylene in this reactor is 95%. The effluent gas from the reactor is cooled to 250°C and then fed into a gas washer. An aqueous stream and an organic liquid, 2-ethyl-1-hexanol, are used to wash the gas stream. The ratio of gas stream/aqueous stream/organic stream is 10.6 : 1.5 : 1. The residual gas leaves the gas washer at 70°C and is introduced to the bottom of a gas cooler. The liquid stream from the gas washer is pumped into a series of distillation columns at 105°C to recover the by-products, acrylic acid and acetic acid. From the gas cooler, the residual gas stream leaves at 19°C and is fed into another gas washer to recover residual acrolein. The organic phase from the bottom of the cooler is recycled to the first gas washer at 45°C. Part of the aqueous phase is combined with the organic phase of the second gas washer, cooled to 16°C, and recycled to the gas cooler. Part of the aqueous phase from the gas cooler is recycled to the first gas washer. The second gas washer uses a water/2-ethyl-1-hexanol mixture to wash the residual gas at 2°C. The aqueous phase is then combined with part of the aqueous phase from the gas cooler to recover acrolein product. 2-Ethyl-1-hexanol is also recovered and combined with makeup 2-ethyl-1-hexanol and water. This stream is cooled to 2°C and fed into the second gas washer. The chemistry involved is:

Primary reaction:

$$\underset{\text{propylene}}{CH_2CHCH_3} + \underset{\text{oxygen}}{O_2} \rightarrow \underset{\text{acrolein}}{CH_2CHCHO} + \underset{\text{water}}{H_2O}$$

Side reactions:

$$\underset{\text{propylene}}{CH_2CHCH_3} + \underset{\text{oxygen}}{1.5\,O_2} \rightarrow \underset{\text{acrylic acid}}{CH_2CHCOOH} + \underset{\text{water}}{H_2O}$$

$$\underset{\text{propylene}}{CH_2CHCH_3} + \underset{\text{oxygen}}{1.5O_2} \rightarrow \underset{\text{acetic acid}}{CH_3COOH} + \underset{\text{formaldehyde}}{CH_2O}$$

$$\underset{\text{propylene}}{CH_2CHCH_3} + \underset{\text{oxygen}}{O_2} \rightarrow \underset{\text{acetaldehyde}}{C_2H_4O} + \underset{\text{formaldehyde}}{CH_2O}$$

(b) Identify and label all the unit operations.

(c) Identify and label all the waste streams.

(d) Are there any waste streams that can be eliminated or reduced?

(e) By looking at the process flow diagram, which green chemistry or green engineering principles can be applied to improve the process? How?

9.3 Complete the mass balance of Example 9.3, estimating the composition of each stream. Remember that because 3-pentanone is slightly soluble in water, the purity of the product obtained from the decanter is about 91.6%. Assume that 156 kg/h of nitrogen is needed at standard temperature and pressure and the concentration of propionic acid in the product obtained from the decanter is negligible.

(a) Can the waste generated by the process be reduced? How?
(b) Which green chemistry or green engineering principles can be applied to improve the process? How?

9.4 Use the information presented in Figure 9.5 as additional data points of the problem.

(a) Complete the mass balance of Example (9.4) estimating the composition of each stream.
(b) What is the per-pass conversion of chlorine in the reactor?
(c) What is the selectivity of hypochlorous acid relative to sodium chlorate?
(d) What are the green engineering issues associated with the waste stream coming out of the dryer?
(e) Is there a way to avoid having chlorine in the gaseous waste stream coming out of the drier?
(f) Which other green chemistry or green engineering principles can be applied to improve the process? How?

9.5 Perform a dimensional analysis of equation 9.4. Which units would the pressure and specific volume need to have? In which units can flow work be expressed?

9.6 Complete the derivation of the energy balance equation for an open steady-state system [equation (9.7)]. Derive equation (9.14) from equation (9.13).

9.7 Solve Example 9.7 now using the thermodynamic table values in Table P9.7 given (NIST[6] and Chase[9]). To find values in between temperatures, interpolation can be used.

TABLE P9.7 Thermodynamic Values

Temperature (K)	C_p (J/mol·K)	$H^\circ - H^\circ_{298.15}$ (kJ/mol)
298	28.91	−0.02
300	28.96	0.03
400	30.56	3.02
500	31.56	6.13

9.8 It was mentioned that average heat capacities can be used to simplify enthalpy calculations.

(a) Derive a generic equation to estimate the average heat capacity at constant pressure for a given range of temperatures, and estimate the average heat capacity for oxygen between 298 and 500 K using the C_p expression given in Example 9.7.
(b) Develop a spreadsheet that calculates the sensible heat between two temperatures for a material with a heat capacity expressed as $C_p = a + bT + cT^2 + dT^3$, once you provide coefficients a, b, c, and d for the heat capacity equation.

9.9 Hydrochloric acid is produced by the absorption of gaseous hydrogen chloride in water. Calculate the heat needed to produce 200 kg/h of hydrogen chloride at 36%. Is the heat transferred to or from the adsorption column? How is this achieved?

9.10 For the typical distillation tower shown in Figure P9.10:

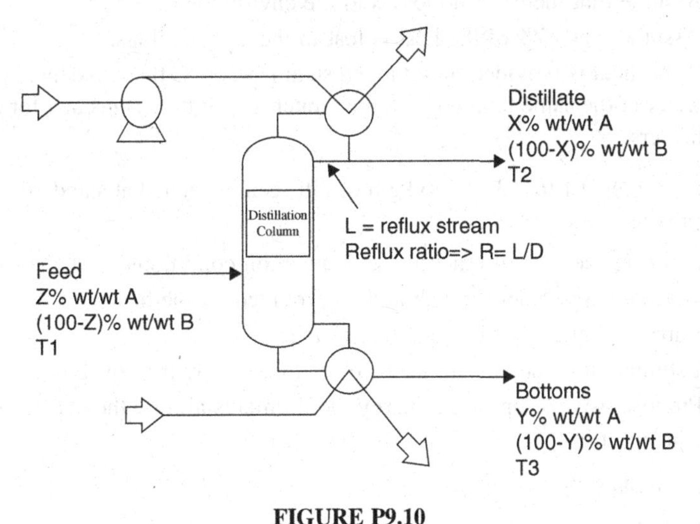

FIGURE P9.10

(a) Write the equation for the global mass balance.
(b) Write the equations for the mass balances for A and B.
(c) Write the equation for the energy balance around the column.
(d) Write the equation for the energy balance around the condenser that would estimate the cooling requirements of the condenser.
(e) Write the equation for the energy balance around the reboiler that would estimate the heating requirements for the reboiler.
(f) What can be done to optimize the energy efficiency for this unit operation?

9.11 If the distillation column of Problem 9.10 is used to produce 90% pure acetone from a 50% acetone–water solution, where the bottoms leave the column as a 5% aqueous acetone waste, estimate:
(a) How much distillate can be produced if the column is fed at a rate of 100 kg/h of total feed
(b) The rate of the reflux stream
(c) The amount of waste produced
(d) The energy requirements for the condenser and reboiler
(e) How much carbon dioxide would need to be emitted to produce the steam if it were produced by the combustion of pure carbon with 100% efficiency (assume a heating value of 15 MJ/kg)

9.12 Estimate the amount of energy that will be required by the reboiler to the distillation column of Examples 9.1 to 9.3. Assume that 20% of the heat is lost to the environment. How would you reduce the heat requirements in this distillation column?

9.13 Calculate how much heat will need to be provided to the three heat exchangers of Example 9.5.

(a) Assume that there are no losses to the environment.

(b) Assume that 15% of the heat is lost to the surroundings.

(c) If this heat is provided by saturated steam at 200°C that condenses to saturated water at the same temperature, how much steam flow is needed for each of the heaters?

9.14 For Examples 9.1 to 9.3, if 156 kg/h of nitrogen is needed at standard temperature and pressure:

(a) Estimate the heat of reaction at the reaction conditions (300°C, 1 atm).

(b) Estimate how much heat should be provided by the heater.

(c) Carry out an energy balance for the reactor.

(d) Estimate how much heat needs to be removed by the condenser.

(e) Propose ways to optimize energy requirements around the heater, reactor, and condenser.

9.15 For Example 9.5:

(a) Calculate the standard heat of reaction for each of the three competing reactions and at the reaction conditions (105°C and 1 atm).

(b) Estimate how much heat needs to be provided to the reactor assuming 15% heat losses.

(c) If the heat needed for the reactor is provided by saturated steam at 200°C that condenses to saturated water at the same temperature, how much steam flow is needed for each heater?

(d) Propose ways to optimize the energy requirements of the heater, reactor, and condenser.

9.16 The enzymatic route for 7ACA involves the conversion of the potassium salt of cephalosporin C into 7-ACA in a three-step process (see below). Two steps are catalyzed by enzymes and the intermediate step is spontaneous. A solution of cephalosporin C is stirred in the presence of immobilized enzyme diamino acid oxidase (DAO), and oxygen is supplied via compressed air. Hydrogen peroxide (which is a by-product of the bioconversion) reacts spontaneously with the keto intermediate to give glutaryl 7-ACA. To ensure satisfactory conversion the reaction is carried out at elevated pressure (5 bar) and the temperature is controlled to 18°C throughout. The pH is also controlled by starting at 7.3 but is increased to 7.7, where it remains until the reaction is completed. Further hydrogen peroxide can be added to promote conversion to glutaryl 7-ACA. When the conversion is complete, the product is separated from DAO. The solution is then stirred in the presence of the enzyme glutaryl 7-ACA acylase (GAC), where the pH and temperature are controlled at 8.4 and 14°C. Some dilution may be necessary to control the concentration, and the resulting 7-ACA is separated from GAC and recycled back into the process.

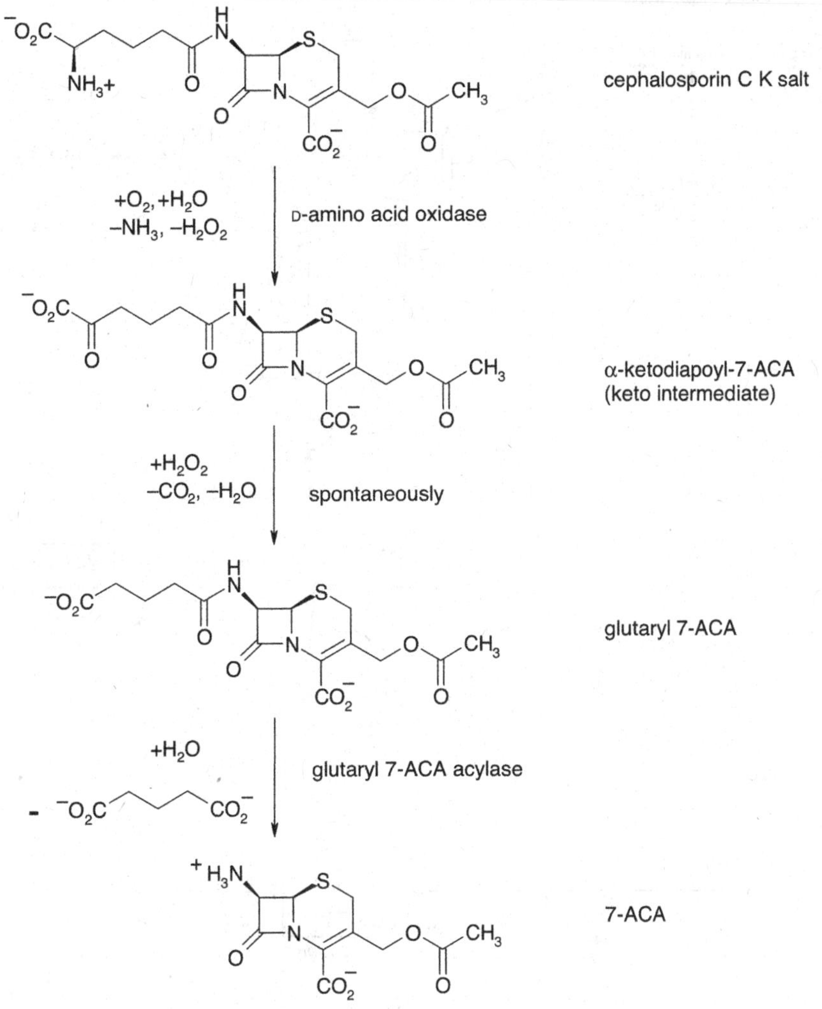

cephalosporin C K salt

$+O_2, +H_2O$
$-NH_3, -H_2O_2$
D-amino acid oxidase

α-ketodiapoyl-7-ACA
(keto intermediate)

$+H_2O_2$
$-CO_2, -H_2O$
spontaneously

glutaryl 7-ACA

$+H_2O$
$- {}^-O_2C\text{-}(CH_2)_3\text{-}CO_2^-$
glutaryl 7-ACA acylase

7-ACA

(a) Draw a process flow diagram for the enzymatic process.
(b) Identify all the waste streams.
(c) Explain your hypothesis of why the enzymatic process fared better than the traditional chemical process.
(d) Proposed green chemistry and green engineering improvements for the enzymatic process.

9.17 Tetra (*n*-butyl)ammonium bromide is a quaternary ammonium salt often used as a phase-transfer catalyst in pharmaceuticals and refinery applications.[17,18] Such tetra-alkylammonium halides also have antimicrobial properties.[19] Synthesis is generally through reaction of a tertiary amine with the desired alkyl halide at moderate temperatures (60 to 100°C) (Kirk-Othmer[4]).

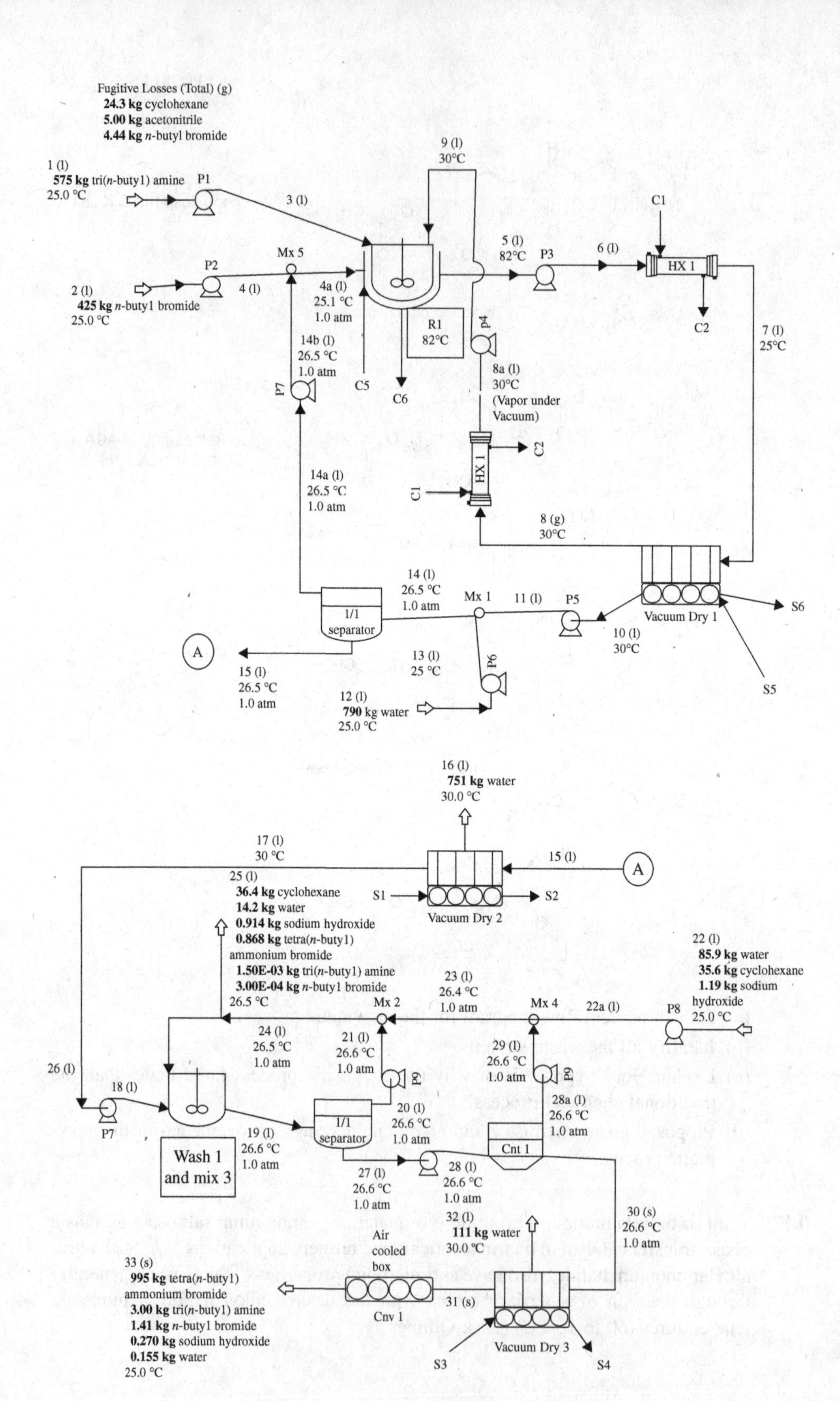

Fugitive Losses (Total) (g)
24.3 kg cyclohexane
5.00 kg acetonitrile
4.44 kg n-butyl bromide
1 (l)
575 kg tri(n-butyl) amine
25.0 °C
P1
3 (l)
9 (l)
30°C
C1
Mx 5
5 (l)
82°C
P3
6 (l)
HX 1
P2
2 (l)
425 kg n-butyl bromide
25.0 °C
4 (l)
4a (l)
25.1 °C
1.0 atm
R1
82°C
p4
C2
7 (l)
25°C
14b (l)
26.5 °C
1.0 atm
P7
C5
C6
8a (l)
30°C
(Vapor under Vacuum)
14a (l)
26.5 °C
1.0 atm
8 (g)
30°C
14 (l)
26.5 °C
1.0 atm
Mx 1
11 (l)
P5
1/1 separator
Vacuum Dry 1
S6
A
10 (l)
30°C
13 (l)
25 °C
P6
15 (l)
26.5 °C
1.0 atm
12 (l)
790 kg water
25.0 °C
S5
16 (l)
751 kg water
30.0 °C
17 (l)
30 °C
15 (l)
25 (l)
36.4 kg cyclohexane
14.2 kg water
0.914 kg sodium hydroxide
0.868 kg tetra(n-butyl) ammonium bromide
1.50E-03 kg tri(n-butyl) amine
3.00E-04 kg n-butyl bromide
26.5 °C
S1
S2
Vacuum Dry 2
22 (l)
85.9 kg water
35.6 kg cyclohexane
1.19 kg sodium hydroxide
25.0 °C
23 (l)
26.4 °C
1.0 atm
Mx 2
Mx 4
22a (l)
P8
24 (l)
26.5 °C
1.0 atm
21 (l)
26.6 °C
1.0 atm
29 (l)
26.6 °C
1.0 atm
P9
26 (l)
18 (l)
P7
20 (l)
26.6 °C
1.0 atm
28a (l)
26.6 °C
1.0 atm
19 (l)
26.6 °C
1.0 atm
1/1 separator
Cnt 1
Wash 1 and mix 3
27 (l)
26.6 °C
1.0 atm
28 (l)
26.6 °C
1.0 atm
32 (l)
111 kg water
30.0 °C
30 (s)
26.6 °C
1.0 atm
Air cooled box
33 (s)
995 kg tetra(n-butyl) ammonium bromide
3.00 kg tri(n-butyl) amine
1.41 kg n-butyl bromide
0.270 kg sodium hydroxide
0.155 kg water
25.0 °C
Cnv 1
31 (s)
S3
Vacuum Dry 3
S4

FIGURE P9.17

$(C_4H_9)_3N$ + $(C_4H_9)Br$ → $(C_4H_9)_4NBr$

tri (*n*-butyl) amine + *n*-butyl bromide tetra (*n*-butyl) ammonium bromide

The reaction mechanism is nucleophilic bimolecular substitution (SN2) and is often conducted in a polar solvent. No side reactions have been identified so far. Using the common synthesis mechanism for tetraalkylammonium halides, tri(*n*-butyl)amine is reacted with *n*-butyl bromide at 82°C for about 21 h. The reaction is conducted in acetonitrile, which is removed and recycled under vacuum. A 5 mol %

TABLE P9.17 Energy Requirements

Process Diagram Label	Unit Operation	Energy requirement (MJ/1000 kg product)
	Electricity	
P1	Pump 1	1.09×10^{-4}
P2	Pump 2	4.84×10^{-5}
P3	Pump 3	0.0883
MxE1	Mixer electricity 1	0.432
P5	Pump 5	5.17×10^{-4}
P4	Pump 4	4.80×10^{-5}
P6	Pump 6	1.62×10^{-4}
P7	Pump 7	1.45×10^{-8}
P8	Pump 8	3.42×10^{-4}
MxE3	Mixer electricity 3	1.18
P9	Pump 9	3.40×10^{-3}
P10	Pump 10	2.75×10^{-4}
P11	Pump 11	2.26×10^{-3}
P12	Pump 12	1.09×10^{-6}
Cnt1	Centrifuge 1	30.0
Cnv1	Conveyer 1	4.20×10^{-3}
	Heating	
Dry1	Dryer 1	378
Dry2	Dryer 2	1835
Dry3	Dryer 3	275
	Cooling	
R1	Reactor 1	−210
Hx1	Heat exchanger 1	−139
Hx2	Heat exchanger 2	−402
	Totals	
	Electricity	31.7
	Heating	2487
	Energy input requirement	2519
	Cooling	−751

excess of *n*-butyl bromide is used and recycled from a decanter to the reactor. The crystallized product is dried to about 95 wt % solids then washed with water–cyclohexane. These washing solvents are separated and reused. Centrifugation and drying are used to produce the final tetra (*n*-butyl) ammonium bromide to a 99.5 wt % purity. The global mass balance is shown in Figure P9.17. The energy requirements were estimated using energy balance principles (Table P.17):

(a) Estimate the overall yield from tri(*n*-butyl)amine.
(b) Calculate the overall mass efficiency for the process.
(c) Calculate the reaction mass efficiency.
(d) Identify good green engineering/green chemistry practices in this process.
(e) Identify green engineering/green chemistry issues in this process.
(f) Propose green chemistry/green engineering improvements based on the process description, mass balance, and process flow diagram.

REFERENCES

1. Felder, R. M., Rousseau, R.W. *Elementary Principles of Chemical Processes*, 3rd ed., Wiley, New York, 2000.
2. Ketones. In *Ullmann's Encyclopedia of Industrial Chemistry*, online edition. Wiley-VCH, New York, 2005.
3. Hussman, G. P., U.S. Patent 4,754,074, to Amoco Corporation, 1988.
4. *Kirk–OthmerEncyclopedia of Chemical Technology*, 5th ed. Wiley, Hoboken, NJ, 2004.
5. Melton, J. K., Shaffer, J. H., Hilliard, G. E., Wojtowicz, J. A. U.S. Patent 5,037,627, to Olin Corporation, 1991.
6. NIST Standard Reference Database 69, June 2005 Release. NIST Chemistry WebBook. http://webbook.nist.gov/chemistry/.
7. Reid, R. C., Prausnitz, J. H., Poling, B. E. *The Properties of Gases and Liquids*. McGraw-Hill, New York, 2007.
8. Perry, R. H., Green, D. W. *Perry's Chemical Engineers' Handbook*, 8th ed. McGraw-Hill, New York, 2007.
9. Chase, M. W., Jr. *NIST-JANAF Themochemical Tables*, 4th ed., Monograph 9. J. Phys. Chem. Ref. Data, 1998, 1–1951.
10. *CRC Handbook of Chemistry and Physics*, 88th ed., CRC Press, Boca Raton, FL, 2007.
11. Henderson, R. K., Jiménez-González, C., Preston, C., Constable, D. J. C., Woodley, J. M. EHS and LCA assessment for 7-ACA synthesis: a case study for comparing biocatalytic and chemical synthesis. *J. Ind. Biotechnol.*, 2008, 4(2), 180–192.
12. Aspen Technology, Inc., Aspen Plus. http://www.aspentech.com/.
13. Gani, R., *ICAS Documentations*. CAPEC Internal Report. Technical University of Denmark, Lyngby, Denmark, 2001.
14. ProSim. Software and Services for Process Simulation. http://www.prosim.net, 2007.
15. Acrolein and methacrolein. In *Ullmann's Encyclopedia of Industrial Chemistry*, online edition, Wiley-VCH, New York, 2005, pp. 6–7.
16. Noll, E., Schaefer, H., Schmid, H., Weigert, W. U.S. Patent 3,926,744, to Deutsche Gold- und Silber-Scheideanstalt vormals Roessler, Dec. 16, 1975.
17. H. Sasaki.U.S. patent 4,948,520, to Lion Corp, Aug. 14, 1990.
18. A. Rogers et al. U.S. patent 2,844,466, to Armour and Co., July 22, 1958.
19. T. Yoneyama et al., U.S. patent 5,015,469, to Shiseido Co., May 14, 1991.

10

THE SCALE-UP EFFECT

What This Chapter Is About In previous chapters we presented several strategies for applying the principles of green chemistry to chemistry and route selection and measuring our success in greening routes and processes. In this chapter we offer an overview of the factors affecting the scale-up of a chemical process from laboratory to full-scale manufacturing, their impact on green engineering and green chemistry design, and the integration of green chemistry and green engineering principles during the scale-up process.

Learning Objectives At the end of this chapter, the student will be able to:

- Understand the typical scale-up process for chemical processes.
- Identify the key factors that affect the scale-up process.
- Understand the impact that the key scale-up issues have in the green chemistry/green engineering design context.
- Identify strategies to address scale-up issues.
- Understand the implications of early chemistry and process choices in scale-up decisions.

10.1 THE SCALE-UP PROBLEM[1,2]

In the last few chapters we described several strategies for incorporating green chemistry and green engineering principles that can be applied to chemistry and route design and

Green Chemistry and Engineering: A Practical Design Approach, By Concepción Jiménez-González and David J. C. Constable

development. Design of a new chemical process almost invariably starts at the bench in a laboratory where we use round-bottomed flasks, beakers, pipettes, and the like, running initial experiments involving only grams or micrograms of a new substance. It takes a great deal of time and effort to discover a specific chemistry or chemical route that renders the desired product with acceptable mass efficiency, yield, selectivity, and reaction mass efficiency, and that has economic promise.

Once a new reaction for a specific process is verified at laboratory scale and there is a robust degree of reproducibility and understanding of the science behind the reaction at this scale, there is still a long way to go in process development. Going directly from laboratory scale to a manufacturing setting is almost never possible. Process development should, among other things, enable one to transfer the production of a few grams of materials from the laboratory scale to full manufacturing, where it is common to have production levels of tons or millions of tons, with even larger amounts of raw materials needed to produce the product.

Scale-up refers to those activities that enable the change in the production scale of a new chemical. Formally, *scale-up* is defined as "the successful start up and operation of a commercial size unit whose design and operating procedures are in part based upon experimentation and demonstration at a smaller scale of operation."[3] Scaling up is, however, rarely a direct and linear procedure. In other words, even though the types of molecules that are reacting are the same, many other factors can and will affect process development and therefore the greenness of the process. For example, imagine that we have discovered a great synthetic route to a new compound that might revolutionize the way in which we live, and then imagine that there is a trace amount of some colored by-products from the first step in our reaction sequence that will need to be removed. Since activated carbon is commonly used in laboratories to remove colored impurities from reaction mixtures, we can certainly try that. We might consider adding a few grams of activated carbon to our reactor (a flask); then perhaps we could stir for a few minutes, remove the agitator, use filter paper and vacuum filtration to separate the carbon (or just rely on gravity to do our filtration), then dispose of the paper with the carbon, and we are ready to continue with the next step. At the laboratory scale, this is a very simple and common step that happens many times, without much thought about the fate of the waste we have generated, or even where to connect the filter to the vacuum.

At the manufacturing scale, however, it is definitely not efficient to linearly (or literally) mimic this process from the laboratory: Imagine for a moment that we did try to build a huge flask (our reactor). How would you add all the activated carbon needed for a scale a million times bigger than the one in the laboratory? How would you ensure that the agitator does sufficient work to achieve good mixing? Maybe a mammoth-size agitator? How much energy would that require? Could you build a big-enough funnel to filter the carbon? How much energy would we need to generate the vacuum? Or if using gravity, how much time would it take to complete the filtration? How would you remove the big carbon filter when it is filled with carbon? How about disposal? How about the costs of this giant laboratory process? How much additional paper and carbon would you need to procure? How about the safety issues of moving mountains of carbon every day? How would you transfer the resulting clear mixture?

In other words, we cannot scale-up linearly what the chemist has just done; it is not practical, it is not cost-effective, and it certainly will be against many green engineering and green chemistry principles, even if we started with the greenest route we could develop. It is then obvious that to achieve a successful and green industrial process, scale-

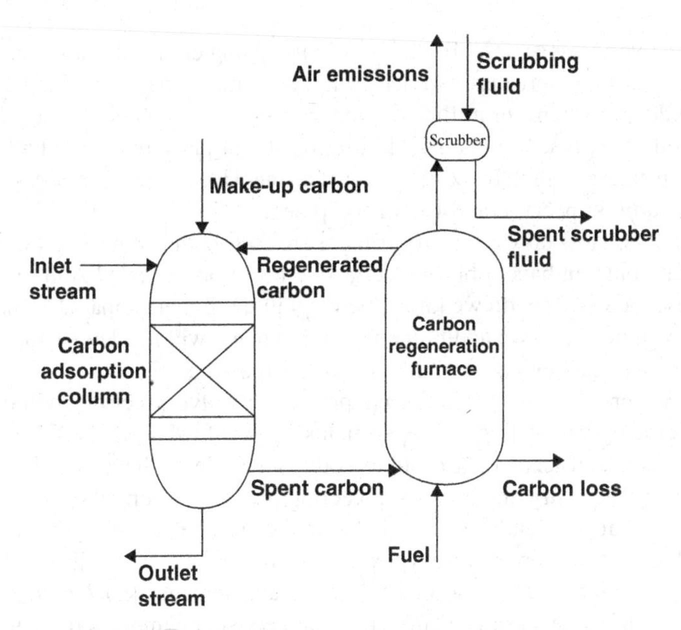

FIGURE 10.1 Activated carbon adsorption process.

up becomes a pivotal part on the successful completion of the development of a new synthesis or product.

Example 10.1 For the activated carbon filtration process described above, provide an example of a feasible large-scale process to achieve the same result.

Solution There are several systems that can perform activated carbon adsorption at the industrial scale. Variations of this process are widely used, both in chemical processing and in water and wastewater treatment. Figure 10.1 is a very simplified flow diagram of a filtration system. In this example the system consists of an adsorption column and a regeneration system. The adsorption system holds the carbon bed while the stream to be treated is passed through. The regeneration of carbon is done through a thermal method in which organic matter within the pores of the carbon is oxidized. A furnace or a rotary kiln can be used, among other methods. About 5 to 10% of the carbon is destroyed or lost and must be replaced with fresh carbon. The regenerated carbon has an adsorption capacity slightly less than that of virgin carbon.[4] Figure 10.1 is indeed simplified and it does not show a series of additional equipment needed for this operation, such as pumps, conveyors, quench tanks, holding tanks, and instrumentation and control.

Additional Points to Ponder What is the probable fate of the spent carbon in this adsorption–regeneration system? Why is there a need for a scrubber? What could be used in its place? What is the probable fate of the spent scrubber fluid?

For proper scale-up we need to think differently than when we are in the laboratory, since what happens is in essence different. Mass and energy transfer become very important players as we move across scales. Mixing a few grams of material in the laboratory under our watchful eye is a simple process, but achieving the same degree of mixing at a scale of hundreds of kilograms, tons, or millions of tons is a significant challenge. To dissipate the heat of an exothermic reaction in a small flask, we might just have an ice bath or even let the environment take care of it. In the manufacturing plant it can become a very serious safety concern that requires special controls and equipment.

The question is: How do we identify if there are factors that were not important at the laboratory scale but can have a big impact as we move up in scale? And, once we have identified these factors, how do we know the magnitude of their impact? To answer these questions we will need to have additional information and will need to perform additional experiments above and beyond our initial chemical reaction.

The scale-up process and the scale-up problem involve precisely definition of the additional information, modeling and experiments that would be needed to fully characterize our reaction system and achieve the desired production scale, with reproducible results and integrating green chemistry and green engineering principles along the way.

It is common that different stages of development are referred to by the scale at which process development experiments are typically performed (i.e., we may hear of bench scale, lab scale, pilot plant scale, half-scale, demonstration unit, and the like). *Laboratory* or *bench scale* normally refers to experiments that use relatively small amounts of material and the chemical synthesis can be carried out at a conventional laboratory bench. The size and throughput of a *pilot plant* varies widely according to the type of industry, but the important aspect is that the industrial constraints need to be taken into account and modeled for these types of experiments. It is during experiments at the pilot plant scale that scale-up issues need to be addressed. Sometimes industrial units are built at a moderate scale, about one-tenth the size of manufacturing. When they are built in general they tend to be very expensive and tend to add time to project time lines. *Manufacturing scale* refers to plants built to a scale that satisfies the market demands of the product. The size of these also varies widely, from thousands of tons per annum for commodity chemicals to hundreds of kilograms per annum for fine and specialty chemicals.

10.2 FACTORS AFFECTING SCALE-UP

In a chemical process there will be variables that are independent of the size and geometry of the process, such as the reactivity of the species involved, the stoichiometry of the main and secondary reactions, kinetics, reaction thermodynamics, and so on. However, there are a series of parameters that can and will be affected by the size and geometry of the equipment in the plant. These aspects are mainly in the realm of physical phenomena and have direct implications not only in the scale-up problem but in the overall green engineering performance of the chemical process that is being developed. In general, the cause of most scale-up issues can be traced back to a series of factors, such as mass and energy transfer, trace chemicals, material properties, and phase equilibrium. For example, Table 10.1 shows various scale-up problems for catalyst scale-up and manufacturing and the linkages to these parameters.[5]

Next, we cover some of the main factors that affect the scale-up process.

Mass Transfer Dispersion and diffusion are greatly affected by the scale and geometry. It is little wonder that in chemical industrial processes, mixing is a fundamental factor used

TABLE 10.1 Reported Catalyst Scale-up Problems and Factors Affecting Scale-up

Unit Operation	Scale-up Problem	Factors Affecting Scale-up
Precipitation/ co-precipitation	Reagent purity, pH, temperature, and concentration control	Impurities Phase equilibrium
Solution/slurry transfer	Clogged lines, contamination	Impurities Fluid dynamics
Filtration	Mixing problems, concentration gradients	Mass transfer
Drying	Temperature gradients, residence time control	Heat transfer Mass transfer/fluid dynamics
Calcination	Temperature gradients, residence time control, sintering	Heat transfer Mass transfer
Washing/ion exchange	Mixing, atmosphere control	Mass transfer Impurities
Particle formation	Control of pH and concentration, particle strength	Phase equilibrium Material properties Mass transfer
Coating	Support residence time, solution concentration, viscosity	Mass transfer Phase equilibrium Material properties
Impregnation	Support residence time, solution concentration, viscosity	Mass transfer Phase equilibrium Material properties

throughout a chemical plant for homogeneous reactions and separations. Mixing is also often cited as one of the main difficulties in scale-up for batch reactors, semibatch reactors, and continuously stirred tank reactors (CSTRs). Small-scale vessels are easy to mix, but agitation is affected by the scale. Mixing can be either macromolecular (*macromixing*), referring to the average composition of the vessel, or micromolecular (*micromixing*), referring to the interactions of the molecules. When working with larger reactors, micromixing becomes more difficult, and this mechanism is important for many processes, such as crystal formation, heat removal, or chemical reactions sensitive to stoichiometry or heat. Turbulence-dominated mixing tends to be scaled up assuming constant mixing, thus leaving the power per volume (P/V) factor constant during scale-up, although there are other strategies, such as using tip speed, the Reynolds number, or mixing time, and the best strategy will depend on the type of mixing that is sought and the type of system to be mixed. More detailed analysis of mixing may be found in the literature.[6,7]

However, mass transfer issues are not only limited to mechanical mixing, and different mass transfer mechanisms might be in play in chemical processes that use heterogeneous systems, such as liquid–liquid, solid–liquid, gas–liquid, and gas–solid. For example, fluid dynamics, such as flow patterns in liquid-phase reactions and separations (hydrodynamics) and the flow patterns in gas reactions and separations (aerodynamics), can make a big difference in separations without mechanical mixing, as in the case of packed columns, absorbers, and gas–liquid and liquid–liquid extractors. Channels that are formed in packed beds or ring columns, or gas passing too fast through a scrubber or a reactor, can impede mass transfer and lower efficiency.

When facing a mass transfer scale-up issue, it is important to identify which mass transfer mechanism is limiting the performance (the step-limiting mechanism), as where temperature and fluid dynamics control the technical characteristics of packed bed reactors (or packed bed columns with adsorbents). At low process temperatures, a fixed-bed catalytic reactor is probably reaction rate–limited because the chemical reaction rate constant is generally small relative to the film mass transfer rate constant, and the chemical reaction rate constant dominates the global rate constant. Changing the catalyst in these cases will improve the conversion rate only if the process is reaction rate–limited. On the other hand, changing the physical properties of a solid catalyst (i.e., shape, pore-size distribution, pore size, surface area), will only increase the conversion rate if the process is pore diffusion rate–limited. Finally, increasing the linear fluid velocity through the bed of solids will raise the conversion rate only if the process is film diffusion rate–limited. A great amount of time and expense can be wasted exploring catalyst structure when in fact the overall conversion rate depends on the fluid flow. In general, the main resistance force in the pilot plant packed bed reactor is film diffusion, while in the commercial unit it will be either pore diffusion or, to a lesser extent, reaction rate.[8]

From a green chemistry/green engineering perspective, ineffective mass transfer leads to a loss of mass efficiency and reaction mass efficiency, potential increase of by-products and impurities, potential localized heating, loss of yield due to side reactions, slow progression of reaction (and consequent loss of plant capacity and profit), and inefficient separations, among other factors.

Example 10.2 It is desirable to scale-up a mixing step from the laboratory to an agitated vessel in the plant. If the same level of turbulence in the mixing step is required:

(a) Should the agitator speed (in rpm) at the plant be faster, equal to, or slower than the speed at the lab?
(b) If this reaction is dependent on turbulent mixing, what green engineering implications would result if the scale-up process was done incorrectly?

Solution

(a) The Reynolds number, N_{Re}, a dimensionless number, is useful in describing fluid flow and conditions. A Reynolds number below 20 indicates laminar flow, Reynolds numbers above 10,000 denote turbulent flow, and Reynolds numbers in between denote transitional flow, flow that combines laminar and turbulent flow. Since $N_{Re} = (D^2 \rho N)/\mu$, where D is the impeller diameter (cm), N the rotational speed (rev/s), ρ the density of the fluid (g/cm^3), and μ the viscosity of the fluid (g/s·cm), to maintain the Reynolds number constant (same degree of turbulence), the impeller speed will have to slow down.
(b) Depending on the reaction and the system, inefficient mixing could result in concentration gradients, which could, in turn, result in a decrease of mass efficiency due to a loss in yield (molecules do not have the opportunity to interact). In addition, there would be increased waste, including unreacted chemicals that would need to be

separated later, and these separations will increase the environmental footprint. Also, if this were a heat-dependent reaction, not all the reaction mixture would be at the same temperature and therefore this is likely to cause an inefficient use of energy.

Additional Points to Ponder Would the answer be the same if using P/V for scaling up? What parameter would you use to scale-up highly exothermic and rapid reactions?

Heat Transfer When conducting an experiment in a test tube or a flask, exothermic reactions are relatively easy to control. At a commercial scale, heat is more difficult to remove. A rule of thumb is that a 1000-gal reactor has 10% of the relative heat-removal surface area of a 1-gal reactor.[9] Heat transfer is very important when scaling up chemical reactors. For example, batch reactors are heat transfer–limited; thus in most instances for batch reactors the heat-transfer rate dictates the capacity of the next-larger-scale reactor.[10] Heat transfer scale-up issues (coupled with mixing issues) are often key stumbling blocks when scaling-up continuous processes. It is a typical axiom of scale-up that "everything runs hotter and takes longer in the plant," and this is caused mainly by the loss of heat transfer efficiency when going to a larger scale.

In heat-sensitive systems such as biological systems, heat transfer issues during scale-up are even more important. In the case of bioprocesses such as fermentations where biomass is grown, the biomass production is highly dependent on the rate of heat removal. For example, with a cylindrical bioreactor the heat balance will show that cell productivity is proportional to $1/V^{1/3}$, where V is the reactor volume. In other words, large-scale fermentation vessels will require a larger heat transfer area to counteract the loss of cell productivity.[11] In addition, long cooling times due to inefficient heat transfer can lead to destruction of thermally sensitive materials in both bioprocesses and traditional chemical processes.

Heat transfer issues coupled with mixing difficulties can also lead to localized heating and the formation of hot spots that could favor secondary reactions. From a green engineering/green chemistry perspective, unresolved heat transfer issues during scale-up can result in an inefficient use of energy (i.e., losing energy resources (heating or cooling) or using them inefficiently), potential formation of impurities, loss of yield/reaction mass efficiency due to secondary reactions, and potential safety issues associated with runaway exothems, gas evolution, and explosions.

Example 10.3 A highly exothermic semibatch reaction between just two reactants was scaled up to a 750 gal plant reactor, running only 5 to 10 degrees hotter than in the laboratory to meet cycle time requirements. At the plant, yield loss to the formation of a dimer by-product was 4 times as large as in the laboratory.

(a) What could have happened?
(b) What are the green engineering implications?
(c) What are other potential alternatives to solve this scale-up issue?

Solution

(a) Leng[9] reported that following laboratory studies, the main cause of dimer formation was suspected to be heat transfer. The main reaction, $A + B \rightarrow C$, was very fast, whereas the competing reaction, A + C → D, was very slow. Temperature was shown to be the most important parameter.

(b) Green engineering impacts include reduced mass efficiency, inefficient use of energy, potential loss of product due to excess impurities, the potential increased environmental footprint due to separations (e.g., a mass separating agent might be needed), and the creation of waste (a dimer by-product that needs to be separated).

(c) In this particular reaction, the development team discovered that if the reaction was carried out at 0 to 5°C for 9 h, the dimer concentration was less than 1%. If the concentration of reactant A was maintained low through metered addition, the percent of the dimer was less than $^1/_2$%, even at a relatively high temperature of 20°C. However, even though this was reported in the literature for this example, one must suspect that heat transfer is not the full story. The fact that the main reaction is fast and the competing reaction is slow may indicate that mass transfer and mixing might also be an issue. The combination of mass *and* energy transfer issues might be creating an opportunity for the formation of areas of localized mixing at relatively high temperatures for long periods of time, which would favor the competing reaction and formation of the by-product. The system would have benefited from good design of experiment studies to better understand the combined effects of residence time, mixing, and kinetics.

Additional Points to Ponder What other types of reactors might have been considered for the reaction described in this example? What is the potential effect of chemical kinetics on the reaction?

Phase Equilibrium Multiphase systems are very common in chemical process, and in many cases they are desirable and necessary. Even when reactions and separations are designed to work homogeneously, different phases can form and coexist without adverse consequences. Understanding the phase equilibrium and interfacial relationships of the fluids involved in the process is in many cases crucial for a successful scale-up.

Crystallization, for example, is achieved by causing a state of supersaturation, either by cooling, evaporation, reaction, or using an antisolvent with a different phase equilibrium profile. Thus, an understanding of the phase equilibrium such as the solubility limit and the supersaturation point is basic for scaling-up crystallization processes. For example, changing the liquid composition can have a large effect on crystallization as a result of how these changes affect the degree of saturation of the mother liquors. Crystallization is indeed a challenging unit operation to scale-up, because of the interactions between mass transfer, heat transfer, and phase equilibrium.[12]

Another important consideration in phase equilibrium is the presence of water, and the interaction of water with other chemicals, such as solvents. In the laboratory, it is easier to run reactions and separations anhydrously; in the plant that is not always easy or even possible. In addition, the degree of miscibility of organic materials in water is important for separations and workups. The degree of water miscibility of a solvent might make separations and purifications easier (if phase separation is ready) or extremely difficult (if azeotropes are formed). Phase equilibrium data for reaction and separation systems is typically modeled using assumptions; therefore, it is important to add experimental data to knowledge about the system as much as possible, and the experiments to obtain this information should be included in the scale-up plan. From a green chemistry/green engineering perspective, phase equilibrium issues during scale-up can cause decreased efficiency in the process, additional waste formation, and increased energy needed for separations and workups.

Impurities Small amounts of impurities can have a huge effect on processes being scaled-up. Impurities in general can complicate purifications and decrease yield and are particularly costly for separations and purifications, especially for high-value products such as fine chemicals and pharmaceuticals. At the laboratory scale, most reactants used are analytical grade, but in the plant the impurities present in raw materials can react adversely with some of the equipment or other materials in the unit operations. It is also important to know what types of impurities may cause a problem, as not all 99%-pure materials are born equal.

When performing experiments in the laboratory, the conditions tend to be controlled to avoid disturbances, and the amounts of material produced are small enough that traces of unreacted material do not make a significant difference. At larger scales, traces of by-products or unreacted materials can poison a catalyst or membrane or can accumulate in the system if recycling loops are present to a point that can hinder the reaction or separation. Purges in the recycling loops to avoid these problems are one strategy that can be used.

From a green chemistry/green engineering perspective, impurities and the associated scale-up problems can lead to decreased mass efficiency, increased energy for separations, additional separation steps (e.g., potential need for a mass transfer agent), potential loss of product, and increased waste, among others.

Example 10.4 A semibatch crystallization process was scaled-up from a 1-L crystallizer to a 2000-gal crystallizer, followed by a filter. In general, the scale-up worked well for the crystallization, but the filtration, which started out just fine in the first half hour, then became very slow, with the added complication of very difficult cake washing and very slow drying. The filtration and drying process ultimately required three to five days to complete.

(a) What could have happened?
(b) What are the green engineering implications?
(c) What are potential alternatives for solving this scale-up issue?

Solution

(a) To solve this problem, samples were taken from the plant and analyzed at the laboratory. It was determined that a flocculent solid was clogging the wet cake pores and slowing down the filtration and the washing and impeding subsequent drying. The flocculent solid was traced back to an impurity in the raw material.

(b) Green engineering impacts include inefficient use of resources due to increased energy to filter, to wash the cake, and to dry, potential loss of product due to excess impurities, increased environmental footprint due to increased energy, increased waste production, and loss of plant capacity.

(c) In this case, the plant personnel found that if they left the slurry to settle, then decanted the mother liquids containing the flocculent material, they could dilute the slurry with fresh water and then proceed to a more rapid filtration. However, this is not an ideal solution of the issue. This engineering problem might also have been solved through supplier development and management to ensure that the flocculent solid was not present in the raw material, perhaps even transferring some green chemistry or engineering practices to the supplier to avoid production of the impurity altogether.

Additional Points to Ponder What other alternative could have been considered in this example? Why is the timing for filtration that important?

Material Properties The ideal reactant or reagent for scale-up produces the desired product at the maximum yield, in the expected time, with no toxicity, no environmental impact, and with minimal effort for workup, isolation, and purification of the product desired. For example, characteristics of the ideal reactant or reagent are likely to include the following:[13]

- *Selective:* specific for the desired transformation
- *Nonhazardous:* to operators, property, or the environment
- *Stable:* does not decompose or react with changes in temperature and pressure, moisture, or air
- *Catalytic:* instead of stoichiometric, and easy to recover
- *Enables separations:* facilitates workup and downstream processing
- *Easy to handle:* easy to transfer and manage at scale
- *Produces nonhazardous by-products:* or produces no by-products at all
- *Consistent quality:* standard quality desired with no impurities
- *Readily available:* easy to manufacture in a green manner
- *Inexpensive:* economically attractive for the desired market

In a similar fashion, ideal construction materials are inert to the reaction mixture, offer the desired heat transfer properties, and do not corrode. However, chances are that the reagents and materials involved in the process are not going to be ideal. Toxic and hazardous materials are easier to control and contain in the laboratory and at small scale, whereas in the manufacturing plant they can represent a large scale-up and operational issue, as we have seen in earlier chapters. The properties of construction materials can affect heat transfer and can cause interferences with reaction mixtures or process fluids, or vice versa; for example, extremely acid or alkaline solutions can corrode the materials of construction. Some physical characteristics of the chemicals used in the reactions might change with temperature and or pressure and might cause significant scale-up effects, and this is particularly important when dealing with non-Newtonian fluids (e.g., fermentation broths), as most design equations assume Newtonian behavior.

For example, one prevalent problem during scale-up and manufacturing is the formation of emulsions, especially in multiphase reaction systems and extraction operations. Emulsions not only decrease the yield of the separation or reaction, but can hinder plant operation. When looking at emulsions, in addition to mass transfer factors such as agitation, material properties such as viscosity and density play important roles. In addition, salts formed from corrosion of the materials used to make the equipment can aid in the formation of an emulsion.

From the viewpoint of green chemistry and green engineering, issues related to material properties during scale-up can affect worker exposure, environmental toxicity, and process safety, can cause lower mass efficiency due to poor separations, and can produce waste resulting from lower efficiencies and reactor or other equipment malfunctions.

Downstream Processing The scale-up of a chemical reaction must be done with a systems mentality, that is, with the entire process in mind. For example, laboratory experiments may have established that the highest mass efficiency and yield of a reaction between an olefin, ammonia, and oxygen are achieved when the olefin is present at a significant excess. However, the separation and recycle of the olefin could involve a large increase in capital and operating investment, a large increase in energy requirements, the potential need for mass separating agents, and additional emissions that could make it so costly that the decrease in mass efficiency for a reaction with a lower ratio of the olefin can be justified as a more cost-efficient and greener alternative.[14] This alternative will also probably result in a larger mass efficiency when the process is looked at in its entirety.

10.2.1 Type of Processing

In addition to the factors mentioned above, the type of operation for the process being scaled-up (e.g., batch, continuous, semibatch) plays an important role in scale-up. Almost all the development work at laboratory scale for liquid-phase processes is performed as a batch operation, primarily because it can be performed easily with typical laboratory equipment and without the need for special rigs. They can also be done with very limited amounts of materials, and the amount of waste produced, if an experiment fails, is limited. However, most industrial commodity processes are carried out as continuous processes, as large volumes are produced and continuous manufacture allows for close monitoring to maximize productivity. Typical examples of continuous processes are found in the traditional refinery and most bulk chemical, food, and agricultural processes. Continuous processes are also very useful for reactions with fast kinetics and those that need rapid heat transfer for heat dissipation. In practice, batch operations are limited primarily to pharmaceutical, fine chemicals, and other operations that require smaller production volumes.

Therefore, one has to consider the implications of scaling-up batch laboratory experiments into a continuous process. Although batch data such as physical properties and phase equilibrium data translate well into continuous processing operations, it is important to perform pilot-scale or mock-up experiments that represent the continuous operation to be set in the plant. These are needed to understand parameters such as mixing, temperature changes, heat transfer, and materials handling that do not translate directly from smaller batch experiments to continuous operations. One must also consider additional equipment requirements for running processes continuously, such as pumps, heat exchangers, and recycling loops.

10.3 SCALE-UP TOOLS

How do we know the real-life implications of our design choices in the laboratory? Ideally, we would like to be able to know the chemical and physical systems involved in a new process well enough to be able to model our system mathematically and translate it into our newly designed plant. However, this fundamental model is impractical, as the basic relationships of each unit operation can be rather complex, and it might take years and much money to achieve such an understanding of the entire process. On the other hand, a purely empirical understanding is obviously impractical, wasteful, and extremely costly; we saw earlier that it is impractical to build a chemical plant that mimics the conditions in the laboratory, and this has not only operational, but economic and green engineering consequences.

From a green engineering/green chemical perspective, it is indispensable to us to understand the scale-up implications of our decisions at each stage of process development. The developer, either the chemist or engineer, needs to strive for a sufficiently robust understanding of the implications resulting from the operational properties, the science, and the market forces for the decisions taken during process design, as these can turn an otherwise green chemistry into a very inefficient process, ultimately defeating the purpose of designing a green process.

Example 10.5 You have invited a very attractive friend out for dinner. She is a chemist and works in a fine-chemical company, and while you are waiting for your table she tells you that in her laboratory they have developed a very mass efficient reaction in a 100-mL flask maintained at 0°C in an ice bath with almost 100% conversion and mass efficiency. However, when the engineers took the reaction to scale it up to a 8000-L jacketed batch reactor, they found a series of by-products in the mother liquors that have significantly reduced the mass efficiency, yield, and conversion that your friend worked so hard to achieve. In addition, traces of some of these by-products are poisoning the catalyst in the next process step, so now the engineers are considering the addition of an intermediate separation operation and need to devise a way to dispose of the by-products, as these also have human and ecotoxicity concerns. She cannot tell you the details of the reaction, but since she knows you are taking a graduate green chemistry and engineering class, she is asking for your help, as she cannot believe that such a clean reaction is causing so many issues. You know you cannot ask about the specifics of the reaction, but since you really want to impress her, what could you point out to her as potential causes of the problem? What potential alternatives could you suggest?

Solution If you are waiting for your table at the bar, you might watch the bartender preparing a drink. There might, for example, be mass transfer issues. In the small equipment used at the bench scale, it is easier to produce close-to-perfect mixing, just as the bartender is doing. In a larger agitation tank, mixing is normally achieved by propellers, and flow patterns could be causing localized mixing and no-mixing areas so that the reactants might not be in contact at all in some parts of the reactor, resulting in decreased mass efficiency and conversion. Some alternatives that she might want to propose to the engineers are to perform mixing studies for the reactor (e.g., modeling, hydraulic mock-ups using markers).

Second, there might be heat transfer issues. The bartender has added some ice to the ingredients in a cocktail shaker, and that and the vigorous mixing are helping to keep the temperature uniform, just as an ice bath can do in a small flask. In the reactor she described, heat removal in this presumably exothermic reaction might not be sufficiently fast, and worsened by inefficient mixing, it might be generating hot and cold areas inside the reactor.

When this happens, there is a possibility of having reacting areas inside the reactor, enabling secondary reactions that would not happen at a lower temperature. If the residence time is longer than required by the reaction, this situation can be aggravated. This might be another cause of the presence of secondary reactions that weren't present at the laboratory scale. Some alternatives that you might suggest to her is to perform kinetic studies to investigate the appropriate residence time required, to study the heat transport inside the reactor (the mixing studies will come in handy here) to investigate potential creations of hot and cold spots, and to look a bit closer at the reaction mechanisms to better understand the potential for side reactions to occur.

Additional Points to Ponder What can you also think about potential safety issues with the problems she is experiencing in the reactor? If you are not old enough to be sitting at a bar, what example would you use to explain your answers?

Obtaining the additional information needed for process scale-up can be done by a combination of mathematical modeling or experiments (which in turn are physical or chemical models). The target of the scale-up is to achieve the optimal combination of fundamental and experimental work to minimize development time, cost, and waste generated during the development process and at the manufacturing scale. One illustration of this for reactor design is *Prater's principle of optimum sloppiness*, in which one seeks to find the optimal net value of a plant (benefit – cost), somewhere in between a purely fundamental model and a purely empirical model. Using models or performing experiments allows us to obtain the data we need, predict the behavior of the process beyond the limits we have studied, design appropriate control mechanisms, and in general understand how the process varies with size. Experiments to facilitate the scale-up process are complementary, and normally involve:

1. *Laboratory experiments.* These are needed to obtain information that is generally independent of the size and geometry of the process, such as chemical kinetics (reaction order, reaction-rate equation), thermodynamics (specific heats, heats of reaction), physical properties (density, viscosity), and phase equilibrium data. Use of statistical design of experiment is highly recommended during scale-up, as it permits us to probe many different variables of interest with the fewest number of experiments.
2. *Mock ups.* These are useful to study the physical mechanisms that will be more sensitive to size and geometry, such as hydrodynamics within a packed column and flow patterns in a mixing vessel.
3. *Pilot-plant experiments.* These reveal representative characteristics of the industrial plant and therefore allow for the simultaneous study of the chemical and physical systems and their interactions: for example, how the heat transfer coefficient in a specific type of vessel could affect the formation of by-products or moderate exothermic reactions. The key challenge in pilot-plant experiments is to achieve a representative model of the process on an industrial scale.
4. *Mathematical models.* These attempt to describe the reality of a chemical plant as closely as possible so that engineers are able to make design decisions as close to fundamental principles as possible, without the need to create potentially vast amounts of waste during large-scale experiments. Mathematical models include

equations that describe mass and energy balances in addition to mass, energy, and momentum transfers for a process in time and space. These models can be either:

a. *Steady state*, mainly for steady-state continuous processes. Steady-state models serve to describe a process in which the variables will have the same values after an intial startup period.
b. *Dynamic*, applicable to batch processes or continuous processes during startup and shutdown. Dynamic models are extremely important for the design of online monitoring and control systems, especially if there is a possibility of runaway reactions.

Simulators, which in essence are mathematical models, are extremely important in understanding the disturbances of a process and to develop online monitoring and appropriate controls for the process, such as exothermic reaction quenching, excess pressure buildup, changes in reactant feeds, and changes in composition or concentrations that if left uncontrolled can result in potential safety issues (e.g., explosions, fires), environmental issues (e.g., spills, rejected batches, out-of-spec wasted product, loss of efficiency), and health issues (e.g., accidental releases, leaks).

The optimal combination is not a trivial process, as illustrated in Figure 10.2. Physical and chemical data from experiments need to feed the mathematical models, the environment, health, and safety assessments, and of course, the scale-up work as the project moves from laboratory to full scale. As the process is better understood, additional experiments might be needed to refine the mathematical models or the EHS assessments. Information from the marketplace will be needed for the economic assessments and also to understand potential impurities of the raw materials that can affect the process. For example, when working with some membrane separation process, even traces of acid in the materials can cause the

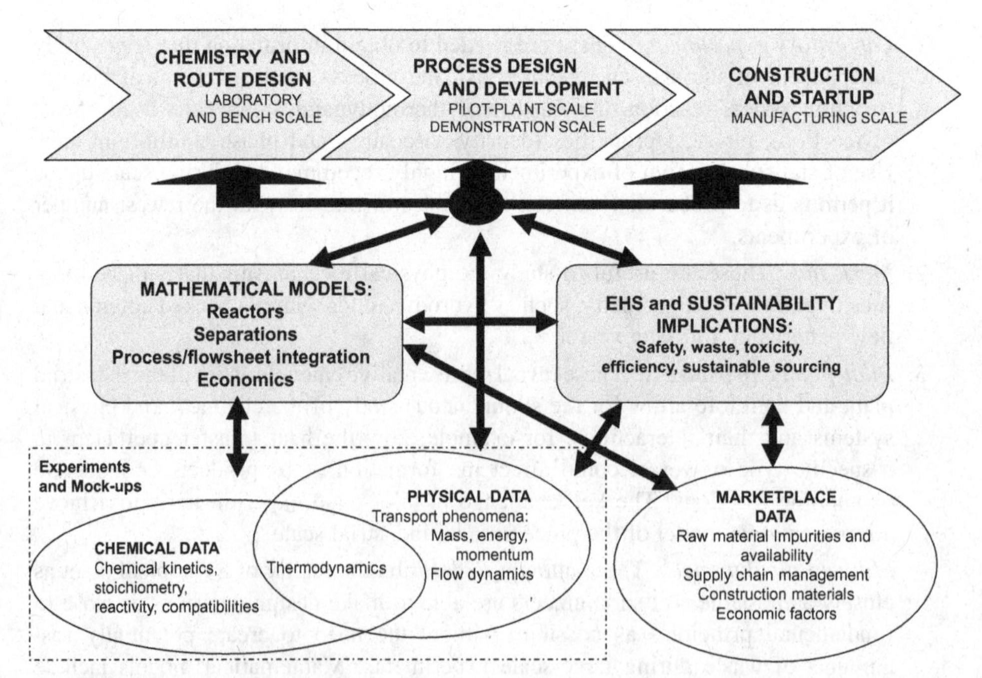

FIGURE 10.2 Scale-up process and how the various tools interact to provide the right information for the process at the right development stage.

membrane to degrade and thereby potentially render the process economically prohibitive given membrane replacement costs.

Additional information from the process, and the physical, chemical, and market characteristics can then be used to address the most common issues in scaling up a chemical process and can help to devise a strategy for manufacturability. Table 10.2 contains a summary of scale-up issues with their negative impacts on green engineering/green chemistry and potential strategies to address these issues.

TABLE 10.2 Potential Strategies to Address Scale-up Issues and Their Negative Impacts on Green Engineering and Green Chemistry

Scale-up Issue	Green Chemistry/Green Engineering Impacts	Potential Scale-up Strategies
Mass transfer	Loss of mass efficiency Decrease reaction mass efficiency Potential increase of by-products and impurities Inefficient separations Potential localized heating Loss of yield due to side reactions Slow progression of reaction Reduction of plant capacity and profit Increased costs	Perform mixing and fluid dynamic studies and modeling (e.g., perform sensitivity to mixing studies) Model fluid dynamics using mock-ups or computer modeling Perform kinetic studies Establish rate-limiting step mechanism Consider different methods of reactant incorporation (reverse addition, different feed place) Consider effect of different phases (e.g., solid formation) Determine critical design parameter for mixing system When scaling agitation, consider the propeller or turbine design Consider other sources of turbulence and mixing
Heat transfer	Potential loss of mass efficiency due to secondary reactions Inefficient use of energy Waste due to lost batches Potential safety issues such as runaway reactions, gas evolution, fire, and explosion Damage to temperature-sensitive products and materials Increased costs	Estimate heats of reaction using reaction calorimetry Evaluate heating and cooling requirements Perform dynamic heat balances to determine exothermic behavior and heat transfer rates Evaluate heat transfer capabilities of equipment If working with existing equipment, fouling might be an issue Perform reactive chemical testing such as differential scanning calorimetry and accelerated rate calorimetry Perform dust explosion testing if solid powders are involved Use computer models to simulate plant conditions Consider different methods of reactant incorporation (reverse addition, different feed place)

(continued)

TABLE 10.2 (*Continued*)

Scale-up Issue	Green Chemistry/Green Engineering Impacts	Potential Scale-up Strategies
Phase equilibrium	Decreased mass efficiency Additional waste formation Increased energy needed for separations and work-ups Increased costs	Obtain phase equilibrium data (e.g., VLE, SLE) among reactants, products, and impurities as much as possible Use computer modeling to evaluate saturation curves and to simulate disturbances Study interactions of equilibrium with mass and energy transfer
Impurities	Decreased mass efficiency Increased energy for separations Additional separation steps (e.g., potential need for a mass transfer agent) Potential loss of product Increased waste Increased costs	Perform experiments with commercial-grade raw materials Consider all sources of impurities though a mass balance, computer modeling, and sampling Analyze process streams for trace impurities Consider recycling studies to evaluate the need for purging Spike laboratory samples with impurities to test robustness Evaluate variation of impurity profiles with variations in yield
Material properties	Worker exposure Environmental toxicity, process safety Lower mass efficiency Increased energy use due to inefficient separations Poor separations Waste related to lower efficiencies and malfunctioning of reactors and other equipment Increased costs	Investigate environment, health, and safety properties of materials and effects on scale-up Conduct experiments to determine how physical properties change with pressure, temperature, concentration, etc. Experiment with large- scale construction materials
Downstream processing	Increased energy requirements Potential need for mass separating agents Potential additional emissions during separations Decreased reaction mass efficiency Increased costs	Evaluate knock-on effects of separation and treatment Perform mass and energy balances of the entire plant Use computer models to simulate plant conditions Perform flowsheet analysis studies Explore mass and energy integration across the entire plant

10.4 NUMBERING-UP VS. SCALING-UP

In previous sections we have covered key scale-up issues and some tools and techniques to circumvent them. However, given that the picture of scale-up and the work required is not rosy, it begs the question of how we might avoid scale-up issues altogether. How might we accomplish this?

One potential answer to that question is the idea of being able to "number-up" instead of scale-up. Imagine that once you have worked hard in the laboratory to produce a very clean reaction and a very efficient separation train, you could just replicate this many times to obtain the amount of product required. Since you will keep the dimensions of the equipment close to those used at the development level, there will be no need to worry about losses in mass transfer and mixing, as the equipment will be roughly the same. Heat transfer will also be as fast as it was in the laboratory, and the systems will behave the same way.

If this is possible, instead of scaling up and having to deal with issues arising from moving the production from a small scale to a larger one, we could think of using enough small-scale units to provide the production level required. This concept, known as *numbering-up*, is able to bypass the issues and problems of scale-up. This means that we would be designing small-scale manufacturing plants (benchtop plants) effectively and replicating the benchtop plants as many times as we need to produce the demand required. This image of replicating smaller manufacturing plants is at the heart of the concept of *process intensification.*

Process intensification is covered in more detail in Chapter 24, but it is important to understand how it relates to the idea of numbering-up and scale-up issues. Process intensification was first defined as a strategy to make dramatic reductions in the size of a chemical plant, either by reducing the size of the equipment or dispensing with steps so that we can achieve a given production goal.[15] The concept of process intensification has since been expanded to include increases in production capacity within a given piece of equipment, significant decreases in energy consumption, a marked cut in waste or by-product formation, and other advantageous outcomes.[16]

Process intensification uses only engineering principles to achieve the desired result; this means that the changes needed to intensify a process will be in the "how" the chemistry performs, not the "what" of the chemistry itself. To achieve the desired intensification, a combination of techniques and equipment can be used. Process intensification techniques are different ways of processing to intensify the production, such as the use of multifunctional reactors (e.g., reaction plus separation in one), hybrid separations (e.g., membrane absorption and stripping), and the use of alternative forms of energy (e.g., centrifugal force instead of gravity for phase separations). Process intensification equipment includes microreactors, static mixers, and compact heat exchanger reactors, among others.

If engineers are successful in intensifying processes, we may perhaps not only avoid scale-up issues but may even improve the performance of traditional equipment. For example, static mixers have a more size- and energy-efficient way of mixing liquids and for that reason can be considered for reactions in which mass and heat transfer need to be fast and efficient, something not easily achieved with traditional agitated vessels at large scale. To exemplify the concept of numbering-up, one could think of a plant in which a train of small static mixers set in parallel could be used instead of one large agitated vessel, after which the fluid can continue toward additional downstream processing (Figure 10.3).

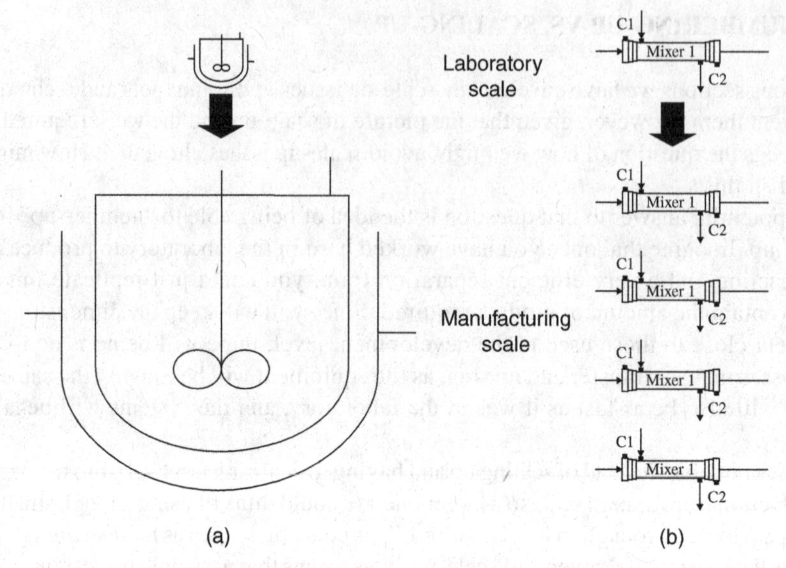

FIGURE 10.3 Scaling-up (a) vs. numbering-up (b).

Of course, the concept of numbering-up in the place of scaling-up works only to the extent that the intensified process meets the final demand. If process intensification is not successful, the possibility of numbering-up might be limited to small-volume production.

PROBLEMS

10.1 Describe the advantages of close collaboration of chemists and engineers during scale-up studies from a green engineering/green chemistry standpoint.

10.2 From a green engineering perspective, what are the advantages and disadvantages of the carbon adsorption system shown in Example 10.1?

10.3 Represent Prater's optimum sloppiness principle in a graph depicting curves for cost, value, and net benefit. What are the drawbacks of this principle? Why is this important from the green engineering viewpoint?

10.4 Provide three examples in which it would be useful to build mock-ups during the scale-up of chemical processes.

10.5 Develop a table contrasting the ranges of applicability of the following scale-up tools: laboratory experiments, computer models, mathematical models, and mock-ups. Describe when each method is useful to achieve an optimal balance of the tools used during scale-up, and when it is not.

10.6 For the following unit operations, describe the type of scale-up issues you might expect to encounter and its green chemistry/green engineering implications:
(a) Distillation
(b) Crystallization
(c) Drying

(d) Filtration
(e) Reactions
(f) Milling

10.7 During the scale-up of a surfactant that was the product of an oxidation reaction, air was used as the oxidant and to manage the heat generated by the reaction. In a 500-mL semibatch autoclave, it took less than an hour for complete conversion and there was no need to throttle the air supply. Thus, in the laboratory, the rate of heat transfer was sufficient to control the process temperature. When this reaction was scaled-up into a 2-gal semibatch autoclave, complete conversion was achieved in about 4 h, but the team was forced to throttle the air supply to maintain a constant process temperature. There was some product in the vent line from this reactor, but the team supposed that it had blown over during the reaction. When a 50-gal semibatch reactor was used, full conversion was achieved, but this required 8 h. After about 50% conversion, the team had to cut the airflow to the reactor to maintain a constant process temperature and keep the product in the reactor. When a 100-gal semibatch reactor was used, it was decided to operate it half full of reactant. The run went well until 50% conversion was reached; then the reactor slowly began to fill with foam, the process temperature started to rise, and the team had to throttle the air supply to control it. The run lasted about 10 h but achieved full conversion. When scaling this up to a commercial campaign, a 5000-gal reactor was filled half-full. At 50% conversion, the reactor filled with foam, and the first run took 24 h to reach completion and another 24 h to depressurize the reactor.

(a) Draw a flow diagram of the process.
(b) What type of scale-up issues might be present in this problem?
(c) What are the green chemistry/green engineering concerns?
(d) What caused the foam?
(e) What would you recommend using to solve the problem?

10.8 When using an agitated vessel for a reaction that has very fast kinetics with competing reactions, which of the following is a more appropriate scale-up factor: constant power per volume (P/V) or constant mixing time? Why?

10.9 Table P10.9 shows a comparison of parameters between one 1000-gal plant reactor and a 2-gal reactor of similar geometry. You have been asked to design mixing experiments in the 2-gal reactor that simulate the behavior of the 1000-gal reactor to solve some scale-up issues.

TABLE P10.9 Comparison of Parameters

	1000-gal Reactor		2-gal Reactor	
Speed (rpm)	Mixing Time (min)	P/V [(ft-lb$_f$/sec)ft^3]	Mixing Time (min)	P/V [(ft-lb$_f$/sec)ft^3]
50	2.3	0.6	1.9	0.01
100	1.2	4.6	1.0	0.10
180	0.7	26.9	0.5	0.60
250	0.5	72.2	0.4	1.57

(a) If the 1000-gal reactor is normally operated at 50 rpm, what speed would you use in your lab reactor to achieve a representative simulation?

(b) Why is this important from a green chemistry/green engineering perspective?

10.10 If the reaction of Problem P10.9 is very fast and very exothermic, would you do anything differently regarding the scale-up?

10.11 A two-step reactive precipitation was performed by first adding acid to a sodium organic compound and then reacting the organic acid with hydrogen peroxide to form the product desired. When scaling up the reaction, the resulting product had very fine particles and very poor centrifugation performance, with very long washing times. A follow-up study showed that the concentration of the organic acid at the end of the addition exceeded the solubility saturation limit threefold and precipitated out of solution.

(a) Write the chemical equations for this reaction.

(b) Draw a process flow diagram.

(c) What type of scale-up issues might be taking place?

(d) What are the green chemistry/green engineering concerns?

(e) What would you recommend to solve the problem?

10.12 The aerobic production of a protein from natural gas using *Methylococus capsultus* has an enthalpy of reaction of 79.6 MJ/kg,[17] represented by the following reaction:

$$CH_4 + 1.618O_2 + 0.0728NH_3 \rightarrow 0.3642CH_{1.8}O_{0.5}N_{0.2} + 0.6358CO_2 + 1.782H_2O$$

In a 1-L reactor, cell productivity was 6 g/m^3·h. If cell productivity is proportional to $1/V^{1/3}$, where V is the reactor volume:

(a) If the heat generation rate is balanced with the heat removal rate, what will be the expected cell productivity in a 50-L reactor?

(b) What is the global warming potential derived from this process?

(c) Which green chemistry/green engineering problems are derived from the scale-up of this process?

10.13 What are the advantages and limitations of numbering-up vs. traditional scale-up principles?

10.14 Monolithic reactors are an example of process intensification equipment. Describe what these reactors are and their most important features and limitations.

10.15 You are considering the use of microreactors for a fine-chemical production process that involves a rapid exothermic reaction, so you want to avoid scale-up issues by numbering-up the reactors needed. To do this, you need to have the idea approved by the head of the development department.

(a) What would be your argument?

(b) You want to bring the EHS director to the meeting as your ally. How would you approach her to sell the idea?

(c) The head of the development department asks you what the risk factors of this strategy are and what aspects other than the reactor need to be considered? What would your answer to him be?

REFERENCES

1. Euzen, J.-P., Trambouze, P., Wauquier, J.-P. *Scale-up Methodology for Chemical Processes*. Editions Technip, Paris, 1993.
2. Peters, M. S., Timmerhaus, K. D., West, R. E. *Plant Design and Economics for Chemical Engineers*, 5th ed. McGraw-Hill, New York, 2002.
3. Bisio, A., Kabel, R. L., *Scaleup of Chemical Processes: Conversion from Laboratory Scale Tests to Successful Commercial Size Design*. Wiley, New York, 1985.
4. U.S. Environmental Protection Agency. *Granular Activated Carbon Absorption and Regeneration.* Wastewater Technology Fact Sheet. EPA 832-F-00-017. U.S. EPA, Washington, DC, Sept. 2000. Available at http://www.epa.gov/OWM/mtb/carbon_absorption.pdf.
5. Turaga, U. T., Engelber, D. R., Beever, W. H., Osbourne, J. T., Wagner, B., Allen, R., Braden, J. Succeed at catalyst scale-up. *Chem. Eng. Prog.* 2006, 102(6), 29–33.
6. Bird, B., Stewart, W. E., Lightfoot, E. N. *Transport Phenomena*, 2nd ed. Wiley, Hoboken, NJ, 2006.
7. Perry, R. H., Green, D. W. *Perry's Chemical Engineer's Handbook*, 8th ed. McGraw-Hill, New York, 2007.
8. Worstell, J. H. Scaling fixed-bed catalyzed processes. *Chem. Eng. Prog.*, 2007, 103(3), 31–35.
9. Leng, R. B. From bench to plant: scale-up specialty chemical processes directly. *Chem. Eng. Prog.*, 2004, 100(11), 37–44.
10. Worstell, J. H. Succeed at reactor scale-up. *Chem. Eng. Prog.*, 2000, 96(6).
11. Doble, M. Avoid the pitfalls of bioprocess development. *Chem. Eng. Prog.*, 2006, 102(8), 34–41.
12. Genck, W. J. Optimizing crystallizer scale-up. *Chem. Eng. Prog.*, 2003, 99(6), 36–44.
13. Anderson, N. G. *Practical Process Research and Development*. Academic Press, San Diego, CA, 2000.
14. Carrberry, J. J. *Chemical and Catalytic Reaction Engineering*. Dover Publications, Mineola, NY, 2001.
15. Ramshaw, C. The incentive of process intensification. *Proceedings of the First International Conference on Process Intensification for the Chemical Industry*. BHR Group, London, 1995.
16. Stankeiwicz, A. I., Moulijn, A. J. Process intensification: transforming chemical engineering. *Chem. Eng. Prog.*, 2000, 96(1), 22–34.
17. Leiba, T. M., et al. Bioreactors: a chemical engineering perspective. *Chem. Eng. Sci.*, 2001, 56, 5485–5497.

11

REACTORS AND SEPARATIONS

What This Chapter Is About In previous chapters we explored how to perform mass and energy balances, how to develop a flowsheet, and the effects of the conditions on the reactions. In this chapter we explore a few of the unit operations that enable greener reactions and separations in chemical processes. We present an overview of reactors and separations (e.g., distillation crystallization, drying, size reduction) and the relationship of these unit operations to the design of greener, more sustainable chemical processes, integrating the concepts of green engineering.

Learning Objectives At the end of this chapter, the student will be able to:

- Understand the importance of reactors and separations for the design of chemical processes in the context of green engineering and green chemistry.
- Identify the key factors that affect the selection of reactors, separators, and their implications on green engineering.
- Identify advantages and disadvantages of reactors and separators in the context of green engineering.
- Understand and contrast the implications of unit operations (reactors and separations) using the principles of green engineering as a guide.

Green Chemistry and Engineering: A Practical Design Approach, By Concepción Jiménez-González and David J. C. Constable

11.1 REACTORS AND SEPARATIONS IN GREEN ENGINEERING

Reactors and separations are the major components of chemical process design; their characteristics and how they are related to each other in the chemical process have direct green engineering impacts. The reactor can be seen as the central unit operation in a chemical process; it is where the main chemical transformations take place and it represents the first step toward the desired product. However, separations are indispensable for obtaining the desired product at the level of purity required, resulting more often than not in the separation train representing the largest part of the chemical process: in design, footprint, investment, and energy requirements.

From a green engineering viewpoint, the reactor design and selection will influence the efficiency of the chemistry taking place in the reactor, creating a series of design questions: How pure is the desired product? How efficient is the reaction? Did secondary reactions take place? Can heat from exothermic reactions be dissipated easily? How much waste will be generated as a result of the reactions? Furthermore, the reactor and reaction will also affect the types of separations that will be needed to recover and purify the desired product with an increased number of design questions: How easy is it to separate the desired product and by-products? How can the desired level of purity be achieved? How can the unreacted species be recovered and reused? How easy is it to recycle by-products? Is a purge needed? How much energy will this separation require?

It is therefore very important to understand the type of unit operation under consideration for the design of industrial processes and the green engineering implications that the selection of a certain unit operation will have.

In this chapter we explore reactors and separations in terms of the green engineering principles commonly accepted to date, but these can be simplified.[1] At the end of the day, applying green engineering principles to your work as an engineer or process chemist means that when selecting reactors or separations, you should strive to maximize resource utilization, minimize hazards and pollution, and evaluate the life cycle implications of your decisions.

11.2 REACTORS

The reactor is the heart of an industrial chemical process. Woods had previously proposed that considering sustainability issues, the impact on the environment, and the hazards associated with the reaction as the first step.[2] The objective is to utilize the types of reactors that maximize conversion and selectivity while minimizing or eliminating secondary reactions. There are, however, a series of challenges that stand in the way of accomplishing this objective:

- Competing secondary and/or parallel reactions that decrease mass efficiency
- Exothermic or endothermic reactions that require temperature control
- Different reaction speeds, which require a designer to address reactions controlled by equilibrium, mass transfer, or kinetics
- Nonideal mixing, which can cause concentration and temperature gradients
- Other mass and energy transfer problems, driven by different phases of the reaction mixture (e.g., homogeneous vs. heterogeneous reactions)

The factors mentioned above and others contribute to reduced efficiency of reactions and potential safety issues. Designing greener chemical processes should involve designing and/or selecting reactors to minimize these issues and maximize efficiencies. Operating conditions such as temperature, pressure, concentrations, and catalysts also need to be taken into account when designing the reaction system to avoid safety hazards. For example, the optimization of catalysts during gasoline and diesel production resulted in increased yield, reduced coking, reduced emissions, less extreme operating conditions, increased throughput, increased catalyst life, increased conversion, improved safety, and improved stability, contributing to a greener process design.[3] A summary of some of the factors affecting the selection of reactors to be used in a chemical process are presented in Table 11.1.

There are two perincipal broad types of reactors based on flow characteristics: plug flow reactors (PFRs) and or stirred tank reactors (STRs). STRs can be operated continuously [also known as continuously stirred tank reactors (CSTRs)] or in batch mode. In general, a PFR is a tube through which the reactants and products flow. These tubes can be in the form of packed columns, loops, tray columns, and others. Stirred tanks are what the name describes, in the form of a relatively small pharmaceutical 100-gal reactor or as a large aeration tank or oxidation pond. There are a series of these reactors, which are a subset of these two general classifications, as can be seen in Figure 11.1.

TABLE 11.1 Some Factors Influencing Reactor Selection

Area	Factor
Chemistry	Reaction kinetics
	Equilibrium (reversible vs. irreversible)
	Parallel and competing reactions
	Thermodynamics (endothermic, exothermic)
	Chemical route
	Heat and mass transfer: mass, rate, or equilibrium controlled
	Intermediates
	Selectivity
	Hazards
Reactants and products	Hazards
	Concentrations
	Impurities
	Phase
	Cost
Catalysts	Particle size and porosity
	Sintering
	Fouling
	Regeneration
	Activity
	Selectivity
	Surface area
	Temperature limitations

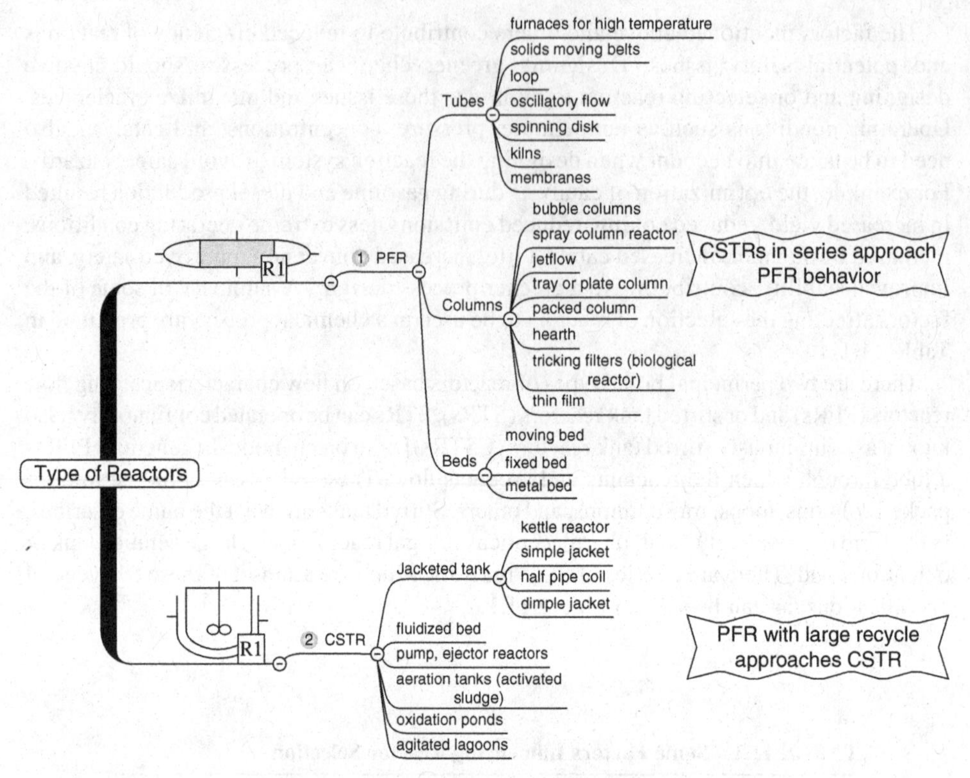

FIGURE 11.1 Reactor configurations.

From a green engineering perspective, the objective is to select the reactor configuration that would contribute to the most efficient process by:

- Maximizing resource utilization
- Minimizing hazards
- Accounting for the life cycle implications or your choice

Selecting the type of reactor should account for the type of separation train that will bring the products and by-products to the desired level of purity, thus maximizing the efficiency of the entire process. Therefore, there is in general not a single reactor that is best, but there is a need to carry out the reactor selection and design process with a good understanding of the type of reaction (e.g., first order, second order), the phases involved (solid, gas, liquid, a combination), the operating regime (continuous, batch), the mechanism that controls the reaction (mass, rate, equilibrium), and the thermodynamics of the reaction (endothermic, exothermic), to name just a few of the many factors to be considered. There will be a series of circumstances in which a CSTR would be more appropriate than a PFR, and vice versa. For example, Figure 11.2 shows how the general area of applicability for different types of reactors varies with a series of parameters, such as the production rate, the residence time of the reaction fluid, and the hourly space velocity.

Example 11.1 In previous experiments it was determined that to have a throughput of 70 L/day, the desired residence time for a first-order reaction was 15 min in a single shift of 8 h. Using Figure 11.2 as general guidance, which type of reactor can be considered for this reaction?

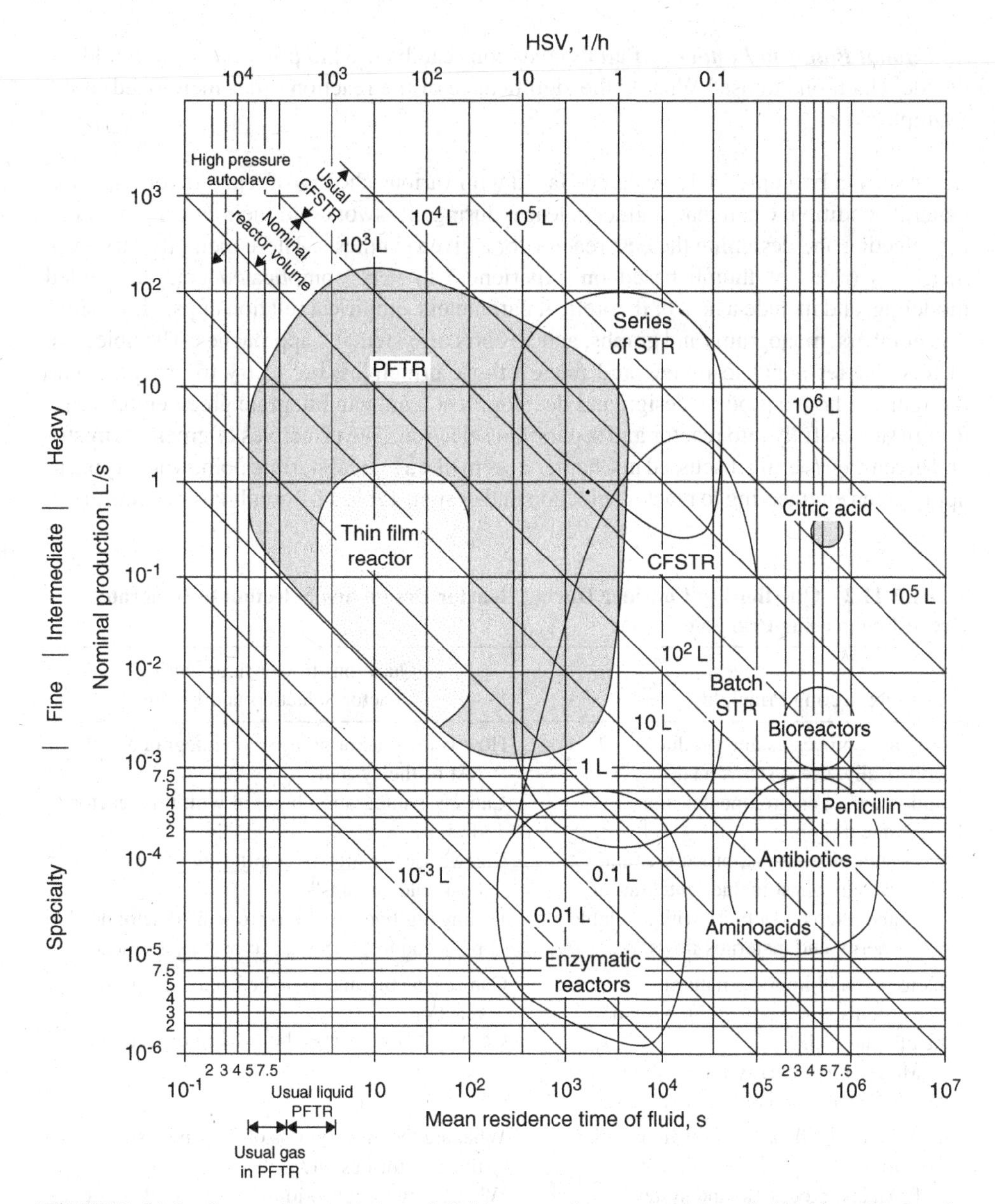

FIGURE 11.2 Some characteristics of fluid processing reactors in relationship with hourly spatial velocity, production, and residence time. (From Woods,[2] with permission. Copyright © 2007 Wiley-VCH, New York.)

Solution To be able to use Figure 11.2, we need to convert the residence time and throughput units to seconds and liters per second.

$$15\,\text{min} \times 60\,\text{s/min} = 900\,\text{s}$$

$$(70\,\text{L/day})(1\,\text{day/8 h})(1\,\text{h/3600 s}) = 0.0024\,\text{L/s} = 2.4 \times 10^{-3}$$

From Figure 11.2, we have that as a general guidance, it can be seen that a CSTR, a batch STR, a PFR, or a series of STRs can, in principle, all be used to perform this reaction.

Additional Points to Ponder If all the reactions can be used in principle, how would we decide which one to use? What is the significance of the reaction order mentioned in the example?

As seen in Example 11.1, we can be faced with various choices of reactors for a specific general condition. Chemical engineers learn, during classwork and their careers, methods for selecting and designing the best reactor for a given synthesis. These methods vary from heuristics (rules of thumb based on experience) to very complicated computer-aided modeling, and include a host of theoretical parameters, empirical relationships, dimensionless numbers, monograms and graphs, and all sorts of systematic approaches. The objective in this chapter is not to review and relearn those principles, but to try to provide some thoughts on how a process design and development team can integrate green engineering design successfully into reactor and separation selection. The principles of green chemistry and green engineering discussed in Chapter 2 seem like a logical starting point when trying to apply green engineering to reactor selection and design. Table 11.2 combines the nine green

TABLE 11.2 Questions to Consider During Reactor Design and Selection to Integrate Green Engineering Principles

Green Engineering Principles	Questions to Consider During Reactor Selection and Design
1. Engineer processes and products holistically, use systems analysis, and integrate environmental impact assessment tools. Design of products, processes, and systems must include integration and interconnectivity with available energy and materials flows.	How is the reactor going to be integrated with the rest of the system? Can mass integration be used with this reactor design? Can we use hot streams coming out of the reactor to heat cold streams? Do any fugitive emissions from the reactor need to be accounted for in an impact assessment?
2. Conserve and improve natural ecosystems while protecting human health and well-being. Material and energy inputs should be renewable rather than depleting.	Can renewable materials be used during the reaction? Can renewable energy be used to operate the reaction?
3. Use life cycle thinking in all engineering activities. Products, processes, and systems should be designed for performance in a commercial "afterlife." Material diversity in multicomponent products should be minimized to promote disassembly and value retention. Targeted durability, not immortality, should be a design goal. Embedded entropy and complexity must be viewed as an investment when making design choices on recycle, reuse, or beneficial disposition.	What are the implications of decommissioning for this reactor design? What are the implications of reactor cleaning for this configuration? How would this reactor affect energy consumption for the process and for the plant? Is this reactor maximizing the overall mass efficiency of the process? Is this reactor configuration contributing to additional separations? What type of energy is being used in this process? Can the catalyst be regenerated or recycled? Can the solvent be recovered?

TABLE 11.2 (*Continued*)

Green Engineering Principles	Questions to Consider During Reactor Selection and Design
4. Ensure that all material and energy inputs and outputs are as inherently safe and benign as possible. Designers need to strive to ensure that all material and energy inputs and outputs are as inherently nonhazardous as possible.	Is this reactor minimizing the risk of runaway reactions? Are we integrating inherent safety principles? Are there any other aspects of this reactor that can pose a hazard to the facility?
5. Minimize depletion of natural resources. Separation and purification operations should be designed to minimize energy consumption and materials use. Products, processes, and systems should be designed to maximize mass, energy, space, and time efficiency. Products, processes, and systems should be "output pulled" rather than "input pushed" through the use of energy and materials.	Is this reactor maximizing mass efficiency? Is this reactor minimizing energy use? Is heat transfer being optimized to maximize energy efficiency? Are mixing and mass transfer being optimized? Can we increase the mass efficiency by driving the equilibrium with water removal as the reaction progresses? Can catalysts or solvents be recycled? Can we apply process intensification? What are the optimal operating conditions (temperature, pressure, concentration) for this reaction and this reactor?
6. Strive to prevent waste. It is better to prevent waste than to treat or clean up waste after it is formed.	Can the reactor help us to run a more concentrated reaction to avoid solvent waste? Can the catalyst be regenerated? Is heat transfer being optimized to maximize energy efficiency? Are mixing and mass transfer being optimized? Can the solvent be recycled? What is the kinetics of the reaction?
7. Develop and apply engineering solutions, while being cognizant of local geography, aspirations, and cultures	Do we have in-house expertise for this type of reactor or do we need to develop the expertise? If we implement the new reactor technology, would training be available in the local language?
8. Create engineering solutions beyond current or dominant technologies; improve, innovate, and invent (technologies) to achieve sustainability. Design for unnecessary capacity or capability (e.g., "one size fits all") solutions should be considered a design flaw.	Have we tried a newer reactor design or a less traditional reactor that would provide better mass efficiency? Can we consider process intensification? Would the reaction be more efficient if we tailor the reactor instead of using existing equipment?
9. Actively engage communities and stakeholders in the development of engineering solutions.	Can we work with our vendors and suppliers to design a reactor that would: Increase mass efficiency? Have fewer fugitive emissions? Better control exotherms?

engineering principles from the San Destin Declaration[4,5] and the 12 green engineering principles of Anastas and Zimmerman (mostly as a subset of the San Destin principles)[6] to provide a series of questions and factors to consider during reactor design and selection in order to integrate green engineering principles. It can be noted that several of these questions appear repetitive, but that reflects the fact that different green engineering principles are extremely closely interrelated.

One important point to highlight from the table is that traditional reactor configurations may be used with increased efficiency, but one can only get so far in achieving a greener engineering vision. For this reason, innovative ways of designing and selecting reactor systems are needed. For example, scientists and engineers have explored new energy sources to activate chemical reactions, such as ultraviolet light, microwaves, and ultrasound[7] as we saw in Chapter 6. However, the industrial application of these types of technologies is limited, in great part, by the lack of suitably developed technology to carry out these reactions at scale.

Example 11.2 Microwave chemistry has the aim of accelerating reactions as a means of avoiding or minimizing the use of organic solvents, reducing reaction times, and reducing excess components, among other objectives. In Table 11.3 some of the microwave chemistry reactors currently available for synthetic applications are shown. What type of optimization or process development would this technology need at the current degree of development?

Solution It can be seen from Table 11.3 that the scale of the reactors for microwave chemistry is limited. Although the scale issue can be circumvented by numbering-up rather than scaling-up the reactors, there is still a considerable amount of reaction engineering innovation to allow for a fully effective application of this type of chemistry. For example, there is a need to optimize control of radiation penetration, to transfer this primarily batch operation to continuous operation, to investigate safety implications and the controls needed, to improve the overall reactor design, and to extend the range of applications to other chemical reactions.

TABLE 11.3 Some Characteristics of Current Microwave Reactors

	Manufacturer			
	Sharp	Personal Chemistry	CEM	MLS/Milestone
Maximum power (W)	800 pulsed	300 unpulsed	300 unpulsed	1000 pulsed or unpulsed
Cavity volume (L)	15.7	<1	<1	42.8
Maximum power density in empty cavity (W/L)	~50	>300	>300	~23
Reaction scale (g)	100	<20	<50	3000

Source: ref. 8.

Additional Points to Ponder What advantages are to be gained with this additional optimization and development? What is the range of applicability of this technology?

Throughout the years, several reaction engineering technologies have been investigated and developed to provide answers in the quest for greener, more sustainable processes. Some examples of these less traditional reaction technologies that intend to contribute to greener, more sustainable reaction engineering options are shown in Table 11.4. These technologies are, however, in need of additional innovation as they are in different degrees of development, availability, and application. As shown in the green engineering principles, the engineer needs to strive to innovate, optimize, and overcome the barriers of the current state of the art to truly apply green engineering principles.

Example 11.3 Your company is exploring the possibility of using a spinning tube-in-tube reactor, which consists of one cylinder (a rotor) turning inside a second cylinder (a stator). A summary of this technology[9] claims that this reactor generates a high-shear zone where reactants can achieve molecular-scale mixing, which is achieved by reducing the annular

TABLE 11.4 Examples of Less Traditional Reaction Technologies and Their Potential Green Engineering Advantages

Reaction Technology	Brief Description	Potential Green Engineering Advantages
Microwave reactors	Using microwave irradiation to carry out chemical synthesis. The microwave radiation passes through the walls of the vessel and heats only the reactants and solvent, not the vessel. The energy transfer is produced by dielectric loss (not convection or conduction).	Rapid heating of reaction mixtures Accurate temperature control at elevated temperatures Microwave enhancement of reaction profiles Solvent-free reactions Increased yields Increased purities
Electrochemistry	An oxidation or reduction reaction that occurs in an electrochemical cell.	Cleaner technology for oxidations or reductions Reactions can be carried out rapidly using relatively simple equipment Avoids the use of potentially hazardous or toxic reagents
Photochemistry	An electronically excited state of a molecule is created via the absorption of electromagnetic energy at the appropriate wavelength. The excited molecule undergoes a direct chemical transformation into a stable product, or becomes an intermediate chemical reactant capable of initiating a secondary reaction.	Reagent-less system for radical chemistry

(*continued*)

TABLE 11.4 (*Continued*)

Reaction Technology	Brief Description	Potential Green Engineering Advantages
Solid-phase chemistry	Synthetic transformations are conducted with one of the reactant molecules attached to an insoluble material (solid support). It was originally developed for peptide synthesis.	Simple workup, just filter and wash No solubility issues of intermediates No handling and drying of intermediates, as is the case with the solution-phase route Coupling with combinatorial chemistry Ease of operation High speed of processing (save as much as 30% of the time taken for a similar route via solution-phase processing)
Particle bed reactor (metal bed reactor)	A reactor alternative for heterogeneous systems. The substrate (usually an aryl or alkyl halide) is passed up through a bed of metal chippings or powder, providing enhanced contact in a heterogeneous reaction. The use of an ultrasonic probe allows reproducible initiation of reactions by cleaning the surface of the metal particles.	Enhanced exotherm control Enhanced process robustness
Oscillatory flow reactors	An oscillatory flow reactor (OFR) is a continuous reactor in which tubes fitted with orifice plate baffles or smooth periodic cavities have an oscillatory motion superimposed on the net flow of the process fluid. It provides highly effective mixing (fluid turbulence) in tube reactors by the combination of fluid oscillations and baffle inserts.	Enhanced heat and mass transfer Efficient dispersion for immiscible fluids Uniform particle suspension Gas-in-liquid dispersions Multiphase mixing Batch and continuous modes of operation
Spinning disk reactor	The reactant streams are mixed in the center of the spinning disk, which gyrates at rotational speeds of up to 5000 rpm. When the fluids are applied to the disk, a thin film is formed (50 to 600 μm, depending on the viscosity of the streams). The film is subject to very high shear stress, which leads to high heat and mass transfer rates.	Improved heat and mass transfer rates Numbering-up instead of scaling-up Continuous operation

TABLE 11.4 (*Continued*)

Reaction Technology	Brief Description	Potential Green Engineering Advantages
Spinning tube-in-tube reactor	A cylinder (rotor) turns inside a second cylinder (stator). Rotor-stators have an annular gap of about 5 mm or more between the outside wall of the rotor and the inside wall of the stator. While the rotor is spinning rapidly, liquids, solids, or gases enter the annular space, where they encounter shear forces that can create mixtures, pastes, slurries, or emulsions, depending on the machine's design.	Precise control over the fluid dynamics of the reaction streamMolecular-scale mixing of the reactants Rapid mass transfer and higher yields Numbering-up instead of scaling-up Modular maintenance without having to shut down operations
Loop reactor	Reactants are introduced via a mixer to create a well-dispersed mixture. This mixture is circulated throughout the volume of the loop, which permits high mass transfer rates. A heat exchanger in the external loop allows for independent optimization of heat transfer. For continuous operation, product is separated by an in-line cross-flow filter which retains the suspended solid catalyst within the loop.	Shorter reaction times Optimized catalyst loading Proven scale-up Reliable control and improved safe operation Rapid translation of process data into operating hardware
Microreactors	A microreactor is a continuous reactor in which the characteristic dimensions are typically less than a millimeter. In its simplest form, it is a channel etched into a silicon or metal plate and sealed by a top flat plate, but more elaborate forms are possible, aiming at improved mixing and contacting.	Improved mass and heat transfer Improved control of residence time and temperature distribution Small diffusion lengths Possibility of higher selectivity, yield, and quality Numbering-up instead of scaling-up High-pressure operations Continuous operation of traditional batch processes

gap to between 0.25 and 1.5 mm. The reactor is made from stainless steel or corrosion-resistant Hastelloy and can operate at temperatures up to 300°C and pressures up to 400 atm. Reactor sizes vary from units processing a few pounds per day to large reactors with throughputs of more than 1 ton of product per hour. The annular gap size and shear rate can be scaled linearly. Reaction rates and thermal control can be carried out efficiently by halving heat dissipation. The reactor design is modular, so it can be used in series or in parallel and connected to other unit operations where a single module can be taken off-line and bypassed for maintenance or repairs without disturbing the remaining operations. Based on the information provided, what would be some advantages of using this type of reactor from a green engineering perspective?

Solution See Table 11.5.

Additional Points to Ponder What are the disadvantages of this type of reactor from a green chemistry/green engineering perspective? What types of companies would consider this technology?

TABLE 11.5 Advantages of Green Engineering Principles

Green Engineering Principles	Advantages
1. Engineer processes and products holistically, use systems analysis, and integrate environmental impact assessment tools. Design of products, processes, and systems must include integration and interconnectivity with available energy and materials flows.	The reactor has a modular design that might make it easier to integrate in a plug-and-play application. Allows for few disturbances to the process for maintenance and repairs. Since the shear rate can be scaled up linearly and the reactor can also be numbered up, fewer scale-up issues are expected.
2. Ensure that all material and energy inputs and outputs are as inherently safe and benign as possible. Designers need to strive to ensure that all material and energy inputs and outputs are as inherently nonhazardous as possible.	Since reaction rates and thermal control can be carried out in this reactor, this could allow for potentially safer handling of exothermic reactions. If micromixing is taking place in the reactor, this could allow for better heat exchange and heat dissipation, making the reaction inherently safer.
3. Minimize depletion of natural resources. Separation and purification operations should be designed to minimize energy consumption and materials use. Products, processes, and systems should be designed to maximize mass, energy, space, and time efficiency. Products, processes, and systems should be "output pulled" rather than "input pushed" through the use of energy and materials.	Enhanced mixing would contribute to enhanced mass efficiency. Enhanced mixing would contribute to enhanced heat transfer (better energy use, less energy-related emissions). Intensified process would contribute to a smaller plant footprint.
4. Strive to prevent waste. It is better to prevent waste than to treat or clean up waste after it is formed.	Enhanced mixing would contribute to enhanced mass efficiency. Enhanced mixing would contribute to enhanced heat transfer (better energy use, less energy-related emissions).
5. Create engineering solutions beyond current or dominant technologies; improve, innovate, and invent (technologies) to achieve sustainability. Design for unnecessary capacity or capability (e.g., "one size fits all") solutions should be considered a design flaw.	Applies non-traditional reactor technologies and improves on current technologies.

11.2.1 Back to Reactor Conditions

The question arises as to how to select the right reactor for the application desired, using traditional CSTRs, PFRs, or batch reactors or by involving newer, less traditional technologies. But selecting the type of technology is just one term in the green engineering equation. Once we have identified a few reactors that can effectively and efficiently achieve the outcome desired, the next step will be to define how best to operate the reactors from a green engineering point of view. As shown in Example 11.1, for roughly the same operating space, several reactors can give us the results we want but how do we determine which is the optimal design? Are there designs that are equivalent from a green engineering perspective? As we noted in Chapter 7, the conditions of the reaction play a pivotal role in green chemistry and green engineering. Just running a reaction a few degrees below reflux, as shown in Example 7.4, will provide energy savings and decrease energy-related emissions significantly. It is not enough to evaluate the type of equipment available; the solution has to be seen in terms of the desired outcome, in combination with the reaction conditions (temperature, pressure, concentration, mixing, etc.) and the specific reactive system.

For example, reaction kinetics plays a major role in the selection of a reactor for green engineering purposes. A first-order reaction can be run very effectively in a PFR if the reaction is isothermal, but if it is adiabatic, a combination of CSTR plus PFR might be the best combination. For multiple reactions, either parallel or sequential, the ideal reactor operation will be dependent on the kinetics of the reaction and the operating conditions. For example, as we have seen previously, extended residence times might promote side reactions and loss of yield.

11.3 SEPARATIONS AND OTHER UNIT OPERATIONS

Once a reaction is completed, the valuable products and perhaps unreacted species will need to be separated so that they can be sold and recycled, respectively. In addition, sometimes we will need to employ other types of unit operations to transform the materials to the right form so that they can be formulated, better managed, or sold (e.g., granulation, milling). In a typical chemical plant, most of the plant footprint and operating costs are related to separations and other unit operations, such as distillation towers, crystallizations, and liquid extractions. Given this, there are many opportunities to apply green engineering principles integrally in the separations and purifications to provide the desired product.

Separations and unit operations themselves represent a very wide topic that can occupy several volumes, let alone trying to cover unit operations that integrate green engineering principles. One common way is to visualize the variety of unit operations and classify them according to separation, transport, size enlargement, and size reduction operations, as shown in Figure 11.3. It is also common to classify separations according to the number of phases involved (e.g., homogeneous, heterogeneous). As can be seen, many types of unit operations are available in the process design toolkit. However, the level of technical understanding varies between operations. For example, in the United States, about 90% of chemical process separations are distillations,[10] so it is not a surprise that considerable effort has been dedicated to optimizing this unit operation. However, this lack of separations diversity has to

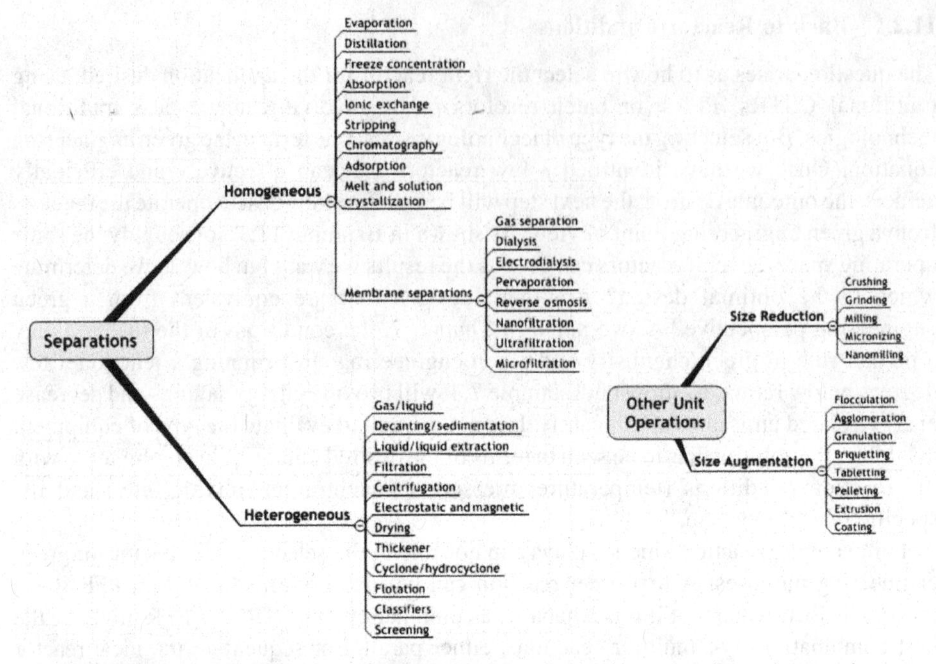

FIGURE 11.3 Examples of separations and size reduction/augmentation unit operations.

a certain point stifled the implementation of new technologies that might contribute to greener, more sustainable manufacturing.

One of the challenges of the engineer and process chemist is to look beyond the status quo toward more innovative chemical processing options that might render benefits from a green engineering viewpoint. The best known option is not always the best alternative. For example, the total energy requirements for concentrating mother liquors using some membrane separation processes (e.g., solvent resistant nanofiltration) are about 20 times smaller than that required for atmospheric distillation. This also translates into energy-related CO_2 emissions that are almost 6000-fold smaller for nanofiltration systems. This is also true when there is a driving force or equilibrium condition that will hinder the separation in the normal setting, as where there are azeotropes or eutectic mixtures. The following example illustrates what options might be available in the case of an azeotropic mixture.

Example 11.4 There is the need to separate an aqueous mixture of isopropyl alcohol. However, isopropyl alcohol forms an azeotrope with water, so normal atmospheric distillation will not provide the purity required. What other options could be used? What are their disadvantages from a green engineering perspective?

Solution There are several ways to "break" the azeotrope and dehydrate alcohols. Examples include:

- *Azeotropic distillation*: will require a mass separating agent
- *Extractive distillation*: will require a mass separating agent

- *Pervaporation*: will require that the membrane in the pervaporation unit be replaced and might not withstand certain solvents or impurities in the solvent mixture

Additional Points to Ponder Are there any advantages from a green engineering perspective? How would you know which option above is best?

As we did with reactors, Table 11.6 provides examples of questions to consider during the design and selection of unit operations. It can be seen that some of the questions for selecting separations unit operations are very aligned with questions for selecting reactors. This alignment occurs because, as with reactors, we are faced with the challenge of selecting the best separation or unit operation that will:

- Maximize resource utilization
- Minimize hazards
- Account for life cycle implications

Example 11.5 Your boss (who really likes to look at data) has asked you to compare two technology options for dehydrating isopropanol so that it may be recycled back into a process. The feed stream is composed of 59 wt% isopropanol and 41% water. Isopropanol with at least 99.5% purity is desired. The options presented to you are:

- Azeotropic distillation using benzene as a mass separating agent
- Extractive distillation using propylene glycol as a mass separating agent

What recommendation would you give to your boss and how would you justify your decision?

Solution Since you know that the green engineering principles would exclude the use of benzene as a mass separating agent given its toxicity, the extractive distillation seems a better alternative. However, you also know that the green engineering principles would advise you to avoid the use of a mass separating agent altogether. You know that pervaporation might be suitable for this task, so you decide to use mass and energy indicators to compare the two proposed options along with your idea of using pervaporation, as follows:

Option 1: Azeotropic distillation. For the design and system proposed, the ratio between the upper and lower layers of the azeotrope leaving the azeotropic column is $93.6/6.4 = 14.6$.[11] This ratio was used in calculations for the material balance. The chemical losses considered for this option are the fugitive emissions of isopropanol and benzene (estimated at 1% according to their boiling points)[12] and the wastewater produced as a waste stream, containing isopropanol and benzene. The composition of the upper and lower layers of the decanter were given in the literature. The overall mass balance is shown in Figure 11.4.

The main energy requirements are the heating requirements to operate the columns, the cooling requirements for condensing, and the electricity for pumping. To calculate the energy requirements, the following assumptions can be made:

TABLE 11.6 Examples of Questions to Consider During Unit Operation Design and Selection to Integrate Green Engineering Principles

Green Engineering Principles	Examples of Questions to Consider During Separation Selection and Design
1. Engineer processes and products holistically, use systems analysis, and integrate environmental impact assessment tools. Design of products, processes, and systems must include integration and interconnectivity with available energy and materials flows.	How is this separation going to fit in the overall process design? Can mass integration be used? What will be the effects if I change the separation sequence (e.g., removing B first instead of A)? Can we use hot streams coming out of this separator to heat cold streams? Are there any fugitive emissions from the separator that need to be accounted for in an impact assessment? Is this separation generating additional waste that I need to treat elsewhere? Can I combine this separation with another, or combine it with a reactor? Do I even need to separate/purify?
2. Conserve and improve natural ecosystems while protecting human health and well-being. Material and energy inputs should be renewable rather than depleting	If additional materials are needed, can renewable materials be used for this separation? Can renewable energy be used to operate the separation?
3. Use life cycle thinking in all engineering activities. Products, processes, and systems should be designed for performance in a commercial "afterlife." Material diversity in multicomponent products should be minimized to promote disassembly and value retention. Targeted durability, not immortality, should be a design goal. Embedded entropy and complexity must be viewed as an investment when making design choices on recycle, reuse, or beneficial disposition.	How would this separation affect energy consumption for the process and for the plant? Is this separation maximizing the overall mass efficiency of the process? Is this unit operation configuration contributing to additional separations? Do we really need it? What type of energy is being used in this process? Can mass separating agents be avoided? If a mass separating agent cannot be avoided, can it be recovered?
4. Ensure that all material and energy inputs and outputs are as inherently safe and benign as possible. Designers need to strive to ensure that all material and energy inputs and outputs are as inherently nonhazardous as possible.	Are the materials used in this separation benign? If not, how can we eliminate or avoid them? Are we integrating inherent safety principles? Can we eliminate occupational exposures by design with this unit operation? Are the operating conditions extreme for this separation? Can any other aspects of this separation pose a hazard to the facility?

TABLE 11.6 (*Continued*)

Green Engineering Principles	Examples of Questions to Consider During Separation Selection and Design
5. Minimize depletion of natural resources. Separation and purification operations should be designed to minimize energy consumption and materials use. Products, processes, and systems should be designed to maximize mass, energy, space, and time efficiency. Products, processes, and systems should be "output pulled" rather than "input pushed" through the use of energy and materials.	Is this separation maximizing mass efficiency? Is it really needed? Is this separation minimizing energy use? Are we purifying beyond the point required? Is heat transfer being optimized to maximize energy efficiency? Is mass transfer being optimized? Can mass separating agents be avoided, minimized, or recycled? Can we apply process intensification?
6. Strive to prevent waste. It is better to prevent waste than to treat or clean up waste after it is formed.	Are we generating additional waste with this separation that could otherwise be avoided? Are we using the easiest separation technique for this system? Can we apply some heuristics for separations? Can mass separating agents be avoided, minimized, or recycled? Is heat transfer being optimized to maximize energy efficiency? Is mass transfer being optimized?
7. Develop and apply engineering solutions while being cognizant of local geography, aspirations, and cultures.	Do we have in-house expertise for this technology, or do we need to develop the expertise? If we implement the new separation, would training be available in the local language? Are human factors being integrated into the design?
8. Create engineering solutions beyond current or dominant technologies; improve, innovate, and invent (technologies) to achieve sustainability. Design for unnecessary capacity or capability (e.g., "one size fits all") solutions should be considered a design flaw.	Are there fewer traditional separations that would provide better mass efficiency? Can this separation be combined with the reactor? Can two different separation techniques be combined?
9. Actively engage communities and stakeholders in the development of engineering solutions.	Can we work with our vendors and suppliers to design a separation system that would: Eliminate the need for a mass separating agent? Reduce energy consumption? Have fewer fugitive emissions? Be inherently safer?

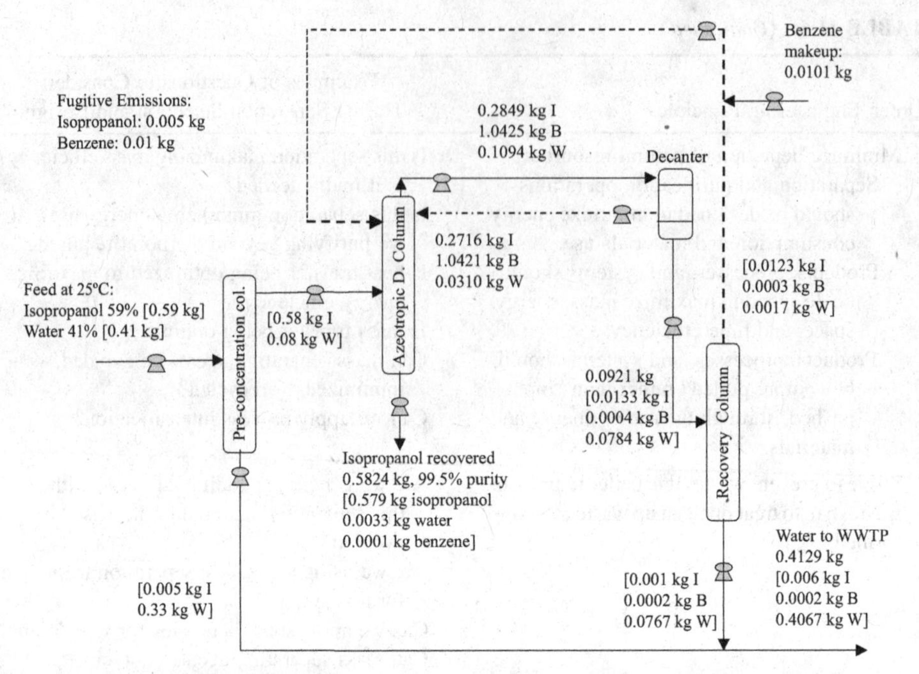

FIGURE 11.4 Mass balance for azeoptropic distillation of Example 11.5.

- Heat losses: 15%
- All distillates and bottoms cooled to 25°C (including the distillate of the azeotropic column before being fed into the decanter)
- Reflux ratio: 1.3
- Operating temperature of the preconcentration column: 80 °C; for the azeotropic column, 67°C, and for the recovery column, is 80°C;

Data used[13]:

- Basis: 1 kg of feed
- C_{pl} of isopropanol (I) = 154 J/mol
- ΔH_v of isopropanol = 45.5 kJ/mol (b.p. 83°C)
- m.w. isopropanol = 60 g/mol
- C_{pl} of water (W) = 75.37 J/mol
- ΔH_v of water = 43.99 kJ/mol (b.p. 100°C)
- m.w. water = 18 g/mol
- C_{pl} of benzene (B) = 136 J/mol
- ΔH_v of benzene = 33.9 kJ/mol
- m.w. benzene = 78 g/mol

Heating requirements of preconcentration column:

$$\Delta H = \text{sensible heat} + \text{vaporization heat} = \frac{1}{\eta}\left[\sum_{i}^{n} m_i C_{pi}\,\Delta T + (1+R)m_v\,\Delta H_v\right]$$

$$= \frac{1}{0.85}\left\{\left[\frac{0.6258\ \text{kg I}}{60\text{g/mol}}(154\ \text{J/mol}\cdot{}^\circ\text{C}) + \frac{0.4875\ \text{kg W}}{18\ \text{g/mol}}(75.37\ \text{J/mol}\cdot{}^\circ\text{C})\right](80-25)^\circ\text{C}\right.$$

$$\left. + 2.3\left[\frac{615\ \text{g I}}{46\ \text{g/mol}}(42.6\ \text{kJ/mol}) + \frac{85\ \text{gW}}{18\ \text{g/mol}}(44\ \text{kJ/mol})\right]\right\}$$

$$= 1928\ \text{kJ} = 1.93\ \text{MJ/kg feed} = 3.3\ \text{MJ/kg isopropanol recovered}$$

Heating requirements of azeotropic column:

$$\Delta H = \text{sensible heat} + \text{vaporization heat} = \frac{1}{\eta}\left[\sum_{i}^{n} m_i C_{pi}\,\Delta T + (1+R)m_v\,\Delta H_v\right]$$

$$= \frac{1}{0.85}\left\{\left[\frac{0.8673\ \text{kg I}}{60\ \text{g/mol}}(154\ \text{J/mol}\cdot{}^\circ\text{C}) + \frac{1.0426\ \text{kg B}}{78\ \text{g/mol}}(136\ \text{J/mol}\cdot{}^\circ\text{C})\right.\right.$$

$$\left. + \frac{0.1127\ \text{kgW}}{18\ \text{g/mol}}(75.37\ \text{J/mol}\cdot{}^\circ\text{C})\right](67-25)^\circ\text{C} + 2.3\left[\frac{284.9\ \text{gI}}{60\ \text{g/mol}}(45.5\ \text{kJ/mol})\right.$$

$$\left.\left. + \frac{1042.5\ \text{gB}}{78\ \text{g/mol}}(42.6\ \text{kJ/mol}) + \frac{109.4\ \text{gW}}{18\ \text{g/mol}}(44\ \text{kJ/mol})\right]\right\}$$

$$= 2757\ \text{kJ} = 2.76\ \text{MJ/kg feed} = 4.73\text{MJ/kg isopropanol recovered}$$

Heating requirements of recovery column:

$$\Delta H = \text{sensible heat} + \text{vaporization heat} = \frac{1}{\eta}\left[\sum_{i}^{n} m_i C_{pi}\,\Delta T + (1+R)m_v\,\Delta H_v\right]$$

$$= \frac{1}{0.85}\left\{\left[\frac{0.0133\ \text{kg I}}{60\text{g/mol}}(154\ \text{J/mol}\cdot{}^\circ\text{C}) + \frac{0.0004\ \text{kgB}}{78\ \text{g/mol}}(136\ \text{J/mol}\cdot{}^\circ\text{C})\right.\right.$$

$$\left. + \frac{0.0784\ \text{kgW}}{18\ \text{g/mol}}(75.37\ \text{J/mol}\cdot{}^\circ\text{C})\right](80-25)^\circ\text{C} + 2.3\left[\frac{12.3\ \text{g I}}{60\ \text{g/mol}}(45.5\ \text{kJ/mol})\right.$$

$$\left.\left. + \frac{0.3\ \text{gB}}{78\text{g/mol}}(33.92\ \text{kJ/mol}) + \frac{1.7\ \text{gW}}{18\ \text{g/mol}}(44\ \text{kJ/mol})\right]\right\}$$

$$= 60\ \text{kJ} = 0.06\ \text{MJ/kg feed} = 0.10\ \text{MJ/kg isopropanol recovered}$$

Total heating requirements:

$$8.13\ \text{MJ/kg isopropanol recovered}$$

Cooling requirements:
Since all the bottoms and distillates are assumed to be cooled to 25°C (the temperature of the feed), the energy required for cooling is

$$-8.13\ \text{MJ/kg isopropanol recovered}$$

Electricity requirements:
Electricity requirements are negligible in comparison to heating and cooling requirements. To simplify the calculations, a rule of thumb using the average standard pumping electricity requirements for the economical pipe size (15 m long) was used, with an average pumping energy of 0.003 kJ/kg pumped:

$$(0.003\ \text{kJ/kg pumped})(1 + 0.66 + 0.335 + 0.5824 + 0.0921 + 0.0778$$
$$+ 0.0139 + 1.4368 + 1.3447)\ \text{kg} = 1.7 \times 10^{-2}\ \text{kJ/kg feed}$$
$$= 2.9 \times 10^{-5}\ \text{MJ/kg IPA recovered}$$

Option 2: Extractive distillation. For the extractive distillation option, the solvent/feed ratio (or simply the feed ratio, on a molar basis) is normally between 1 and 4.[14,15] Two feed ratios were used in the calculations:

(a) S/F = 1 (minimum in the normal range)
(b) S/F = 2.5 (middle point in the normal range)

The overall mass balance for these two cases is shown in Figure 11.5.

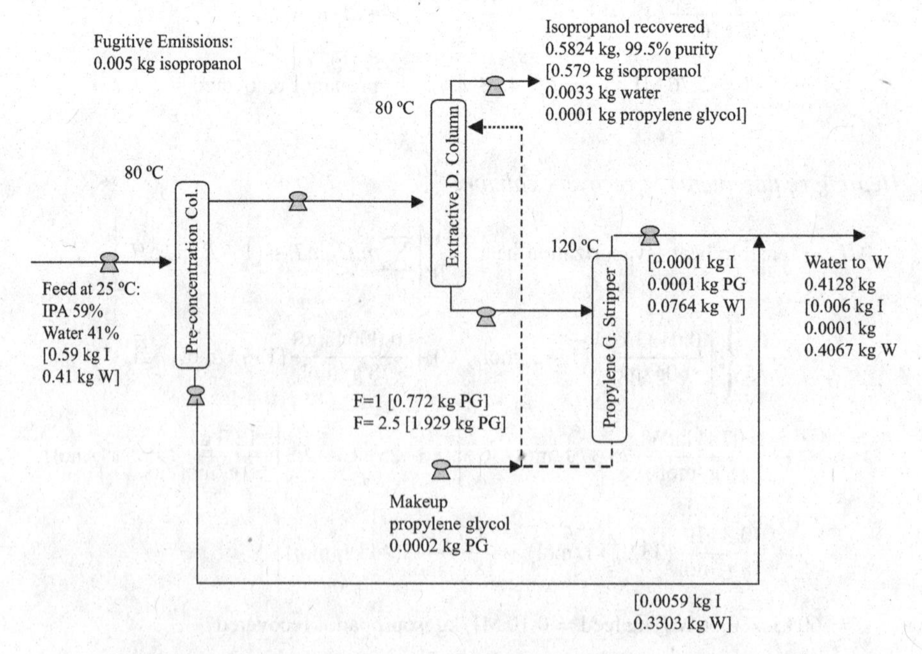

FIGURE 11.5 Mass balance for extractive distillation of Example 11.5.

The main energy requirements are the heating requirements to operate the columns (both columns operate at roughly 80°C) and the stripper (operating at 120°C), the cooling requirements for condensing and cooling, and the electricity for pumping. To calculate the energy requirements, the following assumptions can be made:

- Heat losses: 15%
- Stripper bottoms not cooled before being recycled to the extractive distillation column (remaining distillates and bottoms cooled to 25°C)
- Reflux ratio used for all the columns: 1.3
- Pumping energy: 0.003 kJ/kg pumped

Data used:

- Basis: 1 kg of feed
- C_{pl} of isopropanol = 154 J/mol
- ΔH_v of isopropanol = 45.5 kJ/mol (b.p. 83°C)
- m.w. isopropanol = 60 g/mol
- C_{pl} of water = 75.37 J/mol
- ΔH_v of water = 43.99 kJ/mol (b.p. 100°C)
- m.w. water = 18 g/mol
- C_{pl} of propylene glycol (PG) = 180 J/mol
- ΔH_v of propylene glycol = 64.4 kJ/mol (b.p. 187°C)
- m.w. propylene glycol = 76 g/mol

Heating requirements for preconcentration column:

$$\Delta H = \text{sensible heat} + \text{vaporization heat} = \frac{1}{\eta}\left[\sum_i^n m_i C_{pi}\,\Delta T + (1+R)m_v\,\Delta H_v\right]$$

$$= \frac{1}{0.85}\left\{\left[\frac{0.59\ \text{kg I}}{60\ \text{g/mol}}(154\ \text{J/mol}\cdot{}^\circ\text{C}) + \frac{0.41\ \text{kg W}}{18\ \text{g/mol}}(75.37\text{J/mol}\cdot{}^\circ\text{C})\right](80-25)^\circ\text{C}\right.$$

$$\left. +2.3\left[\frac{579\ \text{g I}}{46\ \text{g/mol}}(42.6\ \text{kJ/mol}) + \frac{79.7\ \text{gW}}{18\text{g/mol}}(44\ \text{kJ/mol})\right]\right\}$$

$$= 1924\ \text{kJ} = 1.92\ \text{MJ/kg feed} = 3.3\ \text{MJ/kg isopropanol recovered}$$

Heating requirements for the extractive distillation column:

(a) S/F = 1

$$\Delta H = \text{sensible heat} + \text{vaporization heat} = \frac{1}{\eta}\left[\sum_i^n m_i C_{pi}\,\Delta T + (1+R)m_v\,\Delta H_v\right]$$

$$= \frac{1}{0.85}\left\{\left[\frac{0.5791\ \text{kg I}}{60\ \text{g/mol}}(154\ \text{J/mol}\cdot{}^\circ\text{C}) + \frac{0.0797\ \text{kg W}}{18\ \text{g/mol}}(75.37\ \text{J/mol}\cdot{}^\circ\text{C})\right](80-25)^\circ\text{C}\right.$$

$$+\frac{0.772\ \text{kg PG}}{76\ \text{g/mol}}(180\ \text{J/mol}\cdot{}^\circ\text{C})(80-120)^\circ\text{C}$$

$$\left.+2.3\left[\frac{582.1\ \text{g I}}{60\ \text{g/mol}}(45.5\ \text{kJ/mol}) + \frac{3.3\ \text{g W}}{18\ \text{g/mol}}(44\ \text{kJ/mol}) + \frac{0.1\ \text{g PG}}{76\ \text{g/mol}}(64.4\ \text{kJ/mol})\right]\right\}$$

$$= 1124\ \text{kJ} = 1.24\ \text{MJ/kg feed} = 1.93\ \text{MJ/kg isopropanol recovered}$$

(b) S/F = 2.5

$$\Delta H = \text{sensible heat} + \text{vaporization heat} = \frac{1}{\eta}\left[\sum_i^n m_i C_{pi}\,\Delta T + (1+R)m_v\,\Delta H_v\right]$$

$$= \frac{1}{0.85}\left\{\left[\frac{0.5791\ \text{kg I}}{60\ \text{g/mol}}(154\ \text{J/mol}\cdot{}^\circ\text{C}) + \frac{0.0797\ \text{kg W}}{18\ \text{g/mol}}(75.37\ \text{J/mol}\cdot{}^\circ\text{C})\right](80-25)^\circ\text{C}\right.$$

$$+\frac{1.929\ \text{kg PG}}{76\ \text{g/mol}}(180\ \text{J/mol}\cdot{}^\circ\text{C})(80-120)^\circ\text{C}$$

$$\left.+2.3\left[\frac{582.1\ \text{g I}}{60\ \text{g/mol}}(45.5\ \text{kJ/mol}) + \frac{3.3\ \text{g W}}{18\ \text{g/mol}}(44\ \text{kJ/mol}) + \frac{0.1\ \text{g PG}}{76\ \text{g/mol}}(64.4\ \text{kJ/mol})\right]\right\}$$

$$= 995\ \text{kJ} = 0.995\ \text{MJ/kg feed} = 1.71\ \text{MJ/kg isopropanol recovered}$$

Heating requirements for propylene glycol stripper:

(a) S/F = 1

$$\Delta H = \text{sensible heat} + \text{vaporization heat} = \frac{1}{\eta}\left[\sum_i^n m_i C_{pi}\,\Delta T + (1+R)m_v\,\Delta H_v\right]$$

$$= \frac{1}{0.85}\left\{\left[\frac{0.0001\ \text{kg I}}{60\ \text{g/mol}}(154\text{J/mol}\cdot{}^\circ\text{C}) + \frac{0.0764\ \text{kg W}}{18\ \text{g/mol}}(75.37\ \text{J/mol}\cdot{}^\circ\text{C})\right.\right.$$

$$\left.+\frac{0.772\ \text{kg PG}}{76\ \text{g/mol}}(180\ \text{J/mol}\cdot{}^\circ\text{C})\right](120-80)^\circ\text{C} + 2.3\left[\frac{0.1\ \text{g I}}{60\ \text{g/mol}}(45.5\ \text{kJ/mol})\right.$$

$$\left.\left.+\frac{76.4\ \text{g W}}{18\ \text{g/mol}},(44\ \text{kJ/mol}) + \frac{0.1\ \text{g PG}}{76\ \text{g/mol}}(64.4\ \text{kJ/mol})\right]\right\}$$

$$= 607\ \text{kJ} = 0.61\ \text{MJ/kg feed} = 1.04\ \text{MJ/kg isopropanol recovered}$$

(b) S/F = 2.5

$$\Delta H = \text{sensible heat} + \text{vaporization heat} = \frac{1}{\eta}\left[\sum_{i}^{n} m_i C_{pi}\,\Delta T + (1+R)m_v\,\Delta H_v\right]$$

$$= \frac{1}{0.85}\left\{\left[\frac{0.0001\ \text{kg I}}{60\ \text{g/mol}}(154\ \text{J/mol}\cdot{}^{\circ}\text{C}) + \frac{0.0764\ \text{kg W}}{18\ \text{g/mol}}(75.37\ \text{J/mol}\cdot{}^{\circ}\text{C})\right.\right.$$

$$\left. + \frac{1.929\ \text{kg PG}}{76\ \text{g/mol}}(80\ \text{J/mol}\cdot{}^{\circ}\text{C})\right](120-80)^{\circ}\text{C}$$

$$\left. + 2.3\left[\frac{0.1\ \text{g I}}{60\ \text{g/mol}}(45.5\ \text{kJ/mol}) + \frac{76.4\ \text{gW}}{18\ \text{g/mol}}(44\ \text{kJ/mol}) + \frac{0.1\ \text{g PG}}{76\ \text{g/mol}}(64.4\ \text{kJ/mol})\right]\right\}$$

$$= 735\ \text{kJ} = 0.74\ \text{MJ/kg feed} = 1.26\ \text{MJ/kg isopopanol recovered}$$

Total heating requirements:

(a) S/F = 1: 6.27 MJ/kg isopropanol recovered

(b) S/F = 2.5: 6.27 MJ/kg isopropanol recovered

This indicates that for each scenario shown, the total heating requirement is not sensitive to the feed ratio.

Cooling requirements:
The energy required for cooling in this option is that needed to condense and cool the product stream to 25°C, condense and cool the overhead stream of the stripper from 120°C to 80°C, and condense the distillate of the pre-concentration column and cool the waste to 25°C. Therefore, it is independent of the S/F ratio.

Total cooling requirements:

$$3.49\ \text{MJ/kg feed} = 5.96\ \text{MJ/kg isopropanol recovered}$$

Electricity requirements:
The electricity requirements are negligible compared to the heating and cooling requirements. To simplify the calculations, a rule of thumb using the average standard pumping electricity requirements for the economical pipe size (15 m long) was used, with an average pumping energy of 0.003 kJ/kg pumped:

(a) S/F = 1

$$(0.003\ \text{kJ/kg pumped})(1 + 0.0002 + 0.5824 + 0.3362 + 0.0766 + 0.6588 + 0.772)\ \text{kg}$$

$$= 1.02 \times 10^{-2}\ \text{kJ/kg feed}$$

$$= 1.8 \times 10^{-5}\ \text{MJ/kg isopropanol recovered}$$

(b) S/F = 2.5

$$(0.003\ \text{kJ/kg pumped})(1 + 0.0002 + 0.5824 + 0.3362 + 0.0766 + 0.6588 + 1.929)\ \text{kg}$$
$$= 1.4 \times 10^{-2}\ \text{kJ/kg feed}$$
$$= 2.4 \times 10^{-5}\ \text{MJ/kg isopropanol recovered}$$

After doing the calculations above, you notice that both options presented to you have green engineering challenges so you decide to add an additional option to the assessment

Option 3: Pervaporation. In this separation process, a multicomponent liquid stream is exposed to a membrane that preferentially permeates one or more of the components. As the feed liquid flows across the membrane surface, the preferentially permeated component passes through the membrane as a vapor. Transport through the membrane is induced by maintaining a lower vapor pressure on the permeate side of the membrane than that of the feed liquid. Since different species permeate the membrane at different rates, a substance at low concentrations in the feed stream can be highly enriched in the permeate. Thus, separation occurs, with the efficacy of the separation effect being determined by the relative volatilities and the physicochemical structure of the membrane. For this comparison, the pervaporation system shown in Figure 11.6 was considered. The feed is equalized and heated to 70°C. The entire process is maintained at 70°C to maximize the vapor pressure difference across the membrane.[11] After the pervaporation module, the resulting isopropanol–water mixture was assumed to have 20% isopropanol.[16] This stream is then distilled to recover and recycle the isopropanol. Therefore, with this option there is no need for external substances, although a distillation column is needed to complete the circle. The chemical losses considered for this option are the fugitive emissions of ethanol (estimated at 1% according to its boiling point) and the wastewater

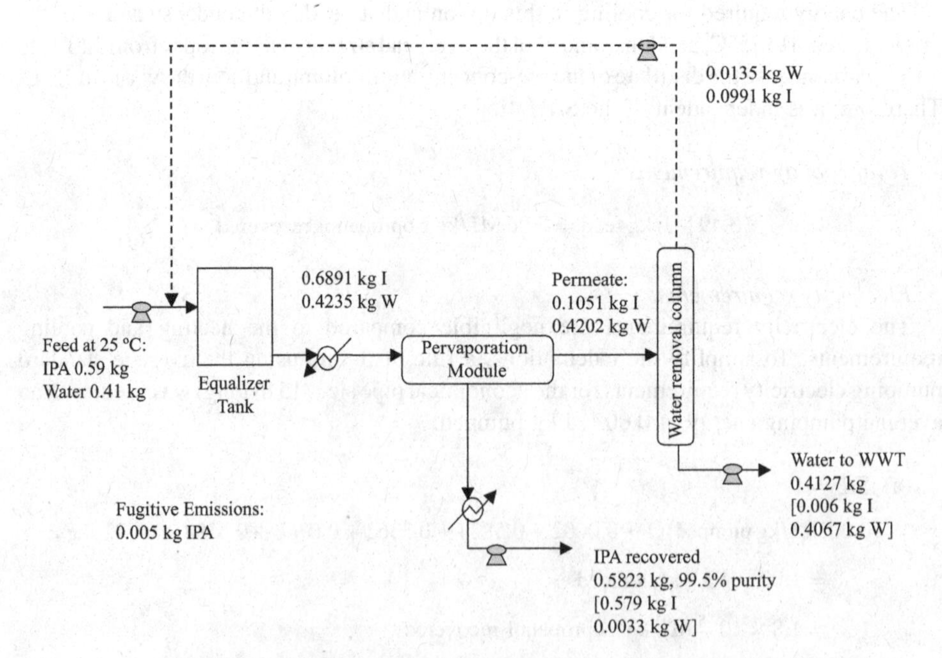

FIGURE 11.6 Mass balance for pervaporation of Example 11.5.

produced as a waste stream, containing some isopropanol. The overall mass balance is also shown in Figure 11.6.

Another issue to take into consideration is the membrane life. A membrane operating under these conditions has an average life of 3 years. A wide range of total fluxes have been reported in the literature, for different initial concentrations. The reported fluxes are mainly between 0.01 and 2.2 kg/m^2·h.[17–20] For the sake of comparison, three cases for membrane life can be considered: at 0.01, 1.1, and 2.2 kg/m^2·h. If the system operates 8 h/day during each day of the 3 years (8760 h), we have:

$$\text{At } 0.5\,\text{kg/m}^2\cdot\text{h} \Rightarrow [(0.01\,\text{kg/m}^2\cdot\text{h})(8760\,\text{h})]^{-1} = 1.1\times 10^{-6}\,\text{m}^2/\text{kg of feed}: = 1.8\times 10^{-6}\,\text{m}^2/\text{kg of IPA recovered}$$

$$\text{At } 2.5\,\text{kg/m}^2\cdot\text{h} \Rightarrow [(1.1\,\text{kg/m}^2\cdot\text{h})(8760\,\text{h})]^{-1} = 1.26\times 10^{-4}\,\text{m}^2/\text{kg of feed}: = 2.16\times 10^{-4}\,\text{m}^2/\text{kg of IPA recovered}$$

$$\text{At } 5.0\,\text{kg/m}^2\cdot\text{h} \Rightarrow [(2.2\,\text{kg/m}^2\cdot\text{h})(8760\,\text{h})]^{-1} = 2.5\times 10^{-4}\,\text{m}^2/\text{kg of feed}: = 4.3\times 10^{-4}\,\text{m}^2/\text{kg of IPA recovered}$$

These fluxes are taken as an average for the purposes of this study, but in practice it is commonly observed that a decrease in the water content results in a progressive decrease in the flux.

The main energy requirements are the heating needed to operate the pervaporation unit at 70°C and the heating required for the distillation column operating at about 80°C. In addition, there are energy requirements for condensing and cooling, and electricity for pumping. To calculate the energy requirements, the following assumptions were made:

- Heat losses: 15%
- Isopropanol stream recovered cooled to 25°C
- Reflux ratio used for the distillation column: was 1.3[12]
- Pumping energy: 0.003 kJ/kg pumped for all the pumps except the permeate pump
- Permeate pump assumed to be a vacuum pump operating at about 10 mbar pressure (1 kPa)

Data used:

- Basis: 1 kg of feed
- C_{pl} of isopropanol = 154 J/mol
- ΔH_v of isopropanol = 45.5 kJ/mol (b.p. 83°C)
- m.w. isopropanol = 60 g/mol
- C_{pl} of water = 75.37 J/mol
- ΔH_v of water = 43.99 kJ/mol
- m.w. water = 18 g/mol

Heating requirements for the preheater:

$$\Delta H = \text{sensible heat} = \frac{1}{\eta}\sum_{i}^{n} m_i C_{pi}\,\Delta T$$

$$= \frac{1}{0.85}\left[\frac{0.51\,\text{kg I}}{60\,\text{g/mol}}(154\,\text{J/mol}\cdot{}^\circ\text{C}) + \frac{0.41\,\text{kg W}}{18\,\text{g/mol}}(75.37\,\text{J/mol}\cdot{}^\circ\text{C})\right](70\text{–}25)^\circ\text{C}$$

$$\frac{1}{0.85}\left[\frac{0.0991\,\text{kg I}}{60\,\text{g/mol}}(154\,\text{J/mol}\cdot{}^\circ\text{C}) + \frac{0.0135\,\text{kg W}}{18\,\text{g/mol}}(75.37\,\text{J/mol}\cdot{}^\circ\text{C})\right](70\text{–}80)^\circ\text{C}$$

$$= 171\,\text{kJ} - 4\,\text{kJ} = 0.17\,\text{MJ/kg feed} = 0.29\,\text{MJ/kg isopropanol recovered}$$

Heating requirements for the distillation column:

$$\Delta H = \text{sensible heat} + \text{vaporization heat} = \frac{1}{\eta}\sum_{i}^{n} m_i C_{pi}\,\Delta T + \frac{1}{\eta}(1+R)\sum_{i}^{n} m_v\,\Delta H_v$$

$$= \frac{1}{0.85}\left[\frac{0.1051\,\text{kg I}}{60\,\text{g/mol}}(154\,\text{J/mol}\cdot{}^\circ\text{C}) + \frac{0.4202\,\text{kg W}}{18\,\text{g/mol}}(75.37\,\text{J/mol}\cdot{}^\circ\text{C})\right](80\text{–}70)^\circ\text{C}$$

$$\frac{1}{0.85}(2.3)\left[\frac{99.1\,\text{g I}}{60\,\text{g/mol}}(45.5\,\text{kJ/mol}) + \frac{13.5\,\text{g W}}{18\,\text{g/mol}}(44\,\text{kJ/mol})\right]$$

$$= 316\,\text{kJ} = 0.32\,\text{MJ/kg feed} = 0.54\,\text{MJ/kg IPA recovered}$$

Total heating requirements:

0.83 MJ/kg IPA recovered

Cooling Requirements:
The energy required for cooling in this option is that needed to condense and cool the product stream to 25°C, and to condense the overhead stream of the distillation column, which would be the same as the energy required for heating. Therefore:

For condensing the distillation column overheads:

$$\Delta H = -\text{vaporization heat} = -\frac{1}{\eta}(1+R)\sum_{i}^{n} m_v\,\Delta H_v$$

$$= -\frac{1}{0.85}(2.3)\left[\frac{99.1\,\text{g I}}{60\,\text{g/mol}}(45.5\,\text{kJ/mol}) + \frac{13.5\,\text{g W}}{18\,\text{g/mol}}(44\,\text{kJ/mol})\right]$$

$$= 292\,\text{kJ} = 0.29\,\text{MJ/kg feed} = 0.50\,\text{MJ/kg IPA recovered}$$

For cooling the product:

$$\Delta H = \text{sensible heat} = \sum_{i}^{n} m_i C_{pi}\, \Delta T$$

$$= \frac{1}{0.85}\left[\frac{0.579\,\text{kg I}}{60\,\text{g/mol}}(154\,\text{J/mol}\cdot{}^{\circ}\text{C}) + \frac{0.0033\,\text{kg W}}{18\,\text{g/mol}}(75.37\,\text{J/mol}\cdot{}^{\circ}\text{C})\right](70-25)^{\circ}\text{C}$$

$$= 79\,\text{kJ} = 0.08\,\text{MJ/kg feed} = 0.14\,\text{MJ/kg IPA recovered}$$

Total cooling requirements:

0.64 MJ/kg IPA recovered

Electricity requirements:

A vacuum pump operating at around 1 kPa requires around 40 kW to operate.[21] If we take a typical feed flow rate for pervaporation of 8 gpm,[22] corresponding to 0.4 kg/s of feed in our system and 0.23 kg of product/s, it requires

$$20\,\text{kJ}/0.23\,\text{kg} = 87\,\text{kJ/kg of isopropanol recovered}$$
$$= 0.087\,\text{MJ/kg of isopropanol recovered.}$$

For the rest of the pumps, the average standard pumping electricity requirements for the economical pipe size (15 m long) have been used. The corresponding pumps are also illustrated in Figure 11.5. Therefore, the estimated electricity requirements for the rest of the pumps are

$$(0.003\,\text{kJ/kg pumped})(1 + 0.4127 + 0.5823)\,\text{kg}$$
$$= 6 \times 10^{-3}\,\text{kJ/kg feed}$$
$$= 1 \times 10^{-5}\,\text{MJ/kg isopropanol recovered}$$

Total electricity requirements:

0.087 MJ/kg IPA recovered

The alternatives in option 1,2, and 3 are compared in Table 11.7.

Therefore, you would recommend exploring the pervaporation option.

Additional Points to Ponder What are the advantages and disadvantages regarding efficiency and safety? How could you get a better estimation of the fugitive emissions? How about the emissions for energy production?

11.4 BATCH VS. CONTINUOUS PROCESSES

Another interesting point from the green engineering perspective is the type of regime under which the process will be operated. Historically, batch processes have been favored for industries with seasonal demands for which an accurate forecast is difficult to obtain, or when the volumes are small, such as agrochemical, fine chemicals, and pharmaceuticals.

Although batch production offers flexibility as to the number of products and generally reduces initial capital expenditures; from a green engineering point of view, there are a series

TABLE 11.7 Comparison of Options

		Extractive Distillation		Pervaporation		
	Azeotropic Distillation	S/F = 1	S/F = 2.5	0.01 $kg/m^2 \cdot h$	1.1 $kg/m^2 \cdot h$	2.2 $kg/m^2 \cdot h$
Mass Indicators						
Mass intensity (not including water), kg/kg usable product	1.030	1.013	1.013	1.013	1.013	1.013
Added solvent intensity, kg added/kg usable product						
Benzene	0.017	—	—	0	0	0
Propylene glycol		3.4 E -4	3.4 E -4			
Wastewater intensity (kg/kg usable product)	0.698	0.698	0.698	0.698	0.698	0.698
Specific compounds released (kg/kg usable product)						
Benzene	1.73 E–2	0	0	0	0	0
Isopropanol	1.89E–2	1.89E–2	1.89E–2	1.89E–2	1.89E–2	1.89E–2
Propylene glycol	0	3.40E–4	3.40E–4	0	0	0
Membrane usage (m^2/kg usable product)	0	0	0	1.8E-6	2.16E-4	4.3E-4
Efficiency (% recovery)	98.6	98.6	98.6	98.6	98.6	98.6
Quality (% purity)	99.5	99.5	99.5	99.5	99.5	99.5
Energy Indicators						
Heating (MJ/kg usable product)	8.13	6.27	6.27	0.83	0.83	0.83
Cooling (MJ/kg usable product)	−8.13	−5.96	−5.96	0.64	0.64	0.64
Electricity (M/kg usable product)	2.9 E-5	1.8E-5	2.4 E–5	0.087	0.087	0.087
Cooling (refrigeration cycle) (MJ refrigeration/ kg usable product)	0	0	0	0	0	0

of inefficiencies related to batch processes. For example, production needs to be stopped, and the equipment cleaned, reconfigured, and validated before changing to another batch or product, thereby producing additional waste (in both materials and time). In addition, the inherent flexibility of batch production equipment to accommodate different processes means that opportunities to maximize the efficiency of any specific process may be missed. When comparing a batch and a continuous process from a green engineering perspective, the assessment needs to be done on the merits of each process. For example, a Dow technical manual[23] cites some of the advantages and disadvantages of a batch process over a continuous reverse osmosis process:

Advantages	Disadvantages
Flexibility when the feedwater quality changes	No continuous permeate flow
Simple automatic controls	No constant permeate quality
Permeate quality can be controlled by termination of the process	Large feed tank required
Permeate quality can be improved by total or partial second-pass treatment	Larger pump required
Favorable operating conditions for single (or low number)-element systems, because the membranes are in contact with the final concentrate only for a short time	Larger power consumption
Rather easy expansion	Longer residence time for feed/concentrate
Lower investment costs	Higher total running costs

As this comparison shows, what is gained by having a flexible batch process affects other areas, in this case the cost, when compared to a continuous process.

11.5 DOES SIZE MATTER?

We saw in Chapter 10 that fast and highly exothermic reactions could benefit from process intensification,[24,25] in which smaller equipment might have some direct sustainability benefits. For example, a pharmaceutical reaction that required 4 h in a 6000-L reactor to dissipate the heat resulting from a large exothermic reaction proceeded in just 10 s in a minireactor (about 300 mL) with better heat dissipation.

Process intensification is covered in more detail in Chapter 24 , however, it is important to understand how it relates to the selection of reactors and unit operations, as it utilizes engineering principles to achieve the desired result; that is, the changes required to intensify a process will be in the "how" the chemistry performs category, not a matter of the "what" of the chemistry itself. To achieve the desired intensification, a combination of techniques and equipment can be used. Process intensification techniques include such approaches as the use of multifunctional reactors (e.g., reaction plus separation in one), hybrid separations (e.g., membrane absorption and stripping), and the use of alternative forms of energy (e.g., centrifugal force instead of gravity for phase separations). Process intensification equipment includes microreactors, static mixers, and compact heat exchanger reactors, among others. As we also discussed in Chapter 10, for relatively low tonnages, when smaller equipment can be utilized, instead of requiring scale-up of the process, the kits can be numbered-up, thereby avoiding most of the scale-up problems.

PROBLEMS

11.1 Do current techniques used to select and design reactors and separations integrate green engineering principles? Which do and which do not? Explain why or why not.

11.2 How are the concepts reviewed in Chapter 7 related to the concepts reviewed in this chapter?

11.3 Create a table showing what additional information you would need to review the remaining green engineering principles for the spinning tube-in-tube reactor of Example 11.2?

11.4 Your boss has asked you to develop a research and development strategy to integrate microwave chemistry into your company's production. You will be presenting this to the head of R&D in an upcoming technical review meeting.

(a) Draw a flow diagram explaining your proposed plan.

(b) Create a table explaining the advantages of this reaction technology based on green engineering principles.

(c) Your boss warns you to be prepared to answer questions about the disadvantages of this reaction technology, and you draft some points in advance of the meeting. What are those points?

11.5 In Problem 11.3 you did a fantastic job of selling the idea of microwave reactors to the head of R&D. He has now asked you to develop a list of questions for the chemistry development department, to be used as a screening mechanism for reactions considered for microwave chemistry. Develop such a list and link each question to a green engineering or green chemistry principle.

11.6 Reactions of aldehydes with small nitrated molecules such as nitromethane and ethyl nitroacetate can be performed in the laboratory under electrochemical conditions.[26] Starting from *meta*- or *para*-benzenedicarboxyaldehydes and nitromethane, it is possible to selectively prepare single or double addition products with an electrogenerated base (EGB). The reaction conditions are very mild, with high yields and selectivities using readily available equipment, simply designed cells, and regular organic glassware. In most cases isolation of the product is simple. A solution of the crude product is filtered through a layer of silica gel, and after evaporation of the solvent, an analytically pure product is obtained.

(a) List some advantages and disadvantages of these syntheses from a green engineering perspective.

(b) What additional development work would this type of technology require?

11.7 Using the reactor described in Example 7.5. employ green engineering principles to show the advantages and disadvantages of changing the operational conditions for the reactor. What other type of reactor would you consider to carry out the same type of reaction?

11.8 You want to select a reactor for the following system of competing reactions:

$$A \rightarrow P \qquad \text{reaction rate equation } r = k_1 C^{a1}$$

$$A \rightarrow W \qquad \text{reaction rate equation } r = k_2 C^{a2}$$

where P is a product and W is a waste, a2 > a1, and the conversion is about 95%. With the information given, which reactor would you propose, and under which operating conditions? Explain your answers in terms of green engineering implications.

11.9 Solve Problem 11.8 if a1 > a2 and the conversion is about 50%.

11.10 Compare the advantages and disadvantages of a 2000-L batch reactor with a traditional impeller against those of a 10-L continuous reactor using a vortex mixer. What are the application ranges of each option for an efficient reaction?

11.11 From a green engineering perspective what are the advantages and disadvantages of batch vs. continuous processes?

11.12 There is a need to recover tetrahydrofuran (THF) from aqueous mother liquors after a reaction so that the solvent can be separated and reused. THF forms an azeotrope with water, so normal atmospheric distillation might not be the most desirable option. The following alternatives are considered:

- Pressure swing distillation
- Atmospheric distillation followed by vapor permeation

(a) What are the advantages and disadvantages of each option from a green chemistry/green engineering perspective?

(b) Draw a flow diagram of each option.

(c) Estimate and compare the mass and energy requirements for these options to recover 99% of the THF from a 50% THF–water feed. For pressure swing distillation, assume a low-pressure column at 0.5 atm and a high-pressure column at 3 atm.

11.13 Solve Problem 11.12 for *tert*-butanol under the same conditions.

11.14 Solve Problem 11.12 for ethanol under the same conditions.

11.15 Phenols are by-products found in wastewater streams of several industrial processes. They are highly toxic and may inhibit the biological wastewater treatment process when present at high concentrations (>200 mg/L). Phenols can be removed using activated carbon adsorption, liquid–liquid extraction with solvents, membranes, and polymeric resins. Descriptions of membrane and resin systems follow.

1. In a membrane aromatic recovery system (MARS), phenol separation is achieved by coupling separation with a reversible reaction to form phenolate salts. Because phenolic compounds are weak acids, they dissociate in base

and associate in acid. This is a reversible reaction that may be manipulated by adjusting the pH of a solution. In general, MARS could be described as follows: The solution to be treated is passed through a nonporous membrane under basic conditions to dissociate the phenols as phenolates. The ionized phenols are collected as a concentrated phenolic solution in what is called the *stripping solution*. Acid is added and the phenols are reassociated, creating a two-phase mixture composed of an organic phenolic phase and an aqueous saline solution. The resulting two phases are separated, the organic phase containing phenol is collected, and the saline solution is recycled to the wastewater feed. The phenols removed from the wastewater into the stripping solution can easily be in the range of 99%, depending on the length of the membrane used. For the specific case study, a solution of 12.5% (w/w) of sodium hydroxide is used as the base and concentrated hydrochloric acid (HCl 37%) is used as the acid. The system temperature is kept at 50°C.[27]

2. Polymeric adsorbent resins are similar in size, shape, and appearance to conventional spherical ion-exchange resins. However, the polymeric adsorbents contain no ionically functional sites. Surface areas for the various polymeric resins normally range from 300 to 800 m^2/g.[28] Phenolic compounds are removed with polymeric resins by physical adsorption. Resin capacity is a function of phenol concentration, the higher the phenol concentration in the feed, the higher the resin capacity for the phenols.[29,30] In general, the adsorption is carried out in two resin columns, one in the loading step while the other is being regenerated. Normally, a 6-h loading cycle is employed, with an average of two cycles per day per resin bed. To regenerate an adsorbent resin loaded with phenols, a solvent system is used. Good organic solvents for phenolics include acetone and methanol. In addition, a dilute aqueous strong inorganic base can also be used.[31] Traditionally, for the regeneration of the resin and recovery of the solvent, a train of three distillation towers had been used. For this specific case study a "superloading" process is taken into consideration for the calculations, as described in the literature.[31,32] In this process the resulting solution from the regeneration cycle is distilled to recover the solvent (in this case, methanol) and the bottoms are cooled and allowed to separate in two phases. The organic phase with the phenol is recovered (about 72 to 75% of phenol in water) and the aqueous phase (with 8 to 10% of phenol) is recycled to the end of the loading cycle of the resin.

Given the information provided for the membrane aromatic recovery system and the polymeric resin adsorption system for the removal of phenolic compounds from wastewater streams:

- Draw a flow diagram of both alternatives.
- Compare them using green engineering principles.

11.16 For Problem 11.15, recommend an option based on mass and energy requirements for an influent concentration of 10 g phenol/L.

11.17 To concentrate a mother liquor containing 2 wt% of a pharmaceutical intermediate (m.w. intermediate = 279.27 g/mol) in methanol to a final concentration of 10 wt%, the options of nanofiltration and direct distillation are being explored.

(a) Draw a process flow diagram of each option.

(b) Contrast the two options using green engineering principles.

11.18 For Problem 11.17, recommend an option based on mass and energy requirements, using the following information:

- For the distillation column:
  - o Heat losses: 15%
  - o Reflux ratio: 1.3
  - o Fugitive losses: 1%
  - o Intermediate in the distillate: negligible concentration
  - o Heat capacity of methanol used as an approximation of the heat capacity of the mixture
  - o Column temperature: 65°C
  - o Pumping of the feed (0.003 kJ/kg pumped)
  - o Bottoms and distillate: are assumed to be cooled to 25°C.
- For the nanofiltration unit:
  - o 10% concentration of the intermediate achievable and desired
  - o Rejection: 99.99% (rejection = $1 - C_p/C_i$)
  - o Fugitive losses: 1%
  - o Intermediate in the permeate: negligible concentration
  - o System total temperature: 25°C (neglecting any cooling needed)
  - o Energy required: only enough to pump and apply pressure to the membrane module
  - o Flow rate of feed: 1200 L/h
  - o Flow rate of permeate: 1200 L/h
  - o Pressure applied 25 bar (360 psi, 2.5×10^6 N/m^2).
  - o Pump efficiency: 50%

REFERENCES

1. Beckman, E. J. Using principles of sustainability to design "leap-frog" products. Keynote presentation at the 11th Annual Green Chemistry and Engineering Conference, June 26–29, 2007.
2. Woods, D. R. *Rules of Thumb in Engineering Practice.* Wiley, Hoboken, NJ, 2007.
3. Allen, D. T., Shonnard, D. R. *Green Engineering: Environmentally Conscious Design of Chemical Processes.* Prentice Hall, Upper Saddle River, NJ, 2002.
4. Abraham, M. A., Nguyen, N. Green engineering: defining the principles – results from the San Destin Conference. *Environ. Prog.*, 2003, 22(4), 233–236.

5. Ritter, S. A green agenda for engineering. *Chem., Eng., News*, July 2003, 81(29), 30–32.
6. Anastas, P. T., Zimmerman, J. B. Design through the twelve principles of Green Engineering. *Env. iron Sci. Technol.*, 2003, 37(5), 94A–101A.
7. Jenck, J. F., Agterberg, F., Droescher, M. J. Products and processes for a sustainable chemical industry: a review of achievements and prospects. *Green Chem.*, 2004, 6, 544–556.
8. Nüchter, M., Ondruschka, B., Bonrath, W., Gum, V. Microwave assisted synthesis: a critical technology overview. *Green Chem.*, 2004, 6, 128–141.
9. Ritter, S. K. A new spin on reactor design. *Chem. Eng. News*, 2002, 80(30), 26–27.
10. Humphrey, J. L. Separation processes: playing a critical role. *Chem. Eng. Prog.*, 1995, 91(10), 31–41.
11. Seiryo, F. U.S. patent 5,207,902 to Ryota Techno Engineering & Construction Co., Mitsubishi Kasei Corporation, Japan, May 4, 1993.
12. Jiménez-González, C., Kim, S., Overcash, M. Methodology to develop a gate-to-gate LCI information. *Int. J. Life Cycle Asses.*, 2000, 5(3), 153–159.
13. Afeefy, H. Y., Liebman, J. F., Stein, S. E. Neutral thermochemical data. NIST Standard Reference Database Number 69. In *NIST Chemistry WebBook*, Mallard, W. G., Linstrom, P. J., Eds. National Institute of Standards and Technology, Gaithersburg MD, Feb. 2000. (Available at http://webbook.nist.gov).
14. Ruthven, D. M., Ed. *Encyclopedia of Separation Technology*. Wiley, New York, 1997.
15. Lee, F. M. Extractive distillation: close-boiling-point. *Chem. Eng.*, Nov. 1998, pp. 112–121.
16. Uramoto, H., Kawabata, N. Separation of alcohol-water mixture by pervaporation through a reinforced polyvinylpyridine membrane. *J. Appl. Polym. Sci.*, 1993, 50, 115–121.
17. Michizuki, A., et al. U.S. Patent 4,944,881, to Agency of Industrial Sciences and Technology, Japan, July 31, 1990.
18. Bartels, C. R. U.S. Patent 5,009,783 to Texaco, Inc., USA, June 4, 1990.
19. Sugai, S., et al. ^{1}H NMR characterization of syndiotactic poly(methacrylic acid). *Makromol. Chem. Rapid Commun.*, 1986, 7, 47–51.
20. Qunhui, G., et al. The effects of the characteristics of support layers on separation of water-alcohol mixtures through Chitosan pervaporation composite membranes. *Membrane*, 1995, 20(3), 229–238.
21. Woods, D. R. *Process Design and Engineering Practice*. PTR Prentice Hall, Englewood Cliffs, NJ, 1995.
22. U. S., Environmental Protection Agency. *Cross-Flow Pervaporation Technology*. Innovative Technology Report, EPA 540/R-95/511, U.S. EPA, Washington, DC, Dec. 1998.
23. Dow Chemical Company. Filmtech Membranes System Design: Batch vs. Continuous Process. Tech manual excerpt. Form 609-02047-1004.
24. Ramshawm, C. The incentive of process intensification. In *Proceedings of the First International Conference on Process Intensification for the Chemical Industry*. BHR Group, London, 1995.
25. Stankeiwicz, A. I., Moulijn, A. J. Process intensification: transforming chemical engineering. *Chem. Eng. Prog.*, 2000, 96(1), 22–34.
26. Niyazymbetov, M. E. Electrochemical methods in organic synthesis of valuable intermediates. *Watts New Q. Newsl.*, 1997, 3(2). Available at http://www.electrosynthesis.com/news/m6watts.html
27. Han, S., Castelo-Ferreira, F., Livingston, A. Membrane aromatic recovery system (MARS): a new membrane process for the recovery of phenols from wastewater. *J. Membrane Sci.*, 2001, 188 (2), 219–233.
28. Crook, E. H., McDonnel, R. P. McNulty, J. T. Removal and recovery of phenols from industrial waste effluents with amberlite XAD polymeric adsorbents. *Ind. Eng. Chem. Product Res. Dev.*, 1975, 14(2), 113–118.

29. Vera-Avila, L. E., Gallegos-Pérez, J. L., Camacho-Frias, E. Frontal analysis of aqueous phenol solutions in amberlite XAD-4 columns, implications on the operation and design of solid phase extraction systems. *Talanta*, 1999, 50, 509–526.

30. Juang, R. S., Shiau, J. Y. Adsorption isotherms of phenols from water onto macroreticular resins. *J. Hazard. Mater.*, 1999, B70, 171–183.

31. Fox, C. R. Plant uses prove phenol recovery with resins. *Hydrocarbon Process.*, 1978, 11, 269–273.

32. Vulliez-Sermet, P. R. E., Fiorentino, E. Adsorption process. U. S. patent 3,979,287, Sept. 7, 1976.

12

PROCESS SYNTHESIS

What This Chapter Is About In Chapter 11 we covered reactors and separators in the context of green engineering. In this chapter we explore the role that process synthesis plays in the context of green engineering: in other words, how we combine the various process components such as reactors and separations to design more optimized, greener, more sustainable chemical processes.

Learning Objectives At the end of this chapter, the student will be able to:

- Understand the concept and importance of chemical process synthesis in the context of green engineering and green chemistry.
- Identify existing process synthesis tools and methods to design more sustainable chemical processes
- Apply process synthesis principles to examples and explain how these relate to green engineering principles.

12.1 PROCESS SYNTHESIS BACKGROUND

In Chapter 11 we covered how green engineering relates to reactors and separators as isolated unit operations. However, chemical processes typically do not consist of a single unit operation that receives reactants and outputs the desired product at the desired purity with no waste [Figure 12.1(a)]. As we saw in Chapter 1, this would be the ideal situation, but it does

Green Chemistry and Engineering: A Practical Design Approach, By Concepción Jiménez-González and David J. C. Constable

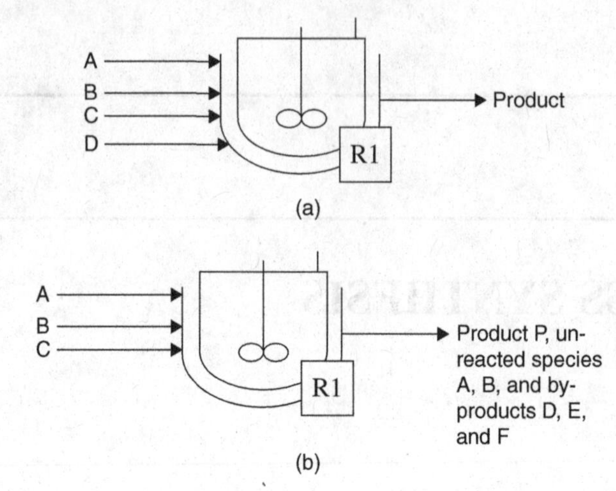

FIGURE 12.1 (a) Ideal reaction situation, where there is no waste and thus no need to separate or purify the product. (b) Typical reaction situation, where the output is a mixture of the desired product and undesired by-products, unreacted species, and impurities in need of separation and purification steps.

not happen in reality. Most chemical processes are complex systems in which a series of reactors and separators are intimately related and interrelated to each other through pumps, transport pipes, valves, utilities, heat exchangers, controls, recirculation loops, waste streams, and pollution control devices, among others. The reactor can be seen as the heart or central unit operation in a chemical process design, but most of the footprint and the resources (e.g., energy) in a chemical process are used in the separation trains, and without them the end product would not be possible. This picture of a chemical process does not occur in a vacuum, and the order of the separation units has in most cases been designed and thought out very carefully to optimize the use of resources and therefore the efficiency of the process. The goal of chemical engineers and process chemists is to design a chemical process that is as close as possible to the idealized process.

Process synthesis, a systematic decision-making process to select the unit operations required for a process, dates back to the early 1970s.[1] The aim of process synthesis is to optimize the sequence of processing steps (e.g., reactions, distillations, extractions) within a chemical process, especially as it relates to the order and interactions among the processing steps (e.g., recycle streams). Process synthesis can be performed to revise an existing process (i.e., retrofitting) or to design a new process (e.g., flowsheet design).

Let's assume that we have a reactor such as the one shown in Figure 12.1(b), with reactants A, B, and C of defined purity, which are fed into the reactor to render product P. In addition to P, several by-products are produced (D, E, and F). Assuming that we need P at a high purity level, how can we obtain our desired result? Do we separate the impurities during the reaction or after? What types of separation do we use to purify the end product? Distillation? Crystallization? Membranes? Once we have selected the separation technology, there are several flowsheet options that can be generated. Which one is best? Figure 12.2 shows two alternatives that may be possible in this very simple hypothetical system.

Process synthesis is very much the fun part of engineering: inventing and optimizing the structure of a new or retrofitted chemical process. Optimizing the chemical process is aligned

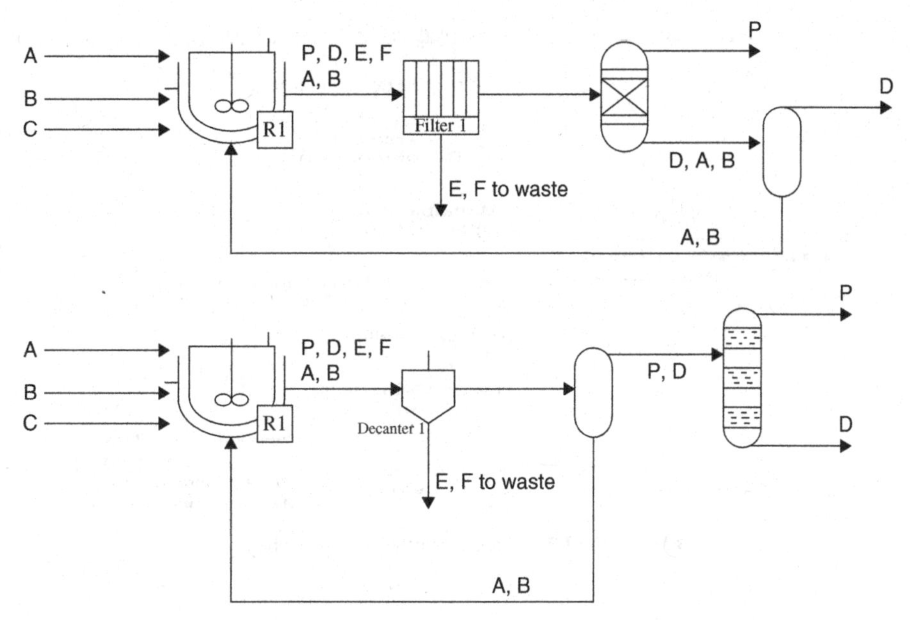

FIGURE 12.2 Separation sequences for the simple hypothetical example of Figure 12.1.

directly with the principles of green engineering, as a more efficient process flowsheet will utilize energy and mass resources more effectively and minimize waste.

12.2 PROCESS SYNTHESIS APPROACHES AND GREEN ENGINEERING

Process synthesis is a complex activity that involves converting an undefined input of desired products into a very concrete outcome for a specific process and system. This involves many decision makers and multiple levels of decision steps. Douglas[2] estimated that for a typical design the number of alternatives that might accomplish the same goal can be over 1 billion. Out of these many potential alternatives the designers want to select the one that best suits the objectives.

As process synthesis is a complex undertaking, a series of different approaches have been developed to facilitate and optimize the design of a process. These techniques and tools have been utilized and tailored directly to produce process designs that will optimize material and energy requirements and minimize environmental impacts. Figure 12.3 shows some examples of process synthesis approaches.[3] These approaches are normally related to tools, primarily software and mathematical algorithms, to simplify and automate the calculations. There are numerous software packages both in academia and industry that are designed to facilitate computer-aided chemical process design. The design of chemical plants is arguably a complex problem, and there are normally several competing objectives with their associated trade-offs.

Although process synthesis techniques have been used as tools to design more sustainable processes, this was not always the case, and there has been an evolutionary approach to process synthesis and design that has increasingly incorporated environmental, health, and safety aspects into process synthesis. For example, initially, chemical

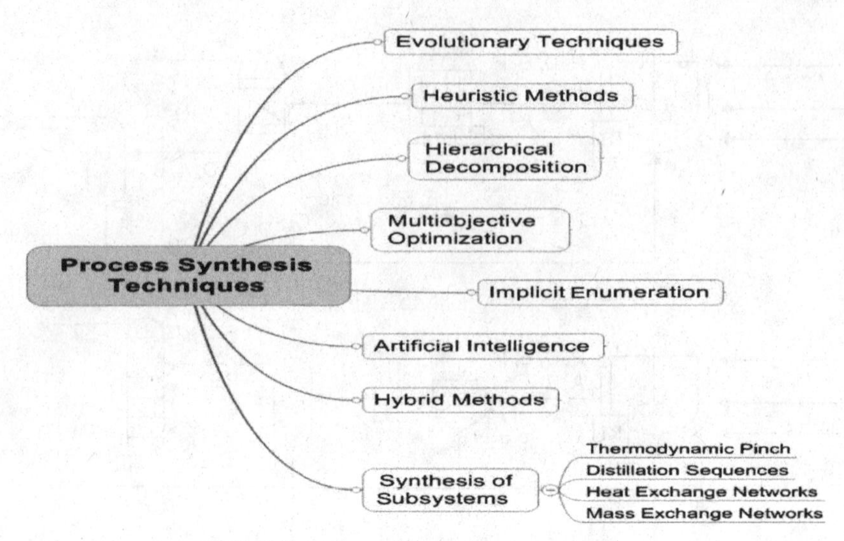

FIGURE 12.3 Process synthesis approaches.

process design was limited to reaction and separation steps. However, with the energy crisis of the 1970s, energy requirements and heat and power integration were included. As environmental costs and impacts became more significant, end-of-pipe treatments were added to the scope for process synthesis, and in recent years the emerging trends are to integrate more holistic approaches into process design, including materials integration, life cycle assessment, and risk–impact assessments. To illustrate the concept, Cano-Ruiz and McRae[4] have presented an evolutionary framework for chemical process synthesis in the last 40 years or so, from the viewpoint of integrating environmental aspects into process design, as presented in Table 12.1.

Next we explore several process synthesis techniques that have been applied to the design of more sustainable chemical processes. These techniques have integrated green engineering to a greater extent by attempting to minimize waste, optimize the use of energy and resources, and optimize the inherent safety of the process, to name only a few of the opportunities explored.

12.3 EVOLUTIONARY TECHNIQUES

In this approach a first approximation flowsheet is generated and then refined incrementally to the desired endpoint. This can be done using computer systems and evolutionary rules. At each step, performance metrics for the flowsheet are calculated and the incremental improvements are repeated until no further significant improvements can be measured. For example, genetic algorithms are used to generate a "population" of alternative flowsheets. New flowsheets ("children") are created by combining the characteristics of some alternative flowsheets ("parents") or by random variations ("mutations"). From a green engineering perspective, the metrics used to evaluate alternatives should include metrics for mass and energy efficiency, toxicity, and life cycle impacts.

TABLE 12.1 Evolution of Chemical Process Synthesis from the Perspective of Integrating Environmental Considerations into Chemical Process Design

Time Line	Chemical Process Synthesis Scope	Description
1960s	Reactions Separations Raw materials Products Utilities	The process design considered only the core reaction and separation processes.
1970s–1980s	Reactions Separations Raw materials Products Utilities Utility plants Heat recovery networks: hot and cold streams Waste	Incorporation of heat and power integration.
1990s	Reactions Separations Raw materials Products Utilities Utility plants Heat recovery networks: hot and cold streams Waste Waste treatment facilities	Incorporation of end-of-pipe alternatives into process synthesis, and some pollution prevention and waste minimization.
21st century	Reactions Separations Raw materials Products Utilities Utility plants Heat recovery networks: hot and cold streams Mass recovery networks: lean and rich streams Waste Waste treatment facilities Mass separating agents Supplier production plants Natural resources	Trends include further heat and mass integration, pollution prevention, and consideration of impacts throughout the production supply chain (life cycle assessment).

Source: Adapted from ref. 4.

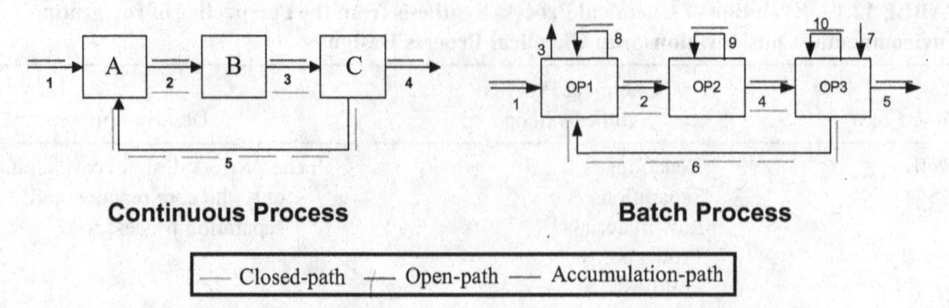

FIGURE 12.4 Flowsheet decomposition for a continuous process and a batch process, identifying closed paths, open paths, and accumulation paths according to the methodology developed by Carvalho et. al.[11] A, B, and C denote continuous unit operations; OP1, OP2, and OP3 denote batch unit operations.

One very good example of integrating sustainability and green engineering into evolutionary process synthesis is the methodology that Carvalho et al. have developed for retrofitting or designing new processes once the first flowsheet is available.[5–7] The steps in this methodology are described briefly below.

1. *Collect process data.* Mass and energy balance data are collected or calculated. The purchase and sales price for each compound are also collected. These data can be provided from real plant data or through model-based simulation.

2. *Flowsheet decomposition.* Decomposition identifies all the open and closed paths for each compound in the process flowsheet. For batch processes, accumulation paths are also identified between unit operations. The open and closed paths for a continuous and a batch process are illustrated in Figure 12.4.

3. *Estimation of metrics.* In this step, any mass, energy, sustainability, and safety indicators that are important parameters for the system are estimated. The mass and energy indicators are calculated for all the paths determined in step 2. Through the values obtained for these indicators, it is possible to identify the locations within the process where the mass and energy "paths" face "barriers" with respect to costs, benefits, or accumulation. Table 12.2 shows some of the main indicators.

TABLE 12.2 Principal Indicators

Indicator	Description	Potential for Improvement of Values That Are Highly Negative or Positive
MVA	Material value added	Negative
EWC	Energy and waste cost	Positive
TVA	Total value added	Negative
RQ	Reaction quality	Negative
AF	Accumulation factor	Positive
REF	Reusable energy factor	Positive
DC	Demand cost	Positive
TDC	Total demand cost	Positive

Analysis of the process impact is determined through the calculation of a set of sustainability metrics, which can include metrics set by the company, sustainability metrics set by groups such as the AIChE or the IChE, metrics contained in the WAR algorithm,[8] life cycle assessment impacts, and others. The safety of the process is measured by the use of a set of inherent safety indices developed by Heikkila[9] (inherent safety is covered in more detail in Chapter 14).

4. *Indicator sensitivity analysis (ISA) algorithm.* To identify the indicators and metrics that have the largest impact on the sustainability performance of the process, an objective function is specified (e.g., gross profit or process, environmental impact). A sensitivity analysis can then be performed to determine those indicators that provide the best improvements in selected indicators and the objective function.

5. *Sensitivity analysis.* A sensitivity analysis is performed to identify the operational variables (e.g., flows, temperatures) that need to be changed to improve progress toward an optimal objective function.

6. *Generation of new design alternatives.* New alternatives are generated using a systematic analysis that allows for the identification of improvement opportunities. Different options for improvement and different process synthesis algorithms can be followed depending on the type of operational variables that have the largest impact on the objective function. This could be related to flow rates in either closed or open paths, the reaction, or the separation train, as shown in Figure 12.5. A new design alternative is considered more sustainable if, and only if, it improves the target indicators without compromising the performance criteria by more than 1% compared to their initial value.

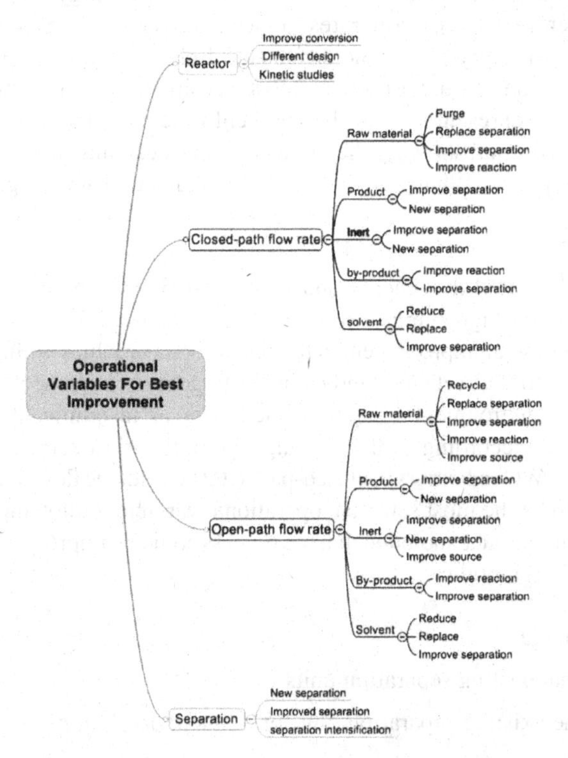

FIGURE 12.5 Operational variables for best improvement during the generation of alternatives.

Carvalho et al. have developed an Excel-based software called Sustain-Pro to follow the algorithm automatically, although the rationale can be followed with other tools and mechanisms. The inputs for Sustain-Pro are the mass and energy balance data as well as the prices of the compounds present in the process. The Sustain-Pro software reads the balance data, automatically performs flowsheet decomposition, calculates the indicators and various metrics, and performs the sensitivity analysis.

Example 12.1 A vinyl chloride monomer (VCM) plant is comprised of 31 unit operations, which include four distillation columns and three reactor systems, six recycle streams, 52 streams, and 35 compounds. The production of VCM can be divided into five sections:

- *Section 1*: Direct chlorination of ethylene to produce 1,2-dichloroethane (EDC), which is the raw material of VCM.
- *Section 2*: EDC is produced in this section by oxychlorination of ethylene with the recycled HCl and O_2.
- *Section 3*: Purification of EDC.
- *Section 4*: Thermal cracking of EDC to form VCM.
- *Section 5*: VCM purification takes place here, allowing the HCl to be recycled back to the oxychlorination reactor.

Carvalho et al. performed an analysis of this VCM plant to identify more sustainable retrofit process synthesis alternatives. Following the methodology described above, they identified 138 open paths with flow rates higher than zero and 252 closed paths in the flowsheet decomposition step. For the entire set of closed and open paths, a full set of indicators was calculated and only the most sensitive indicators were highlighted. Figure 12.6 shows a representation of the vinyl chloride monomer process and indicator sensitivity analysis output, showing the indicator values calculated for the most sensitive closed and open paths. What do these results indicate from a green engineering perspective?

Solution It can be noted that energy and waste cost (EWC) indicators have very large values, which indicates high energy consumption and the need for a reduction in the flow rates for these paths, or an improvement in the operational variables within the process may be preferred. Carvalho et al. set target indicators by targeting simultaneous improvements in the gross profit and environmental impact while having an insignificant effect on all other metrics and indices. According to the ISA algorithm, the most sensitive indicators were identified to be the EWC indicators in closed-path CP219, with the flow rate of a raw material (EDC) in CP219 to be the most sensitive operational variable. Following the mind map of Figure 12.5, Carvalho et al. concluded that the process could be improved by employing one of the following design options:

- Inserting a purge
- Improving the existing separation units
- Replacing the existing separation unit for a new separation process
- Improving the reaction

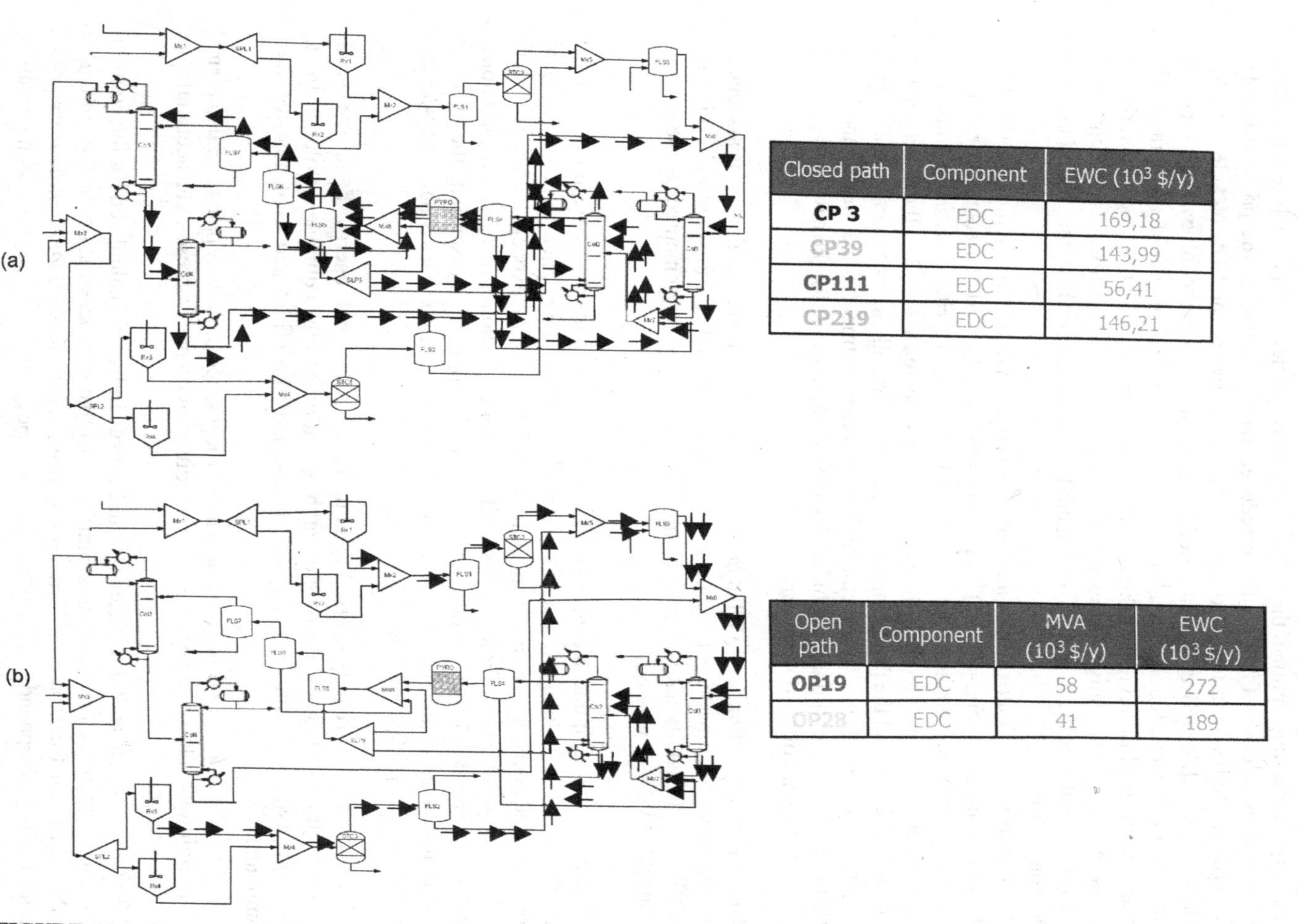

Closed path	Component	EWC (10^3 \$/y)
CP 3	EDC	169,18
CP39	EDC	143,99
CP111	EDC	56,41
CP219	EDC	146,21

Open path	Component	MVA (10^3 \$/y)	EWC (10^3 \$/y)
OP19	EDC	58	272
OP28	EDC	41	189

FIGURE 12.6 Representation of the vinyl chloride monomer process and output of the indicator sensitivity analysis, showing the indicator values calculated for the most sensitive closed paths (a) and open paths (b).

The first two alternatives were tested but did not meet the criteria established. The third alternative was the insertion of a new separation process in closed-path CP219. This could be a potentially interesting option if the investment cost of the new equipment is relatively low and if the flow rate of CP219 could be reduced without reducing the production rate. To determine the best unit to use to separate the impurities in the EDC recycle loop, the methodology of Jaksland et al. was used.[10] Applying this methodology, four types of separation units were found: gas absorption, liquid absorption, liquid membrane, and pervaporation. A close review of the overall process streams identified pervaporation as the most promising alternative. The alternative of improving the reaction was not considered since insufficient data are available on potential replacement catalysts that would improve EDC conversion.

A simulation to validate the new alternative was done and the target indicators (EWC of EDC closed paths in Figure 12.5) were improved by 31%. At the same time, all the performance criteria remained undisturbed whereas other criteria actually improved. Further simulations revealed that the process energy decreased by 2%, the water metrics decreased by 2.8%, and the profit improved by 0.34%, with all the other metrics and indices remaining constant. Based on a calculation of the investment for a membrane-based pervaporation unit, it could be concluded that the investment cost for the new pervaporation unit would be paid off within a year.

Additional Points to Ponder What are safety and occupational health aspects associated with this process? How might life cycle considerations change the solution in this case study? After the additional alternatives for this process are implemented, is the process sustainable? Why or why not?

Example 12.1 is applicable to a continuous process; however, Carvalho et al. have extended their methodology to batch systems,[11] which can be applied to pharmaceutical processes, fine chemicals, batch fermentations, and the like. As expected, a series of data need to be collected that differ from those collected for continuous processes, and additional indicators are needed for batch processes. The rest of the approach is essentially the same and the differences are mainly in the data required, the way the flowsheet is decomposed, and the indicators, as described below.

1. *Process data requirements.* In addition to mass and energy data, the batch process requires us to know the time of each operation and the equipment volume. In the batch process, mass data are expressed in terms of the initial and final mass of each compound.

2. *Flowsheet decomposition.* In the batch process, accumulation paths are also identified in addition to closed and open paths. An accumulation path represents an accumulation of mass and energy.

3. *Indicators.* In addition to the mass and energy indicators calculated for the continuous process, there are two new sets of indicators for batch processes: *operation indicators* (Table 12.3), which compare the operational performance, and *compound indicators*, which indicate which compound is most likely to cause operational problems for each operation. This new set of indicators focuses on the time, volume, and energy utilization per unit operation or per compound.

The application of this methodology to batch processes in a pharmaceutical setting can be illustrated using insulin production as a case study.

TABLE 12.3 Operation Indicators for Batch Processes

Operation Indicator	Formula	Definition
Total free volume	$\text{TFV} = \dfrac{V_{eq} - \sum_j^J M_{AP,j}/\rho_j}{V_{eq}}$	The fraction of free volume in the total equipment volume. High values indicate that the equipment volume is not filled to a high level and indicate a potential for improvements.
Operation time factor	$\text{OTF} = \dfrac{t_j}{\sum_j^J t_j}$	The fraction of total time that a given operation represents for the entire sequence of operations. High values indicate a potential for improvement.
Operation energy factor	$\text{OEF} = \dfrac{E_j}{\sum_j^J E_j}$	The percentage of total energy used for a given operation. High values indicate an opportunity for improvement.

Variables: V_{eq} = equipment volume
j = operation unit
J = total number of operations
ρ = density of the compound
t_j = time spent in operation j
E_j = energy consumed in operation j

Example 12.2 The insulin process is divided into four sections: (1) fermentation, (2) primary recovery, (3) reactions, and (4) final purification. Carvalho et al. demonstrated the application of their methodology in batch processes for the fermentation section only (Figure 12.7).

Solution Carvalho et al.[35] used a SuperPro Designer simulation model of insulin synthesis to obtain the process information. They performed flowsheet decomposition within the fermentation section, involving 21 streams, 14 operations, and eight compounds (Figure 12.8). The indicators and sustainability metrics were then calculated. For this study the most sensitive indicator was the material value added (MVA) for OP 7 and OP10, with very negative values ($\$-7233 \times 10^3$ and $\$-1757 \times 10^3$ per year, respectively). Regarding the batch indicators, the time factor (TF) in AP55 for ammonia was the most sensitive indicator, and its value should be reduced in order to have a lower operational time. Through the ISA algorithm it was found that among the selected indicators, the MVA of OP7 was the most sensitive and consequently the one that should primarily be improved. The process sensitivity analysis found that the most significant variable for improving the MVA of OP7 was the flow rate reduction.

To reduce the OP7 flow rate, Carvalho et al. propose implementing the recycle of unutilized glucose, whose recovery will require an additional separation step, identifying a

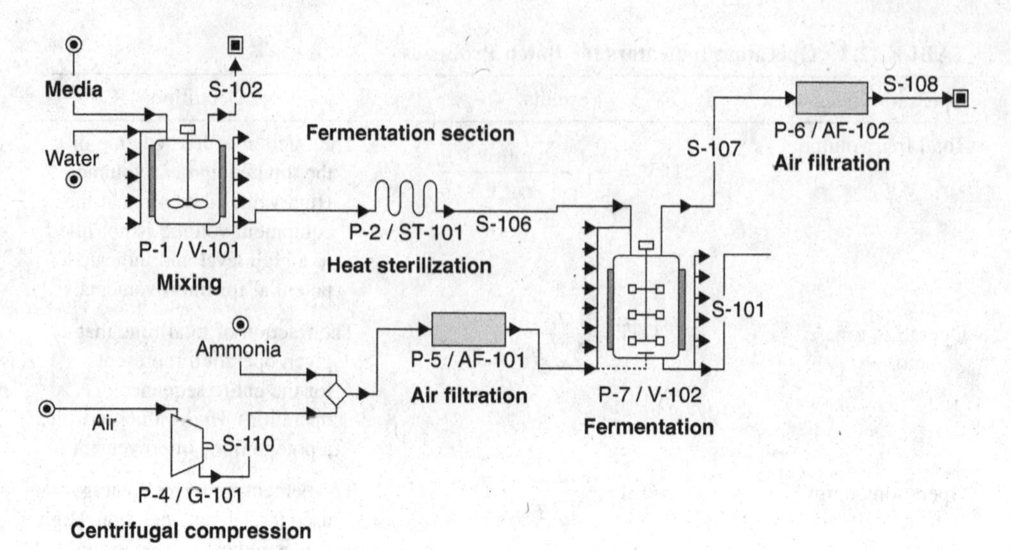

FIGURE 12.7 Fermentation section of the insulin production process. (From Carvalho et al., ref 35. Reprinted with permission. Copyright © 2009, Elsevier.)

filter and an evaporation unit as a suitable activity. From the TF value of AP55 it is known that to reduce fermentation time, the concentration of ammonia needs to be increased. However, in this case the improvement achieved by the ammonia concentration increase is not significant. This fact indicates that the fermentation process is already optimized and that nothing can be done to reduce the operational time.

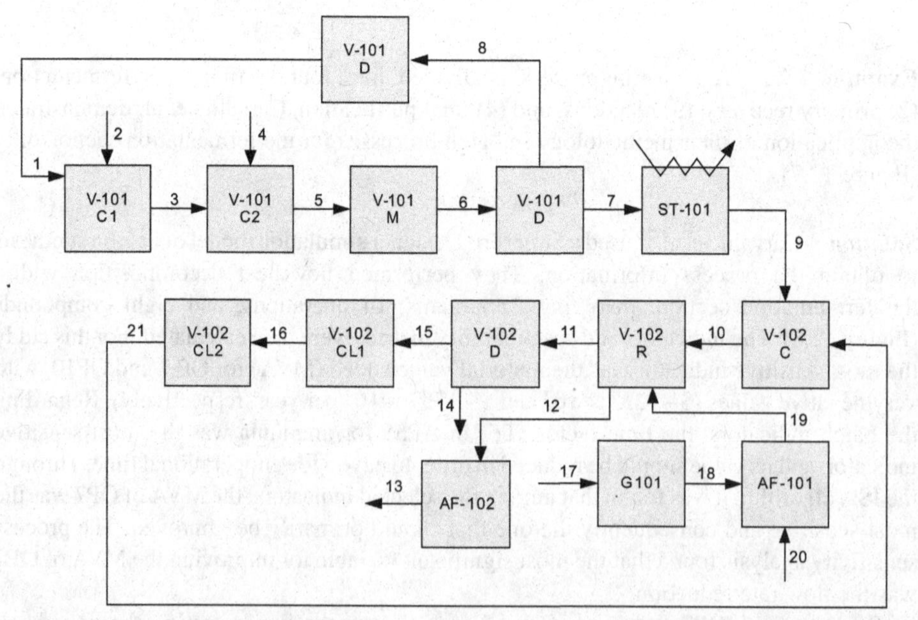

FIGURE 12.8 Insulin fermentation process used for the flowsheet decomposition of a batch process. (From Carvalho et al., ref 35. Reprinted with permission. Copyright © 2009, Elsevier.)

By using a simulation it was estimated that with the addition of glucose recycling, the profit increased 1.04%, the water and the energy metrics per value added improved by 1.03%, and the material metrics improved by 2.55% per kilogram of final product and per value added. All the other parameters remained constant. The target indicators improved by 100%.

Additional Points to Ponder What are the safety and occupational health aspects associated with this process? Is a 1% or 2.5% improvement significant for this type of process? Why or why not?

12.4 HEURISTICS METHODS

Rules of thumb have been used routinely in chemical engineering to select the type of equipment and to generate alternative flowsheets. By observing the behavior of systems when solving many problems of that type, we can devise a series of rules. For example, in distillations of nearly ideal mixtures, we might observe that by removing the most plentiful component, we almost always achieve improved separation. The ideal-distillation literature began with heuristics, and interestingly, most of them seem to work well.[12] Heuristics techniques for process synthesis utilize rules of thumb for the selection of the best unit operation, their sequence, and their recycle streams.

Many rule-of-thumb compendiums have been published to condense the empirical knowledge for chemical processes, such as Woods's *Rules of Thumb in Engineering Practice*,[13] Walas's *Chemical Process Equipment: Selection and Design*,[14] or Branan's *Rules of Thumb for Chemical Engineers*.[15] Heuristics are used broadly in industrial settings in companies such as Shell and BASF[16] and in many cases are combined with other algorithms. It is precisely when the heuristic knowledge is coupled with robust optimization methods that this approach renders the best insights and benefits.[17] For example, heuristic methods have recently been combined with artificial intelligence approaches.

Example 12.3 In 1983, Nadgir and Liu[18] published a heuristic method describing their approach to generating initial flowsheet sequences for multicomponent separations. Some of the heuristics presented in that paper are summarized below, in no particular order (M refers to a method heuristic, S refers to a sequence heuristic, and C refers to a component heuristic).

- M1 (a) All things being equal, select separation methods that use only energy separating agents and avoid using separation techniques that involve using reagents not present in the process. (b) If a mass separating agent is used, remove it immediately after it is used. (c) A method using a mass separating agent cannot be used to isolate another mass separating agent.
- S1 Remove corrosive or hazardous components first.
- S2 Perform difficult separations last.
- S3 Perform easy separations first.
- C1 Remove the most plentiful components first.

How do these heuristics align with green engineering principles and goals?

TABLE 12.4 Heuristic Principles

Heuristic	Green Engineering Principles
M1	This aligns with mass efficiency; by using physical means of separations, there is no need to add additional mass to the system. This heuristic accounts for the potential need for mass separating agents but offers them as an alternative to be used only if other approaches have been exhausted.
S1	This aligns partially with removing hazardous materials. The ideal situation would be to avoid the formation and use of toxic or corrosive materials, but this heuristics attempts to limit the presence of toxic materials to the first few unit operations.
S2 and S3	These align with energy efficiency. Difficult separations require more energy than do easy separations. Additionally, the greater the mass or flow of materials in a given step of the synthesis, the more energy that will be required. By performing the easiest separations first, when the mass of materials is the largest, and the most difficult separation last, when the mass of materials present in the system is the smallest, these heuristics attempt to reduce the energy needed for the separation. This is accomplished by having the most flow with the easiest, more energy-efficient separations and leaving the most energy intensive separation to be run with small volumes.
C1	This aligns with mass and energy efficiency in addition to S2 and S3. Assuming that the separation is achievable and relatively easy, by removing the most plentiful component first, this heuristic is attempting to optimize the remaining processing by the separation train by handling reduced masses of components during that time.

Solution As expected, some of these heuristics were developed to optimize the separation efficiency and minimize energy use, both of which are principles of green engineering. Table 12.4 shows some of the alignment of these heuristics with green engineering principles.

Additional Points to Ponder What would be a better way to express heuristic S1? How could we measure the improvement in the sustainability of a process that would result through application of this heuristic?

Nadgir and Liu proposed an ordered system of heuristics for process synthesis for multicomponent separations, which can be summarized as follows:

1. *Heuristic M1.* Favor ordinary distillation and remove any mass separating agents first. Distillation is feasible if the relative volatility is greater than 1.05; otherwise, use a mass separating agent (MSA) such as a solvent. The use of a MSA already present in the stream is preferable.
2. *Heuristic M2.* Avoid vacuum distillations and refrigeration. Avoid excursions from ambient temperatures and pressures. Use vacuum distillation if the boiling point is high or to improve volatility. Use vacuum to keep temperatures low and thereby

prevent product degradation and/or side reactions. Refrigeration may be avoided by using higher column pressures.

3. *Heuristic D1.* Favor the smallest product set and take out the product as distillate if possible. Avoid separating and then recombining materials.
4. *Heuristic S1.* Remove corrosive and hazardous materials first.
5. *Heuristic S2.* Perform difficult separations last.
6. *Heuristic C1.* Remove the most plentiful component first.
7. *Heuristic C2.* If component compositions do not vary widely, favor a 50 : 50 split (i.e., separate streams into equimolar splits). If it is not easy to determine which split is closest to 50 : 50, carry out the split with the highest *coefficient of ease of separation* (CES) first. CES $= f\Delta T$, where f is the ratio of molar flow of products (smaller than 1) and ΔT is the boiling-point difference. The claim is that by following the heuristics in this order, the separation cost can be minimized, as energy requirements are minimized.

Example 12.4 The following mixture of paraffins and olefins needs to be separated into pure components.

TABLE 12.5 Paraffin/Olefin Data

Compound	Mole Fraction	Relative Volatility, α	CES
A: Ethane	0.20	3.50	62.5
B: Propylene	0.15	1.20	10.7
C: Propane	0.20	2.70	139.1
D: Isobutene	0.15	1.21	9.0
E: *n*-Butane	0.15	3.0	35.3
F: *n*-Pentane	0.15		

The engineering team is considering the best separation sequence to use to perform this separation. Using Nadgir and Liu's ordered heuristics, offer a proposal for the separation sequence.

Solution Following the heuristics, we have:

1. *Heuristics M1 and M2:* Use ordinary distillation with refrigeration at high pressure.
2. *Heuristics D1 and S1:* Not applicable.
3. *Heuristics S2:* Splits B/C and D/E have low relative volatility; do these last.
4. *Heuristic C1:* Not applicable, as the molar ratios are very close.
5. *Heuristic C2:* Split ABC/DEF represents a 55 : 45 split with the largest coefficient of ease of separation, so it should be done first.

The separation proposed will therefore be as represented in Figure 12.9.

Additional Points to Ponder How can this separation sequence be improved further? What are the energy requirements for this proposed separation?

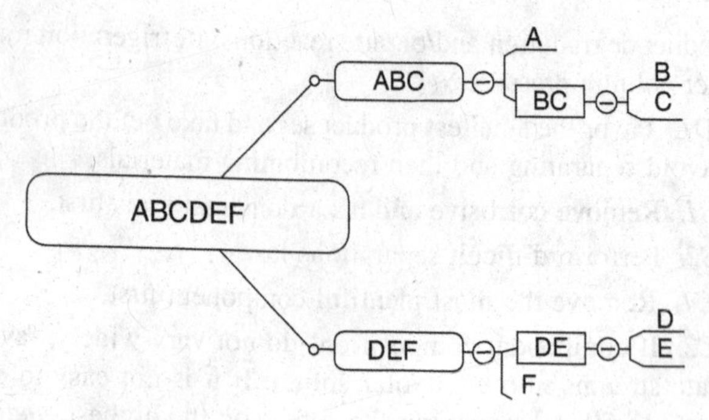

FIGURE 12.9 Proposed separation sequence.

12.5 HIERARCHICAL DECOMPOSITION

Hierarchical decomposition is probably the most widely used semiformal design approach. It breaks the design down in such a way that "broad strokes" decisions are taken first in the design phase (e.g., identifying reaction–separation schemes and recycle streams). After the broad strokes decisions are made decisions on each unit operation are refined. For example, Linnhoff et al.[19] presented the onion diagram, which has been used to decompose the design procedure conceptually into distinct steps (Figure 12.10) starting with the selection of a reactor path, followed by separators, compressors, and finally, the heat exchanger network. This concept has been used together with simulators to build the process flowsheet in an iterative manner starting with the reactor as the center of the chemical plant.[20] This approach does not help in the integration of an entire chemical plant, as decisions cannot be taken simultaneously; however, the use of simulators helps to speed the iterative nature of this technique. In addition, as Smith has pointed out in his book *Chemical Process Design*,[21] use of this hierarchy very often relegates environmental, health, safety, and, in general, green engineering aspects to the

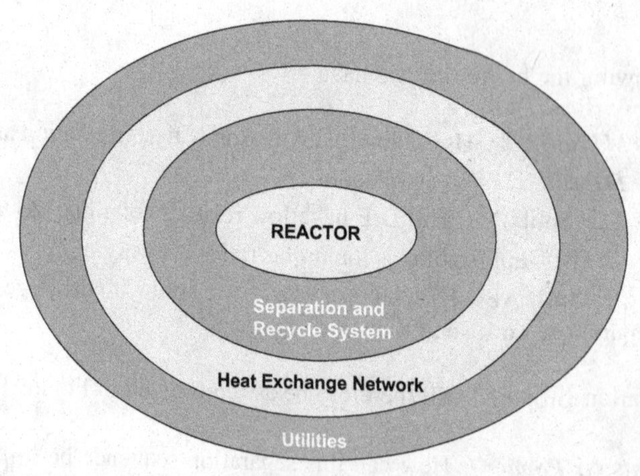

FIGURE 12.10 Linnoff's onion diagram used to decompose the design process.

very final steps of the synthesis instead of integrating green engineering concepts into the core of the process design.

Douglas developed a hierarchical decomposition approach, which is now widely accepted, and in 1992 he proposed to modify his hierarchy to account for waste minimization in the process synthesis.[22] The levels of decision proposed were:

Level 1: input information—type of problem

Level 2: input–output structure of the flowsheet

Level 3: recycle–structure of the flowsheet

Level 4: specification of the separation system

- *4a:* general structure—phase splits
- *4b:* vapor recovery system
- *4c:* liquid recovery system
- *4d:* solid recovery system

Level 5: energy integration

Level 6: evaluation of alternatives

Level 7: flexibility and control

Level 8: safety

Douglas's proposal is that waste minimization would be achieved by decisions or information given for levels 1 to 4 only (Table 12.6). Therefore, the items related to heat integration and safety considerations were left outside the scope of the proposal.

Example 12.5 Many chemical intermediates are produced by air oxidation of hydrocarbons. In these reactions, some carbon dioxide and water are formed, with some nitrogen-producing nitrogen oxides. Using the questions offered by Douglas on level 2, which alternatives do we have to minimize waste formation?

Solution The questions at level 2 are related to the purity of the reactants and whether one should recycle some unreacted species, by-products, and catalysts. The questions can prompt alternatives to reduce the formation of nitrogen oxides and carbon dioxide that, once generated, can be evaluated further to determine suitability. Some of the alternatives that can be generated are listed in Table 12.7.

Alternative 1 has the drawback that as the nitrogen in the air modulates the reaction temperature, using pure oxygen will require additional heat transfer controls. Alternative 4 would help in controlling reaction temperature, as the carbon dioxide would moderate the temperature. Alternatives 2 and 3 point to a closed-loop design vs. a one-pass flowsheet, which could increase the efficiency of the process.

Additional Points to Ponder What other alternatives can be generated for this problem? How can we capture other options for temperature control given that oxides of nitrogen are formed between certain temperature ranges? How can we measure the degree to which these alternatives contribute to the principles and goals of green engineering?

TABLE 12.6 Summary of Decisions or Information Needed for Waste Minimization Using Douglas's Hierarchical Decomposition Approach

	Decisions or Information for Waste Minimization Purposes
Level 1: input information—type of problem	1. Desired product, production rate, product value, and product 2. Reactions and reaction conditions 3. Raw material streams, conditions, and costs 4. Data on the product distribution 5. Data concerning the reaction rates and catalyst deactivation 6. Any processing constraints 7. Plant and site data and physical property data 8. Data concerning the safety, toxicity, and environmental impact 9. Cost data for the by-products produced
Level 2: input–output structure of the flowsheet	1. Purify the feed streams? 2. Do not recover and recycle some reactants? 3. Use a gas recycle and purge stream or vent gaseous reactants? 4. Recover and recycle or remove a by-product formed by a secondary reversible reaction?
Level 3: recycle structure of the flowsheet	1. Excess reactant at the reactor inlet? 2. Reactor heat effects: adiabatic, isothermal, heat carrier? 3. Shift equilibrium conversion? How? 4. Diluent to improve the product distribution? 5. Reactor or product solvents? 6. Complete conversion to avoid a separation? 7. Reactor–separator (distillation, extraction, etc.)
Level 4: specification of the separation system 4a: general structure—phase splits 4b: vapor recovery system 4c: liquid recovery system 4d: solid recovery system	Vapor systems 1. Condensation: low temperature and/or high pressure 2. Gas absorption 3. Gas adsorption 4. Reactive absorption 5. Membranes Liquid systems 1. Stripping 2. Distillation 3. Azeotropic, extractive, and reactive distillation 4. Extraction 5. Crystallization 6. Adsorption 7. Membranes

TABLE 12.7 Alternatives at Level 2

	Questions	Alternatives
Level 2: input–output structure of the flowsheet	1. Purify the feed streams? 2. Do not recover and recycle some reactants? 3. Use a gas recycle and purge stream, or vent gaseous reactants? 4. Recover and recycle or remove a by-product formed by a secondary reversible reaction?	1. Use pure oxygen instead of air to eliminate nitrogen oxide production. 2. Recycle back unreacted species. 3. Purge recycled streams to reduce accumulation of impurities in the system. 4. Use pure oxygen to eliminate nitrogen oxide formation and recycle back carbon dioxide.

12.6 SUPERSTRUCTURE AND MULTIOBJECTIVE OPTIMIZATION

The process synthesis problem can also be formulated and solved as an optimization problem. In this case, all the possible flowsheets are combined into one, normally called a *superstructure*. The superstructure contains more streams and unit operations than would be needed in reality, but by selectively removing unit operations and specific streams, the superstructure can be reduced to a practical flowsheet. The optimization problem is reduced to continuous design variables, such as pressure and temperature and to integer variables that indicate if a specific stream or unit operation is or not included. Figure 12.11 shows a very simple example of a superstructure generated to represent two alternative reactors, where the squares represent logical choices (e.g., it is either reactor 1 or reactor 2) and the triangle represents the convergence of the hypothetical choices (e.g., after the reactor, the process continues to the next unit operation through the exit stream).

The objective of the optimization is to remove those aspects of the flowsheet that are not going to render the best options. To solve this optimization problem, computer-aided and mathematical approaches have been used. Given the type of problem, the mathematical approaches utilize algorithms that solve mixed-integer nonlinear problems (MINLP).

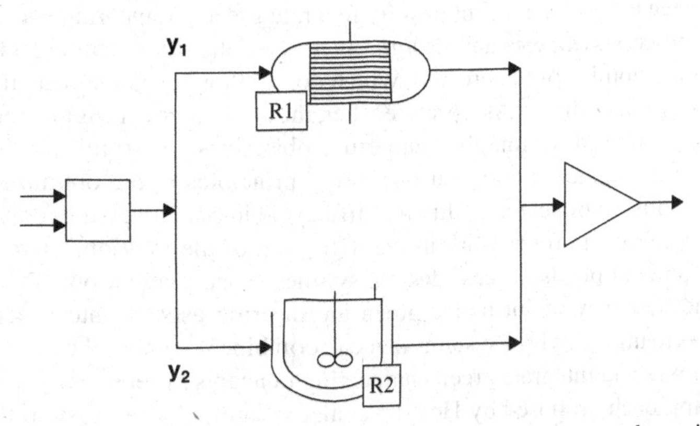

FIGURE 12.11 Example of a superstructure generated to represent two alternative reactors. If $y_1 = \text{true}$, then $y_2 = \text{false}$ and reactor 1 is used. If $y_2 = \text{true}$, then $y_1 = \text{false}$ and reactor 2 is used.

To illustrate the optimization problem, we can loosely follow Grossmann's MINLP definition.[23] Let's assume that we want to minimize the total environmental impact of our process superstructure flowsheet; we can describe the optimization problem as follows:

$$
\begin{aligned}
F_{\mathrm{obj}} = \mathrm{Min}\{C^T y + f(x)\} \\
\text{subject to:} \quad & (1)\ Cx + Dy \leq d \\
& (2)\ Ey \leq e \\
& (3)\ x \geq 0 \\
& (4)\ y \in \{0, 1\}p
\end{aligned}
$$

What this set of mathematical equations means is that the total environmental impact of our flowsheet would depend on a combination of process-driven equations [or $f(x)$], while the definition of the flowsheet would be guided by which combination of unit operations are considered [$C^T y$]. Our objective function (F_{obj}) would be minimized through the relationship, as that would minimize the total environmental impact. In constraint (3) we state that x represents a vector of continuous variables, such as temperature and pressure. In constraint (4) we state that y represents the logical variables that will tell us if a unit operation in the superstructure is either kept or not (0 value if it is not kept, 1 if it is). Logical constraints are set in constraint (2), and constraint (1) represents the overall types of process constraints, such as mass and energy balances and process design specifications.

A number of approaches have been developed in this area,[24,25] with a series of accompanying computer tools, algorithms, solvers, and simulators to expedite the solution of these types of optimization problems, which can be rather complex for some chemical systems. Early in the state of the art of process synthesis, the role of superstructure optimization was not as prominent, but with improvements in computer capabilities and the development of better algorithms, both by chemical engineers and by mathematical programmers, the approach has become more practical. One of the limitations of this method, however, is that the initial superstructure in most cases is limited by the imagination of the engineer who prepares the first draft. In recent times, engineers have used other techniques, such as artificial intelligence and thermodynamic pinch to generate the first superstructure.

Multiobjective optimization, alone or in combination with other approaches in hybrid methods, has been used in the recent past to integrate green engineering and sustainability concepts into process synthesis and design, as most of the optimization techniques have focused on the economics of the process, which don't allow for implementation of the full aspects of green engineering.[26] As we saw earlier, the areas of green engineering represent a balance between several seemingly competing objectives. A promising theory for the inclusion of green engineering and sustainability principles in the optimization of alternatives is the inclusion of sustainability and life cycle impact assessment metrics into the design, using a life cycle framework in the definition of the multiobjective optimization problem for chemical plants process design, synthesis, and integration.[27–29]

Perhaps the best way to introduce green engineering aspects into process synthesis methodology is to utilize hybrid systems that can combine several of the best aspects of the various techniques and integrate green engineering concepts as seamlessly as possible. For example, an approach proposed by Hostrup et al. explores a hybrid system that integrates mathematical optimization techniques, heuristic approaches, and thermodynamic insights into process synthesis to determine a flowsheet that optimizes separation efficiency, energy and material costs, and environmental aspects.[30,31] In this approach, the nonoptimal flowsheet generated by heuristics and thermodynamic insight techniques is taken as the first iteration of the superstructure, and this superstructure is optimized with the mathematical technique to determine the optimal flowsheet. The optimal flowsheet is obtained by solving the optimization problem through a suitable MINLP model.

Let's follow an example of a hybrid system proposed by Hostrup et al. to illustrate how solving process synthesis through superstructure optimization would work. In this case, we seek to find the optimal flowsheet and the best solvent to separate a mixture. The proposed steps for the methodology are:

1. Determine the number and type of separation techniques based on solvent property differences.
2. Identify separation alternatives using external materials (e.g., membranes).
3. Define the initial superstructure based on pure-component properties analysis to determine which separation techniques are feasible and which are not.
4. Perform a binary mixture property analysis to eliminate unfeasible separation techniques from the superstructure.
5. Generate solvent alternatives based on process and environmental constraints.
6. Perform a multicomponent mixture property analysis with the solvents identified to eliminate unfeasible separation techniques from the superstructure.
7. Solve the remaining optimization problem. The problem statement can be formulated mathematically as follows:

$F_{\text{obj}} = \text{Max}\{C^T y + f(x)\}$	sets optimization function
subject to	
$h_1(x) = 0$	process design specifications
$h_2(x) = 0$	mass and energy balances
$h_3(x) = 0$	solvent design constraints
$l_1 \leq g_1(x) \leq u_1$	process design specifications
$l_2 \leq g_2(x) \leq u_2$	environmental and property constraints
$l_3 \leq By + Cx \leq u_3$	enforces logical conditions
$x \geq 0$	sets continuous variables
$y \in \{0, 1\}p$	sets logical variables

Example 12.6 Benzene is used to separate acetone from a well-known binary azeotropic mixture between acetone and chloroform. Given the health and safety issues associated with benzene, the desire is to replace benzene and find the optimal design to accomplish this. In this instance the problem can be stated: "Given a chemical species that must be separated from a mixture, determine the flowsheet design that optimizes separation efficiency, cost of energy requirements, process issues, and environmental constraints."

Solution A detailed solution can be found in a publication by Hostrup et al.[31] Note that once the problem is solved, the "optimal" flowsheet design will not have a minimal environmental impact, but it will satisfy a set of environmental constraints in that the solvent selected will not have the same concerns as benzene.

Step 1. The separation techniques considered as a starting point were adsorption, absorption, pervaporation, filtration, crystallization, distillation, distillation plus decanter, extractive distillation, azeotropic distillation, liquid–liquid extraction, and supercritical extraction.

Step 2. No membrane system is known for this mixture, so membrane-based techniques are eliminated.

Step 3. There is no interest in a solid-state product, so crystallization is eliminated.

Step 4. Binary mixture properties analysis validates the presence of an azeotrope and confirms the feasibility of pressure swing distillation, since the location of the azeotrope changes with the pressure. High-pressure distillation is feasible but would probably require very large amounts of energy, so it is eliminated as an alternative. Miscibility analysis also eliminates distillation plus decanter as an alternative, since there are no miscibility gaps.

Step 5. For this step the desired outcome is to replace benzene as a separating agent. For this problem, the solvent design problem is set first, thus reducing the number of alternatives for subsequent optimization. The design criteria for the solvent search is to include acyclic alcohols, aldehydes, ketones, acids, ethers, and esters with the following property constraints:

$$g_2(x)\begin{cases} 340\,\text{K} < T_b < 420\,\text{K} \\ \beta(\text{selectivity}) > 3.5 \\ S_p(\text{solvent power}) > 2.0 \\ S_1(\text{solvent loss}) < 0.9 \end{cases}$$

$$h_3(x)\begin{cases} \text{No azeotrope with any component in the binary feed mixture} \\ \text{Binary feed mixture molar composition : acetone 34.4\%, chloroform 65.6\%} \end{cases}$$

After designing the solvent, Hostrup et al. found two potential solvents from the molecular design exercise: 1-hexanal and amyl methyl ether. They checked the EHS data on both solvents and found that 1-hexanal did not have the carcinogenicity problems that benzene has. There was, however, very limited data on methyl amyl ether, but they decided not to exclude this solvent from the analysis at this point, although they indicated that additional exploration was necessary. Note that if the solvent selection criteria are changed, a different set of solvents or even more solvents might be found as alternatives. However, for the purpose of illustration, two alternatives plus the baseline (benzene) for comparison would be sufficient.

Step 6. The two solvents identified form homogeneous systems with the acetone–chloroform mixture; therefore, liquid–liquid extraction and azeotropic distillation are eliminated. Since carbon dioxide is found to have low solubility with acetone and poor selectivity in relation to chloroform, supercritical fluid extraction is also eliminated. Therefore, the reduced superstructure will only consider pressure swing distillation and extractive distillation with the three solvent alternatives (the two additional solvents plus benzene as a baseline for comparison). Note that both separation

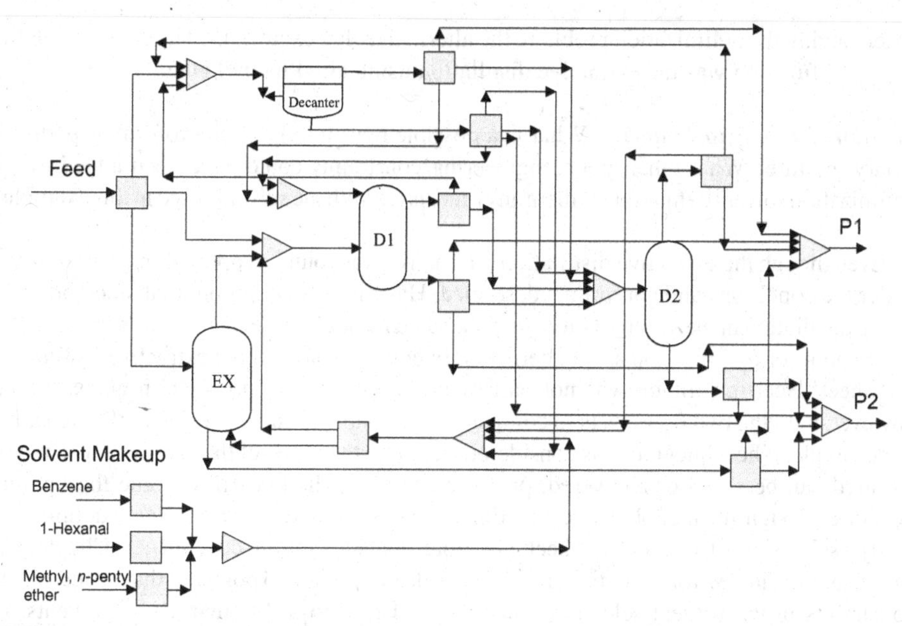

FIGURE 12.12 Simplified representation of the reduced superstructure for Example 12.6. (From Hostrup and Gani, ref 30. Copyright © 1999, with permission from Elsevier.)

alternatives include two distillation columns. A simplified version of the superstructure is presented in Figure 12.12.

Step 7. The optimization problem was based on maximizing profits. The optimization problem can be expressed as follows:

$$\begin{aligned} F_{\text{obj}} &= \text{Max}\{\text{profit}\} \\ &= \text{Max}\{\text{sales-cost of solvent-cost of steam-cost of electricity-cost of cooling}\} \end{aligned}$$

subject to

$$h_1(x)\begin{cases} \text{column 1 pressure} = \begin{cases} \text{1 atm if extractive distillation} \\ \text{10 atm if pressure swing distillation} \end{cases} \\ \text{outlet pressure of pump} = \text{column 1 pressure} \\ \text{Feed} = 5\ \text{kmol/h acetone} + 5\ \text{kmol/h chloroform} \end{cases}$$

$$h_1(x)\begin{cases} \text{mass and energy balances for the mixer} \\ \text{energy balance for the pump} \\ \text{energy balance for the heat exchanger} \\ \text{mass and energy balances for columns 1 and 2} \\ \text{total mass and energy balances for the entire flowsheet} \end{cases}$$

$$g_1(x)\begin{cases} \text{composition of acetone in the distillate of column 1} > 0.99 \\ \text{recovery of acetone in column 1} > 99\% \\ \text{composition of chlorofrom in the distillate of column 2} > 0.98 \\ \text{recovery of chloroform in column 2} > 90\% \end{cases}$$

After solving the optimization problem, the alternative that exhibits the largest value for the objective function was the extractive distillation using amyl methyl ether.

Additional Points to Ponder What is a possible reason to have chloroform as part of a binary mixture? What other green engineering constraints could have been added to the optimization solvent definition? Is there any other issue with the solvents used in this example?

Even though the extractive distillation alternative was found as optimal in Example 20.6 under the conditions and constraints described, Hastrup et al. point out that since pressure swing distillation in the superstructure flowsheet did not include heat integration, there is still some room for optimization as an alternative in comparison with the extractive distillation flowsheet. Heat integration was not part of the scope of the optimization at that point; however, with the two flowsheets to compare it is possible to run some "what if" scenarios. For example, if heat integration is considered and more than 25% of the energy requirements required can be saved or recovered, pressure swing distillation will become the optimal flowsheet design for this objective function. The system under study in the example was really a subsystem of a larger flowsheet; therefore, heat integration (covered next Chapter 13) should be considered for the entire process. It is also important to point out that changing the constraints in the solvent selection criteria would lead to a different set of solvents, or changing the conditions of the separation techniques might have changed the reduced superstructure (e.g., by adding a cosolvent to carbon dioxide to increase the solubility of the substrate might have made supercritical fluid extraction an attractive alternative).

Finally, it is very important to point out that the formulation of the optimization function in this case was based purely on an economic objective that was maximized while attempting to satisfy environmental, health, and safety constraints. In this case the objective function was not minimization of the environmental impact, and this provides a very interesting contrast with the methodology presented by Carvalho et al. If the objective function had been written to minimize the environmental impact using one or a series of green engineering metrics, it might have been possible to obtain another solution for the optimal flowsheet.

12.7 SYNTHESIS OF SUBSYSTEMS

Most process synthesis work has concentrated on subsystems in which the main objective has been the reduction and optimization of energy. Reduction and optimization of energy use is an important green engineering objective. The synthesis and integration of subsystems is dealt with in more detail in Chapter 13, but mentioning some of the most common subsystems syntheses briefly warrants inclusion in this chapter as a matter of introduction.

1. *Distillation sequencing.* The synthesis of sequences for simple distillation based on heuristics is very well developed; some examples of these heuristics were covered earlier in the chapter. There have also been enumeration methods, evolutionary search procedures, and thermodynamic analyses of the separation of mixtures.
2. *Thermodynamic pinch.* The pinch method is applied directly to energy optimization to determine the possibility for heat integration given thermodynamic constraints. Pinch can be applied to driving forces other than temperature: for example, in the case of contaminant concentrations in water.

3. *Reactor networks.* Heuristics-based approaches are generally limited to simple reactions. Some of the reactor networks have been synthesized using superstructure representation with serial recycle reactors without bypass.

4. *Heat exchange networks (HENs).* Heat exchange network synthesis is by far the best developed technique, and a wide variety of methods and software packages are available to optimize heat integration, based primarily on thermodynamic pinch.

5. *Mass exchange networks (MENs).* These were initially motivated by the application to waste recovery systems. Interesting analogies can be drawn from the HENs in terms of using the pinch point concept and the ways to solve the optimization problems.

PROBLEMS

12.1 You have a process with five components, A, B, C, D, and E, and you want to estimate how many different flowsheets can be drawn to achieve that.

(a) How many different options would you have, assuming that the only separation technique available to you is distillation? Show these options in a diagram.

(b) How many different options would you have, assuming that you could use distillation, extraction, and membranes interchangeably?

(c) How would the number of choices affect your decisions when striving for a final flowsheet with a green engineering design?

12.2 Methyl *tert*-butyl ether (MTBE) is manufactured by catalytically reacting isobutylene and methanol. The isobutylene is fed to the reactor containing several impurities, such as *n*-butane, isobutane, 1-butene, and *cis*- and *trans*-2-butene. Water is used to recover and recycle methanol. Figure P12.2 shows the initial flow diagram for MTBE.

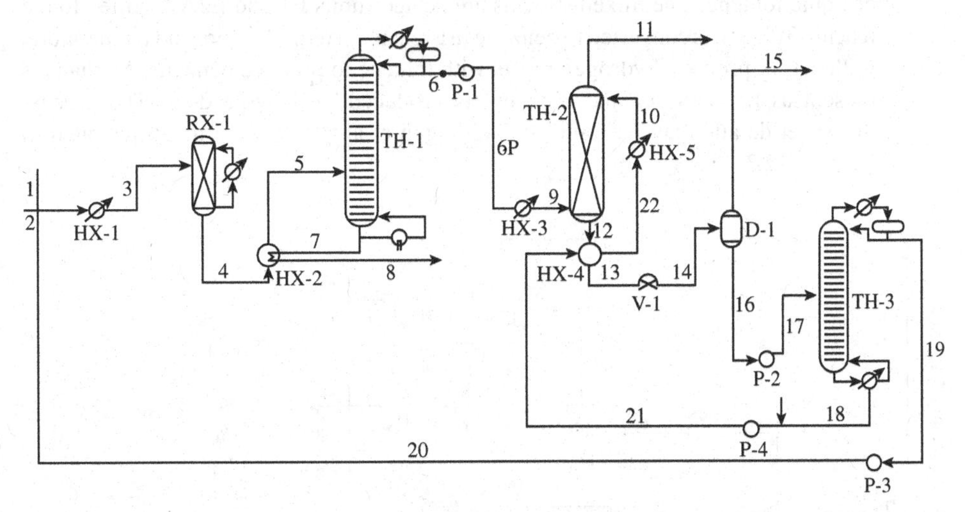

FIGURE P12.2 (From Carvalho et al., ref 6. Copyright © 2006, with permission from Elsevier.)

In the modeling that Calvalho et al. performed using the ISA methodology explained previously, the information in Table P12.2 is obtained for several compounds in either open paths (i.e., mass and energy come into the process and go out

TABLE P12.2 Data for Compounds in Open and Closed Paths

Compounds in Open Paths	Flow Rate (kg/h)	Material Value Added (MVA, 10^3 \$/yr)	Energy and Waste Costs (EWC, 10^3 \$/yr)
n-Butane	4,446.3	−376.4	9.1
Isobutane	20,255.5	−1714.8	260.7
1-Butene	3,338.4	−282.6	43.4
cis- and *trans*-2-Butene	4,769.1	−403.7	63.9
Compounds in Closed Paths	**Flow Rate (kg/h)**	**Accumulation Factor (AF)**	**Energy and Waste Costs (EWC, 10^3 \$/y)**
Water	6,744.9	626.5	313.3

from the process through these paths) or closed paths (recycles). It was found that the indicators with the greatest potential for improvement were the AF and EWC of the water stream.

(a) Identify the open and closed paths in the MTBE flowsheet.

(b) Assuming that the remaining metrics and indicators will remain unchanged, what do these modeling results mean in terms of process synthesis and green engineering?

12.3 In the hydrodealkylation of toluene or HDA process, toluene (S1) is converted into benzene by reaction with hydrogen (S2) in a reactor (unit R1) at high pressure and temperature. Hydrogen containing approximately 5% methane and toluene feed are mixed (unit M1) with recycled gaseous hydrogen and methane and a recycle stream of liquid toluene. The mixed streams are heated (units E1 and E2A) and fed to the reactor. Toluene is converted to benzene in a reactor at high pressure and temperature. In the HDA process, hydrogen reacts with toluene to produce benzene. Methane is present as an impurity and biphenyl is produced as a by-product. The process flowsheet details may be found in Uerdingen et al.[32,33] but are also illustrated in Figure P12.3.

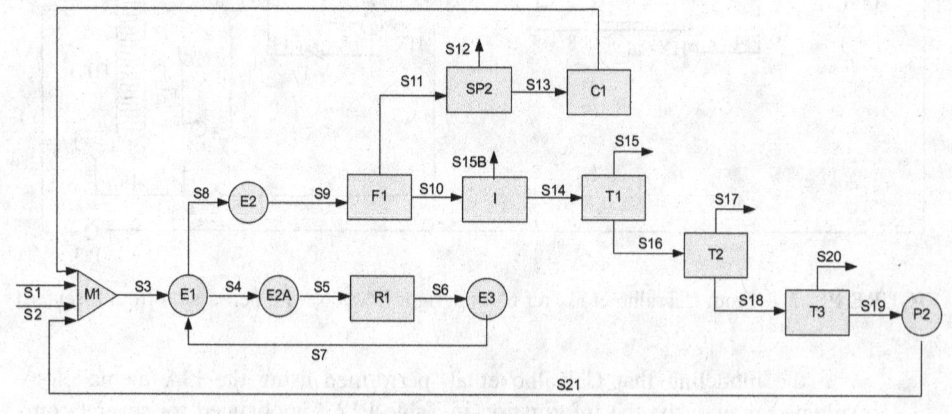

FIGURE P12.3

TABLE P12.3 Indicator Values for Various Paths

Path	Compound	Path Type	Flow Rate (kg/h)	Total Value Added (TVA, 10^3 $)
02	Hydrogen	Open	244.3	−2,981.7
06	Methane	Open	1,887.7	−8755.7
012	Methane	Open	103.9	−482.3
C1	Methane	Gas cycle	10,926.7	−484.5
C2	Hydrogen	Gas cycle	1,281.2	−179.4
C10	Toluene	Liquid cycle	3,695.4	−1,828.8

The mass and energy indicators have been calculated by Gani et al.[34] for the initial flowsheet. There are five compounds and 20 open paths (i.e., mass and energy come into the process and go out from the process through these paths) and 15 closed paths (i.e., recycle streams). The most important paths and the corresponding indicator values are given in Table P12.3.

(a) What does a negative value of TVA indicate from the viewpoint of process synthesis?

(b) If a sensitivity analysis concludes that the greatest positive impact can be obtained by reducing the methane in gas cycle path C1, what retrofit options would you recommend for the existing flowsheet?

(c) How would your suggested changes affect the process synthesis from a green engineering prospective? Explain the potential advantages and disadvantages.

12.4 How do the heuristics M2, D1, and C2 from Nadgir and Liu align with green engineering principles?

12.5 Table P12.5 shows a mixture of light hydrocarbons.

TABLE P12.5 Hydrocarbon Data

Compound	Fraction	Relative Volatility at 37.8°C and 1.72 MPa
A: Propane	0.05	
B: Isobutane	0.15	2.00
C: *n*-Butane	0.25	1.33
D: Isopentane	0.20	2.40
E: *n*-Pentane	0.35	1.25

(a) Use the heuristics of Nadgir and Liu in the order given to propose a flowsheet for the separation of a mixture.

(b) Generate a second flowsheet without following the heuristics.

(c) Estimate the energy required for each flowsheet generated. Assume perfect separations (efficiency 100%).

(d) Which flowsheet is closer to the principles of green engineering?

12.6 A stream resulting from thermal cracking consists of hydrogen (18 mol/h), methane (5 mol/h), ethylene (24 mol/h), ethane (15 mol/h), propylene (14 mol/h), and propane

(6 mol/h). It is desired to keep the methane and hydrogen in a single stream and the rest of the components separated into single streams.

(a) Use the heuristics of Nadgir and Liu in the order given to propose a flowsheet for the separation of the stream.

(b) Generate a second flowsheet without following the heuristics.

(c) Estimate the energy required for each flowsheet generated. Assume perfect separations (efficiency 100%).

(d) Which flowsheet is closer to the principles of green engineering?

12.7 Allyl chloride is produced by the reaction of propylene and chlorine. The reactor effluent consists of a mixture of unreacted species, allyl chloride, hydrochloric acid, 1,3-dichloropropane, and acroleindichloride, as shown in Table P12.7.

TABLE P12.7 Effluent Data

Compound	Moles	Boiling Point (°C)
Allyl chloride	9.3	50
Chlorine	3.0	−34
Hydrochloric acid	93	−85
Propylene	105	−85
1,3-Dichloropropane	0.2	112
Acroleindichloride	1.8	84

(a) Use the heuristics of Nadgir and Liu in the order given to propose a flowsheet for the separation of the stream.

(b) What would you say about this system from a green engineering prospective?

12.8 For Example 12.4:

(a) Generate a different alternative flowsheet.

(b) Estimate the energy required to perform the separation in each of the flowsheets generated.

(c) What would this mean in terms of green engineering?

12.9 For Example 12.5:

(a) Provide a generic example of the reactions involved.

(b) Generate alternatives using the level 3 questions proposed by Douglas's hierarchical decomposition.

(c) Draw the flowsheet that represents the processes you are proposing.

(d) Describe the advantages and disadvantages of the various alternatives from a green engineering prospective.

12.10 Ethylene can be produced by means of ethane cracking. A simplified set of reactions is as follows.

$$C_2H_6 \rightarrow C_2H_4 + H_2 \quad \text{main reaction}$$

$$C_2H_6 + H_2 \rightarrow 2CH_4 \quad \text{side reaction}$$

To reduce the production of methane, a by-product stream is normally added as a diluent, as it shifts the product distribution, acts as a heat carrier, and is easy to condense and separate from the gaseous mixture. However, this produces a wastewater stream that needs to be treated. Use the level 3 questions proposed by Douglas's hierarchical decomposition to generate alternative flowsheets. Describe the advantages and disadvantages of the alternatives in terms of green engineering.

12.11 For the optimization problem of Example 12.6

(a) Propose additional constraints that could have been added to the optimization problem.

(b) Rewrite the objective function to minimize the environmental impact of the separation.

12.12 A chemical product (P) is formed by the hydrogenation of the raw material (R). The reaction takes place in a solvent (S) under excess of R. Hydrogen and R enter the reactor and the reaction mixture is sent to a flash distillation unit where S is separated in the light fraction and returned to the reactor. The unreacted R and P are separated to obtain the product P at the desired purity.

(a) Provide an example of an industrial process that follows the description above.

(b) Draw the superstructure for this flowsheet considering two different reactors and three different separation techniques.

(c) Propose an objective function that integrates green engineering principles into the process synthesis.

12.13 We want to develop a flowsheet for the separation of propane, isobutane, and *n*-butane that are in equimolar compositions in a stream with a total flow of 300 kmol/h at 350 K.

(a) Which separation techniques can be considered for this flowsheet?

(b) Select two of the separation techniques you have identified and draw a superstructure.

(c) Write a conceptual objective function and its constraints in such a way that the flowsheet design would minimize the environmental impact of the separation.

REFERENCES

1. Rudd, D. F., Powers, G. J., Siirola, J. J. *Process Synthesis*. Prentice Hall, Englewood Cliffs, NJ. 1973.
2. Douglas, J. M. *Conceptual Design of Chemical Processes*. McGraw-Hill, New York, 1988.
3. Johns, W. R. Process synthesis: poised for a wider role. *Chem. Eng. Prog.*, 2001, 97(4), 59–65.
4. Cano-Ruiz, J. A., McRae, G. J. Environmentally conscious chemical process design. *Ann. Rev. of Energy Environ.*, 1998, 23, 499–536.
5. Carvalho, A., Gani, R., Matos, H., et al. Design of sustainable chemical processes: systematic retrofit analysis generation and evaluation of alternatives. *Process Saf. Environ. Prot.*, 2008, 33(12), 2075–2090.

6. Carvalho, A., Gani, R., Matos, H. Design of sustainable processes: systematic generation and evaluation of alternatives. In *16th European Symposium on Computer Aided Process Engineering and 9th International Symposium on Process Systems Engineering*, Vol. 2, Marquardt, W., Pantelides C., (Eds.) Elsevier, Amsterdam, The Netherlands, 2006, pp. 817–822.
7. Carvalho, A., Gani, R., Matos, H. Systematic methodology for process analysis and generation of sustainable alternatives. In *18th European Symposium on Computer Aided Process Engineering (ESCAPE 18)*, Braunschweig, B., Joulia, X., Eds. Elsevier, Amsterdam, The Netherlands, 2008.
8. Cabezas, H., Bare, J., Mallick, S. Pollution prevention with chemical process simulators: the generalized waste reduction (WAR) algorithm– full version. *Comput. Chem. Eng*, 1999, 23(4–5), 623–634.
9. Heikkila, A. M. Inherent safety in process plant design: an index-based approach. Ph.D. dissertation, VTT Automation, Espoo, Finland, 1999.
10. Jaksland, A. C., Gani, R., Lien, K. M. Separation process design and synthesis based on thermodynamic insights. *Chem. Eng. Sci.*, 1995, 50(3), 511–530.
11. Carvalho, A., Matos, H. A., Gani, R. Systematic methodology for continuous/batch processes: analysis and generation of sustainable alternatives. In, *Proceedings of the Foundations of Computer-Aided Process Operations*, 2008.
12. Westerberg, A. W. *Comput. Chem. Eng.*, 2004, 28(4), 447–458.
13. Woods, D. R. *Rules of Thumb in Engineering Practice*. Wiley-VCH, Weinheim, Germany, 2007.
14. Walas, S. *Chemical Process Equipment: Selection and Design*. Butterworth-Heinemann, Oxford, UK, 2002.
15. Branan, C. *Rules of Thumb for Chemical Engineers*. Gulf Professional Publishing, Houston, TX, 2005.
16. Harmsen, J. G. Industrial best practices of conceptual process design. *Chem. Eng. Process.*, 2004, 43, 677–681.
17. Butner, R. S. A heuristic design advisor for incorporating pollution prevention concepts in chemical process design. *Clean Technol. Environ. Policy*, 1999, 1(3), 164–169.
18. Nadgir, V. M., Liu, Y. A. Studies in chemical process design and synthesis: V. A simple heuristic method for systematic synthesis of initial sequences of multicomponent initial sequences of multicomponent separations. *AIChE J.*, 1983, 29, 926–934.
19. Linnhoff, B., Townsend, D. W., Boland, D., Hewitt, G. F., Thomas, B. E. A., Guy, A. R., Marsland, R. H. *A User Guide on Process Integration for the Efficient Use of Energy*. Institute of Chemical Engineers, Rugby, UK, 1982.
20. Foo, D. C. Y., Manan, Z. A., Selvan , M., McGuire, M. L. Process synthesis by onion model and process simulation. *Chem. Eng. Prog.*, Oct. 2005, 101(10), 25–29.
21. Smith, R. *Chemical Process Design*. McGraw-Hill, New York, 1995.
22. Douglas, J. M. Process synthesis for waste minimization. *Ind. Eng. Chem. Res.*, 1992, 31, 238–243.
23. Grossmann, I. E. A modelling decomposition strategy for MINLP optimisation of process flowsheets. *Comput. Chem. Eng.*, 1989, 13, 797.
24. Grossmann, I. E. Daichendt, M. M. Comput. Chem. Eng., 1996, 20(6–7), 665–683.
25. Grossmann, I. E., Biegler, L. T., Westerberg, A. W. Retrofit design of processes. In *Foundations of Computer Aided Process Operations*, Reklaitis, G. V., Spriggs, H. D. Eds. Elsevier, Amsterdam, 1987, p.403.
26. Jackson, T. Clift, R. Where's the profit in industrial ecology? *J. Ind. Ecol.* 1998, 2, 3–5.
27. Azapagic, A., Clift, R. The application of life cycle assessment to process optimisation. *Comput. Chem. Eng.*, 1999, 23, 1509–1526.
28. Korevaar, G. Sustainable chemical process and products: New Design Methodology and Design Tool. Ph.D. dissertation. Technische Universiteit Delft, Delft, The Netherlands, 2004.

29. Kaggerud, K. H. Chemical and Process Integration for Environmental Assessment. Ph.D. dissertation. Norwegian University of Science and Technology. Trondheim, Norway, 2007.
30. Hostrup, M., Harper, P. M., Gani, R. Design of environmentally benign processes: integration of solvent design and separation process synthesis. *Comput. Chem. Eng.*, 1999, 23, 1395–1414.
31. Hostrup, M., Gani, R., Kravanja, Z., Sorsak, A., Grossmann, I. Integration of thermodynamic insights and MINLP optimization for the synthesis, design and analysis of process flowsheets. *Comput. Chem. Eng.*, 2001, 25, 73–83.
32. Uerdingen, E. Retrofit Design of Continuous Chemical Processes for the Improvement of Production Cost-Efficiency. Ph.D. dissertation. ETH–Zürich, Zürich, Switzerland, 2002.
33. Uerdingen, E., Gani, R., Fisher, U., Hungerbuhler, K. A new screening methodology for the identification of economically beneficial retrofit options for chemical processes. *AIChE J.*, 2003, 49(9), 2400–2418.
34. Gani, R., Jørgensen, S. B., Jensen, N. Design of sustainable processes: systematic generation and evaluation of alternatives. Presented at the 7th World Congress of Chemical Engineering, 2005.
35. Carvalho, A., Matos, H. A., Gani, R. Design of batch operations: systematic methodology for generation and analysis of sustainable alternatives. *Comput. Chem. Eng.*, 2009, 33(12), 2075–2090.

13

MASS AND ENERGY INTEGRATION

What This Chapter Is About In Chapter 12 we covered how green engineering and sustainability concepts can be integrated into chemical process synthesis. In this chapter we continue to cover the design and retrofit of chemical process with the integration of green engineering principles, in this case how to design processes in such a way that the energy and mass requirements are optimized. Mass and energy integration seeks to productively utilize energy (e.g., heat) and materials (e.g., trace products in waste stream) that otherwise would have been wasted.

Learning Objectives At the end of this chapter, the student will be able to:

- Understand the concept of energy integration in the context of green engineering.
- Understand the concept of mass integration in the context of green engineering.
- Identify opportunities for mass and energy integration in a chemical process flowsheet to design more sustainable chemical processes.
- Apply mass and energy integration principles to examples and explain how these relate to green engineering principles.

13.1 PROCESS INTEGRATION: SYNTHESIS, ANALYSIS, AND OPTIMIZATION

We have seen how a chemical process is a system of interrelated units, inputs, outputs, and recycle streams. A chemical process should therefore not be treated as separated groups

Green Chemistry and Engineering: A Practical Design Approach, By Concepción Jiménez-González and David J. C. Constable

of elements, but as a complete unit. For example, when a chemical process has two or more reactions in series, the efficiency of one reaction should not be considered in isolation from the others but as an integrated system that includes the two or more reactions and all the necessary separation steps. Process integration is best understood as a holistic, systematic, methodical framework for process synthesis, retrofit, and operation. Process integration methodologies have three key components: process synthesis, process analysis, and process optimization.

In Chapter 12 we studied process synthesis and process optimization and considered how green engineering principles can be incorporated into process synthesis and retrofit. While process synthesis intends to combine the various elements (e.g., unit operations) of a process to form the general flowsheet, process analysis would involve the decomposition of the process to study and optimize individual components. Process synthesis and analysis are therefore complementary techniques. Once the flowsheet is defined through process synthesis, specific characteristics of the process, such as temperatures, flow rates, and compositions, and the process performance are defined using analysis techniques. The principal means by which process analysis is carried out range from mathematical modeling, computer-aided tools, and graphical techniques to laboratory and pilot-plant experiments, among others.

Once the process characteristics and the performance of a process have been evaluated and defined, we can engage in process optimization. The process engineer can determine if the targets and objectives of the process have been met. These objectives can be economic (e.g., profit, total cost), technical (e.g., purity, yield), environmental (e.g., waste, impact), or a combination of any of these. Although these objectives may be met, that does not necessarily mean that the flowsheet would be the optimal one, so the next step is to find the "best" combination of process parameters with respect to one or more targets that are either maximized or minimized (i.e., the objective function of Chapter 12) under a series of equality and inequality restrictions (i.e., the constraints of Chapter 12).

In general, process integration involves the following activities:

- *Task identification.* In other words, describe why the process intensification work is being conducted (goal setting) and then define a specific task to achieve the goal. One example is to have the goal of improving overall energy efficiency, and the specific task is to reduce net energy requirements which might involve energy integration.
- *Target setting.* In this step the desired performance levels are identified ahead of the detailed design: for example, a 25% reduction of net energy requirements.
- *Alternatives generation.* In other words, this is step 1 of process synthesis (see Chapter 12). In engineering, there are often multiple solutions to achieve a certain target; therefore, it is necessary to follow a methodology that would allow identifying the potential alternatives that would achieve the desired target.
- *Alternative analysis and selection.* This involves evaluating the alternatives based on a set of desired performance metrics (e.g., efficiency, economic, environmental, safety) that would allow determination of the best alternative for the given target.

Process integration is very tightly aligned with green engineering principles, as it attempts to optimize the resources required for a given process within a holistic framework. A chemical process can, to a point, be described as an entity with the objective of converting mass into a more valuable product. This conversion will, of course, require energy, and mass and energy are intimately related in any production process. Also, mass and energy, as we have seen elsewhere in the book, are cornerstone concepts of green chemistry and green engineering. The green chemistry/green engineering metrics that have been proposed are different conceptualizations and measurements of mass and energy, either how much mass or energy (e.g., mass productivity, mass intensity, cumulative energy demand, solvent intensity), a type of mass or energy (e.g., mass of materials of concern, heating requirements), or a characteristic of mass or energy (e.g., impact factors of chemicals, energy from renewable sources).

Given the primal importance of mass and energy in the realm of green engineering, it warrants dedicating an entire chapter to mass and energy integration. Mass and energy integration may be considered as a logical continuation of process synthesis revision in the context of designing greener processes. Chemical engineers over the past few decades have used a varied arsenal of techniques for mass and energy integration as a means of optimizing resources in chemical process design, and there is a direct link to green engineering that can be found in these two more traditional process design approaches.

13.2 ENERGY INTEGRATION

As we have seen earlier, chemical processes require energy to carry out chemical and physical transformations. As a result, it is typical to have several hot streams that need to be cooled and several cool streams that need to be heated. To achieve this, one could cool all the hot streams with cooling water or refrigerants and then heat all the cold streams with stream and heating fluids; but by doing this the total energy requirements of the chemical process would be maximized. To minimize the amount of utilities that need to be used, and therefore minimize costs and environmental impacts, heat integration is used.

The core concept of energy integration is to use hot streams to heat cold streams, and vice versa, before any additional utilities are used, with the result of reducing the overall use of utilities. In a more formal way, energy integration has been defined as "a systematic methodology that provides a fundamental understanding of energy utilization within the process and employs this understanding in identifying energy targets and optimizing heat-recovery and energy-utility systems."[1] Energy integration comprises all forms of energy that can potentially be used in a chemical process, such as heating, cooling, electricity, pressure, and fuel consumption.

Process integration technology for energy efficiency has generally experienced significant uptake around the world. There are now many examples of successful heat integration in a wide variety of industries, such as refineries, petrochemical, chemical, food and drink, pulp and paper, and metallurgical.[2] The net results of heat integration applied successfully are cost savings due to energy cost reductions and de-bottlenecking for increased throughput, with corresponding reductions in emissions and environmental impacts.

13.2.1 Heat Exchange Networks

Industrial heat exchange networks (HENs) are one or more heat exchangers that collectively satisfy the energy conservation requirements for a process by taking advantage of the heat carried by hot process streams and transferring it to cold process streams, thereby reducing the overall heating and cooling requirements for the process. In principle the concept is simple: heat the cold process streams with the hot streams. However, to achieve an optimal or even suboptimal design where the right streams are paired in the right sequence can be a large problem in practice. Therefore, these HEN optimization and sequencing problems often require systematic approaches to arrive at the best solution.

Some of the questions that need to be answered when designing HENs might include[3]:

- How many heat exchangers are needed?
- How should the heat exchangers be arranged?
- Should any streams be mixed?
- Which heating and cooling utilities (steam, cooling water, refrigeration) should be employed?
- What is the optimal heat load of each utility?
- Which hot and cold streams should be matched?

13.2.2 Thermodynamic Pinch

Probably the best known tool for energy integration is *thermodynamic pinch analysis*. Also known as *pinch technology*, this tool emerged during the energy crisis of the 1970s to help design heat exchange networks. Since then, this methodology has become more broadly based and now includes systems such as distillation and distillation networks, operability, emissions, and a general approach to process synthesis and analysis, not only for heat exchange networks but also for process flowsheet synthesis and design. Pinch technology is probably the most successful process synthesis tool, it has been embedded in many optimization software packages, and it has been applied widely industrially.[4,5]

Pinch analysis or pinch technology is based on basic thermodynamic principles of heat and power, with the key strategy of setting heat integration targets at the beginning of the design. Pinch technology uses the first law of thermodynamics to calculate the enthalpy changes in the streams and the second law to determine the direction of heat flow. In other words, thermodynamics will tell us that heat will flow only in the direction of hot to cold, and a hot stream cannot be cooled below the temperature of the cold stream, and vice versa. There has been a great deal of literature devoted to this method[6–8] but there are a couple of graphic tools important to pinch analysis:

- *Composite curves,*shown in Figure 13.1, are based on heat and mass balances and show all the hot and cold streams in the process, including the total amount of heat required as a function of the stream's temperatures. This allows for the identification of hot and cold utility targets, the heat transfer driving forces, and the heat recovery pinch. Since enthalpy is relative, the objective is to try to obtain as much heat from the hot streams to transfer to the cold streams, which can only happen when the hot stream is at a higher temperature than the cold stream. Graphically, this would be achieved by moving the hot composite curve as far to the right as possible without touching the cold curve but still being above it. If there is no limit to how far it can be moved, either all heating or all cooling requirements can be met using process streams. Otherwise, the lines touch at a point known as the *pinch*, which represents the theoretical maximum heat transfer. In reality, it is, of course, not possible to operate at this point, as some minimum temperature difference is needed as a driving force for heat transfer. The temperature difference is called the *temperature approach.*

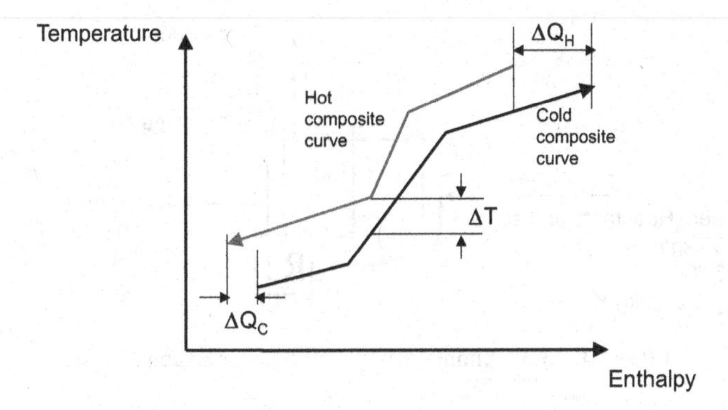

FIGURE 13.1 Pinch diagram with hot and cold composite curves, ΔQ_H is external heating, ΔQ_C is the external cooling, and ΔT is the temperature approach in the pinch.

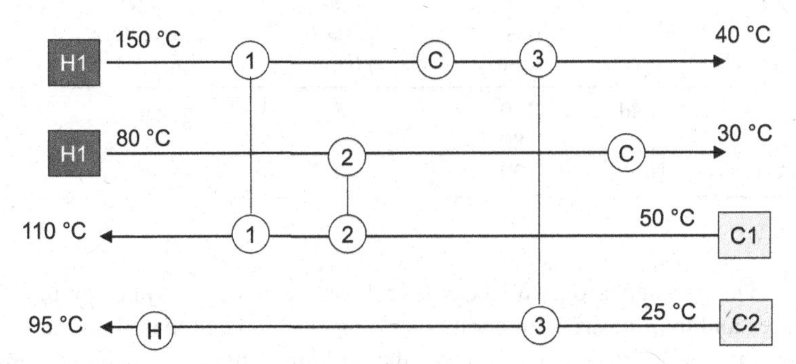

FIGURE 13.2 Grid diagram representing hot streams, cold streams, heat exchanges (numbered pairs), and external utilities such as heating (H) or cooling (C).

- *Grid diagrams,* shown in Figure 13.2, allow us to devise heat recovery networks in conjunction with engineering judgment. In this diagram the hot streams are represented from left to right at the top, and the cold streams from right to left at the bottom. The matches and utilities are also represented, and the pinch point can be identified in the diagram. This diagram also provides a good way of screening for how complex the heat exchange network can be.

Example 13.1 We have a small reaction step in a pharmaceutical process, as illustrated in Figure 13.3. In this small diagram we are looking not at a full process, only at a reactor system that is very common in industry. The feed, at 100 kg/h, enters the reaction at room temperature (25 °C), the reaction takes place at 150 °C, and therefore the product stream leaves the reactor at about the same temperature. For simplicity in this example, the heat capacity of all materials would be assumed to be approximately the same, 2 kJ/kg · K. The reaction would proceed more efficiently if the feed is preheated to the reaction temperature. The product stream is needed at 60 °C for the next operation, and the by-product needs to be cooled to ambient temperature. How much heat can be integrated in this example using pinch analysis and a minimum temperature approach of 25 °C?

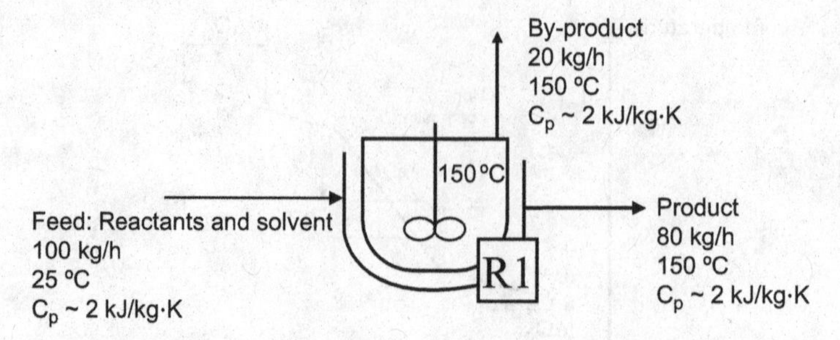

FIGURE 13.3 Simple reaction system for Example 13.1.

TABLE 13.1 Description of Streams

Stream	Type	Mass (kg/h)	Initial Temperature, T_i (°C)	Target Temperature, T_f (°C)	Enthalpy Change Needed, ΔH (MJ/hr)
Feed 1, C_1	Cold	100	25	150	25
Product, H_1	Hot	80	150	60	−14.4
By-product, H_2	Hot	20	150	25	−5

Solution The first step is to list all the cold and hot streams in the system with their current temperatures and their desired temperatures (Table 13.1). Then, based on the temperature ranges and the heat capacity, we estimate the enthalpy change that the streams need to undergo to achieve the desired temperature. To build a composite curve, we first need to "add" the hot streams and the cold streams graphically or algebraically to form a composite curve. In Figure 13.4 you can see the graphs for the hot streams. Since enthalpy is relative, we have assigned the lowest temperature of the hot streams an enthalpy value of zero. The point farthest right in the graph represents the total amount of heat that needs to be removed from

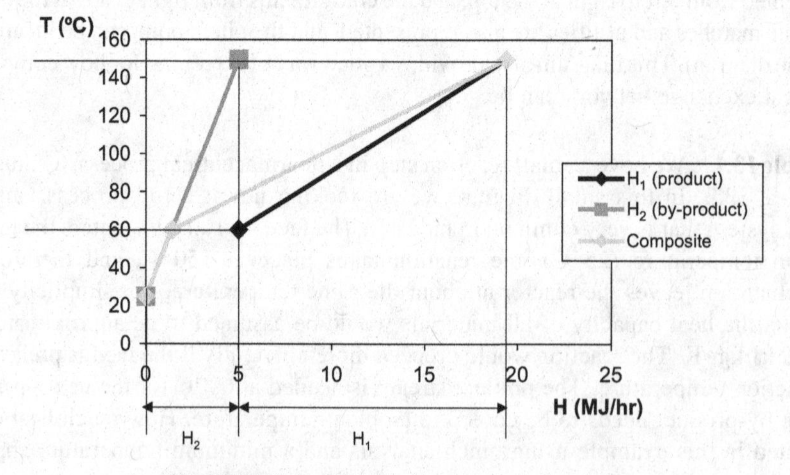

FIGURE 13.4 Building hot composite curves.

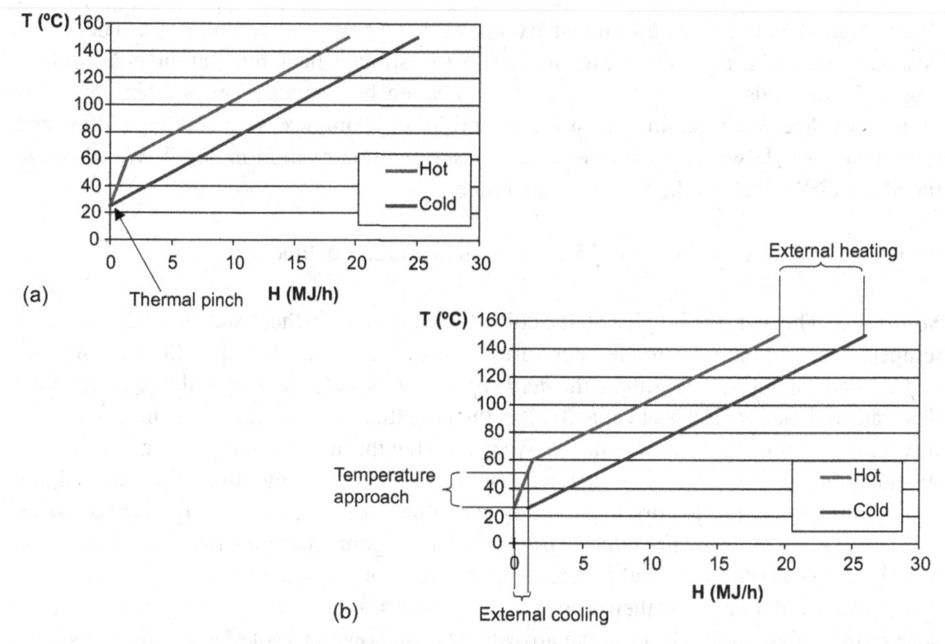

FIGURE 13.5 Pinch diagram for Example 13.1: (a) thermal pinch (b) the graph once the cold composite curve has moved toward the right to allow for the minimum temperature approach of 25° C.

the hot streams, in this case $H_1 + H_2 = 14.4 + 5 = 19.4$ MJ/h. The composite curve is drawn using the diagonal rule for adding lines.

We would do something similar if we had more than one cold stream, but in this case we have only one, so the composite graph is comprised of that single stream. The pinch diagram is shown in Figure 13.5 (a). As can be seen, there is an obvious pinch point at 25 °C. In other words, a hot stream at 25 °C will not be able to transfer heat to a cold stream at that same temperature. Since the enthalpy scale is relative, we can slide both curves to the right and left until we have the approach temperature of 25 °C, which in the graph is the minimum vertical distance between the cold and hot streams, as shown in Figure 13.2(b). It can be seen that at that approach temperature, some external heating would be required to heat the cold stream (e.g., steam, heat exchange fluids), and some external cooling will be needed to cool the hot streams (e.g., cooling water), with the rest being integrated. Reading the temperatures from the graph, we have:

$$\text{external cooling required} = mC_p\Delta T = 20\,\text{kg/h}\,(50\text{–}25\,°\text{C})\,2\,\text{kJ/kg}\cdot\text{K} = 1000\,\text{kJ/h} = 1\,\text{MJ/h}$$

$$\text{external heating required} = mC_p\Delta T = 100\,\text{kg/h}\,(150\text{–}117\,°\text{C})\,2\,\text{kJ/kg}\cdot\text{K} = 6600\,\text{kJ/h} = 6.6\,\text{MJ/h}$$

$$\text{total heat integrated} = mC_p\Delta T = 100\,\text{kg/h}\,(117\text{–}25\,°\text{C})\,2\,\text{kJ/kg}\cdot\text{K} = 18{,}400\,\text{kJ/h} = 18.4\,\text{MJ/h}$$

Additional Points to Ponder Can you explain how we determined the required temperature for heating? How might this result be affected by the other parts of the process before and after this reactor? Is there any advantage to analyzing a specific unit operation, apart from the other parts of the process?

Example 13.1 was solved almost exclusively by graphical calculations. There are, however, algebraic methods available to use to estimate heat integration possibilities. Algebraic methods are very useful, especially when heat integration problems become more complex, with additional streams and more complex settings. Papoulias and Grossman,[9] El-Halwagi,[10] and others have described this method in detail. We illustrate the algebraic method in the following example.

Example 13.2 Solve Example 13.1 using an algebraic method.

Solution The first step is to list all the cold and hot streams in the system with their current temperatures and their desired temperatures (Table 13.2). Then, based on the stream flows and the heat capacity, we estimate the *heat capacity flow rate*, defined as the product of the flow rate and the specific heat capacity. It is the potential enthalpy that a stream can give or receive for every change in temperature. We saw earlier that heat can only be transferred from the hot to the cold streams when there is some temperature driving force. A useful tool that shows whether or not this driving force is available is the *temperature interval diagram* (also known as a *temperature interval table*). In this diagram, the temperature scales for both hot (T) and cold (t) are shifted by the temperature approach determined for the network. The hot and cold streams are then represented by arrows. Temperature intervals are set by the start of the arrows and the end of the arrowheads, as shown in Table 13.3. This helps us to visualize the energy, or heat load, that is available for transfer to or from each stream within

TABLE 13.2 Heat Capacity Calculations

Stream	Type	Initial Temperature, T_i (°C)	Target Temperature, T_f (°C)	Heat Capacity Flow Rate, C_P, (kJ/hr · °C)
Feed 1, C_1	Cold	25	150	200
Product, H_1	Hot	150	60	160
By-product, H_2	Hot	150	25	40

TABLE 13.3 Temperature Interval Diagram

Interval	Hot Streams T	Cold Streams t
	175	150
1	150	125
2	60	35
3	50	25
4	25	0

H_2 (Cp=40): arrow from 150 down to 25; H_1 (Cp=160): arrow from 150 down to 60; C_1 (Cp=200): arrow from 25 up to 150.

TABLE 13.4 Heat Loads

Interval	C_1 Heat Load (kJ/h)	H_1 Heat Load (kJ/h)	H_2 Heat Load (kJ/h)	Total Hot Heat Load (kJ/h)	Total Cold Heat Load (kJ/h)
1	5,000	—	—	0	5,000
2	18,000	14400	3,600	18,000	18,000
3	2,000	—	400	400	2000
4	0	—	1,000	1,000	0

each interval and is calculated by the heat capacity flow rate (C_P) times the temperature interval. Table 13.4 shows the heat loads for each of the four intervals.

Now, the energy integration can be looked at as an energy balance where each interval is a process, or a part of a process, and energy is transferred from the hot streams to the cold streams with the leftover energy carried over to the next interval. For each interval, the energy balance can be drawn as in Figure 13.6(a). Therefore, the residual heat leaving the interval (r_z) is equal to the heat available from the hot streams (HH_z), minus the heat removed by the cold streams in that interval (HC_z), plus any residual heat entering the interval (r_{z-1}):

$$r_z = HH_z - HC_z + r_{z-1}$$

This formula allows us to create a spreadsheet with all the intervals linked across the process. Since there is no residual heat entering the first interval, the residual heat is zero. Next, we estimate the heat residuals for the system and represent them in a cascade diagram in Figure 13.6(b). When residual heats r_z are negative, this means that the heat is flowing in

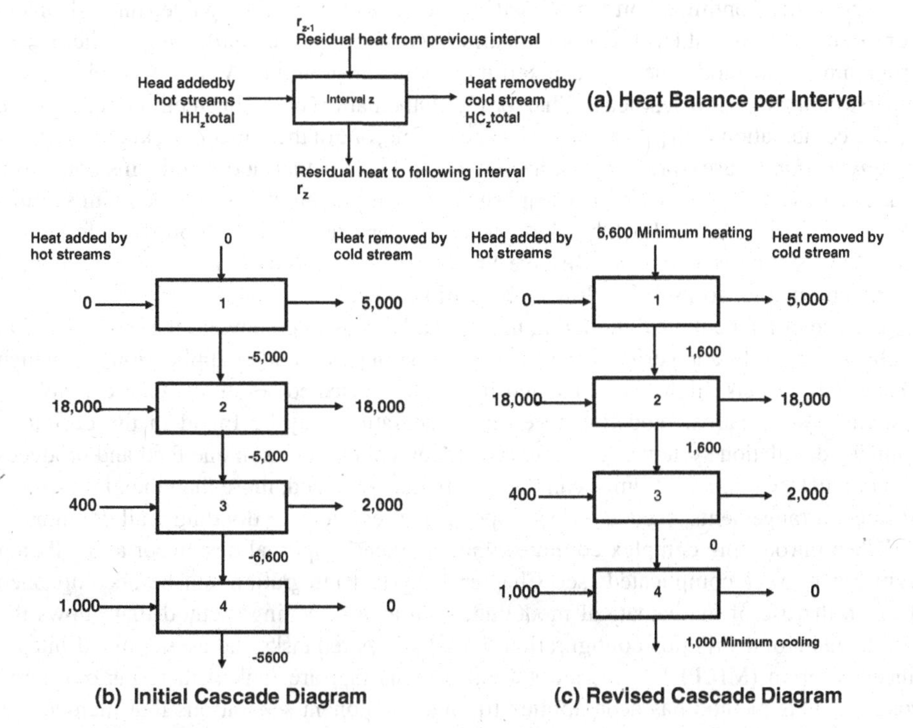

FIGURE 13.6 Cascade Diagrams for Example 13.2. All amounts in kJ/h.

the wrong direction, which is thermodynamically impossible. Thermodynamic compliance can be achieved by adding an amount of heat equal to the most negative of the values, or the minimum heat requirements. In this case, this is 6600 kJ/h (or 6.6 MJ/h, as we estimated graphically in Example 13.1). We can recalculate the energy balance after adding that amount of heat at the beginning of the interval, as in the revised cascade diagram shown in Figure 13.6(c). A zero residual heat determines the thermal pinch, and the residual leaving the last interval represents the minimum cooling requirements (1000 kJ/h or 1 MJ/h, as we also determined graphically).

Additional Points to Ponder What probable impact would this heat integration have from a green engineering principles perspective? Why do we say that when the residual heats are negative, the heat is flowing in the "wrong direction"? Why is this not possible?

An algebraic methodology for solving heat integration problems permits us to use a computer program containing mathematical algorithms for optimizing energy integration for any given process.

13.2.3 Mathematical Optimization Modeling: Energy Integration Beyond HEN

Even with the power of pinch technology, following a pinch method rigidly can result in extremely complex systems. This is best explained by the difficulty in including equipment limitations and the need for applying engineering judgment and common sense to the analysis. Efforts to enhance the pinch methodology have therefore led to the application of mathematical optimization modeling (MOM).

Mathematical optimization can be used independent of the insights gained through pinch. For example, we might try direct optimization of energy requirements using mathematical programming methods such as a mixed-integer linear program (MILP) to optimize the superstructures,[11] as we learned in Chapter 12. Mathematical modeling methods may also be used in combination with pinch analysis. For example, one of these methods, known as *block decomposition*,[12] uses process synthesis techniques and thermodynamic insights from pinch analysis to break down the composite curves into a number of blocks with streams of similar characteristics. Each block becomes a superstructure that is optimized following the process synthesis techniques we have already covered. This helps us generate optimal or nearly optimal networks with a low number of heat exchange units.

The use of mathematical modeling methods allows us to go beyond the design of heat exchange networks and perform energy integration in several other applications for which traditional analysis might have resulted in very complicated solutions. One example of applying mathematical modeling to energy integration may be found in the design of complex distillation systems. A simple distillation column takes in one feed and produces two products. Complex columns can have more than one feed, more than one product and different arrangements, such as side strippers, side rectifiers, or dividing wall columns.

When introducing complex columns, finding a nearly optimal design for a distillation train can be very complicated (see Chapter 12). Heat integration can be accomplished through the use of mathematical modeling, such as a screening method that allows the development of distillation configurations based on desired tasks and uses a mixed-integer linear program (MILP) to generate several designs that are ranked in terms of energy cost.[13,14] This method has been applied to multicomponent separations that include less conventional settings (e.g., a side rectifier, a prefractionator) and provide a lower energy requirement. It has also been applied to low-temperature systems, where it provided significant reductions in refrigeration requirements. Heat integration with mathematical modeling has also been applied to dividing-wall distillation columns. These columns combine a prefractionator with the main column in the same shell. A design method developed at UMIST[15] optimizes for minimum energy consumption while allowing for additional potential heat integration, thereby achieving energy and capital savings up to 30% above the base case. Additional examples of heat integration beyond heat exchange networks and pinch analysis include design of adsorption systems, design of azeotropic distillation, energy integration of site utility and cogeneration systems, design of cooling water systems, and design of refrigeration systems.[16]

13.3 MASS INTEGRATION

Mass integration is also based on chemical engineering principles that integrate process synthesis and optimization techniques. Similar to energy integration, mass integration has been defined as "a systematic methodology that provides a fundamental understanding of the global flow of mass within the process and employs this understanding in identifying performance targets and optimizing the generation and routing of species throughout the process." The concept of targeting means that the design is driven to a specific goal, such as minimum cost, maximum yield, minimum amount of mass separating agents, and minimum emissions. Very important drivers for mass integration techniques have been environmental implications and pollution prevention concepts. Therefore, from the green engineering/ green chemistry perspective, mass integration, together with energy integration, allows us to optimize our processes by paying close attention to the resources we employ for separations.

13.3.1 Mass Exchange Networks

The concept of a heat exchanger is generally easier to understand directly from process design, but the term *mass exchanger* sometimes requires a little more explanation. To understand what mass exchange networks are, we need to start by understanding mass exchange units. Let's assume that we have a stream of a process that contains the product of a reaction with some unreacted species. One of the flowsheet design options we explored in Chapter 12 was to recover the unreacted species and recycle them back to the reactor. A mass exchanger is a unit operation that allows us to separate the unreacted species from the stream and recycle it back to the reactor. This is sometimes easy, sometimes not. Mass exchanger units very often employ mass separating agents to help separate the species that need to be recovered. Some examples of mass exchange units are shown in Table 13.5.

13.3.2 Solving MENs by Using Their Similarities to HENs

One way to illustrate the MEN design problem is to use a hypothetical process having one or several process streams with a relatively high concentration of a material that needs to be recovered through a separation train. These are generally referred to as *rich streams*, or *sources*, similar to the hot streams in energy integration. We also have one or several process streams with relatively low concentrations of the same material, also known as *lean streams*, or *sinks*, equivalent to the cold streams in energy integration. The idea is to utilize as much of

TABLE 13.5 Example of Mass Exchange Units

Mass Separation Unit	Examples
Absorption	Removal of ammonia using water Desulfurization of gases using amines
Adsorption	Removal of aromatic compounds using activated carbon Removal of hexavalent chromium using zeolites
Extraction	Separation of pharmaceutical ingredients from mother liquors Removal of phenol from wastewater
Ion exchange	Sweetening of potable water Removal of heavy metals from water
Leaching	Recovery of precious metals from mining tails Removal of metals from sludge
Membrane units (nanofiltration, reverse osmosis)	Desalinization of seawater
	Recovery of fine-chemical intermediates from process streams
Stripping	Regeneration of spent activated carbon using steam Removal of volatile organic compounds using air

the process lean streams to recover (and perhaps recycle) the material to reduce the amounts of mass separating agent (MSA) required for the same purpose (which will be equivalent to the utilities in energy integration).

Since there are easily drawn similarities between HEN and MEN, we can use these similarities to synthesize networks of separators that could, for example, optimally recover a compound that otherwise would be a pollutant and recycle it back to the production process. El-Halwagi and Manousiouthakis[17] introduced this concept for synthesis of MEN based on the HEN pinch analysis.

In the same way that the thermal pinch point was located with the cold and hot composite curves, the mass transfer pinch point can be located with the lean and rich composite curves. One can also estimate the minimum MSA required for the network as we estimated the minimum utility in the HEN design. These lean and rich composite curves can be constructed by plotting the compositions of the lean and rich streams vs. the amount of mass transferred. In a more recent application to recycling problems they have also been drawn by plotting the flow rate against the concentration load of the material in the stream.

As we saw previously for heat integration, where pinch analysis requires a minimum temperature difference as a driving force, the design of MEN requires a minimum composition difference, ε. Similar to the HEN design, mass composite curves can be drawn for rich streams and lean streams. In MENs the minimum composition difference is governed by mass equilibrium limits to ensure thermodynamic feasibility. For example, let's assume that we are trying to remove the pollutant j from a waste stream by using a mass separating agent (MSA). The waste stream is carrying pollutant j with a composition y_j. When this waste stream is put into contact with the MSA, there is a maximum concentration of j that can be contained in the solvent, x_j, which is also known as the *equilibrium composition*. Mass transfer stops beyond this point. By having a composition difference between the equilibrium composition and the operating composition, we can ensure that the rate of mass transfer is

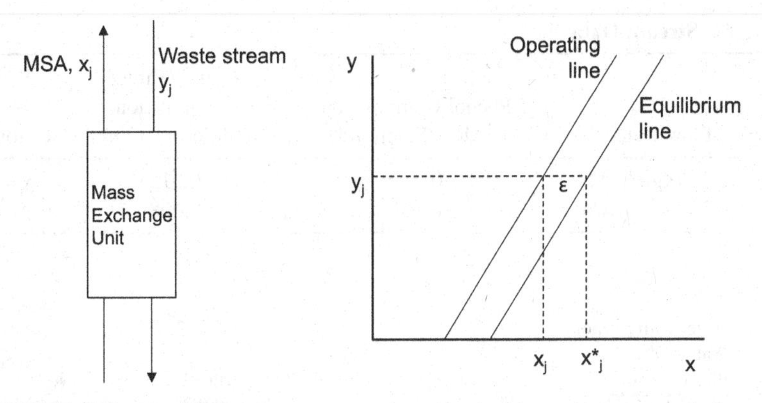

FIGURE 13.7 Equilibrium in a mass exchange unit using a mass separating agent.

feasible and practical for that mass exchange unit (see Figure 13.7). This principle is widely used by chemical engineers to design mass separation units such as adsorbers and absorbers.

One can estimate the operating line from equilibrium equations. Because many of the mass integration applications have been centered on pollution prevention and recovery of substances in small amounts, most of the equilibrium ranges can be assumed to behave ideally; thus, we can use the same scale to represent compositions across several streams (which are very often rich streams). From the equilibrium equation we have

$$y_j = m_j x_j + b_j \tag{13.1}$$

where m_j is the slope and b_j is the abscissa. Since the composition of the MSA in the operating line is

$$x_j = x_j - \varepsilon \tag{13.2}$$

the operating line equation is

$$y_j = m_j(x_j + \varepsilon) + b_j \tag{13.3}$$

The amount of material lost from a rich stream can be calculated as

$$m = F(y_i^s - y_i^t)$$

where m is the mass lost and F is the flow.

Example 13.3 A company removes phenolic compounds from wastewater using adsorption with polymeric resins in a two-column setting. Typically, the first column is operated while the second is being regenerated. To regenerate a column loaded with phenols, a solvent system is used, consisting of one or more distillation columns. In this example the stream from the regeneration cycle is distilled to recover the solvent, and the bottoms are cooled and allowed to separate into two phases. The phenol-rich phase is recovered, and the aqueous phase is recycled to the end of the loading cycle of the resin, as shown in Figure 13.8. This is a mass- and energy-intensive way to recover solvent, so we want to explore the possibility of performing mass integration by using two by-product streams from elsewhere in the process to capture and recover the phenols prior to their being distilled and sent back to the process. The data for these two streams are given in Table 13.6. Assuming a minimum composition difference of 0.001, is there any possibility that these streams can be used to recover phenol instead?

TABLE 13.6 Stream Data

Stream	Flow Rate	Phenol Composition (Mole Fraction)	Phenol Target Composition (Mole Fraction)	Equilibrium Equation for Phenol
S1	200 kg · mol/h	0	0.012	$y = 0.4 \times 1$
S2	300 kg · mol/h	0.001	0.008	$y = 0.5 \times 2$

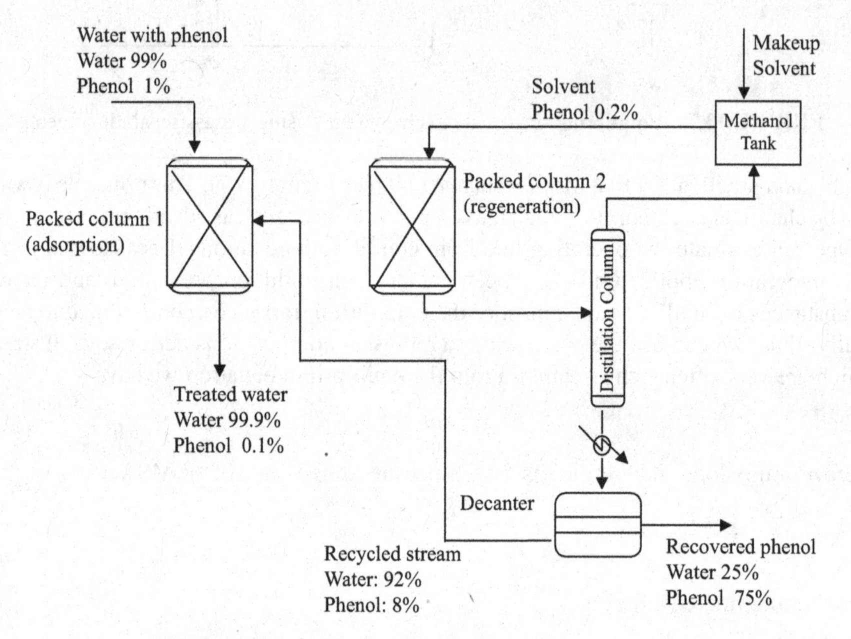

FIGURE 13.8 Phenol recovery using polymeric resins.

Solution Similar to the HEN problem, we can use either a graphical approach or a mathematical approach. Since the data for the waste stream are on a mass percent basis, the first thing we need to do is to convert the data in Figure 13.8 from mass percent to mole fraction. Thus for 1 L of the source stream, we have:

For 1000 g of wastewater in			moles	mole fraction
10	g phenol in	m.w. = 94.1	0.106	0.002
990	g water in	m.w. = 18	55.000	0.998
1000	g total		55.106	

For 1000 g of wastewater out			moles	mole fraction
1	g phenol in	m.w. = 94.1	0.011	0.0002
999	g water in	m.w. = 18	55.500	0.9998
1000	g total		55.511	

TABLE 13.7 Stream Data

Stream	Flow Rate	Phenol Concentration (Mole Fraction), y_i^s	Phenol target Concentration (Mole Fraction), x_i^t	Supply Conc., y_i^s	Target Conc., y_i^t	Equilibrium Equation for Phenol
S1	200 kg · mol/h	0	0.012	0.0004	0.0052	$y = 0.4(x_1 + \varepsilon)$ or $x_1 = y/0.4 - \varepsilon$
S2	250 kg · mol/h	0.001	0.008	0.001	0.0045	$y = 0.5(x_2 + \varepsilon)$ or $x_2 = y/0.5 - \varepsilon$

TABLE 13.8 Interval Data

	Waste Streams		Process Streams			
Interval		y	$x_2 = (y - b_2)/m_2 - \varepsilon$		$x_1 = (y - b_1)/m_1 - \varepsilon$	
		0.0052	0.0094		0.012	
1		0.0045	0.008		0.01025	
				S2 300 kmol/h ↑		S1 200 kmol/h ↑
2		0.002	0.003		0.004	
3		0.001	0.001		0.0015	
	W 1000 kmol/h ↓					
4		0.0004			0	
5		0.0002				

For this example we will solve the problem algebraically, and for this particular problem it will be useful to express the compositions in terms of the operating line using equation (13.3), as in Table 13.7. Table 13.8, the composition interval table for this problem, allows us to calculate a table of exchangeable loads, similar to what we did for heat integration, showing the available phenol loads that can be exchanged between streams within thermodynamic limits (Table 13.9).

Figure 13.9 shows the cascade diagram and illustrates the pinch point. It can be seen that there is indeed some opportunity for mass integration. As a matter of fact, there is 3.2 kg · mol/h excess capacity in the process streams that could remove phenol. However, there is still a need to use some external means (e.g., the resin system) since there is an interval in which there is no suitable process stream for removing phenol (interval 5).

Additional Point to Ponder Would the result be different if the resin in the adsorption column had been included in the mass integration exercise?

TABLE 13.9 Exchangeable Loads Data

Interval	Waste Mass Load (kmol/h)	Total Waste Load (kmol/h)	S1 Mass Load (kmol/h)	S2 Mass Load (kmol/h)	Total MSA Load (kmol/h)
1	0	0	0.35	0	0.35
2	0	0	1.25	1.5	2.75
3	1	1	0.5	0.6	1.1
4	0.6	0.6	0.3	0	0.3
5	0.2	0.2	0	0	0
Total load		1.8	2.4	2.1	4.5

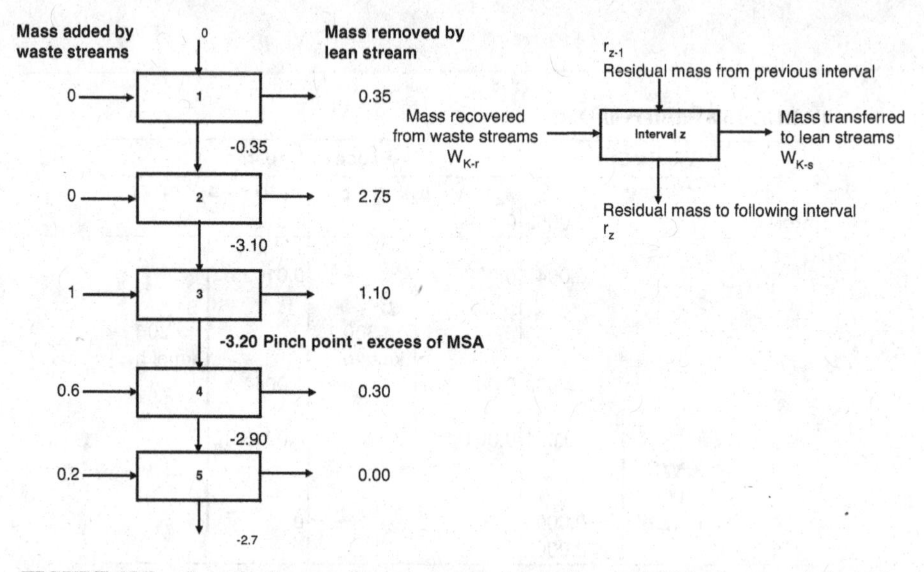

FIGURE 13.9 Cascade diagram for mass integration in Example 13.3. Amounts in kmol/h.

13.3.3 Other Applications of MEN

As was the case for HEN design, MEN design has been expanded beyond composite curves into mathematical programming and numerical analysis to facilitate optimization of the mass exchange network by, for exmple, designing it for the minimum number of mass exchange units for a specific task. Thanks to mathematical modeling and linear programming, the application of MEN synthesis and optimization has been widened to solve an array of problems, with the special case of water minimization having thus far received special attention. Some applications of MEN synthesis are shown in Table 13.10.

Another useful application of MENs is in the design of material recycle and reuse networks to identify the option that would allow for the minimum use of fresh resources for the recovery and recycle of materials and by-products. There is a graphical approach developed by El-Halwagi et al. [43] to target the minimum freshwater consumption and wastewater discharged by the transfer of contaminants from process streams to water streams. This graphical approach is based on defining an optimization problem in which the

TABLE 13.10 Examples of Mass Exchange Network Applications

Applications of MEN	Illustrative References
Simultaneous synthesis of mass exchange and regeneration networks	18
Synthesis of waste-interception networks	19
Pervaporation networks	20
Reverse osmosis networks	21
MENs in batch processes	22–24
Synthesis of reactive MENs	25,26
Heat-induced separation networks:	27
VOC recovery	28,29
Crystallization	30
Evaporation and Crystallization	31
Water minimization	32–42

objective is to minimize the consumption of fresh resource subject to the system constraints of sources and sinks, following two optimization rules:

1. If a sink requires the use of a fresh source, the inlet composition to the sink and the inlet pollutant load to the sink should be maximized.
2. Maximize the recycle and reuse of the available amount of source (i) until it is fully consumed; then maximize the recycle and reuse of the next source in ascending order of composition ($i + 1$); and so on.

The optimization problem and rules are translated into the steps of a very simple graphical approach:

1. Rank the sinks in ascending order of maximum admissible composition.
2. Rank sources in ascending order of pollutant composition.
3. Plot the maximum load of each sink vs. its flow rate. Create a sink composite curve by superimposing the sink arrows in ascending order.
4. Plot the load of each source vs. its flow rate. Create a source composite curve by superimposing the sources in ascending order.
5. Move the source composite stream until it touches the sink composite stream, with the source composite below the sink composite in the overlapped region. This will identify the material recycle–reuse pinch point. As with heat transfer, the flow rate of sinks below any source is the target for minimum fresh discharge, and the flow rate of the sources above any sinks is the target for waste discharge.

This procedure will ensure that we start utilizing the first sink and recycle or reuse the first source until it is fully consumed or until the maximum inlet load to the sink is met and then move to the following source, and so on. We then move to the second sink and repeat the process. The procedure above is for single-component systems, but it can also be applied to multicomponent systems where there is a limiting component and the solution strategies for the other components are consistent with those of the limiting component. For other multicomponent problems, a more general approach should be developed.

TABLE 13.11 Sink/Source Data

Sink	Flow (mol/s)	Maximum Inlet Impurity Concentration (mol %)	Load (mol/s)
1	2495	19	484
2	180	21	38
3	554	22	124
4	721	25	179

Source	Flow (mol/s)	Impurity Concentration (mol %)	Load (mol/s)
1	624	7	44
2	416	20	83
3	1802	25	451
4	139	25	35
5	347	27	94
6	457	30	137

Example 13.4 El-Hawagi et al. presented an example of using the recycle or reuse strategy for the optimization of a hydrogen distribution system in a refinery presented by Alves and Towler.[44] This system is comprised of four sinks and six sources as shown in Table 13.11. Using this information, construct the shifted source–sink composite curves and determine the minimum hydrogen required and waste hydrogen to be discharged.

Solution To solve this problem we only need to follow the steps noted above. To plot the cumulative sink curve, we only need to add the flow rates in ascending order and plot them against their corresponding load. The points to plot are given in Table 13.12.

We do a similar thing for the sources. When we plot both cumulative curves (Figure 13.10) the lines will intersect and we will have to move the source composite line to the right until we find the pinch. We have to move the line to the point that corresponds to a flow Rate of 244 mol/s and the minimum amount of fresh hydrogen that is required. As we do this, the difference between the source and the sink line above the pinch is 79 mol/s, which is the minimum amount of waste to be generated after mass integration.

Additional Points to Ponder Would the result be different if the fresh resource included a certain degree of impurity? How is this different from Example 13.3?

TABLE 13.12 Plot Point Data

Cumulative Sink Flow Rate	Sink Load
0	0
2495	484
2495 + 180 = 2675	522
2675 + 554 = 3229	646
3229 + 721 = 3950	825

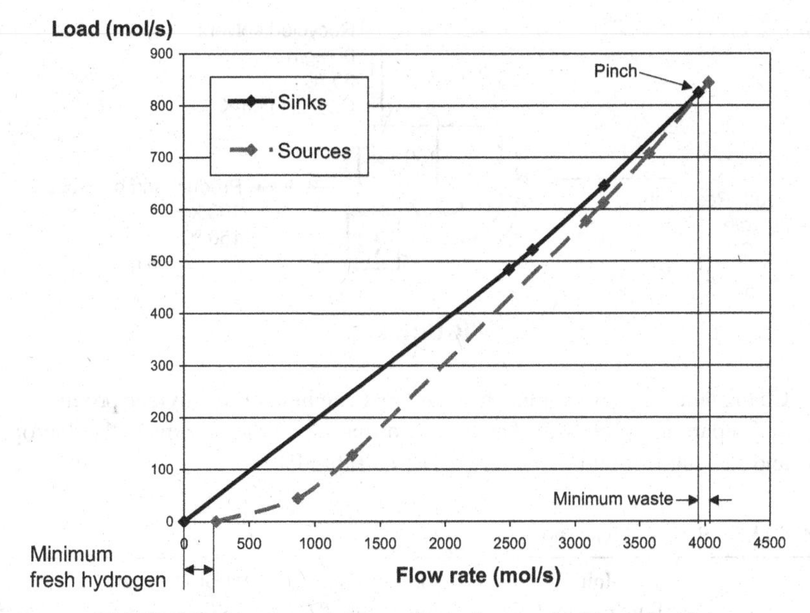

FIGURE 13.10 Recycle/reuse pinch diagram for Example 13.4.

PROBLEMS

13.1 In process optimization we maximize or minimize an objective function subject to certain constraints expressed as equality or inequality relationships.

(a) Provide two examples of an objective function related to green engineering principles.

(b) Provide two examples of equality constraints and two examples of inequality constraints related to green engineering principles.

(c) Explain the relationship of your examples in parts (a) and (b) with green engineering principles.

13.2 What would have been the total energy requirements (heating and cooling) of Example 13.1 if energy integration hadn't been applied? What percentage of savings in utilities has the energy integration brought?

13.3 How would you propose to implement the heat integration for the process in Problem 13.1?

(a) Draw a process flow diagram and a grid diagram to explain your proposal.

(b) Estimate the input and output temperature of the streams.

13.4 The process in Example 13.1 has now been modified in such a way that the product, by-product, and solvent leave the reactor in the same stream, and the solvent used is recovered in the next unit operation and recycled back to the reactor, as shown in Figure P13.4. If the product stream has to be cooled to 60 °C:

(a) Estimate how much heat can be integrated using pinch analysis and a minimum temperature approach of 15 °C.

(b) How much heat would be needed in external utilities?

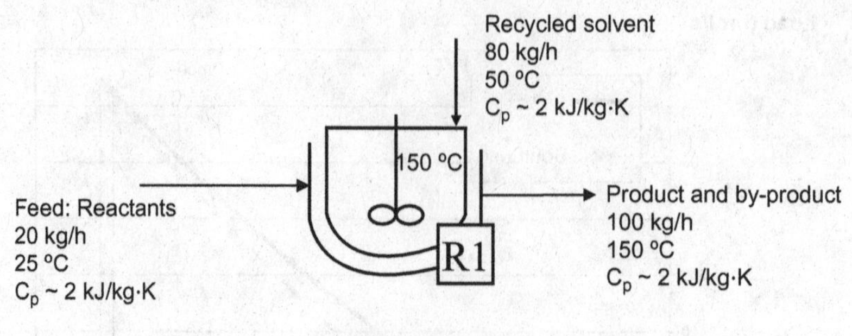

FIGURE P13.4

13.5 Chang and Guo have studied heat energy in an industrial ethylene product recovery and separation process.[45] The heat information on the streams in the depropanizer and debutanizer subsystem are given in Table P13.5.

TABLE P13.5 Stream Heat Data

Stream	Type	Initial Temperature (°C) (Phase)	Target Final Temperature (°C) (Phase)	Heat Capacity Flow Rate (MJ/h · °C)	Heat Duty (MJ/h)
1	Hot	43.7 (V)	43.6 (V-L)	23,170	2,317
2	Hot	5.2 (V)	4.2 (V-L)	31,138	31,026
3	Cold	91.5 (L)	91.6 (V-L)	185,759	18,576
4	Cold	82.9 (L)	87.8 (V)	885	4,345
5	Cold	61.6 (L)	65.2 (V-L)	3,612	12,931
6	Cold	55.8 (L)	66.6 (V)	463	5,018

(a) Draw a pinch diagram that includes the hot and cold composite curves.

(b) What heat integration suggestions would you make for this system?

13.6 For Problem 13.5:

(a) Estimate the minimum amount of heat that needs to be removed or added through external utilities.

(b) Using a temperature approach of 10 °C, estimate the total heat that needs to be removed or added through external utilities.

(c) Draw a grid diagram of the streams and propose a heat integration configuration.

(d) What environmental savings would you gain through heat integration?

13.7 Table P13.7 contains information for hot and cold streams in a process.[46]

(a) What is the minimum amount of heat that needs to be removed or added through external utilities?

(b) Using a temperature approach of 15 °C, estimate the amount of heat that needs to be added or removed from the system by external utilities.

(c) You have available in-plant cooling water with an inlet temperature of 38 °C and a maximum outlet temperature of 82 °C. Saturated steam at 270 °C is available for heating, with a minimum outlet temperature of 140 °C. What would be the most effective outlet temperature for your cooling water and steam?

TABLE P13.7 Stream Heat Data

Stream	Initial Temperature (°C)	Target Temperature (°C)	Heat Capacity Flow Rate (kW/K)
Hot 1	249	138	10.6
Hot 2	60	93	8.8
Cold 1	160	160	7.6
Cold 2	116	260	10.1

13.8 Figure P13.8 is a flow diagram of an ethyl ether plant with no energy integration. The stream temperature and compositions are given in Table P13.8, where the negative numbers for the stream flows and compositions denote exits from the system.

(a) Estimate the current heat requirements of the plant.

(b) What energy integration retrofits would you suggest for the plant?

13.9 For Example 13.2, Assume that the price for cooling water is $0.2 (per gigajoule) and the price for steam is $3.2 (per gigajoule). How would you write the optimization problem (i.e., the optimization function and the constraints) so that linear programming could be used to minimize the overall:

(a) Utilities cost?

(b) Global warming potential?

(c) Acidification potential?

13.10 Table P13.10 below gives some selected resources and emissions associated with the production of 1 MJ of steam and the removal of 1 MJ of heat by cooling water. What is the reduction in environmental impact from the energy integration of Example 13.1?

13.11 For mass integration, one can plot a pinch diagram with the compositions (y) along the horizontal axes and the mass exchanged along the vertical axes. The composite lean and rich curves are drawn based on the mass exchanged vs. the composition, with the slope of the curves corresponding to the flow rate of the stream. The vertical scale (mass lost or gained) is relative. For example, each rich stream is represented as an arrow starting at the supply composition and finishing at the target composition. Solve Example 13.3 graphically using a pinch diagram.

13.12 In Example 13.3 there is an excess of process streams available to remove the phenol from the wastewater without having to use the resin system. What might the designers do in this situation? Explain your design with a cascade diagram.

13.13 Table P13.13 has information on four water streams containing a by-product and four process streams that can be used as sinks to capture this contaminant and recycle the water, thereby minimizing freshwater needs and wastewater discharge (as presented by Polley and Polley[39] and El-Halwagi et al.[43]).

(a) Estimate the minimum amount of wastewater that can be discharged by the process and the minimum amount of freshwater required by the system to recover and recycle this by-product.

(b) What is the material recycle pinch concentration?

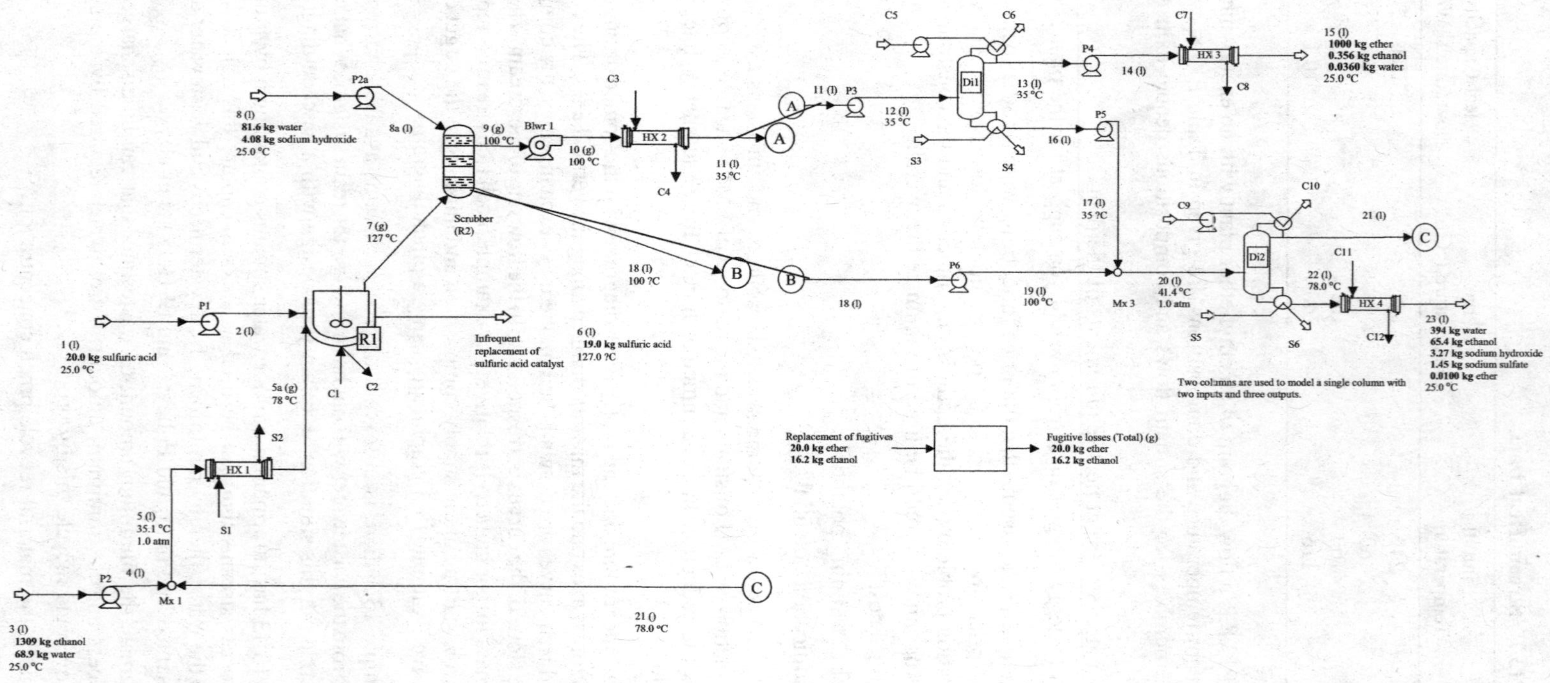

1 (l)
20.0 kg sulfuric acid
25.0 °C

3 (l)
1309 kg ethanol
68.9 kg water
25.0 °C

8 (l)
81.6 kg water
4.08 kg sodium hydroxide
25.0 °C

P1 2 (l) P2 4 (l) Mx 1 P2a 8a (l)

5 (l)
35.1 °C
1.0 atm

HX 1 S1 S2

5a (g)
78 °C

R1 C1 C2

7 (g)
127 °C

Scrubber
(R2)

Infrequent
replacement of
sulfuric acid catalyst

6 (l)
19.0 kg sulfuric acid
127.0 °C

9 (g)
100 °C

Blwr 1

10 (g)
100 °C

HX 2 C3 C4

11 (l)
35 °C

11 (l) P3

12 (l)
35 °C

18 (l)
100 °C

18 (l) P6

19 (l)
100 °C

Di1 C5 C6 S3 S4

13 (l)
35 °C

P4 14 (l) HX 3 C7 C8

15 (l)
1000 kg ether
0.356 kg ethanol
0.0360 kg water
25.0 °C

P5 16 (l)

17 (l)
35 °C

Mx 3

20 (l)
41.4 °C
1.0 atm

Di2 C9 C10 S5 S6

21 (l)

22 (l)
78.0 °C

HX 4 C11 C12

23 (l)
394 kg water
65.4 kg ethanol
3.27 kg sodium hydroxide
1.45 kg sodium sulfate
0.0100 kg ether
25.0 °C

Two columns are used to model a single column with two inputs and three outputs.

Replacement of fugitives
20.0 kg ether
16.2 kg ethanol

Fugitive losses (Total) (g)
20.0 kg ether
16.2 kg ethanol

21 (l)
78.0 °C

FIGURE P13.8

TABLE P13.8 Stream Temperature/Composition Data

Comments	Streams	Temp. (°C)	P	Phase	Total Flow	Water	Sulfuric Acid	Ethanol	Ether	Sodium Hydroxide	Sodium Sulfate
Input	1	25.0	1.00	l	20.0	—	20.0				
	2	25.0	1.00	l	20.0	—	20.0				
Input	3	25.0	1.00	l	1378	68.9	—	1309			
	4	25.0	1.00	l	1378	68.9	—	1309			
	5	35.1	1.00	l	1703	85.1	—	1617	1.00		
	5a	78.0	1.00	g	1703	85.1	—	1617	1.00		
Waste	6	127	1.00	l	−19.0	0	−19.0	0	0	0	0
	7	127	1.00	g	1704	328	1.00	374	1001	0	0
Input	8	25.0	1.00	l	85.7	81.6	—	—	—	4.08	
	8a	25.0	1.00	l	—	81.6	—	—	—	4.08	
	9	100	1.00	g	1716	360		356	1001		
	10	100	1.00	g	1716	360	0	356	1001	0	0
	11	35.0	1.00	l	1716	360	0	356	1001	0	0
Feed	12	35.0	1.00	l	1716	360	0	356	1001	0	0
Distillate	13	35.0	1.00	l	1000	0.0360	0	0.356	1000	0	0
Bottoms	16	35.0	1.00	l	716	360	0	355	1.00	0	0
	14	35.0	1.00	l	1000	0.0360	0	0.356	1000	0	0
Main product	15	25.0	1.00	l	−1000	−0.0360	0	−0.356	−1000	0	0
	17	35.0	1.00	l	716	360	0	355	1.00	0	0
	18	100	1.00	l	73.4	50.0	0	18.7	0	3.27	1.45
	19	100	1.00	l	73.4	50.0	0	18.7	0	3.27	1.45
Feed	20	41.4	1.00	l	790	410	0	374	1.00	3.27	1.45
Distillate	21	78.0	1.00	l	326	16.2	0	308	0.991	0	0
Bottoms	22	78.0	1.00	l	464	394	0	65.4	0.0100	3.27	1.45
Waste	23	25.0	1.00	l	−464	−394	0	−65.4	−0.0100	−3.27	−1.45

TABLE P13.10 Resource and Emission Data

Level 2	Unit	Steam (1 MJ)	Cooling Water (1 MJ of cooling water potential)
Energy Resources			
Coal	MJ	9.12×10^{-3}	4.65×10^{-3}
Natural gas	MJ	7.50×10^{-1}	1.75×10^{-3}
Oil	MJ	6.96×10^{-1}	3.88×10^{-4}
Hydro power	MJ	9.50×10^{-4}	4.85×10^{-4}
Nuclear power	MJ	4.75×10^{-3}	2.42×10^{-3}
Total		1.461	9.697×10^{-3}
Air emission			
CH_4	g	1.45×10^{-1}	1.86×10^{-3}
CO	g	5.53×10^{-2}	1.44×10^{-4}
CO_2	g	7.91	5.28×10^{-1}
NMVOC	g	4.02×10^{-1}	1.22×10^{-4}
NO_x	g	2.60×10^{-1}	1.12×10^{-3}
SO_x	g	3.46×10^{-1}	1.57×10^{-3}
Water emissions			
BOD5	g	1.93×10^{-2}	6.40×10^{-5}
COD	g	6.22×10^{-3}	1.60×10^{-3}
TDS	g	3.60×10^{-1}	4.93×10^{-2}
Solid waste			
Total solid waste	g	1.70×10^{-1}	2.40×10^{-2}

TABLE P13.13 Wastewater and Process Stream Data

Wastewater Streams	Flow (metric tons/h)	Concentration (ppm)	Load (kg/h)
1	50	50	2.5
2	100	100	10
3	70	150	10.5
4	60	250	15
Process Streams	**Flow (metric ton/hs)**	**Maximum Inlet Concentration (ppm)**	**Load (kg/h)**
A	50	20	1
B	100	50	5
C	80	100	8
D	70	200	14

13.14 Jacob et al. presents a study of a thermomechanical pulp and newsprint mill consisting of 54 sinks and 10 sources. The contaminants here are present as fines in the wastewater. Using the source and sink information in Table P13.14, construct the shifted source–sink composite curves, and determine the minimum amount of fresh water required and the waste to be discharged after mass integration.

TABLE P13.14 Source/Sink Data

Source	Flow (L/min)	Fines Concentration (%)	Load (L/min)
TMP clear water	25,000	0.07	17.5
TMP cloudy water	39,000	0.13	50.7
Screen water	5,980	0.5	29.9
Press header water	2,840	0.49	13.9
'Save-all' clear water	6,840	0.08	5.5
	3,720	0.1	3.7
Silo water	73,000	0.39	284.7
Machine chest white water	8,585	0.34	29.2
Vacuum pump overflow	2,570	0	0
Residual showers	1,940	0.13	2.5

Sink	Flow (L/min)	Maximum Allowable Fines Concentration (%)	Load (L/min)
1	200	1	2
2	400	1	4
3	355	0.02	0.1
4	150	1	1.5
5	13,000	1	130
6	4,250	1	42.5
7	2,800	1	28
8	4,580	1	45.8
9	1,950	1	19.5
10	500	1	5
11	1,000	1	10
12	3,000	1	30
13	435	1	4.4
14	310	1	3.1
15	60	1	0.6
16	1,880	1	18.8
17	4,290	1	42.9
18	9,470	1	94.7
19	6,500	1	65
20	620	1	6.2
21	55	1	0.6
22	70	1	0.7
23	320	1	3.2
24	1,050	1	10.5
25	73,000	1	730
26	1,765	1	17.7
27	235	1	2.4
28	95	1	1.0
29	20	1	0.2
30	180	0	0
31	160	0.018	0.03
32	30	0.018	0.005
33	20	0.018	0.004
34	315	0	0

(continued)

TABLE P13.14 (*Continued*)

Sink	Flow (L/min)	Maximum Allowable Fines Concentration (%)	Load (L/min)
35	315	0	0
36	930	0.018	0.2
37	460	0.018	0.1
38	30	0.018	0.005
39	30	0.018	0.005
40	315	0	0
41	315	0	0
42	110	0.018	0.02
43	110	0.018	0.02
44	190	0	0
45	190	0	0
46	100	0	0
47	20	0	0
48	15	0	0
49	60	0.018	0.01
50	30	0.018	0.005
51	100	0	0
52	20	0	0
53	100	0	0
54	20	0	0

REFERENCES

1. El-Halwagi, M. M. Process integration. In *Process Systems Engineering*, Vol. 7, Academic Press, San Diego, CA, 2006.
2. Smith, R. State of the art in process integration. *Appl. Therm. Eng.*, 2000, 20, 1337–1345.
3. Dunn, R. F., El-Halwagi, M. M. Process integration technology review: background and applications in the chemical process industry. *J. Chem. Technol. Biotechnol.*, 2003, 78, 1011–1021.
4. Johns, W. R. Process synthesis: poised for a wider role. *Chem. Eng. Prog.*, 2001, 97(4), 59–65.
5. Koufos, D., Retsina, T. Practical energy and water management through pinch analysis for the pulp and paper industry. *Water Sci. Technol.*, 2001, 43(2), 327–332.
6. Linnoff, B. Pinch analysis:a state of the art overview. *Chem. Eng. Res. Des., Trans. IChemE, A.*, 1993, 71, 503–522.
7. Linnoff, B., Mason, D. R., Wardle, I. Understanding heat exchangers networks. *Comput. Chem. Eng.*, 1979, 3, 295–302.
8. Linnoff, B., Hindmarsh, E. The pinch design method for heat exchange networks. *Chem. Eng. Sci.*, 1983, 38(5), 745–763.
9. Papoulias, S. A., Grossmann, I. E. A structural optimization approach in process synthesis:—II. Heat recovery networks. *Comput. Chem. Eng.*, 1983, 7(6), 707–721.
10. El-Halwagi, M. M. *Pollution Prevention Trough Process Integration: Systematic Design Tools.* Academic Press, San Diego, CA, 1997.
11. Grossman, I. E. Mixed-integer programming approach for the synthesis of integrated process flowsheets. *Comput. Chem. Eng.*, 1985, 9, 463.

12. Zhu, X., et al. A method for automated heat exchanger network synthesis using block decomposition and non-linear optimization. *Trans. IChemE*, 1995, 73(8),919–930.
13. Shah, P., Kokossis, A. Systematic optimization technology for high-level screening and scoping of complex distillation systems. *Comput. Chem. Eng.*, 1999, 23(Suppl.), S137–S142.
14. Shah, P., Kokossis, A. Novel designs for ethylene cold-end separation using conceptual programming technology. *Comput. Chem. Eng.*, 1999, 23(Suppl.), S895–S898.
15. Amminudin, K.Design and Optimisation of the Dividing Wall Column. Ph.D. dissertation. UMIST, Manchester, UK, 1999.
16. Hallale, N. Burning bright: trends in process integration. *Chem. Eng. Prog.*, 2001, 97(7), 30–41.
17. El-Halwagi, M. M., Manousiouthakis, V. Synthesis of mass exchange networks. *AIChE J.*, 1989, 35(8), 1233–1244.
18. El-Halwagi, M. M., Manousiouthakis, V. Simultaneous Synthesis of Mass Exchange and Regeneration Networks. *AIChE J.*, 1990, 36,(8), 1209–1219.
19. El-Halwagi, M. M., Hamad, A. A., Garrison, G. W. Synthesis of waste interception and allocation networks. *AIChE J.*, 1996, 42,(11), 3087–3101.
20. Srinivas, B. K., El-Halwagi, M. M. Optimal design of pervaporation systems for waste reduction. *Comput. Chem. Eng.*, 1993, 17,(10), 957–970.
21. Zhu, M., E1-Halwagl, M. M., A1-Ahmad, M. Optimal design and scheduling of flexible reverse osmosis networks. *J. Membrane Sci.*, 1997, 129, 161–174.
22. Foo, C. Y., Manan, Z. A., Yunus, R. M., Aziz, R. A. Synthesis of mass exchange network for batch processes: I. Utility targeting. *Chem. Eng. Sci.*, 2004, 59, 1009–1026.
23. Foo, C. Y., Manan, Z. A., Yunus, R. M., Aziz, R. A. Synthesis of mass exchange network for batch process systems: 2. Batch network design. In *Proceedings of the International Conference on Chemical and Bioprocess Engineering,* San Francisco, Nov. 16–21, 2003.
24. Foo, C. Y., Manan, Z. A., Yunus, R. M., Aziz, R. A. Synthesis of mass exchange network for batch process systems: 3. Targeting and design for network with mass storage system. In *Proceedings of the International Conference on Chemical and Bioprocess Engineering,* San Francisco, Nov. 16–21, 2003.
25. El-Halwagi, M. M., Srinivas, B. K. Synthesis of reactive mass exchange networks. *Chem. Eng. Sci.*, 1992, 47(8), 2113–2119.
26. Srinivas, B. K., El-Halwagi, M. M. Synthesis of reactive mass exchange networks with general non-linear equilibrium functions. *AIChE J.*, 1994, 40(3), 463–472.
27. El-Halwagi, M. M., Srinivas, B. K., Dunn, R. F. Synthesis of optimal heat-induced separation networks. *Chem. Eng. Sci.*, 1995, 50(1), 81–97.
28. Dunn, R. F., Srinivas, B. K., El-Halwagi, M. M. Optimal design of heat-induce separation networks for VOC recovery. *AIChE Symp. Ser.*, 1995, 90(303), 74–85.
29. Richburg, A., El-Halwagi, M. M. A graphical approach to the optimum design of heat-induced separation networks for VOC recovery. *AIChE Symp. Ser.*, 1995, 91(304), 256–259.
30. Dye, S. R., Berry, D. A., Ng, K. M. Synthesis of crystallisation-based separation scheme. *AIChE Symp. Ser.*, 1995, 91(304), 238–241.
31. Parthasarathy, G., Russell, F. D. Development of heat-integrated evaporation and crystallization networks for ternary wastewater systems: 1. Design of the separation system. *Ind. Eng. Chem. Res.*, 2001, 40, 2827–2841.
32. Wang, Y. P., Smith, R. Wastewater minimisation. *Chem. Eng. Sci.*, 1994, 49, 981–1006.
33. Dhole, V. R., Ramchandani, N., Tainsh, R. A., Wasilewski, M. Make your process water pay for itself. *Chem. Eng.*, 1996, 103, 100–103.
34. Olesen, S. G., Polley, G. T. A simple methodology for the design of water networks handling single contaminants. *Trans. IChemE, A.*, 1997, 75, 420–426.

35. Hallale, N. A new graphical targeting method for water minimisation. *Adv. Environ. Res.*, 2002, 6 (3), 377–390.

36. Foo, C. Y., Manan, Z. A., Tan, Y. L. Synthesis of maximum water recovery network for batch process systems. *J. Cleaner Prod.*, 2005, 13, 1381–1394.

37. Xiao, F., Seider, W. D. New structure and design methodology for water networks. *Ind. Eng. Chem. Res.*, 2001, 40, 6140–6146.

38. Sorin, M., Bédard, S. The global pinch point in water reuse networks. *Trans. IChemE, B.*, 1999, 77, 305–308.

39. Polley, G. T., Polley, H. L. Design better water networks. *Chem. Eng. Prog.*, 2000, 96(2), 47–52.

40. Bagajewicz, M. A review of recent design procedures for water networks in refincries and process plants. *Comput. Chem. Eng.*, 2000, 24, 2093–2113.

41. Dunn, R., Wenzel, H. Process integration design method for water conservation and wastewater reduction in industry: 1. Design for single contaminants. *Clean Prod. Process.*, 2001, 3, 307–318.

42. Dunn, R., Wenzel, H. Process integration design method for water conservation and wastewater reduction in Industry: 2. Design for multiple contaminants. *Clean Prod. Process.*, 2001, 3, 319–329.

43. El-Halwagi, M. M., Gabriel, F., Harell, D. Rigorous graphical targeting for resource conservation via material recycle/reuse networks. *Ind. Eng. Chem. Res.*, 2003, 42, 4319–4328.

44. Alves, J. J., Towler, G. P. Analysis of refinery hydrogen distribution systems. *Ind. Eng. Chem. Res.*, 2002, 41, 5759.

45. Chang, H., Guo, J. J. Heat exchange network design for an ethylene process using dual temperature approach. *Tamkang J. Sci. Eng.*, 2005, 8(4), 283–290.

46. Jezowski, J. M., Jezowska, A. Some remarks on heat exchanger networks targeting. *Chem. Pap.*, 2002, 56(6), 362–368.

14

INHERENT SAFETY

What This Chapter Is About In this chapter we discuss the principles of inherent safety. In Chapter 2 we discussed principles of green chemistry and green engineering that have been developed over the past 20 years or so. One of the green chemistry principles states that "substances and the form of a substance used in a chemical process should be chosen so as to minimize the potential for chemical accidents, including releases, explosions, and fires." Inherent safety is a way of thinking about synthetic route and process design in such a way as to avoid chemical accidents. It is very similar to pollution prevention, and the two approaches are complementary, so the reader is advised to review pollution prevention methodologies.

Learning Objectives At the end of this chapter, the student will be able to:

- Understand basic principles of chemical process safety and how to apply them to a chemical reaction and/or process.
- Describe the basic elements of inherent safety and how it is complementary to the pollution prevention hierarchy.
- Apply several approaches to inherently safer process design.

14.1 INHERENT SAFETY VS. TRADITIONAL PROCESS SAFETY

When we talk about safety, we are typically concerned with protecting people, property, and the environment. Historically, safety practices have generally involved features we need to add on to our processes or procedures to prevent some type of accident or harm to people,

Green Chemistry and Engineering: A Practical Design Approach, By Concepción Jiménez-González and David J. C. Constable

equipment, or both, and these have been known as passive, active, and procedural actions. A brief description of these traditional safety program elements follows:

- *Passive*: includes things like designing equipment to withstand a certain pressure, or putting a dike around a storage tank.
- *Active*: includes systems such as safety interlocks or having alarm systems that would have multiple active elements, such as a sensor to detect a hazardous condition, a logic device to decide what to do, and a control element that would implement a corrective action. All of these things are designed to prevent incidents or to mitigate the consequences of incidents. Everyday examples of active safety systems include a child lock in a car, a lock on a front-loading washer that prevents the door from opening when the washer is in a middle of a wash cycle, and in a chemical plant, a low-level alarm that triggers a refill or shuts down the reactor's heating elements.
- *Procedural*: standard operating procedures, safety rules, emergency response procedures, and, of course, a lot of training.

Traditional safety activities have also focused on *mitigation measures*, things we might do in the event of an accident. Such things as sprinkler or deluge systems, water curtains, emergency response systems, and catastrophic event planning would fall into the category of mitigation measures. All these mitigation measures are reactive or responsive activities; that is, they occur following a process upset or in response to an accident or unexpected event and are designed to minimize the overall risk of a catastrophic loss. The combining of elements of these loss-prevention strategies, generally known as *layers of protection*, is illustrated in Figure 14.1.

Over a period of years and in response to some major accidents, such as the 1974 accident at the Nypro plant in Flixborough, UK,[1] the 1989 explosion at a petrochemical facility in Pasadena, Texas,[2] or the 1995 Lodi, New Jersey explosion during a blending operation that

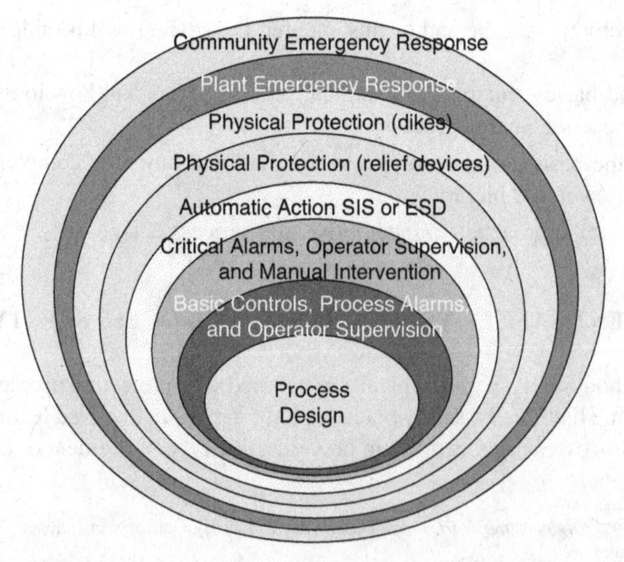

FIGURE 14.1 Layers of protection in classic chemical plant safety.

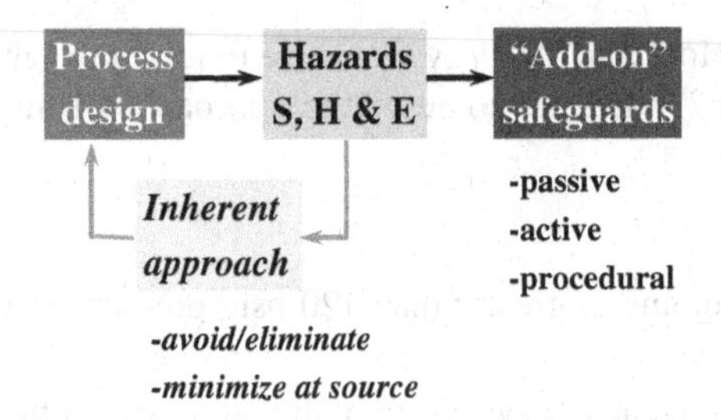

FIGURE 14.2 Traditional vs. inherent safety approach.

resulted in five fatalities and destruction of much of the manufacturing facility,[3] people began to think about safety in a different way. Perhaps the earliest pioneer in inherent safety was Trevor Kletz, who worked for ICI in the UK and following the Flixborough disaster began to work on specific design principles for the chemical industry. Additional notable work has been carried out by Preston and Turney, also of ICI, Hendershot at Rohm & Haas, Englund and Rogers at Dow, and many others. The point here is that inherent safety as a concept in many respects preceded green chemistry and engineering and to a certain extent, pollution prevention, and industry has always provided the intellectual and practical development of the discipline.

Clearly, all of the mitigation practices cited above generally work, work well, and continue to be used to this day. We also know that we can safely handle a variety of extremely hazardous materials on a daily basis; indeed, society would not have as many of its modern conveniences if this were not the case. But why not eliminate the hazard from the manufacturing process altogether rather than managing or controlling the risk? As we have noted throughout this book, avoiding a hazard and its attendant risk is always cheaper, safer, and provides less of an impact. Inherent safety is all about designing-out potential issues at the earliest possible moment in the development of a chemical synthetic route and its associated process.

Inherent safety fits very well into any discussion about green chemistry and green engineering. In each case, we are interested in avoiding the short- and long-term consequences of fires, explosions, toxic material releases, and so on (see Figure 14.2). We can accomplish this in ways that are a natural fit with what we already do; evaluate reaction pathways, evaluate the chemistries and the materials we use, carefully consider our reaction kinetics and thermodynamics, use technology that ideally eliminates the need for hazardous substances and/or activities, and so on. We can also think about where we site our plants, the ways in which we transport our raw materials to and from those facilities, and how we store things. Finally, we can think about the details of equipment design and how our processing approach can eliminate risk even if there is a need to use a hazardous material.

Example 14.1 As part of a chemical synthesis you are designing, there are several stages where you are isolating powders and drying them. What are classic and inherent safety approaches that you might take to ensure safe handling of these powders?

Solution First, you need to think about the potential issues. The powder is going to be organic, and most powders will support combustion. Are they explosible? When they explode, what would the force of the explosion be? How quickly would the explosion propagate? How much energy would be required to ignite the powder: a low- or a high-energy spark? What are some of the approaches I might take to reduce the risk?

- Passive
  Build stronger equipment: greater than 120 psig pressure rating
- Active
  Inerting (assuming an active system to maintain an inert atmosphere)
  Explosion venting
  Explosion containment
  Explosion suppression systems
- Procedural
  Procedures to maintain electrical grounding and bonding of equipment
  Procedures to avoid getting metal objects into the system (spark hazard)
  Procedures to avoid the generation of dust clouds
- Inherent
  Use a different (noncombustible) material
  Larger particles: granules or pellets (eliminate, or more likely, reduce the dust explosion hazard)
  Eliminate the drying step: pass to the next stage as a wet cake

Additional Points to Ponder What other hazards and risks might there be in this example? Will any of the solutions introduce or increase the magnitude of other hazards and risks?

14.2 INHERENT SAFETY AND INHERENTLY SAFER DESIGN

It is important to remember that as much as we might like to, we can't get something for nothing. So, as in the case of pollution prevention, it is generally impossible to eliminate all waste and it is equally impossible to make everything perfectly safe. Consequently, we should talk about something being inherently safer rather than inherently safe; a certain relativity needs to be understood as we move forward. But what do we really mean when we say that something is inherently safer? If we dissect the term, *inherent* means that there is a feature that exists in something, or there is an attribute of something which is a permanent or inseparable part of that thing. So the phrase "an inherently safer process" means that safety is an inseparable or permanent part of a process from first principles or by design; it is not added or "bolted" on, but built in. Ideally, this means that hazard is eliminated in a process so that the process is risk-free. This is clearly something that exists only as an ideal; it is very difficult to attain in practice.

Having said that, there is much that we can do to make our chemical synthetic routes and their associated processes inherently safer, and there are a great many resources available to assist us.[4–9] As with pollution prevention, we speak of a hierarchy of controls or approach, as illustrated in Figure 14.3. Let's look a bit more closely at each of the steps of the inherent safety hierarchy.

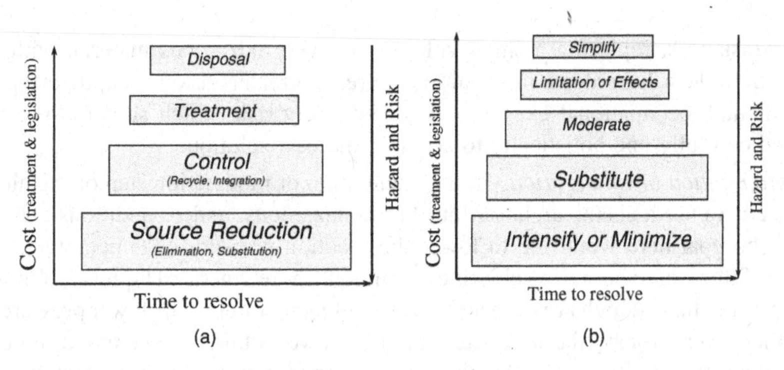

FIGURE 14.3 (a) Pollution prevention hierarchy; (b) inherent safety hierarchy.

1. *Intensification or minimization.* Intensification or minimization can be understood in several ways: from a materials perspective, from a combined equipment and materials perspective, and from an energy perspective. From a materials perspective, the objective is to reduce the amount of a hazardous material or materials in a process or a plant. A useful way of looking at this is: "What you don't have can't leak." This objective can be met through such strategies as in situ generation, or using less of a hazardous material at a given time through logistical planning and just-in-time delivery. It also has to do with simple things such as storage, intermediate storage, piping, and the types of process equipment in use. Alternatively, process intensification strategies such as those described in Chapter 15 can be used to probe what can be done to run reactions at a higher concentration or by using less material through a change in reactor design. For process intensification strategies that rely on increasing reaction concentrations, we must ensure that doing so will not result in an uncontrolled event. This is clearly where alternative reaction technology can and should play a roll.

As with many things in green chemistry, it is imperative that you understand what controls any chemical reaction so that process equipment can be designed to optimize the reaction. In general, you will want to develop reactions that occur very rapidly but with the proviso that they are not highly exothermic. One also needs to pay attention to energy and mass transfer or transport phenomena to optimize mixing, chemical equilibrium, and other molecular phenomena across phases and surfaces. Reactions that occur at room temperature and pressure, reactions that use single-phase systems of low viscosity, and reactions that are not sensitive to variations in reaction conditions are all desirable. Ideally, changes in temperature, pressure, concentration, or the presence of trace quantities of contaminants such as water, oil, or particulate matter such as rust would not create problems for the system. There is also the possibility of employing phase-transfer catalysts as a means of overcoming phase separation issues, or permitting the use of biphasic solvent systems that enhance the separability and isolation of the desired product or that can be used to control reaction kinetics.

2. *Substitution.* As with pollution prevention, the strategy here is to use a safer material or a set of reactions to replace more hazardous materials: replacing highly flammable substances with those that are less flammable, or toxic substances with those that are less toxic. If possible, complete elimination of a hazardous substance or set of conditions is ideal but is often not possible. It is also important to evaluate a substance and the volume required.

In some instances, replacing a small volume of a very hazardous material with a large volume of a less hazardous material may create greater risk of another type (e.g., environmental, occupational exposure, or other safety risk). Each situation needs to be evaluated carefully and holistically to arrive at the best solution.

3. *Attenuation or moderation.* In the attenuation or moderation step of the hierarchy, the desire is to use a chemical, hazardous or nonhazardous, under conditions that are less severe. The goal here would be to lower the reaction pressure or temperature. A good example of this would be storing chlorine or ammonia as refrigerated liquids at atmospheric pressure rather than at high pressure and at ambient temperature. The lower pressure means that if there were a leak, the leak rate would be lower while the decreased temperature reduces the rate of vaporization. Another possible approach would be to employ catalysts. As catalysts lower the activation energy of a reaction, reactions can often be run at lower temperatures and pressures. As with intensification and minimization, phase-transfer catalysts may be used to good effect to moderate reactions or control reaction kinetics.

4. *Limitation of effects.* Simple changes in the way that chemicals are introduced into a reaction vessel can have profound effects on the overall safety of a process. So changing reactor designs, changing process conditions, or relying on protective equipment is desirable. For example, we can lower the final temperature of reaction liquors if we introduce a reactive chemical at a reduced temperature, or we might change the order or rate of addition of a very reactive substance rather than relying on a control system to sense a problem and attempt to control it.

5. *Simplicity.* The acronym KISS, "keep it simple, stupid," is in view here. While we may need to build complex molecules, complexity in synthetic and process design is generally not a good thing. Simpler synthetic designs and simpler plants are generally safer than complex syntheses, processes, and plants because they generally provide fewer opportunities for human error and equipment failure. They are also easier to control if there are deviations from normal conditions before the situation turns into a catastrophic accident.

Now that we have the broad outline of the inherent safety hierarchy, you might ask: How do we know if a system or process is inherently safer? A reasonably simple way to look at this would be to say that if a system or process is perturbed yet remains within or returns to a safe and stable condition in the absence of human intervention or automatic controls, it is inherently safe. Getting to the point where we can use the hierarchy effectively will, of course, require a strong fundamental knowledge of the physical and chemical processes that are underlying and governing our reactions and manufacturing processes. It is important to remember that inherently safer design, like green chemistry, is first and foremost a way of thinking about the problems facing us and the tools and methods we use to solve those problems. It is how we think about things: from the conception of a route during retrosynthetic analysis to the final aspects of how we design our processes.

Example 14.2: Carbaryl Process Many of you may be familiar with the first process shown below as the process used by the former Union Carbide in Bhopal, India to make the pesticide carbaryl. Contamination of the methyl isocyanate with water resulted in the overpressurization of a tank holding methyl isocyanate and the rupture of a pressure relief valve. Because over 100 tons of the intermediate was stored on site, and the emergency safety systems were either inactivated or ineffective, over 2000 people were killed, many more were injured, and ultimately, the company went out of business. A second process was then proposed and is shown below. Which process is inherently safer, and why?

Old process:

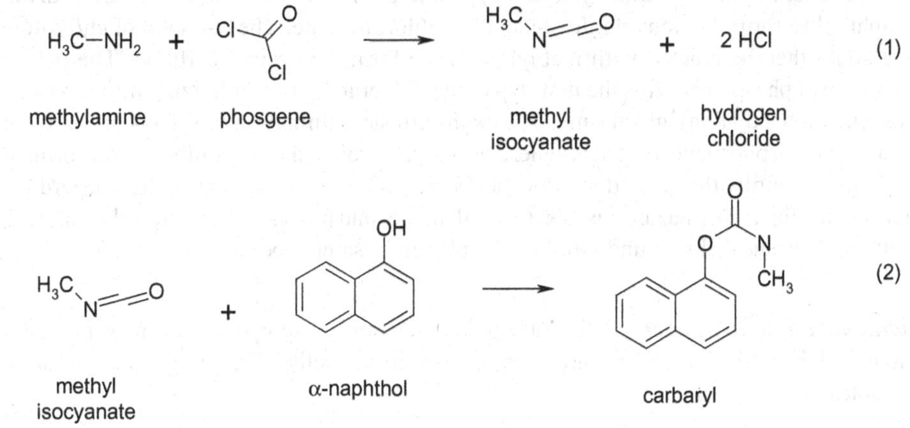

$H_3C{-}NH_2$ + Cl–C(=O)–Cl → $H_3C{-}N{=}C{=}O$ + 2 HCl (1)

methylamine phosgene methyl isocyanate hydrogen chloride

$H_3C{-}N{=}C{=}O$ + OH–naphthalene → carbaryl (O–C(=O)–N–CH_3) (2)

methyl isocyanate α-naphthol carbaryl

New process:

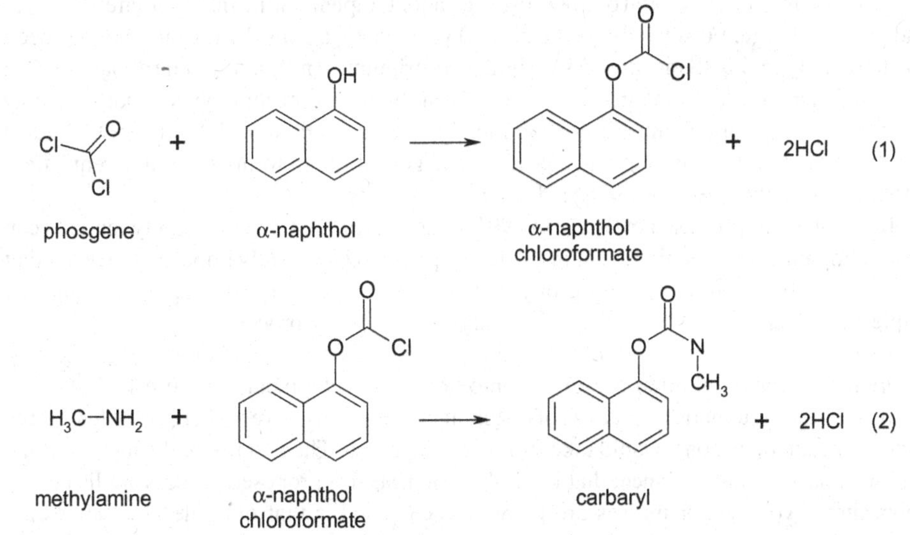

Cl–C(=O)–Cl + OH–naphthalene → naphthyl O–C(=O)–Cl + 2HCl (1)

phosgene α-naphthol α-naphthol chloroformate

$H_3C{-}NH_2$ + naphthyl O–C(=O)–Cl → naphthyl O–C(=O)–N–CH_3 + 2HCl (2)

methylamine α-naphthol chloroformate carbaryl

Solution First, it should be noted that phosgene and methylamine are both toxic, hazardous compounds and have a certain degree of associated risk in their manufacture, handling, and storage. More recent synthetic strategies for making carbaryl have eliminated the use of both compounds, and these satisfy the second step in the IS hierarchy. In the Bhopal case, the methyl isocyanate (MIC) intermediate was a denser-than-air but still very volatile gas that is highly reactive with water. Storing 100 tons of MIC was not necessary, as the process could have been run close to continuously with only a few kilograms of MIC needed at any time. Had the plant operators made and used only the precise amount of MIC that was needed, they would have fulfilled the first step of the IS hierarchy and intensified or minimized the hazard and resulting risk of an adverse event.

In the second synthetic strategy, in situ generation of phosgene could be reacted with the α-naphthol to form the nonvolatile α-naphthol chloroformate. The α-naphthol chloroformate could then be reacted with methyl amine to form the desired carbaryl. The in situ generation of phosgene fulfills the first step of the IS hierarchy by minimizing the amount of phosgene on hand at any given time, and the fourth step, limitation of effects, as only the exact amount of phosgene needed is generated. Going through the α-naphthol chloroformate intermediate fulfills the second step of the IS hierarchy by substituting a less hazardous intermediate for a more hazardous one. Overall, the second process, although still exhibiting safety hazards and risks, is undoubtedly an inherently safer process.

Additional Points to Ponder What other changes would you propose to the newer process to make it inherently safer? What are other less tangible benefits of applying inherent safety principles?

14.3 INHERENT SAFETY IN ROUTE STRATEGY AND PROCESS DESIGN

All that remains is a discussion of how best to go about implementing inherent safety in route and process design. Fortunately, in the past 30 years or so, a considerable amount has been written on integrating IS into process design. It is intriguing to note a comment about inherent safety design, and that is that while inherent safety as a term may be well known, how inherent safety is implemented is not common knowledge.[10] Figure 14.4 is a very high level flow diagram that outlines the basic process one goes through to integrate inherent safety thinking into either route or process design.

Table 14.1 was proposed by Kletz in 1991[11] and shows what inherent safety features can best be implemented at which point in the design phase. It is critically important to recall that inherent safety, as with everything in green chemistry and green engineering, should be implemented as early as possible in the design of a route or process.

Recently, Palaniappan et al. developed a material-centric methodology for integrating inherent safety and pollution prevention into process development.[12] Figure 14.5 shows a process for the systematic evaluation of different process alternatives. There are at least three very important points one should take from their approach. The first is that the methodology is based on a materials balance; that is, all the materials in a proposed process are listed and approximate volumes or masses are known. The second is that multiple alternatives are screened. It is too often the case that a single route is evaluated, (or perhaps two or so) before the route is tested. It should be understood that a green chemistry/green engineering approach is most successful if more time is spent in options generation. The third point is that options evaluation is carried out in a systematic and objective manner. Very frequently, "conventional wisdom" or "chemists' intuition" is used to justify a proposed approach, only to find that the route becomes unworkable on scale-up because of a processing, environmental, safety, or health problem. Bringing forward a systematic approach to process generation enables us to bring the scientific method into process design, limiting the possibility of making decisions based on erroneous intuition or personal bias.

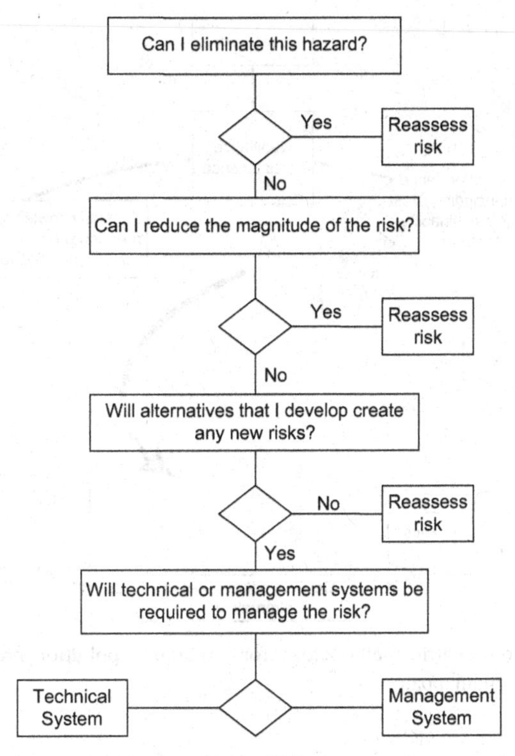

FIGURE 14.4 Flowchart for integrating IS into route strategy and process design.

TABLE 14.1 Inherent Safety Principles Considered in First Project Stages

Feature	Conceptual Stage	Flowsheet Stage	PI Stage
Intensification	×	×	
Substitution	×	×	
Attenuation	×	×	
Limitation of effects			
By equipment design			×
By changing reaction conditions	×	×	
Avoiding knock-on effects			
By layout	×	×	×
In other ways		×	×
Making incorrect assembly impossible			×
Making status clear			×
Simplification	×	×	
Tolerance			×
Ease of control	×	×	
Software			×

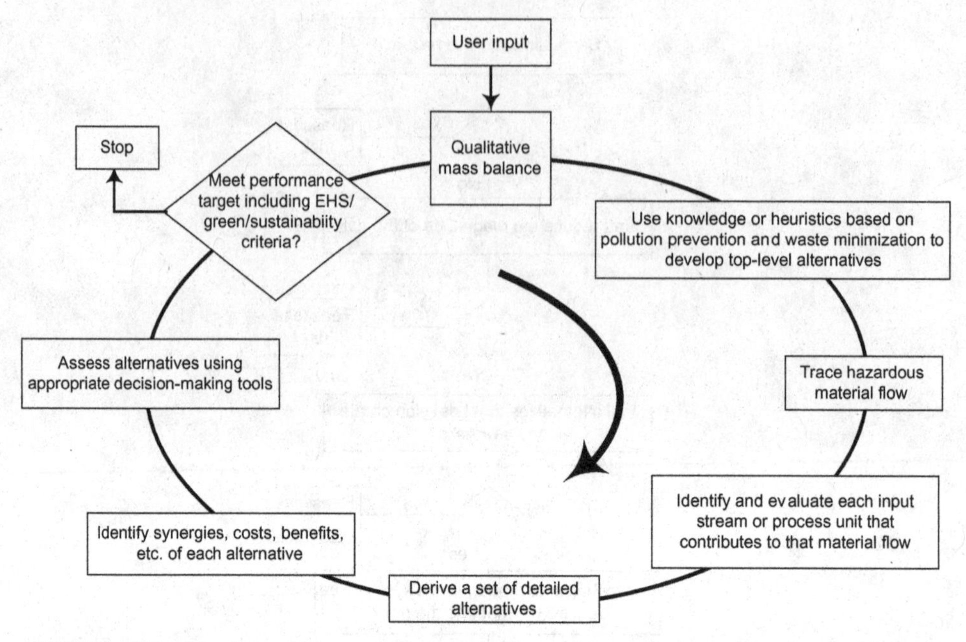

FIGURE 14.5 Material-centric methodology for integrating pollution prevention and inherent safety. (Adapted from Palaniappan et al.[12])

In the materials-centric approach of Palaniappan et al., four aspects are considered: process materials, material–process condition interactions, material–material interactions, and material–process unit interactions.

1. *Process materials.* All the materials in a process, including potential by-products, intermediates, and processing aids need to be identified. EHS issues may then be identified and assessed according to the key physical properties described in Chapter 3.

2. *Material–process condition interactions.* These interactions highlight the potential for materials to behave differently as process conditions change. For example, as the pH, temperature, or pressure is increased or decreased, a solvent may vaporize, a chemical may become unstable, or there may be a phase change—any of which might lead to an unsafe process condition. Knowledge of a material's thermal decomposition temperature and their freezing or boiling points all need to be evaluated carefully under the process conditions expected. A good checklist to better understand material–process interactions has been drawn up by Hendershot[13] and is shown in the accompanying box. However, many tools and checklists are available to step you systematically through a review of a chemical route and process to determine the suspected interactions.

Chemical Reaction Hazard Identification

1. Know the heat of reaction for the intended and other potential chemical reactions.
2. Calculate the maximum adiabatic temperature for the reaction mixture.
3. Determine the stability of all individual components of the reaction mixture at the maximum adiabatic reaction temperature.

4. Understand the stability of the reaction mixture at the maximum adiabatic reaction temperature.
5. Determine the heat addition and heat removal capabilities of the pilot plant or production reactor.
6. Identify potential reaction contaminants.
7. Consider the impact of possible deviations from intended reactant charges and operating conditions.
8. Identify all heat sources connected to the reaction vessel and determine their maximum temperature.
9. Determine the minimum temperature to which the reactor cooling sources could cool the reaction mixture.
10. Consider the impact of higher temperature gradients in plant-scale equipment compared to a laboratory or pilot-plant reactor.
11. Understand the rate of all chemical reactions.
12. Consider possible vapor-phase reactions.
13. Understand the hazards of the products of both intended and unintended reactions.
14. Consider preparing a chemical interaction matrix and/or a chemistry hazard analysis.

3. *Material–Material interactions.* Material incompatibilities are a key component of any process safety evaluation, as they can lead to uncontrolled exothermic excursions, rapid and uncontrolled gas evolution, rapid polymerizations, and/or combinations of such factors. Process parameters may also facilitate or retard these interactions, so the materials interactions need to be evaluated in light of these process conditions.

4. *Material–process unit interaction.* Unit operations common to chemical processing have been described throughout the book. As you have learned, these unit operations can profoundly affect such things as mixing rates, mass and energy transport phenomena, and other chemical equilibrium phenomena. In addition, materials can interact with the unit operation materials of construction and cause corrosion or other damaging effects to reactors, separators, and other equipment. Once again, these types of interactions need to be looked at very carefully and evaluated as to their potential for unsafe or environmentally damaging effects.

Many analytical tools and methodologies have been developed to identify potential areas of concern for both materials and chemical processes. Many of these are listed in Table 14.2.

A notable variation on the materials-centric approach is the hierarchical approach to evaluating chemical processes taken by Shah et al.[15] These authors take a layered approach to the assessment of hazards and risks associated with a given route or process alternative. The first layer is known as the substance assessment layer, the second is the reactivity assessment layer, the third is the equipment assessment layer, and the fourth is the safety technology assessment layer. As you can see, once a particular route or process alternative has been chosen, this is very similar to the previous methodology. However, it is worth describing each layer in a bit more detail. The flowchart in Figure 14.6 illustrates the decision logic associated with this approach. As with the previous approach, the process begins with a chemical or process flowsheet where all the materials in the synthesis or process have been identified.

TABLE 14.2 Tools for Developing Better Process Safety Understanding

Theoretical and Computational Screening	Chemical Hazard Identification: Experimental Screening for Thermal Stability	Process Hazard Identification: Screening Tools for Reaction Rate and Kinetics	Emergency Relief Systems (ERS) Design, Screening, and Direct Scale-up	Process Design and Optimization
MSDS	Blasting cap test	Isothermal storage test (IST)	RSST	Reaction calorimetry (RCI)
Chemical compatibility matrices	Flame test	Accelerating rate calorimetry (ARC)	SuperChems Expert, for DIERS, QuickSize	Contalab
Literature reactivity data, such as Bretherick's handbook, NFPA, hazard ratings	Gram-scale heating test	Vent sizing package (closed test cell) (VSP)	Simple nomographs	Atomic pressure-tracking adiabatic calorimetry (APTAC)
Incident data	Shock sensitivity test	RC1 (Pressure vessel only, after screening tests)		Computational fluid dynamics, SuperChems, Expert/DIERS
Chemical structure	Drop weight test	APTAC		Specialized large-scale test (mixing limited reactions, injection of reaction killers, chemical rollover, reactions at interface, etc.)
Formation energies	Thermogravimetric analysis (TGA)			
Heats of reaction, decomposition, solution	Differential thermal analysis (DTA)			
Computed adiabatic reaction temperature at constant pressure and/or volume, (CART)	Reactive systems screening tool (RSST)			
Oxygen balance	Differential scanning calorimetry (DSC)			
Software tools such as the ASTN CHETAH, NASA CET89, SuperChems, TIGER, etc.				

Source: Adapted from ref. 14.

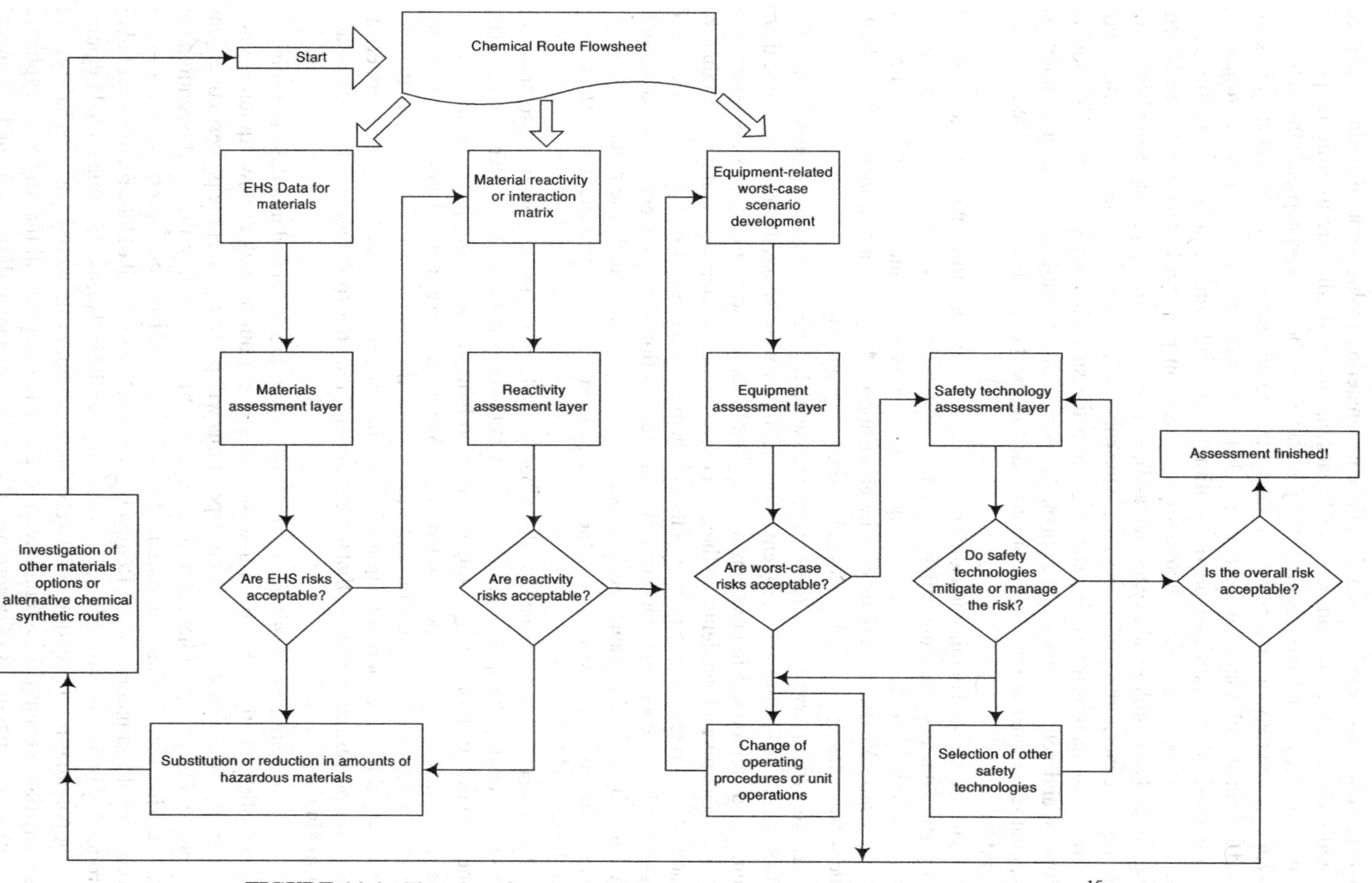

FIGURE 14.6 Flowchart for process design decision making. (Adapted from Shah et al.[15])

1. *Substance assessment layer*. In the first layer, the substance assessment layer, the first objective is to substitute less hazardous or benign materials for hazardous materials. Next, we would seek to reduce the amounts of hazardous materials that are used in the process. Substances are evaluated by looking at key physical and chemical properties of the materials, along with a variety of environmental, safety, and health parameters. Eleven categories are used to evaluate substances: mobility (i.e., where the compound ends up in the environment), fire, explosion, reaction and decomposition, acute and chronic toxicity, irritation, air-mediated effects (e.g., photochemical ozone creation potential), water-mediated effects (e.g., hydrolysis, biological oxygen demand), solid waste, persistence, and degradability. In a sense, the exact number of categories and the way in which they are scored or assessed is not as critical as having the right categories from the sustainability perspective of looking globally and acting locally. Consequently, there may be some variation in parameters, depending on whether or not you are considering local conditions and how these may be weighted.

As has been noted in this book and elsewhere, EHS and sometimes physical and chemical property data are often unavailable for new materials in early development or recent commercialization. In these cases, one needs to rely on quantitative structure–activity relationship (QSAR) models, nearest-neighbor approaches, or other means to arrive at best-guess estimates of physical, chemical, and/or EHS properties.

2. *Reactivity assessment layer*. This is essentially the same as the material–material assessment level in the material-centric methodology we discussed earlier. As noted then, compound–compound interactions are frequently the cause of many process upsets and safety accidents and incidents, so the main objective of this step is to make certain that we know of any potential for process upsets. As chemists, we frequently make use of energetic materials because we value their reactivity; faster reactions that are kinetically and thermodynamically more favorable are less likely to result in unwanted side reactions and by-product formation. However, using these materials at scale is often difficult at best and they are better avoided.

Mosley et al.[16] and others have described procedures for the systematic assessment of reactive chemical hazards. Figure 14.7 is a fictitious example of an interaction matrix. The intention of tools such as this matrix is to help in visualizing potential interactions between materials and process conditions. While there are many databases containing reactivity data, it takes some time and effort to compile accurate data from scratch. It is also necessary through experimentation to determine potential reactivity concerns under the proposed process conditions so that there are no surprises due to changes in pH, temperature, or mixing.

3. *Equipment assessment layer*. Once substances are evaluated and their potential for uncontrolled reactivity has been assessed, it is important to look at how equipment and collections of equipment used in unit operations will perform under the process conditions proposed. The objective here is to determine what might occur in the event of equipment failure. This is an extremely important step, as many industrial accidents occur as a result of some sort of equipment failure. Remember that risk is a function of hazard, the potential or probability for an event to occur (i.e., exposure potential in the case of occupational hygiene, or the potential for a pressure relief valve to stick shut in the case of process safety), the severity of the event (i.e., a temperature rise of a few degrees all the way to an explosive event), and the frequency of occurrence. By substituting materials or changing the chemical reactions, we reduce the severity of the event and the inherent hazard of the process.

	Reactant A	Reactant B	Solvent A	Solvent B	Reactant A / Solvent A Mixture	150° C steam	Operators	Etc.
Reactant A	★	★	★	★	★	★	★	
Reactant B		★	★	★	★	★	★	
Solvent A			★	★	★	★	★	
Solvent B				★	★	★	★	
Reactant A / Solvent A Mixture					★	★	★	

FIGURE 14.7 Interaction matrix: example of all interactions safe.

The equipment assessment layer seeks to reduce the frequency of occurrence of an adverse event. Through the use of worst-case scenarios that are related to equipment failures in different unit operations used for a process, risk can be characterized and prevention and protection measures modeled using unit operation models. Once again, there are systematic ways of doing this for any given process scenario. For example, Stoessel[17] has developed a systematic approach to assessing exothermic runaway reactions that simplifies the analysis of risk.

The equipment assessment layer is important, as it enables the consideration of vent sizing, vent locations, different reactor configurations, reactant feeds, mixing rates, and other factors. Hendershot[13] has also drawn up a list of potential options for process design considerations that involve process and equipment:

- Rapid reactions are desirable.
- Avoid batch processes in which all of the potential chemical energy is present in the system at the beginning of the reaction step.
- Use gradual addition or semi batch processes for exothermic reactions.
- Avoid using control of the reaction mixture temperature as the only means of limiting the reaction rate.
- Account for the impact of vessel size on the heat generation and heat removal capabilities of a reactor.
- Use multiple temperature sensors in different locations in the reactor for rapid exothermic reactions.
- Avoid feeding a material to a reactor at a higher temperature than the boiling point of the reactor contents.

Given the history of process safety catastrophes and the critical part that equipment failure has played in these events, equipment assessment is critical in the design of a safe process and plant.

4. *Safety Technology Assessment Layer*. Despite our best efforts to remove process hazards associated with materials, to reduce the probability of reactive chemical upsets and prevent equipment malfunctions or limit the frequency of their occurrence, every process will always have a certain amount of associated risk. This means that there will invariably be a need for certain bolt-on pieces of safety equipment, control operations, or control technologies to help mitigate or manage the residual risk. Once again, there are a variety of recommendations that can be made, based on the particular scenario being evaluated for each unit operation in the overall process configuration.

14.4 CONCLUSIONS ON INHERENT SAFETY

Inherently safer design is an extremely important aspect of green chemistry and green engineering that is still often underappreciated. Clearly, the benefits of implementing pollution prevention and inherent safety approaches throughout the process and product development cycles have been demonstrated repeatedly in many industries. Failure to implement such strategies and methodologies often leads to catastrophic events that result in loss of life or a significant impacts on health and well-being, environmental degradation, destruction of property, and potentially a significant loss of income and perhaps even the eventual demise of a company. It is in your hands to move inherent safety from something that you may have heard about to something that you do as a normal part of your work.

PROBLEMS

14.1 A semibatch nitration process is shown in Figure P14.1(a), compared to the continuous stirred tank reactor shown in Figure P14.1(b).

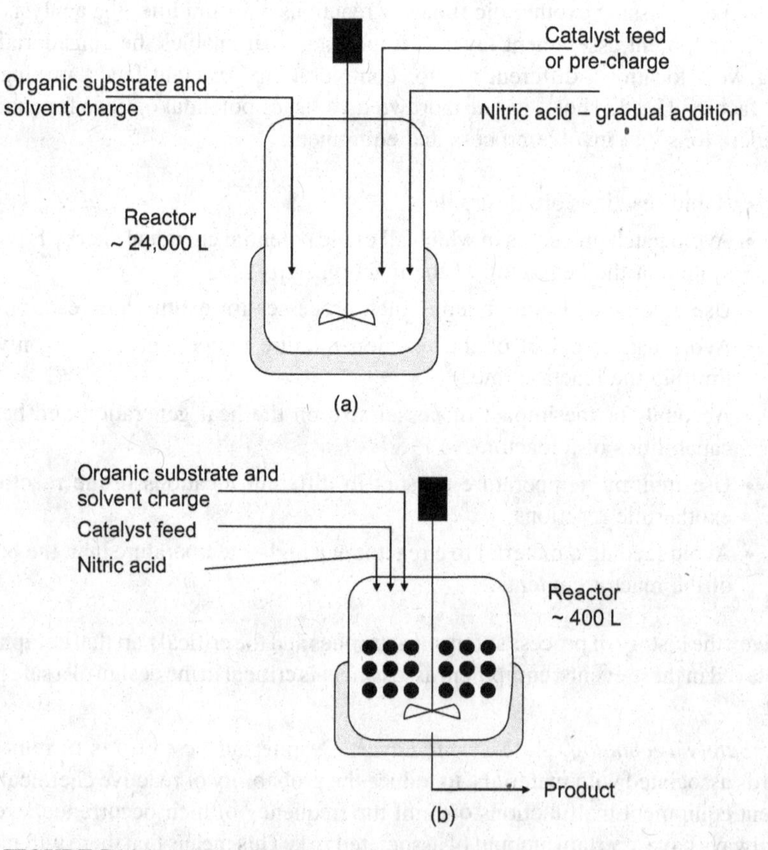

FIGURE P14.1 (a) Batch reactor; (b) continuous tank reactor.

(a) What controls the rate of reaction?
(b) Does control over the rate of the reaction change with the size of the reactor?

(c) Evaluate the inherent safety of each reactor. Is one reactor configuration inherently safer than the other? Why or why not?

(d) How might you make the reaction and both processing options more inherently safe?

14.2 Refrigeration has changed over the last 100 years. Prior to the 1930s, refrigeration was accomplished using several kilograms of such gases as ammonia, light hydrocarbons (e.g., propane, butane), and sulfur dioxide. In the 1930s these gases were generally replaced by new materials known as chlorofluorocarbons (CFCs). In the 1980s, continued use of CFCs became a matter of widespread concern, and replacements were developed.

(a) Compare and contrast the inherent safety and environmental advantages and disadvantages of various approaches to refrigeration gases, including current refrigerants.

(b) Which refrigerants are best, and why?

(c) What technology options existed to make each alternative inherently safer?

14.3 One of the processes used most commonly to synthesize acrylic esters is the Reppe process. An alternative process is available through a propylene oxidation process. These processes are depicted below.

Reppe process:

$$HC\equiv CH + CO + ROH \xrightarrow[HCl]{Ni(CO)_4} H_2C{=}CH{-}C({=}O){-}OR$$

Propylene oxidation process:

$$H_2C{=}CH{-}CH_3 + 1.5\ O_2 \xrightarrow{\text{catalyst}} H_2C{=}CH{-}C({=}O){-}OH + H_2O$$

$$H_2C{=}CH{-}C({=}O){-}OH + ROH \xrightarrow{H+} H_2C{=}CH{-}C({=}O){-}OR + H_2O$$

(a) Using a structured methodology such as one of those described in this chapter, compare and contrast the green aspects of each synthesis, including the inherent safety aspects of each. Which synthesis is greener, and why? Which synthesis is inherently safer?

(b) What are the advantages and disadvantages of the starting materials for each synthesis? Is one inherently safer than the other?

14.4 A traditional batch emulsion polymerization reaction is shown in Figure 14.4(a). This is compared to a newer approach that employs a loop reactor in the place of the batch reactor, shown in Figure 14.4(b).

(a) Which of the process designs is inherently safer?

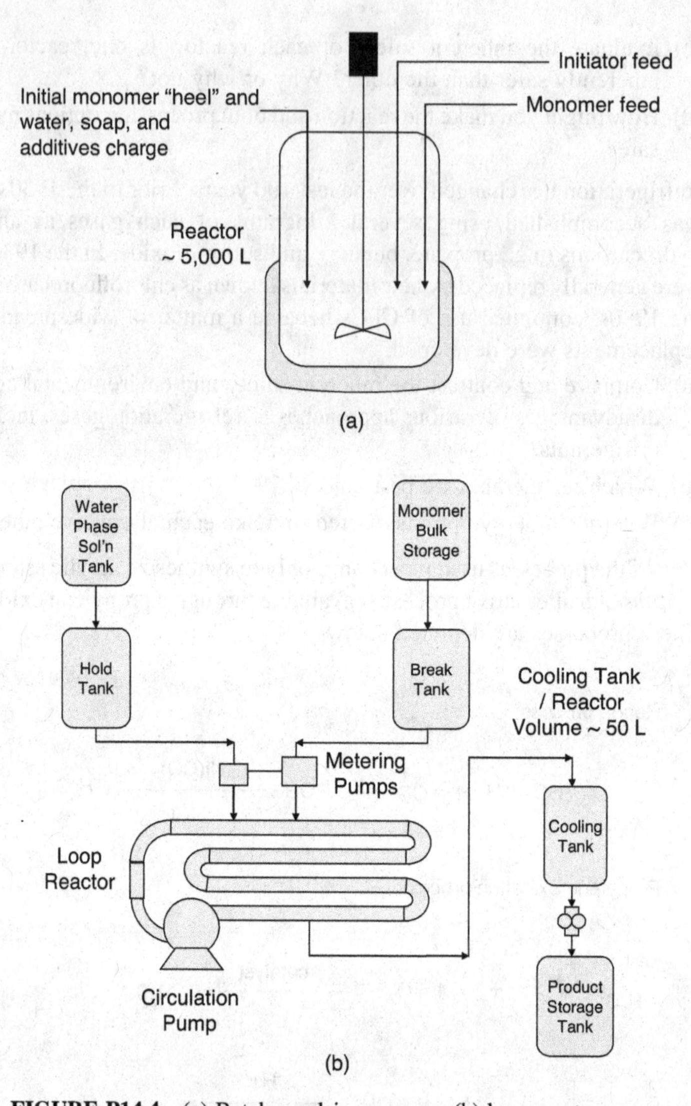

FIGURE P14.4 (a) Batch emulsion process; (b) loop reactor.

(b) Imagine that the product demand for this emulsion reaction suddenly skyrockets. Which reactor design would allow you to meet that demand more easily? What might be some of the trade-offs?

14.5 What might be some inherently safer chemical choices for each of the following commonly used substances? Defend your answers.

(a) Ammonia gas

(b) HCl gas

(c) Oleum

(d) Benzoyl peroxide

(e) Dimethylformamide

(f) Chloroform
(g) Diethyl ether
(h) Sodium borohydride
(i) LiAlH

14.6 Many reactions in industry are exothermic. If a reaction is exothermic enough and heat is not removed quickly enough, a runaway reaction, may result.

(a) Describe what might happen in a batch reactor during an exothermic reaction when the temperature begins to rise. Keep the Arrhenius equation in mind:

$$k = Ae^{-E_a/R_g T}$$

(b) What might you do to prevent a runaway reaction from a materials standpoint?

(c) What inherently safer designs might you consider to prevent a runaway reaction from an engineering standpoint?

(d) If you couldn't control the reaction, what other actions might you take to prevent the runaway reaction from becoming critical?

14.7 Complete Table P14.7.

TABLE P14.7

Process Unit	Condition	Safety or Waste Issue	Inherent Safety or P2 Principle	Design Recommendations
Equipment	Temperature > 150°C or pressure > 25 bar			
Reactor	Low yield			
Reactor	Reaction temperature > autoignition temperature			
Packed bed reactor	Gas-phase-catalyzed exothermic reaction			
Separation	Use of hazardous mass separating agent			
Distillation	Boiling point between solvents below 5°C and one solvents very toxic			
Inlet stream	Raw material contaminated with catalyst poison			
Heat exchanger	Heat of reaction > heat exchanger medium decomposition temperature			
Fluid-bed reactor	Loss of primary air jets			

14.8 In the processing train shown in Figure P14.8, two substances react in the presence of a catalyst, the reaction mixture is cooled, and the product is extracted in an absorber unit operation. Reactant A is the limiting reagent and both toxic and highly reactive. Reactant B is present at a stoichiometric excess of 1.5. The catalyst can be poisoned by exposure to an excess of oxygen. The mass separating agent is toxic and must be removed from the final product to less than 2 ppm, and although the catalyst may be regenerated, this is possible only if it is removed from by-product D.

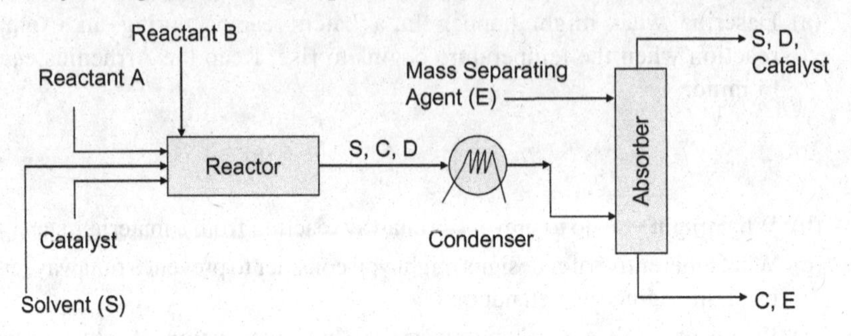

FIGURE P14.8

(a) Write a balanced equation for the reaction.

(b) Use a systematic approach to assess the IS and P2 issues associated with this process.

(c) How can the process be made inherently safer?

14.9 A variety of solvent alternatives have been investigated in recent years as greener alternatives. Develop a table to compare the inherent safety issues associated with the following proposed solvent replacements:

(a) Supercritical CO_2

(b) Polyethylene glycols

(c) Fluorous biphasic solutions

(d) Ionic liquids

14.10 Explain how catalysts might contribute to making a process inherently safer.

14.11 In Chapter 10 we discussed scale-up issues and their effect on green chemistry and green engineering. Windhorst and Koen[18] make the case that whereas many designs for larger plants are linear scale-ups, the associated process safety risks increase not linearly, but exponentially. Consider the following scale-up issues and describe why the process safety risk might increase exponentially.

(a) Larger equipment size

(b) Larger nozzle size

(c) Increased adiabaticity (i.e., larger equipment has a slower rate of heat loss per unit of material in the processing equipment)

(d) Increase in number of support systems

(e) Duplicated, triplicated, etc. piping, valves, flanges, and welds

14.12 A company wants to manufacture an energetic material and has a glass-lined stainless steel reactor available for use. This particular energetic material would be present in the reactor as a solution and once it is made, most of the solvent used for the reaction would need to be removed. The energetic material can detonate if it is overheated or if it is in the presence of iron contaminants.

(a) What are the process safety risks associated with a conventional batch evaporation process?

(b) What inherently safer approaches might the company take to reduce the safety risk? Describe thêse and defend your answer.

REFERENCES

1. Health and Safety Executive. Flixborugh (Nypro, UK) Explosion 1st June 1974, Case Study. http://www.hse.gov.uk/comah/sragtech/caseflixboroug74.htm.
2. Yates, J. Phillips Petroleum Chemical Plant Explosion and Fire Pasadena, Texas (October 23, 1989). Report 35. FEMA, U.S. Fire Administration National Fire Data Center, Technical Report Series. http://www.interfire.org/res_file/pdf/Tr-035.pdf.
3. U.S. Environmental Protection Agency and U.S. Occupational Safety and Health Administration. *EPA/OSHA Joint Chemical Accident Investigation Report: Napp Technologies, Inc., Lodi, New Jersey*. EPA 550-R-97-002. U.S. EPA, Washington, DC, Oct. 1997.
4. American Institute of Chemical Engineers, Center for Chemical Process Safety. *Safety Alert: Reactive Material Hazards*. AIChE, New York, 2001.
5. *Bretherick's Handbook of Reactive Chemical Hazards*. Butterworth-Heineman, Oxford, UK, 1999.
6. U.S. National Oceanic and Atmospheric Administration. Chemical Reactivity Worksheet. http://response.restoration.noaa.gov/chemaids/react.html.
7. American Institute of Chemical Engineers, Center for Chemical Process Safety. *Guidelines for Safe Storage and Handling of Reactive Materials*. AIChE, New York, 1995.
8. Health and Safety Executive. *Guidelines for Chemical Reactivity Evaluation and Application to Process Design: Designing and Operating Safe Chemical Reaction Processes*, HSE Books, Norwich, UK, 2000.
9. Barton, J., Rogers, R. *Chemical Reaction Hazards: A Guide to Safety*. Gulf Publishing Company, Houston, TX, 1997.
10. Gupta, J. P., Edwards, D. W. Inherently safer design—present and future. *Trans. IChemE B*, May 2002, 80, 115–125.
11. Kletz, T. A. *Plant Design for Safety*. Taylor & Francis, Bristol, PA, 1991.
12. Palaniappan, C., Srinivasan, R., Halim, I. A material-centric methodology for developing inherently safer environmentally benign processes. *Comput. Chem. Eng.*, 2002, 26, 757–774.
13. Hendershot, D. A checklist for inherently safer chemical reaction process design and operation. Presented at the Center for Chemical Process Safety International Conference and Workshop on Risk and Reliability, Jacksonville, FL, Oct. 8–11, 2002.
14. Melhem, G. A. Systematic evaluation of chemical reaction hazards. In *Proceedings of the 2nd AIChE International Symposium on Runaway Reactions: Pressure Relief Design and Effluent Handling*, Feb. 1998, pp. 399–443.
15. Shah, S., Fischer, U., Hungerbuhler, K. A hierarchical approach for the evaluation environmental protection. *Trans. IChemE B*, 2003, 81, 430–443.
16. Mosley, D. W., Ness, A. I., Hendershot, D. C. Screen reactive chemical hazards early in process development. *Chem. Eng. Prog.*, 2000, 96(11), 51–63.
17. Stoessel, F. What is your thermal risk? *Chem. Eng. Prog.*, 1993, 89(10), 68–75.
18. Windhorst, J. C. A., Koen, J. A. Economies of scale of world-scale plants and process safety. In *Proceedings of Asia-Pacific Safety*, Kyoto, Japan, 2001.

15

PROCESS INTENSIFICATION

What This Chapter Is About In this chapter we cover some general principles of process intensification and its potential advantages in the framework of green engineering. We will also explore examples of some of the technologies and techniques used for process intensification in the context of green technology.

Learning Objectives At the end of this chapter, the student will be able to:

- Understand the concept of process intensification and its green engineering impacts.
- Identify some of the technologies and techniques used for process intensification and their applications.
- Identify the advantages and disadvantages of process intensification technologies and techniques in the context of green engineering.
- Understand some of the trade-offs of process intensification regarding sustainability.

15.1 PROCESS INTENSIFICATION BACKGROUND

As we saw in Chapter 10, process intensification was first defined by Colin Ramshaw as a strategy to make dramatic reductions in the size of a chemical plant (100 to 1000-fold), either by reducing the size of the equipment or combining multiple operations into fewer pieces of equipment.[1] The concept of process intensification has since been expanded to include increases in production capacity within a given piece of equipment, multi-functional systems, significant decreases in energy consumption, a marked cut in waste or by-product formation, and other advantageous outcomes.[2] Process intensification can be

Green Chemistry and Engineering: A Practical Design Approach, By Concepción Jiménez-González and David J. C. Constable

TABLE 15.1 Potential Green Engineering Advantages and Disadvantages of Process Intensification

Area	Potential Green Engineering Advantages	Potential Green Engineering Disadvantages
Improved mass transfer	Reduction of by-product formation and thus a reduction in waste	
	Increased throughput	
	Faster reactions: equipment tailored to intrinsic kinetic rate	
	Possibility of higher selectivity, mass efficiency, yield, and quality	
Improved energy transfer	Reduction of hot spots, improved control of temperature distribution	
	Reduction of energy requirements and related life cycle impacts (see Chapter 18)	
	Potential reduction in secondary reactions	
Reduced solvent use	Increased mass efficiency	Potential for decreased inherent safety (see Chapter 14)
		Potential for reaction runaway if heat transfer is not sufficient
Reduced size	Reduced inventories of potentially hazardous materials	Potential for blockages during operation
	Reduction or elimination of scale-up issues (see Chapter 10)	Less incentive to eliminate hazardous material
	Might not need frequent cleaning if equipment is dedicated	Potentially more difficult to clean and maintain

achieved through the use of large forces, such as increased pressure, smaller geometry, microfluidic interactions, high pressures, and different types of energy (e.g., magnetic fields, ultrasound, oscillatory forces).

Table 15.1 shows some of the potential advantages of process intensification from a green engineering perspective. It can also be seen from the table that although there are many advantages, there are some aspects that will invariably need to be managed.

Process intensification uses only engineering principles to achieve the desired result, which means that the changes needed to intensify a process will be in "how" the chemistry performs, not in the "what" of the chemistry itself. To achieve the desired intensification, a combination of techniques and equipment can be used. Process intensification techniques use different ways of processing to intensify production, such as the use of multifunctional reactors (e.g., reaction plus separation in one reactor), hybrid separations (e.g., membrane absorption and stripping), use of alternative forms of energy (e.g., centrifugal force instead of gravity for phase separations). Process intensification technology includes microreactors, static mixers, and compact heat exchanger reactors, among others. Figure 15.1 illustrates some areas of process intensification. Of course, there will be interrelations between methods and equipment, as new methodologies will require new equipment, and vice versa.

Although it is true that process intensification leads to smaller, cleaner, safer, more energy-efficient processing plants, and therefore has significant green engineering advantages, the main drive for process intensification has been the need for innovation, either

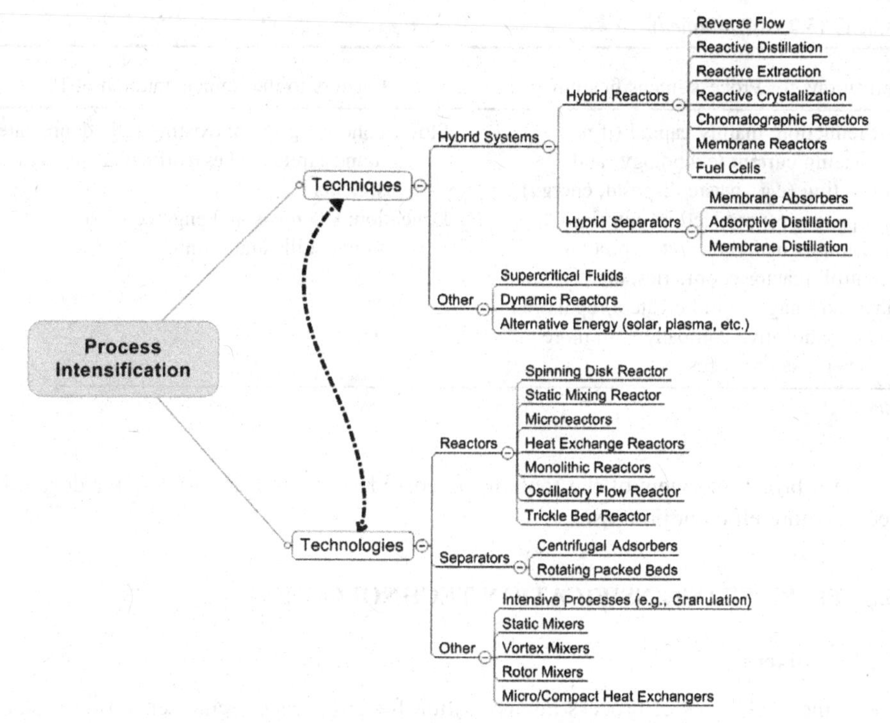

FIGURE 15.1 Some areas of process intensification.

through new technologies or by optimizing existing technologies closer to their full potential. Table 15.2 shows some of the motivations for using process intensification and some of the barriers to adopting this type of engineering.

In the following sections we explore examples of process intensification techniques and technologies and how they may be applied to a given process to make it more sustainable. The technologies and techniques described are by no means comprehensive, but attempt to

TABLE 15.2 Motivations and Barriers to the Implementation of Process Intensification

Motivations for Process intensification	Barriers to the implementation of PI
Novel or better products that cannot be produced through conventional methods	Resistance to change: conservatism among senior engineers and management might require full-scale demonstration before adopton
Improved chemistry: higher mass efficiency, higher yield and quality, lower energy	PI portfolio limitations: main focus of the PI portfolio has been on reactors, heat exchangers, and combination units
Enhanced processing: going from batch to continuous, better control, reduced scale-up issues, just-in-time processing	Existing batch mentality: in some industries, chemist will need to use different equipment for R&D
Low inventories: just-in-time processing, smaller volumes of hazardous materials	Issues with fouling: in some industrial applications, microdevices might be subject to fouling

(continued)

TABLE 15.2 (*Continued*)

Motivations for Process intensification	Barriers to the implementation of PI
Cost reduction: mainly capital (if not replacing current technology) and operating (e.g., waste disposal, energy)	Slow economic growth: existing fully depreciated in-ground capital makes it difficult to justify new investments
Safer processes: reduced inventories of hazardous materials, better process control, reactor geometries, etc.	Education: Chemists and engineers need additional skills and training
Enhanced image: could create a reputation as an innovative company with more efficient, safer processes	

Source: ref. 3.

provide a broad overview of the tools to be considered within the toolkit for designing greener, more efficient processes.

15.2 PROCESS INTENSIFICATION TECHNOLOGIES

15.2.1 Mixers

One of the backbones of process intensification for both reactive and separation systems is mixing. As we saw in Chapters 7 and 10, mixing is a very important factor in chemical reactivity and process conditions within a green engineering/green chemistry context because it helps us to avoid:

- Concentration gradients (reactants uniformly distributed)
- Temperature gradients (desired and uniform temperature)
- Runaway reactions (due to hot spots, localized concentration or temperature gradients)
- Secondary reactions (with by-product formation and reduction in mass efficiency)

Mixing technology has evolved steadily over the last 30 years or so and has been intensified to enable micro- and molecular mixing. We can start with a general view of fluidic mixers, which need no mechanical energy to be input within the mixer itself (although they can have external energy applied to them). In practical terms, this translates to there not being any rotating shafts, seals, or blades. The generally small size and rapid mixing of fluidic mixers offers significant advantages for exothermic and fast reactions compared with mixing in a larger reaction vessel, where the feed rate for reactants and reagents is limited by the heat and mass transfer capacity of the vessel and its contents. Some examples of fluidic mixers are vortex, static, and jet mixers.

Static mixers are small tubes (as small as 0.8 inch in external diameter and 6 inches in length)[4] that have stationary mixing elements positioned such that reaction mixtures are mainly blended through a radial movement [see Figure 15.2(a)]. The mixer design also has a high surface area/volume ratio, which increases heat transfer. These mixers can be used for highly exothermic reactions requiring intensive heat removal and instantaneous mixing, such as nitration or neutralization, since these reactions can be "quenched" instantaneously with rapid and efficient heat removal. In fact, for some static mixers, such as the Sulzer SMR static mixer reactor, the mixing elements are made out of heat transfer tubes, thereby

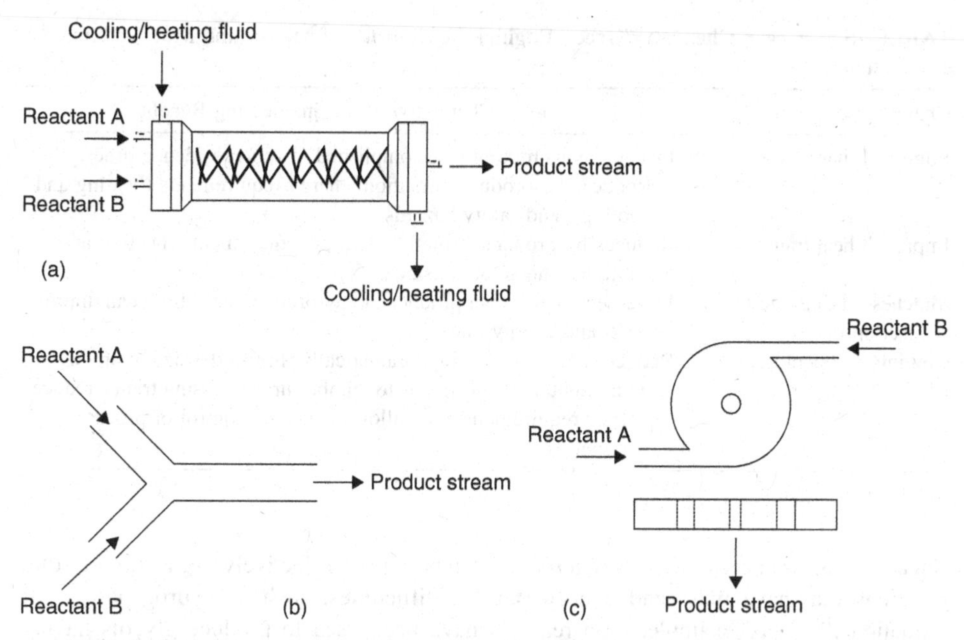

FIGURE 15.2 Illustrative schemes for static (a), Y-shaped jet (b), and vortex (c) mixers.

ensuring rapid heat removal. The mixing elements can also be constructed from catalytic materials, and in that way one may avoid potential clogging that would be encountered for reactions that use slurry catalysts. A good example of where this type of static mixer may be employed is in some gas-phase exothermic reactions that are normally performed in packed bed reactors. However, this type of static mixer array has a relatively low specific area compared with other process-intensification technologies (e.g., monolithic catalytic reactors).[5]

A vortex mixer is constructed as a cylindrical chamber into which liquid or gaseous reagents are injected tangentially, both in the same direction. The reagents, conserving angular momentum and mass flow, accelerate toward the axial outlet [see Figure 15.2(c)]. Because they are accelerating, the streams thin out. This provides excellent conditions for molecular mixing due to the dissipation of turbulent energy and diffusion across concentration gradients.[6,7] Vortex mixers are also used where enhanced temperature control and mass transfer are required, especially where mass transfer is the rate-limiting step.[8,9] As with fluidic mixers in general, one application of vortex mixers will be exothermic and fast reactions. Other examples of applications for vortex mixers are precipitation reactions (where the mixing rate can have a profound effect on the morphology and particle size),[10] highly exothermic reactions such as neutralizations using acid or base addition,[7] fast reactions such as the iodide–iodate system,[11,12] and hydrolysis reactions.[13]

There are, of course, other ways to enhance or intensify mixing besides the use of static and vortex mixers. Some additional approaches, used mainly on the meso- and microscales, take advantage of different geometries using T- or Y-shaped structures and a simple jet injector principle so that when the two jet streams collide, they are mixed rapidly and effectively [see Figure 15.2(b)]. In general, the contact chamber is very small, and this produces short residence times with high flow rates. Such a reactor works well for very fast reactions, but for reactions with rates longer than the residence time in jet mixers, the reactants can be recirculated to better align the reaction kinetics with the residence time.

TABLE 15.3 Green Chemistry/Green Engineering Benefits of Micro- and Macromixing Technologies

Characteristic	Green Chemistry/Green Engineering Benefit
Improved mass transfer	Instantaneous mixing is accomplished in short residence times. Reduces by-product formation, energy requirements (heating and cooling), and safety hazards.
Improved heat transfer	Reduces by-product formation, energy requirements (heating and cooling), and safety hazards.
Matches mixing rate to reaction rate	Increased control over product and by-product formation; can improve mass and energy intensity.
Continuous operation	Reduces need for cleaning, dramatically shrinks the size of processing equipment, and allows one to number-up processing trains, reduces energy requirements, and allows real-time control of process parameters.

Such a reactor is known as a *loop reactor* and can be applied effectively to gas–liquid–solid reactions that have mass and energy transfer difficulties, such as hydrogenations or oxidations.[14,15] For example, loop reactors have been used to produce glycols having specific chain lengths through the reaction of ethylene oxide with the corresponding alcohols and to produce polyaniline through the oxidation of aniline with ammonium persulfate.[22] In this type of high-energy environment, one note of caution when using jet mixers is the impact of collisions on the product or streams (i.e., they can cause agglomeration or deagglomeration in particulate-forming reactions).

Mixing can also be enhanced through the use of external forces, such as mechanical vibrations or vibrations caused by sound (e.g., ultrasound), electromagnetic fields, and others. The main challenge with mixers used for process intensification (and with many types of microprocessing equipment) is the presence of solids. Clogging or blockages can create havoc during operation, and cleaning can be more difficult. A very good summary of the principles of micromixing and the associated technologies has been given by Hessel et al.[16] From a green chemistry/green engineering perspectives some of the advantages of using the mixers described above are described in Table 15.3.

Example 15.1 A Darzens' reaction to convert pharmaceutical intermediate A to pharmaceutical intermediate B is limited by the transfer rate (diffusion) of a hydroxide ion from the aqueous phase into the organic phase. Although the reaction is an extremely rapid one, the mass transfer exchange rate between the aqueous and organic phases in a batch reactor configuration is not maximized, so the reaction kinetics are constrained. A phase-transfer catalyst (PTC) is used to facilitate the diffusion of the hydroxide ion into the organic phase and to increase the reaction rate. Experiments performed without the PTC showed that the conversion to B was extremely low (<10%).[17] Two processes were explored:

Batch process. A PTC is used in a batch process at the manufacturing scale, and the general process is illustrated in Figure 15.3(a). However, carrying out Darzens' reaction in a batch reactor with the PTC results in the formation of an undesirable by-product, since intermediate B continues to react with an excess of the hydroxide ion. The desired product, B, is isolated as a wet cake, washed, and dried prior to the next reaction.

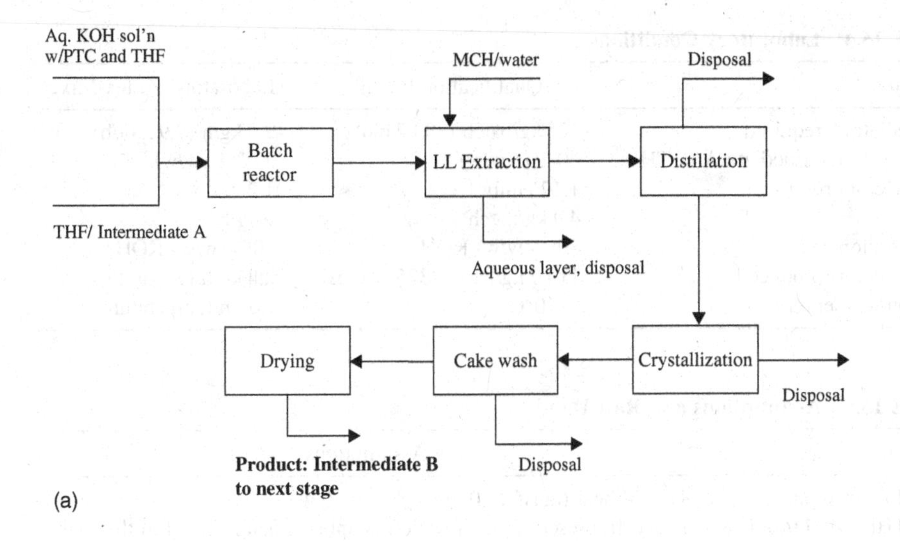

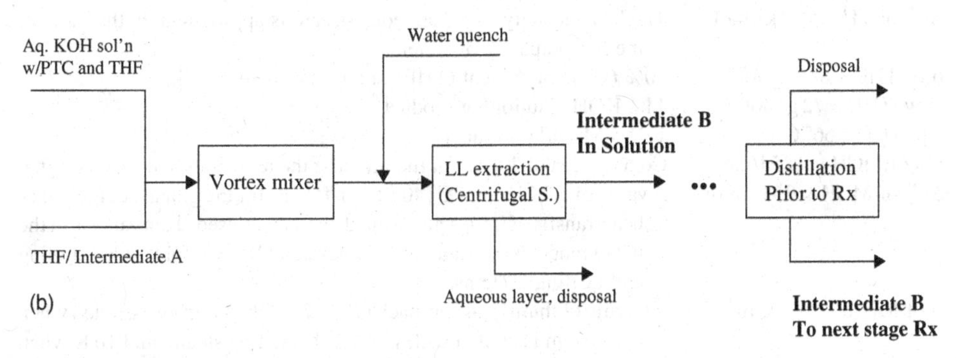

FIGURE 15.3 Batch process (a) and continuous process using a vortex mixer (b).

Continuous process with vortex mixer. A continuous process using a vortex mixer was also investigated. The block flow diagram for a continuous process using a vortex mixer is shown in Figure 15.3(b). No methylcyclohexane (MCH) is added in this process because of the faster kinetics resulting from the improved phase separation encountered in the vortex mixer. To make the process continuous, a continuous centrifugal separator was used for the extraction of product with tetrahydrofuran (THF). In addition, the THF must be removed before the next reaction can take place in the following stage.

Laboratory conditions for the vortex mixer and batch conditions are given in Table 15.4. Estimate and compare the energy requirements for the two systems.

Solution

General data used and assumptions. The input mass streams were estimated from the information in Table 15.4 and process descriptions [8,9] and the THF/KOH and THF/MCH ratios. The THF/KOH solution ratio is 4.8 and THF/MCH ratio is 2.6. The data and assumptions used appear on Table 15.5.

TABLE 15.4 Laboratory Conditions

Parameter	Qualification Batch	Laboratory Vortex Mixer
Intermediate A required	125 kg/batch (393.7 mol)	2.5 kg/h (7.9 mol/h)
Weight % of intermediate A in THF	30.2% (w/w)	27.3% (w/w)
Chloroacetonitrile use	1.12 equiv.	1.2 equiv.
PTC use	4.4 kg/batch	90 g/h
Concentration of base	30% (w/w) KOH	30% (w/w) KOH
Intermediate B produced	114.7 kg/batch (325.8 mol)	2.4 kg/h (7.1 mol/h)[2]
Reaction temperature	0–10°C	Room temperature

TABLE 15.5 Assumptions and Raw Data

Raw Data	Assumptions
Basis: 1 kg of product	85% of heat transfer efficiency
C_{pl} of THF = 124 J/mol	The THF present in the reaction is approximately equal to the THF distilled in the current batch reaction.
ΔH_v of THF = 32 kJ/mol	The heat capacity of the aqueous stream is approximately the same as the heat capacity of water.
ρ of THF = 886 kg/m^3	30% (w/w) of solvent (THF) in the intermediate cake
m.w. THF = 72 g/mol	1 kg KOH solution/kg product
b.p. THF = 66° C	1.9 kg MCH/kg product
C_{pl} of MCH = 185 J/mol	Contact area of 3.8 m^2 for maintaining the temperature during cooling.
ΔH_v of MCH = 35 kJ/mol	A value of 5.77 J/s·m^2·°C [40 Btu/hr-ft^2-°F] for the global coefficient of heat transfer (U) for a glass-lined vessel was used. This value is in the mean range of the values of U reported in Perry[18] for various cooling and heating systems.
ρ of MCH = 769 kg/m^3	As a rule of thumb, an approach ($T_w - T$) of 20 K can be assumed when the system is cooled with water or heated by steam, and 10 K when the system is cooled with refrigerant.[19]
m.w. MCH = 98 g/mol	
b.p. MCH = 101 °C	
C_p water = 4.18 J/g·°C	

For current batch reactor system

Cooling requirements: The reaction mixture is cooled to 0 °C for 2 h (1 h charging and 1 h reaction).

(a) To cool the reaction mixture:

$$\Delta H = \text{sensible heat} = \frac{1}{\eta}\left(\sum_i^n m_i C_p \, \Delta T\right)$$

$$= \frac{1}{0.85}\left[1\,\text{kg KOH sol'n}(4.18\,\text{J/g}\cdot°\text{C}) + \frac{4900\,\text{g THF}}{72\,\text{g/mol}}(0.124\,\text{kJ/mol}\cdot°\text{C})\right](0\text{–}25)°\text{C}$$

$$= -371\,\text{kJ/kg product} = -0.37\,\text{MJ refrigeration/kg product}$$

(b) To maintain the temperature for 2 h:

$$Q = \frac{UA(T_w - T)t}{\eta}$$

where:
Q = heat added (kJ)
A = energy transfer area (m^2)
T_w = temperature of cooling or heating medium (K)
T = temperature of the system (K)
t = time (h)
U = global coefficient of heat transfer ($kJ/h \cdot m^2 \cdot K$)
η = heat transfer efficiency (85%)

$$Q = UA(T_w - T)t \qquad \text{(in this case refrigeration)}$$

$$= (5.77 \times 10^{-6}\ \text{MJ/s} \cdot \text{m}^2 \cdot {}^\circ\text{C})(3.8\ \text{m}^2)(-10^\circ\text{C})(7200\ \text{s})\left(\frac{1\ \text{batch}}{114.7\ \text{kg product}}\right)$$

$$= -0.02\ \text{MJ/kg product}$$

Condense the THF distilled:

$$\Delta H = -(\text{sensible heat} + \text{vaporization heat}) = -\frac{1}{\eta}\left(\sum_i^n m_i Cp_i\ \Delta T + m_v \Delta H_v\right)$$

$$= \frac{1}{0.85}\left\{\frac{4900\ \text{g THF/kg product}}{72\ \text{g/mol}}(0.124\ \text{kJ/mol THF})(65-25)^\circ\text{C}\right.$$

$$\left. + \left[\frac{4900\ \text{g THF/kg product}}{72\ \text{g/mol}}(32\ \text{kJ/mol THF})\right]\right\}$$

$$= -2558 - 406\ \text{kJ/kg product} = -2954\ \text{kJ/kg product} = -2.95\text{MJ/kg product}$$

Cool the crystallization feed from the distillation temperature to 0 °C:

$$\Delta H = \text{sensible heat} = \frac{1}{\eta}\left(\sum_i^n m_i Cp_i\ \Delta T\right)$$

For cooling water (down to 25°C)

$$\Delta H = \frac{1}{0.85}\left[\frac{1900\ \text{g MCH}}{98\ \text{g/mol}}(0.185\ \text{kJ/mol} \cdot {}^\circ\text{C}) + \frac{100\ \text{g THF}}{72\ \text{g/mol}}(0.124\ \text{kJ/mol} \cdot {}^\circ\text{C})\right](25-66)^\circ\text{C}$$

$$= -181\ \text{kJ/kg product} = -0.18\ \text{MJ cooling water/kg product}$$

For refrigeration (from 25°C to 0°C)

$$\Delta H = \frac{1}{0.85}\left[\frac{1900\text{ g MCH}}{98\text{ g/mol}}(0.185\text{ kJ/mol}\cdot{}^\circ\text{C}) + \frac{100\text{ g THF}}{72\text{ g/mol}}(0.124\text{ kJ/mol}\cdot{}^\circ\text{C})\right](0-25)^\circ\text{C}$$

$$= -110\text{ kJ/kg product} = -0.11\text{ MJ refrigeration/kg product}$$

Total cooling requirements

$$\text{cooling water} = -2.95-0.18 = -3.23\text{ MJ/kg product}$$

$$\text{refrigeration} = -0.37-0.02-0.11 = -0.50\text{ MJ/kg product}$$

Heating requirements

Distill the THF from the reaction mixture, which is equivalent to condensing.

$$\Delta H = \text{sensible heat} + \text{vaporization heat} = \frac{1}{\eta}\left(\sum_i^n m_i C_{pi}\,\Delta T + m_v\,\Delta H_v\right)$$

$$= \frac{1}{0.85}\left\{\frac{4900\text{ g THF/kg product}}{72\text{ g/mol}}(0.124\text{ kJ/mol THF})(65-25)^\circ\text{C}\right.$$

$$\left. + \left[\frac{4900\text{ g THF/kg product}}{72\text{ g/mol}}(32\text{ kJ/mol THF})\right]\right\}$$

$$= 2558 + 406\text{ kJ/kg product} = 2954\text{ kJ/kg product}$$

$$= 2.95\text{ MJ/kg product}$$

Dry the intermediate cake. Assume that the energy required = the energy required to evaporate the solvent in the cake.

$$\Delta H = \text{vaporization heat} = \frac{1}{\eta}(m_v\,\Delta H_v)$$

$$= \frac{1}{0.85}\left[1000\text{ g product in cake}\left|\frac{30\text{ g THF in cake}}{70\text{ g product in cake}}\right|\frac{32\text{ kJ/mol THF}}{72\text{ g/mol}}\right]$$

$$= 223\text{ kJ/kg product} = 0.22\text{ MJ/kg product}$$

$$\text{total heating requirements} = 2.95 + 0.22 = 3.17\text{ MJ/kg product}$$

Electricity requirements

Agitate the batch reactor. Using the heuristics of 5 hp/1000 gal (9.85×10^{-4} kJ/L·s), a reaction time of about 2 h (1 h charging, 1 h reaction), 114.7 kg product/batch, and 750 L of reaction volume per batch, we have the energy for mixing:

$$(9.85 \times 10^{-4}\ \text{kJ/L}\cdot\text{s})(7200\ \text{s})\frac{750\ \text{L}}{\text{batch}}\left|\frac{1\ \text{batch}}{114.7\ \text{kg product}}\right|$$

$$= 46.4\ \text{kJ/kg product} = 0.05\ \text{MJ/kg product}$$

Operate the centrifuge. Assuming a 30-inch basket centrifuge of 10 hp given in *Perry's Handbook* with a centrifugation time of 2 h per batch and a production rate of 114.7 kg product/batch, we have

$$(7.46\ \text{kJ/s})\left|\frac{7200\ \text{s}}{1\ \text{batch}}\right|\frac{1\ \text{batch}}{114.7\ \text{kg product}} = 468\ \text{kJ/kg product} = 0.47\text{MJ/kg product}$$

$$\text{total cooling requirements} = 0.05 + 0.47 = 0.53\ \text{MJ/kg product}$$

For vortex mixer system

Heating requirements: To distill THF prior to the next reaction step:

$$\Delta H = \text{sensible heat} + \text{vaporization heat} = \frac{1}{\eta}(m_i C_{pi}\Delta T + m_v \Delta H_v)$$

$$= \frac{1}{0.85}\left\{\frac{900\ \text{g THF/kg product}}{72\ \text{g/mol}}(0.124\ \text{kJ/mol THF})(65-25)^{\circ}\text{C}\right.$$

$$\left. + \left[\frac{900\ \text{g THF/kg product}}{72\ \text{g/mol}}(32\ \text{kJ/mol THF})\right]\right\}$$

$$= 444 + 70\ \text{kJ/kg product} = 514\ \text{kJ/kg product} = 0.51\ \text{MJ/kg product}$$

Cooling requirements: To condense the distilled THF, assume that the energy required is equivalent to the heating requirements = 0.51 MJ/kg product.

Electricity requirements: According to information provided in *Perry's Handbook*, a centrifugal separator of 16 × 12 × 30 inches operating continuously at its normal capacity will require about 35 kJ/L processed, so approximating the kg processed/kg product yields

$$(35.5\ \text{kJ/L})(10.2\text{L/kg product}) = 362\frac{\text{kJ}}{\text{product}} = 0.36\ \text{MJ/kg product}$$

In summary, the energy requirements estimated for both systems are shown in Table 15.6.

Additional Points to Ponder How is refrigeration achieved? What other process configurations might be used to avoid the problems that we encountered with a batch reactor? Why was the distillation of the solution containing intermediate B included in the energy calculations?

TABLE 15.6 Energy Requirements (MJ/kg Intermediate B)

Energy Requirements	Vortex Mixer	Batch Reactor
Heating	0.51	3.17
Cooling with cooling water	−0.51	−3.23
Refrigeration	0	−0.5
Electricity	0.36	0.53

15.2.2 Reactors

As we saw in Example 15.1, some of the mixing elements used for process intensification can be used directly for reaction technology. We will now cover some additional reactors used for process intensification as a means of illustrating some of the technologies shown in Figure 15.1. However, our discussion here is not meant to be comprehensive, as process intensification is an ever-expanding field, and new technologies and geometries are being developed constantly. It would be helpful to consult a few of the very useful summaries of process intensification technologies that have been published to date.[20–23]

Microreactors Microreactors are probably the most common reactors used for process intensification. Microreactor devices are generally defined as miniaturized reaction systems fabricated by microtechnology and precision engineering. The reactions take place in channels or chambers whose dimensions are on the order of micrometers (see Figure 15.4). Several types of geometries are used in microreactor designs, two of the most commonly used being the microchannel and the microplate structures. If necessary, these can also be used as direct heat exchangers. In either case, the geometries for both the microplate and the microchannel are stacked cross-flow structures that have channels, slots, grooves, or other desired flow pathways for the reactants and heat exchange media. Given this configuration, they may also be used for serial endothermic and exothermic reactions in the same plate reactor.[22]

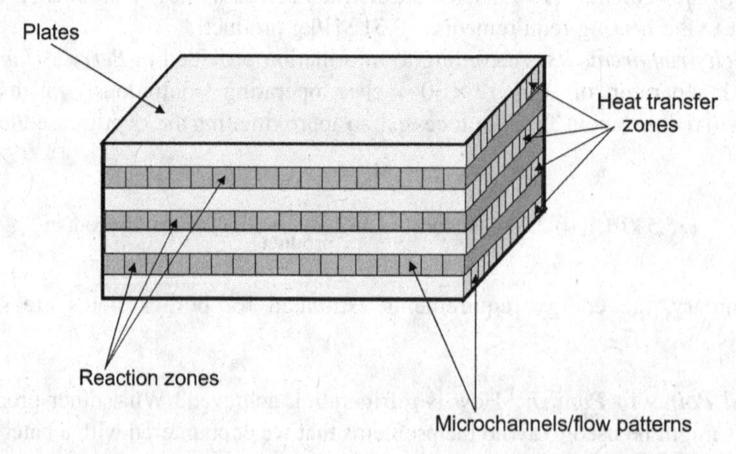

FIGURE 15.4 Microchannel reactor. The reaction zones are darker than the heat transfer zones. This is also known as a typical HEX reactor.

Microfabrication techniques have fueled advances in microreactor designs in recent years. Some of these reactors are fabricated using processes very similar to those used to manufacture computer silicon wafers, and include such approaches as deep-x-ray lithography, nonlithographic micromachining, and silicon micromachining. Although the world market for microreactors was still small at the time this chapter was written, about 20 to 30 chemical plants are running with micro- or minireactions: with stand-alone microreactor modules comprising about 50% of the market and the remaining half being integrated systems bought mainly by fine-chemical companies. Major known microtechnology players at this time include the Institute für Mikrotechnik Mainz (IMM) and its spin-offs—CPC, Mikroglas, Ehrfeld BTS—plus some others.[23]

The high surface/volume ratio of the mixing section in combination with the temperature-controlled reaction channels allows heat transfer coefficients up to 20,000 $W/m^2 \cdot K$, which suggests that these systems work best with fast reactions, primarily on the order of minutes, seconds, or less.[24–26] This would potentially allow for more aggressive reaction conditions at higher yields than those of conventional larger reactors.[27,28] Some of the mixers we saw above, such as the vortex or static mixers, can be used as the microreactor or as part of a microreactor system.

A few microreactor examples include the production of ethylene oxide (highly exothermic reaction) in a poly(methyl methacrylate)–nickel microreactor with a silver catalyst,[29] Pd-catalyzed hydrogenation of alkenes and alkynes,[30] photoreactions,[31] enzymatic reactions,[32] synthesis of cyanobiphenyls using a modified Suzuki reaction,[33] high-temperature HCN generation,[34] catalytic partial oxidation of methane to syngas,[35] and selective oxidations and hydrogenations.[36] Table 15.7 provides some potential advantages and disadvantages of microreactors from a green engineering perspective.

Example 15.2 The following reaction is carried out between a carbonyl compound and an organometallic agent to produce a fine chemical:[28]

TABLE 15.7 Potential Advantages and Disadvantages of Microreactors

Advantages	Disadvantages
Improved heat transfer	Unwanted precipitation of solids may lead to clogging and fouling
Improved heat removal: less extreme temperatures	
Improved control of residence time and temperature distribution	Low velocities (laminar flow with parabolic profiles)
High operating pressures (~20 bar)	Micromixing may not occur
Alignment with the kinetics of the reaction	
Reduced possibility for side reactions, higher selectivity and yield	
Modular nature: numbering-up instead of scaling-up (see Chapter 10)	
Compact, continuous operation.	

TABLE 15.8 Reactor System Characteristics

Reactors Type	T (°C)	Residence Time	Yield (%)	Volume/ Area (m^2/m^3)	Dimensions
Flask	−40	0.5 h	88	80	0.5 L
Stirred vessel (production)	−20	5 h	72	4	6000 L
Microreactor	−10	<10 s	95	10,000	2 × 16 channels of w × h = 40 × 220 μm

This reaction proceeds in the liquid phase and it is exothermic (standard heat of reaction ca. −300 kJ/mol), the main reaction and most side reactions are fast (<10 s), some parallel and consecutive reactions can occur, and the compounds are sensitive to temperature. Experiments were carried out in microreactors and were compared with laboratory- and full-scale operations. The characteristics of the reactor systems evaluated are shown in Table 15.8.

Given the limited data in the table:

(a) Compare green mass and energy metrics for the three reactor systems above.
(b) Compare the reactors from a green engineering perspective.

Assume that the reactants are fed into the system at an equimolar ratio.

Solution

Mass metrics. The main aspects to be considered from a mass perspective are the mass efficiency and the mass of unwanted products. Since there were not enough data to estimate the relative solvent use, and there is no basis to sustain an assumption at the present time, the mass intensity will need to exclude solvent use and will be expressed on a molar basis instead of on a mass basis, as no molecular weights are given. From the yield information given, we obtain Table 15.9. Unfortunately, the information given does not include the mechanism of by-product formation (e.g., concurrent vs. subsequent reactions), so it would not be fair with the information at hand to declare the nonformation of by-products as a general characteristic or as a benefit of microreactors. Furthermore, since no conversion of reactants is given, it is not possible to estimate how much by-product is

TABLE 15.9 Metric Comparison

Mass Metrics	Flask 0.5-L	6000-L Vessel	Micro-Reactor	Theoretical Minimum
Mass intensity (does not include solvent use) (total moles in/moles product)	2.27	2.78	2.10	2.00
Efficiency (yield, %)	88	72	95	100

formed in each case. The concept of less production of by-products is considered important, but because there is insufficient information to establish a reasonable assumption, it is only treated qualitatively.

Energy metrics. To compare the energy requirements of the various systems, we may use the electricity for refrigeration, since there are no heating requirements (exothermic reaction) and we know that the temperatures of one of the reactors goes below −23°C. The following formula is used to estimate the electricity required for refrigeration:

$$W = \frac{Q}{\eta}\frac{T_1 - T_2}{T_1}$$

where

W = work (electrical energy), MJ

Q = heat to be dissipated by the refrigeration system, MJ

T_1 = outlet temperature of the system to be cooled, K

T_2 = inlet temperature of the system to be cooled, K

η = efficiency of the system (common value = 50%)

It is expected that the major contribution to the refrigeration requirements is the dissipation of the reaction energy, thus the standard heat of reaction is taken as an approximation of the total heat to be dissipated. Therefore, for the systems analyzed we have:

$$\text{0.5-L flask:}\quad W = \frac{Q}{\eta}\frac{T_1 - T_2}{T_1} = \frac{0.3\ \text{MJ}}{0.5}\left(\frac{25+40}{233}\right) = 0.167\ \text{MJ/mol product}$$

$$\text{6000-L vessel:}\quad W = \frac{Q}{\eta}\frac{T_1 - T_2}{T_1} = \frac{0.3\ \text{MJ}}{0.5}\left(\frac{25+20}{253}\right) = 0.107\ \text{MJ/mol product}$$

$$\text{Microreactor:}\quad W = \frac{Q}{\eta}\frac{T_1 - T_2}{T_1} = \frac{0.3\ \text{MJ}}{0.5}\left(\frac{25+10}{263}\right) = 0.080\ \text{MJ/mol product}$$

Based on the mass and energy metrics evaluated and the data given, one can a priori conclude that micro-reactors have more advantages than full-size reactors because:

- The reaction is performed at less extreme temperatures, reducing the energy requirements.
- The reaction is more mass efficient.
- The residence time is reduced from hours to seconds, which reduces the baseload energy and could explain the increase in mass efficiency, as it reduces the potential for secondary reactions.

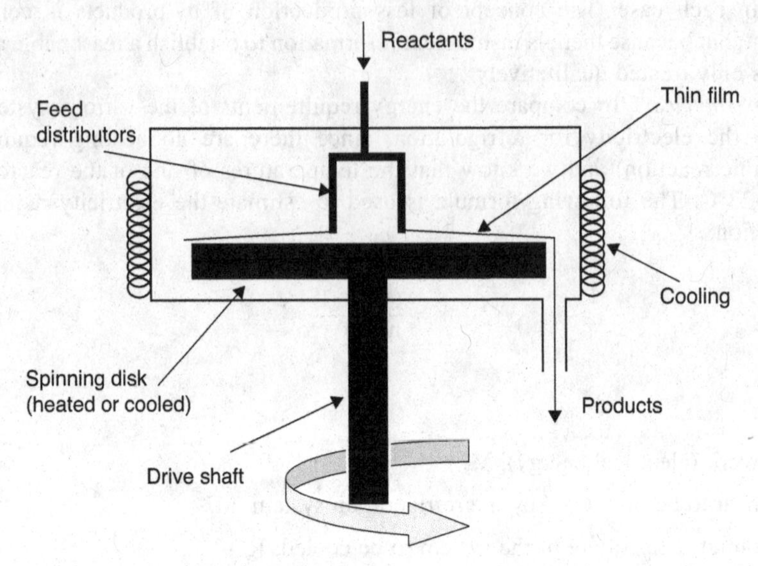

FIGURE 15.5 Spinning disk reactor.

- The reactor volume is reduced considerably, which reduces scale-up issues, and could be inherently safer with smaller hold-up volumes of hazardous chemicals (see Chapter 10).

Additional Points to Ponder What disadvantages might you imagine with this design? What additional data would you need to complete the assessment fully?

Spinning Disk Reactor This reactor is one of the original process intensification technologies developed by Ramshaw and has undergone several modifications and improvements over the years. The general concept of a spinning disk reactor is that of a disk rotating at high speed with the reactants fed at the center of the disk, from where they flow outward due to centrifugal force. The exiting fluid is either cooled or heated through the chamber wall (see Figure 15.5). As the disk is spinning rapidly (in the neighborhood of 1000 rpm), the reactant solutions are dispersed as a very thin, highly sheared film. The disk can have a smooth surface or one can engineer a given flow path that is designed to the target fluid dynamic for the specific mixing desired.

The design of a spinning disk reactor leads to enhanced mass and energy transfer and is of special interest when performing gas–liquid–solid contacting operations, as the reactants and phases are in intimate contact during operation. This design also has the advantage over microreactors in that it can handle viscous fluids or slurries and is resistant to clogging given the high shear of the moving disk.[37] Some of the characteristics, advantages, and disadvantages of a spinning disk reactor are presented in Table 15.10. The spinning disk reactor therefore has been targeted primarily for use with very fast reactions needing large heat dissipation, such as nitrations, sulfonations, phase-transfer Darzen processes, crystallizations, exothermic condensations,[38] and polymerizations.[39]

TABLE 15.10 Some Characteristics, Advantages, and Disadvantages of Spinning Disk Reactors

Characteristics	Advantages	Disadvantages
Able to handle heat fluxes of up to 100 kW/m^2·K	Variable rotational speed	Low throughputs
Typical mass transfer coefficients of 0.01–0.03 cm/s for low-viscosity fluids	Enhanced mass transfer	Rotating parts at high speed might pose design and basic safety challenges
Typical liquid residence times 1–5 s	Enhanced heat transfer: ability to cope with very exothermic reactions	
Liquid film thickness 50–200 nm	Low inventory hold-up: potentially inherently safer (see Chapter 14)	
Radial speed 600–1200 rpm	Improved control due to short residence time	
	Relatively easy to clean	
	Good resistance to clogging	

Example 15.3 Oxley et al.[38] reviewed several pharmaceutical reactions to test the suitability of using a spinning disk reactor. One of the reactions reviewed was Darzens' reaction using a phase-transfer catalyst (benzyltriethylammonium chloride):

$$R_1C(O)R_2 + ClCH_2CN \xrightarrow[BnEt_3NCl]{KOH,\ THF} \text{epoxide}(CN, R_1, R_2)\ (\text{desired}) + \text{epoxide}(CONH_2, R_1, R_2)\ (\text{impurity})$$

desired | impurity

Oxley et al. performed tests in batch reactors and spinning disk reactors. They estimated that a 15-cm-diameter spinning disk reactor will have a throughput of about 8 tons/yr, an amount that is equivalent to a batch reactor but with a reduction in reactor volume of 99%. Heat transfer coefficients of about 4 kW/m^2·K were modeled. The results obtained are shown in Table 15.11. Identify advantages from a green chemistry/green engineering perspective.

TABLE 15.11 Comparison Test

Parameter	Batch Process	Spinning Disk Reactor
Reaction/contact time	30 min to 2 h	1 s
Reaction temperature	0°C	20°C
Impurity level	~ 1.5%	~ 0.1%

Solution In comparison to the batch reactor, the spinning disk reactor offers significant advantages from a green engineering perspective:

- The reaction is performed at ambient temperature, which will reduce the energy requirements.
- The reaction time is reduced by 99.9%, which reduces complexity and baseload energy, and potentially increases mass efficiency as the potential for secondary reactions is reduced.
- Relatively high heat transfer coefficients mean effective cooling and lower energy needs.
- The reactor volume is reduced by 99%, which reduces or eliminates scale-up issues, reduces the complexity of the reaction, and is inherently safer, as it will hold smaller volumes of potentially hazardous materials (see Chapter 10).
- The impurity level is reduced by 93%, which indicates an increase in mass efficiency. It may also reduce or eliminate the need for separations and purifications (at less than 0.1% there might not be a need for additional purifications).

Additional Points to Ponder What disadvantages are there with this design? How could the spinning disk reactor be further optimized?

Spinning Tube-in-Tube Reactor The spinning tube-in-tube reactor (STT), Figure (15.6) was developed by Holl Technologies Co. In this reactor a cylinder (a rotor) turns inside a second cylinder (a stator). Rotor–stators have an annular gap of about 5 mm or more between the outside wall of the rotor and the inside wall of the stator. While the rotor is rapidly spinning, liquids, solids, or gases enter the annular space, where they encounter shear forces that can create mixtures, pastes, slurries, or emulsions, depending on the machine's design.

The concept on which the STT is based, the rotor–stator concept, is not new. Rotor–stators have been used to incorporate ingredients for food processing, the production of paints, in the deagglomeration of solids, and in premixing chemicals. However, the key feature of the STT is to be able to control the reaction fluid dynamics by shifting them from a volume-based flow

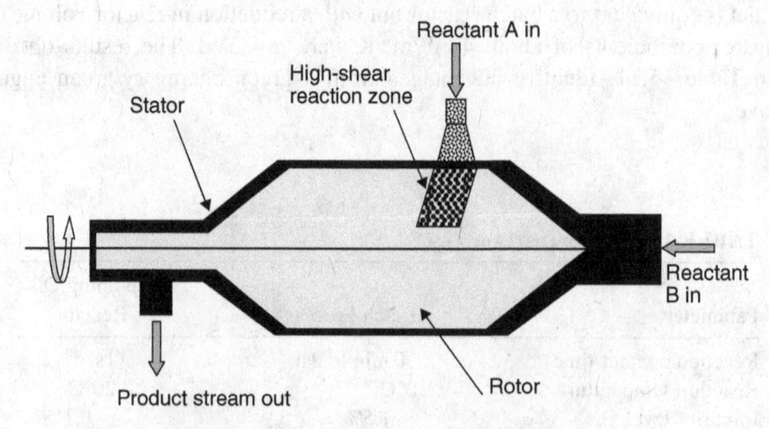

FIGURE 15.6 Spinning tube-in-tube reactor.

to an area-based flow by reducing the annular gap to a smaller dimension (about 0.25 mm). This allows the reactor to achieve almost instantaneous micromixing of the reactants at the molecular level as they enter the very high shearing area while the rotor spins with an angular velocity of several thousand rpm. The size of these reactors can vary from desktop units to units that are about 15 ft long.[40] Heat transfer for this reactor is also enhanced, as heat transfer coefficients of up to 10 $kW/m^2 \cdot K$ have been reported.[22] The reactor may be designed to have separate temperature zones, including different temperature profiles across the zones to aid in overall temperature control. The residence time is controlled by the feed rates and the rotor speed.[41]

The STT can be used in liquid–liquid, gas–liquid, gas–gas, and solid–liquid reactions (solids need to be fed in slurries). This reactor can work with highly viscous liquids such as in polymerization reactions. Also, when the stator is made out of glass, photocatalyzed reactions such as photo-brominations and photo-chlorinations can take place. In some instances, phase-transfer catalysts can be eliminated completely given the high shear mixing rates. Some additional applications reported are biodiesel production,[42] fermentation, and reactions such as condensation, addition–elimination, substitution and rearrangement. Some of the disadvantages of this reactor are the potential fouling of the surfaces and the maintenance of the rotating equipment.[20]

Oscillatory Flow Reactor An oscillatory flow reactor (OFR) is a continuous reactor in which tubes are fitted with orifice plate baffles or smooth periodic cavities, and an oscillatory motion is superimposed externally upon the net flow of the process fluid (see Figure 15.7). The flow patterns have a relatively complicated eddy mixing pattern provided by the baffles and the oscillatory motion. An external oscillator provides the movement at a range of 0.5 and 15 Hz at an amplitude of 1 to 100 mm.[22] This frequency and amplitude control the intensity of the movement within the reactor. The combination of fluid oscillations and baffle inserts promotes highly effective mixing due to the fluid turbulence within the reactor.

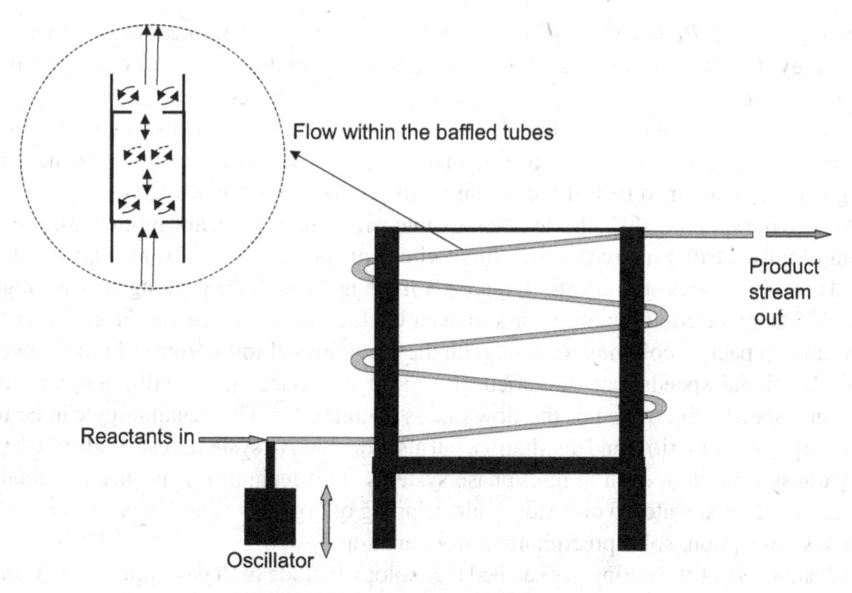

FIGURE 15.7 Oscillatory flow reactor.

This type of reactor tends to be more flexible than tubular continuous reactors. In the case of a tubular continuous reactor, the residence time and net mass flow rate are optimized, which in turn optimizes mixing conditions (i.e., fluid turbulence). In the case of an OFR, the residence time and plug flow performance are optimized, hence heat and mass transfer are totally independent of the net mass flow rate through the tubes. This means that ideal mixing conditions can be achieved at very low net mass flow rates, thereby providing a wider range of operating conditions and residence times while maintaining a very small footprint and high throughputs.

Some additional advantages of the OFR from a green engineering perspective are that this reactor is highly compact, has a continuous parallel tube reactor configuration, has good control over mixing, temperature profiles, solids suspension and size distribution, and has very good heat transfer characteristics. A disadvantage of this type of reactor is the potential for backmixing given the type of fluid dynamics, and this can lead to reduced reaction rates.

Some applications for OFRs have included the laboratory-scale production of biodiesel,[43] biotechnological applications,[44] isolation of active pharmaceutical ingredients by crystallization (i.e., cooling, antisolvent addition, reaction precipitation), where high throughput is desired, and continuous reactions in multiphase reaction systems, such as liquid–liquid, solid–liquid, and gas–liquid systems.[45]

15.2.3 Separators

So far we have covered a few examples of different reactor technologies. As can be seen in Figure 15.1, there are still many opportunities to develop process intensification-specific separation technologies, although most of the advantages in the separation area come not so much from the technology, but from the techniques and methodologies we cover in the next section. We now cover two examples of separation technology designed and used for process intensification: the rotating packed bed and continuous liquid–liquid extractors.

Rotating Packed Bed The use of rotational energy is not limited in use to reactor technology. One of the most developed separation technologies at reduced scale is the rotating packed bed, also known as high-gravity (HIGEE) contactor technology. This technology was originally developed at ICI Imperial Chemical Industries by Ramshaw as a spinoff from a NASA research project in microgravity. It is not terribly surprising that its design is very similar to that of the spinning disk reactor (see Figure 15.8)

As with the spinning disk, this technology intensifies mass, heat, and momentum transfer through high centrifugal forces in the neighborhood of 1000g.[2,20,46,47] In gas–liquid systems, liquid residence times are very short, typically moving through the packing at an average of 1 m/s.[48] The reported volumetric mass transfer coefficients are up to 100 times larger than conventional packed columns, varying with the gravitational force from 0.14 to a power of 0.54. Rotational speeds can be varied and mean residence times will decrease as the rotational speed is increased or the flow rate is decreased.[49] This technology can be used for absorption, extraction, and distillation, but also for reactive systems, especially if they are mass transfer–limited, even in multiphase systems. This technology has been applied for deaeration of flood water in oil fields,[50] air stripping of organics,[20] and in general reactions with gas absorption, solid precipitation, or ozonation.[22]

Advantages of the rotating packed bed technology include high gas–liquid mass transfer rates and fast residence times. As flow through the packed bed is regulated through

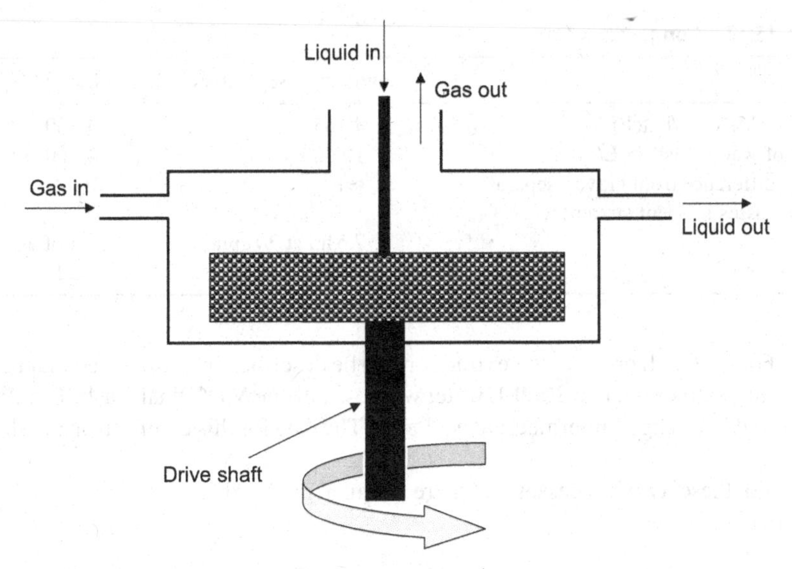

FIGURE 15.8 Rotating packing bed.

differences in rotational speed, and packed beds can sometimes suffer from nonuniform packing; the combination can lead to lower separation efficiencies.

Centrifugal Liquid–Liquid Extractors In centrifugal extractors, contact time between different phases is reduced and phase separation is accelerated through the application of centrifugal forces, which greatly exceed gravitational forces. CLLE units are normally compact with relatively high throughput per unit volume. For example, *Perry's Handbook* has a model that operates at 30 gpm (about 113.6 L/min) for an extractor that is 55 inches wide, 33 inches high and 67 inches long.[18]

Centrifugal separators are particularly useful in systems where the compounds being extracted are chemically unstable (e.g., in the extraction of antibiotics), or for systems in which the phases are slow to settle by gravity. Also, in contrast with the gravity separators, centrifugal separators can be operated continuously, they require less space, handle emulsified material relatively easy, and work with systems with little liquid density difference.[51–53]

CLLE are useful for liquid-phase systems that are prone to forming emulsions, since the centrifugal force reduces the loss of water in solvent and solvent in water. Although there will be advantages to using CLLEs with most any solvent system, the advantages will be less dramatic for systems that include water-miscible solvents (e.g., ethyl acetate). CLLE are useful in processes having many extractions or in cases where the extraction is a bottleneck in the process, as CLLE can improve throughput. The limitations of CLLE are that they cannot handle solids or slurries and have a limited ability to handle liquid–gas separations. CLLEs have been used extensively for commercial pharmaceutical fermentation processes to extract the active ingredient from the fermentation broth. They have also been used successfully for a few chemical synthesis processes; in one instance the use of CLLE resulted in an annual waste reduction of about 66% and an annual cost reduction of £500,000.[45]

Example 15.4 A comparison of centrifugal and gravity separators was completed for the extraction of a pharmaceutical intermediate. In this process, CLLE is used to separate mother liquors containing methylene chloride (MDC, specific gravity of 1.33) and water

TABLE 15.12 Comparison Data

	Centrifugal Separators	Gravity Separators
Volume of MDC (L/batch)	4,165	4,900
Volume of water washes (L/batch)	10,000	4,000
Yield (% difference from gravity separators due to losses in spent stream)	≈10	0
Power use	7.5 hp at 30 gpm	2 h of agitation
Process time (h)	≈12	≈22

washes. For the batch process, the extraction can be described in a simplified manner as an MDC–water extraction, two 2000-L water washes, and one MDC final wash. The standard batch output is 179 kg of intermediate per batch. The data for this comparison are shown in Table 15.12.

How do these two processes compare from a green chemistry green engineering perspective?

Solution A quick calculation of the mass intensity gives the data in Table 15.13. In addition, these two alternatives require electrical energy to operate (in the case of the gravity separator, the energy for the mixer). For the gravity separators, around 2 h of agitation is needed, and the holdup volume in each wash is about 6500 L (including mother liquors and each 2000 L of wash water). To estimate mixing power, a rule of thumb of 10 hp per 10,000 gal mixed is used. Therefore, we have:

For the gravity separator:

$$6500\,\text{L}\left(\frac{10\,\text{hp}}{10,000\,\text{gal}}\right)\left(\frac{1\,\text{gal}}{3.875\,\text{L}}\right)\left(\frac{0.746\,\text{kJ/s}}{1\,\text{hp}}\right)\left(\frac{3600\,\text{s}}{1\,\text{h}}\right)\left(\frac{2\,\text{h}}{\text{batch}}\right)\left(\frac{1\,\text{batch}}{179\,\text{kg stage product}}\right)$$

$$= 50.3\frac{\text{kJ}}{\text{stage product}} = 0.050\,\text{MJ/kg stage product}$$

For the CLLE:

$$7.5\,\text{hp}\left(\frac{0.746\,\text{kJ/s}}{1\,\text{hp}}\right)\left(\frac{3600\,\text{s}}{1\,\text{h}}\right)\left(\frac{12\,\text{h}}{\text{batch}}\right)\left(\frac{1\,\text{batch}}{179\,\text{kg stage product}}\right)$$

$$= 1350\frac{\text{kJ}}{\text{stage product}} = 1.35\,\text{MJ/kg stage product}$$

TABLE 15.13 Mass Intensity Data

Mass Indicators	Centrifugal	Gravity
Mass intensity (kg/kg intermediate)	36.5	58.8
Wastewater intensity (kg/kg stage product)	5.6	22.4
Reduction of chemical losses, (% reduction compared to the traditional method)	10	0

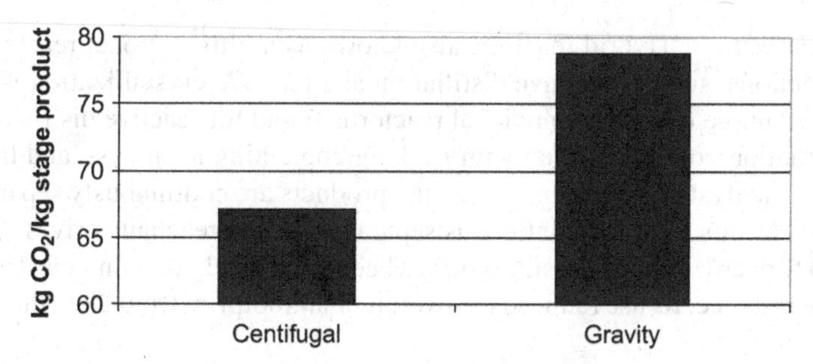

FIGURE 15.9 Carbon dioxide emissions for Example 15.4, estimated using a streamlined life cycle assessment approach (see Chapters 16 and 17).

This is an interesting case, as so far it seems that CLLEs are more efficient than the gravity separators (this translates to about 10% higher yield). Due to a more efficient separation, they require less solvent for stream washing and workup, shorter process cycles (about half!), handle emulsions well, and have the added advantage of continuous operation. Also, the spent aqueous streams contain less solvent, which means that they could be treated in a wastewater treatment plant and thereby avoid the use of an incinerator. Finally, a shorter contact time implies that there is a possibility of less evaporative solvent loss. However, gravity separators require considerably less energy to operate than centrifugal separators (about 27-fold!). So, which one is better?

At this point it would be useful to explore a streamlined life cycle approach (more to come in Chapter 16). This will enable us to evaluate the overall differences between the two options, including emissions for production of methylene chloride, emissions for waste treatment, and emissions for electricity generation. When this was done, it was clear that the environmental life cycle savings in emissions achieved by avoiding the manufacture of solvents (in this case, MDC) surpass the emissions derived for larger energy use. As an example of the results from the simplified life cycle assessment, the carbon dioxide emissions for the two systems are shown in Figure 15.9.

Additional Points to Ponder Why is an extraction needed in the first place? What problems might we encounter when using methylene chloride?

15.3 PROCESS INTENSIFICATION TECHNIQUES

In this section we cover some general aspects of process intensification techniques (see Figure 15.1). As shown earlier, most process intensification technologies focus on reaction technology, or mixing for achieving better reactions. Although there are some technologies for process intensification that are specifically designed for separations, most process intensification focus has been on methods and techniques that allow us to do things differently by combining what would traditionally be two or more unit operations into a single piece of equipment to perform the functions more effectively within a more compact structure. Therefore, to intensify our processes we use hybrid reactors, hybrid separations, and other methods. In the paragraphs below we provide cover a brief description and an example of each.

1. *Hybrid reactors.* Hybrid reactors, also known as multifunctional reactors perform multiple functions, such as reactive distillation and reactive crystallization. Perhaps the best known example of a multifunctional reactor is found for reactive distillation. In this case, a distillation column is filled with packing containing a catalyst, and the reaction occurs on the catalyst of the column while the products are continuously separated in the fractionators. In this way, the continuous separation of the reactants drives the reaction equilibrium forward. Reactive distillation has been extensively used in industrial applications. In one instance, its use reduced the overall plant footprint from 28 to three pieces of equipment.

Good candidate reactions for reactive distillation are acetylations, aldol condensations, alkylations, aminations, dehydrations, esterifications, etherifications, hydrolysis reactions, and isomerizations. There are, of course, many more examples of hybrid reactors that have been applied industrially, and Stankiewicz has published a comprehensive and very good review of multifunctional reactions that is worth reading.[54]

2. *Hybrid separations.* Similar to what we found with multifunctional reactors, when we consider hybrid separators we find separations that combine two types of separation mechanisms within a single piece of equipment, such as in the case of membrane absorbers, adsorptive distillation, and heat-integrated distillation. Most of the advances in this area are derivations on membrane work combined with another unit operation. Hollow-fiber membranes present an opportunity to create large mass transfer areas. Membrane distillation consists of passing the vapor of the more volatile component of a liquid mixture through a porous membrane and condensing it on the permeate side. The driving force for the process is the temperature difference. Membrane distillation was originally designed as a lower-energy alternative to distillation or reverse osmosis. Some of the advantages of membrane distillation are close to a 100% rejection of ions, macromolecules, cells, and other nonvolatiles; lower operating temperatures than distillation, lower operating pressures than conventional membranes, less membrane fouling, and more compact processing than traditional distillation. The main limitations of membrane distillations are related to the way the separation is performed, as the solutions need to be aqueous and sufficiently dilute to prevent wetting the membrane, which limits the application to processes such as desalination and the removal of aromatic organics or other nonvolatile components from aqueous solutions.[55,56]

3. *Other methods.* We can also intensify processes through the use of a variety of energy sources, such as solar, plasma, microwave, or electric fields; or by operating the traditional technologies in a different way, such as operating reactors in the dynamic state or using supercritical fluids.

One application that is based on the use of less traditional energy sources is sonocrystallization. It is well known that cavitation of a supersaturated solution causes nucleation. Acoustic power applied to continuous tank reactors in series is exploited to manipulate the nucleation rate independent of the growth rate, thereby allowing control of the product particle size while operating at low supersaturation levels to minimize encrustation. This also delivers a greater level of purification and more predictable size distributions. The equipment is generally smaller compared to that of conventional continuous crystallizers, thereby providing opportunities for better hydrodynamic conditions.

Sonocrystallization is recommended for high-throughput crystallizations, where a very tight control of the particle size is required and the desired particle-size distribution is very small by first intent. This will avoid the need for size-reduction processes such as milling or micronization. The use of sonocrystallization is limited by the fact that adequate mixing conditions are difficult to achieve in each vessel, there is a risk of particle segregation, and there are scale-up issues with this technique.

15.4 PERSPECTIVES ON PROCESS INTENSIFICATION

In previous sections we have seen that process intensification has several advantages for green chemistry and engineering, including increased mass productivity and reduced waste generation; enhanced mass transfer and efficiency, enhanced energy transfer with reduced energy use, lower plant profile (height) and smaller plant footprint (area), reduced residence times; increased reaction selectivity and rates, decreased volumes of hazardous materials for an inherently safer operation; minimized side reactions, and reduced energy-intensive workup steps, such as distillation and extraction. Smaller unit operations made possible by process intensification translate into a more-compact plant.[57] The BHR group, for example, reported that the benefits of their process intensification projects include:[58]

- A 60% reduction in capital cost
- A 99% reduction in impurity levels
- A 70% reduction in energy use
- Better yields than a fully optimized batch process
- A 99.8% reduction in reactor volume for a potentially hazardous process, leading to inherently safer operation

In general, process intensification represents a great opportunity for enhancing existing and new chemical processes by taking advantage of new technologies and processing methods. With all these benefits, why isn't the use of process intensification more widespread? In part, this is because process intensification is still a relatively immature area of research and development compared with more traditional processing approaches. In addition, when we look at the barriers presented in Table 15.2, we can perhaps see that the main two barriers may be the invested capital and knowledge in the existing technologies, but most important, the lack of expertise and understanding of process optimization and the constraints and limitations of the existing and new technologies.

PROBLEMS

15.1 In this chapter we did not cover all the process intensification technologies.
(a) Investigate the operating principles of the trickle bed reactor.
(b) Provide an example of its application.
(c) Identify its advantages and disadvantages from a green chemistry/green engineering perspective.

15.2 In this chapter we did not cover all the process intensification techniques.
(a) Investigate the operating principles of a membrane reactor.
(b) Provide an example of its application.

(c) Identify its advantages and disadvantages from a green chemistry/green engineering perspective.

15.3 In Example 15.4 the evaluations of the centrifugal liquid–liquid extractors involved a streamlined life cycle assessment. Explain what this is.

15.4 Estimate the mass intensity, solvent intensity, and waste intensity of the two processes of Example 15.1. Additional data on the streams are given in Table P15.4.

TABLE P15.4 Stream Data

	Vortex Mixer (kg/kg product)	Qualification Batch (kg/kg product)
Aqueous layer	8.3	6.0
THF distillate	—	4.9
THF/product layer	1.9	—
Mother liquor	—	2.0
Cake wash	—	2.0

15.5 Brechtelsbauer and Ricard performed comparisons between a batch process and a static mixer in the protection of an amine group by BOC-anhydride in IPA/water with KOH acting as the catalyst[59]:

$$H_2\text{–N–R} + (Boc)_2O \xrightarrow[\text{IPA/water}]{\text{KOH}} \text{BocHN–R} + \text{dimers (undesired by-products)} \qquad Boc = (C_4H_9)CO_2$$

This is a biphasic reaction system, which requires good mixing to facilitate mass transfer. It also has a high reaction exotherm of −213 kJ/mol for the reactant amine with typical heat transfer–related issues when run in semibatch and batch mode, resulting in thermally induced impurities. These impurities were mainly dimers of the starting material, as shown in the scheme above. Some the results of the study are given in Table P15.5.

TABLE P15.5 Study Results

Parameter	Static Mixer Reactor	Batch Reactor
Residence time	13 min	4 h
Selectivity	99.9%	97%
Conversion	96%	98%
Reactor volume	350 mL	1000 L

(a) How would you compare the two processes from a green chemistry/green engineering perspective?

(b) Why was the selectivity improved?

(c) Identify the advantages and disadvantages of the static mixer reactor from a green chemistry/green engineering perspective.

15.6 Another alternative for the system of Example 15.1 is a nonisolation batch system.

(a) Propose a block flow diagram for the nonisolation process.

(b) Estimate the energy requirements of the nonisolation system.

(c) How does that compare with the vortex and the batch reaction?

15.7 A conventional batch reactor for the production of nitroglycerine may have between 10 and 50 kg of TNT at a given time.

(a) Would you recommend the use of microreactors for this process?

(b) Why or why not? Defend your answer.

15.8 The minireactor concept was developed in the synthesis of Example 15.2 after some problems with blockage in the initial microreactor experiments. The minireactor has the same design as the microreactor, but with wider channels, small enough to keep the desired features of a microreactor and large enough to avoid blockage. The characteristics of the minireactors are given in Table P15.8.

TABLE P15.8 Minireactor Characteristics

Reactor Type	T (°C)	Residence Time	Yield (%)	Volume/ Area (m^2/m^3)	Dimensions
Minireactor	−10	<10	92	4000	Capacity of 3×10^{-5} m^3/s (30 mL/s)
Five minireactors	−10	<10	92	4000	Number-up of the minireactor described above

(a) Validate the assertion that five minireactors will be enough to substitute for the full-scale batch reactor presented in Example 15.2.

(b) Identify the advantages and disadvantages of the minireactors compared to the micro- and batch reactors of Example 15.2, from a green chemistry/green engineering perspective.

15.9 Mitake, et al.[60] have described the preparation of ZnS particles by the reaction

$$ZnSO_4\text{(solution)} + Na_2S\text{ (solution)} \rightarrow ZnS \downarrow + Na_2SO_4\text{ (solution)}$$

Table P15.9 compares the batch reactor with a two-jet mixer configuration.

TABLE P15.9 Comparison Characteristics

Batch Reactor	Two-Jet Mixer Configuration
Agitator speed 5000 rpm	Reactant streams speed 25.5 m/s
Mean particle diameter 1.4 μm (10% diameter/90% diameter is 0.31/4.52)	Reactant streams flow 600 mL/s
	Mean particle diameter 0.07 μm (10% diameter/90% diameter is 0.03/0.09)

(a) What are the green chemistry/green engineering advantages and disadvantages of the jet mixers compared to the batch reactor?

(b) What other applications would you see for this type of mixing technology?

15.10 2-Butene-1,4-diol is used in the pharmaceutical (e.g., endosulfan, vitamin B_6), insecticides, resins, paper, and textile industries. Commercially, 2-butene-1,4-diol is obtained via the selective Pd-catalyzed hydrogenation of 2-butyne-1,4-diol in an aqueous solution at elevated pressures, as shown in the reaction sequence below:[61–63]

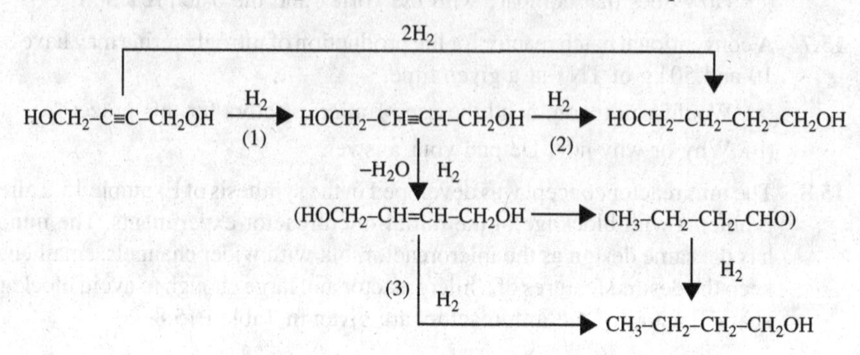

As can be seen, 2-butene-1,4-diol can undergo further hydrogenation to give saturated alcohols. In general, the variation in selectivity relates to the rate of hydrogenation relative to the rate of mass transfer, and the rate of triple- to double-bond hydrogenation.

A loop reactor was employed in this reaction using filaments of activated carbon fibers as catalyst support. The fabrics and filaments are woven from the long threads about 0.5 mm in diameter. The loop reactor setup is shown in Figure P15.10. The reaction in the loop reactor showed a selectivity of up to 97% toward 2-butene-1,4-diol with conversions of 80%.[64]

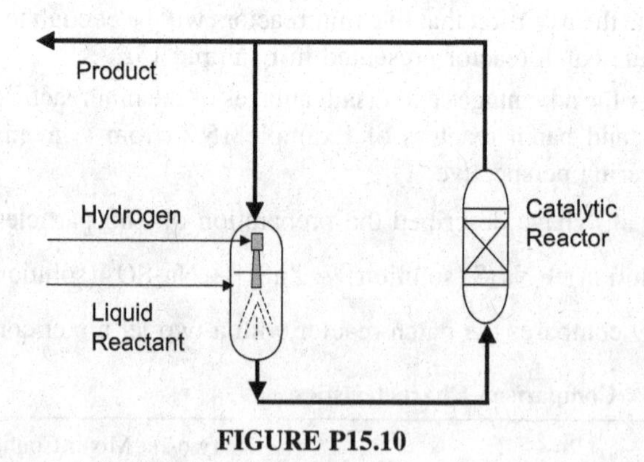

FIGURE P15.10

(a) Explain the high selectivity obtained.

(b) Describe advantages and disadvantages of this approach from a green engineering perspective.

15.11 Phenyl boric acid can be prepared with improved purity using a microchannel microreactor similar to the one shown in Figure 15.4. The phenyl product is used in pharmaceutical, agricultural, and polymerization applications. Production of the unwanted diphenyl by-product can be significant, as the monophenyl intermediate reacts with the phenyl magnesium bromide. The side reaction is rapid at higher temperatures. In the batch process, formation of the desired product has around 71% yield, with about 14% of the yield lost to the by-product. When using the microchannel reactor, the yield is about 94% and the yield lost to the by-product is less than 1%.[65]

$$B(OCH_3)_3 \xrightarrow{C_6H_5\text{–}MgBr} C_6H_5\text{–}B(OCH_3)_3 \xrightarrow{C_6H_5\text{–}MgBr} (C_6H_5)_2\text{–}BOCH_3$$

$$C_6H_5\text{–}B(OCH_3)_3 \xrightarrow{H^+,\ H_2O} C_6H_5\text{–}B(OH)_2$$

$$(C_6H_5)_2\text{–}BOCH_3 \xrightarrow{H^+,\ H_2O} (C_6H_5)_2\text{–}BOH$$

(a) Explain the difference in performance between a microchannel reactor and a batch reactor.

(b) Identify the green chemistry/green engineering advantages of a microchannel reactor.

(c) Identify the green chemistry/green engineering disadvantages of a microchannel reactor.

15.12 Figure P15.12 is a flow diagram of a process that is used for alcohol dehydration. The process employs a distillation column followed by a pervaporation unit to break an azeotrope instead of using azeotropic or extractive distillation. The feed to the pervaporation unit can be the distillate at the azeotropic composition.

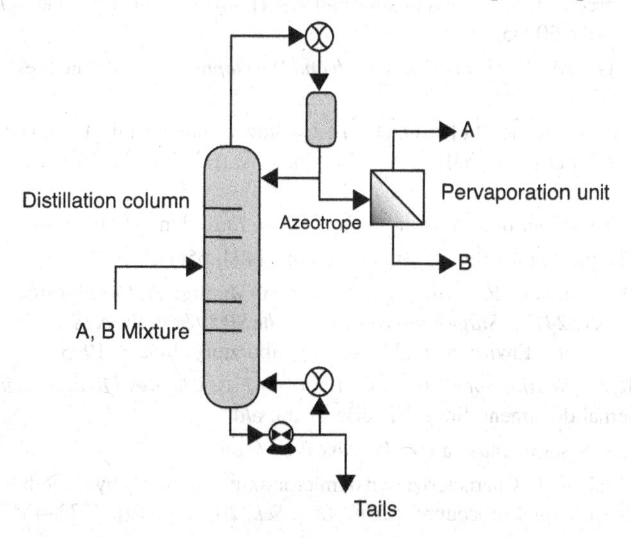

FIGURE P15.12

(a) Is this an example of process intensification? Why or why not?

(b) What advantages or disadvantages in terms of green chemistry and green engineering would this process have?

15.13 The production of sterols by ester saponification is currently performed in batch mode using large stirred tank reactors [50-m^3 reactor at 115°C and a pressure of 2 bar (gauge), on a 24-h batch cycle (only 2 h for the reaction)]. The ester saponification reaction system is very complex, with many esters in the feed from a natural source, but in simple terms the reaction desired is between steryl esters reacting with an base to render sterols. An oscillatory flow reactor was used successfully with an expected residence time for the industrial scale of 12 min, with a reactor of only 4.1 m^3 operating at 85°C, with equivalent selectivities and conversion. In both cases the process takes place in IMS (industrial methylated spirits).[66] Explain the difference in temperature and the size of the process.

15.14 We studied heat and mass integration in Chapter 13. One concept that has been developed is the heat-integrated distillation column, which uses intrinsic heat integration in the design.

(a) Is this an example of a process intensification technique?

(b) What advantages or disadvantages would a heat-integrated distillation column have from a green chemistry/green engineering perspective?

REFERENCES

1. Ramshaw, C. The incentive of process intensification. In *Proceedings of the First International Conference on Process, Intensification for the Chemical Industry*. BHR Group, London, 1995.
2. Stankeiwicz, A. I., Moulijn, A. J. Process intensification: transforming chemical engineering. *Chem. Eng. Prog.*, 2000, (96) 1, 22–34.
3. Tsouris, C., Porcelli, J. V. Process intensification—Has its time finally come? *Chem. Eng. Prog.* Oct. 2003, 99(10): 50–55.
4. Anderson, N. G. *Practical Process Research and Development*. Academic Press, San Diego, CA, 2000.
5. Stringaro, J.-P, Collins, P., Bailer, O. Open cross-flow channel catalysts supports. In *Structured Catalysts and Reactors*, Cybulski, A., Moulijn, J. A., Eds. Marcel Dekker, New York, 1998, pp. 393–416.
6. Redman, J. Getting out of a mess in mixing. *Chem. Eng.*, Jan. 31. 1991, 489, 13.
7. Brown, S. The mixing revolution. *Manuf. Chem.*, 1994, 65(1), 19–21.
8. Herrmann, R., Hannah, R. *Evaluation of Vortex Mixing: A Comparative Case Study for Production of SB 224773, Stage 8 Intermediate in the SB 207499 Process*. Internal memorandum, SmithKlineBeecham, Environmental Research Laboratory, June 8, 1998.
9. Herrmann, R. *Applications of Motionless Mixers for Fast Chemical Reactions in the SB-207499 Process*. Internal document. SmithKlineBeecham, n/d.
10. Knott, M. Mixers set to cause a stir. *Process Eng. (London)*, 1992, 73(3), 29–30.
11. Guichardon, P., Falk, L. Characterization of micromixing efficiency by the iodide–iodate reaction system: I. Experimental procedure. *Chem. Eng. Sci.*, 2000, 55(19), 4233–4243.
12. Guichardon, P., Falk, L., Villermaux, J. Characterization of micromixing efficiency by the iodide-iodate reaction system: II. Kinetic study. *Chem. Eng. Sci.*, 2000, 55(19) 4245–4253.
13. Brokers, O. S. Power fluidics or harnessing hydrodynamics. *Chem. Eng. (London)*, 1992, 521, 19–20.
14. Cihonski, J., Dolbear, G. *Process Intensification from a Chemist's Perspective*. Dolbear and Associates, Inc., http://www.gedolbear.com/processintensification/.

15. Dautzenberg, F. M., Mukherjee, M. Process intensification using multifunctional reactors. *Chem. Eng. Sci.*, 2001, 56, 251–267.
16. Hessel. V., Löwe, H., Schönfeld, F. Micromixers: a review on passive and active mixing principles. *Chem. Eng. Sci.*, 2005, 60, 2479–2501.
17. Herrmann, R. Internal communication, GlaxoSmithKline, Mar. 21 2001.
18. Perry, R. H., Green, D. W. *Perry's Chemical Engineers' Handbook*, 8th ed. McGraw-Hill, New York, 2007.
19. Walas, S. M. Rules of thumb. *Chem. Eng.*, Mar. 16, 1987.
20. Costello, R. Process intensification—Think small. *Innov. Pharm. Technol.*, June 2002, 126–132.
21. Ranshaw, C. Process intensification and green chemistry. *Green Chem.*, 1999, 1, G15–G17.
22. Doble, M. Green reactors. *Chem. Eng. Progr.*, 2008, 104(8), 33–42.
23. Yole development Web site. Microreaction Technologies. http://www.yole.fr/pagesAn/products/mrt.asp, accessed Jan. 4, 2009.
24. Bender, M., Knocks, A., Hover, S., Freund, H. High temperature micro catalysis: a new design for comparative reaction studies in technical and model systems. In *Microreaction Technology: Industrial Prospects (IMRET 3: Proceedings of the 3rd International Conference in Microreaction Technology)*, Ehrfeld, W. Ed. Springer-Verlag, New York, 2000, pp. 54–61.
25. Green, A. Process intensification: The key to survival in global markets? *Chem. Ind.* Mar. 2. 1998, 5, 168–172.
26. Srinivasan, R., Hsing, I.-M., Berger, P., Jenssen, E. K. F., Firebaugh, S. L., Schmidt, M. A., Harold, M. P., Lerou, J. J., Ryley, J. F. Micromachined reactor for catalytic partial oxidation reactions. *AIChE J.*, 1997, 43(11), 3059–3069.
27. Jenssen, K. F. Micromechanical systems: status, challenges and opportunities. *AIChE J.* 1999, 45 (10), 2051–2054.
28. Krummradt, H., Koop, V., Stoldt, J. Experience with the use of microreactors in organic synthesis. In *Microreaction Technology: Industrial Prospects (IMRET 3: Proceedings of the 3rd International Conference in Microreaction Technology)*, Ehrfeld, W. Ed. Springer-Verlag, New York, 2000, pp. 181–186.
29. Becht, S., Franke, R., Geißelmann, A., Hahn, H. Micro process technology as a means of process intensification, *Chem. Eng. Technol.*, 2007, 30(3), 295–299.
30. Kobayashi, J. et al. Micro-reactor: multi-phase organic synthesis reactions using micro-space. *Chem. Chem. Ind.*, 2006, 59(3), 248–251.
31. Kerby, M. B., et al. Measurements of kinetics in a micro-fluidic reactor. *Anal. Chem.*, 2006, 78 (24), 8273–8280.
32. Schwalbe, T., et al. Chemical synthesis in microreactors. *Chimia*, 2008, 56, 636–646.
33. Skelton, V., et al. Microreactor synthesis of cyanobiphenyls using a modified Suzuki coupling of an aryl halide and aryl boronic acid. In *Microreaction Technology: Industrial Prospects (IMRET 3: Proceedings of the 3rd International Conference in Microreaction Technology)*, Ehrfeld, W., Ed. Springer-Verlag, NewYork, 2000, 235–242.
34. Hessel, V., et al. High temperature HCN generation in an integrated microreaction system. In *Microreaction Technology: Industrial Prospects (IMRET 3: Proceedings of the 3rd International Conference in Microreaction Technology)*, Ehrfeld, W., Ed. Springer-Verlag, NewYork, 2000, pp. 151–164.
35. Mayer, J., et al. A microstructured reactor for the catalytic partial oxidation of methane to syngas. In *Microreaction Technology: Industrial Prospects (IMRET 3: Proceedings of the 3rd International Conference in Microreaction Technology)*, Ehrfeld, W., Ed. Springer-Verlag, NewYork, 2000, pp. 187–196.

36. Kursawe, A., et. al. Selective reactions in microchannel reactors. In *Microreaction Technology: Industrial Prospects (IMRET 3: Proceedings of the 3rd International Conference in Microreaction Technology)*, Ehrfeld, W.,Ed. Springer-Verlag, NewYork, 2000, pp. 213–223.
37. Aoune, A., Ramshaw, C. Process intensification: heat and mass transfer characteristics of liquid films on rotating discs. *Int. J. Heat Mass Transfer*, 1999, 42, 2543–2556.
38. Oxley, P., Brechtelsbauer, C., Ricard, F., Lewis, N., Ramshaw, C. Evaluation of spinning disk reactor technology for the manufacture of pharmaceuticals. *Ind. Eng. Chem. Res.*, 2000, 39, 2175.
39. Boodhoo, K. V. K., Jachuck, R. J. Process intensification: spinning disk reactor for styrene polymerisation. *Appl. Thermal. Eng.*, 2000, 20, 1127–1146.
40. Ritter, S. A new spin in reactor design. *Chem. Eng. News*, 2002, 80(30), 26–27.
41. Costello, R. C. Costello Consulting Engineers Web site. http://www.rccostello.com/STT.html, accessed Jan. 6, 2009.
42. Kriedo Laboratories Web site. http://www.kreidobiodiesel.com/, accessed Jan. 6, 2009.
43. Harvey, A. P., Mackley, M. R., Seliger, T. Process intensification of biodiesel production using a continuous oscillatory flow reactor. *J. Chem. Technol. Biotechnol.*, 2003, 78, 338–341.
44. Fernandes-Reis, N. M. Novel Oscillatory Flow Reactors for Biotechnological Applications. Ph.D. dissertation, Universidade do Minho, Apr. 10, 2006.
45. GlaxoSmithKline. *Green Technology Guide*. Internal report. 2007.
46. Ramshaw, C. "Higee" Distillation: an example of process intensification, *Chem. Eng.*, 1983, 389 (2), 13–14.
47. Ramshaw, C., Mallinson, R. H. Mass transfer apparatus and its use, Eur. patent. 0,002,568, June 20, 1984.
48. Burns, J. R., Jamil, J. N., Ramshaw, C. Process intensification: operating characteristics of rotating packed beds — determination of liquid hold-up for a high voidage structured packing. *Chem. Eng. Sci.*, 2000, 55, 2401–2415.
49. Keyvani, M., Gardner, N. C. *Operating Characteristics of Rotating Beds.* Technical Report, DOE/PC/79924-T3. U.S. Department of Defence, Washington, DC, Jan 1988.
50. Zheng, C., Guo, K., Song, Y., Zhou, X., Al, D., Xin, Z., Gardner, N. C. Industrial practice of HIGRAVITEC in water deaeration. In *Proceedings of the 2nd International Conference on Process Intensification in Practice*. BHR Group, London, 1997, pp. 273–287.
51. Ruthven, D. M.Ed., *Encyclopedia of Separation Technology*, Wiley, New York, 1997.
52. Letki, A. G. Know when to turn to centrifugal separation. *Chem. Eng. Prog.*, 1998, 94(9), 29–44.
53. SmithKlineBeecham Environmental Research Laboratory, *Green Technologies.* Internal document. 2000.
54. Stankiewicz, A. Reactive separations for process intensification: an industrial perspective. *Chem. Eng. Process*, 2003, 42, 137–144.
55. Lawson, K. W., Lloyd, D. R. Membrane distillation. *J. Membrane Sci.*, 1997, 124(1), 1–25.
56. Foster, P. J., Burgoyne, A., Vahdati, M. M. A new rationale for membrane distillation processing. Presented at the *International Conference on Process Innovation and Intensification,* Manchester, UK. Institute of Chemical Eengineers, Rugby, UK., 1998.
57. Costello, R. Tiny Reactors Aim for Big Role. http://www.chemicalprocessing.com/articles/2006/176.html, accessed Jan. 6, 2009.
58. BHR, Group. http://www.bhrgroup.co.uk/pi/aboutpi.htm, accessed Jan. 6 2009.
59. Brechtelsbauer, C., Ricard, F. Reaction engineering evaluation and utilization of static mixer technology for the synthesis of pharmaceuticals, *Organ. Proc. Res. Dev.*, 2001, 5, 64.
60. Mitake, K., Miyake, F., Yasuda, F., Yazaki, T., Toda, M. High speed collision reaction method. U.S. patent 6,227,694, Genus Corporation and Hakusui Tech. Co., 2001.

61. Hort, E. V., Graham, D. E., Verfahren zur partiellen Hydrierung von Butin-(2)-diol-(1,4) zu buten-(2)-diol-(1,4). General Aniline and Film Corporation, New York, 1962, 1, 139–832.

62. Hoffmann, H., Boettger, G., Bör, K., Wache, H., Kräfje, H., Körning, W. *Katalysator zur partiellen Hydrierung*, Vol. 1. BASF, Ludwigshafen, Germany, 139–832.

63. Winterbottom, J. M., Marwan, H., Viladevall, J., Sharma, S., Raymahasay, S. Influence of dispersity on the activity, selectivity and stability of raney nickel during the hydrogenation of 1,4-butynediol into 1,4-butanediol. *Stud. Surf. Sci. Catal.* 1997, 108, 59–66.

64. Kiwi-Minsker, L., Joannet, E., Renken, A. Loop reactor staged with structured fibrous catalytic layers for liquid-phase hydrogenations. *Chem. Eng. Sci.*, 2004, 59, 4919–4925.

65. Yoshida, J.-I., Nagaki, A., Iwasaki, T., Suga, S. Enhancement of chemical selectivity by microreactors. *Chem. Eng. Technol.*, 2005, 28, 259.

66. Harvey, A. P., Mackley, M. R., Stonestreet, P. Operation and optimization of an oscillatory flow continuous reactor. *Ind. Eng. Chem. Res.,* 2001, 40, 5371–5377.

PART IV

EXPANDING THE BOUNDARIES

16

LIFE CYCLE INVENTORY AND ASSESSMENT CONCEPTS

What This Chapter Is About In this chapter we explore the wider impacts of industrial activities across the supply chain. For a process or activity to be considered green, it is necessary to assess the cumulative environmental impacts across the entire life cycle of the product: raw material extraction, manufacturing, use, maintenance, reuse, transportation, and final fate. This is done using life cycle inventory and assessment concepts. The intent of this section is to provide background and context for placing a particular chemical process within the broader chemical enterprise. It will also explain the methodology used to assess the environmental impacts of a product from raw materials extraction to recycle and reuse or end-of-life considerations. Both macro- and microeconomic considerations of these life cycle impacts are explained.

Learning Objectives At the end of this chapter, the student will be able to:

- Understand the concepts and methodology of life cycle inventory and assessment.
- Apply the methodology for evaluating life cycle impacts through inventory and assessment.
- Identify opportunities for improvements through life cycle assessment evaluations.
- Apply life cycle inventory and assessment methodology to assess processes.

Green Chemistry and Engineering: A Practical Design Approach, By Concepción Jiménez-González and David J. C. Constable

16.1 LIFE CYCLE INVENTORY AND ASSESSMENT BACKGROUND

In previous sections of the book we have explored how green chemistry and green engineering principles are used to design greener, more environmentally friendly processes. However, to measure if a new or retrofitted process is actually green, it is necessary to measure the true impacts of the process throughout its entire life cycle. There are many examples of changes in technology or chemistry that on the surface appear to be an improvement in the local environmental performance when looking at a single stage of a process. However, when expanding the boundaries beyond what is done in the plant, one might discover that they are merely transferring environmental impacts from one medium to the other (i.e., ground to water or water to air) instead of solving the problem. One of the typical examples of this is removing a contaminant from a waste stream using a solvent. On the surface we have prevented pollution, but when we account for the impacts related to generating the solvent and disposing of the pollutant-rich stream, we have to ask if that is really an overall gain for the environment. But how can we estimate the overall environmental performance of a process?

Life cycle inventory and assessment (LCI/A) is a methodology that allows one to more precisely estimate the cumulative environmental impacts associated with manufacturing all the chemicals, materials, and equipment used to make a product or deliver a service; thereby providing a comprehensive view of the potential trade-offs in environmental impacts associated with a given process or product choice. The term *life cycle* refers to the major activities in the life span of the process, product, or activity, from the extraction of all raw materials, to the final manufacture, transportation, use, maintenance, reuse, to its final fate.[1,2] This is typically called a *cradle-to-grave approach* and is the ultimate aim of life cycle approaches reported in the literature, as represented in Figure 16.1.

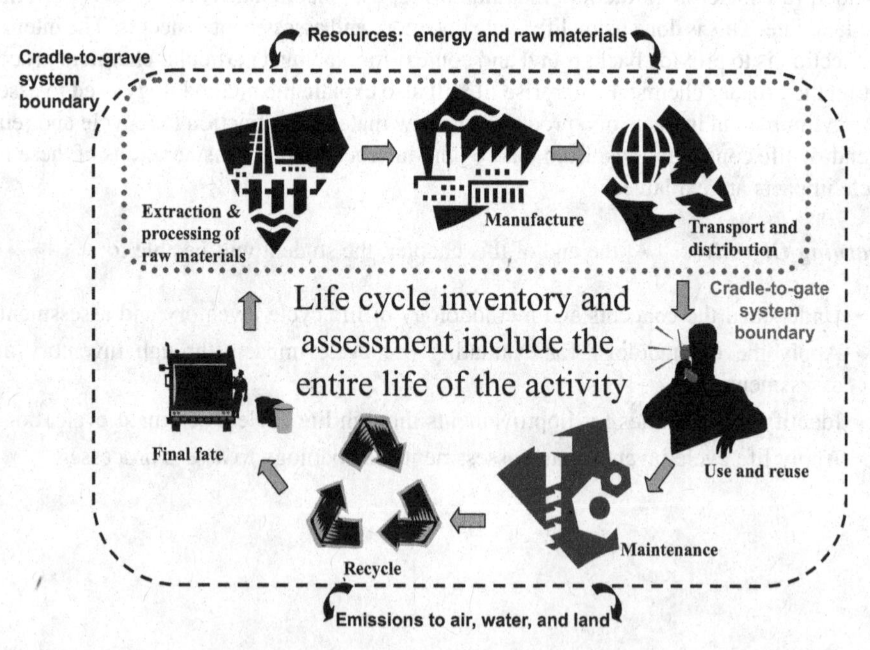

FIGURE 16.1 Life cycle inventory and assessment. Including all the phases in the life cycle of an activity aims at cradle-to-grave evaluation.

Life cycle assessment traces its roots back to the 1960s, where the main interest was in cumulative energy consumption. Later, global modeling studies published in *The Limits to Growth*[3] and "A Blueprint for Survival"[4] were focused on estimations of population change in the face of limited resources. The predicted depletion of fossil fuels caused more calculations of energy use but the foundations for modern life cycle methods began in 1969 with the initiation of an internal study for the Coca-Cola Company to compare the environmental releases associated with two different types of beverage containers. Raw material use was quantified as were the environmental emissions associated with the manufacture of each container. Other companies performed similar inventories in the early 1970s. At that time, most of the data came from publicly available sources, and industry-specific data were hard to find. From 1975 to the late 1980s, interest in performing these types of assessments decreased as the influence of the oil crises waned and environmental issues focused on hazardous waste management.

When solid waste became a global issue, life cycle reemerged as a tool for analyzing environmental problems. Still there was the need for a standardized approach, and there were many questions about the usefulness and appropriateness of the tool. These shortcomings drove the standardization of life cycle practices, which was led by the Society of Environmental Toxicology and Chemistry[5–7] and culminated with the development of international LCA standards through the International Standards Organization (ISO) 14000 series (14040, 14044).[8,9]

One of the principles of LCI/A methodologies is that all the phases of a process life cycle need to be evaluated from the perspective that the phases are interdependent and that changes in one can and do affect the other phases. Therefore, the results of a LCA are an estimate of the cumulative impacts resulting from all stages in a product life cycle, often including effects not considered in more traditional analyses, and providing a comprehensive view of the environmental aspects of the product or process with a more accurate picture of the true environmental trade-offs in product and process selection.[10]

Using a life cycle approach to estimate environmental impacts of processes and products provides the practitioner with the following advantages:

- Evaluates the true greenness of a process (e.g., avoids shifting impacts).
- Provides a set of commonly accepted metrics to measure specific impacts.
- Identifies trade-offs between impacts and phases in the life cycle and highlights hidden costs and impacts.
- Identifies "hot spots," the phases or areas that contribute most to the environmental footprint of an activity to drive opportunities for resource optimization.
- Systematically identifies key environmental impacts at each of the life cycle stages of a system.
- Provides information to decision makers for strategic planning, priority setting, and product or process design or redesign.
- Identifies information gaps.
- Provides scientific data that can be applied to recognized marketing schemes (e.g., ecolabeling, environmental claims, environmental product declarations).

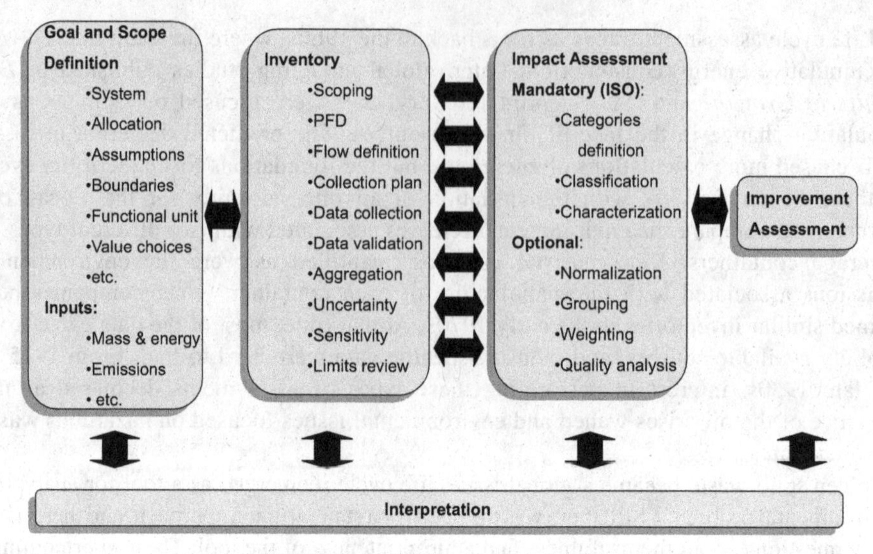

FIGURE 16.2 Phases of an LCI/A.

16.2 LCI/A METHODOLOGY

The LCA process is a systematic, phased approach and consists of four components: goal definition and scoping, inventory analysis, impact assessment, and interpretation, as illustrated in Figure 16.2.

16.2.1 Goal and Scope Definition

In this stage we define and describe why we are performing a life cycle inventory, life cycle assessment, or both. According to ISO, the following items should be stated:

- The intended application
- The reasons for carrying out the study
- The intended audience
- Whether the results will be used in comparisons to be disclosed to the public

Example 16.1 What might be potential goals for life cycle inventories?

Solution

- Support environmental assessments used for environmental submissions sent to regulators.
- Collect baseline process information while a product moves through R&D.
- Estimate contributions to environmental releases by specific steps in a process so that chemists and chemical engineers may improve the environmental footprint of a process or product.

- Identify data gaps in a process required to reduce the environmental footprint of R&D.
- Use external reporting to support product certification for a "blue angel" committee.
- Provide guidance to company-based scientists and engineers for product or process development and greening.

Additional Points to Ponder How would the answer change for a life cycle assessment? Are all of the requirements from ISO fulfilled in those statements?

During the scope definition phase of life cycle assessment, methods, boundaries, and limitations of life cycle assessment need to be described. Table 16.1 shows the scope definition requirements as set by ISO.

There are several aspects that need special consideration when setting goals and scopes for a life cycle inventory/assessment:

1. *Functional unit.* In other words, why and how is this product used? This is particularly important when comparing two or more products or systems. If the comparison includes products that are used at different rates, with different efficiencies, those aspects need to be integrated. For example, if we are performing a comparative life cycle assessment between two liquid detergents, one potential functional unit could be the mass or volume of clean clothes washed during a month in 15 standard laundry loads. If detergent A is sold in containers that can be used to wash 20 loads and detergent B is sold in containers that can be used to wash 15 loads, this needs to be taken into account.

2. *System boundaries.* In other words, do we have to cover the entire life cycle? The traditional life cycle assessment covers a system from raw materials and energy extractions through the material's final fate (or a cradle-to-grave boundary). A decision to use a boundary other than a cradle-to-grave boundary needs to be understood and documented. If the boundary is too limited, significant elements may be excluded; if the boundary is too open, resources may be wasted unnecessarily. It is also possible to exclude some stages or activities and still have a well-developed life cycle assessment, but it will depend on the goal of the study and the functional unit. For example, if the goal of the study is to assess different packaging materials for the same product, it may be justified to exclude the raw material acquisition and product production for the main product and limit the assessment to packaging. On the other hand, if the goal is to compare two different synthetic routes to the same material, it may be possible to include raw material extraction and the synthesis, but exclude anything that follows the production of the material (also known as a *cradle-to-gate boundary*).

3. *Allocation.* In other words, the manufacture of a functional unit for any given product may produce other valuable secondary outputs, so when a process produces more than one material, which life cycle impacts belong to which product? There is, therefore, a need for an objective way to assign environmental burdens to each co-product. This method is known as the *allocation problem.* ISO standard 14040/44 dictates that allocation be avoided whenever possible, either by more clearly delineating the details of the assessment for each part of the process used for each product, or by broadening the system boundaries (known as the *system expansion method*). If allocation cannot be avoided, one could use physical relationships (e.g., mass, heat value) to allocate the impacts appropriately. When

TABLE 16.1 ISO Requirements During Scope Definition of an LCI/A

ISO Requirement	Examples
Product system	Manufacture of the active ingredient in medications to treat depression Biofuels used in private car transportation Production of paint
Functions of the product system(s)	Manufacture of the active ingredient in medications to treat depression Biofuels used in private car transportation Protection of houses from the elements
Functional unit	1 kg of active ingredient 100 miles by car Effective paint protection for five years
System boundary	Cradle to grave Cradle to gate
Allocation procedures	By mass By heating value By system expansion No allocation
LCI/A methodology and types of impacts	EcoIndicator 99 and related impacts EDIP and related impacts TRACI and related impacts
Data requirements	Manufacturing logs Transportation distances and modes Commercial forecasts
Assumptions	Data used to fill gaps Foreground and background Calculations of physical properties
Value choices and optional elements	Would the LCA impacts be weighted? How? Would the LCA impacts be normalized? How?
Limitations	Data not found Phases not included Materials excluded
Data quality requirements	Geographic coverage Industry-specific Precision Completeness Consistency Reproducibility
Type of critical review, if any	Journal peer-review panel Internal review Third-party review
Type and format of the report required	Internal report Journal publication Web application

allocation or the use of physical properties cannot be avoided, other relationships, such as economics, may be used. For more information on this topic, see the book by Baumann and Tillman.[11]

16.2.2 Inventory Analysis

Once the goal and scope are defined, the backbone of a life cycle study is a life cycle inventory. The objective of this part of life cycle assessment is to identify, quantify, and produce a detailed accounting of the resources used (e.g., energy, water and materials) and the environmental emissions associated with producing the product or service (e.g., air emissions, solid waste disposal, wastewater discharges). The following are the recommended steps for performing an LCI:

1. *Goal and scope definition.* This will clarify the reasons for performing the LCI, under which conditions, and following which methodologies.
2. *Prepare a process flow diagram.* This will provide an understanding of the system under study and the process flows (e.g., are there any recycling loops?).
3. *Prepare a data collection plan.* This will determine how we are planning to obtain the necessary information for the LCI: from industrial surveys, from databases, using estimation methodologies, and so on. A decision has to be made whether to use average data, measured data, or marginal data. Different collection plans might be needed for the foreground data (data of the main system under study) and background data (data of systems farther back in the supply chain).
4. *Data collection.* This is the actual step in which the practitioner(s) obtain the data by calculations, surveys, and other means.
5. *Data validation.* The objective is to confirm that the data obtained fulfill the data quality requirements for the study.
6. *Relating data to unit processes and functional unit and data aggregation.* This is the equivalent to performing a large mass and energy balance across the system boundary, in such a way that the functional unit is the basis for calculation.
7. *Refining system boundaries.* LCI is an iterative process. After the first calculations for an LCI, the boundaries, significance, and relevance of the data need to be questioned. This could result in the exclusion of some unit processes (reducing the boundaries), the inclusion of new unit processes (expanding the boundaries), or the inclusion or exclusion of inputs or outputs that may be considered to have negligible significance to the assessment.

Any calculation or data acquisition methods that are used in the assessment need to be documented, and all calculation procedures need to be applied consistently throughout the study. There are several documented methodologies that can be used to calculate life cycle inventory information, either in full, on a gate-to-gate basis, or for specific systems, such as energy production, waste treatment, and others.[12–17] There are also commercially available software packages and databases, such as Eco-Invent,[18] SimaPro,[19] DEAM,[20] UMBERTO,[21] GABI,[22] and others that contain life cycle inventory information for generic processes, energy production, waste treatment, and some other subsystems. The literature also contains life cycle information that can be used in LCIs.

Transparency in inventory development is critically important since it is very unlikely that information can be obtained from a single source and that calculated data will be derived from multiple inputs. It is important to clearly document the sources of all information, assumptions, data gaps, and so on. Data validation is tremendously important at this step, and one must keep in mind that data obtained from a database are not necessarily relevant, transparent, nor applicable to the system under study. Comparative studies of data obtained for the same system from different LCI databases have found a lack of transparency, double-counting of primary sources, and significant discrepancies.[23]

Example 16.2 Estimate the life cycle inventory for the production of 3-pentanone to be used in a series of pharmaceutical and fine-chemical production processes.

Solution The life cycle inventory for 3-pentanone was estimated by Environmental Clarity.[24] The methodology utilized was the modular gate-to-gate methodology developed by Jiménez-González et al.,[12] where gate-to-gate life cycle information is generated for each chemical in the supply chain using mass and energy balances, and applying chemical engineering principles to estimate the process flows.

The first step of this methodology is to define the production process for the chemical. There are several ways to define it, but the most important guidance in selecting the production process is the relevance to the goal of the LCI, as the type of process utilized needs to provide insight to the question the LCI is intending to answer. In this case, 3-pentanone produced by the catalytic ketonization of propionic acid was chosen as the process.

The next step is to develop a data collection plan. In this case, the methodology calls for the utilization of chemical engineering principles to estimate gate-to-gate inventories for all the substances involved in the supply chain for the manufacture of 3-pentanone. Since the scope of this assessment is cradle-to-gate, the next step is to identify all the manufacturing processes that are involved in the supply chain of 3-pentanone until we have located the primary materials taken from the Earth or from an agreed cutoff point (e.g., less than 0.5% contribution to the overall life cycle mass). The graphical representation of this is sometimes called the *chemical pedigree* or a *chemical tree*. Figure 16.3 shows a chemical tree for 3-pentanone for the process we selected, the catalytic ketonization of propionic acid.

Level 1	Level 2	Level 3	Level 4	Level 5	Level 6
3-Pentanone 1,000	Ethanecarboxylic acid 1,734	**Carbon monoxide 797**	**Carbon dioxide 398**	Natural gas 81.6	Natural gas (unprocessed) 83.2
				Nitrogen from air 153	Air (untreated) 153
				Oxygen from air 67.9	Air (untreated) 67.9
				Water for reaction 106	Water (untreated) 106
			Natural gas 268	Natural gas (unprocessed) 274	
			Water for reaction 163	Water (untreated) 163	
		Ethylene 702	Naphtha 716	Crude Oil 737	
		Water for reaction 790	Water (untreated) 790		
	Nitrogen 156	Air (untreated) 216			

FIGURE 16.3 Chemical tree for the production of 3-pentanone. All figures in kilograms.

From the tree you can see that propionic acid (ethanecarboxylic acid) and nitrogen are needed to produce the 3-pentanone. Propionic acid is produced using carbon monoxide, ethylene, water, and so on, until you arrive at resources such as natural gas or water that are obtained from the biosphere. The numbers in the cells correspond to the amount of each chemical needed to produce 1000 kg of 3-pentanone. The next step is to gather detailed information, which in this case, we obtained through chemical engineering design estimations. In other methodologies, this could be done through databases or supplier surveys if the data are available and reliable. For each one of the chemicals in the chemical tree, it is necessary to estimate the raw material consumption, energy requirements, and emissions to air, water, and land.

We will use the 3-pentanone process to show the way a gate-to-gate life cycle inventory is estimated. As you might remember, in Chapter 9 we covered process flow diagrams, and in Example 9.1 described this process for producing 3-pentanone. The process flow diagram is shown in Figure 16.4. Based on the unit operations and the process descriptions, a mass and energy balance is performed around the process to estimate energy requirements and material flows (inputs and emissions). This is performed following the methods described in Chapter 9. You might want to revisit Example 9.1 to remind yourself about the details in the calculations, but Table 16.2 is a summary of the mass balance, including the composition of each process stream in the process flow diagram. Table 16.3 shows the energy requirements for each unit operation, cumulative energy requirements, cooling requirements (exotherms), and assumed heat recovery from hot streams being cooled.

After you obtain the desired information from the mass and energy balances for the 3-pentanone process, you repeat the process for all the chemicals within the chemical tree. In addition to collecting the resource requirements and emissions related to energy production, you must also include estimates for transportation and waste treatment for the entire tree. For processes that produce more than one valuable product, a mass allocation algorithm is used to ensure that you are not overburdening a product inventory. A description of potential impacts from transportation is given in Chapter 17, a description of potential impacts derived from energy production is given in Chapter 18, and a description of the potential impacts from waste treatment is given in Chapter 19. Table 16.4 is a final summary of the estimated cradle-to-gate life cycle inventory for 3-propanone.

Additional Points to Ponder What effect would different allocation rules have on the results? What effect would selecting a different process have on the results?

16.2.3 Life Cycle Impact Assessment

At this stage, the life cycle inventory results are used to assess the potential human and ecological effects of the process or activity under study. However, you should be cautioned that estimating environmental impacts is easier said than done. Most types of emissions represented by our inventory outputs are not associated with one and only one direct impact, but they are usually involved in a variety of cause–effect chains and networks that make the overall assessment challenging.

For example, as we saw in Chapter 3, green house gases contribute to a change in radiative forcing in the atmosphere by absorbing energy as heat (the primary effect); this energy absorption is expected to change the temperature of the Earth's atmosphere (the secondary effect, although there will be some regional or geographic differences as the

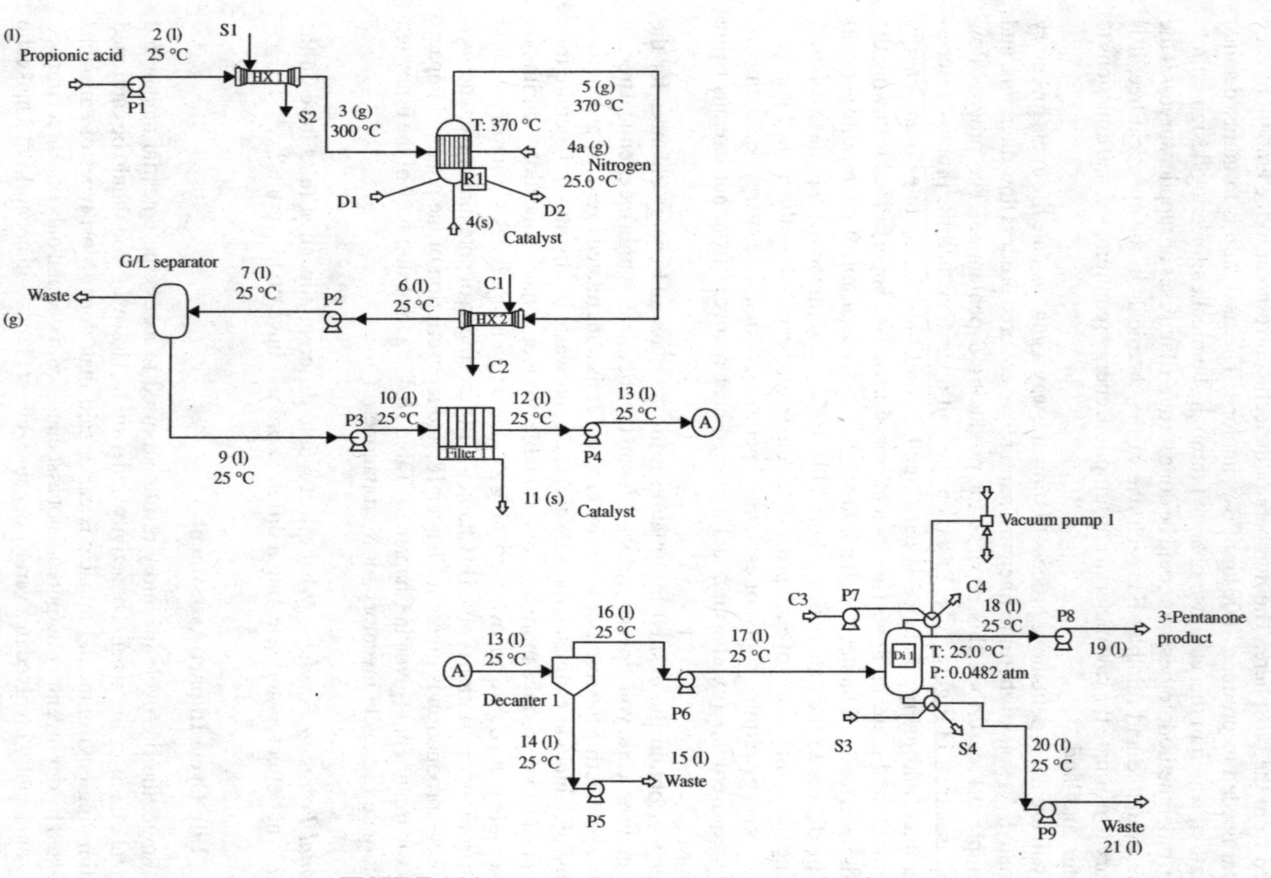

FIGURE 16.4 Process flow diagram for Example 16.2.

TABLE 16.2 Mass Balance of Chemicals in Each Process Stream[a]

	Comments	Streams	Temp (°C)	P	Phase	Total flow	Propionic acid	3-Pentanone	Carbon dioxide	Water	Catalyst	Nitrogen	Steam	Water
		0		1.00		0	0	0	0	0	0	0	0	0
Input		1	25.0	1.00	l	1734	1734							
		2	25.0	1.00	l	1734	1734							
		3	300	1.00	g	1734	1734							
		4	25.0	1.00	s	1.00					1.00			
		4a	25.0	1.00	g	156						156		
R1	1725	kg	Propionic acid			is converted in reaction 1 (99.5 % of reactor input)								
		kg				is lost in reaction 2								
		kg				is lost in reaction 3								
		Input to reactor				1891	1734	0	0	0	1.00	156		
		R1 Reaction coefficient 1					−2.00	1.00	1.00	1.00				
		R1 Conversion 1 (kg/h)				0.233	−1725	1003	512	210				
		R1 Conversion 1 (kgmol/h)				11.6	−23.3	11.6	11.6	11.6				
		R1 Reaction coefficient 2												
		R1 Conversion 2 (kg/h)												
		R1 Conversion 2 (kgmol/h)				Non-number in input								
		R1 Reaction coefficient 3												
		R1 Conversion 3 (kg/h)												
		R1 Conversion 3 (kgmol/h)												
		Flow out of reactor				1891	8.67	1003	512	210	1.00	156		
		Primary product				3-Pentanone								
		Total conversion					99.5	NA	NA	NA	−0	−0		
		Per-pass conversion					99.5	NA	NA	NA	−0	−0		
	Total yield from reactor						100	NA						

(*continued*)

TABLE 16.2 *(Continued)*

Comments	Streams	Temp (°C)	P	Phase	Total flow	Propionic acid	3-Pentanone	Carbon dioxide	Water	Catalyst	Nitrogen	Steam	Water
	5	370	1.00	g	1891	8.67	1003	512	210	1.00	156		
	6	25.0	1.00	l	1891	8.67	1003	512	210	1.00	156		
	7	25.0	1.00	l	1891	8.67	1003	512	210	1.00	156		
g/l separation <1>	Percentage of input in vapor phase					0.100	0.100	100	0.100	0.100	100		
	Percentage of input in liquid phase					99.9	99.9	0	99.9	99.9	0		
	Boiling temperature (°C)					141	102	−78.7	99.9		−194		
Waste	8	25.0	1.00	g	−670	-8.67×10^{-3}	−1.00	−512	−0.210	-1.00×10^{-3}	−156	0	0
	9	25.0	1.00	l	1221	8.66	1002	0	209	0.999	0		
	10	25.0	1.00	l	1221	8.66	1002	0	209	0.999	0		
Filter	Percentage of input in solid phase					0	0	0	0	100	0		
	Percentage of input in liquid phase					100	100	100	100	0	100		
	Boiling temperature (°C)					141	102	−78.7	99.9		−194		
Waste	11	25.0	1.00	s	−0.999	0	0	0	0	−0.999	0	0	0
	12	25.0	1.00	l	1220	8.66	1002	0	209	0	0		
	13	25.0	1.00	l	1220	8.66	1002	0	209	0	0		
Decanter <1>	Percentage in oil phase					80.0	99.5	0	40.0	0	0		
	Percentage in aqueous phase					20.0	0.500	100	60.0	100	100		
	14	25.0	1.00	l	132	1.73	5.01	0	126	0	0		
Waste	15	25.0	1.00	l	−132	−1.73	−5.01	0	−126	0	0	0	0
	16	25.0	1.00	l	1087	6.93	997	0	83.7	0	0		
	17	25.0	1.00	l	1087	6.93	997	0	83.7	0	0		
Feed	17	25.0	1.00	l	1087	6.93	997	0	83.7	0	0		

D_i <1>	Percentage of input in distillate					0.100	99.9	100	5.00	0	100		
	Percentage of input in bottoms					99.9	0.1000		95.0				
	Boiling temperature (°C)					141	102		99.9				
	18	25.0	1.00	l	1000	6.93×10^{-3}	996	0	4.19	0	0		
Main product	19	25.0	1.00	l	−1000	-6.93×10^{-3}	−996	0	−4.19	0	0	0	0
	20	25.0	1.00	l	87.5	6.92	0.997	0	79.6	0	0		
Waste	21	25.0	1.00	l	−87.5	−6.92	−0.997	0	−79.6	0	0	0	0
	Product purity (%)				0.996								
	Main product			3-Pentanone									
	Overall reaction coefficients					−2.00	1.00	1.00	1.00				
	Total yield of process (from reactant)					98.8	NA	NA	NA				
Waste	Fugitive losses (total)			g	−12.6	0	−10.0	−2.56	0	0	0	0	0
	Input sum				1891	1734	0	0	0	1.00	156		
	Fugitive replacement of reactants				0	0							
	Total input (input + fugitive replacement)				1891	1734	0	0	0	1.00	156		
	Product sum				1000	6.93×10^{-3}	996	0	4.19	0	0		
	Main product flow				1000	6.93×10^{-3}	996	0	4.19	0	0		
	Net input (in − out, omitting fugitives)				0.233								
Input	C1	20.0	1.00	l	1.39×10^{4}								1.39×10^{4}
Cooling out	C2	50.0	1.00	l	-1.39×10^{4}	0	0	0	0	0	0	0	-1.39×10^{4}
Input	C3	20.0	1.00	l	3341								3341
Cooling out	C4	50.0	1.00	l	−3341	0	0	0	0	0	0	0	−3341
Input	S1	207	1.00	g	808							808	
Steam out	S2	207	1.00	l	−808	0	0	0	0	0	0	−808	0
Input	S3	207	1.00	g	304							304	
Steam out	S4	207	1.00	l	−304	0	0	0	0	0	0	−304	0

[a]All flow rates are given in kg/h. NA, not available.

TABLE 16.3 Energy Input for Each Unit Process, Cumulative Energy Requirements, Cooling Requirements (Exotherms), and Assumed Heat Recovery from Hot Streams Receiving Cooling

Energy Input (MJ/batch)						Cooling Requirements (MJ/batch)								
Process Diagram Label	Unit	Energy Input (MJ/1000 kg)	Cumulative Energy (MJ/1000 kg)	To (°C) (Used to determine energy type)	Energy Type	Process Diagram Label	Unit	Energy Loss	Cumulative Cooling Water	Tef (°C) (for Recovery)	Recovery Efficiency	Energy Recovered	Cumulative Recovered (MJ/1000 kg)
P1	Pump 1	0.102	0.102		E	Hx2	Heat exchanger 2	−2046	−2046	370	0.600	−1227	−1227
Hx1	Heat exchanger 1	1312	1312	300	D	Di1	Distillation condenser 1	−493	−2539	25.0	0	0	−1227
R1	Reactor 1	516	1829	370	D								
P2	Pump 2	2.23×10^{-3}	1829		E								
P3	Pump 3	0.174	1829		E								
P4	Pump 4	6.66×10^{-4}	1829		E								
P5	Pump 5	1.34×10^{-6}	1829		E								

P6	Pump 6	4.86×10^{-4}	1829		E		
Di1	Distillation reboiler 1	493	2323	25.0	S		
P7	Pump 7	0.206	2323		E		
P8	Pump 8	3.86×10^{-4}	2323		E		
P9	Pump 9	4.50×10^{-7}	2323		E		
	Vacuum pump 1	54.0	2377	0	0		
	Potential recovery	−1227	1149				
	Net energy		1149			Potential recovery:	−1227
	Electricity	0.486	E	(MJ/h)			
	DowTherm	1829	D	(MJ/h)			
	Heating steam	493	S	(MJ/h)			
	Direct fuel use	0	F	(MJ/h)			
	Heating natural gas	0	G	(MJ/h)			
	Energy input requirement	2323		(MJ/h)			
	Cooling water	−2539		(MJ/h)			
	Cooling refrigeration			(MJ/h)			
	Potential heat recovery	−1227		(MJ/h)			
	Net energy	1095		(MJ/h)			

TABLE 16.4 3-Pentanone Cradle-to-Gate Summary Including Energy-Related Emissions[a]

	3-Pentanone				
	Total	Process	Energy-Related	Transportation 3-Pentanone	Total 3-Pentanone GTG
Raw Material (kg)					
Air (untreated)	437	437	0		0
Coal	229	0	229		0.0233
Crude oil	968	737	169	62.2	0.169
Natural gas	314	0	314		33.1
Natural gas (unprocessed)	357	357	0		0
Water (untreated)	1058	1058	0		0
Energy (MJ)					
Electricity	4190	4190		0	0.486
DowTherm	1829	1829		0	1829
Heating steam	7507	7507		0	580
Direct fuel	6954	6954		0	0
Natural gas	8124	8124		0	0
Natural gas, refinery	2390	2390		0	0
Coal	0	0		0	0
Coal, refinery	107	107		0	0
Diesel	2693	0		2693	440
Undefined	1194	1194		0	0
Heavy oil, refinery	1009	1009		0	0
Hydro power, refinery	7.16	7.16		0	0
Nuclear power: refinery	7.16	7.16		0	0
Energy input	3.60×10^4	3.33×10^4		2693	2850
Cooling water	-2.24×10^4	-2.24×10^4		0	−2539
Refrigeration	36.0	36.0		0	0
Potential recovery	−9,396	−9,396		0	−1227
Net energy (Input, potential recovery)	2.66×10^4	2.39×10^4		2693	1622
Air Emissions (kg)					
1,3-Butadiene	0.139	0.139	0		0
3-Pentanone	11.0	11.0	0		11.0
Acetylene	0.0186	0.0186	0		0
Ammonia	3.29	3.29	0		0
Argon	2.43	2.43	0		0
Butane	0.0695	0.0695	0		0
Butene	0.241	0.241	0		0
C5 and higher	1.90	1.90	0		0
CO_2	2883	771	1907	205	0
CO	110	108	0.332	1.13	0.175
Catalyst	1.00×10^{-3}	1.00×10^{-3}	0		1.00×10^{-3}

TABLE 16.4 *(Continued)*

	Total	3-Pentanone Process	3-Pentanone Energy-Related	Transportation 3-Pentanone	Total 3-Pentanone GTG
Ethane	27.9	27.9	0		0
Ethanecarboxylic acid	8.67×10^{-3}	8.67×10^{-3}	0		8.67×10^{-3}
Ethylene	11.1	11.1	0		0
Ethylene carbonate	46.6	46.6	0		0
High alkanes	0.0707	0.0707	0		0
Hydrogen	1.03	1.03	0		0
CH_4	13.6	9.17	4.27	0.202	0.216
Naphtha	14.0	14.0	0		0
Nitrogen dioxide	0.984	0.984	0		0
Nitrogen monoxide	0.643	0.643	0		0
NMVOC	12.5	6.31	4.83	1.32	0.640
NO_x	10.8	2.43	4.52	3.84	0.808
Propane	0.0708	0.0708	0		0
Propylene	2.30	2.30	0		0
Propyne	0.0167	0.0167	0		0
SO_x	6.39	1.69	4.44	0.261	−0.138
Total air emissions	3160	1023	1925	212	617
Water Emissions (kg)					
3-Pentanone	6.01	6.01	0		6.01
BOD	0.116	4.59×10^{-5}	0.104	0.0119	-2.21×10^{-3}
BOD5	0.144	0.144	0		0
C5 and higher	0.0151	0.0151	0		0

[a]Level: 4 (cradle-to-gate data after energy requirement and transportation); basis: 100 kg; allocation: 1.00; date: 7/21/07. Not all inputs are raw materials.

temperature would not be expected to change in a uniform pattern); the higher atmospheric temperature is then expected to lead to such effects as climate change, ice-cap melting, and increased sea levels (the tertiary effects); and these changes would then cause higher-order effects such as biodiversity changes and an increase in disease vectors such as malaria (see Figure 16.5).

As the order of the effects increases, so does the complexity and uncertainty of the impact assessment. For example, although radiative forcing changes can be measured in the laboratory, and increased sea levels can be modeled under a series of assumptions, a question arises as to how far one must go to estimate all the environmental impacts of any given pollutant. This is known as the *midpoint vs. endpoint debate* when performing an LCA. There have been many attempts to estimate the endpoint impacts within life cycle assessment; however, the price that is paid is in terms of uncertainty. It is currently general practice to estimate midpoint impacts, as they provide useful and comprehensive information on the effects of an activity and simplify communications while minimizing extrapolations, assumptions, and the complexity of the model.

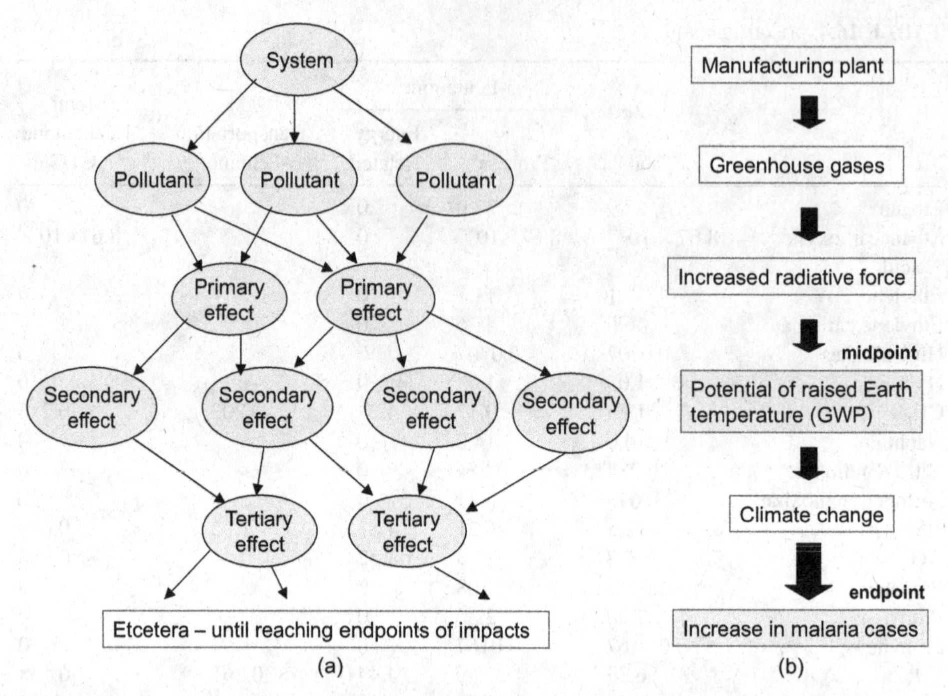

FIGURE 16.5 Representation of a chain/network of effects (a) and an example (b).

Nonetheless, using midpoint impacts still involves a certain level of uncertainly, and the level of uncertainty increases as one moves from midpoints to endpoints. Therefore, most impacts are expressed in terms of potentials (e.g., global warming potential, acidification potential) instead of using actual effects or outcomes resulting from the specific emission or pollutant. In addition, the global effects are better estimated than are local or regional effects, since it is more difficult to account for or explicitly attribute specific outcomes at the local or regional level (i.e., what the final fate of pollutants is, or to untangle particular diseases from several potential causes).

16.2.4 Performing an LCI/A

ISO standard 14044 includes defined guidelines for each of the elements that comprise the impact assessment (Figure 16.2).

1. *Category selection and definition.* This element is about determining which impacts would be evaluated (e.g., cumulative energy demand, acidification potential), the models to be used (e.g., CML, EDIP), the framework chosen (e.g., Nordic guidelines, SETAC, corporate guidelines), the cause–effect chains, and the methods to measure them (e.g., use of midpoints or endpoints). Selecting and defining the right categories will depend on the goal and scope of the overall LCI/A, although it is common practice to select categories that measure the effects of the activity on resources, humans, and ecosystems. Another point to consider is the debate between endpoints and midpoints described above. Table 16.5 includes some examples of category definitions commonly used.[25–29]

TABLE 16.5 Some Examples of Commonly Used Life Cycle Impact Assessment Methods and Guidelines

Guideline	Impact Categories
Nordic Guidelines on Life Cycle Assessment[25]	Resources
	Energy and materials
	Water
	Land (including wetlands)
	Human health
	Toxicological impacts (excludes work environment)
	Nontoxicological impacts (excludes work environment)
	Impacts in work environment
	Ecological consequences
	Global warming
	Depletion of stratospheric ozone
	Acidification
	Eutrophication
	Photochemical ozone creation
	Ecotoxicological impacts
	Habitat alterations and impacts in biodiversity
	Inflows that are not traced back to the "cradle" (not an impact category, but it should be included)
	Outflows that are not traced back (not an impact category, but it should be included)
SETAC – Europe[26]	Input-related
	Abiotic resources
	Biotic resources
	Land
	Inputs and outputs from and to other technical systems
	Input-related (energy, materials, plantation wood, etc.)
	Output-related (solid waste, etc.)
	Output-related
	Global warming
	Depletion of stratospheric ozone
	Human toxicological impacts
	Ecotoxicological impacts
	Photochemical ozone creation
	Acidification
	Eutrophication
	Odor
	Noise
	Radiation
	Casualties
CML[27]	Baseline categories
	Impact on land use
	Ecotoxicity
	Freshwater aquatic
	Marine aquatic
	Terriestrial
	Human toxicity
	Climate change

(continued)

TABLE 16.5 (*Continued*)

Guideline	Impact Categories
	Stratospheric ozone depletion
	Photochemical ozone formation
	Acidification
	Eutrophication
EPS[28]	Safeguard subjects
	Human Health
	Biological diversity
	Ecosystem production capacity
	Abiotic resources
	Cultural and recreational values
TRACI[29]	Environmental
	Global warming
	Depletion of stratospheric ozone
	Photochemical ozone creation
	Acidification
	Eutrophication
	Ecotoxicity
	Human health
	Cancer
	Noncancer
	Criteria air pollutants (initial focus on particulates)
	Resources
	Fossil fuel use
	Land use
	Water use

2. *Classification.* This element requires that we match the inventory results (e.g., total kilograms of carbon dioxide emissions) with specific impact categories (e.g., global warming potential).

3. *Characterization.* This element is concerned with estimating the environmental impact for each category using science-based conversion factors. You may recall that in Chapter 3 we covered in some detail most of the impact categories used in LCAs. Characterization factors for global warming potential, ozone depletion potential, photochemical ozone creation potential, acidification potential, and eutrophication potential may be found in Tables 3.1 through 3.5. Example 3.1 showed how to estimate those impacts.

4. *Normalization.* This element is about expressing impact results in such a way that meaningful comparisons may be drawn (e.g., normalizing by regional impact, benchmarking by process)

5. *Grouping.* For this element we think about how to sort the results, as, for example, into local, regional, and global impacts. This may be done either for presenting the results or as part of completing the assessment.

6. *Weighting.* This element can be understood in two ways. First, we may aggregate the results for several pollutants across different categories that may be related to give them greater importance. Second, we may actually give an impact category greater importance

subjectively because of its potential to cause a particularly bad outcome (e.g., as in weighing a human cancer outcome more highly than crop damage).

7. *Quality assessment.* This element is where we work with the data to obtain a better understanding of the reliability of the results by applying appropriate statistical and other analytical methods (e.g., sensitivity analysis, data quality indicators, uncertainty analysis).

Figure 16.6 shows how most elements of the life cycle assessment might be applied in an industrial setting. An example of a sensitivity assessment is shown in Figure 16.7.

In Example 16.2 the impact categories that were used follow GlaxoSmithKline's corporate life cycle assessment and sustainability guidelines and metrics. These guidelines and metrics were developed in-house based on state-of-the-art life cycle methodology and science. It should be noted, however, that there are a variety of ready-to-use life cycle impact assessment methods that have been designed to simplify the work of an LCA practitioner, and these have been embedded in the most common software packages used for LCI/A, including the EcoIndicator 99 model,[30] EDIP, input–output LCA,[31] environmental themes, constant-volumes method, EPS, CML, and TRACI, among others.

With these ready-to-use LCI/A methods and software packages the characterization methods are integrated into characterization indicators or even a single LCI/A index that is derived from previously established approaches to weighting, normalization, and groupings that are normally based on societal, corporate, or government values. Since these different methods reflect different values and priorities, it is not worth one's time debating whether one is more correct than the other, but it is essential for you to be aware of the underlying assumptions, values, priorities, and potential biases (i.e., everyone has a different idea as to what is or is not important).

It is worth a bit more time to reflect on this. If one is considering using one of these methods, it is important to understand that the priorities and assumptions behind them may cause the conclusions to vary depending on what impact is considered to be more important and how much weight is going to be given to different impacts. In addition, one should be aware that there may be a lack of impact data, or the impact data may be a bit suspect, so the developers of these methods by necessity have resorted to the use of default values and nearest-neighbor approaches to use in the indexes. Consequently, it is essential that you understand how the underlying assumptions will affect the sensitivity of the conclusions when using preset impact assessment methodologies.

Example 16.3 Estimate the environmental impacts of a chiral active pharmaceutical ingredient using life cycle assessment.

Solution A full life cycle inventory and assessment was undertaken to estimate the environmental footprint of a chiral API.[32]

Goal and scope. For this assessment the goal was to provide life cycle insight into the process selection and development of complex drugs within the pharmaceutical industry. This study was a comparative cradle-to-gate life cycle assessment of different synthetic routes (production processes) to manufacture a chiral pharmaceutical product [CPP; an active pharmaceutical ingredient (API)]. The analysis included eight variations of the production process for the regulatory starting material (RSM) and five different processes for production of the CPP from the RSM (Route 5, Route 6, EtOH process, THF process, TOL process). The packaging and distribution of the drug once manufactured were not

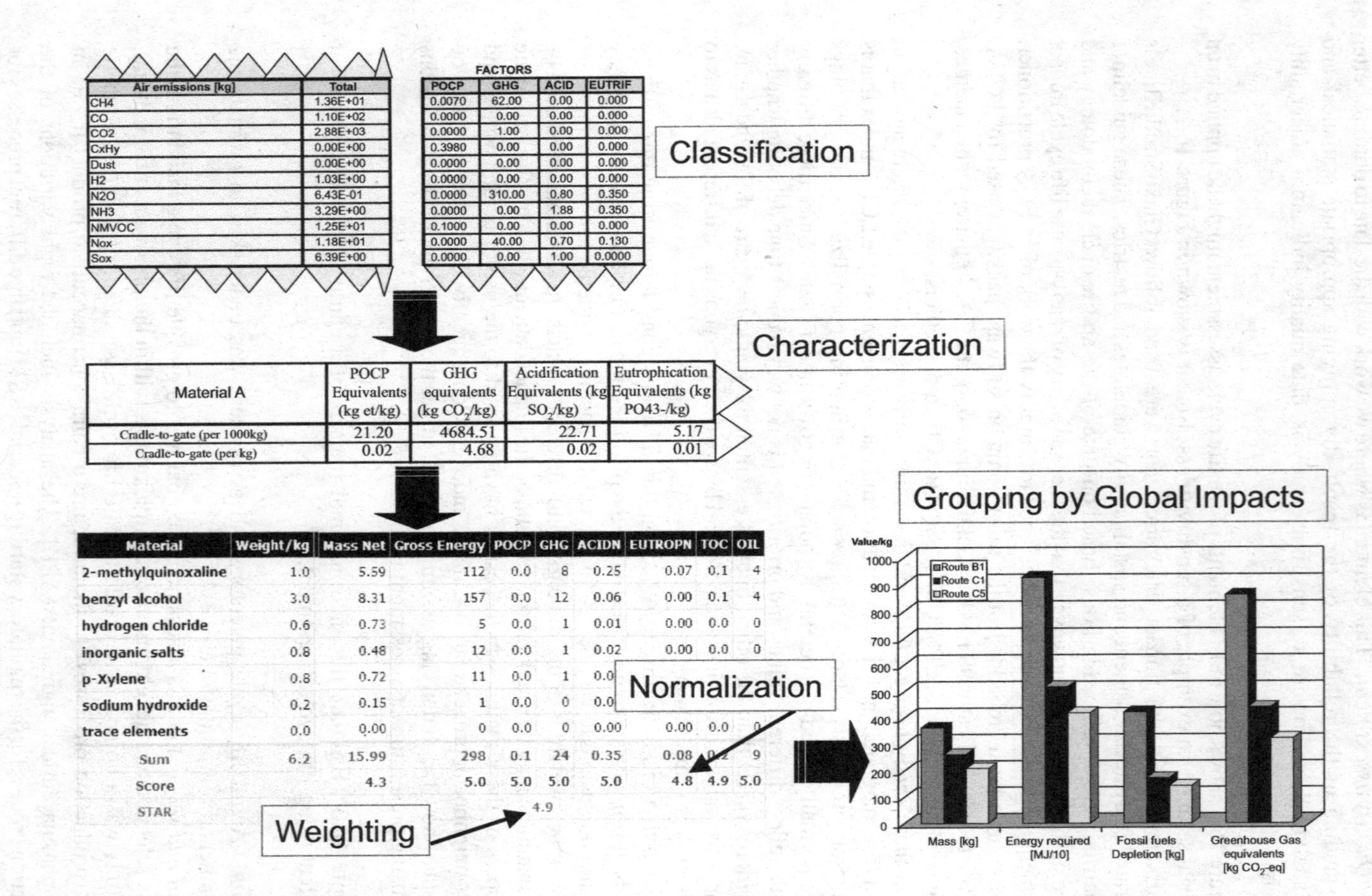

FIGURE 16.6 Examples of an industrial application of most of the elements of a life cycle impact assessment. In addition, sensitivity analysis was performed.

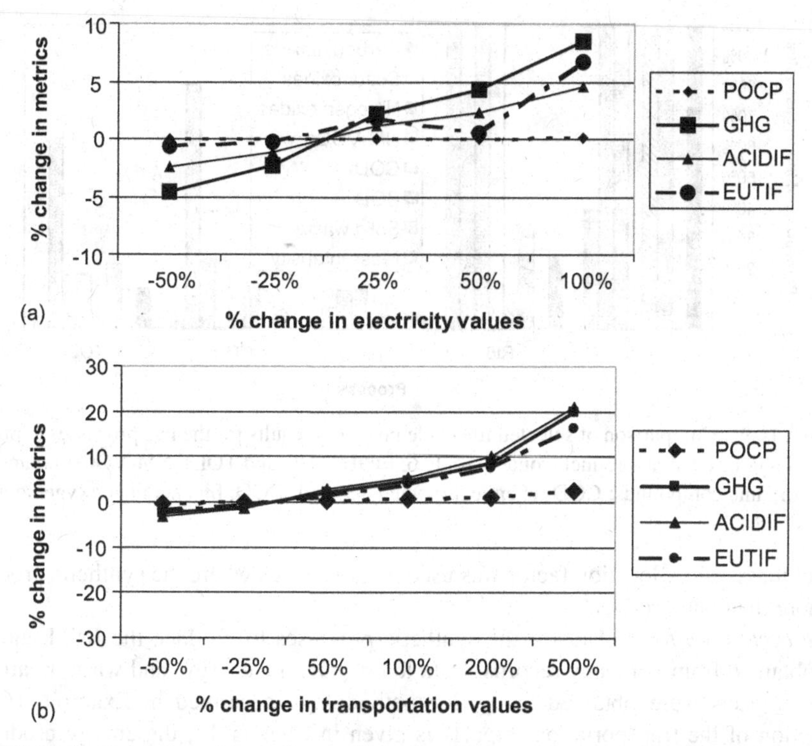

FIGURE 16.7 Examples of a quality analysis. Sensitivity analysis on the effects of changes in electricity (a) and transportation (b) values in a cradle-to-gate LCIA. (From Jiménez-González, ref. 39. Reproduced with kind permission from Springer Science and Business Media. Copyright © 2004, Springer Science and Business Media.)

considered, as shown in Figure 16.8. The functional unit was defined as 1000 kg of CPP produced. This functional unit was chosen since the exact amount of CPP (API) dosages that would fulfill the function of the drug will vary depending on the patient and the severity of the ailment. In addition, since this research was conducted using the cradle-to-gate approach, a mass-based functional unit makes the comparison between routes and processes easier. A

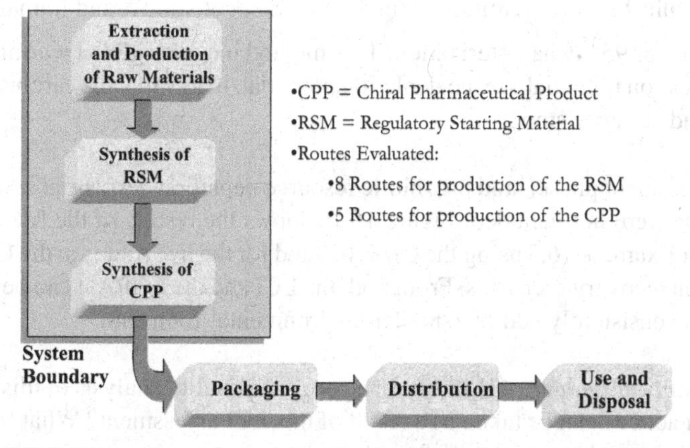

FIGURE 16.8 System boundaries for the life cycle assessment of an API.

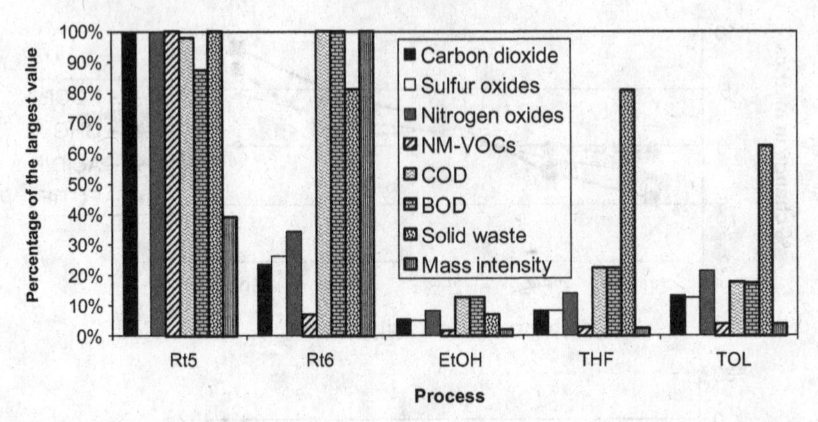

FIGURE 16.9 Comparison of selected life cycle inventory results for the five processes to produce the chiral pharmaceutical product (route 5, route 6, EtOH, THF, and TOL). NM-VOCs, nonmethane volatile organic compounds; COD, chemical oxygen demand; BOD, biochemical oxygen demand. (From ref 32.)

product mass ratio allocation factor was used in those cases where the synthetic processes had more than one product.

Life cycle inventory. Data for the synthetic processes to produce the RSM and CPP were obtained from company records. Data to estimate the energy- and waste treatment–related impacts were obtained using the methodology described in Example 16.2. A description of the transportation impacts is given in Chapter 17, the energy production impacts are described in Chapter 18, and the impacts from waste treatment are described in Chapter 19. Figure 16.9 shows a comparison of selected life cycle inventory results for the five processes to produce the chiral pharmaceutical product.

Life cycle impact assessment. Three different impact assessment methodologies were used:

- Critical volumes method, BUWAL 132.[33] This method includes factors for air, water, and energy.
- CML (Centrum voor Milieukunde in Leiden, The Netherlands).[34] This method includes global warming potential, eutrophication potential, ozone depletion potential, photochemical ozone creation potential, aquatic ecotoxicity, and human toxicity.
- Eco-Indicator 95[35]. Characterization. This method includes global warming potential, eutrophication potential, ozone depletion potential, heavy metals, carcinogens, winter smog, and summer smog.

For the assessment phase, land use, biotic resource depletion, terrestrial ecotoxicity, and landfill volume were not included. Figure 16.10 shows the results of the life cycle impact assessment for Example 16.3 using the CML method for the five routes to the CPP, with and without solvent recovery scenarios. From both the LCI and the LCI/A it can be seen that the EtOH process consistently had the smaller environmental footprint.

Additional Points to Ponder How can one perform a quality analysis of this assessment? What types of actions can be taken as a result of this LCI assessment? What happens if the results of the impacts have trade-offs?

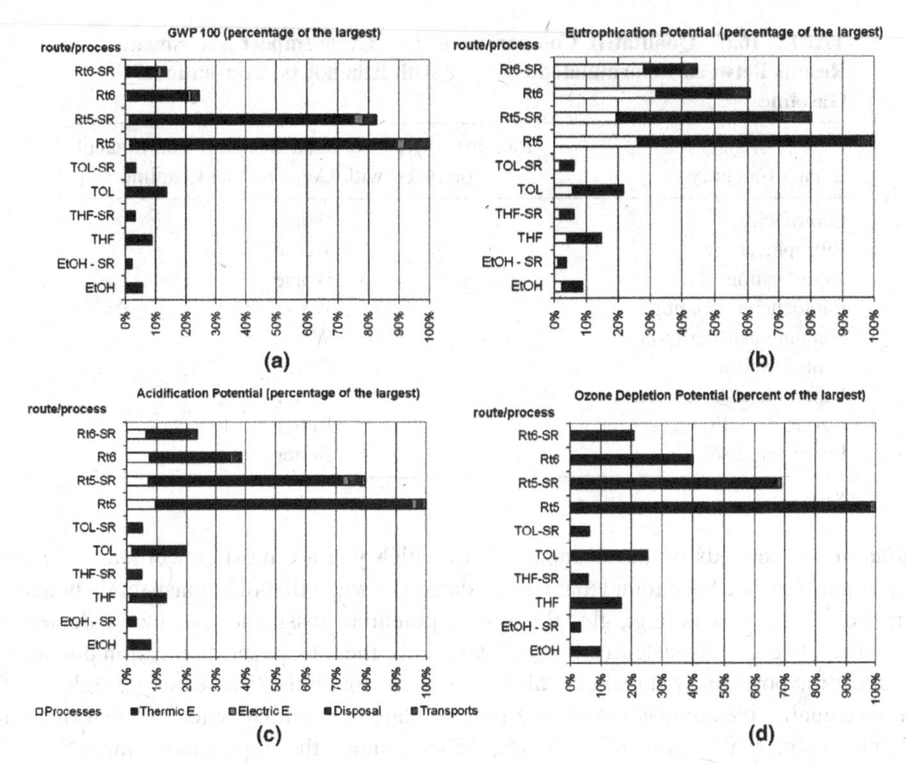

FIGURE 16.10 Results of the life cycle impact assessment for Example 16.3 using the CML method for global warming potential (a), eutrophication potential (b), acidification potential (c), and ozone depletion potential (d). (From ref 32.)

16.3 INTERPRETATION: MAKING DECISIONS WITH LCI/A

During this part of the life cycle assessment, the results of the inventory analysis and impact assessment are evaluated for the products, processes, or services under study. This phase should include a clear understanding of any uncertainty in the data and the assumptions used to generate the results. In principle, this may sound simple, and in theory we would have a system such as the one described in Example 16.3, where it was clear from both the inventory and the impact assessment which route has the lowest impacts. But what happens when that is not the case? It is relatively common that with a long list of impacts to evaluate, one can find that process A might have a better profile than process B in some impacts, but worse in the remaining impacts. One must therefore come to terms with these trade-offs between environmental impacts during the decision-making process. Whether the trade-offs are important or not generally depends on the specific priorities of the decision makers, how large a difference there is among the trade-offs, and how much uncertainty there is in the data quality or overall assessment.

Example 16.4 Which is better: reformulated gasoline containing ethanol or conventional gasoline?

TABLE 16.6 Qualitative Comparison of Life Cycle Impact Assessment Results Between Reformulated Gasoline with Ethanol vs. Conventional Gasoline

Impact Category	Ranking of Reformulated Gasoline with Ethanol Compared with Conventional Gasoline
Ecotoxicity	Worse
Eutrophication	Worse
Acidification	Worse
Photochemical smog	Worse
Human health criteria	Worse
Global warming	Better
Human health noncancer	Better
Ozone depletion	Better
Fossil fuel use	Better

Source: ref. 36.

Solution It depends on those impacts with which you are most concerned. Different assessments to date have found that reformulated gas with ethanol is consistently better in terms of some impacts (e.g., global warming potential, fossil fuel use, ozone depletion potential), but it is consistently worse in terms of others (e.g., eutrophication potential, acidification potential, photochemical ozone creation potential). For example, Table 16.6 shows a qualitative comparison between reformulated gas with ethanol and conventional gasoline using a 5% cutoff rule and TRACI to estimate the impact assessment.[36]

Additional Points to Ponder What would be the effect of a change in land use to grow biofuel and biofeedstock materials? What happens when people report only one impact?

As can be seen from the biofuel example above, transparency once again plays an important role in the decision-making process. Some people tend to report only one or two impact categories, either because they desire to simplify the results of the life cycle assessment, lack experience in LCI/A, or lack a sufficient understanding of the life cycle and its nuances. This can be extremely misleading if there are some environmental impact trade-offs that are not being identified or considered, or if the environmental impact categories do not render the same results using different assessment methods.

Here are some important considerations to keep in mind when making decisions based on life cycle impact assessments or life cycle inventory assessments:

- *The goal and scope of the LCA.* One cannot easily extrapolate from one system to another unless the data and science support it.
- *Boundaries and boundary conditions are extremely important.* In comparative LCI/As these are fundamental, but it is even more important when talking about stand-alone LCAs, as the differences in boundaries can mean different outcomes.
- *The degree of uncertainty in the data and the associated results.* This is especially important when comparing outputs. How significant are the differences?
- *Sensitivity of the results to variations in specific parameters.* If other databases or assessment methods were used, would the results remain valid?
- *Trade-offs.* Are all the important impacts considered, or are there hidden trade-offs in the assessment results?

- *Applicability of the data.* In some instances, spatial and temporal differences in the data might cause important differences.
- *Organizational values.* Which impacts are valued above others, and do these align with the inherent priorities of the impact assessment methods employed?
- *Micro- and macroeconomic factors.* From a holistic viewpoint, would the total cost of the activity change the results? Are market forces affecting the assessment in some way?
- *Societal values.* Are there cultural impacts that might not be immediately apparent in the metrics used for the assessment?

Another example of life cycle concepts being used in industry is found in the ecoefficiency and SEEBalance assessments developed by BASF.[37] In the BASF methodology, the ecoefficiency assessment attempts to measure both environmental and economic factors within a life cycle assessment framework. BASF has been using these assessments since 1996 on numerous products and processes as a means to achieve constant improvement in environmental and economic performance.

Some examples where the tool has been applied include the selection of the best synthetic route to indigo based on an assessment of four different technologies for indigo production; the assessment of two methods for cereal crop preservation to serve as a basis for discussing chemical additives in animal feeds, the analysis of two furniture board processes, which aided in making technology investment decisions; and others.[38] According to the BASF Web site, results of the ecoefficiency analysis are intended to inform business decisions about:

- Whether or not further improvements are possible or if the assessment can be used as a selling point when the analysis confirms a high degree of ecoefficiency.
- Whether and to what extent the economic or ecological footprint can be improved when the analysis shows that a product or process has inadequate ecoefficiency but it is possible to improve it.
- Whether to abandon or substitute a product or process when the analysis shows that it has poor ecoefficiency and cannot be improved with a reasonable investment.
- Whether to direct further research and development to the most ecoefficient alternative when a product or process is in the early stages of research and there are several credible paths for further development to follow.

16.3.1 Presenting LCI/A Results

There are certainly many different ways to present the results of an LCI/A, depending on the type of information to be presented, the goal of the LCI/A study, and the stakeholders for the assessment. In a few words, the results and the corresponding information should strive to answer the question posed in the goal and scope definition and be easy to understand by the intended audience. In many situations, the same results need to be presented in several different ways to convey the appropriate and desired message. This can be done:

- *Quantitatively.* The results may be presented in terms of the LCI (Figure 16.9) or LCI/A results (Figure 16.10). Depending on what question is being answered,

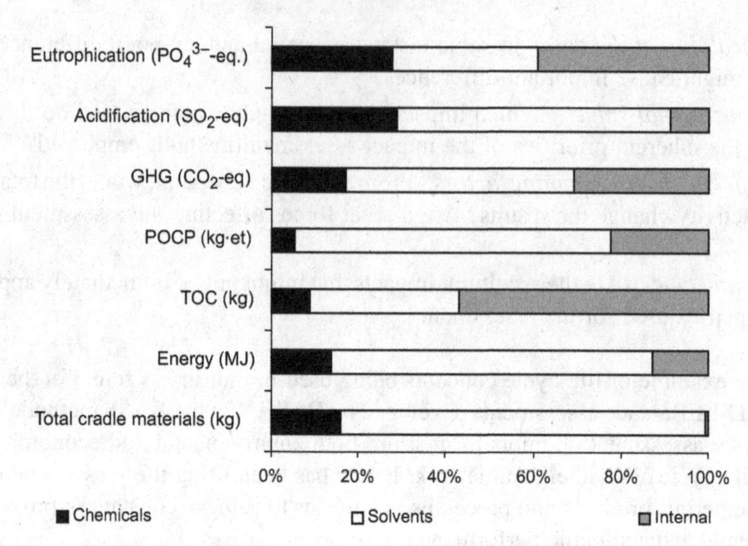

FIGURE 16.11 Cradle-to-gate LCA pretreatment contributions of solvent manufacturing, production of nonsolvent chemicals, and internal active pharmaceutical ingredient manufacturing. (From Jiménez-González et. al, ref 39. Reproduced with kind permission from Springer Science and Business Media. Copyright © 2004, Springer Science and Business Media.)

the information may be presented in such a way that additional information is easy to obtain. For example, one might choose to show the results in terms of a:

Dominance analysis. This can be used to ascertain which of the life cycle phases or activities contribute most to the environmental footprint (Figures 16.11 and 16.12).[39]

Contribution analysis. Similar to the dominance analysis, but instead of determining those activities with the largest contributions, it focuses on environmental loads or emissions (Figure 16.13).[40]

Break-even analysis. This serves to determine environmental impact trade-offs or to set goals for product development [41] (Figure 16.14).

Data quality analysis. This can be expressed in terms of sensitivity or uncertainty analysis (Figure 16.7). Data quality indicators can also be used as a semi-quantitative assessment uncertainty of the LCI/A.[42]

- *Qualitatively.* Some times it is desired to maintain a very high level of detail as a means of showing contrast or identifying hot buttons on an LCA (Table 16.6).

Example 16.5 An ecoefficiency analysis was performed to compare the life cycle environmental impacts and costs for BASF and several competitors to produce ibuprofen. The basis for the assessment was the production of 1 kg of ibuprofen for use in over-the-counter pain relief medication sold in North America. The processes being compared were:

- Ibuprofen produced by BASF in the United States
- Acetylchloride and cyanohydrin chemistry in the United States
- Acetylchloride, glycidester, and Cr(VI)-oxidation chemistry in India
- Chlorpropionylchloride and rearrangement chemistry in China

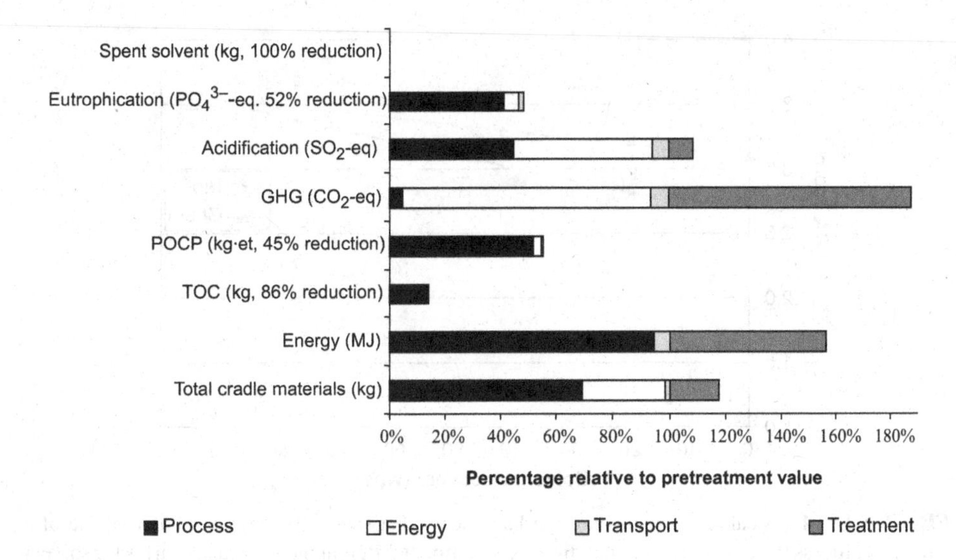

FIGURE 16.12 Cradle-to-gate LCA posttreatment contributions of energy, production processes, transportation, and treatment systems for pharmaceutical active ingredient manufacturing. (From Jiménez-González et. al, ref 39. Reproduced with kind permission from Springer Science and Business Media. Copyright © 2004, Springer Science and Business Media.)

The results of the assessment are presented in Figures 16.15 and 16.16. What conclusions can be drawn from the assessment?

Solution In general, the ibuprofen produced by BASF is more ecoefficient, followed closely by the U.S. competitor. The ibuprofen produced in the United States costs more

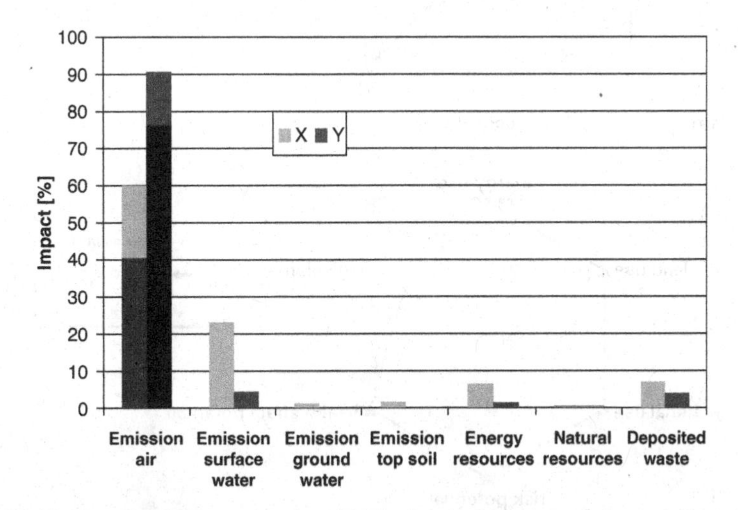

FIGURE 16.13 Impact of X compared to Y. The hatched portion of the emission air segment represents the combustion gases (CO_2, NO_x, SO_2) from energy production. (Source: ref 40, Reproduced with permission.)

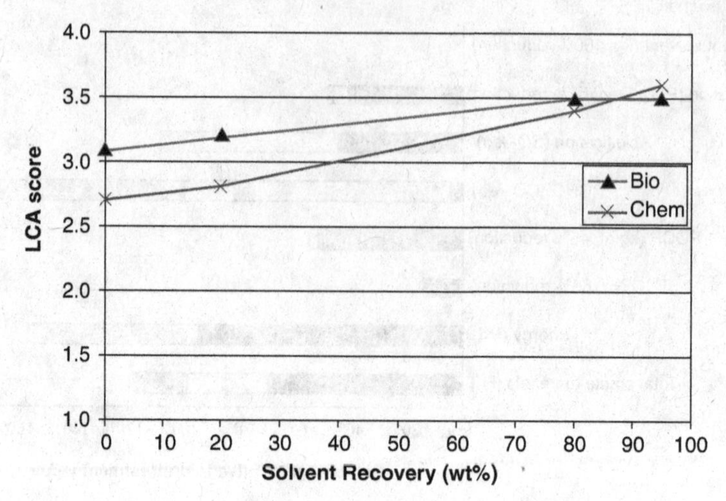

FIGURE 16.14 Break-even analysis of an LCA score that shows the environmental profile of a chemical process (Chem) and a biocatalytic process (Bio) at different theoretical scenarios for solvent recovery. (Source: ref 41.)

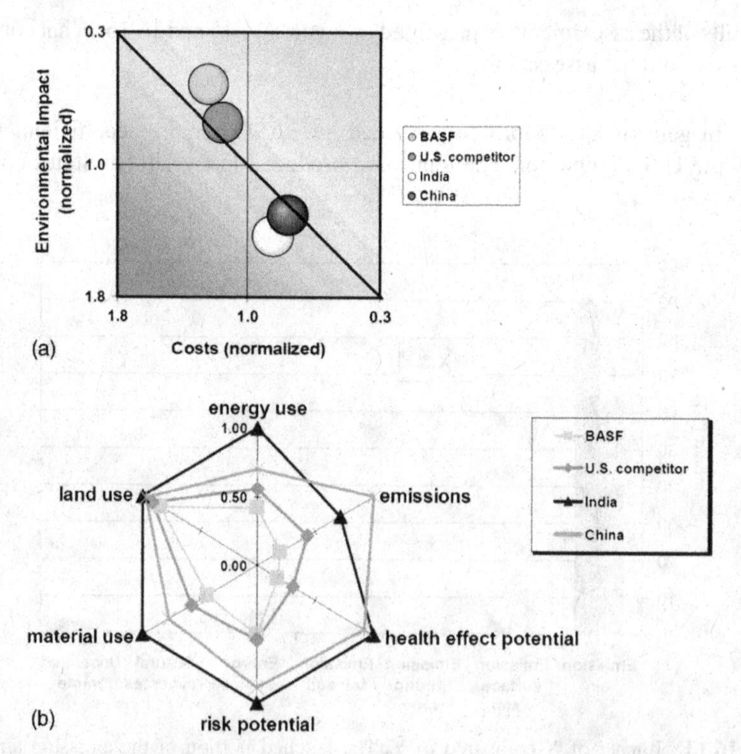

FIGURE 16.15 Ecoefficiency and footprint comparisons for the BASF example. (Source: ref 38, Reproduced with permission from BASF.)

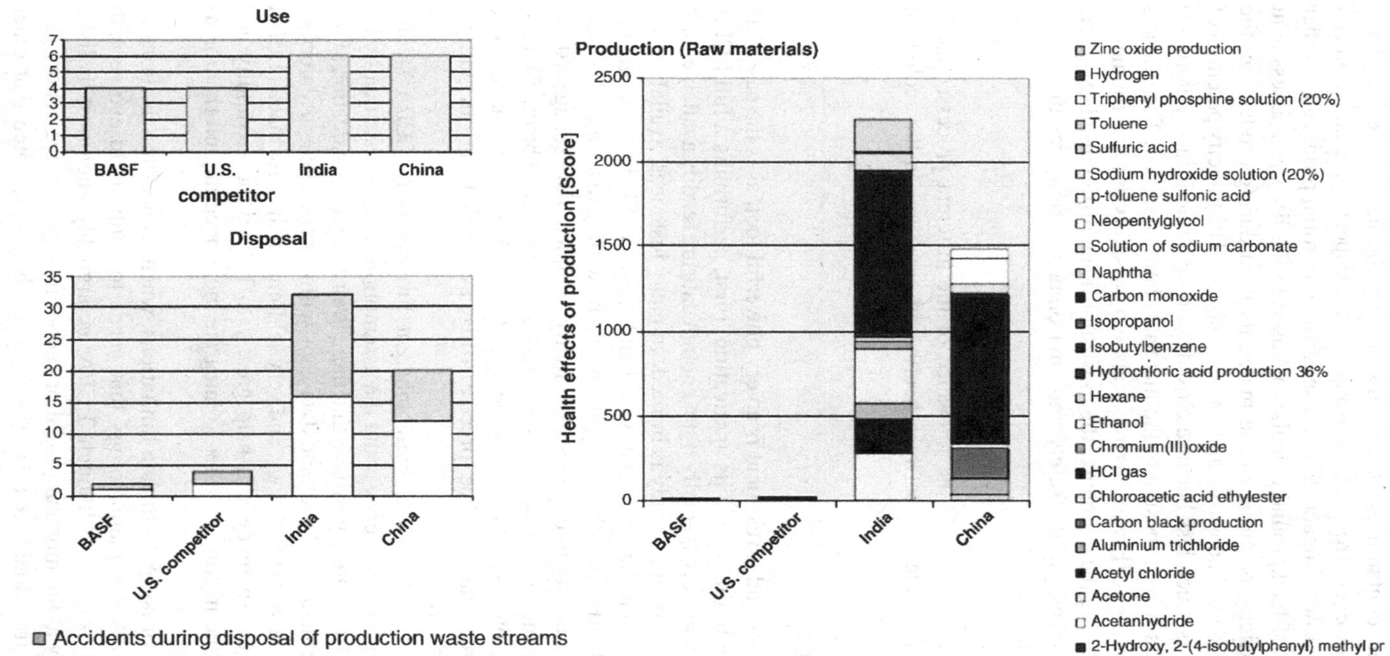

FIGURE 16.16 Some of the results of the life cycle assessment for the ibuprofen comparisons during the production, use, and disposal phase. (Source: ref 38. Reproduced with permission from BASF.)

than ibuprofen produced in India or China at the time of the assessment. These costs are somehow a result of the resources used to optimize their processes, maintain and operate their equipment to prevent contamination, and ensure that the environment and the workers are protected. However, from an environmental perspective, U.S. produced ibuprofen has significantly lower impacts than those of the ibuprofen produced in other countries. Some examples of background calculations used in the life cycle assessment (Figure 16.5) show the toxicity potential of the materials used during production, the disposal impacts due to accidents and waste, and the use impacts. The toxicity potential of the U.S.-based production is much less because closed-loop manufacturing systems are used, which protect workers from exposure to the materials being handled. The impacts for disposal reflect the quantities of hazardous materials and the safety and environmental controls that are present. During the use phase, the most significant differentiating factors are the transportation distances and the safety and environmental practices that are in place at the various manufacturers.

Additional Points to Ponder What is the relevance of the different processes used? How would you interpret the trade-offs between health and emissions in the ecological footprint?

16.3.2 Limitations of LCI/A

Although life cycle inventory and assessment is a very powerful tool, as is the case with any methodology, there are limitations in its application. First, performing a full LCI/A can be resource and time intensive, especially if not enough data are readily available, or if one's knowledge of the system under study is limited. In many instances, available data are limited, and for those data that are readily at hand, there is likely to be greater uncertainty associated with these data than what one might be willing to accept. One may therefore wish to weigh the ability to obtain a sufficiently complete data set against the desired degree of accuracy that one believes is necessary for the study to be credible. In almost all LCI/A studies one will inevitably be faced with having to make some assumptions about the source data or will be forced to utilize averaged data from multiple sources (i.e., background data).

As has been stated repeatedly above, these inherent limitations in LCI/A studies necessitate that one maintains transparency about the assumptions and the data quality for any data used in the study. Transparency ensures that there is an explicit demarcation of the sources, assumptions, information utilized, and data gaps within a life cycle inventory or assessment. LCI/A is an iterative methodology, and very seldom is it the case that an LCI/A result is fixed; thus, data transparency will help greatly to bolster the credibility and acceptance of the study while making it easier to integrate more accurate information as new data become available.

In addition to problems or issues with data limitations, when seen within the broader sustainability context, LCI/A is a methodology that needs to continue to evolve with the state of the art and science. Although LCI/A is arguably the most holistic sustainability methodology at the moment, social or economic aspects have historically been out of the scope of traditional LCI/As. However, it should be noted that other tools, such as social LCA and total cost accounting (life cycle costing), tend to cover these aspects (see Chapter 20).

16.3.3 Critical Review

As we saw in Section 16.3.2, LCI/A methodology has several limitations, many of them linked to the transparency of the underlying data, the assumptions, and the methodology used to undertake the assessment. In some LCAs, particularly those used to assess high-profile commercial, societal, or policy questions, there have been concerns that the practitioners might be trying to demonstrate a specific point of view instead of looking for a scientifically defensible explanation. In an attempt to ensure the credibility of LCI/A methodology and to answer concerns about objectivity, there has been a movement toward increased standardization of LCI/A methodologies and critical peer reviews.

Besides enhancing credibility, critical peer review helps to maintain data confidentiality and integrity by allowing a reviewer to validate the study as they review the entire assessment. This allows the final results to be communicated with confidence without having to show the confidential information. In addition, as one delves more deeply into LCI/A and develops greater experience with it, it becomes clear that performing a high-quality assessment is more complex than one might at first think upon learning the conceptual idea of the cradle-to-grave assessment. A critical reviewer can therefore help the practitioner to choose the appropriate and relevant methodology to achieve the goal of the study or might help to ensure that the ISO standards are maintained. Finally, when the results of the LCA may possibly affect policy or market decisions, inviting interested parties and stakeholders as part of the critical review is an advisable option.

The ISO standard contemplates three types of reviews: review by internal experts, review by external experts, and review by interested parties. Reviews performed by interested parties are normally done as a panel of experts, whereas internal and external expert reviews tend to be done by a single reviewer. Not all LCAs are subject to critical review, but it is definitely advisable to consider it in the planning. In addition, not all critical reviews are the same. The ISO standards recognize two comprehensive and formal critical reviews:

- The ISO standard full review, which is mandatory in cases where comparative assessments are done for a public audience in mind. The primary focus of this review is on the methodology, its consistency with the ISO standard, and its scientific validity.
- The review of environmental product declarations, where the primary focus is on the underpinning data. In these cases, customers are expected to draw conclusions and comparisons based on potentially nontransparent (commercially sensitive or proprietary) aggregated data.

There are very specific guidelines for these two types of mandatory critical reviews. In contrast to these reviews, the focus of the internal expert or external expert reviews will need to be decided as part of the goal and scope of the LCA in question.

16.4 STREAMLINED LIFE CYCLE ASSESSMENT

As we saw in Section 16.4, performing a full LCI/A can be very time consuming, and such a study normally requires an extensive collection of data and a large database. When the results of LCI/A are intended for process or product development decision making, producing a full LCI/A is not always feasible due to limited data availability during process development. Waiting to obtain a complete and well-validated data set that could stand up to the full rigor of an LCA would very likely mean that assessment results would be produced long after any potential recommendations could be put to good use. At the same time, the most efficient and cost-effective time to introduce sustainability considerations into a process, product, or activity is precisely during development. In addition, the more holistic these considerations are (i.e., with life cycle thinking integrated), the greater the likelihood that the process, product, or activity will be aligned to the principles of sustainability.

One way to address this dilemma is to use a simplified or streamlined life cycle assessment method. According to SETAC, a streamlined LCA applies LCA methodology and covers the same aspects, but at a higher level. Thus, instead of using site-, time-, mode-, and technology-specific data throughout the assessment, a streamlined LCA would tend to use generic background data or standard modules for transportation, waste treatment, and energy production to conduct a simplified assessment that focuses on the most important phases or impacts. One would then assess the reliability of the results with uncertainty and sensitivity analysis.

Most simplified LCAs consist of three phases: an initial screening to determine the main points of focus; a simplification phase that uses the results of the screening to determine which areas can be simplified and which need more in-depth assessments, and a quality analysis phase that assesses the reliability of the results. As is common with a full LCA, a streamlined LCA can be used iteratively to identify those parts of the system that require further assessment, or it can be used to determine indicators, as a prior step to performing a full LCA, as a complementary study to a full LCA, or as a basis to evaluate similar systems in a faster manner. As might be expected, the quality analysis phase is a very important part of any streamlined LCA because as one streamlines the study the uncertainty increases, and there is an increased chance of obtaining significantly different results than would be obtained with a full LCA. However, there are many occasions where the risk associated with a streamlined LCA is acceptable, as in the case of process and product development, when there are insufficient data to perform a full LCA, when the time lines required to complete a full LCA would hinder the integration of any results into the improvement assessment, when the effort is better spent on the most important subsection of the LCA, and so on. Given the increased importance of streamlined LCAs, some general guidance has been proposed by Hunt et al. for how to conduct streamlined LCAs and how to select a streamlining method so as to reduce the potential for error.[43]

One example of the application of a streamlined LCA is given in Chapter 6. In Figure 6.8 a company-specific solvent selection guide is shown. It can be seen that there is a column that ranks the life cycle profiles of the solvents against each other. Although there is underlying rigorous LCI data behind the one-digit comparative rankings, the final user is given a streamlined way to quickly assess and compare the life cycle effects of the solvent choice.

Another industrial example of a streamlined LCA tool is shown in Figure 16.17. This streamlined LCA tool is used to compare the life cycle impacts of materials used in synthetic chemistry routes and was designed based on a full LCA of an active pharmaceutical ingredient. The scores are shown on a scale of 1 to 5 (1, lowest impact; 5, highest impact), and the user is able to obtain a quick, high-level benchmark of the impacts of the materials used in the synthetic route. This assessment is typically performed in less than 30 minutes.[44]

Example 16.6 Ionic liquids have been of great academic interest as a means of avoiding some of the environmental impacts associated with organic solvents, such as those having high volatilities or high POCP. However, there have been many questions about whether or

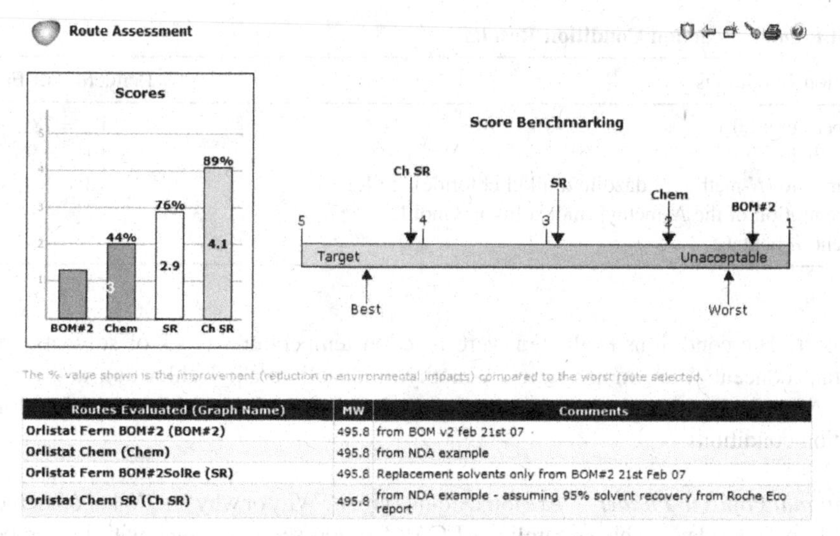

Routes Evaluated (Graph Name)	MW	Comments
Orlistat Ferm BOM#2 (BOM#2)	495.8	from BOM v2 feb 21st 07
Orlistat Chem (Chem)	495.8	from NDA example
Orlistat Ferm BOM#2SolRe (SR)	495.8	Replacement solvents only from BOM#2 21st Feb 07
Orlistat Chem SR (Ch SR)	495.8	from NDA example - assuming 95% solvent recovery from Roche Eco report

FIGURE 16.17 Output of a streamlined life cycle assessment tool used to compare the impacts of materials of synthetic chemical routes. The scores shown are on a 1 to 5 scale (1, lowest impact; 5, highest impact). The percentages over the bar charts show the relative average improvement on the environmental life cycle impacts of the materials.

not and to what extent they are truly green, due to toxicity and life cycle footprint concerns. Use a streamlined life cycle assessment method to aid in optimization of the alkylation step of the production of 1-hexyl-3-methylimidazolium chloride from *N*-methylimidazolium:

$$H_3C{-}\text{(imidazole)} \xrightarrow[\text{X = halogen, Br, Cl}]{C_6H_{13}X} H_3C{-}\text{(imidazolium}^{+}\text{)}{-}C_6H_{13}\;\; X^-$$

Solution This reaction has been studied within a life cycle framework to help design an alkylation step having the smallest environmental footprint and the example is reported in the literature.[45] The researchers utilized the ECO method, a streamlined life cycle assessment method that seeks to combine environmental impact and cost optimization of chemical synthesis at the R&D stage. It utilizes three objective functions to account for the effects of materials, reaction, separation, use of products, and disposal:

- *Energy factor* (EF): assessing cumulative energy demand
- *Environmental and human health factors* (EHF): assessing environmental and toxicological aspects
- *Cost factor* (CF): intending to assess life cycle costing aspects (see Chapter 20 for more details)

The three objective functions are minimized to obtain the best parameters for optimizing the synthetic route. For this reaction, the initial experimental parameters were evaluated to obtain the conditions that minimized energy use, environmental and human health impacts,

TABLE 16.7 Reaction Condition Results

Reaction Conditions	Reduction of Factors
Temperature: 80°C	EF = 78%
Time: 30 h	EHF = 98%
Molar ratio (*N*-methylimidazolium/alkyl chloride): 1 : 1.2	CF = 87%
Concentration of the *N*-methylimidazolium: 3 mol/L	
Solvent: *n*-heptane	

and cost. The conditions evaluated were reaction temperature, types of solvents, initial reactant concentration, molar ratio (*N*-methylimidazolium/alkyl chloride), and reaction time. After the optimization procedure, the results given in Table 16.7 were found for the reaction conditions.

Additional Points to Ponder Are ionic liquids green? Why or why not? What other factors would you consider in this streamlined LCA? Do you see any issues with the proposed optimal reaction parameters?

The importance of streamlined LCAs has increased over the past decade and the development and application of reliable streamlined methods is one of the main areas in the LCA arena that would improve the uptake of LCA significantly in the research, development, and industrial sectors in general. Many streamlined life cycle assessment methods have been developed in industry and academia,[46–55] reflecting the generalized need for adequate LCA screening methods and indicators.

In Chapter 17 we explore the application of several examples of streamlined LCA to evaluate the environmental footprint of materials in the supply chain and to help us understand the impacts associated with the transportation of goods. In the following chapters we present additional examples of applying streamlined (simplified) LCA to cover several of the big building blocks of a life cycle inventory, such as energy production (Chapter 18), waste treatment (Chapter 19), and life cycle costing (Chapter 20). These separate LCA assessments can be used in a modular form as building blocks that simplify the work of conducting a life cycle assessment either through streamlining it or having ready-to-use modules that would reduce the time needed to perform a full life cycle assessment.

PROBLEMS

16.1 Provide three examples in which a full cradle-to-grave boundary might be unnecessarily wide.

16.2 Provide three examples in which a cradle-to-gate boundary might be too limited for a life cycle assessment.

16.3 What functional unit would you propose to compare the life cycle impacts of two types of paint used for car painting and water protection? Explain why.

16.4 What functional unit would you propose to compare two light bulbs? Explain why.

16.5 Define your data quality goals for the life cycle inventory to be developed for Problem 16.3.

16.6 Define your data quality goals for the life cycle inventory to be developed for Problem 16.4.

16.7 Describe what type of issues can arise from allocation methods when conducting a life cycle inventory in a multiproduct facility.

16.8 Develop a chemical tree to produce hypochlorous acid, following the process described in Example 9.5.

16.9 Develop a chemical tree to produce ethanol:
(a) Using bioprocesses
(b) Using a synthetic route

16.10 What would you suggest doing when characterization factors are lacking? Provide examples.

16.11 Estimate the global warming potential, ozone depletion potential, photochemical ozone creation potential, acidification potential, and eutrophication potential for the results of Example 16.2 using the characterization factors in Tables 3.1 through 3.5.

16.12 Table P16.12 shows the cradle-to-gate life cycle inventory results for the production of 1000 kg of acetaldehyde. Estimate the global warming potential, ozone depletion

TABLE P16.12 Cradle-to-Gate Results

	Total
Raw Material (kg)	
Air	6.18×10^2
Alum	3.74×10^{-2}
$BaCO_3$	7.14×10^{-2}
Chlorine	4.38
Crude oil	8.59×10^2
Ethylene	7.41×10^2
Hydrofluosilicic acid	5.31×10^{-3}
Hydrogen chloride	1.74×10
Lime	2.43×10^{-2}
Na_2CO_3	2.41×10^{-2}
Naphtha	8.37×10^2
Oxygen	4.48×10^2
Salt rock	6.04
Sodium chloride	9.66
Sodium hydroxide	1.81×10
Water for reaction	2.99
Water, including water for reaction	1.33×10^3
Energy [MJ]	
Coal	1.25×10^2
Cooling water	2.10×10
Diesel	1.35×10^3
Electricity	2.49×10^3
Heating fuel	1.46×10

(*continued*)

TABLE P16.12 *(Continued)*

	Total
Heavy oil	1.18×10^3
Hydro power	8.37
Natural gas	2.80×10^3
Nuclear power	8.37
Potential energy recovery	-4.65×10^3
Refrigeration	5.63×10^2
Steam	1.18×10^4
Total	7.77×10^3
Air Emissions (kg)	
1,2-Dichlorethane	9.55×10^{-2}
Acetic acid	1.15
Acetaldehyde	4.22×10
C_2H_6	5.93×10^{-1}
CH_4	6.08
Cl_2	2.44×10^{-1}
CO	1.93×10
CO_2	1.43×10^3
C_xH_y	1.85
Ethylene	6.72
Ethylene chloride	2.96×10
H_2	7.90
H_2S	2.50×10
HCl	9.86×10^{-2}
HOCl	1.73×10^{-4}
NMVOC	1.04×10
NO_x	7.22
SO_x	5.59
Vinyl chloride	1.93×10^{-2}
Water Emissions (kg)	
Acetaldehyde	4.22×10
Acetic acid	1.09×10^2
$BaSO_4$	4.18×10^{-2}
BOD	2.74×10^{-1}
$CaCO_3$	1.78×10^{-2}
COD	1.95×10^2
Hg	3.12×10^{-6}
$Mg(OH)_2$	2.98×10^{-3}
Na_2S	6.30
TDS	4.73
TOC**	6.68×10
Wastewater	1.31×10^3
Solid Waste (kg)	
s_$BaSO_4$(g)	$4.34 \times \times 10^{-2}$
s_$CaCO_3$(g)	1.93×10^{-2}
s_$Mg(OH)_2$(g)	4.82×10^{-4}
Solid waste	2.87×10

potential, photochemical ozone creation potential, acidification potential, and eutrophication potential for this system using the characterization factors in Tables 3.1 through 3.5

16.13 Table P16.13 shows the results of a gate-to-gate estimation for a plant producing ethylene glycol, diethylene glycol, and triethylene glycol. They are produced from

TABLE P16.13 Gate-to-Gate Estimation

UID	CAS	Chemical	Amount	Purity (%)	Units
		Inputs			
750-21-8	750-21-8	Ethylene oxide	799		kg/h
7732-18-5	7732-18-5	Water	310		kg/h
		Total	1109		kg/h
		Products			
107-21-1	107-21-1	Ethylene glycol	1000	99.7	kg/h
111-46-6	111-46-6	Ethylene glycol,di	94.6	89.3	kg/h
112-27-6	112-27-6	Ethylene glycol, tri	10	91.1	kg/h
		Total	1104.229894		kg/h

UID	CAS	Chemical	Amount: Gas	Liquid	Solid	Solvent	Units
		Chemical Emissions					
75-21-8	75-21-8	Ethylene oxide	3.97				kg/h
107-21-1	107-21-1	Ethylene glycol					kg/h
111-46-6	111-46-6	Ethylene glycol,di-					kg/h
112-27-6	112-27-6	Ethylene glycol, tri-	—	0.0893			kg/h
UID higher glycols		Higher glycols	—	0.434			kg/h
		Total	3.97	0.523	0	0	kg/h
		Mass balance difference	0				kg/h

Source	Amount	Units	Comments
	Energy Use		
Electricity	412	MJ/h	
DowTherm	0	MJ/h	
Heating steam	1.51×10^4	MJ/h	85% efficiency has been included to determine how much steam is needed for heating process fluid
Direct fuel use in high-temperature heating	0	MJ/h	
Heating natural gas	0	MJ/h	
Energy input requirement	1.56×10^4	MJ/h	Electricity + steam + direct fuel oil + DowTherm
Cooling water	-1.45×10^4	MJ/h	

the reaction of ethylene oxide and water and then separated in a separation train that produces, first, ethylene glycol, with the di- and tri- products separated from the higher glycols in two additional distillation columns. Propose a way to allocate the gate-to-gate results.

16.14 The cradle-to-gate life cycle assessment of an active pharmaceutical ingredient was undertaken and reported in the literature. The pretreatment life cycle impact assessment results are shown in Figure 16.11.

(a) What conclusions would you draw from these results?

(b) What additional questions would you have?

16.15 The post-treatment life cycle impact assessment results of the life cycle impact assessment of Problem 16.11 are shown in Figure 16.12.

(a) What conclusions would you draw from these results?

(b) How would you couple these with the results from Figure P16.10?

(c) What additional insights do post- and pre-treatment results bring to the interpretation of the assessment?

16.16 Figure 16.12 shows some results from an LCA comparing two complex compounds.

(a) What interpretations can be drawn from the data in the figure?

(b) What additional questions would you have?

16.17 You are the president of a corporation that develops, manufactures, and sells electronic equipment globally.

(a) How would you propose to include life cycle assessment in your business decisions?

(b) How would you propose to include life cycle assessment in your marketing plans?

16.18 Investigate the details of the ecoefficiency method developed by BASF and list its advantages and disadvantages.

16.19 Describe, compare, and contrast the considerations for the two mandatory LCA critical reviews: comparative assessments and environmental product declarations.

16.20 Describe the advantages and disadvantages of using streamlined life cycle assessment methods.

REFERENCES

1. U.S. Environmental Protection Agency. *Life Cycle Assessment: Principles and Practice.* EPA 600-R-06-060. National Risk Management Reseach Laboratory, Cincinnati, OH, 2006. Available at http://www.epa.gov/ORD/NRMRL/lcaccess/pdfs/600r06060.pdf, accessed, Nov. 1, 2008.
2. Wenzel, H., Hauschild, M., Alting, L. *Environmental Assessment of Products*, Vol. 1, *Methodology, Tools and Case Studies in Product Development*. Chapman & Hall, London, 1997.
3. Meadows, D. H. , et al. *The Limits to Growth: A Report for the Club of Rome's Project on the Predicament of Mankind.* Universe Books, New York, 1972. p. 205.

4. Goldsmith, E., Allen, R. A blueprint for survival. *Economist*, 1972, 2 (1).
5. Consoli, F., Allen, D., Boustead, I., Fava, J., Franklin, W., Jensen, A. A., Oude, N., Parrish, R., Perriman, R., Postlethwaite, D., Quay, B., Seguin, J., Vigon, B., Eds. *Guidelines for Life Cycle Assessment: A "Code of Practice."*Society of Environmental Toxicology and Chemistry, Brussels, Belgium, 1993.
6. Barnthouse, L., Fava, J., Humphreys, K., Hunt, R., Laibson, L., Noessen, S., Owens, J. W., Todd, J. A., Vigon, B., Wietz, K., Young, J.,Eds. *Life Cycle Impact Assessment: The State-of-the-Art*. Society of Environmental Toxicology and Chemistry, Brussels, Belgium, 1997.
7. Hunkeler, D., Rebitzer, G., Finkbeiner, M., Schmidt, W.-P., Jensen, A. A., Stranddorf, H., Christiansen, K. *Life-Cycle Management*. Society of Environmental Toxicology and Chemistry, Brussels, Belgium, 2004.
8. International Organization for Standardization. *Environmental Management: Life Cycle Assessment—Principles and Framework*, ISO 14040.ISO, Geneva, Switzerland, 1997.
9. International Organization for Standardization. *Environmental Management: Life Cycle Assessment—Requirements and Guidelines*. ISO 14044.ISO, Geneva, Switzerland, 2006.
10. Fava, J., Denison, R., Jones, B., Curran, M. A., Vigon, B., Selke, S., Barnum, J. Eds. *A Technical Framework for Life Cycle Assessment*. Society of Environmental Toxicology and Chemistry, Brussels, Belgium, 1991.
11. Baumann, H., Tillman, A.-M. *The Hitch Hiker's Guide to LCA*. Studentlitteratur, Lund, Sweden, 2004.
12. Jiménez-González, C., Kim, S., Overcash, M. Methodology of developing gate-to-gate life cycle analysis information. *Int. J. Life Cycle Assessm.*, 2000, 5(3), 153–159.
13. Jiménez-González, C., Overcash, M. Energy sub-modules applied in life cycle inventory of processes. *J. Clean Products Process.*, 2000, 2, 57–66.
14. Jiménez-González, C., Overcash, M., Curzons, A. Treatment modules: a partial life cycle inventory. *J. Chem. Technol. Biotechnol.*, 2001, 76, 707–716.
15. Capello, C., Hellweg, S., Badertscher, B., Betschart, H., Hungerbühler, K. Environmental assessment of waste-solvent treatment options: 1. The ecosolvent tool. *J. Ind. Ecol.*, 2007, 11(S), 26–38.
16. Doka, G. *Life Cycle Inventories of Waste Treatment Services*. Ecoinvent report 13. Swiss Centre for Life Cycle Inventories, Dec. 2003.
17. Kim, S., Dale, B. Life cycle inventory information of the United States electricity system. *Int. J. Life cycle Assess.*, 2005, 10(4), 294–304.
18. Eco-Invent: Swiss Centre for Life Cycle Inventories. 2008. http://www.ecoinvent.org, accessed Nov. 10, 2008.
19. SimaPro. PRé Consultants. http://www.pre.nl/simapro/default.htm.
20. DEAM™ Database. EcoBilan. http://www.ecobalance.com/uk_deam.php.
21. UMBERTO software. Institute for Environmental Informatics, Hamburg, Germany. http://www.umberto.de/en/
22. GaBi Software. PE International, http://www.gabi-software.com/
23. Jiménez-González, C., Overcash, M. Life cycle inventory of refinery products: review and comparison of commercially available databases. *Environ. Sci. Technol.*, 2000, 34(22), 4789–4796.
24. Environmental Clarity. http://www.environmentalclarity.com/
25. Lindfors, L.-G., Christiansen, K., Hoffman, L., Virtanen, Y., Juntilla, V., Hanssen, O.-J., Rønning, A., Ekvall T., Finnveden, G. *Nordic Guidelines on Life Cycle Assessment*. Nordic Council of Ministers, Copenhagen, Denmark, 1995.
26. Udo de Haes, H. A., Ed. *Towards a Methodology for Life Cycle Impact Assessment*. Report from the SETAC–Europe Working Group on Impact Assessment. Society of Environmental Toxicology and Chemistry, Brussels, Belgium, 1996.

27. Gunée, J., Ed. *Life Cycle Assessment: An Operational Guide to the ISO Standards*. Centrum voor Milieukunde Leiden, Leiden University, Leiden, The Netherlands, 2002.
28. Steen, B. *A Systematic Approach to Environmental Priority Strategies in Product Development* (*EPS*), Version 2000, *General System Characteristics*. CPM Report 1999:4. Center for Environmental Assessment of Products and Material Systems, Chalmers University of Technology, Göteborg, Sweden, 1999.
29. U.S., Environmental Protection Agency. *Tool for the Reduction and Assessment of Chemical and Other Environmental Impacts* (*TRACI*): *User's Guide and System Documentation*. EPA 600-R-02-052. U.S. EPA, National Risk Management Research Laboratory, Cincinnati, OH, 2003.
30. Eco-Indicator, 99.*A Damage Oriented Method for Life Cycle Impact Assessment*. PRé Consultants, Amersfoort, The Netherlands, 2001.
31. Hendrickson, C. T., Lave L., Matthews, H. S. *Environmental Life Cycle Assessment of Goods and Services: An Input–Output Approach*. Resources for the Future,Washington, DC, 2006.
32. Jiménez-González, C. Life cycle assessment in pharmaceutical applications. Ph.D. dissertation, North Carolina State University, Raleigh NC, 2000.
33. Bundesamt für Umwelt, Wald und Landschaft. *Ökobilanzen von Packstoffen*. Schriftenreihe Umwelt 132. BUWAL, Bern, Switzerland, 1991.
34. Heijungs, R. *Environmental Life Cycle Assessment of Products: Guide and Backgrounds*. NOH report 9266 and 9267, commissioned by the National Reuse of Waste Research Programme (NOH), in collaboration with CML, TNO, and B&G, Leiden, The Netherlands, 1992.
35. Goedkoop, M. J., Demmers, M., Collignon M. X. *The Eco-Indicator 95: Manual for Designers*. NOH Report 9524. PRé Consultants, Amersfoort, The Netherlands, July 1995.
36. U.S., Environmental Protection, Agency. U.S. EPA, Washington, DC, 2006.
37. Saling, P., Kicherer, A., Dittrich-Krämer, B., Wittlinger, R., Zombik, W., Schmidt, I., Schrott, W., Schmidt, S. Eco-efficiency analysis by BASF: The method. *Int. J. Life Cycle Assess.*, 2002, 7(4), 203–218.
38. BASF Web site. http://www.basf.com/group/sustainability_en/index, accessed Nov. 2008.
39. Jiménez-González, C., Curzons, A. D., Constable, D. J. C., Cunningham, V. L. Cradle-to-gate life cycle inventory and assessment of pharmaceutical compounds: a case-study. *Int. J. Life Cycle Assess.*, 2004, 9(2), 114–121.
40. Conradt, S. Life Cycle Assessment of a Pharmaceutical Compound. Master's thesis in Environmental Science. Safety and Environmental Technology Group, ETH, Zürich, Switzerland, Aug. 2008.
41. Henderson, R. K., Jiménez-González, C., Preston, C., Constable, D. J. C., Woodley, J. EHS and life cycle assessment of biocatalytic and chemical processes: 7ACA production. *Ind. Biotechnol.*, 2008, 4(2), 180–192.
42. Weidema, B. P., Wesnaes, M. S. Data quality management for life cycle inventories: an example of using data quality indicators. *J. Cleaner Prod.*, 1996, 4, 3–4.
43. Hunt, R. G., Boguski, T. K., Weitz, K., Sharma, A. Examining LCA streamling techniques. *Int. J. Life Cycle Assess.*, 1998, 3(1), 36–42.
44. Curzons, A., Jiménez-González, C., Duncan, A., Constable, D., Cunningham, V. Fast life-cycle assessment of synthetic chemistry tool, FLASC(TM) tool. *Int. J. Life Cycle Assess.*, 2007, 12(4), 272–280.
45. Kralisch, D. Application of LCA in process development. In *Green Chemistry Metrics*, Lapkin, A., Constable, D., Eds.Wiley-Blackwell, Hoboken, NJ, 2008.
46. Weidenhaupt, A., Hungerbüler, K. Integrated product desubg in chemical industry: a plea for adequate life cycle screening indicators. *Chimia*, 1997, 51(5), 217–221.

47. Fleischer, G., Gerner, K., Kunst, H., Lichtenvort, K., Rebitzer, G. A semi-quantitative method for the impact assessment of emissions within a simplified life cycle assessment. *Int. J. Life Cycle Assess.*, 1997, 6(3), 149–156.
48. Legarth, J. B., Åkesson, S., Ashkin, A.Imrell, A. M. A screening level life cycle assessment of the ABB EU 2000 air handling unit. *Int. J. Life Cycle Assess.*, 2000, 5(1), 47–58.
49. Nakaniwa, C., Graedel, T. E. Life cycle and matrix analyses for re-refined oil in Japan. *Int. J. Life Cycle Assess.*, 2002, 7(2), 95–102.
50. Hochschorner, E., Finnveden, G. Evaluation of two simplified life cycle assessment methods. *Int. J. Life Cycle Assess.*, 2003, 8(3), 119–128.
51. Graedel, T. E. Weighted matrices as product life cycle assessment tools. *Int. J. Life Cycle Assess.*, 1996, 1(2), 85–89.
52. Curran, M. A., Young, S. Report from the EPA conference on streamlining LCA. *Int. J. Life Cycle Assess.*, 1996, 1(1), 57–60.
53. Hochschorner, E., Finnveden, G. Life cycle approach in the procurement process: the case of defence materiel. *Int. J. Life Cycle Assess.*, 2006, 11(3), 200–208.
54. Mori, Y., Huppes, G., Udo de Haes, H., Otoma, S. Component manufacturing analysis: a simplified and rigorous LCI method for the manufacturing stage. *Int. J. Life Cycle Assess.*, 2000, 5(6), 327–334.
55. Mueller, K., Lampérth, M. U., Kimura, F. Parameterised inventories for life cycle assessment: systematically relating design parameters to the life cycle inventory. *Int. J. Life Cycle Assess.*, 2004, 9(4), 227–235.

17

IMPACTS OF MATERIALS AND PROCUREMENT

What This Chapter Is About In this chapter we review how the concepts of life cycle assessment covered in Chapter 16 can be applied to estimate the impact of materials sourced for a particular activity. We also cover application examples of streamlined LCA to evaluate the environmental footprint of materials in the supply chain, and explore to what extent the transportation of goods affects the life cycle impact profile of production processes or activities.

Learning Objectives At the end of this chapter, the student will be able to:

- Understand the concept of life cycle management.
- Understand the potential impacts of production chains.
- Understand the impacts that procurement choices of materials and services have in the sustainability of companies.
- Identify the differences between environmental declarations and greenwashing.
- Understand the role of transportation on the environmental footprint of products.

17.1 LIFE CYCLE MANAGEMENT

In Chapter 16 we reviewed the concepts of life cycle inventory and life cycle impact assessment and saw that LCI/A is an important tool for decision making that strives to give us

Green Chemistry and Engineering: A Practical Design Approach, By Concepción Jiménez-González and David J. C. Constable

a holistic picture of the cradle-to-cradle environmental impacts of a product or process. However, an environmental LCI/A represents only one element of what we need to make a more sustainable decision. That is, the information obtained from an LCI/A should be a part of a more comprehensive decision process assessing the trade-offs with total cost, performance, and social aspects.

Applying life cycle thinking to modern business practices is called life cycle management (LCM). LCM seeks to move an organization's products and services toward more sustainable consumption and production, thereby minimizing the environmental and socioeconomic impacts of products across their entire life cycle and supply chain.[1] Life cycle management is the operational side of life cycle thinking and provides an integrated framework to address environmental, economic, technological, and social aspects of products, services, and organizations. As with most any other management techniques, companies who choose to apply LCM do so on a voluntary basis and adapt it to their specific needs.[2]

It is also true that LCM is not a methodology, but a collection of tools that are used to collect, analyze, structure, and communicate information about the environmental, economic, and social programs that are associated with products, product portfolio, and services across their entire life cycle. Some of the potential elements that comprise the toolkit of a life cycle management system are shown in Figure 17.1. As you can see, LCM operationalizes the concepts and areas on the left through the tools, methodologies, and systems shown on the right. All these tools are not necessarily used at the same time, and within the LCM system, one can use several of them selectively, depending on the goals of the organization, the areas that the organization wishes to highlight, changes in the market and/or society, and so on.

Introducing LCM into a company is a strategic decision that will require support from senior management, especially during the first phases of implementation, since integrating LCM practices into the business will require time, resources and probably, a considerable culture change. There are many external and internal drivers for a company to seek a culture of sustainability through an LCM system. Examples of external drivers include business strategies, market changes and pressures, financial pressures, and regulatory requirements. Internally, organizations might look to LCM for a variety of reasons, including a desire to increase resource efficiency, reduce cost, increase operational efficiency, change or align with company values and culture, increase product or service innovation or market share, enhance brand value, achieve competitive advantage, improve public reputation and stakeholder relations, develop criteria for product development, and identify improvement opportunities and risks.

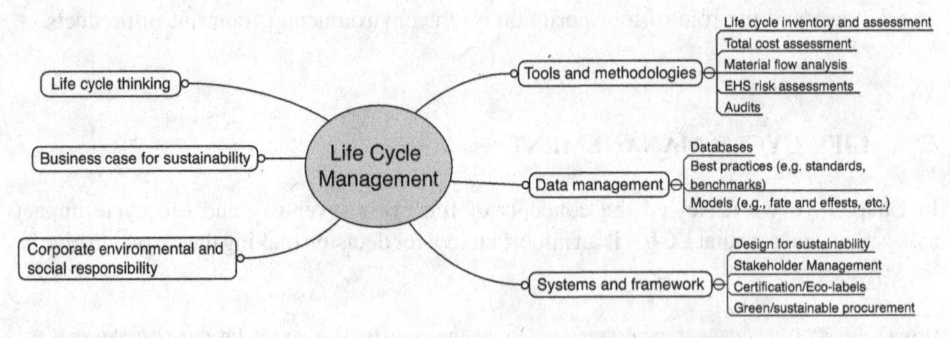

FIGURE 17.1 Life cycle management as building the operational side of life cycle thinking.

One of the main challenges in implementing a LCM system is to be able to embed the concepts and tools in the normal business context so as to avoid this being an additional requirement on top of the normal functions of the company. In small and medium-sized companies, this might not be as complicated as with large companies, where there are multiple departments (e.g., you might have several procurement departments servicing different parts of the organization) and different functional areas, sometimes with seemingly divergent goals.

Thus, it is critically important to clearly, consistently, and repeatedly communicate the goals of an LCM system and the overall concept of life cycle thinking. One should not assume that there is a shared view of what these mean until LCM is embedded into the everyday practice of a company. In addition to internal communications, external communications about life cycle information are an important part of LCM, as these help to increase the transparency and visibility of company activities with the stakeholders and to improve the level of trust with the public. Some avenues to disseminate information and obtain feedback on a company's LCM program are:

- Annual corporate social responsibility report
- Stakeholder panels
- Life cycle environmental product declarations
- Life cycle data
- Peer-review publications
- Presentations at scientific conferences
- Benchmarks at industry organizations
- Green procurement guidelines
- Supplier development programs
- Product brochures
- Ecolabels and certifications
- Audits
- Supplier and contract manufacturers requirements

In Chapter 2 we covered the principles of green chemistry and green engineering in the context of sustainability, and in Chapter 16 we explored the general concepts of life cycle thinking and life cycle inventory and assessment, both of which are crucial elements of LCM. In this chapter we expand our horizon by thinking about how we evaluate and manage the impacts of supply chains and networks in a more operational sense.

17.2 WHERE CHEMICAL TREES AND SUPPLY CHAINS COME FROM

To use life cycle management, it is necessary to understand the supply chain and how it works in an increasingly interconnected world. We saw in Chapter 16 how to construct a chemical tree. The trees we constructed there depended on the type of manufacturing process used to produce a given chemical or product. Figures 17.2 and 17.3 show chemical trees for the production of tetrahydrofuran (THF). As you can see, the different chemical trees presented in the figures represent two different processes for the production of THF. For a

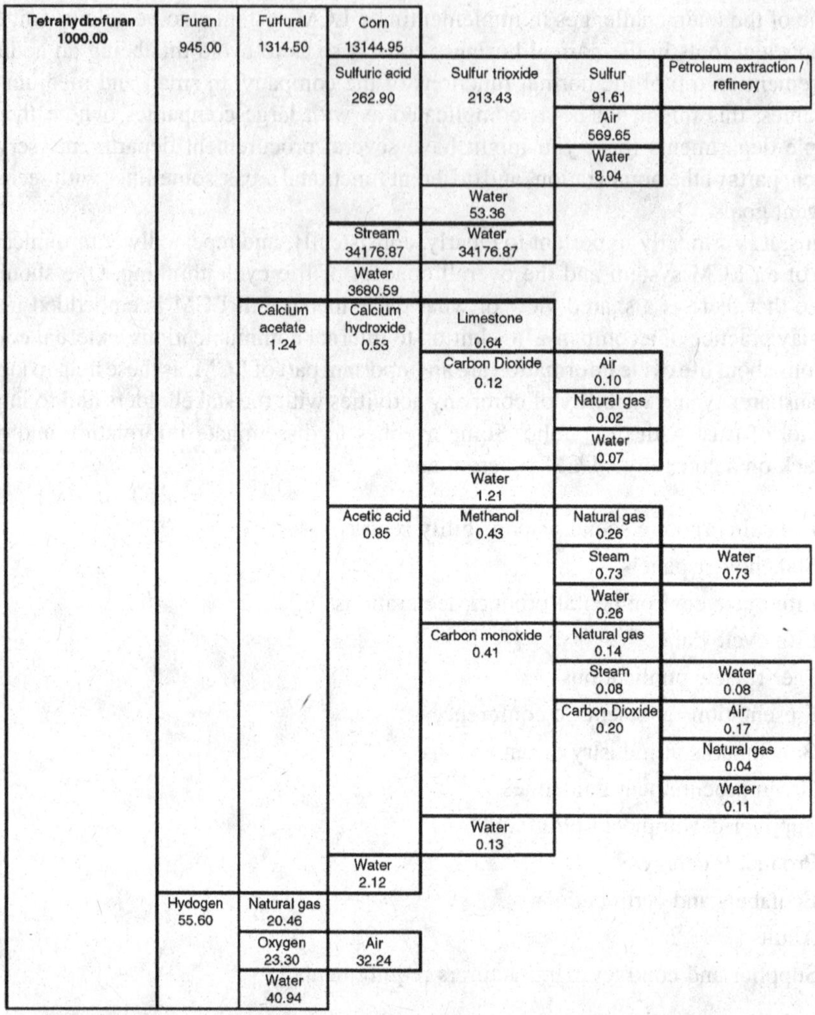

- Tetrahydrofuran 1000.00
  - Furan 945.00
    - Furfural 1314.50
      - Corn 13144.95
      - Sulfuric acid 262.90
        - Sulfur trioxide 213.43
          - Sulfur 91.61
            - Petroleum extraction / refinery
          - Air 569.65
          - Water 8.04
        - Water 53.36
      - Stream 34176.87
        - Water 34176.87
      - Water 3680.59
    - Calcium acetate 1.24
      - Calcium hydroxide 0.53
        - Limestone 0.64
        - Carbon Dioxide 0.12
          - Air 0.10
          - Natural gas 0.02
          - Water 0.07
        - Water 1.21
      - Acetic acid 0.85
        - Methanol 0.43
          - Natural gas 0.26
          - Steam 0.73
            - Water 0.73
          - Water 0.26
        - Carbon monoxide 0.41
          - Natural gas 0.14
          - Steam 0.08
            - Water 0.08
          - Carbon Dioxide 0.20
            - Air 0.17
            - Natural gas 0.04
            - Water 0.11
        - Water 0.13
      - Water 2.12
  - Hydrogen 55.60
    - Natural gas 20.46
    - Oxygen 23.30
      - Air 32.24
    - Water 40.94

FIGURE 17.2 Chemical tree of tetrahydrofuran derived from natural sources.

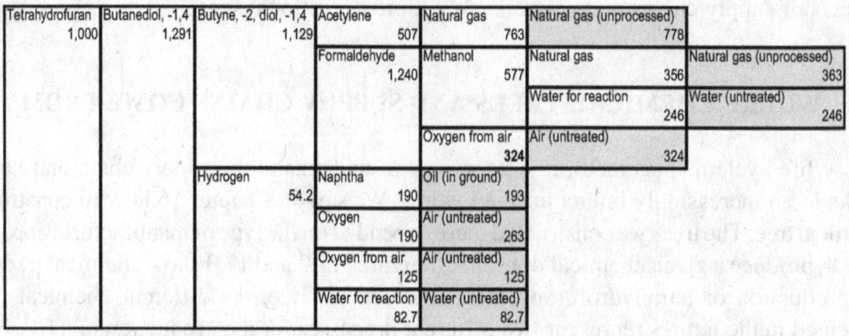

- Tetrahydrofuran 1,000
  - Butanediol, -1,4 1,291
    - Butyne, -2, diol, -1,4 1,129
      - Acetylene 507
        - Natural gas 763
          - Natural gas (unprocessed) 778
      - Formaldehyde 1,240
        - Methanol 577
          - Natural gas 356
            - Natural gas (unprocessed) 363
          - Water for reaction 246
            - Water (untreated) 246
        - Oxygen from air 324
          - Air (untreated) 324
    - Hydrogen 54.2
      - Naphtha 190
        - Oil (in ground) 193
      - Oxygen 190
        - Air (untreated) 263
      - Oxygen from air 125
        - Air (untreated) 125
      - Water for reaction 82.7
        - Water (untreated) 82.7

FIGURE 17.3 Chemical tree of tetrahydrofuran derived from a synthetic route.

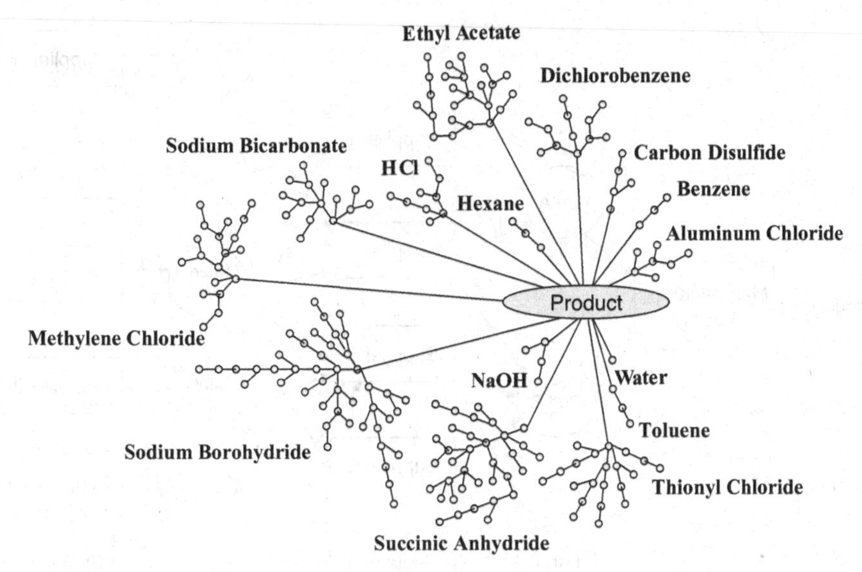

FIGURE 17.4 One route to producing a fine chemical product used in the manufacture of active pharmaceutical ingredients. Each shaded dot represents a manufacturing or extraction process. (Source: ref 3.)

given chemical or product, there could be several possible processes to evaluate, and the chemical trees are likely to look very different.

When talking about more complex systems, the chemical trees of all the substances or materials that are combined to give a specific product can be quite large and cumbersome. For example, the components required to produce an automobile are on the order of thousands, and almost every one of them requires a separate production process. Figure 17.4 shows a representation of the complexity of the chemical tree for one development route to produce a single fine chemical, which also happens to be the regulatory starting material of a commercial active pharmaceutical ingredient.[3] As can be seen, the complexity of the chemical tree can expand rather quickly.

There is still another layer of detail and complexity that can be added to the chemical tree and that involves identifying which particular manufacturing plant would be producing the chemical or the part. This is what a supply chain is. In the current economic environment, supply chains very frequently become supply networks, as companies tend to use several suppliers for a given good, and in turn the suppliers may have several suppliers for each part or chemical. These suppliers can be located anywhere in the world, and this usually increases the complexity of the overall supply chain. Figure 17.5 represents the potential pathways for a simple supply chain (network) for a product that requires three parts, A, B, and C. Part A can be procured from either supplier 1 or 2, part B can be procured from supplier 2, 3, or 4, and part C can be procured from supplier 3 or 5. Each parts supplier in turn needs raw materials for manufacture, which can come from one or more suppliers. This is a simple example, but some production systems can be rather complicated.

Example 17.1 In Example 16.5 we saw a comparison among four suppliers of ibuprofen: two in the United States, one in India, and one in China. Provide an example where the differences in the supply chain affected the results of the ecoefficiency evaluation.

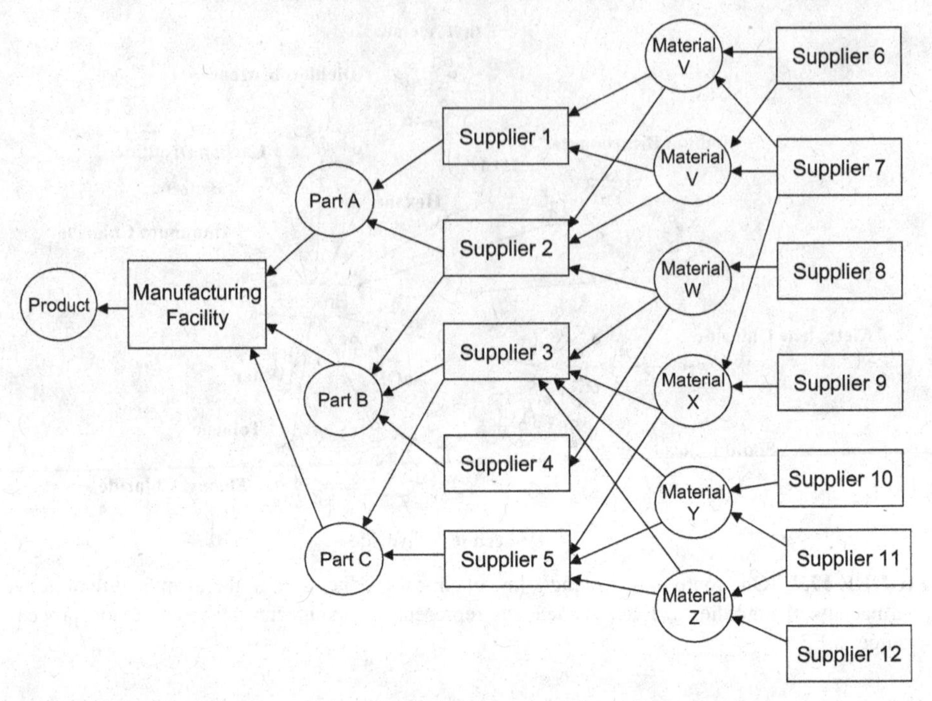

FIGURE 17.5 Potential pathways for a simple supply chain.

Solution In Example 16.5 we saw that the toxicity potential is heavily related to the type of process that is used for the production of ibuprofen. After accounting for those materials having toxicity potentials, we found that the Indian and Chinese syntheses use materials such as dichloroethane, aluminum trichloride, triphenyl phosphine, and other substances of significant toxicity. We also found an increase in the environmental impacts from the waste streams associated with these processes. As more hazardous materials are used, the more difficult it is to control emissions and potential accidents. Consequently, the Indian and Chinese supply chains were found to be less ecoefficient than those of their U.S.-based counterparts.

Additional Points to Ponder How would this be different if the production processes in India and China were contained? How would that affect the price structure?

In terms of measuring, managing, and minimizing the environmental or sustainability footprint of production systems, the more complex the supply chain (network), the more complex it is to estimate the environmental or sustainability footprint of the system. We can add to that complexity the fact that suppliers might change from time to time and that market fluctuations might affect the way that products and processes get to our homes.

One way to obtain the environmental footprint information for a supply chain is to obtain it directly from the first layer of suppliers. Generally, this is an approach that would require more resource and time to complete, and the results would depend on the market share that the buyer has for a given supplier. For example, large retailer chains such as Tesco and Wal-Mart have recently started to measure the environmental impacts related to the packaging used for products sold at their stores. Wal-Mart has, for example, begun requesting packaging data from its suppliers to determine a sustainability scorecard and produces a normalized score between 1 and 10 (10, good; 1, bad). This score is a normalized weighted average of the following packaging metrics:

- Greenhouse gas emissions from packaging production
- Sustainable materials content
- Average transportation distance
- Package/product ratio
- Cube utilization (volume efficiency)
- Recycled content
- Recyclability
- Renewable energy to power each facility
- Innovation (different from energy)

The suppliers provide the information to the retailer for each of these nine metrics and a raw score is calculated from weight-based formulas. The raw scores for each metric provide a means of "ranking" one product's packaging against another product's packaging for a given segment of products (e.g., beauty products). This ranking is used with products having similar packaging to calculate a final normalized score. It should be noted that the score will vary over time, as it is based on the number and type of packaging data that is obtained from other suppliers.

As you can imagine, this is not a simple task to set up and coordinate. It is not always possible to request and capture information from suppliers within the time frame required for decision making. In this case, another way to estimate the environmental impacts of complex supply chains is to use streamlined life cycle methodologies. As we saw in Chapter 16, a streamlined LCA covers the same aspects as a full LCA, but at a higher level. One can use streamlined life cycle methodologies to get quick estimations of supply chain impacts for a given activity, provided that the groundwork to understand the system boundaries, goal, and scope of the study has been completed. There are also collections of life cycle impact data that may contain generic data for the production of materials and chemicals. These databases can be used directly for a streamlined assessment, followed by some quality analysis that provides an idea of the uncertainty and sensitivity of the numbers.

Example 17.2 Estimate the environmental footprint of the materials needed to produce a fine chemical with a formula weight of 174. The raw materials needed are listed in Table 17.1.

Solution If life cycle inventory data are available from commercially available databases, that would be one way to estimate the footprint of the materials. However, for this example it was not possible to obtain information about the di-*N*-oxide or the sodium dithionite. In view of that, the environmental life cycle impacts of the materials needed to produce this specialty chemical were estimated using GlaxoSmithKline's FLASC tool (fast life cycle assessment for synthetic chemistry).[4] As discussed in Chapter 16, this streamlined LCA tool was designed based on a full LCA, and it is intended for use in making quick comparisons of different synthetic chemical routes. The scores are shown on a scale of 1 to 5 (1, lowest

TABLE 17.1 Raw Materials

Material	Mass (kg/kg product)
Di-*N*-oxide	3.9
N,*N*-Dimethylacetamide	33.7
Sodium dithionite	5.7
Chloroform	211.6
Ethanol	14.2
Hydrogen chloride	2
Hydrogen peroxide	3.2
Sodium hydroxide	1.2

impact; 5, highest impact), and the user can get a quick, high-level benchmark of the impacts of the materials used in the synthetic route. This assessment is typically performed in less than 30 minutes (see Figure 16.13). In this case, the bill of materials (material name and mass required) and the molecular weight were entered into the tool. For the materials that were not in the database, life cycle environmental profiles were estimated from averaged data.

As can be seen in Table 17.2, the environmental life cycle impacts were estimated for each chemical needed to manufacture the desired substance. These estimations include manufacturing, energy production, waste treatment, and transportation. The numbers provided in each cell represent the estimated contribution of that chemical to the corresponding impact. The totals at the end of each column represent the overall estimated environmental impact for the materials in the production of this chemical.

Those totals were then normalized by an internal benchmark and the normalized scores rolled up into a single high-level score using the geometric average of the scores of individual impacts, thus giving the same weight to all impacts. The high-level output is shown in Figure 17.6.

Additional Points to Ponder Which chemicals make the largest contribution to the environmental footprint? What other impacts currently not included would you think of adding to this assessment?

17.3 GREEN (SUSTAINABLE) PROCUREMENT

As we saw in Section 17.2 and Example 17.2, the way and the process by which goods are made determine the environmental profile of processes. In addition, a company's own waste and overall environmental performance are linked to the materials it buys and their quality. The responsibility for acquiring these goods normally lies with procurement departments, which, depending on the corporation, may or may not be aligned with other activities and strategies, such as sustainability efforts. The effort to align procurement departments with a corporation's overall goal to reduce its environmental profile comes under such names as green procurement, green purchasing, environmentally preferred purchasing (EPP), affirmative procurement, eco-procurement, and environmentally responsible purchasing. Green procurement should be a normal part of supply chain management, and attempts to identify and reduce environmental impact and maximize resource efficiency by working closely with new and existing suppliers. Some of the potential advantages in implementing a green procurement (green purchasing) strategy or program are shown in Table 17.3.

TABLE 17.2 FLASC estimations for the Chemical Route of Example[a]

Material	Mass (kg/kg product)	Classification	Net Mass (kg/kg)	Gross Energy (MJ/kg)	POCP (ethene-eq/kg)	GHG (CO_2-eq/kg)	Acid (SO_2-eq/kg)	EUTRO ($(PO_4)^{3-}$-eq/kg)	TOC (kg/kg)	Oil (kg/kg)
Di-*N*-oxide	3.9	Average complex organic	22.48	449	0	31	1	0.29	0.5	15
***N,N*-Dimethyl-acetamide**	33.7	Solvent	95.12	1214	0.6	218	3.45	0.31	12.1	88
Sodium dithionite	5.7	Average metal cation inorganic	8.61	90	0	9	0.08	0.02	0	3
Chloroform	211.6	Solvent	243.85	6326	0.6	2098	3.79	0.57	3.9	221
Ethanol	14.2	Solvent	0.03	387	0	36	0.22	0.02	0	9
Hydrogen chloride	2	General inorganic	2.66	18	0	5	0.05	0	0	0
Hydrogen peroxide	3.2	General inorganic	0.08	2	0.1	0	0	0	0	0
Sodium hydroxide	1.2	Metal cation organic	1.06	10	0	2	0.01	0	0	0
Sum	**275.5**		**373.88**	**8496**	**1.4**	**2398**	**8.59**	**1.22**	**16.5**	**337**
Score			**1**	**1**	**1.5**	**1**	**1**	**1.3**	**1**	**1**
FLASC score			**1.1**							

[a]The molecular weight is 174.

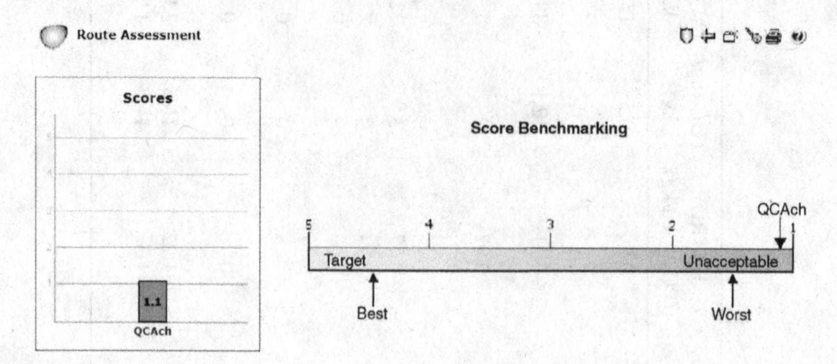

FIGURE 17.6 High-level output of the streamlined LCA for Example 17.2 (FLASC score). The score provides a benchmark of the LCA impacts of the materials used for a chemical synthesis (1, bad; 10, good).

In concept this probably sounds simple, but as we have already seen, supply chain networks can be extremely complicated and can expand into several continents. This invariably adds complexity to the implementation of a green or sustainable procurement program, as obtaining sufficient collaboration from suppliers often requires persistence and a concerted effort. In addition, there might be some resistance to change within a company's procurement department and other functional areas, especially as there are many definitions for green procurement, which tends to create confusion and barriers to implementation.

When a new or modified product is introduced to the market, it is not uncommon to see marketing claims labeling that product as green or environmentally friendly; these claims may or may not be true. For example, in some instances you might see a specific

TABLE 17.3 Potential Benefits of Implementing a Green Procurement Strategy

Potential Benefit	Examples
Reduced total cost	Reducing hazardous waste management costs by procuring materials that are inherently less toxic or that require the use of less of a hazardous material
	Reducing operational cost by using more energy-efficient equipment
	Reducing repair cost by buying more durable and reliable equipment
	Identifying hidden costs in an organization and developing strategies to reduce them
Improved market presence	Capturing market segments that are interested in environmentally friendly products
	Increasing institutional trust with stakeholders
	Product differentiation through product and service offerings that are of equivalent quality with fewer environmental impacts
Increased operational efficiency	Simplification of supply chains
	Comprehensive procurement strategy that integrates environmental, health, safety, and sustainability aspects

Source: ref. 5

characteristic of the product (e.g., low VOC emissions) highlighted to make it appear friendly, whereas other characteristics (e.g., toxicity) of the product might not be favorable or might even be unknown. On the other hand, some products, especially if they are novel products with environmentally superior characteristics, might not be as readily available as traditional products, or they might require further optimization to meet certain specifications.

All of these factors, of course, complicate the selection of sustainable alternatives and point to the need for some general guidance. One example of such guidance may be found in the U.S. EPA's five guiding principles for environmentally preferable purchasing. These principles were established for the U.S. federal government but are general enough to be applied in other settings. They are also flexible enough to accommodate a variety of factors, such as the type and complexity of the product, the commercial availability, the type of procurement strategy (e.g., negotiated contract, bid), the time frame, and the total purchasing value. Table 17.4 shows the EPA principles on environmentally preferable purchasing.

The five principles are based on three simplified green engineering principles (see Chapter 2):

TABLE 17.4 EPA Principles on Environmentally Preferable Purchasing

No.	Principle	General Area
1	Environmental considerations should become part of normal purchasing practice, consistent with such traditional factors as product safety, price, performance, and availability.	Environmentally preferable purchasing = environment + price + performance
2	Consideration of environmental preferability should begin early in the acquisition process and be rooted in the ethic of pollution prevention, which strives up-front to eliminate or reduce potential risks to human health and the environment.	Pollution prevention
3	A product or service's environmental preferability is a function of multiple attributes from a life cycle perspective.	Life cycle perspective
4	Determining environmental preferability might involve comparing environmental impacts. In comparing environmental impacts, one should consider: the reversibility and geographic scale of the environmental impacts, the degree of difference among competing products or services, and the overriding importance of protecting human health.	Comparison of environmental impacts
5	Comprehensive, accurate, and meaningful information about the environmental performance of products or services is necessary in order to determine environmental preferability.	Environmental performance information

Source: ref. 6

1. Resource efficiency: which includes giving preference to reusable content, recycled materials over virgin materials, and conservation of water and energy.
2. Minimize hazards and pollution: with the goal of avoiding waste creation from the start of a process, favoring source reduction, and eliminating toxicity and emissions.
3. Life cycle thinking: aiming to look at costs beyond the purchase price (total cost assessment) and assessing environmental impacts over the entire life cycle.

One example of a company that has established green procurement practices is General Motors of Canada, which works with suppliers and dealers to establish pollution prevention practices. By using reusable crates and pallets the company has eliminated the need for disposable crates. Several cities and counties are also implementing green procurement strategies, such as the requirement from the city of Santa Monica, California, to include environmental and health specifications in bids for cleaning products,[7] while King County in Seattle, Washington, has an environmental purchasing program that provides information used to identify, evaluate, and purchase environmentally preferable products. In 2007, King County purchased $41 million of these products with an estimated savings of $875,000 compared to the cost of conventional products.[8]

A natural extension of green procurement is to embed environmental aspects into supply chain management. The Pacific Northwest Pollution Prevention Resource Center has published a list of case studies and examples of companies that have worked with green procurement and greening the supply chain. Table 17.5 shows some examples.

17.3.1 Environmental Declarations

As we saw earlier, the central question of a green procurement strategy would be how one knows if one product is preferred over another from an environmental perspective. One strategy used by manufacturers to communicate specific environmental characteristics of products is through the use of environmental declarations. The ISO 14020 standards series establish principles and procedures for the development and use of environmental labels and declarations, and the other standards of the series are intended to be used in conjunction with it.[10] According to the ISO standards, there are three types of environmental declarations:

1. *Environmental labels*, or *ecolabels*, also called *type I environmental declarations*, provide companies with an opportunity to promote the environmental performance of their products and services using symbols. These are based on criteria established by third parties based on the product's life cycle impacts. Figure 17.7 shows some ecolabels.[11]
2. *Auto-declarations*, also called *type II environmental declarations*, are self-declarations by manufacturers. Type II declarations can be made on a single attribute (e.g., energy efficiency, use of recycled materials), or they may include several environmental attributes.[12]
3. *Ecoprofiles*, also called *type III environmental declarations*, provide information based on life cycle assessment studies according to the ISO 14040 series. These declarations are intended primarily for business-to-business communications.[13]

A reflection of the growing importance of environmental and sustainability aspects in the current marketplace is the proliferation of environmental labels. For example, 56 global ecolabels have been reported to date: 118 in Europe, 85 in North America, and 32 in Asia.[14]

TABLE 17.5 Some Examples of Green Procurement and Greening the Supply Chain in Companies

Company	Description
Ben and Jerry's	Work with suppliers to develop chlorine-free packaging and to reduce berry packaging.
BMS	Partner with suppliers in Egypt, Venezuela, and Greece to eliminate the use of methylene chloride. Education of internal procurers on considering the environmental costs of products rather than just the product cost.
Eli Lilly	Collaborate with a supplier to research and find an alternative raw material for a lead-bearing catalyst.
Levi Strauss	Work with suppliers worldwide on water quality. More than 35 water treatment systems were built or upgraded at laundry and product-finishing facilities.
SC Johnson	Work with suppliers on VOC reduction and packaging reduction
Xerox	Work with suppliers to make "smarter" parts and products for their office equipment, especially to facilitate remanufacture of Xerox copiers and other equipment at the end of their useful lives.
Volvo	Provide customers with environmental information on products. Requirement that suppliers certify that they do not use, in products or in manufacturing, any substances on Volvo's list of banned substances.
Clorox	Work with suppliers on new process developments to reduce wastes and inefficiencies.
Eastman Kodak	Evaluate life cycle impacts of components and raw materials from suppliers.
Home Depot	Educate suppliers and themselves about forestry issues. The first major U.S. home improvement retailer to offer certified sustainable wood products.
Hewlett-Packard	Require contract and purchasing agreements with suppliers to reflect environmental and social expectations. Expectation from suppliers to formally include health, safety, labor, and human rights practices in their performance standards. Implementation of a supply chain social and environmental policy and a supplier code of conduct.
Anheuser-Busch	Have suppliers manage on-site storage/recycling of plastic banding. Encourage suppliers to use reusable packaging.
Motorola	Have suppliers manage chemical inventory and offer alternative selection, optimized containers, and efficient logistics.

Source: ref. 9.

There are ecolabels for a wide range of applications and geographical areas, ranging from energy to food to forestry products and the like.

There are a series of publications, lists, and databases that are intended to help the user or the procurement expert to navigate, compare, and procure products within the guidelines of sustainable or green purchasing. For example, the Responsible Purchasing Network has issued a series of responsible purchasing guides for such products as computers, cleaners, bottled water, green power, and paint.[15,16] The USEPA has issued comprehensive procurement guidelines[17] that include product resource guides on cars, paper products, office supplies, and construction products. They have also created a series of green purchasing tools for electronics and construction, among other industries. As of this date, there is very little

FIGURE 17.7 Some environmental labels.

centralized information about bulk substances and chemicals that one might normally encounter in a chemical process: the ecoefficiency assessments of BASF covered in Chapter 16 being a notable exception. However, it is expected that more information will become available, led primarily by changes in the regulatory landscape of chemical management strategies and regulations, such as REACH (European Community regulation on chemicals and their safe use, dealing with the registration, evaluation, authorization and restriction of chemical substances).[18]

17.3.2 Implementing a Green (Sustainable) Procurement Initiative

In general, a green procurement strategy should be compatible with existing procurement programs. Sometimes the least problematic approach may be to make minor changes in existing procurement procedures which ensure that the long-term costs of each contract or purchase are considered along with the purchase price; this type of small change can have a large impact on an organization. Sometimes the main barrier to establishing a green procurement program is in selling the concept, as it might initially be seen as adding an additional large burden to the work the procurement department is already doing. One of the ways to overcome this barrier is to focus on the long-term total cost savings that this type of strategy can bring, since at times the lowest buying price can bring a series of additional expenses to the operation in the form of greater energy consumption, more waste to treat, and more regulations to comply with. It might help to set the stage in simple terms of eliminating unnecessary purchases and reducing total costs.

However, the degree of difficulty in implementing green procurement strategies will vary with the relationship a company has with its suppliers and its market position. Large companies with a very strong market share will generally have less difficulty influencing their suppliers than will smaller companies that have a smaller share of the market. However, when strategies such as green procurement become mainstream (e.g., it becomes a more or less standard practice for many small, medium-sized, and large companies), the degree of influence with suppliers increases, and the possibilities for influencing increases as well.

17.3.3 Greenwashing

One of the challenges in using environmental declarations is to ensure that useful information is provided without deceiving the customer. *Greenwashing* is a term used to describe the perception of consumers that they are being misled by a company about its environmental practices or the environmental benefits of a product or service. These type of practices have significant consequences. First, consumers who desire to reduce their environmental footprint may be misled and may thus be missing an opportunity to obtain an environmental benefit through their purchase. Second, illegitimate claims could be used to compete against products possessing more legitimate environmental benefits. Finally, and perhaps most damaging, is the fact that greenwashing can create skepticism in the minds of consumers regarding the legitimacy of environmental claims. If products having legitimate benefits are not sold, the financial incentives for these products and services will not be there, and ultimately the drivers for innovation and investment in environmentally responsible products will not be present.

In 2007 the environmental marketing firm TerraChoice Environmental Marketing Inc. released a study called "The Six Sins of Greenwashing."[19] TerraChoice conducted a survey of six leading "big-box" stores to describe, evaluate, and quantify the pervasiveness of greenwashing. In this survey they identified 1018 consumer products with 1753 environmental claims. Of all the products identified in the survey, all but one made claims that were either false or misleading. In 2009, Terrachoice released a report adding a seventh sin to the list, the "Sin of worshiping false labels," releasing an updated report[20] as outlined in Table 17.6. The frequency with which each sin was committed was also

TABLE 17.6 Six Sins of Greenwashing

The Sin of...	Description	Example
Hidden trade-offs	Suggesting that a product is "green" based on a single environmental attribute or an unreasonably narrow set of attributes without attention to other environmental issues	Claiming recycled content in paper products without attention to other impacts
No proof	Any environmental claim that cannot be substantiated by easily accessible supporting information, or by a reliable third-party certification	Household lamps and lights that promote their energy efficiency without any supporting evidence or certification
Vagueness	Any claim that is so poorly defined or broad that it is likely to be misunderstood by the consumer	Claiming that a product is "green" or "environmentally friendly" without further elaboration
Irrelevance	Any environmental claim that may be truthful but is unimportant for consumers who want to buy environmentally preferable products	Claiming CFC-free window cleaners, even if CFCs have been banned and the product wouldn't require the use of CFCs

(continued)

TABLE 17.6 (*Continued*)

The Sin of...	Description	Example
Fibbing	Making environmental claims that are false	A product that claims to be packaged in "100% recycled paper," yet whose container is plastic
Lesser of two evils	Any claim that may be true within the product category but that may distract consumers from the greater impacts of the category as a whole	Green ammunition
Worshiping false labels	A product that gives the impression of third-party endorsement where no such endorsement exists	Fake labels; certification-like labels used with phrases such as "eco-safe"

Source: ref. 20.

reported in both studies. In the 2009 update study, the sin of hidden trade-offs was the most frequent, followed by the sin of vagueness, the sin of no proof, and the worship of false labels, in that order. This trend in frequency appeared to be consistent across the United States, Canada, the UK, and Australia.

Example 17.3 What type of greenwashing sin is being committed in these environmental claims, if any, and why is this happening?:

1. "**Roughie** facial tissues are made with post-consumer recycled content."
2. "**There's something 4 Mary**—all natural hair gel."
3. "**Smogie air freshener**—non-toxic scent for your car."
4. "Our color printer/copier **EcoImage** has minimized its carbon footprint."
5. "**Aqualite**—water efficient light bulbs."
6. "Try **Puffs**—the 100% organic cigarettes from sustainably harvested tobacco."

Solution These environmental claims are all committing one of the sins of greenwashing:

- "**Roughie** facial tissues are made with post-consumer recycled content." This is committing the sin of hidden trade-off since it is focusing on only one aspect and ignoring others, such as the use of sustainable harvesting or manufacturing impacts. This claim is also committing the sin of vagueness, as it does not specify how much recycled content goes into the paper, (i.e., is a 1% recycled content good enough?).
- "**There's something 4 Mary**—all natural hair gel." This claim is committing the sin of vagueness, as it does not specify what is meant by *natural*. Not all natural things are good for the environment or for health. As, Hg, and formaldehyde are all found in nature, as well as many poisonous plants.

- "**Smogie air freshener**—non-toxic scent for your car." This claim is committing the sin of vagueness, as it does not specify what is meant by *non-toxic*.
- "Our color printer/copier **EcoImage** has minimized its carbon footprint." This claim is committing the sin of hidden trade-offs, as it seems to ignore other impacts, such as potential health or environmental hazards from the toner or metallic parts. It is also committing the sin of no proof, as it does not seem to provide any manner to verify what the minimization consisted of.
- "**Aqualite**—water efficient light bulbs." This claim is probably committing the sin of irrelevance, as it does not seem to be related to use of the product. Arguably, all light bulbs are water efficient, unless additional explanation is provided that the water efficiency is related to a part of the life cycle process (e.g., manufacturing).
- "Try **Puffs**—the 100% organic cigarettes from sustainable harvested tobacco." This commits two sins—although this might represent a better choice if you smoke, shouldn't we question the environmental and health effects of smoking in the first place?

Additional Points to Ponder Could any of these claims be culpable of the sin of fibbing? How can you confirm if this is the case?

Terrachoice also provided recommendations on how to avoid greenwashing, both as a consumer and as a marketer, as summarized in Table 17.7. At the end of the day these

TABLE 17.7 Suggestions as to How to Avoid the Sins of Greenwashing

How to Avoid the Sin of...	When You Are Buying	When You Are Selling
Hidden trade-offs	Look for multiattribute environmental claims that consider several impacts throughout every phase of a product's life cycle (raw materials, manufacturing, use, and end of life). If there is a single-attribute claim, search other information with a more complete description of the environmental profile.	Understand and communicate the environmental impacts of your product across its entire life cycle. Don't make a claim about a single impact without understanding product performance in other impact categories. It is fine to emphasize one message, but it is not fine to use without first understanding the potential trade-offs or to use it to distract the consumer from other impacts. Consider using a multiattribute system to make environmental claims. Some multiattribute eco-labels are already available, such as Green Seal or EcoLogo.

(continued)

TABLE 17.7 (*Continued*)

How to Avoid the Sin of...	When You Are Buying	When You Are Selling
No proof	Look for evidence of environmental certifications and ecolabels. Look for evidence and additional information on company Web sites, from third-party certifiers, and by calling toll-free phone numbers.	Understand and confirm the scientific and technical basis for any claims. Provide consumers and the public with the means to corroborate the evidence easily.
Vagueness	Look at the explanation that is provided to elaborate on what the claim means in terms of "green" or "environmentally friendly."	Be as specific as possible when defining an environmental claim. Avoid using terms such as "eco-friendly" or "green" without specifying what it actually means.
Irrelevance	Ask yourself if the claim is relevant to the product or if most products in the same category can make a similar claim.	Do not make claims that are irrelevant to the product. Do not make environmental claims that are not a legitimate competitive differentiation.
Fibbing	The most common use of fibbing has been misrepresentation of third-party certifications, so look in the public databases of third-party certifiers and check for fraud alerts as well.	Do not misrepresent information. Always be truthful and transparent.
Lesser of two evils	Ask if the claim is trying to make you feel "greener" about a product that has questionable impacts on health or the environment.	Do not attempt to put a "green spin" to products that are harmful or unnecessary.
Worshiping false labels	Look for products with reliable ecolabels and third-party certifications In the absence of a reliable ecolabel, choose products that offer transparency, information, and education.	Favor ecolabels that are accredited and follow a life cycle approach. If a third-party endorsement is important, get one; do not fake one.

Source: ref. 20.

recommendations are based on the same principle, but looked at from different sides of the marketplace. Avoiding greenwashing is not the same as not issuing environmental claims, but it does mean that sound science and transparency need to be used within an established framework.

17.4 TRANSPORTATION IMPACTS

Transportation-related impacts have garnered considerable attention from the press in recent years, driven by environmental concerns and the desire to foster local economies. Virtually every motorized means of transportation will have environmental impacts related to:

- *Energy consumption*: the energy required to fuel and operate each form of transportation
- *Land use*: land used to construct airports, roads, and highways, with a potential for loss of habitat and other environmental impacts
- *Air emissions*: mainly from internal combustion engines, with emissions from privately owned vehicles tending to be the largest contributors to air pollution in cities
- *Water and land pollution*: arising from leaks of oil, use of road salt, discarded tires, and other sources
- *Noise*: from airports, ports, roads, and highways

There are several efforts to account for and reduce the impacts of transport throughout the supply chain. One of these efforts is known as the SmartWay system, the USEPA's program for reducing emissions in the transportation sector.[21] Companies with trucking fleets enrolled in the SmartWay program are required to measure emissions and demonstrate overall reductions. However, in most cases it is not possible to obtain transportation-related emissions data directly from the supply chains, and in these cases it is necessary to estimate these emissions.

There are two main drivers of transportation-related impacts: transport distances and the modes of transportation. For complex supply networks, it is difficult to know precisely what the transportation modes and distances between specific factories are in order to make an accurate estimation. Unless these are known by the company, or the system being studied is small enough to follow or model, it is common practice in LCA studies to use average or aggregated data for transportation distances and/or modes. Aggregate U.S. transportation data by sector are shown in Table 17.8.

The other part of the equation is the emission profile for the mode of transportation. In this case, standard transportation-related emissions are reported in LCI databases. For example, Table 17.9 shows data from the NREL database. These data were provided by Franklin Associates and are derived from data in the GREET model. The data shown in Table 17.9 are selected life cycle inventory data points for transport by truck, rail, and ship at 100% capacity. The inventory emissions will vary depending on the capacity utilization rate for the transportation mode in question. One way to account for the normal decreased capacity utilization rate is to assume a linear relationship to account for operating at partial capacity. Thus, for a truck capacity utilization rate of 50% we divide the emissions for diesel trucks by 0.5. Similarly, for a ship capacity utilization rate of 70% the ship emissions are divided by 0.7.

Example 17.4 Estimate the typical transportation-related emissions of moving sylvinite by diesel truck for a total distance of 350 miles, based on the data given in Table 17.9. The truck will be operated at 70% of its capacity.

TABLE 17.8 Aggregate U.S. Transportation Data by Industry, 1997

	Basic Chemicals	Metallic Ores and Concentrates	Nonmetallic Ores and Concentrates	Cereal Grains	Coal	Fuel Oils
Average distance (miles)	332	303	174	125	81	28
Contribution of Modes (% of ton-miles)						
Truck	26.4	4.6	31.2	9	1.7	50.7
Rail	50.8	76.8	39.3	58	81.2	1.4
Water	19.3	18.6	19.7	29	4.2	11.2
Multiple modes	2.2	—	9.8	1.5	12	1.2
Pipeline	—	—	—	—	—	34.2
Other/unknown	1.3	—	—	2.5	0.9	1.3

Source: ref. 22.

Solution In this case we can take 1000 kg of sylvinite as the basis to calculate the total ton-miles:

$$(1000\text{ kg})\left(\frac{1\text{ ton}}{2000\text{ lb}}\right)\left(\frac{1\text{ lb}}{0.454\text{ kg}}\right)(350\text{ mi}) = 385.5\text{ ton-mi}$$

Then it is just a matter of multiplying the ton-miles times the emission factors for a diesel truck and accounting for the 70% capacity, as shown in Table 17.10.

TABLE 17.9 Selected Life Cycle Inventory Emissions for Transport by Truck, Rail, and Ship at 100% Capacity Utilization

	Diesel Truck	Barge	Train
Emissions (g/ton-mile)			
CO_2	117	41	28
CO	0.16	0.05	0.07
CH_4	1.88×10^{-3}	9.48×10^{-4}	1.32×10^{-3}
NO_x	0.78	0.52	0.73
N_2O	2.91×10^{-3}	0	0
PM10	0.013	0.013	0.018
SO_x	0.026	9.06×10^{-3}	6.12×10^{-3}
VOC	0.038	0.019	0.027
Diesel consumption			
Gallons	0.0105	3.70×10^{-3}	0.0025
MJ	1.53	0.54	0.37

Source: ref. 23.

TABLE 17.10 Estimated Emissions

	Diesel Truck, 100% Capacity (g/ton-mi)	Diesel Truck, 70% Capacity (g/ton-mi)	Diesel Truck, 70% Capacity, to Move 1000 kg for 350 mi (g)
Emissions per ton-mile			
CO_2	117	167	64,434
CO	0.16	0.23	88
CH_4	1.88×10^{-3}	2.69×10^{-3}	1.04
NO_x	0.78	1.11	430
N_2O	2.91×10^{-3}	4.16×10^{-3}	1.60
PM10	0.013	0.019	7.159
SO_x	0.026	0.037	14
VOC	0.038	0.054	21
Diesel consumption			
Gallons	0.0105	0.015	5.78
MJ	1.53	2.19	843

Additional Points to Ponder How would you account for transportation-related emissions when a truck is empty during its return trip to its origin? How would you estimate transportation-related impacts other than emissions?

17.4.1 Transportation Impact Contribution: A Life Cycle Viewpoint

In recent years, more people seem to be concerned about transportation-related environmental impacts associated with the goods we consume, in part due to concerns about climate change and in part due to concerns about globalization. However, even though the transportation of goods in the current economy is global, and transportation does have environmental impacts, in many systems the transportation-related environmental impacts associated with consumer goods do not play a significant role compared to other impacts (e. g., production phase).

For example, in recent years the term *food miles*,[24] the distance that food travels from farm to fork, has been used as a metric to indicate potential environmental impacts associated with the food we consume. Buying food with low food miles is currently a growing movement among environmentally conscious shoppers, which has been prompted by the increasing distances that food travels before it reaches our plates. For example, half the vegetables and 95% of the fruit eaten in the UK is sourced from overseas.[25] In the United States the average distance covered by food increased from 6760 km to 8240 km from 1997 to 2004. With those numbers in hand, it might seem logical to think that the global warming potential from food might be much higher as a result. However, in terms of distance, because ocean shipping is the largest mode of transportation used for food transport (more than 99%) and requires less energy than normal transport by truck, the total increase in greenhouse gas emissions between 1997 and 2004 from transporting food to the United States was estimated by a life cycle assessment study to have increased by only 5%.[26]

The same life cycle assessment study found that transporting food items from the farm to the marketplace creates only a small percentage of the total global warming potential impact

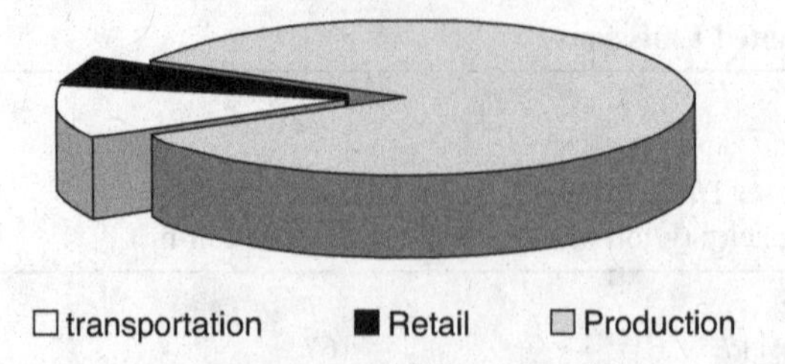

FIGURE 17.8 Contribution of total GHG emissions for food consumption in an average American house. The average annual GHG emission was estimated to be 8.1 tons of CO_2-eq./household. The contribution of transport will vary depending on the type of product.

associated with food, and that globalization of the food markets has increased global greenhouse gas emissions by only 5%. The study traced about 83% of the average U.S. household food items and found that transportation, on average, contributes only about 11% of the greenhouse gas emissions, as shown in Figure 17.8 (transportation from producer to the marketplace accounted for only about 4%). The transportation contribution varies with the type of food group, from 6% for red meat (1% due to delivery from producer to market) to 18% for fruits and vegetables (11% due to delivery from producer to market).

The researchers concluded that if consumers wanted to reduce their food-related greenhouse gas emissions, they would need to consider changing their diets to foodstuffs that require less resource consumption (e.g., energy) to produce. For example, even a relatively small change, such as a 21 to 24% reduction in red meat consumption shifted to chicken, would result in the same reduction in greenhouse gas emissions as the total reduction that could be achieved by switching to locally sourced meats. That is not to say that buying locally is not a good choice, as there may be many good reasons to source food (and other materials) locally, such as supporting local agriculture, increased freshness, better control of the supply chain, taste, and loyalty to the local economy. However, what this study proves is that for the average consumer, buying locally does not have as significant an environmental impact as does the choice of food.

This research provides just one example of why food miles is an inappropriate environmental impact metric. The food-miles concept has become popular because it is a single measure that may be easy for people to understand (similar to carbon footprinting), but it should not be used as a measure of environmental impact because it does not represent a significant environmental impact associated with food production. In addition, metrics such as food miles or carbon footprint include only a single aspect, and as we saw previously, impacts other than greenhouse gas emissions need to be considered.

Additional Points to Ponder The study we have just discussed considered only greenhouse gas emissions. How would transportation fare if we considered other environmental impacts?

Transportation-related environmental impacts beyond greenhouse gas emissions have been assessed as part of life cycle studies applied to active pharmaceutical ingredient manufacturing. As can be seen from the data in Figure 17.9, these studies have shown that transportation accounts for only a small contribution to the pretreatment cradle-to-gate impacts: less than 8% across the impact categories of greenhouse gases, eutrophication, acidification, photochemical ozone creation potential, total organic carbon, and resource consumption (energy and material). When waste treatment impacts are included in the

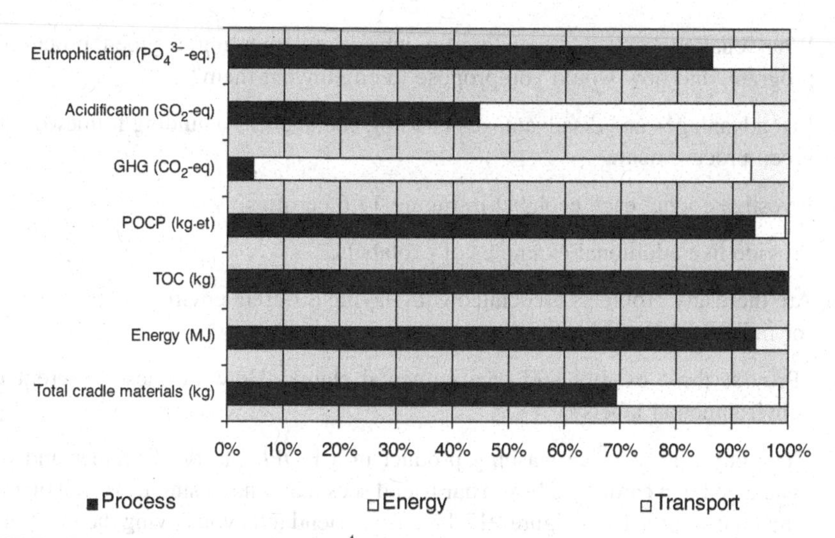

FIGURE 17.9 Relative pretreatment contributions of process-, energy-, and transport-related environmental cradle-to-gate life cycle assessment impacts for the production of an active pharmaceutical ingredient. (From Jiménez-González et al., ref. 27. Reproduced with kind permission of Springer Science and Business Media. Copyright © 2004, Springer Science and Business Media.)

assessment, the contributions from transport are even less significant.[27] This is consistent with another LCA of API production across five different routes,[3] which included all the categories of the previous example, and in addition, ozone depletion potential, and human health impacts (carcinogens and heavy metals). This study estimated the maximum contribution of transport to posttreatment environmental impacts to be 3.3% (ozone depletion potential).

The conclusion of this discussion is that we always need to look at the overall life cycle impacts of an activity or service. Although it is true that in most systems transportation will not play a major role, this should not be construed as a blanket statement. It all depends on the type of system that one is analyzing and the upstream environmental impacts of the system. For example, the production of wood products, in general, tends not to be energy-intensive, and as a result, transportation impacts could be a significant portion of wood furniture production. In contrast, steel production from iron ore tends to be very energy intensive; therefore, it would be expected that transportation would not be a significant contributor to the overall life cycle environmental impacts associated with steel.

In Chapter 18 we look more closely at the impacts derived from energy production and its use in industrial systems.

PROBLEMS

17.1 Develop a chemical tree for the production of sodium hydroxide.

17.2 Develop a chemical tree for the production of acetic acid.

17.3 Develop a chemical tree for the production of isopropanol.

17.4 What steps would you suggest to implement a green procurement program?

17.5 What challenges would you expect when implementing a green procurement program, and how would you propose to circumvent them?

17.6 List advantages and disadvantages of using ecolabels as a guiding framework for a green procurement.

17.7 Investigate what each ecolabel in Figure 17.6 certifies.

17.8 Provide five additional examples of ecolabels.

17.9 Are there any problems associated with having different environmental labels used for individual countries?

17.10 Provide three examples of environmental claims. How are these different from environmental labels?

17.11 You buy a liquid face washing product in a HDPE plastic container and pump packaged in a cardboard box. Your friend sees that it has a single symbol on one of the flanks of the box (Figure P17.11). Your friend tells you having the symbol as it appears is committing a greenwashing sin.

FIGURE P17.11

(a) Is this true? Why or why not?
(b) Would you propose any changes to the claim? Which ones?

17.12 Look at the environmental claims in Example 17.3.

(a) Identify the ones that could actually be supporting real information.
(b) Rewrite the claims you identified previously in such a way that they are appropriate environmental claims.

17.13 Estimate the global warming potential, acidification potential, and eutrophication potential for Example 17.4.

17.14 Redo Example 17.4 estimating the typical transportation-related emissions associated with moving 1000 kg of sylvinite in the United States based on the data given in Tables 17.8 and 17.9.

(a) Assume that all the transportation modes are utilized at 100% capacity.
(b) Assume that the train and truck modes operate at 50% capacity and that the barge mode operates at 70% capacity.

17.15 Estimate the global warming potential, acidification potential, and eutrophication potential for Problem 17.14. How would you explain the differences from Example 17.4?

17.16 Estimate the typical transportation-related emissions of moving 1000 kg of basic chemicals in the United States based on the data given in Tables 17.8 and 17.9.
(a) Assume that all the transportation modes are utilized at 100% capacity.
(b) Assume that the train and truck modes operate at 50% capacity and that the barge mode operates at 70% capacity.

REFERENCES

1. Remmen, A., Astrup, A., Frydental, J. *Life Cycle Management: A Business Case for Sustainability*. United Nations Environment Program, Life Cycle Initiative, 2007.
2. Hunkeler, D., Rebitzer, G., Finkbeiner, M., Schmidt, W.-P., Jensen, A. A., Stranddorf, H., Christiansen, K. *Life-Cycle Management*. Society of Environmental Toxicology and Chemistry, Brussels, Belgium, 2004.
3. Jiménez-González, C. Life Cycle Assessment in Pharmaceutical Applications. Ph.D. dissertation, North Carolina State University, Raleigh, NC, 2000.
4. Curzons, A., Jiménez-González, C., Duncan, A., Constable, D. J. C., Cunningham, V. Fast life cycle assessment of synthetic chemistry tool, FLASC(TM) tool. *Int. J. Life Cycle Assess.*, 2007,12 (4), 272–280.
5. U.S. Environmental Protection Agency, Pollution Prevention Resource Exchange. *Green Procurement: Reasons for Change.* U.S. EPA, Washington, DC, Available at http://www.p2ric.org/TopicHubs/index.cfm?page=subsection&hub_id=13&subsec_id=3, accessed Jan. 27, 2010.
6. U.S., Environmental Protection Agency. *Final Guidance on Environmentally Preferable Purchasing*.U.S. EPA,Washington, DC, Aug. 20,1999. Available at http://www.epa.gov/oppt/epp/pubs/guidance/finalguidance.htm##GuidingPrinciple2, accessed Dec. 21, 2008.
7. GreenBiz.com. Green Procurement.http://www.greenbiz.com/resources/resource/green-procurement, accessed Dec. 21, 2008.
8. King County Web site. Environmental Purchasing.http://www.kingcounty.gov/operations/procurement/Services/Environmental_Purchasing.aspx, accessed Dec. 21, 2008.
9. Pacific Northwest Pollution Prevention Resource, Center. Supply Chain Management for Environmental Improvement.http://www.pprc.org/pubs/grnchain/casestud.cfm#eval, accessed Dec. 22, 2008.
10. International Organization for Standardization.*Environmental Labels and Declarations: General Principles*. ISO 14020:2000.ISO, Geneva, Switzerland, 2000.
11. International Organization for Standardization.*Environmental Labels and Declarations—Type I Environmental Labelling: Principles and Procedures*. ISO 14024:1999.ISO, Geneva, Switzerland, 2000.
12. International Organization for Standardization. *Environmental Labels and Declarations—Type II environmental labelling: Self-Declared Environmental Claims*. ISO 14021:1999.ISO, Geneva, Switzerland, 2000.
13. International Organization for Standardization. *Environmental Labels and Declarations—Type III Environmental Declarations: Principles and Procedures*. ISO 14025:2006.ISO, Geneva, Switzerland, 2006.

14. Ecolabelling.org.http://ecolabelling.org/, accessed Dec. 21, 2008.
15. Responsible Purchasing Network Web site.http://www.responsiblepurchasing.org/publications/index.php, accessed Dec. 22, 2008.
16. Godwin, C., Tilford, D. Bottled water. In *Responsible Purchasing Guide*, O'Brien, C., Ed. Responsible Purchasing Network, Center for a New American Dream, Takoma Park, MD, 2007.
17. U.S. Environmental Protection Agency. Comprehensive Procurement Guidelines.http://www.epa.gov/epawaste/conserve/tools/cpg/index.htm, accessed Dec. 22, 2008.
18. European commission. Regulation 1907/2006.
19. TerraChoice Environmental Marketing, Inc. *The "Six Sins of Greenwashing,™" A Study of Environmental Claims in North American Consumer Markets*, TerraChoice, Ottawa, Ontario, Canada, Nov. 2007.
20. TerraChoice Environmental Marketing, Inc. "*The Seven Sins of Greenwashing,™" Environmental Claims in Consumer Markets Summary Report: North America*. Apr. 2009.
21. U.S. Environmental Protection Agency. *Smart Way Transport Partnership Carrier FLEET Model User Guide*. EPA 420-B-06-011. U.S. EPA,Washington, DC, 2006.
22. U.S. Department of Commerce, Economics and Statistics Administration; U.S. Census Bureau; U.S. Department of Transportation, Bureau of Transportation Statistics. 1997 Commodity Flow Survey, in *1997 Economic Census: Transportation* (issued Dec. 1999). EC97TFC-US.
23. NREL Life Cycle Inventory database. http://www.nrel.gov/lci/
24. Paxton, A. *The Food Miles Report: the Dangers of Long Distance Food Transport*. Safe Alliance, London, 1994.
25. Stacey, C. Food Miles. BBC Web page.http://www.bbc.co.uk/food/food_matters/foodmiles.shtml, accessed Dec. 26, 2008.
26. Weber, C. L., Matthews, H. S. Food-miles and the relative climate impacts of food choices in the United States. *Environ. Sci. Technol.*, 2008, 42(10), 3508–3513.
27. Jiménez-González, C., Curzons, A. D., Constable, D. J. C., Cunningham, V. L. Cradle-to-gate life cycle inventory and assessment of pharmaceutical compounds: a case-study. *Int. J. Life Cycle Assess.*, 2004, 9(2), 114–121.

18

IMPACTS OF ENERGY REQUIREMENTS

What This Chapter Is About In this chapter we continue building on life cycle assessment elements with a review of the life cycle impacts for energy production. We also use examples of streamlined LCA to evaluate the environmental footprint of energy production and explore to what extent the transportation of goods affects the life cycle impact profile of production processes or activities.

Learning Objectives At the end of this chapter, the student will be able to:

- Understand the concepts of energy, primary energy carrier, and energy requirements.
- Understand the potential impact of energy production.
- Understand the impact that energy choices have on the environmental footprint of products.
- Apply streamlined techniques to evaluate the life cycle impact of energy production and delivery.

18.1 WHERE ENERGY COMES FROM

In Chapter 9 we saw how to estimate the mass and energy required to make products through chemical processes. The energy required is mostly thermal or mechanical, with the latter provided by electricity in a great number of cases. These requirements translate into heating requirements and electricity use, respectively. At the same time, heat has to be removed from the system, thereby creating cooling requirements.

Green Chemistry and Engineering: A Practical Design Approach, By Concepción Jiménez-González and David J. C. Constable

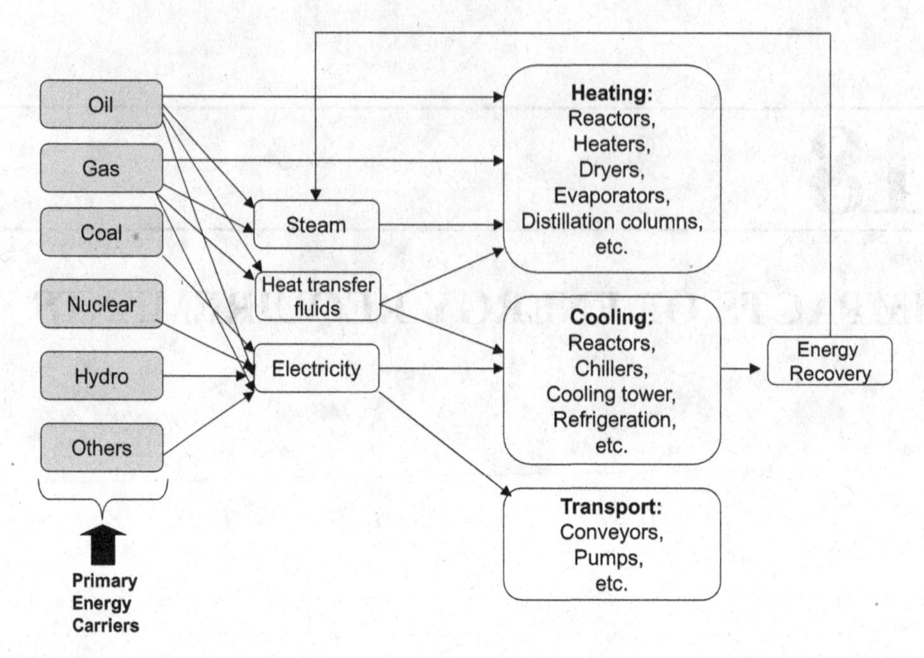

FIGURE 18.1 Potential energy paths in a chemical process. Primary energy carriers are shaded.

To satisfy the electricity, heating, and cooling requirements, other processes have to take place to provide the energy sources or sinks. These energy processes will have material inputs and outputs that can readily be traced back to their extraction from the Earth (e.g., natural gas, coal, potential energy of water). A representative example of a heating requirement is the process of producing steam, and a representative example of a cooling requirement is the use of a cooling tower. Figure 18.1 is a graphical representation of energy production for a chemical process.

As can be seen in the figure, different energy requirements are fulfilled through the use of different direct energy sources (i.e., the energy used directly in chemical processes) and include such sources as petroleum fuels, steam, electricity, and heat transfer fluids. There are also miscellaneous direct energy sources, such as the use of coal or biomass combustion, but these are not significant contributors to direct energy sources at this time. These direct energy sources are produced by different means, and each means of producing the energy will have a different environmental impact profile. For example, producing 1 MJ of steam has a different set of emissions and resource requirements than those producing 1 MJ of electricity. In the same vein, the environmental impacts from energy production will vary depending on the efficiency of the energy generation and/or distribution processes, or on the typical energy sources used in a specific region. For example, in 2005, the production of 1 kWh of electricity in Alaska came from about 9.5% bituminous coal, in contrast with Texas electricity production, where 37% of the energy was derived from bituminous coal. On average, about half of the electricity in the United States comes from bituminous coal. These figures include line-loss factors during energy distribution.[1,2] See Figure 18.2 for a general illustration of the 2005 electricity mix in the United States according to the North American Electric Reliability Corporation.[3] This type of energy mix information can significantly affect the outcome of an LCA, so it is critical to obtain representative data, especially when dealing with averages.

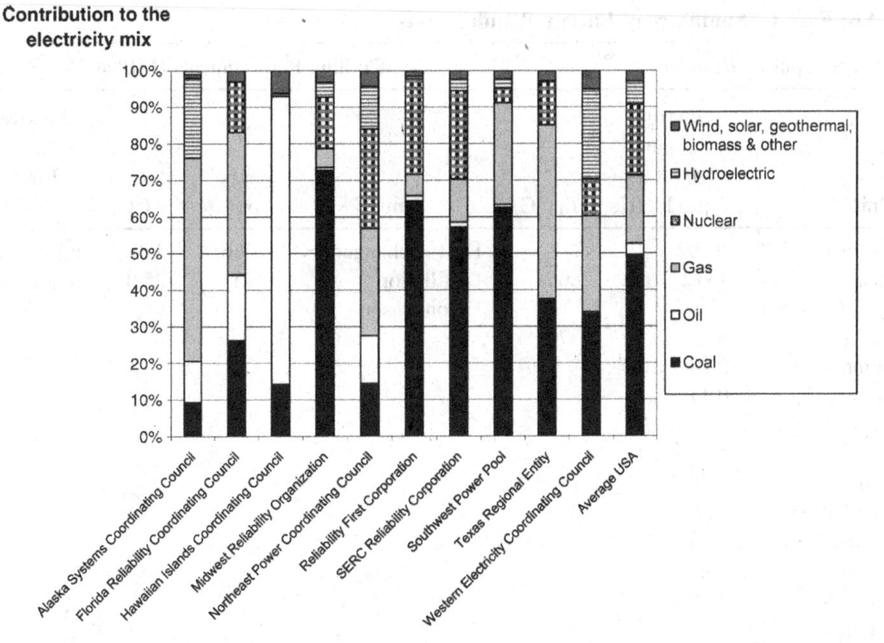

FIGURE 18.2 Contributions to the U.S. electricity grid for 2005 by geographical zone. SERC, Southeast Electric Reliability Council.

Example 18.1 In Example 16.2 we provided gate-to-gate and cradle-to-gate summaries for the production of 3-pentanone. Table 18.1 summaries the energy requirements for that process.

(a) Where does the energy to satisfy the heating requirements come from?
(b) How would you satisfy the cooling requirements?
(c) What might you do with any excess energy from the process?

Solution Table 18.2 shows the direct energy sources to be used for the process. All conveyance equipment (e.g., pumps, vacuum systems) will use electricity. Steam or heat exchange fluid will be used for heating. For this particular example we propose using DowTherm for heating to higher temperatures, and the vacuum distillation will require the use of some steam. The distillation condenser will probably take cooling water (or brine) at around 20°C. Any excess energy could be used to reduce the DowTherm requirement for reactor heating and in the first heat exchanger.

Additional Points to Ponder How was the potential energy recovery estimated? How do you know that a vacuum distillation is taking place?

Since both mass and energy are normally included in a typical LCA, that is, we normally tally both total energy requirements and crude oil for energy production in LCIs, one question that may come to mind is: Are we double counting? In essence, from the viewpoint of the inventory, this is not the case, as it is equivalent to reporting the results for two different units. Nonetheless, one should be mindful of the potential for double counting during the

TABLE 18.1 Summary of Energy Requirements

Energy Input (MJ/batch)			Cooling Requirements (MJ/batch)			
Unit	Energy Input (MJ/1000 kg product)	T_0 (°C)	Unit	Energy Loss	T_{eff} (°C)	Recovery Efficiency/ Energy Recovered
Pump 1	0.102		Heat exchanger 2	−2046	370	60%/−1227
Heat exchanger 1	1312	300	Distillation condenser 1	−493	25.0	0
Reactor 1	516	370				
Pump 2	2.23×10^{-3}					
Pump 3	0.174					
Pump 4	6.66×10^{-4}					
Pump 5	1.34×10^{-6}					
Pump 6	4.86×10^{-4}					
Distillation reboiler 1	493	25.0				
Pump 7	0.206					
Pump 8	3.86×10^{-4}					
Pump 9	4.50×10^{-7}					
Vacuum pump 1	54.0	0				

assessment phase, so no additional weight is given to energy, except for weighting that is desired expressly as part of the value judgment. Another potential for a double-counting error is when a portion of the direct energy (e.g., MJ of steam) is reported along with the primary energy carrier (e.g., MJ of fuel oil to produce steam). As noted in Chapter 17, the key to ensuring that this does not happen would be in maintaining the transparency of the LCI data. Besides avoiding potential double counting when reporting the energy and mass associated with energy production, there are several good reasons to provide separate figures for the resources required for energy production and consumption in general: that energy is normally well understood and it is easy to communicate, measure, and monitor (e.g., it is easier to obtain measured energy requirements than measured emissions).

Another interesting point regarding energy is that some raw materials can be used to produce energy or products. An example of this is when petroleum products are used to produce plastics. The total energy embedded in the materials being manufactured is called *feedstock energy* and the energy content for energy production is not included in feedstock energy. One must be careful when using feedstock energy in LCIs, as it is easy to double count if it is not used appropriately.[4]

When feedstock energy values are reported, it is important to specify the type of heat value that is used. It is common to see either the high heat value (HHV) or the low heat value (LHV), or both, reported. To avoid double counting one must know the distinction between these two values. The HHV represents the energy content as the full intrinsic energy, that is, the gross calorific value, and captures the combustion energy plus the phase transfer heat for the water formed in the combustion reaction. The LHV does not capture the phase-transfer heat of the water formed in the combustion reaction. For example, the HHV of natural gas is in general between 53 and 54 MJ/kg of natural gas, while its LHV is about 48 MJ/kg of natural gas. The

TABLE 18.2 Direct Energy Sources

Energy Input (MJ/batch)				Cooling Requirements (MJ/1000 kg batch)				
Unit	Energy Input (MJ/1000 kg Product)	T_0 (°C)	Direct Energy Type	Unit	Energy Loss	T_{eff} (°C)	Recovery Efficiency/Energy Recovered	Direct Energy Type
Pump 1	0.102		Electricity	Heat exchanger 2	−2046	370	60%/−1227	Heat exchangers
Heat exchanger 1	1312	300	DowTherm	Distillation condenser 1	−493	25.0	0	Cooling water
Reactor 1	516	370	DowTherm					
Pump 2	2.23×10^{-3}		Electricity					
Pump 3	0.174		Electriciy					
Pump 4	6.66×10^{-4}		Electricity					
Pump 5	1.34×10^{-6}		Electricity					
Pump 6	4.86×10^{-4}		Electricity					
Distillation reboiler 1	493	25.0	Steam					
Pump 7	0.206		Electricity					
Pump 8	3.86×10^{-4}		Electricity					
Pump 9	4.50×10^{-7}		Electricity					
Vacuum pump 1	54.0	0	Electricity					

difference between these two figures is the latent heat of the water formed in the combustion reaction.

18.1.1 Energy Requirements in a Chemical Plant

In a chemical plant, energy is required for the reaction (reactor, preheater, compressor, cooler, etc.) or for the separation train (filtration, dryer, heater, cooler, distillation, etc.). An article published by Kim and Overcash analyzed the gate-to-gate process energy for 86 chemical manufacturing processes,[5] where the energy requirements were estimated using a design-based methodology[6] following general chemical engineering principles such as those described in *Perry's Chemical Engineers' Handbook*.[7] In this paper it was found that the net energy used in half of the organic chemicals analyzed was between 0 and 4 MJ/kg, while the net energy requirement for inorganic chemicals was between −1 and 3 MJ/kg. The zero or negative net energy values were due to recovering the surplus energy (e.g., exothermic reactions) from steam production or direct heating and were regarded as an energy credit. In addition, the study found that about half of the process energy used in the organic and inorganic chemical processes was used for separation and purification processes and indicates the potential for general improvements. Table 18.3 shows the contrast between energy requirements for organic and inorganic chemical production found in the study.

As can be seen from the table, all these data have large associated standard deviations and suggest that reliable estimations for chemical production process energy is not possible at this time. However, these values could be used in screening life cycle assessments to establish if a specific chemical process contributes significantly to the cumulative energy demand when no primary information is available. One would, of course, want to do a sensitivity analysis to determine the impact of the estimations on the overall LCI.

Example 18.2 Is the energy required to produce 3-pentanone, as found in Example 16.2, a typical example of the production of this type of chemical?

Solution It will all depend on what "typical" means. Table 18.4 summarizes the energy requirements from Examples 16.2 and 18.1. From Table 18.1 we see that the electricity required and the potential energy recovery are within the expected ranges but just below average. However, given the large variation in energy requirements, this would be expected. Steam requirements seem to be at the lower end of the spectrum, but since

TABLE 18.3 Energy Requirements for Gate-to-Gate Production of Chemicals

Direct Energy	Organic Chemicals Average ± Standard Deviation (MJ/kg)	Inorganic Chemicals Average ± Standard Deviation (MJ/kg)
Electricity[a]	0.6 ± 0.98	1.9 ± 5.1
Steam	7.7 ± 14	3.6 ± 8.2
Heating fuel	0.15 ± 0.5	1.5 ± 3.2
Potential recovery	−1.6 ± −1.9	−2.0 ± −5
Total (range)	0 to 4	−1 to 3

Source: ref. 5.

[a] Heat transfer fluid is not included, due to its infrequent use.

TABLE 18.4 Energy Requirements

Type of Energy	Amount (MJ/kg)
Electricity	0.49
DowTherm	1829
Heating steam	493
Energy input requirement	2323
Cooling water	−2539
Cooling–refrigeration	0
Potential heat recovery	−1,227
Net energy	1095

steam use is so variable, it is still within the standard deviation for organic chemical production processes.

Additional Points to Ponder From an LCA viewpoint, why would it be important to know if the energy requirements are within certain ranges? Would the conclusions be different if the process did not have any energy integration?

18.2 ENVIRONMENTAL LIFE CYCLE EMISSIONS AND IMPACTS OF ENERGY GENERATION

18.2.1 Process Energy: The Concept of Energy Submodules

From a life cycle viewpoint, it is customary to report the cumulative energy demand. In addition, to increase the transparency of the results, it is useful to keep track of the type of energy requirements that are needed for the various production processes. It is also useful to have the environmental life cycle profile of energy production characterized so that one can understand the impacts and improvement opportunities of any given process. A set of life cycle inventory data containing the energy requirements for a unit operation or a complete process has been called an *energy submodule*. The energy submodule contains all the material input and output data for that submodule, as well as data for the primary sources of energy and their associated emissions. One can therefore think of energy production submodules for electricity, steam, cooling water, refrigeration, waste treatment, and the like.

Energy submodules can in turn be coupled with the energy requirements for a specific process, as in Example (18.2), to estimate the life cycle energy-related emissions for any given process. As mentioned above, these impacts will depend on the specific characteristics of the process, such as the efficiency of the boiler, the types of materials used, and the geographic area. When primary data (i.e., specific measured data) are available, it is better to utilize that information. However, this is not always the case, and when primary data cannot be obtained, there are methodologies and general heuristics that can be used to estimate the impacts for energy production. One methodology has been proposed as a way of estimating energy submodule information based on chemical engineering principles.[8] A general rule of thumb for the application of thermal energy submodules is given in Table 18.5.

In the following paragraphs we explore the impacts of producing several common forms of energy through energy submodules.

TABLE 18.5 Energy Submodules: Rule of Thumb for Operating Temperature Ranges

Energy Submodule	Outflow Temperature Range (°C)	Thermal Efficiency (%)
Heating	$T < 207$	~85[9]
Steam		
Heat transfer fluids	$207 < T < 400$	~85
Furnace	$T > 400$	70 to 80
Cooling water at 20°C	$T \geq 25$	~85
Refrigeration systems	$T < 25$	~65[10]

18.2.2 Electricity

Electricity must be generated and then used immediately. Figure 18.3 shows the basic flow of electricity: how it is generated and transmitted through distribution lines (high voltage and lower voltage, respectively) to reach the final user. Transformers at substations step the electric voltage up and down to deliver power to the customers. The generation and transmission components make up the *bulk power system*. There is normally about a 40% loss of energy due to the transmission and distribution of electricity.

Much work has been done in the life cycle community to characterize the environmental life cycle impacts of electricity production. There is information available for the national grid on a regional basis in commercially available LCI databases, such as Ecoinvent and the like,[11] although there is still much work to be done in this area. Maintaining the transparency of these estimations and making it apparent and clear if the life cycle inventory accounts for transmission and distribution losses, or if the boundaries are set at a different point, is of paramount importance.

The life cycle inventory profile will depend on the energy mix that is used to produce the electricity, the efficiency with which it is transmitted and distributed to the final customer, and the voltage at which the electricity is delivered. A sample of these differences can be seen in Table 18.6, where we show a 2001 compilation of life cycle inventory emissions data that

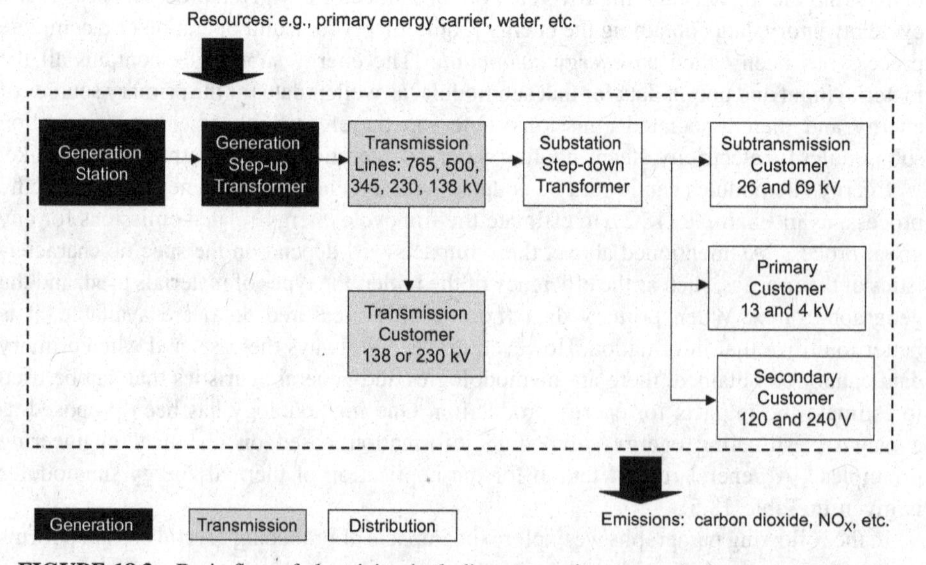

FIGURE 18.3 Basic flow of electricity, including generation, transmission, and distribution.

TABLE 18.6 Selected Life Cycle Inventory Parameters for Electricity Production

	National Grid								
Energy source	UCTE[13]	UCTE[14]	EU, High Voltage[15]	EU, Medium Voltage[15]	EU, Low Voltage[15]	UK[14]	UK, Low Voltage[15]	U.S.A.[16]	U.S.A.[17]
Coal	0.19	0.17	—	—	—	0.59	—	0.57	0.57
Liginite	0.11	0.08							
Fuel oil	0.10	0.11	—	—	—	0.08	—	0.03	0.03
Natural gas	0.10	0.07	—	—	—	0.03	—	0.10	0.10
Hydropower	0.15	0.16	—	—	—	0.03	—	0.09	0.09
Nuclear power	0.36	0.40	—	—	—	0.26	—	0.22	0.22
Other solid fuel									
Sun and wind									

		Major Emissions (per MJ)								
Emission Source	Unit	UCTE[13]	UCTE[14]	EU, High Voltage[15]	EU, Medium Voltage[15]	EU, Low Voltage[15]	UK[14]	UK, Low Voltage[15]	U.S.A.[16]	U.S.A.[17]
					Air Emissions					
CO	g	2.54×10^{-2}	2.17×10^{-2}	2.70×10^{-2}	2.70×10^{-2}	3.30×10^{-2}	2.84×10^{-2}	4.40×10^{-2}	2.91×10^{-2}	2.84×10^{-1}
CO_2	g	1.33×10^{2}	1.19×10^{2}	1.39×10^{2}	1.41×10^{2}	1.58×10^{2}	1.88×10^{2}	2.14×10^{2}	1.82×10^{2}	1.97×10^{2}
Methane	g	3.03×10^{-1}	2.82×10^{-1}	3.63×10^{-1}	3.68×10^{-1}	4.18×10^{-1}	7.42×10^{-1}	7.99×10^{-1}	7.27×10^{-1}	1.24×10^{-3}
Dust	g	2.09×10^{-1}	1.42×10^{-1}	—	—	—	2.95×10^{-1}	—	2.79×10^{-1}	3.30×10^{-1}
NMVOC[a]	g	6.34×10^{-2}	7.39×10^{-2}	6.20×10^{-2}	6.30×10^{-2}	7.30×10^{-2}	6.81×10^{-2}	7.80×10^{-2}	3.70×10^{-2}	2.43×10^{-1}
NO_x	g	2.83×10^{-1}	2.56×10^{-1}	3.07×10^{-1}	3.11×10^{-1}	3.54×10^{-1}	4.74×10^{-1}	5.35×10^{-1}	4.54×10^{-1}	8.65×10^{-1}
SO_x	g	6.61×10^{-1}	6.27×10^{-1}	6.69×10^{-1}	6.81×10^{-1}	7.93×10^{-1}	8.77×10^{-1}	9.38×10^{-1}	7.15×10^{-1}	1.72×10

(*continued*)

TABLE 18.6 *(Continued)*

		Major Emissions (per MJ)								
Emission Source	Unit	UCTE[13]	UCTE[14]	EU, High Voltage[15]	EU, Medium Voltage[15]	EU, Low Voltage[15]	UK[14]	UK, Low Voltage[15]	U.S.A.[16]	U.S.A.[17]
					Water Emissions					
Chlorides	g	5.58×10^{-1}	5.55×10^{-1}	6.38×10^{-1}	6.48×10^{-1}	7.43×10^{-1}	1.22	1.28	1.08	
COD	g	5.64×10^{-4}	5.56×10^{-4}	1.00×10^{-3}	1.00×10^{-3}	1.00×10^{-3}	9.74×10^{-4}	1.00×10^{-3}	7.86×10^{-4}	1.61×10^{-4}
Phosphate	g		2.96×10^{-3}	4.00×10^{-3}	4.00×10^{-3}	5.00×10^{-3}	9.68×10^{-3}	1.00×10^{-2}	9.36×10^{-3}	
Sulfates	g	5.92×10^{-1}	5.68×10^{-1}				8.78×10^{-1}		8.27×10^{-1}	1.45×10^{-1}
TDS	g	1.75×10^{-1}	4.56×10^{-1}	2.90×10^{-2}	2.90×10^{-2}	3.30×10^{-2}	6.91×10^{-1}	7.00×10^{-2}	5.95×10^{-1}	3.35×10^{-2}
					Solid Waste					
	g	1.79×10								2.40×10

Source: ref. 12.

[a] Nonmethane volatile organic compound.

compares electricity profiles from Europe (UCTE, Union for the Co-ordination of Transmission of Electricity), the UK, and from various sources in the United States.[12] Of course, the results of Table 18.6 are not comprehensive and are subject to regional variations and revision as more accurate data are available over time, but it provides a rough idea of the environmental impacts for electricity production in several regions of the world.

Example 18.3 Estimate the LCI emissions profile for electricity production that is required to produce 1000 kg of 3-pentanone.

Solution From Example 18.2 we know that the electricity required to produce 1 kg of 3-pentanone is 0.49 MJ, so 490 MJ will be required to produce 1000 kg of 3-pentanone. The electricity emissions will depend on the geographical area in which the manufacturing plant producing 3-pentanone is located. Using three of the columns from Table 18.3, we may make a very streamlined comparison between the European, UK, and U.S. profiles (Table 18.7).

The variability of the results is illustrated by the large standard deviations associated with each of the emissions factors. For some of the emissions factors in this very limited data set, there is more variation than in others. The differences in these data will also depend on the type of voltage being used, the data source, and the quality of the data. Given this variability, you might ask yourself which data set is more appropriate for the case in question. Are the differences a result of technological differences, a lack of data, a different set of parameters, different definitions, or typos? These types of issues need to be evaluated before deciding to use any given life cycle inventory data set, not only for electricity, but in general.

Additional Points to Ponder How might you determine when differences in inventory data are a result of mistakes or due to other differences? What other parameters would you expect to see in this list?

TABLE 18.7 Major Emissions Related to Electricity (g/1000 kg 3-pentanone)

	UCTE[13]	UK, Low Voltage[13]	U.S.A.[16]	Standard Deviation
Air emissions				
CO	12	22	14	5
CO_2	65,170	104,860	89,180	19,990
Methane	148	392	356	131
Dust	102	0	137	71
NMVOC	31	38	18	10
NO_x	139	262	222	63
SO_x	324	460	350	72
Water emissions				
Chlorides	273	627	529	183
COD	0	0	0	0
Phosphate	0	5	5	3
Sulfates	290	0	405	209
TDS	86	34	292	136
Solid waste	8,771	0	0	5,064

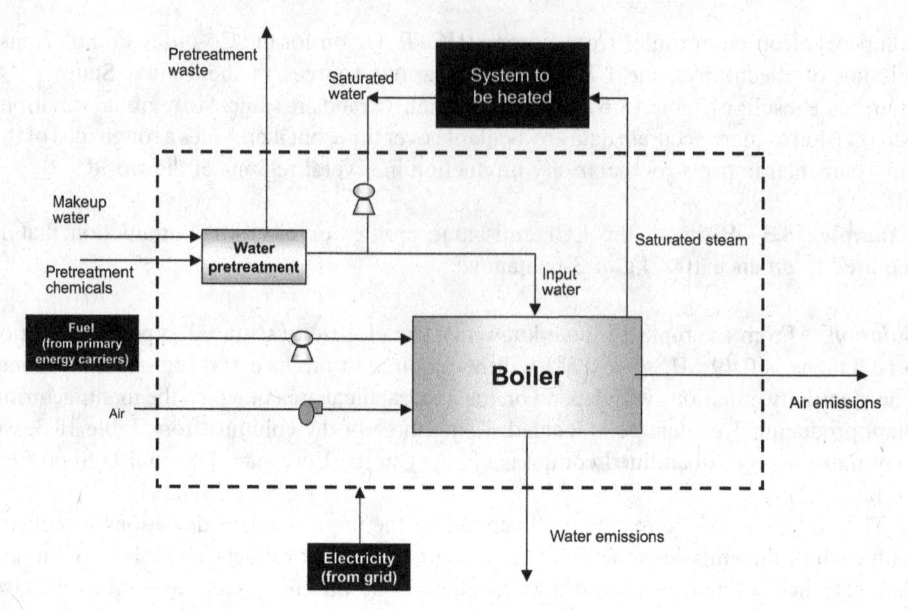

FIGURE 18.4 Typical mass and energy flows to produce steam. (Source: ref 8, with kind permission from Springer Science and Business Media. Copyright © 2000, Springer Science and Business Media.)

18.2.3 Heating: Steam and Furnaces

The energy source for steam production in industry is generally natural gas, fuel oil, or a combination of both. As can be seen in Figure 18.4, the environmental life cycle profile for steam production utilizing natural gas or fuel oil in a combustion chamber will include:

- *Resource requirements*: includes fuel, oxygen (air), makeup water, and chemicals for water pretreatment.
- *Direct air emissions*: from the combustion of the fuel.
- *Direct water emissions*: from the boiler itself, usually a closed-loop water system to generate steam, and hence there is blowdown.
- *Electricity requirements*: for treating and pumping the water and transporting the steam.

The environmental profile for steam production not only varies with the type of fuel but also with the pressure at which the steam is being generated. Table 18.8 shows the selected gate-to-gate parameters for a steam production submodule, and Table 18.9 shows selected cradle-to-gate life cycle inventory parameters for steam production, assuming 50% natural gas and 50% fuel oil consumption.

To estimate these data it was assumed that during the heat exchange process, the vapor enters as saturated steam and leaves as saturated liquid. Water entering the boiler is assumed to be at 71°C, which accounts for the heat losses during piping and storage, and the mixing with makeup water at 20°C. A boiler efficiency of 80% (MJ steam produced/MJ fuel used) was assumed, which accounts for the possibility of heat recovery from the boiler stack through the use of a preheater for gas and air.[18,19] The combustion gas profiles were calculated using material and energy balances, and the heat values of the natural gas and oil were 54.2 and 47.6 MJ/kg, respectively. An air excess of 25%, on a molar basis, was used in

TABLE 18.8 Selected Gate-to-Gate Life Cycle Inventory Parameters for Steam Production Using Two Types of Fuel at Two Different Pressures

	Steam Produced Using No. 2 Fuel Oil		Steam Produced Using Natural Gas as Fuel	
	135 psia	50 psia	135 psia	50 psia
	Inputs			
Air (kg/MJ heating potential)	0.56	0.52	0.60	0.55
Fuel				
kg/MJ heated	0.032	0.03	0.029	0.026
MJ/MJ heated	1.53	1.41	1.53	1.41
Water Makeup (kg/MJ heated)	0.063	0.060	0.063	0.060
Electricity (kWh/MJ heated)	1.92×10^{-3}	1.81×10^{-3}	1.93×10^{-3}	1.81×10^{-3}
11.2% NaCl solution for water pretreatment (kg/MJ heated)	2.97×10^{-3}	2.80×10^{-3}	2.97×10^{-3}	2.80×10^{-3}
	Outputs			
Blowdown waste (kg/MJ heated)	0.024	0.023	0.024	0.023
Pretreatment waste $CaCl_2/MgCl_2$ 10% sludge (kg/MJ heated)	2.97×10^{-3}	2.80×10^{-3}	2.97×10^{-3}	2.80×10^{-3}
Carbon dioxide (kg/MJ heated)	1.03×10^{-1}	9.52×10^{-2}	8.0×10^{-2}	7.0×10^{-2}
Sulfur oxides (kg/MJ heated)	1.99×10^{-4}	1.84×10^{-4}	0	0
Nitrogen oxides (kg/MJ heated)	8.99×10^{-5}	8.33×10^{-5}	3.70×10^{-4}	3.43×10^{-4}

TABLE 18.9 Selected Cradle-to-Gate Life Cycle Inventory Parameters for Steam Production, Assuming 50% Natural Gas and 50% Fuel Oil Consumption

Parameter	Units	Per 1 MJ of Steam Produced
Energy resources		
Coal	MJ	9.12×10^{-3}
Natural gas	MJ	0.750
Oil	MJ	0.696
Hydropower	MJ	9.50×10^{-4}
Nuclear power	MJ	4.75×10^{-3}
Air emissions		
Carbon dioxide	G	79.1
Carbon monoxide	G	0.0553
Methane	G	0.145
NMVOC	G	0.402
NO_x	G	0.260
SO_x	G	0.346
Water emissions		
COD	G	0.0193
BOD	G	6.22×10^{-3}
TDS	G	0.360
Solid waste	G	0.170

the case of natural gas and 22% in the case of oil.[20] Blowdown was estimated as 5% of boiler capacity and evaporative losses as 8% of boiler capacity. The electricity mix used was representative of the average U.S. emission profile.

When it is required to heat processing streams to very high temperatures (e.g., higher than 400°C, as seen in Table 18.2), furnaces are used. Furnaces are generally used for heating in cracking reactors, as part of the boilers/reboilers, coking, reformers and polymerization reactors, and cracking processes (e.g., to produce ethylene). Depending on the application, the heat fluxes would vary between 10 and 60 kW/m^2. In furnaces the heat is transfer by two mechanisms: radiative heating (through the flames) and convection (through the hot gases). The convection section of the furnace can be used to preheat the feed, superheat the output stream, or heat another process stream to increase the thermal efficiency, which can vary from 70%[21] to 80%.[22] In a furnace, the main environmental impacts would be the air emissions (including heat) and the resulting ash, which will depend on the type of fuel that is used. Table 18.10 shows selected cradle-to-gate life cycle inventory parameters for a furnace operated entirely with natural gas, which, incidentally, represents the optimistic scenario.

Example 18.4 Estimate the steam emissions profile to produce 1000 kg of 3-pentanone.

Solution From Example 18.2 we know that the steam required to produce 1 kg of 3-pentanone is 0.493 MJ, so 493 MJ will be required to produce 1000 kg of 3-pentanone. It is interesting to note that the energy requirement for steam production is essentially the same as the electricity requirements for the same process. Using Table 18.10, we can estimate the emission profile for this process shown in Table 18.11. As we saw before, this

TABLE 18.10 Selected Cradle-to-Gate Life Cycle Inventory Parameters for Heating Using a Furnace, Assuming that the Fuel Is 100% Natural Gas

Parameter	Units	Per 1 kg of Natural Gas Combusted
Energy Resources		
Coal	MJ	2.91×10^{-3}
Natural gas	MJ	54.3
Oil	MJ	2.42×10^{-4}
Hydro power	MJ	3.03×10^{-4}
Nuclear power	MJ	1.52×10^{-3}
Air emissions		
Carbon dioxide	g	2,765
Carbon monoxide	g	0.67
Methane	g	6.67
NMVOC	g	16.3
NO_x	g	8.97
SO_x	g	1.11
Water emissions		
COD	g	2.81×10^{-3}
BOD	g	1.71×10^{-4}
TDS	g	1.13×10^{-1}
Solid waste	g	3.66

^{a} *Source:* ref. 22.

TABLE 18.11 Emission Profile Estimate

	g/1000 kg of 3-pentanone
Air emissions	
CO_2	38,980
CO	27
Methane	71
NMVOC	198
NO_X	128
SO_X	171
Water emissions	
COD	10
BOD	3
TDS	177
Solid waste	84

emissions profile already includes the emissions associated with electricity required for pumping and the like.

Additional Points to Ponder What conclusions can you draw from the numbers above and from the preceding example? Why are the profiles different?

18.2.4 Cooling: Cooling Towers and Refrigeration

Heat removal from chemical processes is widely achieved through the use of cooling towers and refrigeration systems, depending on the temperature required for the process, as shown in Table 18.5. Cooling water or brine is used widely to remove heat at near-ambient temperature. Cooling towers are most often used to remove the heat gained by the cooling water. Cooling towers use the latent heat of vaporization of about 8 to 9% of the water in the cycle to remove the sensible heat of the "hot" water; that is, about 9% of the water evaporates.[23] The typical size of cooling towers accommodates flows ranging from 60 to about 1500 L/s. Figure 18.5 illustrates the typical mass and energy flows for a cooling water system.

Refrigeration is used for various applications, such as heat removal from reactions, keeping temperatures below the operating range of cooling water or brine, and purifying products by "freezing-out" one component of a liquid mixture. Many substances can be used as refrigerants, including ammonia, ethylene, some specialty hydrochlorofluorocarbons, and others. In the basic refrigeration cycle there is an area of high pressure and an area of low pressure. In the low-pressure side of the system, a liquid refrigerant boils as heat passes from the fluid being cooled into the refrigerant. The low-pressure vapor refrigerant is then compressed to increase the temperature and pressure to a point where the condenser can liquefy it in the high-pressure portion of the system. In the condenser, heat is transferred from the refrigerant to cooling water or brine. The refrigerant is pumped through the expansion valve where the temperature and pressure of the refrigerant are again reduced on the low-pressure side of the system. Figure 18.6 illustrates the typical mass and energy flows for a refrigeration cycle.

As shown in Figures 18.5 and 18.6, a majority of the impacts from the cooling tower and refrigeration cycle come from the electricity used to operate the cooling tower and the compressor in the refrigeration system. There are additional impacts as well, related to water

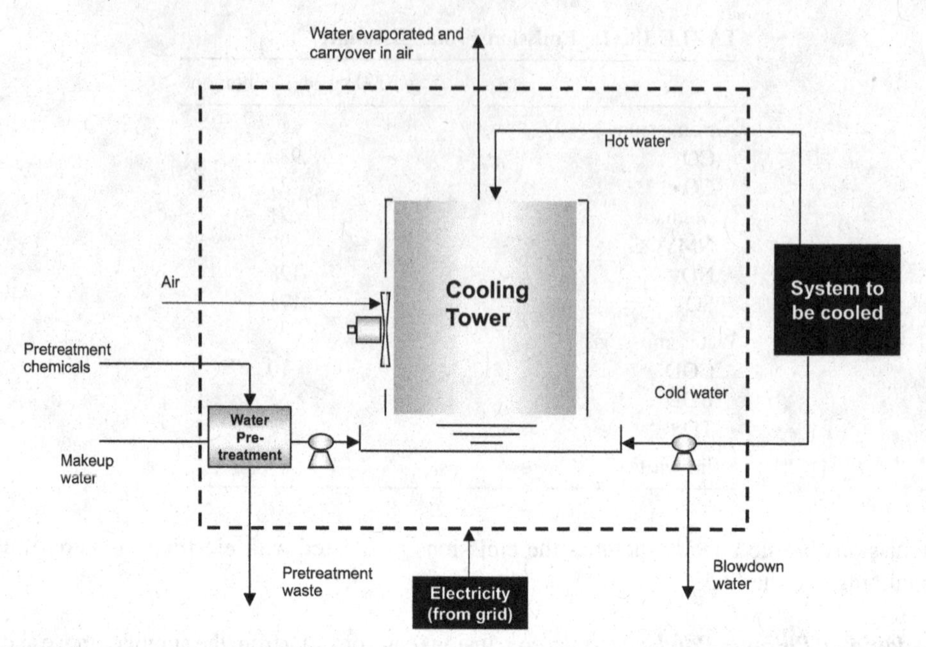

FIGURE 18.5 Mass and energy flows for a typical cooling water cycle. (Source: ref 8, with kind permission from Springer Science and Business Media. Copyright © 2000, Springer Science and Business Media.)

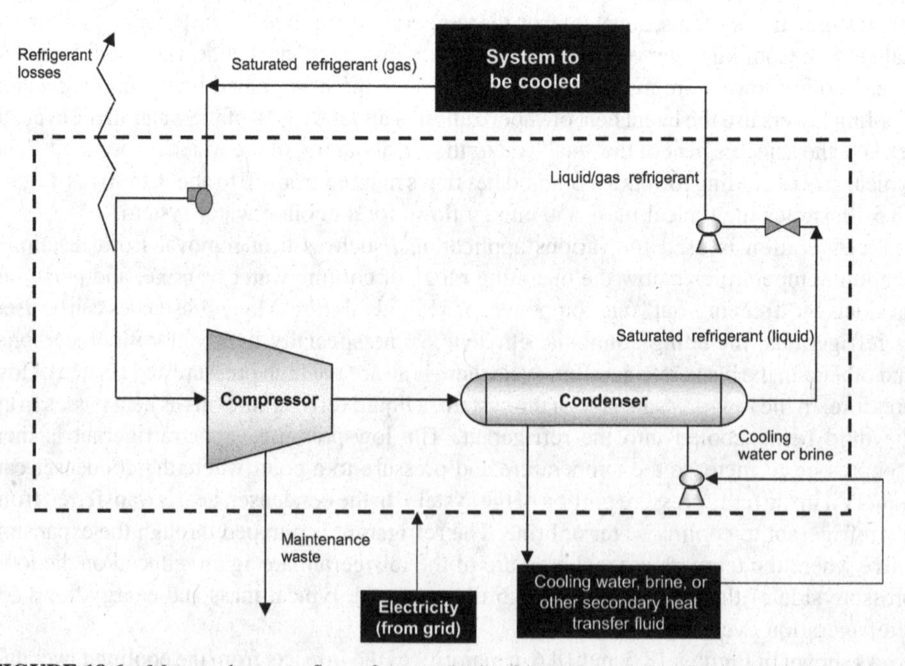

FIGURE 18.6 Mass and energy flows for a typical refrigeration cycle. (Source: ref 8, with kind permission from Springer Science and Business Media. Copyright © 2000, Springer Science and Business Media.)

pretreatment, maintenance waste, and refrigerant losses. In the case of the cooling tower, these impacts are generally minimal. However, in the case of the refrigeration cycle, the impacts can be very significant. Electricity usage for refrigeration cycles is usually rather intensive, as the electricity requirement is governed by the limits of thermodynamics, such as the work (electricity) required for a Carnot cycle:[24]

$$W = \left(Q\frac{T_1 - T_2}{T_2}\right)\frac{1}{\eta}$$

where W is the work, Q the heat removed and η the efficiency. Heat is removed from the refrigerant (absorbed by the cooling water) at the hot refrigerant temperature, T_1. Heat is transferred from the process stream to the refrigerant at the cold refrigerant temperature, T_2 (temperatures must be in terms of an absolute system, either K or °R). Therefore, the electricity requirements and the associated environmental impacts will increase as the temperature to which it is necessary to cool the system (T_2) decreases. You can compare the various life cycle environmental impact profiles for cooling and refrigeration by looking at Table 18.12. In addition to these impacts, fugitive losses of refrigerants have their own environmental and safety impacts. For example, some refrigeration systems, particularly in the developing world where much chemical manufacturing has shifted, still use Freon or other ozone-depleting substances, and some refrigerants, such as ammonia, pose significant safety hazards.

18.2.5 Heat Transfer Fluids

Steam or cooling water are not the only means by which vessels or unit operations are heated or cooled in chemical plants. Heat transfer fluids such as DowTherm, DowFrost, Syltherm,

TABLE 18.12 Selected Cradle-to-Gate Life Cycle Inventory Parameters for Cooling Through Cooling Tower and Refrigeration

Parameter	Units	Cooling Water (1 MJ of cooling potential)	Refrigeration (1 MJ of cooling potential)
Energy resources			
Coal	MJ	4.65×10^{-3}	0.471
Natural gas	MJ	1.75×10^{-3}	0.177
Oil	MJ	3.88×10^{-4}	0.0393
Hydro power	MJ	4.85×10^{-4}	0.0491
Nuclear power	MJ	2.42×10^{-3}	0.245
Air emissions			
Carbon dioxide	g	0.528	53.4
Carbon monoxide	g	$1.44E \times 10^{-4}$	0.0146
Methane	g	1.86×10^{-3}	0.188
NMVOC	g	1.22×10^{-4}	0.0123
NO_x	g	1.12×10^{-3}	0.113
SO_x	g	1.57×10^{-3}	0.159
Water emissions			
COD	g	1.60×10^{-3}	0.162
BOD	g	6.40×10^{-5}	6.48×10^{-3}
TDS	g	0.0493	0.194
Solid waste	g	0.0240	2.43

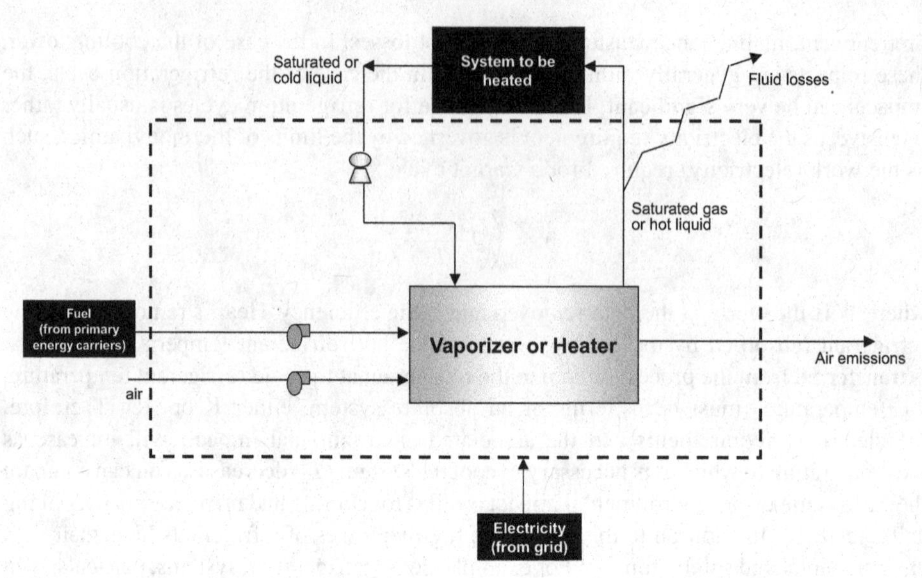

FIGURE 18.7 Mass and energy flows for a typical heat transfer fluid that is heating a system. (Source: ref 8, with kind permission from Springer Science and Business Media. Copyright © 2000, Springer Science and Business Media.)

DowCal, Norkool, Ambitrol, and others are widely used for heating and cooling. They are also used as secondary heat transfer fluids in refrigeration cycles. These fluids are used as either sensible heat carriers (i.e., cooling or heating without changing their phase) or through use of their latent heat of vaporization (i.e., using their phase change heat). Figure 18.7 illustrates this concept.

In addition to the general temperature guidelines of Table 18.5, Woods has provided a very useful monograph on selecting different heating and cooling systems, including the temperature and pressure of application for the different heat transfer systems.[25] In addition, manufacturers and distributors normally provide information about the temperature ranges over which the heat transfer fluids would be most applicable. For instance, Table 18.13 summarizes some of this type of information. As would be expected, most of

TABLE 18.13 Typical Temperature Ranges of Some Heat Transfer Fluids

Heat Transfer Fluid	Typical Composition	Temperature Use Range, Liquid (°F)
DowTherm A	Diphenyl oxide/diphenyl blend	60 to 750
DowTherm G	Mixture of di- and triaryl ethers	20 to 680
DowTherm J	Alkylated aromatic compounds	−110 to 600
DowTherm HT	Partially hydrogenated terphenyl	25 to 650
DowTherm Q	Mixture of diphenylethane and alkylated aromatics	−340 to 625
DowTherm RP	Diaryl alkyl	30 to 660
DowTherm MX	Mixture of alkylated aromatics	−10 to 625
DowTherm T	C14 to C30 alkyl benzene derivatives	14 to 550
SylTherm 800	Poly(dimethylsiloxane)	−40 to 750
SylTherm XLT	Poly(dimethylsiloxane)	−150 to 500
SylTherm HF	Poly(dimethylsiloxane)	−100 to 500

Source: ref. 26.

TABLE 18.14 Selected Cradle-to-Gate Life Cycle Inventory Parameters Using a Heat Transfer Fluid for Heating

Parameter	Units	Heating Fluid (1 MJ of heating potential)
Energy resources		
Coal	MJ	9.45×10^{-4}
Natural gas	MJ	1.25
Oil	MJ	7.88×10^{-5}
Hydropower	MJ	9.85×10^{-5}
Nuclear power	MJ	4.92×10^{-4}
Air emissions		
Carbon dioxide	g	59.2
Carbon monoxide	g	0.0144
Methane	g	0.154
NMVOC	g	0.374
NO_x	g	0.192
SO_x	g	0.0259
Water emissions		
COD	g	3.67×10^{-4}
BOD	g	1.60×10^{-5}
TDS	g	2.86×10^{-3}
Solid waste	g	0.0886

the energy-related impacts from the use of heat transfer fluids will be associated with the electricity required to operate the system, and in the case of heat transfer fluids that are used for heating, the additional fuel that might be used in the heater/vaporizer (Table 18.14).

18.3 FROM EMISSIONS TO IMPACTS

In the last few paragraphs we have discussed several of the commonly used direct sources of energy required to operate a chemical plant and have provided a very high level description of some of the major life cycle emissions. But how does that translate into impacts? Figure 18.8 shows a comparison of impacts for several direct sources of energy, including global warming potential (GWP), acidification potential, eutrophication potential, and smog formation potential.[8]

It can be seen from these data that the generation of steam by oil combustion has the largest GWP, followed by refrigeration using SUVA 123 refrigerant (2,2-dichloro-1,1,1-trifluoroethane), and finally, steam generation using natural gas. For heating, the major contribution to the GWP is associated with fuel combustion, while for cooling it is mostly from electricity. Lower environmental impacts are associated with refrigeration using ammonia, in comparison to refrigeration using SUVA 123. The potential environmental impacts from cooling with a cooling tower are negligible compared with the impacts of the other direct sources of energy.

The next question to be asked is how significant those emissions and impacts are when set in the context of the overall chemical plant. They can be rather significant on both counts. For

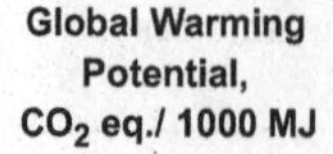

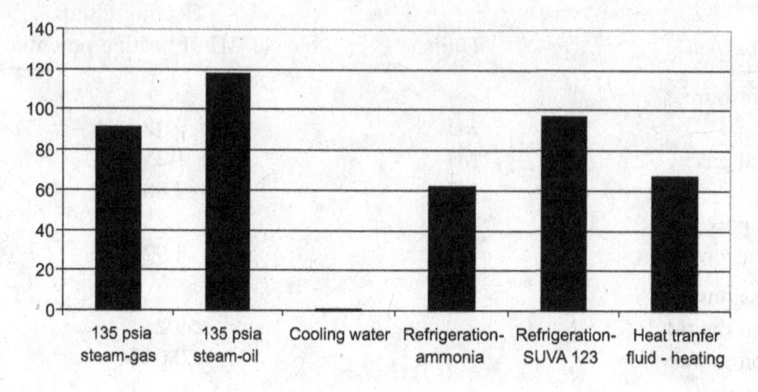

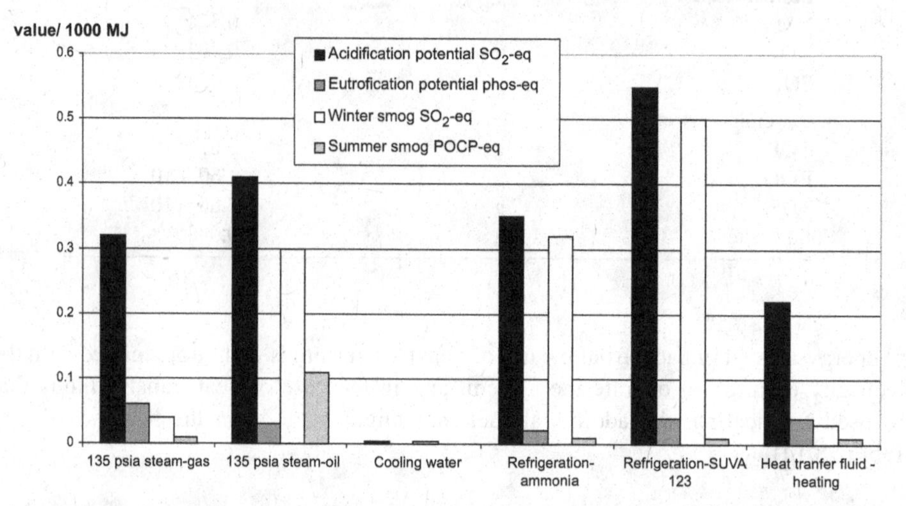

FIGURE 18.8 Life cycle assessment impacts for energy requirements for each 1000 MJ of heat exchanged. (Source: ref. 8.)

example, a review of several life cycle inventory databases for refinery products was undertaken to assess the contributions of heating, electricity, and chemical processing to the emissions from the refinery.[27] As shown in Figure 18.9, the major energy-related air emissions include CO_2, CO, SO_x, and NO_x. In addition, there are some very significant water emissions and a few other important air emissions.

How does this translate into impacts? As we saw in Figure 17.9, energy-related impacts are dominated by GWP and acidification potentials, and can contribute significantly to resource consumption and eutrophication. In summary, life cycle emissions and impacts from energy production and use can be very significant in the chemical processing industries. The techniques we introduced in Chapter 13 for energy integration, energy-efficient technologies and processes, and the use of alternative energy sources (e.g., solar, geothermal) are extremely important elements in our overall strategy to minimize the impacts resulting from energy consumption.

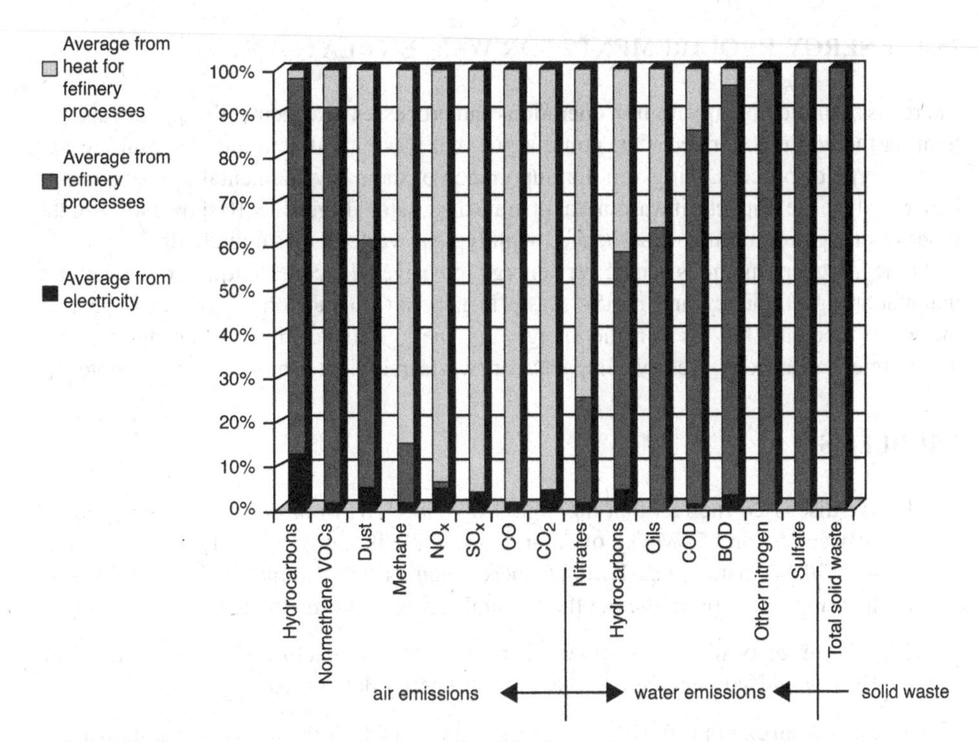

FIGURE 18.9 Contributions to air, water, and solid life cycle inventory emissions in the average refinery process, obtained from several life cycle inventory databases. (Source: ref. 27. Reproduced with permission. Copyright © 2000, American Chemical Society.)

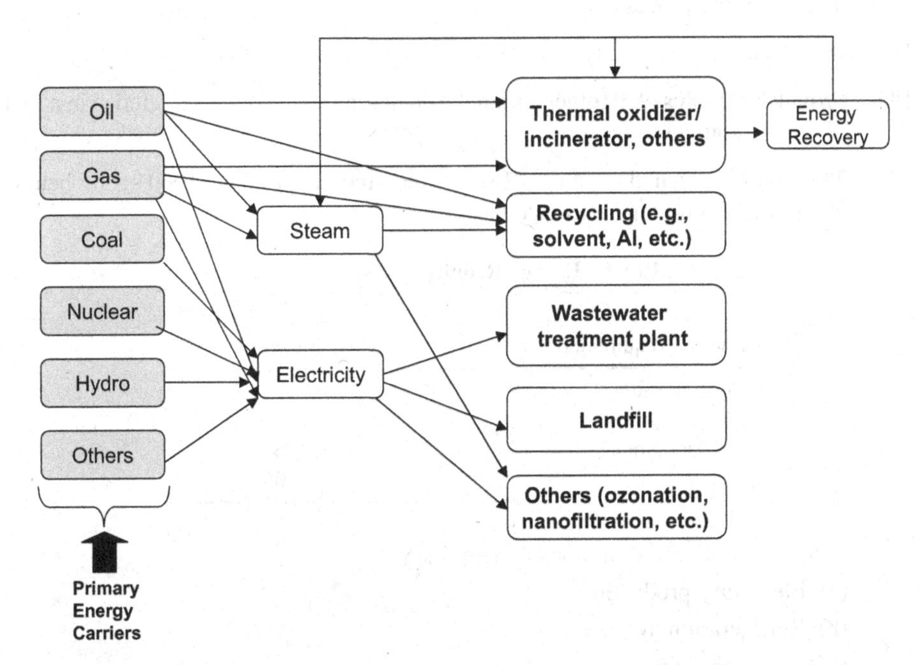

FIGURE 18.10 Graphic representation of some potential energy paths in a chemical process. Primary energy carriers are shaded.

18.4 ENERGY REQUIREMENTS FOR WASTE TREATMENT

Energy is required not only for unit operations and processes in a chemical plant, but also for treating the waste generated. One could argue that waste treatment and disposal are just another type of process, but given the importance of their environmental impacts, and to better analyze the impacts, it warrants a separate discussion. Figure 18.10 shows some of the types of energy requirements associated with typical waste disposal methods.

Some treatment methods can be very energy intensive. Hence techniques to increase the manufacturing efficiency and reduce waste in general will not only have an impact on the waste amounts, but also on the energy and energy-related impacts generated during the treatment operations. We cover impacts from waste treatment in more detail in Chapter 19.

PROBLEMS

18.1 We discussed the fact that the high heat value (HHV) of natural gas is in general between 53 and 54 MJ/kg of natural gas, while its low heat value (LHV) is about 48 MJ/kg of natural gas. Estimate these amounts using combustion enthalphy. As a first approximation, assume that natural gas is 100% methane.

18.2 You are performing a life cycle inventory for the manufacture of a specific chemical. How would you decide on what type of energy data to use?

18.3 Provide an example of when the energy required for purification or separation might be the dominant contributor to the total energy requirements.

18.4 Provide an example of a chemical process that utilizes fuels (oil, natural gas, coal) directly in the process.

18.5 Investigate the electricity mix for the area in which you live.

18.6 Provide examples of 10 other parameters that typically would be included in a LCI of energy sources.

18.7 The energy given the Table P18.7 is required to produce 1000 kg of benzyl chloroformate on a gate-to-gate basis.

TABLE P18.7 Energy Requirements

Energy Requirement	MJ/1000 kg Benzyl Chloroformate
Cooling water	192
Electricity	127
Refrigeration	564
Steam	106

Estimate the major emissions related to:

(a) Electricity production

(b) Refrigeration systems

(c) Steam production

(d) Cooling water

(e) Energy requirements

18.8 Estimate the following environmental impacts for Problem 18.7:
(a) Global warming potential
(b) Acidification potential
(c) Eutrofication potential
(d) Photochemical ozone creation potential
(e) Energy-related resource consumption

18.9 It has been estimated that there is a potential to recover a total amount of energy equivalent to 40.8 MJ/1000 kg of benzyl chloroformate. Repeat Problem 18.7 accounting for this energy recovery.

18.10 Repeat Problem 18.8 accounting for the 40.8 MJ/1000 kg of benzyl chloroformate of potential energy recovery.

18.11 The energy given in Table P18.11 is required to produce 1000 kg of benzyl chloroformate on a cradle-to-gate basis.

TABLE P18.11 Energy Requirements

Energy Requirement	MJ/1000 kg Benzyl Chloroformate
Cooling water	1.01×10^4
Electricity	3.63×10^3
Refrigeration	5.78×10^2
Steam	8.82×10^3

How much energy is required for the benzyl chloroformate on a gate-to-gate basis?

18.12 Repeat Problem 18.11 but account for the 40.8 MJ/1000 kg potential energy recovery from benzyl chloroformate on a gate-to-gate basis and a potential energy recovery of 7690 MJ/1000 kg of benzyl chloroformate on a cradle-to-gate basis.

18.13 Estimate the following environmental impacts of a typical organic chemical:
(a) Global warming potential
(b) Acidification potential
(c) Eutrofication potential
(d) Photochemical ozone creation potential
(e) Energy-related resource consumption

18.14 Estimate the following environmental impacts of a typical inorganic chemical
(a) Global warming potential
(b) Acidification potential
(c) Eutrofication potential
(d) Photochemical ozone creation potential
(e) Energy-related resource consumption

REFERENCES

1. U.S. Environmental Protection Agency. *eGRID2007 Version 1.0 Year 2005 Summary Tables.* U.S. EPA, Washington, DC, 2008.
2. U.S. Environmental Protection Agency. *Emissions and Generation Resource Integrated Database (eGRID) for 2007.* Prepared by E.H. Pechan & Associates, Inc. Contract EP-D-06-001. U.S. EPA, Washington, DC, Sept. 2008.
3. North American Electric Reliability Corporation. http://www.nerc.com/, accessed Dec. 30, 2008.
4. Baumann, H., Tillman, A.-M. *The Hitch Hiker's Guide to LCA.* Studentlitteratur, Lund, Sweden, 2004.
5. Kim, S., Overcash, M. Energy in chemical manufacturing processes: gate-to-gate information for life cycle assessment. *J. Chem. Technol. Biotechnol.*, 2003, 78, 995–1005.
6. Jiménez-González, C., Kim, S., Overcash, M. Methodology for developing gate-to-gate life cycle inventory information. *Int. J. Life Cycle Assess.*, 2000, 5, 153–159.
7. Perry, R. H., Green, D. W., Maloney, J. O. *Perry's Chemical Engineers' Handbook*, 7th ed. McGraw-Hill, New York, 1997.
8. Jiménez-González, C., Overcash, M. Energy sub-modules applied in life cycle inventory of processes. *J. Clean Products Process.*, 2000, 2, 57–66.
9. Griffin, E., Overcash, M. *Unit Process Life Cycle Inventory (LCI) Heuristic 201: Heating and Cooling.* Chemical and Biomolecular Engineering Department, North Carolina State University, Raleigh, NC, 2005.
10. Griffin, E., Overcash, M. *Unit Process Life Cycle Inventory (LCI) Heuristic 117: Refrigeration Systems.* Chemical and Biomolecular Engineering Department, North Carolina State University, Raleigh, NC, 2004.
11. Ecoinvent database, http://www.ecoinvent.com/, accessed Dec. 30, 2008.
12. Kim, S., Overcash, M. *Unit Process Life Cycle Inventory (LCI) Heuristic 01: Electricity Emissions.* Chemical and Biomolecular Engineering Department, North Carolina State University, Raleigh, NC, 2001.
13. *SIMAPRO, Life Cycle Analysis Software.* PRé Consultants, Amersfoort, The Netherlands, 1998.
14. Bundesamt für Umwelt, Wald und Landschaft. *Ökobilanzen von Packstoffen.* Schriftenreihe Umwelt 132. BUWAL, Bern, Switzerland, 1991
15. PEMS, *Life Cycle Analysis Software.* Pira International, Surrey, UK, 1998.
16. ECOPRO, Life Cycle Analysis Software. EMPA (Swiss Federal Laboratories for Material Testing and Research), St. Gallen, Switzerland, 1996.
17. Dumas, R. D. *Energy Usage and Emissions Associated with Electric Energy Consumption as Part of a Solid Waste Management Life Cycle Inventory Model.* Department of Civil Engineering, North Carolina State University, Raleigh, NC, 1997.
18. Griffin, E., Overcash, M. *Unit Process Life Cycle Inventory (LCI) Heuristic 08: Steam for Industrial Sources from Fuel* (50% using natural gas and 50% using fuel oil). Chemical and Biomolecular Engineering Department, North Carolina State University, Raleigh, NC, 2002.
19. Goodwin, K. Griffin, E. *Unit Process Life Cycle Inventory (LCI) Heuristic 100: Boiler Efficiency.* Chemical and Biomolecular Engineering Department, North Carolina State University, Raleigh, NC, 2002.
20. Cohen-Hubal, E. A. Net Waste Reduction Analysis Applied to Air Pollution Control Technologies and Zero Water Discharge Systems. M.S. thesis, North Carolina State University, Raleigh, NC, 1992.

21. Woods, D. R. *Rules of Thumb in Engineering Practice*. Wiley-VCH, Weinhem, Germany, 2007.
22. Overcash, M. *Unit Process Life Cycle Inventory (LCI) Heuristic 05: Natural Gas Combustion*. Chemical and Biomolecular Engineering Department, North Carolina State University, Raleigh, NC, 2002.
23. Walas, S. M. Rules of thumb selecting and designing equipment. *Chem. Eng.*, 1987, 75–81.
24. Smith, J. M., Van Ness, H. C. *Introduction to Chemical Engineering Thermodynamics*. McGraw-Hill, New York, 1975, p. 515.
25. Woods, D. R. *Process Design and Engineering Practice*. Prentice Hall, Upper Saddle River, NJ, 1995.
26. DowTherm and SylTherm Heat Transfer Fluids. Form No. 176-01-545. http://www.dow.com/heattrans/app/chem.htm, accessed Dec. 31, 2008.
27. Jiménez-González, C., Overcash, M. LCI of refinery products: review and comparison of commercially available databases. *Environ. Sci. Technol.*, 2000, 34(22), 4789–4796.

19

IMPACTS OF WASTE AND WASTE TREATMENT

What This Chapter Is About In this chapter we cover the environmental fate and effects data needed to assess environmental risks appropriately. We describe the different test for obtaining environmental information and what can be inferred from the results regarding ecotoxicity, persistence, and mobility. In this chapter we also review some common treatment and recovery technologies and their life cycle environmental impacts. The technologies covered here include wastewater treatment plants and incinerators, and given the prevalence of solvents in the chemical processing industries, we include solvent recovery.

Learning Objectives At the end of this chapter, the student will be able to:

- Understand the meaning of environmental fate and effects.
- Understand how environmental fate and effects data are used to perform an environmental risk assessment.
- Apply streamlined techniques to evaluate the life cycle impacts of waste treatment.
- Apply the concepts of life cycle assessment, energy submodules, and treatment submodules to estimate and understand the environmental impacts for recovering vs. treating waste.

Green Chemistry and Engineering: A Practical Design Approach, By Concepción Jiménez-González and David J. C. Constable

TABLE 19.1 Fate and Effects Parameters and Typical Expected Ranges

Parameter	High	Low
S = water solubility	> 1000 mg/L	<10 mg/L
H = Henry's constant	> 10^{-7} atm · m^3/mol	< 10^{-7} atm · m^3/mol
Log K_{ow}, log D_{ow}, log K_d, log K_{oc}	> 3	< 3
UV/visible = absorption	> 290 nm	< 290 nm
Hydrolysis half-life, $t_{1/2}$	Short: minutes	Long: weeks
Photolysis half-life, $t_{1/2}$	Short: minutes	Long: weeks
Biodegradation half-life, $t_{1/2}$	Short: minutes	Long: weeks
Toxicity	LC_{50} < 1 mg/L	LC_{50} > 100 mg/L

19.1 ENVIRONMENTAL FATE AND EFFECTS DATA

Environmental risk assessment is a structured quantitative and qualitative approach to the estimation of environmental hazard potential, environmental exposure potential, adverse environmental impact potential, and overall environmental risk. The cornerstone for performing an environmental risk assessment is to obtain environmental fate and effects data.[1–5] Fate and effects parameters are presented in Table 19.1 with typical expected ranges for these compounds.[6] These data can be broadly divided into three groups:

1. *Mobility* describes the probable movement of a chemical among and between the various environmental compartments, such as water, air, soil, sediment, biomass, organisms.
2. *Persistence* describes various environmental transformation processes that chemicals may undergo and which may result in the removal of those chemicals from the environment, including such processes as hydrolysis, photolysis, and biodegradation.
3. *Ecotoxicity* describes various effects on environmental organisms that chemicals may have. Organisms used typically for ecotoxicity testing include algae, daphnids, other microorganisms, fish, and other invertebrates, such as earthworms.

19.1.1 Assessing Environmental Mobility and Persistence

If a chemical is introduced into the environment, there is a certain probability that it may move from the point where it is released. Eventually, the compound may be distributed over a broad geographical area as the original parent molecule or as different degradants and/or metabolites. The different environmental compartments and their relationships are illustrated in Figure 19.1, where:

- K_{oc} is the organic carbon/water distribution coefficient, which can be measured directly or estimated from the octanol/water distribution coefficient.
- K_w is the water/air distribution coefficient, the reciprocal of Henry's constant (H), which is the air/water distribution coefficient. H can be measured directly or estimated from the water solubility and the vapor pressure.
- BCF is the bioconcentration factor, which is the organism/water distribution coefficient and can be measured directly or estimated from the octanol/water partition coefficient, log K_{ow} (log P).

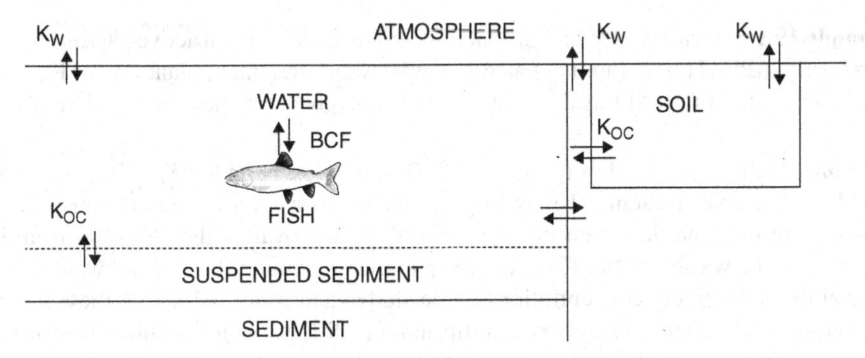

FIGURE 19.1 Model ecosystem.

Assigning volumes or sizes to the various compartments allows calculation of the predicted equilibrium concentrations in each compartment as a function of the input load. Thus, the primary compartments that chemicals will tend to migrate toward, or accumulate in, can be identified. The next step in assessing the environmental fate of a chemical is to consider the transformation or degradation processes that it may undergo. Such processes include hydrolysis, photolysis, and biodegradation. All of these will serve to decrease the concentration of a chemical in a particular compartment and allow estimation of the persistence of the chemical in the environment. Examples of inferences that may be made based on fate data are shown in Table 19.2.

TABLE 19.2 Examples of Inferences Based on Fate Data[a]

Log K_{ow}	H	Depletion Mechanism(s) Exist	Inference
< 3	$< 10^{-7}$	Yes	Chemical will distribute primarily to the water compartment and will degrade over time.
> 3	$< 10^{-7}$	Yes	Chemical will distribute to biomass, soils, and sediments and will degrade over time.
> 3	$> 10^{-7}$	Yes	Chemical will distribute to air, biomass, soils, and sediments and will degrade over time.
< 3	$> 10^{-7}$	No	Chemical will distribute to the air and water compartments and may be persistent.
< 3	$< 10^{-7}$	No	Chemical will distribute to the water compartment and may be persistent.
> 3	$< 10^{-7}$	No	Chemical will distribute to biomass, soils, and sediments, may bioconcentrate in organisms, and may be persistent.

[a] In all cases, acute and subchronic toxicity may be an issue.

Example 19.1 A wastewater stream contains certain amounts of an active pharmaceutical ingredient[7] (API). This stream is treated in a wastewater treatment plant (WWTP). What is the likely fate of this API as the waste stream undergoes treatment in this WWTP?

Solution We may use models to assess WWTP scenarios. High log K_{ow}, D_{ow}, K_d, or K_{oc} would suggest that a chemical may tend to undergo sorption to the activated sludge biomass and that one depletion mechanism may be removal of the chemical from the system with the waste sludge. Knowing the parameters around the WWTP would allow calculation of the likely concentrations in the sludge and water. Although these models are equilibrium models and assume equilibrium conditions, which would generally not be the case in practice, they are useful first approximations to the fate of a chemical in the environment or in a WWTP. For this example, we used the activated sludge model WW-Treat, developed by Cowan et al.,[8] with the results shown in Table 19.3.

TABLE 19.3 Wastewater Treatment Model for an API

	Code	Units	Value
Parameter			
Influent concentration	Ci	g/m^3	1.7
Sludge/water partition coefficient	Kp	—	871
Fraction sludge removed in primary clarifier	Rp	—	0.6
Total suspended solids in influent	S	g/m^3	220
Gas flow rate	G	m^3/h	0.45
Henry's low constant	H	$m^3 \cdot atm/mol$	7.00×10^{-9}
Hydraulic retention time	HRT	h	8
Biodegradation rate of dissolved substance	K1	h^{-1}	0
Biodegradation rate of adsorbed substance	K2	h^{-1}	0
Mixed liquor suspended solids	MLSS	g/m^3	2500
Water flow rate	Q	m^3/h	250
Gas constant	R	$m^3 \cdot atm/mol \cdot K$	0.0821
Fraction of solids removed in reactor	Ra	—	0.95
Sludge retention time	SRT	h	216
Temperature	T	K	293
Primary treatment module			
Concentration on primary sludge	Cps	g/m^3	2.73×10^{-1}
Concentration in effluent from primary settling	Ce1	g/m^3	1.54
Activated sludge module			
Volatilization loss term	Kv	—	3.03×10^{-10}
Biodegradation loss term	Kb	—	0.00
Dissolved + absorbed compound concentration	Cr	g/m^3	1.54
Compound concentration absorbed to sludge	Cs	g/m^3	1.05
Final effluent concentration	Ce2	g/m^3	5.36×10^{-1}
Final distribution			
Total removed			68%
Volatilization			0%
Biodegradation			0%
Adsorption			62%
Effluent			32%

Additional Points to Ponder What do the numbers mean in a broad sense? What does it mean that the biodegradation percent is zero?

19.1.2 Assessing Ecotoxicity

Aquatic toxicity tests are used to detect and evaluate the potential toxicological effects of chemicals on aquatic organisms. Since these effects are not necessarily harmful, a principal function of the tests is to identify chemicals that can have adverse effects on aquatic organisms at relatively low exposure concentrations or body residues. These tests provide a database which can then be used to assess the risk associated with a situation in which the chemical agent, the organisms, and the exposure conditions are defined. Organisms commonly used in aquatic toxicity tests include the following:

- *Microorganisms*, used in tests such as microbial respiration inhibition and Microtox, are useful as toxicity screening tools and are particularly applicable to chemicals in waste streams intended for treatment by activated sludge or other biological treatment systems.
- *Algae*, such as green algae and blue-green algae, are simple photosynthetic organisms found in many terrestrial and aquatic habitats. Algae are extremely important in the functioning of aquatic ecosystems because they serve as a foundation of most aquatic food chains.
- *Daphnids*, such as *Daphnia magna*, are freshwater microcrustaceans commonly referred to as water fleas. They are ubiquitous in temperate fresh waters and are an ecologically important species because they convert phytoplankton and bacteria into animal protein and form a significant portion of the diet of numerous fish species.
- *Fish*, such as warm-water fathead minnow and bluegill and cold-water rainbow trout and brook trout, are ecologically and economically important organisms that are widely distributed throughout most aquatic environments. Fish fill diverse ecological roles and represent an essential link in the food chain by converting aquatic matter into protein that is able to be harvested as human or animal food.
- *Other organisms*, such as *Hyallela azteca*, a benthic amphipod, and earthworms, are used in tests for chemicals likely to concentrate in sediments and soils.

The most common aquatic toxicity tests are short-term (acute) tests, using lethality or immobility (in the case of daphnids) as the endpoint. These tests provide a means of comparing substances whose mechanisms of action may be quite different and indicate whether further toxicity studies should be conducted. An acute toxicity test is conducted to estimate the median lethal concentration (LC_{50}) of the chemical in water to which the test organisms are exposed. The LC_{50} is the concentration estimated to produce 50% mortality of a test population over a specific time period. The length of exposure is usually 24 to 96 h, depending on the species. When effects other than mortality are measured, the general expression EC_{50} is used. The EC_{50} (median effective concentration) is the concentration of a chemical estimated to produce a specific effect (e.g., behavioral or physiological) in 50% of a population of test species after a specified length of exposure (e.g., 24 or 48 h). Typical effect criteria include immobility, a developmental abnormality or deformity, loss of equilibrium, failure to respond to an external stimulus, and abnormal behavior. The general guidelines shown in Table 19.4 may be useful in evaluating acute toxicity data.

TABLE 19.4 Acute Aquatic Toxicity Data and Their Interpretation

LC_{50}	Inference
< 1 mg/L (ppm)	Very toxic
1–10 mg/L	Toxic
10–100 mg/L	Harmful
>100 mg/L	Not toxic

19.2 ENVIRONMENTAL FATE INFORMATION: PHYSICAL PROPERTIES

19.2.1 Water Solubility

Water solubility, an important parameter in determining the fate of a compound in the environment, is defined as the maximum amount of a chemical in solution and at equilibrium with excess chemical in pure water at specified ambient conditions (temperature, atmospheric pressure, and pH). Water solubility is usually expressed as weight/weight (ppm, ppb) or weight/volume (mg/L). Water solubility is generally not useful for gases, because their solubility in water is measured when the gas above the water is at a partial pressure of 1 atm. Thus, the solubility of gases does not usually apply to environmental assessment, as their partial pressure of a gas in the environment is extremely low. Temperature and the solution pH may have a significant effect on water solubility, so these variables should be recorded with the water solubility value. If not specified, a pH of 7 and a temperature of 25°C are assumed.

In general, highly soluble chemicals are more likely than poorly soluble chemicals to be distributed by the hydrological cycle; desorb from biomass, soils, and sediments; have relatively low bioconcentration potential; be more readily biodegradable by microorganisms; be less toxic to aquatic organisms; be less persistent; and be less likely to volatilize from water. Highly water-soluble chemicals are more likely to be transported and distributed by the hydrologic cycle than are relatively water-insoluble chemicals. Water solubility, along with vapor pressure, can be used to predict the extent to which a chemical will volatilize from water into air. Water solubility can also affect possible transformation by hydrolysis, photolysis, oxidation, reduction, and biodegradation in water. Finally, the design of most chemical tests and of many ecological and health tests requires precise knowledge of the water solubility of chemicals.

19.2.2 Dissociation Constant

The significance of the dissociation constant is the relationship between pK_a and pH, and the resulting distribution of a substance in the environment. The degree of ionization of a substance at a particular pH will affect its availability to biological organisms; chemical and physical reactivity, and ultimate environmental fate. For example, an ionized molecule will generally have greater water solubility and will be less likely to partition to lipophilic substances than to its nonionized form. Ionic charge will also affect the potential of a molecule to participate in environmental ion-exchange processes such as those found in soil and sludge systems. Knowledge of the pK_a will assist in designing appropriate adsorption and ecotoxicity studies, and in accurately interpreting the results from these studies.

19.2.3 Vapor Pressure and Volatility

Water solubility, along with vapor pressure, can be used to predict the extent to which a chemical will volatilize from water into air. Vapor pressure is the force per unit area exerted by a gas that is in equilibrium with its liquid or solid phase at a specific temperature. Equilibrium vapor pressure can be thought of as the solubility of a chemical in air. The vapor pressure increases with an increase in temperature; thus, values are meaningful only if accompanied by the temperature at which they were measured.

Although the potential volatility of a chemical is related to its inherent vapor pressure, actual volatilization (or vaporization) rates will depend on environmental factors. *Volatility* is the evaporative loss of a substance to the air from the surface of a liquid or solid. Volatilization is an important source of material for airborne transport and may lead to the distribution of a chemical over wide areas and into bodies of water (e.g., in rainfall) far from the site of release. Chemicals with relatively low vapor pressures, high adsorptivity onto solids, or high solubility in water are less likely to vaporize and become airborne than chemicals with high vapor pressures or less affinity for solution in water or adsorption to solids and sediments.

In addition, chemicals that are likely to be gases at ambient temperatures and that have low water solubility and low adsorptive tendencies are less likely to transport and persist in soils and water. Such chemicals are less likely to biodegrade or hydrolyze but are prime candidates for photolysis and for involvement in adverse atmospheric effects (such as smog formation and stratospheric alterations). On the other hand, nonvolatile chemicals are less frequently involved in significant atmospheric transport, so concerns regarding them should focus on soils and water.

For chemicals such as pharmaceuticals, there may be some concern about volatility from water. Although many pharmaceuticals are used in their nonvolatile salt form, in water this salt will dissociate and some of the nonionized species will be formed, depending on the compound's pK_a value. While vapor pressure is a measure of the volatility of a chemical from itself, a more useful measure for environmental fate predictions is the Henry's law constant. Volatility is thus assessed through a determination of either Henry's law constant for water (a volatility limit test) or vapor pressure (from solution or surface) for a volatile compound.

Some chemicals that have very low vapor pressure and low water solubility, such as DDT and polychlorinated biphenyls, volatilize to a significant extent. This phenomenon is considered responsible for the global spread of these materials. Vapor pressure and volatilization half-lives of selected chemicals are shown in Table 19.5.

The volatility of chemicals in aqueous or soil systems is also influenced by the chemical's rate of movement through water to the water–air interface, the chemical's water solubility,

TABLE 19.5 Vapor Pressure and Volatilization Half-Lives of Selected Chemicals

Compound	Vapor Pressure at 20°C (torr)	Volatilization Half-Life (min)
Chloromethane	3700	27
Dichloromethane	362	21
Carbon tetrachloride	90	29
1,1,2-Trichloroethane	19	21
1,1,2,2-Tetrachloroethane	5	56
Hexachloroethane	0.4	45

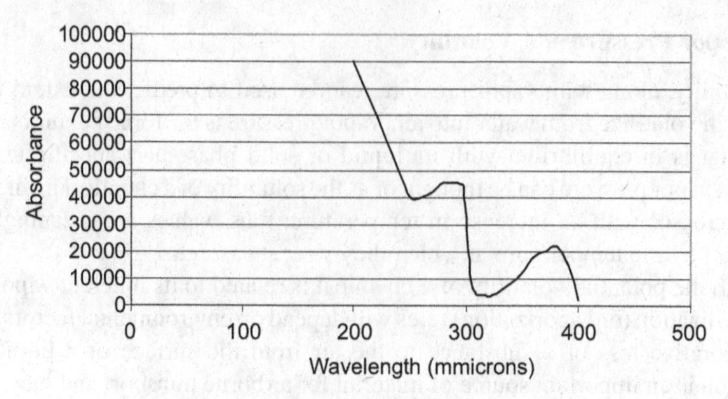

FIGURE 19.2 Sample UV/visible spectrum.

the chemical's tendency to sorb to soil and sediments, the amount of soil water; the evaporation rate of soil water; the depth to which chemical is incorporated into soil; and the wick effect that brings water and dissolved chemicals to soil surfaces.

19.2.4 UV/Vis Spectrum

The ultraviolet–visible (UV/Vis) absorption spectrum is a quantitative measure of the ability of a substance to absorb radiation in the electromagnetic spectral region between 290 and 800 nm. It is generally measured with a spectrophotometer and presented as a function of wavelength or wavenumber.

The UV/Vis spectrum of a compound may be used to evaluate a compound's susceptibility to being degraded by light. In general, a compound possessing absorbance maxima above 300 nm may be susceptible to direct photodegradation. The UV/Vis spectrum for a particular compound will not definitively determine the photodegradability potential, and additional photolysis testing would be required to determine the extent and the rate at which a compound will be degraded. Because UV/Vis absorbance maxima may shift with changes in pH for ionizable compounds, the UV/Vis spectra for these materials are determined at different pH levels. A typical UV/Vis spectrum is shown in Figure 19.2.

19.2.5 Partitioning

Knowledge of the distribution or partitioning behavior of a chemical between phases in the environment is essential to evaluating its potential environmental fate. Partition coefficients of various sorts are determined and used to estimate how a chemical may partition into lipids, fats, or organisms; sorb to particulates such as soils, sediments, biomass, and sludges; partition into air; or distribute among the various environmental compartments.

In addition, they can be used to predict the bioconcentration potential in aquatic and terrestrial organisms and to estimate the amount of sorption to soils, sediments, biomass, and sludges. These processes are major factors in determining the movement of chemicals in the biosphere. The principal distribution coefficients are:

1. The octanol/water distribution (or partition) coefficient—K_{ow} and D_{ow}
2. The organism/water distribution coefficient—the bioconcentration factor
3. The soil (sediment)/water distribution coefficient—K_{oc}

4. The biomass/water distribution coefficient—K_d
5. The air/water distribution coefficient—Henry's law constant (**H**).

Octanol/Water Distribution Coefficient, K_{ow} Although other systems have been used to measure distribution (or partition) coefficients between a chemical and an organic solvent, such as hexane/water or benzene/water, it has become customary in environmental fate assessment to use the *n*-octanol/water system. *n*-Octanol is considered to be a good medium for simulating natural fatty substances. The *n*-octanol/water system is widely used as a reference system, and many data using this system have been reported. The coefficient is designated as P, P_{ow}, K, or K_{ow}. The logarithmic values of the coefficient generally range from about 1 to 6 and are designated as log P, log P_{ow}, log K, or log K_{ow}. Log K_{ow} values > 3 may indicate the propensity of the chemical to absorb or adsorb to sediments and soil and to bioaccumulate in fatty tissue.

The octanol/water distribution coefficient, D_{ow}, is defined as the ratio of the concentration of a chemical in two phases, *n*-octanol and water, when the phases are in equilibrium with one another and the test chemical is in dilute solution in both phases. The *n*-octanol/water distribution coefficient is given by dividing the concentration of the compound in *n*-octanol by the concentration in water.

For ionizable compounds, D_{ow} is usually determined at pH values of 5, 7, and 9. For environmental risk assessments, the value at pH 7 is generally used. Log D_{ow} values less than 1 indicate that the chemical is unlikely to bioconcentrate or adsorb significantly onto organic particles. Log D_{ow} values equal to or greater than 4 indicate that the chemical may bioconcentrate or adsorb significantly. The log D_{ow} values for selected chemicals and the way these values are interpreted is shown in Table 19.6.

If the molecular structure of a chemical is known, it is often possible to estimate this parameter. This is because the partition phenomenon exhibits a reasonable additive-constitutive property. That is, the partition coefficient can be considered as an additive function of the partition coefficients of component parts of the molecule, particularly if the components are nonpolar. This has led to the development of models that can be used to estimate the log K_{ow}, which in turn can be used in environmental fate models for predicting other physical properties or distribution coefficients. The log K_{ow} can also be estimated from the water solubility of a chemical. For example, one such relationship cited in the U.S. Food and Drug Administration's *Environmental Assessment Technical Assistance Handbook*[9] is given by

$$\log K_{ow} = 5.00 - 0.67 \log S$$

TABLE 19.6 Log D_{ow} of Selected Chemicals and Fate Inferences

Compound	Log *Dow*	Inference
Amoxycillin	1.56	Will not sorb significantly
Granisetron	0.15 (pH 7)	Will not sorb significantly
Cimetidine	0.198 (pH 7)	Will not sorb significantly
Paroxetine	1.32 (pH 7)	Will not sorb significantly
Pranlukast	2.69 (pH 7)	Will sorb moderately
Nabumetone	3.13	Will sorb strongly

where S is the aqueous solubility in μmol/L. The n-octanol/water distribution coefficient is corrected for the ionization of the compound so that only the concentration of the nonionized species is considered. This corrected coefficient is designated as K_{ow}, and is given by

$$K_{ow} = D_{ow}(1 + 10\text{abs pH-p}K_a)$$

Log K_{ow} is often represented as log P. For nonionizable chemicals, log P = log D_{ow}. For large, ionizable chemicals such as pharmaceuticals, log P discounts the potentially significant solubility of the ionized species in the octanol layer. For these molecules, use of log D_{ow} is preferred:

- D_{ow}: partitioning from water into octanol, corrected for degree of ionization
- K_{ow}: partitioning from water into octanol, uncorrected for degree of ionization

Bioconcentration Factor The tendency of some chemicals to move through the food chain, resulting in higher and higher chemical concentrations in the organisms at each level of the food chain has been termed *biomagnification* or *bioconcentration*. From an environmental point of view, this phenomenon becomes important when the acute toxicity of the chemical is low and the physiological effects go unnoticed until the chronic effects become evident. For this reason, prior knowledge of the bioconcentration potential of new or existing chemicals is desirable. The bioconcentration factor (BCF) is given by

$$\text{BCF} = \frac{\text{concentration in organism}}{\text{concentration in water}}$$

$$\log \text{BCF} = \log_{10} \text{BCF}$$

BCFs can be estimated from the log K_{ow} using a number of regression equations. One such relationship cited in the *Environmental Assessment Technical Assistance Handbook*[9] is given by

$$\log \text{BCF} 0.79 \log K_{ow} - 0.40$$

where BCF is the bioconcentration factor and K_{ow} is the octanol/water partition coefficient. This relationship is most appropriate for relatively inert compounds that do not undergo rapid biotransformation in the body. Degradable chemicals usually have a lower BCF value because elimination is enhanced.

Biomass/Water Partition Coefficient, K_d Since many organic chemicals are treated in wastewater treatment plants, the tendency of the chemical to sorb to the biosolids in such plants is an important factor that needs to be evaluated. The biosolids/water distribution coefficient, K_d, is the ratio of the concentration of a chemical in two phases, biosolids and water, when the phases are in equilibrium with one another and the test chemical is in dilute solution in both phases. Biomass or sludge adsorption studies are generally run at a biomass concentration of 2500 mg/L. This approximates the biomass concentrations found in typical wastewater treatment plants. The K_d value can be used to estimate the extent of adsorption of the compound during treatment:

$$\text{Fraction adsorbed} = \frac{K_d S}{[1 + K_d S]}$$

where S is the solids/water ratio (g/mL). In some instances, a value for K_d may be estimated from a sorption/desorption isotherm study using activated sludge (sorption of a compound onto the biomass in sludge). A controlled biomass adsorption study may be conducted, monitoring depletion of parent compound as a function of initial biomass concentration [as measured by total suspended solids (TSS)] and of time. The data are then fit to a Freundlich equation:

$$\log(x/m) = \log K + (1/n)\log C_e$$

where

$\log(x/m)$ = logarithm of the amount of chemical sorbed per amount of adsorbent at equilibrium
$\log C_e$ = logarithm of the amount of chemical in solution at equilibrium
K = Freundlich adsorption coefficient
n = a constant describing the degree of nonlinearity of the isotherm (when $n \cong 1$, the F reundlich constant K_f can be used as an adsorption distribution coefficient, K_d)

If a plot of log (x/m) vs. log C_e gives a straight line, the slope of the line is the $1/n$ linearity term and the intercept is log K_d.

Soil/Water Partition Coefficient, K_{oc} Methodology similar to that described above may be used to determine the soil/water partition coefficient. A K_d is determined from isotherms generated using soils or sediments. The K_{oc} may then be calculated from

$$K_{oc} = (K_d/\%\text{ organic carbon}) \times 100$$

or

$$K_{oc} = \mu g \text{ chemical dissolved at equilibrium/g solution}$$

In general, the rate of movement of organic chemicals through soil is inversely correlated with sorption. Compounds having a K_{oc} value of around 1000 (log $K_{oc} = 3$) are quite tightly bound to organic matter in soil and are considered immobile, whereas those with a K_{oc} value below 100 (log $K_{oc} = 2$) are moderately to highly mobile.

Air/Water Distribution Coefficient, H The Henry's law constant **H** represents the ratio of the equilibrium concentration of a chemical in air to its concentration in water. **H** can be measured directly or estimated from the equilibrium vapor pressure and the water solubility:

$$\mathbf{H} = \frac{16.04 \times P \times M}{T \times S}$$

where P is the equilibrium vapor pressure of pure chemical in mmHg, M the gram molecular weight of the chemical, T the temperature in K, and S the solubility in water in mg/L.

Chemicals with values of **H** less than $\sim 10^{-7}$ are less volatile than water and would be considered nonvolatile in the environment. Rates of evaporation of organic chemicals from water can be estimated from

$$K_1 = \frac{221.1}{(1.042/\mathbf{H} + 100.0)M^{1/2}}$$

where **H** is the Henry's Law constant; and M is the molecular weight of the test chemical. K_1 can then be used to determine the evaporation half-life ($t_{1/2}$) in minutes:

$$t_{1/2} = 0.6931(d/K_1)$$

where d is the solution depth.

Thus, a chemical such as *p,p*-DDT, which has a very low vapor pressure (7.16×10^{-7} torr), also has very low solubility in water (0.0017 mg/L). The combination of these properties leads to a not-insignificant Henry's law constant and indicates that the chemical is likely to have measurable volatilization from water in the environment. The global dissemination of *p,p*-DDT is attributed to this mechanism. Under environmental conditions, the actual volatility rates will be influenced by a number of factors, including the rate of dispersion of the chemical away from the evaporative site by wind; the degree of persistence of the chemical in the environment, that is, the extent to which the chemical will undergo photodegradation; oxidation; hydrolysis; or biodegradation before volatilization.

Example 19.2 A new compound in development has the attributes listed in Table 19.7. What can you say about the fate of this compound?

Solution From Table 19.1 we can see that the compound has comparatively low water solubility, and together with the hydrolysis data, we might conclude that it would be stable for reasonably long periods of time in water. Together with a D_{ow} of 2.9, we would also conclude that the compound is likely to sorb to biomass in a wastewater treatment plant, or to partition to fatty substances in environmental organisms, including humans. However, based on its UV/Vis absorbance, we might want to see if photolysis may play a role as a depletion mechanism.

Additional Points to Ponder If you were a synthetic organic chemist, what might you do to this compound to make it less stable in the environment? What are the drawbacks in doing this?

TABLE 19.7 Compound Attributes

Property	Result
Water solubility	0.5 mg/L
UV/Vis absorbance	350 nm and 480 nm
Hydrolysis	Stable for 3 months
Log D_{ow}	2.9

19.3 ENVIRONMENTAL FATE INFORMATION: TRANSFORMATION AND DEPLETION MECHANISMS

19.3.1 Hydrolysis

Hydrolysis is a common reaction occurring in the environment and therefore represents a potentially important degradation and depletion pathway for many classes of compounds. *Hydrolysis* refers to the reaction of an organic chemical (RX) with water:

$$RX + HOH \rightleftharpoons ROH + HX$$

In aqueous systems, rates of hydrolysis usually depend only on the concentration of the organic chemical because water is present in such excess that its concentration does not change during the reaction and thus does not affect the reaction rate. The half-life ($t_{1/2}$) is defined as

$$t_{1/2} = \frac{0.6931}{k}$$

where k is the rate constant observed. Rates of hydrolysis depend on a number of factors that change seasonally and slowly in the aquatic environment, such as the pH, temperature, and concentration of the chemical. However, hydrolysis rates are independent of many rapidly changing factors that normally affect other degradative processes, such as the amount of sunlight, the presence or absence of microbial populations, and the oxygen supply. Hydrolysis data are important in the design and interpretation of other environmental fate and effects tests. If a substance is extremely susceptible to hydrolysis, loss of the compound must be taken into account in tests such as aquatic toxicity and photodegradation. While many hydrolysis products are more polar than the parent compound, this is not always the case. Toxicity tests may need to be carried out on hydrolysis products.

19.3.2 Photolysis

Photolysis represents a second potentially important depletion mechanism, but in water it is generally a less common reaction than hydrolysis. Photolysis is a process whereby chemicals are altered directly as a result of irradiation, or indirectly through interaction with products of direct irradiation. Photolysis experiments may be carried out using a single- or multiple-wavelength light source (e.g., a mercury vapor light) or in direct sunlight between April and October. Tests carried out in direct sunlight are preferred over single-wavelength light studies.

Direct photolysis involves the absorption of light in the UV/Vis region by a molecule with a resultant increase in the molecular energy level. The increased energy then transforms the molecule chemically into one or more products. Molecules whose UV/Vis spectrum does not show absorption in the UV-Vis region (290 to 800 nm) will not undergo direct photolysis.

Indirect photolysis results from a chemical that may either receive energy for degradation from another chemical that has absorbed sunlight (sensitizer) and functions as a catalyst; or may react with products formed through direct photolysis (reactive species). Indirect photolysis in the aquatic compartment is not well documented. In the atmospheric compartment, the hydroxy radical has been implicated as the most important reactive species. In general, chemicals containing C–H and C–C bonds will be susceptible to hydroxy radical attack in the troposphere and will be photolysed indirectly. Volatile, fully halogenated compounds are not likely to be completely photodegraded in the troposphere and are likely to be transported to the stratosphere, where they pose a threat to the stratospheric ozone layer.

Phototransformations have to be taken into account for gases and compounds that occur in the gas phase in environmentally significant quantities. The probability that a compound occurs in the gas phase depends not only on its vapor pressure but also on its water solubility and adsorption/desorption behavior. Therefore, even substances that have relatively low vapor pressure (down to 10^{-3} Pa) can be found in the atmosphere in measurable quantities.

19.3.3 Aerobic Biodegradation

The *biodegradation* of an organic chemical refers to the reduction in complexity of the chemical through the metabolic activity of organisms, particularly bacteria and fungi. Knowledge of the potential biodegradability of an organic chemical is often critical in an assessment of the environmental exposure and impact of the chemical. Biodegradation is the most important depletion mechanism for organic chemicals in the aqueous compartment. Chemicals that biodegrade readily and rapidly are not likely to reach concentrations in the environment at which adverse impacts occur. However, adverse impacts of a biodegradable chemical might still occur if the chemical is introduced at levels that are acutely toxic or contribute to significant levels of inorganics such as nitrogen and phosphorus.

Readily biodegradable chemicals are those that are mineralized, or converted to carbon dioxide, water, and inorganic compounds in the presence of oxygen (aerobic biodegradation) and microorganisms. These chemicals:

1. Will have significant biological oxygen demand (BOD), and thus can exert a significant load on a wastewater treatment plant
2. Have shown significant (>60% of theoretical) carbon dioxide production during standard stringent biodegradation tests (e.g., a Sturm test).

Inherently biodegradable chemicals are those that show biotransformation in activated sludge test systems at biomass concentrations typical of wastewater treatment plants (~2500 mg/L). Typical aquatic biodegradation data and their interpretation are shown in Table 19.8.

Multiple tests may be carried out, depending on what is being assessed. The tests may include any, or a combination of, the following: a Sturm or modified Sturm test, a batch activated sludge (BAS) test (e.g., the Zahn–Wellens test is a specific type of BAS test), a semicontinuous activated sludge (SCAS) test, and a continuous activated sludge (CAS) test.

1. *Sturm or modified Sturm test.* This is a stringent biodegradation test of low biosolids concentration used primarily to assess "ready" biodegradation. The results of this test are considered positive if the actual amount of carbon dioxide produced during the test period is 60% or greater of the amount of carbon dioxide that could theoretically be produced based on the molecular formula of the compound. However, compounds that fail to achieve the extent of biodegradation required under the conditions of the Sturm test may be inherently

TABLE 19.8 Aquatic Biodegradation Data and Their Interpretation

Data	Inference
More than 80% degradation in 28 days	Readily biodegradable
More than 20% degradation in 28 days	Not readily but inherently biodegradable
Less than 20% degradation in 28 days	Not readily or inherently biodegradable
No degradation after extensive testing	Recalcitrant to biodegradation
$BOD_5/COD \geq 0.5$	Likely to be biodegradable
$BOD_5 > 25\%$ ThOD	Likely to be biodegradable
$BOD_{20} > 25\%$	Likely to be biodegradable
$BOD_5/COD < 0.5$	Not likely to be biodegradable
$BOD_5 < 25\%$ ThOD	Not likely to be biodegradable
$BOD_{20} < 25\%$	Not likely to be biodegradable
Biodegradation half-life on the order of minutes	Readily biodegradable
Biodegradation half-life on the order of days	Biodegradable
Biodegradation half-life on the order of weeks	Slowly biodegradable

biodegradable. Negative results might instead mean that the chemical may be biodegraded at a slow rate, partially biodegraded (i.e., the identity of the compound is changed but the compound is not completely changed to carbon dioxide and water), inhibitory or toxic at the concentration tested, and biodegradable under different test conditions (e.g., at a lower concentration of chemical and/or a high concentration of biomass).

2. *Batch activated sludge* (BAS) *test.* In this test the compound is introduced into an activated sludge matrix. The sludge is obtained from local wastewater treatment plants used to treat mixed domestic and industrial wastewater and is composed of a mixed consortium of microorganisms. These microorganisms are present at fixed concentrations on the order of 2500 mg/L total suspended solids (TSS). Degradation of the parent may be assessed by following the disappearance of the parent molecule by using high-performance liquid chromatography or by a more general method such as dissolved organic carbon and total organic carbon.

3. *Semicontinuous and continuous activated sludge* (SCAS *and* CAS) *tests.* This test is essentially a BAS, but the test compound is introduced into an activated sludge matrix that operates either continuously or semicontinuously. The CAS test most closely simulates a wastewater treatment plant and generates the most useful data. Compounds that do not readily mineralize under these conditions but do biotransform into other usually more polar degradants are considered to be "inherently" biodegradable. If the kinetics of the biotransformation reaction is studied, a rate constant can be calculated. This rate constant can be used in environmental fate models to predict depletion, either in wastewater treatment plants or in the environment. These data may be used to estimate a chemical's persistence.

19.4 ENVIRONMENTAL EFFECTS INFORMATION

19.4.1 Microbial Respiration Inhibition Test

Respiration inhibition measures the extent to which respiration (oxygen uptake) is affected following introduction of the test chemical into a mixed consortium of microorganisms. For chemicals that are biodegradable, this test gives some idea of a "safe" concentration for

TABLE 19.9 Respiration Inhibition IC_{50} Interpretation

Inhibition Concentration, IC_{50}	Interpretation
> 50 mg/L	Not toxic to activated sludge microorganisms
$> 1 < 50$ mg/L	Harmful to activated sludge microorganisms
$>0.05 < 1$ mg/L	Toxic to activated sludge microorganisms
< 0.05 mg/L	Very toxic to activated sludge microorganisms

discharge into a wastewater treatment plant (WWTP) and thus allows determination of whether or not waste streams containing these chemicals will have to be pretreated or diluted prior to treatment in the WWTP. It also gives some indication of whether or not the compound will exert an acute toxic effect on microorganisms in a spill scenario. Solutions of varying concentrations of the chemical are added to vessels containing activated sludge samples and aerated. At a specified time, the oxygen uptake rates are determined. The concentration at which there is a 50% inhibition of the sludge respiration is termed the IC_{50}. Toxicity value interpretation is shown in Table 19.9

19.4.2 Microtox Test

The Microtox test uses a sensitive photoluminescent marine microorganism, *Photobacterium phosphoreum,* to assess the potential for a compound to cause mutations in that microorganism. The bacteria are subjected to varying concentrations of a chemical, and the diminution in light output as a function of concentration is measured. The concentration at which the light output is reduced by 50% after a specified number of minutes of contact time is termed the *effective concentration* (EC_{50}). The Microtox results are generally good indicators of acutely toxic effects to aquatic organisms at higher levels of the food chain, although these results are not generally accepted by regulatory agencies. These results are useful, however, because the Microtox test can be carried out rapidly and the results may identify the potential for a compound to cause environmental impacts. Microtox test results should always be confirmed with additional aquatic toxicity testing.

19.4.3 Aquatic Toxicity Tests

Acute Aquatic Toxicity Tests Adverse or toxic effects can be produced in the laboratory or in the natural environment by acute (short-term) or chronic (long-term) exposure to chemicals or other potentially toxic agents. Adverse effects on aquatic ecosystems may result from exposures to toxic substances that directly cause the death of organisms (acute effects), or produce sublethal effects on an organism's ability to develop, grow, and reproduce in the ecosystem (chronic effects). In acute exposure, organisms come in contact with the chemical delivered either in a single event or in multiple events that occur within a short period of time, generally hours to days. Acute exposures to chemicals that are rapidly absorbed generally produce immediate effects, but they may also produce delayed effects similar to those caused by chronic exposure.

To first identify and then control the introduction of toxic chemicals to aquatic systems, laboratory toxicity tests are performed and "safe" pollutant concentrations for aquatic ecosystems are extrapolated from these laboratory-derived data. These toxicity tests are the initial "tools" that provide qualitative and quantitative data on potential adverse or toxic effects in the environment. In particular, acute aquatic toxicity tests are required to determine the scope and degree of acute toxic effects that a compound might exert on organisms at different levels of the food chain. In general, it is desirable to select at least one organism at three different levels of the food chain, most commonly with;

- Algae (usually a form of green algae such as *Selenastrum capricornutum*)
- Aquatic fleas that eat algae (*Daphnia magna* or *pulex*)
- Fish [bluegill sunfish (*Lepomis machrochirus*) or rainbow trout (*Oncorynchus mykiss* or *Salmo gairdineri*], which are generally at the top of the aquatic food chain (nonmammalian or avian)

By including acute tests at three levels of the food chain, the accuracy of the final risk assessment increases, and the safety factors that need to be applied are decreased. Details of applying these safety factors are discussed in Section 19.5. Toxicity value interpretations were shown in Table 19.4.

Chronic Aquatic Toxicity Tests The fact that a chemical does not adversely affect aquatic organisms in acute toxicity tests does not necessarily indicate that it is not toxic to these species. Chronic toxicity tests are used to evaluate possible adverse effects of a chemical under conditions of long-term exposure at sublethal concentrations. In a full chronic toxicity test, the test organism is exposed for an entire reproductive life cycle (e.g., egg to egg) to at least five concentrations of the test material. Partial life cycle toxicity tests involve only several sensitive life stages.

From the data obtained in partial or full life cycle tests, the maximum acceptable toxicant concentration (MATC) can be estimated. This is the estimated threshold concentration of a chemical within a range defined by the highest concentration tested at which no significant deleterious effect was observed (NOEC) and the lowest concentration tested at which some significant deleterious effect was observed (LOEC). Life cycle data were also used to derive the assessment factors (AFs) used to predict permissible no-effect levels (PNECs), which are used in environmental risk assessments.

Relationship Between Acute and Chronic Toxicity Endpoints For much of the history of aquatic toxicology testing, the bulk of the data have been generated by acute testing where exposure duration is short and mortality is used as the response endpoint. It was recognized, however, that information about longer exposure durations and nonlethal response endpoints was needed for appropriate evaluation of the ecological significance of the impacts that a given chemical might have on the environment. Safety factors or assessment factors were often used to generate water quality criteria regulations from laboratory-based acute toxicity data. As more chronic toxicity data became available, models were developed to allow the prediction of potential chronic effects from acute toxicity test data. In general, it was found that chronic effects could be observed at concentrations that were an order of magnitude lower than those producing acutely toxic effects.

However, these relationships have been established based in large part on data generated for industrial chemicals. These chemicals may not be representative of pharmaceutical compounds, which are designed to be pharmacologically active in addition to exerting a toxicological effect. There is currently considerable discussion within the scientific community about the use of these assessment factors in risk assessments for pharmaceutical compounds. Current practice within industry is to use these assessment factors when assessing the acute toxicity of the material, but to base risk assessments concerned with chronic exposures on case-, site-, and situation-specific information so that the appropriate safety factors may be incorporated.

Bioaccumulation Tests Chemicals with low solubility in water usually have an affinity for fatty tissues and thus can be stored and concentrated in tissues with a high lipid content. Such hydrophobic chemicals may persist in water and demonstrate cumulative toxicity to organisms. Chemicals with these characteristics are usually considered for bioconcentration tests, which are designed to determine or predict the bioconcentration factor (BCF), the ratio of the average concentration of test chemical accumulated in the tissues of the test organisms under steady-state conditions to the average concentration measured in the water to which the organisms are exposed. The BCF may also be estimated from the log of the octanol/water distribution coefficient, as discussed before.

Other Tests Other types of toxicity tests using different species may also be carried out, depending on the nature of the chemical and probable environmental release scenarios. For example, for highly sorptive compounds that are likely to partition strongly to soils and sediments, benthic organisms such as *Hyallela azteca* may be used. Another common test organism is the earthworm (*Lumbricus terrestris, Lumbricus rubellus,* and *Eisenia foetida*). The earthworm is generally used when release of a compound to the terrestrial compartment is likely, either directly, or through land application of wastewater treatment plant sludge to which the chemicals may have sorbed.

As discussed above, with compounds such as pharmaceuticals, which tend not to be readily biodegradable, acute aquatic toxicity tests may be inadequate for predicting environmental risk. Careful consideration of the various likely release scenarios may suggest that other types of tests may be advisable. There is currently considerable concern about chemicals in the environment that may have effects at low levels on endocrine systems, leading to developmental and reproductive effects.

19.5 ENVIRONMENTAL RISK ASSESSMENT

As we saw in Section 3.4, risk assessment is the characterization of the potential adverse effects resulting from human and ecological exposures to environmental hazards. Environmental risk assessment can then be defined as the process of estimating and characterizing the likelihood that adverse effects of human actions on the nonhuman environment will occur, are occurring, or have occurred. Environmental risk assessment can be categorized according to the problems addressed:

- *Predictive risk assessments* estimate risks from proposed future actions such as marketing a new medicine, operating a new process, or emitting a new aqueous or atmospheric contaminant.
- *Retrospective risk assessments* estimate the risks posed by actions such as disposal of hazardous wastes, spills, existing effluents, and landscape modifications that occurred in the past and may have ongoing consequences.

Predictive risk assessments involve three major phases: hazard definition, risk quantification, and risk management, as shown in Table 19.10.

TABLE 19.10 Phases of Environmental Predictive Risk Assessments

Hazard Definition	Risk Quantification	Risk Management
Source description (e.g., pollutant fate properties)	Exposure assessment	Decision as to course of action
Environment description (e.g., fate and transport modeling)	Effects assessment	Mitigation solutions
Endpoints definition (e.g., potentially affected organisms)	Risk characterization	Risk vs. benefit

Figure 19.3 illustrates the predictive ecological risk assessment. The solid arrows represent the sequential flow of the procedure. The dashed arrows represent feedback and other constraints on one assessment process by other processes.

Briefly, the sophistication of the fate and transport models depends on the needs of the risk manager. For example, a complete environmental assessment (EA) or environmental impact statement (EIS) may require very sophisticated models; whereas "quick and dirty" assessments may only need simple models, such as the quantitative risk equations described below. Risk managers take the information from the first two phases and combine this information with possible mitigation solutions to arrive at a final course of action. Risk vs. benefit issues are also part of the risk management phase.

For example, the simplest quantitative technique for estimating ecological risks is the *quotient method*. The quotient method is simply estimating a safety factor (SF) by dividing the predicted environmental concentration (PEC) by a toxicological benchmark concentration, normally the predicted no effect concentration (PNEC).

$$SF = \frac{PEC}{PNEC}$$

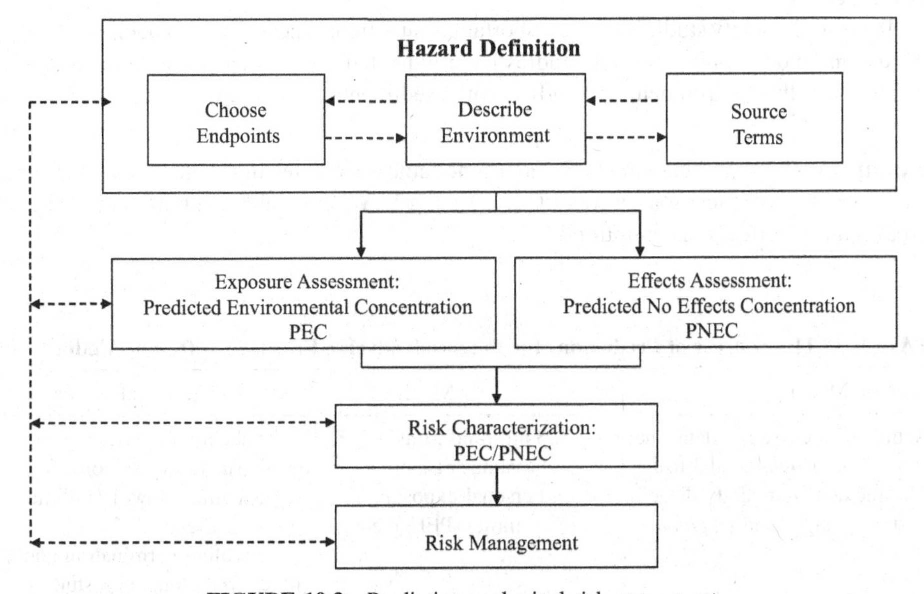

FIGURE 19.3 Predictive ecological risk assessment.

- If SF < 1, the discharge is considered an acceptable risk to the environment.
- If SF > 1, additional fate and/or toxicity tests may be needed to better define the PEC and PNEC values.

Values of SF < 1 do not guarantee that adverse environmental impacts won't occur, but it means that the probability (risk) of adverse impacts occurring is low. Conversely, values of SF > 1 do not guarantee that adverse environmental impacts will occur. To estimate the safety factor we need, therefore, to estimate the PEC and the PNEC. For example, the predicted environmental concentration (PEC) for water may be obtained as follows:

$$\text{PEC} = \frac{A(100 - \sum R_j)}{VD}$$

where A is the amount of chemical entering an environmental compartment, R_j the removal rate by depletion mechanisms, V the volume or flow, and D the dilution factor into receiving media. The predicted no-effect concentration (PNEC) of a chemical in an environmental compartment may be obtained as follows:

$$\text{PNEC} = \frac{\text{L(E)C}_{50}}{\text{assessment factor}}$$

where L(E)C_{50} is the 50% lethal or effect concentration on a test organism and the assessment factor is intended to take into account inter- and intraspecies uncertainties as well as the ratio of acute to chronic effects. The assessment factor is normally 1000 if only one acute test was carried out; 100 if three acute tests on organisms from different trophic levels were performed; 10 if chronic tests were carried out, and 1 if field monitoring studies were done. Table 19.11 shows some sources of predictions for potential adverse effects in different media.

Terrestrial toxicity studies may be important in situations where sludge is being evaluated for use in land farming, and fate studies have indicated a tendency for a chemical to be persistent in the environment and sorb to soils, sediments, and sludges.

Example 19.3 A new chemical has undergone acute toxicity testing in bluegill sunfish, and the lowest lethal concentration was $\text{LC}_{50} = 1.6$ mg/L. What would be a reasonable way to predict the no-effect concentration?

TABLE 19.11 Sources of Predictions for Potential Adverse Effects in Different Media

Aquatic Media	Air Media	Terrestrial Media
Acute aquatic toxicity data, such as algae, *Daphnia*, and fish	Threshold limit values (TLVs)	Subchronic toxicity data, such as earthworm
Chronic aquatic toxicity data, such as algae and *Daphnia*	Personal exposure limits (PELs), etc.	Seedling growth inhibition studies
		Seedling germination and root elongation studies

Solution Since the LC_{50} is an acute toxicity result and no additional data are available for this new chemical, we have

$$\text{PNEC} = \frac{LC_{50}}{AF} = \frac{1.6}{1000} = 0.0016\,\text{mg} = 1.6\,\mu\text{g/L}$$

Additional Points to Ponder What does this result represent? What might an environmental manager at the manufacturing plant designated to produce this chemical have to do?

19.6 ENVIRONMENTAL LIFE CYCLE IMPACTS OF WASTE TREATMENT

In the first part of this chapter we discussed environmental fate and effects data and how they are used to assess environmental risk. Once a risk assessment is performed and it is determined that there is a need to treat waste streams, we need to remember that from an environmental life cycle viewpoint there are additional impacts related to treating the waste, as represented in Figure 19.4.

In Chapter 18 we discussed how to use energy submodules to estimate the environmental life cycle impacts of energy production. In a similar fashion, we can estimate the environmental life cycle impacts of treating waste. Although it is true that as we treat the waste we are working to reduce one or more hazards, there are additional impacts that need to be taken into account. For example, when we treat wastewater in an activated sludge plant, we reduce the organic content of the wastewater, but in so doing, we also incur impacts associated with energy production, transportation, and the production of chemicals to operate the plant. Previous research has explored the impacts from waste treatment, has characterized these impacts, and has compared different waste treatment technologies; the results may be found in the literature.[10–12] In this section we explore briefly some common waste treatment and recovery technologies and their associated environmental impacts through use. The technologies we cover include the wastewater treatment plant, the incinerator, and solvent recovery. There are, of course, many more technologies used for

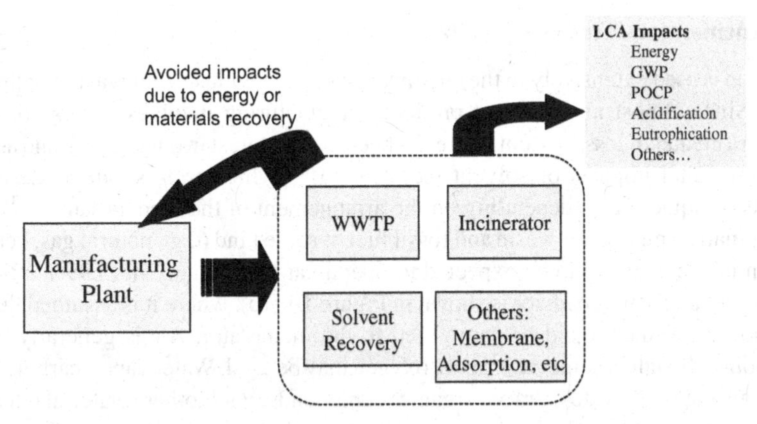

FIGURE 19.4 Environmental impacts of waste treatment and recovery technologies. WWTP, wastewater treatment plant.

treatment and recovery, but they are outside the scope of this chapter. However, the principles of estimating life cycle impacts are generic and applicable to any other type of waste treatment and processing technologies, as we will see in Chapter 23.

19.6.1 Wastewater Treatment Plant

As we described above, operating a waste treatment plant results in environmental life cycle impacts associated with the operational energy consumed in running the plant and from the substances used to treat the water. A general flow diagram is presented in Figure 19.5(a). Many models are used to estimate the biodegradation mechanisms in the waste treatment plant, and some of these have been published widely. For example, the following reactions describe the main biodegradation mechanisms taking place in activated sludge treatment[13]:

Cell autooxidation:

$$C_5H_7NO_2(\text{cells}) + 5O_2 \rightarrow 5CO_2 + 2H_2O + NH_3$$

Cellular energy generation and cell production:

$$C_6H_{12}O_6(\text{substrate}) + O_2 + NH_3 \rightarrow C_5H_7NO_2(\text{cells}) + CO_2 + 4H_2O$$

The substrate cells use for their energy source the organic chemical mixture generated during manufacturing that is suitable for discharge to a wastewater treatment plant (see Section 19.1). The glucose substrate in the second equation is only one example of cellular respiration, and it is common to characterize the potential for all organic chemicals to be biodegraded by calculating the theoretical oxygen demand (ThOD) or the TOC. Eckenfelder[14,15] developed engineering design methods based on the above equations, but generalize for a wide variety of substrates and microbial consortia, which can in turn be used to estimate life cycle impacts. For example, Table 19.12 shows selected LCI emissions associated with treating 1 kg of TOC in a wastewater treatment plant. The negative values for some of the parameters denotes their reduction due to water treatment.

19.6.2 Incinerator (Thermal Oxidizer)

Incineration is used extensively in the chemical processing industry for waste treatment and disposal. Since almost all chemical product manufacturing involves the use of organic solvents to a greater or lesser extent, there has been great interest in quantifying and assessing the environmental impacts of solvent incineration.[16,17] Incinerators can be designed to accept gas or liquid feeds, depending on the arrangement of the feed stream,[18–20] and the feed is normally a mixture of waste and fossil fuel of some kind (e.g., natural gas, kerosene). A modern incinerator would be expected to operate at a minimum of 99.99% efficiency.

A model of a thermal oxidizer is shown in Figure 19.5(b), where it is assumed that spent organic solvent (either liquid or gas) is fed to the incinerator. Air is generally used for combustion, although in some cases, pure oxygen may be used. Water vapor, carbon dioxide, nitrogen and sulfur oxides, trace organic compounds, and other materials leave the incinerator as part of the flue gas with part of the combustion energy usually recovered from the hot stream. Several models, most of which are based on stoichiometric relationships

TABLE 19.12 Example of Selected LCI Emissions (kg) from Treating 1 kg of TOC in a Wastewater Treatment Plant

	Total from WWTP	From WWTP Process	Electricity-Related
Air Emissions			
CH_4	0.01	—	6.47×10^{-3}
CO	0.00	—	5.02×10^{-4}
CO_2	4.33	2.49	1.84
NMVOC	0.00	—	4.24×10^{-4}
NO_x	0.00	—	3.91×10^{-3}
SO_x	0.01	—	5.47×10^{-3}
Water Emissions			
TOC	−0.86	-8.60×10^{-1}	1.95×10^{-3}
BOD	−1.30	−1.30	2.23×10^{-4}
COD	−2.45	−2.46	5.58×10^{-3}
TDS	0.00	—	4.69×10^{-3}
Solid waste			
Biosolids	1.14	1.14	
Solid waste, other	0.08	—	8.37×10^{-2}

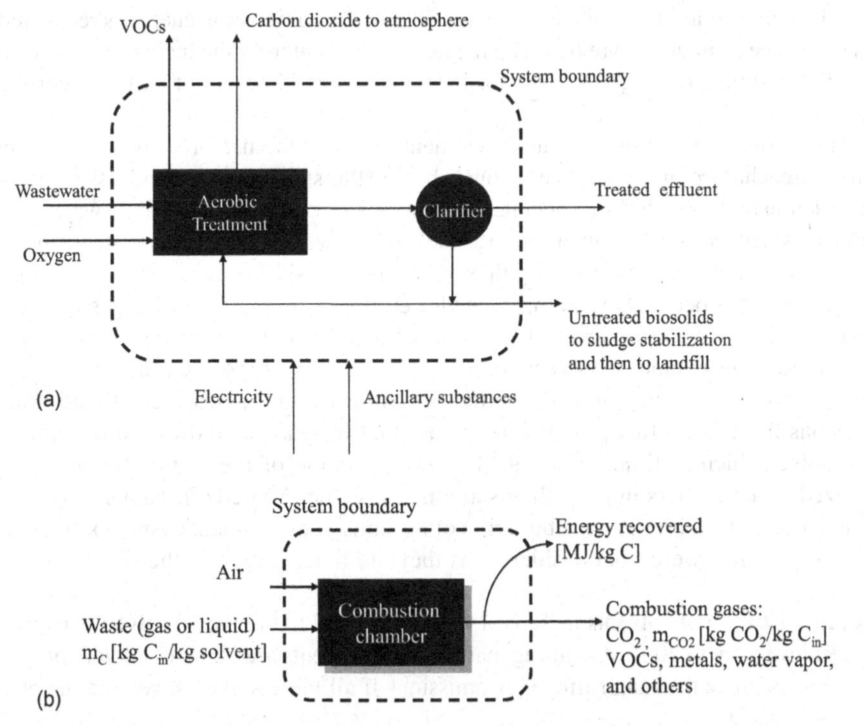

FIGURE 19.5 Examples of inputs and outputs of a wastewater treatment Plant (a) and a thermal oxidizer or incinerator (b). (*Source:* ref. 10. Reproduced with permission from John Wiley and Sons. Copyright © 2001. John Wiley and Sons, Inc.)

TABLE 19.13 Example of Selected LCI Emissions from Treatment in 1000 kg of Toluene

	Total in Incinerator Without Energy Recovery	Total in Incinerator with 30% Energy Recovery
Net energy (MJ)	39,515	32,125
Air emissions (kg)		
CH_4	6.32	5.25
CO	2.12	1.71
CO_2	6,090.38	5,506.02
NMVOC	15.63	12.66
NO_x	8.82	6.90
SO_x	1.06	−1.50
Water emissions (kg)		
TOC	0.00	0.02
BOD	0.00	−0.14
COD	0.00	−0.04
TDS	1.08×10^{-1}	−2.55
Solid waste (kg)	3.47	2.21

or direct industrial measurements, have been used to estimate the environmental impacts of an incinerator. Table 19.13 contains two examples of selected LCI emissions associated with incinerating 1000 kg of toluene, based on a model developed for an industrial thermal oxidizer. The first column contains the emissions when there is no energy recovery, and the second column contains the emissions when 30% of the combustion energy is recovered. In this model the estimations are based on the total carbon entering the incinerator, with about 43 MJ of additional energy required for incineration per kilogram of carbon entering the incinerator.[10]

One incineration option for waste treatment has been the use of kilns, mainly cement kilns, to treat hazardous waste. The rationale behind this strategy is to spend less money than that when using fossil fuels to operate the kiln and, instead, use hazardous waste (which companies will pay to have taken away) as fuel. The benefit of hazardous waste incineration in cement kilns is that most of the flue gases are trapped in the cement matrix in this countercurrent process. For example, Seyler et al.[21] estimated the changes in environmental impacts and emissions when using cement kilns to incinerate waste for four systems: toluene, ethanol with traces of heavy metals, mixtures of ethyl acetate and water, and a mixture of 1-butanol with methylene chloride. The assessment found that the emissions from fossil fuel production and use that were avoided dwarfed the emissions from solvent incineration. Table 19.14 shows a sample of the results for the systems analyzed. The numbers in parentheses are the gross fuel-related emissions from solvent incineration in the kiln before subtraction of the emissions avoided. Cement kiln process-related emissions were not considered, as they are independent of the fuel used.

Example 19.4 The solvents in Table 19.15 correspond to the top 10 solvents reported in the TRI inventory in 2007. Assuming that all of the solvents are incinerated with no energy recovery, estimate the major life cycle emissions if all these solvents were incinerated.

Solution To estimate the emissions we use the incineration model developed by Jiménez-González et.al.[10] The model estimations are based on the total carbon entering the

TABLE 19.14 Change in Selected LCI Parameters for the Use of 1 Metric Ton of Waste Solvent as Alternative Fuel for Clinker Production

Fuel	Toluene	Ethanol with Traces of Heavy Metals	Ethyl Acetate and Water	1-Butanol with Methylene Chloride
		Resource Avoided		
Coal (tons)	−1.22	−0.81	−0.67	−1.00
Oil (tons)	−0.22	−0.15	−0.12	−0.18
		Change in Emissions		
CO_2 (kg)	−610 (3350)	−715 (1910)	−261 (1900)	−877 (2360)
NO_x (kg)	−10.2	−6.77	−5.56	−8.33
As (mg)	−3.69	−2.45	−2.02	−3.02
Cu (mg)	−4.60	6.95 (10)	−2.51	−3.76
Ni (mg)	−0.65	0.57 (1.0)	−0.35	−0.53
Hg (mg)	−129	−85.5	−70.3	−105

TABLE 19.15 Top Ten Solvents

Solvent	Amount to Be Incinerated (kg)
Methanol	9,706,905
Dichloromethane	12,781,416
Toluene	9,032,140
Acetonitrile	2,178,311
n-Hexane	2,041,914
N-Methyl-2-pyrrolidone	1,008,886
N,*N*-Dimethylformamide	941,262
n-Butyl alcohol	931,721
Methyl *tert*-butyl ether	1,075,252
Xylene	907,587

incinerator, with about 43 MJ of additional energy required per kilogram of carbon incinerated. Since the water content of the waste–solvent mixture will dramatically increase the energy required for incineration, a 50% water/solvent ratio was assumed for miscible solvents, and for non-water-miscible solvents the water/solvent ratio was estimated using water solubility data. The use of natural gas as a fuel was assumed, which is a conservative assumption. The results of this assessment are shown in Figure 19.6.

Additional Points To Ponder How different would the emissions profile be if energy recovery were used in the incinerator? How accurate would the water assumptions be in terms of the effect on emissions?

19.6.3 Solvent Recovery Through Distillation

As we have seen in Chapter 6 and other chapters, solvents comprise the largest portion of the mass of materials used in the chemical, fine chemicals, and pharmaceutical industries; in other words, solvents comprise the largest proportion of the mass intensity for many

EPA TRI Solvents				
Calculation basis = 122,336,620 pounds total solvent				
Information needed (enter in the yellow cells)				
Name of the organic substances to be incinerated	Molecular weight	Number of Carbon atoms in formula	Amount to be incinerated [kg]	Organic carbon to incinerator [kg]
methanol	32	1	9,706,905	3,640,089
dichloromethane	84.9	1	12,781,416	1,806,561
toluene	92.13	7	9,032,140	8,235,100
acetonitrile	41	2	2,178,311	1,275,109
n-hexane	87.17	6	2,041,914	1,686,564
N-methyl-2-pyrrolidone	99	5	1,008,886	611,446
N,N-dimethylformamide	73	3	941,262	464,184
n-butyl alcohol	74.12	4	931,721	603,381
methyl tert-butyl ether	88.14	5	1,075,252	731,962
xylene	106.16	8	907,587	820,727
				0
		Total Carbon to incinerator [kg] =		19,875,123
		Total Organics to incinerator [kg] =		40,605,394
		Total aqueous to incinerator [kg]		14,987,006
			Energy Recovered (%)	0%
Energy needed = [MJ of Natural gas]	960,373,068		Energy Recovered (MJ of Steam)	0

	Total from incinerator	From incineration process	Energy usage-related	Energy recovery-related
Air emission [kg]				
CH4	153,660.81		1.54E+05	0.00E+00
CO	47,770.73	3.18E+04	1.60E+04	0.00E+00
CO2	139,728,324.46	7.21E+07	6.76E+07	0.00E+00
NMVOC	379,460.91	3.18E+03	3.76E+05	0.00E+00
NOx	214,328.19		2.14E+05	0.00E+00
SOx	25,672.75		2.57E+04	0.00E+00
Water emission [kg]				
TOC	22.97		2.30E+01	0.00E+00
BOD	3.98		3.98E+00	0.00E+00
COD	65.70		6.57E+01	0.00E+00
TDS	2.61E+03		2.61E+03	0.00E+00
Solid waste [kg]	84,270.51		8.43E+04	0.00E+00

FIGURE 19.6 Results of TRI modeling for incineration with no energy recovery.

industries. Moreover, once solvents are used in a chemical process, there is often resistance to reusing the solvent, based on fears of lowering product quality, concentrating an unwanted impurity, or causing a change in chemical reactivity. Although this is observed in many industries, it is especially a barrier in the fine chemicals and pharmaceuticals industries. As a consequence, in addition to recovering solvents through distillation, a considerable volume of solvent is either incinerated in-house or sent to external incinerators or cement kilns.[21] However, we discussed in previous chapters that most environmental benefit from a life cycle viewpoint will come through recovering the solvent and reusing it in the same process (internal recycling) or in another process (external recycle). To do this, we need to purify and recover the solvent.

Distillation is most commonly used to purify solvents. Solvents with different vapor pressures can be separated from one another by fractional distillation. Azeotropic mixtures can be separated by extractive or azeotropic distillation (e.g., addition of benzene to a water–ethanol mixture), by chemical reaction of a component (e.g., addition of acetic anhydride to an ethanol–ethyl acetate mixture), or by altering the pressure during distillation.[22] Distillation is by far the most commonly used technology for purifying solvents and/or separating solvent mixtures. Several assessments have been made to estimate the environmental impacts of solvent recovery vs. disposing of them through incineration with energy recovery or in cement kilns.[23,24]

It has been found that in contrast to incineration, the reduction in environmental impacts through solvent recovery by distillation (or any related technology, for that matter) is driven by the credits obtained through avoided resource consumption and emissions from the recovery of the solvent, as shown in Figure 19.7.[25] This has been corroborated by several studies, including one by Capello et al.,[26] who found that recovering solvents through distillation is in general the environmentally optimal option, especially when a majority of the solvent is recovered and when the solvents recovered have large associated environmental life cycle profiles. They also found that when atmospheric distillation is very difficult, with low solvent recoveries (e.g., methanol, azeotropic distillation), distillation is not necessarily superior to incineration with solvent recovery. They went as far as to provide a series of rules of thumbs based on solvent type, recovery efficiencies, and the associated life cycle profile to select the type of solvent recovery or treatment, as shown in Table 19.16.

They also performed a solvent-specific assessment for solvent recovery by distillation. This assessment revealed that there were no specific solvents for which incineration is environmentally the better choice. The solvent-specific assessment also revealed that:

- For acetic anhydride, butylene glycols, dichloromethane, formic acid, methyl isobutyl ketone, and tetrahydrofuran, distillation is the most environmentally favorable option in most cases, regardless of the recovery efficiency and at almost minimal solvent recoveries. Some of these solvents (acetic anhydride, butylene glycol, methyl isobutyl ketone, tetrahydrofuran[27]) have large life cycle credits because they have elaborate chemical trees or have low net calorific value, such as formic acid (4.6 MJ/kg[28]), or both, such as dichloromethane.
- For heptane, methyl *tert*-butyl ether (MTBE), and pentane, distillation and incineration had similar environmental profiles given their high calorific values.
- For cyclohexane, ethanol, ethyl benzene, formaldehyde, isohexane, methanol, toluene, and xylene, distillation was only marginally better under the best-case scenario of high

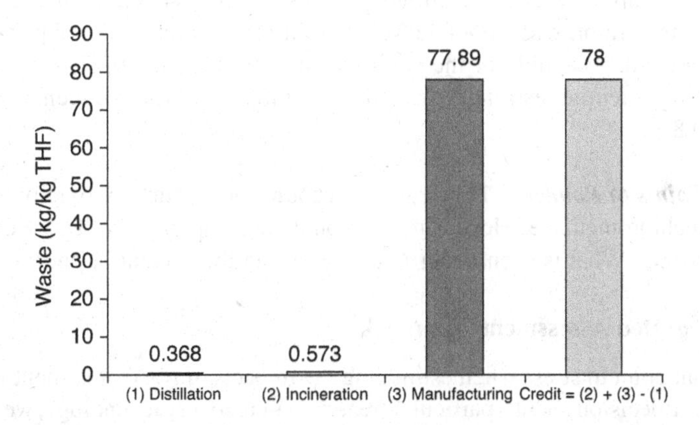

FIGURE 19.7 Total waste credits estimated by recovering 1 kg of THF.

TABLE 19.16 Rules of Thumb for Selecting the Treatment/Recovery Technology with the Fewest Environmental Impacts

Solvent Mixture Has:	Option with Fewest Environmental Impacts
High net calorific value	Incineration with energy recovery (e.g., cement kilns)
Heteroatoms	Recovery through distillation
Possibility of high recovery	Recovery through distillation
High water content	Recovery through distillation
Complex production chain	Recovery through distillation
One of these solvents as main component:	Recovery through distillation
Acetic anhydride	
Butylene glycols	
Dichloromethane	
Formic acid	
Methyl isobutyl ketone	
Tetrahydrofuran	

solvent recovery. This is explained by their high net calorific values (>40 MJ/kg); low environmental credits due to simple chemical trees, difficulty in separating azeotropes, or a combination of these factors.

Example 19.5 It was mentioned above that recovering dichloromethane by distillation is always the best option. Prove that point.

Solution For this case we can assume three scenarios:

- Incineration with no energy recovery
- Incineration with 50% energy recovery
- Recovery by distillation with 75% efficiency and the tails incinerated with energy recovery

For these scenarios we use the same model as in Example 19.5, which is different from the one used by Capello et al. As a reminder, the estimations are based on the total carbon entering the incinerator, with about 43 MJ of additional energy required per kilogram of carbon incinerated. A sample of the results is shown in Table 19.17, and the resulting global warming potentials estimated from these numbers for the three scenarios are shown in Figure 19.8.

Additional Points to Ponder How can we account for the human toxicity hazard associated with dichloromethane? How different would the comparison be if 80% of the energy can be recovered? What is even better than recovering the solvent or energy?

19.6.4 Integrated Assessment Approach

One important point to stress when estimating the impacts of waste treatment is that every time we make a decision about a particular treatment or recovery technology, we must assess the systems from a life cycle viewpoint. For example, some of the generic rules of thumb,

TABLE 19.17 Sample of Results

	75% Solvent Recovery Using Distillation	Incineration with 50% Energy Recovery	Incineration without Energy Recovery
Net energy (MJ)	−15,482	6,212	3,149
Air emissions (kg)			
CH_4	−4.83	0.50	0.99
CO	−2.90	0.28	0.33
CO_2	−1,349.58	734.65	950.19
NMVOC	−7.33	1.26	2.46
NO_x	−5.67	0.70	1.39
SO_x	−5.96	0.08	0.17
Water emissions (kg)			
TOC	−13.80	0.00	0.00
BOD	−0.13	0.00	0.00
COD	−56.62	0.00	0.00
Leachate	0.01	0.01	0.02
TDS	−5.73	0.01	0.02
Solid waste (kg)			
Other solid waste	−27.55	0.28	0.55

methodologies, and what we learned earlier about solvent recovery can be applied to other types of hazardous and nonhazardous waste. It is a matter of assessing the system with expanded boundaries. Many of the solutions will be found when looking outside the plant, as we will see in Chapter 24 when we cover industrial ecology.

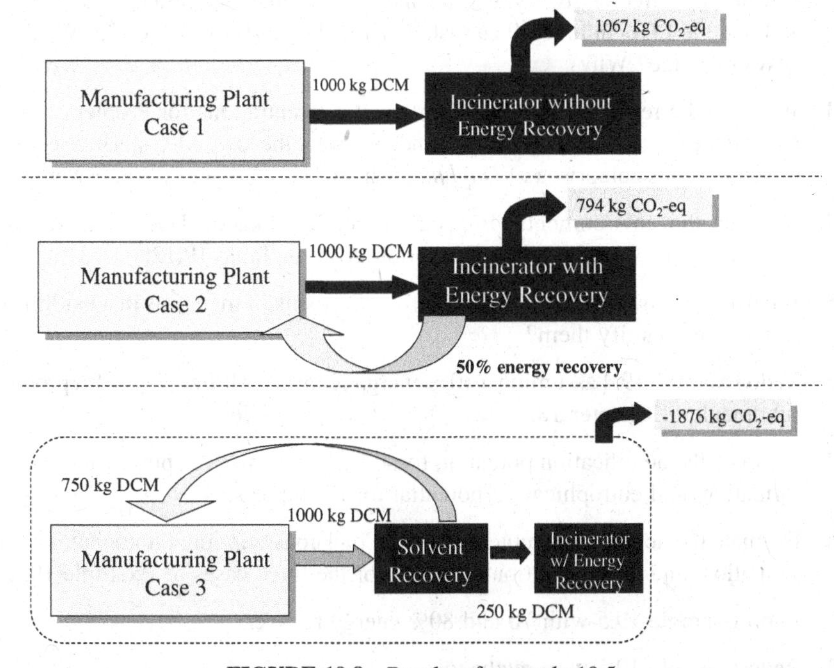

FIGURE 19.8 Results of example 19.5.

The other important point to cover is that although it is important to understand and assess impacts of waste treatment and recovery, it is even more important to design chemical and manufacturing processes in such a way that the generation of waste is eliminated or minimized, and that is precisely the role of green engineering. Having said that, even the most creative and clever engineers will be faced with the laws of physics and thermodynamics, and as we get closer to thermodynamic limits, there will be an increased price to pay. In most cases, the cost will be greater for designing inefficient processes that generate unnecessary waste. And similar to what we have seen previously with life cycle assessment principles, we must estimate and assess the costs by expanding the boundaries beyond the plant's fence, as we cover in Chapter 20.

PROBLEMS

19.1 A manufacturer is planning to introduce a new product to the market that releases chemical A as a major by-product of the manufacturing process. Given current forecasts, it is expected that the average concentration of chemical A in the site's effluent will be 16.8 mg/L. The plant's effluent is discharged into the wastewater treatment plant in the industrial park, which, unfortunately, is not expected to remove or degrade any portion of chemical A before discharging the treated water into the river. You are the EHS manager and have been asked by the new product director if the introduction of this new product will raise environmental concerns, especially since she heard that the only acute toxicity data available is an LC_{50} of 25 mg/L in bluegill sunfish, and that sounded like it was too low. What answer do you give her?

19.2 You have just finished the assessment required to answer Problem 19.1 when the new product director gives you a call and tells you that there is a mistake in the data and that the LC_{50} in bluegill sunfish is really 4.1 mg/L instead of 25. Would your answer change? Why?

19.3 What are the respective predicted no-effect concentrations for Problems 19.1 and 19.2? If the LC_{50}'s in Problems 19.1 and 19.2 were the lowest LC_{50}'s in a set of acute toxicity studies comprised of *Daphnia*, fish, and algae, would your answer change?

19.4 Estimate the environmental impacts of treating 1 kg of TOC in a wastewater treatment plant based on the information given in Table 19.12.

19.5 What are the environmental impacts from disposing of materials in a landfill? How would you quantify them?

19.6 Redo Example 19.4 assuming a 30% energy recovery. Utilize the steam production submodule of Chapter 18.

19.7 Estimate the acidification potential, fossil fuel consumption, photochemical smog formation, and eutrophication potential for Example 19.4.

19.8 Estimate the acidification potential, fossil fuel consumption, photochemical smog formation, and eutrophication potential for the three cases of Example 19.5.

19.9 Redo Example 19.5 with 70 and 80% energy recovery.

19.10 Redo Example 19.5 with methanol.

19.11 A manufacturing plant has a mixed stream of 50% dichloromethane, 25% methanol, and 25% water. Compare the environmental impacts of incinerating the stream without energy recovery versus segregating the streams within the plant to recover the solvents as much as possible, as shown in Figure P19.11.

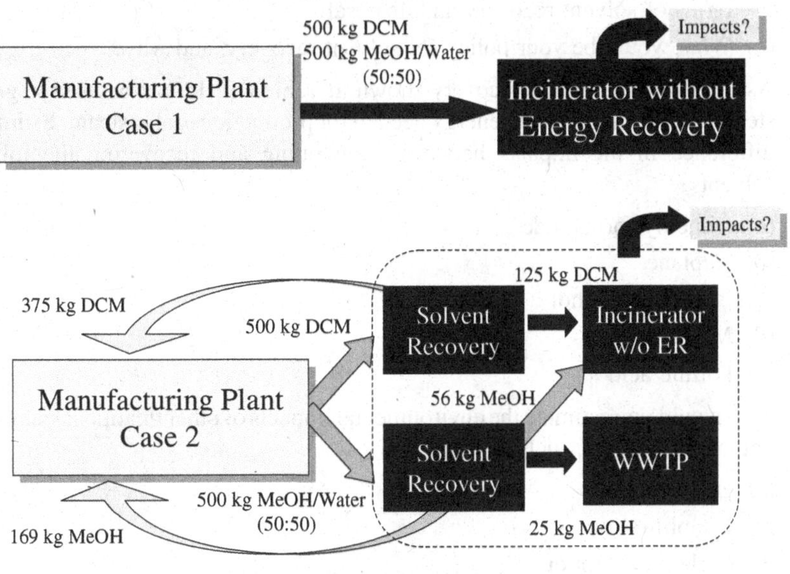

FIGURE P19.11

19.12 Capello et al. published the results of a life cycle cumulative energy demand study that includes production, credits for recovering the solvent through distillation, and credit for recovering energy through incineration for several solvents. The data for some solvents are shown in Table P19.12.

TABLE P19.12 Solvent Data

		Life Cycle Energy for:	
Solvent	Production (MJ/kg solvent)	Distillation (MJ/kg solvent)	Incineration (MJ/kg solvent)
Acetic acid	55.9	−34.9	−15.5
Acetone	74.6	−34.9	−33.9
Acetonitrile	88.5	−79.6	−29.7
1-Butanol	97.3	−74.6	−39.9
Cyclohexane	83.2	−63.4	−53.5
Cyclohexanone	124.7	−99.7	−40.4
Dimethylformamide	91.1	−67.6	−25.9
Dioxane	86.6	−63.8	−27.6
Ethanol	50.1	−31.2	−31.7
Formic aid	73.9	−50.1	−4.7
Heptane	61.5	−43.7	−54.5
Methanol	40.7	−21.7	−22.2
Isopropyl alcohol	65.6	−46.1	−36.5
n-Propyl alcohol	111.7	−87.3	−36.5
Tetrahydrofuran	270.8	−230.7	−37.5

(a) How would you show for which solvents incineration with energy recovery is the option with the fewest environmental impacts?

(b) You are the EHS manager of a plant that handles these types of solvents. Devise a graphical way to guide your process engineers in making the best decisions in terms of solvent recovery and disposal.

(c) What would be your policy for solvent recovery, and why?

19.13 Assume that the energy recovery shown in Table P19.12 can be used to generate steam, and that most of the energy needed for production is also steam. Estimate the difference in the impacts between incinerating and recovering the following solvents

(a) Dimethylformamide

(b) Heptane

(c) Isopropyl alcohol

(d) Methanol

(e) Formic acid

19.14 How would you estimate the environmental impacts of other treatment technologies not evaluated here, such as:

(a) Ozonation

(b) Nanofiltration

(c) Carbon absorption

(d) Scrubbing

REFERENCES

1. Neely, W. B., Ed. *Chemicals in the Environment: Distribution, Transport, Fate, Analysis.* Marcel Dekker, New York,1980.
2. Rand, G. M., Ed. *Fundamentals of Aquatic Toxicology: Effects, Environmental Fate, and Risk Assessment*, 2nd ed. Taylor & Francis, Washington, DC, 1995.
3. Swann, R. L., Eschenroeder, A., Eds. *Fate of Chemicals in the Environment*. American Chemical Society, Washington, DC, 1980.
4. Suffet, I. H., Ed. *Fate of Pollutants in the Air and Water Compartments*, Part I. Wiley, New York, 1977.
5. Verschueren, K., Ed. *Handbook of Environmental Data on Organic Chemicals*, 2nd ed. Van Nostrand Reinhold, New York, 1983.
6. U.S. Food and Drug Administration, Center for Drug Evaluation and Research. *Guidance for Industry for the Submission of an Environmental Assessment in New Drug Applications and Supplements*. U.S. FDA, Washington, DC, Nov. 1995.
7. SmithKline Beecham. *Environmental Assessment: Paroxetine*.
8. Cowan, C.E., Larson, R.J., Feijtel, T.C., Rapaport, R.A. An improved model for predicting the fate of consumer product chemicals in wastewater treatment plants. *Water Res.*, 1993, 27, 561–573.
9. U.S. Food and Drug Administration. *Environmental Assessment Technical Assistance Handbook*. U.S. FDA, Washington, DC, 1987.
10. Jiménez-González, C., Overcash, M. R., Curzons, A. Waste treatment modules: a partial life cycle inventory. *J. Chem. Technol. Biotechnol.*, 2001, 76, 707–716.

11. Muñoz, I. I., Peral, J., Ayllón, J. A., Malato, S., Martin, M. J., Perrot, J. Y., Vincent, M., Domènech, X. Life-cycle assessment of a coupled advanced oxidation-biological process for wastewater treatment: comparison with granular activated carbon adsorption. *Environ. Eng. Sci.*, 2007, 24(5), 638–651.
12. Capello, C., Hellweg, S., Badertscher, B., Betschart, H., Hungerbühler, K. Environmental assessment of waste-solvent treatment options: I. The ecosolvent tool. *J. Ind. Ecol.*, 2008, 11 (4):26–38.
13. Rozich, A. F. *Design and Operation of Activated Sludge Processes Using Respirometry.* Lewis Publishers, Boca Raton, FL, 1992.
14. Eckenfelder, W. *Industrial Water Pollution Control.* McGraw-Hill, New York, 1989.
15. Eckenfelder, W. *Activated Sludge Treatment of Industrial Wastewater.* Technomic, Lancaster, PA, 1995.
16. Seyler, C., Hofstetter, T. B., Hungerbühler, K. Life cycle inventory for thermal treatment of waste solvent from chemical industry: a multi-input allocation model. *J. Cleaner Prod.*, 2005, 13, 1211–1224.
17. Seyler, C., Hellweg, S., Monteil, M., Hungerbühler, K. Life cycle inventory for use of waste solvent as fuel substitute in the cement industry. *Int. J. Life Cycle Assess.*, 2005,10(2), 120–130.
18. Cheremisinoff, P. L., Young, R. A. *Air Pollution Control and Design Handbook*, Part I. Marcel Dekker, New York, 1977.
19. Cooper, C. D., Alley, F. C. *Air Pollution Control: A Design Approach.* Waveland Press, Prospect Heights, IL, 1994.
20. Katari, V. S., Vatavuk, W. M., Wehe, A. H. Incineration techniques for control of volatile organic compound emissions: I. Fundamentals and process design consideration. *J. Air Pollut. Control Assoc.*, 1987, 37(1), 91.
21. Seyler, C., Capello, C., Hellweg, S., Bruder, B., Bayne, D., Huwiler, A., Hungerbühler, K. Waste solvent management as an element of green chemistry: a comprehensive study on the Swiss chemical industry. *Ind. Eng. Chem. Res.*, 2006, 45(22), 7700–7709.
22. Ullmann, F. *Ullmann's Encyclopedia of Industrial Chemistry*, 5th ed. Wiley, New York, 1985.
23. Capello, C., Hellweg, S., Badertscher, B., Hungerbühler, K. Life-cycle inventory of waste solvent distillation: statistical analysis of empirical data. *Environ. Sci. Technol.*, 2005, 39, 5885–5892.
24. Capello, C., Fischer, U., Hungerbühler, K. What is a green solvent? A comprehensive framework for the environmental assessment of solvents. *Green Chem.*, 2007, 9, 927–934.
25. Jiménez-González, C. Life Cycle Assessment in Pharmaceutical Applications. Ph.D. disssertation, North Carolina State University, Raleigh, NC, 2000.
26. Capello, C., Fischer, U., Hungerbühler, K. Environmental assessment of waste-solvent treatment options: II. General rules of thumb and specific recommendations. *J. Ind. Ecol.*, 2008, 12(1), 111–127.
27. Stoye, D. Solvents. In *Ullmann's Encyclopedia of Industrial Chemistry.* Wiley-VCH, New York, 2000.
28. Yaws, C. L., Ed. *Chemical Properties Handbook.* McGraw-Hill, New York, 1999.

20

TOTAL COST ASSESSMENT

What This Chapter Is About This chapter is a brief introduction to total cost assessment (TCA) and about how TCA can be used in conjunction with green chemistry and green engineering as a decision-making tool. Whether you are in academia, government, or business, an understanding of cost drivers from a holistic point of view is essential when considering how to move toward more sustainable practices. Because total cost assessment is a decision-making methodology that attempts to assign monetary value to environmental, safety, and human health impacts, it has traditionally been resisted within businesses and by governments. However, if one does assign economic value to nontraditional services, businesses and governments benefit from it (e.g., environmentally related services such as the nitrogen in air, or the fact that we use river water for cooling in electricity generation). It is more likely that people will pay attention and seek to preserve these services that are currently provided by the environment at little or no cost.

Learning Objectives At the end of this chapter, the student will be able to:

- Understand what total cost assessment is.
- Understand how it can be used in an overall assessment of what is green.

20.1 TOTAL COST ASSESSMENT BACKGROUND

In the late 1980s through the middle to late-1990s, there was a considerable degree of interest in and work on trying to increase awareness of a range of environmental issues. Following from many efforts and considerable success in focusing attention on pollution prevention

Green Chemistry and Engineering: A Practical Design Approach, By Concepción Jiménez-González and David J. C. Constable

TABLE 20.1 Various Names for Total Cost Assessment

Accounting for externalities	Full-cost environmental accounting
Comprehensive accounting	Full-cost accounting
Environmental cost accounting	Green accounting
Environmental accounting	Real accounting
Total cost accounting	Strategic-enviroeconomics
Life cycle costing	

through source reduction instead of waste treatment at the end of the pipe, one effort that saw considerable activity was in the area of costing environmental activities or issues in the hopes of making the environment more tangible to business decision makers[1–10] Table 20.1 contains a brief list of the variety of names that have been used to describe total cost assessment.

Notably, during this period, the U.S. Environmental Protection Agency worked extensively with the Tellus Institute and developed a substantial body of documentation and ultimately a computer-based tool known as P2 Finance that could be used to perform rudimentary environmental cost accounting.[11] The desire was to expose costs that were associated with managing and treating wastes, since many of these costs were buried in overhead costs or spread across multiple product lines. As long as these costs are generally hidden and not allocated specifically to any given product, companies will not necessarily make the most informed decisions about which products to produce or which manufacturing processes to use to make a given product less impactful or truly less costly.

In the latter half of the 1990s a group of companies operating under the now-defunct Center for Waste Reduction Technologies, an organization that operated under the auspices of the American Institute of Chemical Engineers, began to meet to discuss how they might develop a greater understanding of total cost assessment as a means of making business decisions that would move companies toward more sustainable outcomes and business practices. Over the course of about four years, and with the considerable support of the U.S. Department of Energy, a decision-making methodology was developed for total cost assessment.[12] Because one author of this book was intimately involved in leading the CWRT effort, much of what follows in this chapter is based on the experience gained in developing the methodology and a reasonably sophisticated tool to perform routine total cost assessment. The CWRT effort is not the only effort in terms of TCA, but many people have worked to develop TCA up to the present.[13–23] In addition, there have been many who have adapted environmental cost accounting more broadly to general environmental management[24–26] and have applied it to specific case studies.[27,28]

20.2 IMPORTANCE OF TOTAL COST ASSESSMENT

As stated above, total cost assessment and related terms have been studied by many and variously defined, so there are minor differences that have led to mild confusion about exactly what TCA is used for. In the CWRT project, TCA was defined as the identification, compilation, analysis, and use of internal and external cost information on environmental, safety, and environmentally related human health issues. Table 20.2 contains a summary of the cost types that were considered to be a part of a total cost assessment.

TABLE 20.2 Total Cost Assessment Cost Types

Type I costs	Direct costs: traditional capital investment and operational costs, including: Equipment Labor Raw materials Waste treatment
Type II costs	Indirect and hidden costs: Reporting Monitoring Regulatory (e.g., operating permits and fees)
Type III costs	Contingent costs: Liabilities (e.g., cleanup, Workers' Compensation) Lawsuits (e.g., personal injury claims, toxic torts) Damage to resources Accidents
Type IV costs	Internal intangible costs: Company or brand image Consumer loyalty Worker morale Worker relations Community relations
Type V costs	External intangible costs: Increase in housing costs Degradation of resources Cost of increasing emissions (e.g., CO_2, NO_x, SO_x)

If anyone in business or government spends just a few moments looking at Table 20.2 and thinking about it, it will become obvious that even today most of these cost types are not considered routinely by a majority of businesses or governments in routine decisions. While the Dow Chemical company succeeded for a time in embedding an adaptation of the CWRT TCA methodology known as total business cost analysis, even this included only the first three cost types. Although there is generally recognition by businesses that type IV costs are important, there is in general reluctance and a lack of know-how in how to assign type IV costs to a specific or proposed product. The reluctance in assigning these types of costs is due only in part to a lack of know-how and is clearly linked to a lack of trust that the costs are in fact accurate.

As difficult as type IV costs are for most to come to terms with, type V costs are even more difficult to integrate into routine business decisions. The CWRT group did, in fact, develop the means to cost intangibles such as the cost of an odor, or the cost of losses on property value, but these cost types were not explored further or used by any company as, unfortunately, they were seen as having a very real potential of becoming a type III cost. The recent rise in litigation associated with natural resource damages is just one example where these concerns have been in the minds of some, quite justified.

Example 20.1 Describe some situations where total cost assessment might be applied most effectively and explain why.

Solution See Table 20.3.

TABLE 20.3 Effective Use of TCA

Decision	Rationale
Estimating baseline costs	Would provide an objective, comprehensive comparison between two different syntheses beyond the simple cost of materials.
Internal/external benchmarking	Benchmarking allows us to see how we are doing on a comparative basis. By performing a total cost assessment, one is less likely to make an "apples to oranges" comparison between different parts of a company or between different companies.
Process development	As has been seen in previous chapters, process alternative selections can be quite difficult. It is easy to miss key issues if one focuses solely on the technology while excluding other process parameters.
Product mix	Every company is faced with having to choose which products to keep in its portfolio. A TCA across the entire portfolio is likely to reveal strategic risks that may not be evident from a simple consideration of type I and II costs.
Waste management decisions	One is faced with many issues in waste management. Often there is media switching (i.e., we shift the impact from air to water or to land) and in switching media, there may be unintended consequences (e.g., choosing to incinerate waste rather than landfill may increase the production of dioxins and require an excessive amount of abatement equipment, increased energy use, etc.)

Additional Point to Ponder Imagine that you were trying to make a decision about which synthetic chemical process you should use. In one process, a significant amount of energy is used, whereas in the other, a platinum group metal catalyst is used in combination with a natural product derived from a rare but harvestable sea plant. Could TCA be used to choose the most sustainable route?

20.3 RELATIONSHIP BETWEEN LIFE CYCLE INVENTORY/ASSESSMENT AND TOTAL COST ASSESSMENT

In Chapter 16 we learned about life cycle inventory/assessment and how it could be applied to answering questions about which chemistry or technology option might be preferred. The astute student should readily see that there is considerable opportunity to cost the elements of a life cycle inventory as part of the assessment phase of the study. For example, while performing R&D for a new product, a life cycle inventory (LCI) could be used to influence the choice of process options that minimize pollutant discharges cross the life cycle, thereby avoiding the additional cost of control technology and the associated cumulative emissions and/or impacts.

The LCI output may also be used to steer the selections of raw materials toward materials that produce fewer emissions or smaller volumes of waste from the finished product. An additional very important benefit derived from using LCI inputs in an economic assessment of a project, product, or process is the inclusion of the cost types that are typically excluded. By translating the LCI outputs into economic assessments of the contingent risk and liability, and the intangible and external impacts, a more comprehensive view of current and future impacts and costs is possible. From a sustainability perspective, that is undoubtedly a good thing.

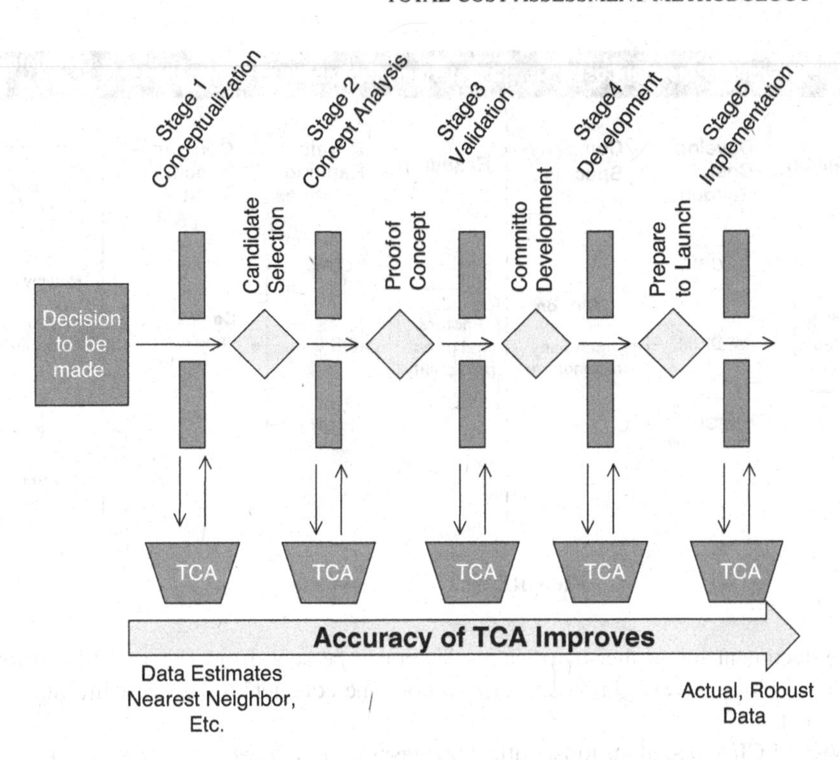

FIGURE 20.1 Timing of a total cost assessment.

20.4 TIMING OF A TOTAL COST ASSESSMENT

As has been stated many times throughout this book, most decisions about how to green a given chemistry or technology are best made as early as possible. Figure 20.1 shows a typical stage gate approach to product development. The point of the figure is to show that TCA can and should be applied at each stage of development. In the early phases of product development, a typical approach to cost estimation is shown in Figure 20.2. As can be seen, one will have to base decisions on estimated data, QSAR or nearest-neighbor relationships, or best judgment to fill in the unknowns. As the product moves forward into development and production, the materials of construction and the process to make the product are clearer, and the estimates become actual data. The TCA becomes more robust and the discipline in conducting the analysis is likely to help people to steer clear of potential showstoppers to getting a product or service to market. If one does identify risks, a very likely outcome for any TCA, the TCA results will help to identify and quantify the potential risk. If these risks are unavoidable, early identification will allow time for risk mitigation and management strategies to be developed.

20.5 TOTAL COST ASSESSMENT METHODOLOGY

Figure 20.3 is adapted from the TCA methodology developed by the AICHE CWRT focus group. It is not in the scope of this book to go into the details of how the TCA methodology might be applied to a particular decision; rather, the student is referred to the original

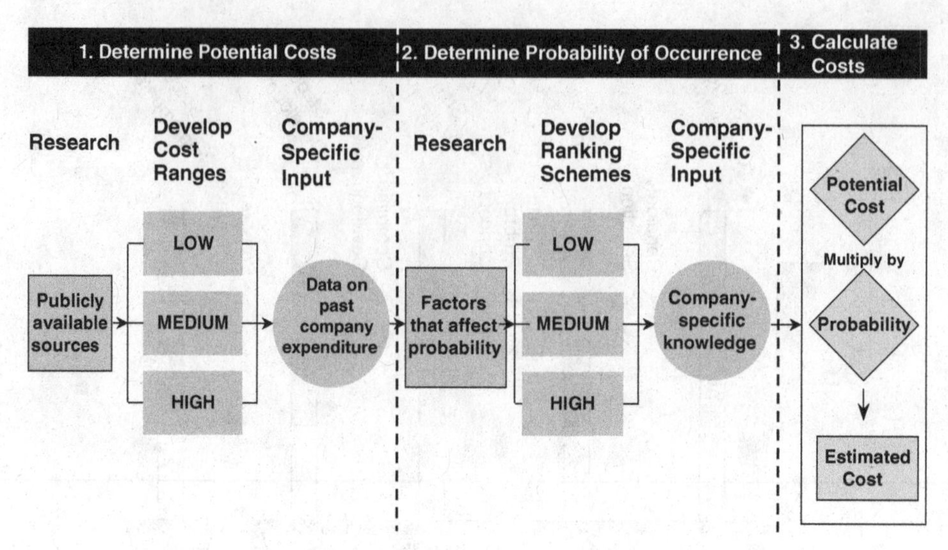

FIGURE 20.2 Cost estimation.

source document for further clarification.[10] As can be seen from Figure 20.1, however, the methodology is very similar in form to how one conducts a life cycle inventory and assessment.

As with LCI/A, one of the most critical first tasks is *project definition and scoping*. Many times the tendency is to rush through this first step too hastily, only to be frustrated as one

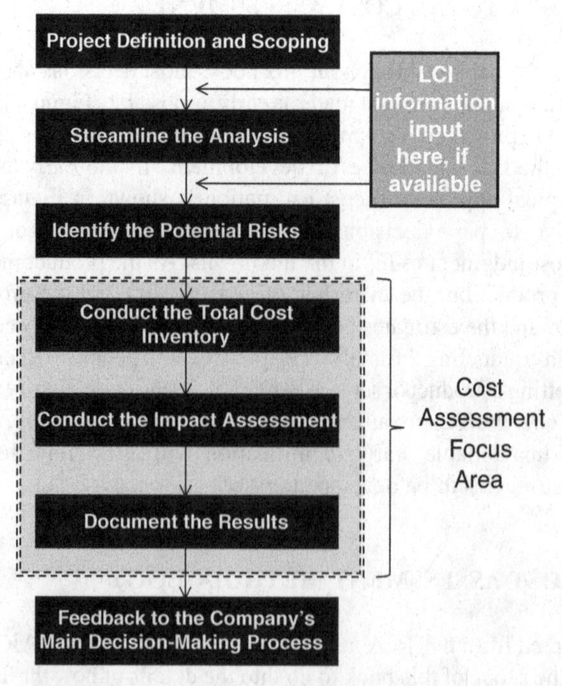

FIGURE 20.3 Total cost assessment methodology.

proceeds through the other portions of the assessment and arrives at a less than satisfactory conclusion. This step therefore requires one to be very clear about the project definition and overall purpose of the TCA analysis, including certain goals and constraints on the analysis.

During the second step, *streamline the analysis*, one refines the first step, including the potential for adding the results of a life cycle inventory or a brainstorming session or two. The main point of this step is to associate sustainability metrics and impact categories with the objectives and other elements of the decision that you are trying to make.

In the third step, *identify potential risks*, one identifies and then evaluates the relative importance of the impact categories and the associated costs for each alternative or project flowing out of the project definition and scoping. Each alternative or option may have a number of unique risk/cost scenarios based on the cost drivers (e.g., compliance obligations, remediation costs) associated with that alternative or cost driver. It is important to note that this third step may entail several iterations to develop an accurate understanding of the most relevant risk/cost scenarios that are passed on to the next step to be costed accurately.

Once potential risks are identified and evaluated, one proceeds to the fourth step and *conducts a financial inventory*. The first and easiest task of this step is to collect the type I and II costs from existing internal cost accounting systems. Once these are known, the next task is to evaluate the potential type III, IV, and V costs. It should be noted that this group of cost types is likely to present the greatest obstacles to performing a total cost assessment. One must first struggle to understand which costs drivers of each type are relevant, and once decided, these costs will undoubtedly have significant associated uncertainties, due to the difficulty of establishing their magnitude and probability of occurrence. A considerable amount of effort was taken to provide reliable cost estimates for these drivers as part of the CWRT effort, and these should be consulted and used as appropriate.

For the fifth step, *conduct impact assessment*, the data collected are analyzed and reviewed. Since the assigned costs that compose the TCA have been collected from very different sources of varying reliability and uncertainty, it is critical that the largest cost contributors in each category are determined and an assessment is made of how that information may best be incorporated into the overall decision-making process. This is something of an iterative process and is, in a sense, a sensitivity analysis for how cost drivers may be affecting the overall results. Although one may have great confidence in type I and II costs, some type III, IV, or V costs, once included, may have a dramatic effect and potentially change the analysis or decision. This will obviously depend on the project or the range of alternatives that are being evaluated and the degree of confidence that one has in the costs that have been collected in these categories. As with any assessment where there is a higher degree of uncertainty, it is useful to include ranges of type III, IV, and V costs for different risk profiles. These different risk profiles might generally represent the high, medium, and low risks for different future scenarios and thereby give decision makers greater comfort with the final cost assessment.

Once the assessment has been completed, it is imperative that you *document results*. As with LCI/A, there are a great many assumptions that are made, and these, together with the final results of the TCA, should be documented carefully. Any use of life cycle information and/or other company-specific data used in the assessment are likely to be critical components of any defense of the assessment. This also provides an opportunity for identifying best practices and permits accurate data sharing between assessments carried out at different times. It also permits one more easily to update the assessment with new cost elements or revised cost data if required.

Ultimately, one must provide *feedback to the company's main decision loop*. It should be recognized that conducting a TCA and obtaining a result from it is not the decision itself, but only one element of the overall decision-making process, which includes many types of information. A TCA is intended to enable a richer discussion of all the potential issues and should, if applied diligently, improve the sustainability profile of any organization that makes use of it regularly.

Example 20.2 A company is interested in making a decision about shipping a phosphorus derivative either by rail or in cans by truck. They have done a life cycle assessment of both means of transportation. What is the cost differential for each mode? What course of action would you recommend?

Solution Because the manufacture of the phosphorus derivative is the same in both instances, a fully burdened cost assessment for that portion of the product is not necessary. Consequently, you are free to focus on the incremental difference between bulk shipment via rail and shipment in cans via truck. Figures 20.4, 20.5, and 20.6 contain graphical representations of the inputs and outputs, including the additional inputs and outputs via truck. This information has also been put into tables for ease of comparison, beginning with a financial inventory in Table 20.4.

At this point of looking at the inventory, it is difficult to conclude that the differences between the two options are particularly substantial, with the exception of the additional

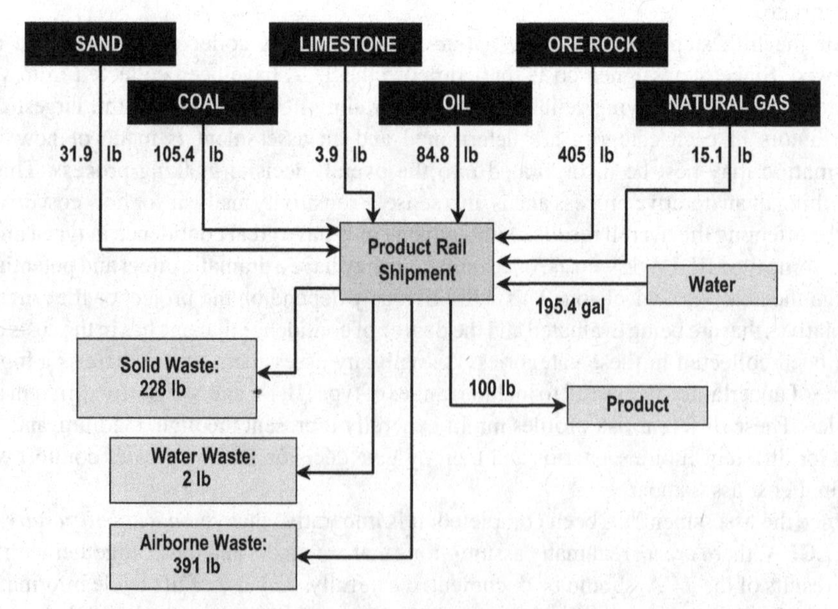

FIGURE 20.4 Life cycle evaluation for rail shipment of a phosphorus derivative.

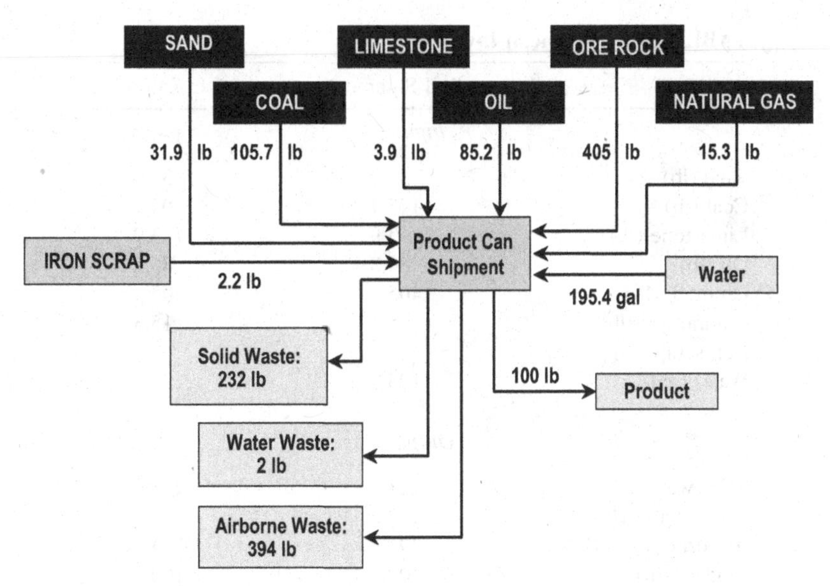

FIGURE 20.5 Life cycle evaluation for can shipment of a phosphorus derivative.

iron scrap required for the cans. When additional information for cans is collected, you find the total differences given in Table 20.5. By scouring the literature and government sources, you find estimates for externalities costs (Table 20.6), and finally, you are able to calculate the total incremental cost for cans (Table 20.7)

These data show that there is an additional societal cost to shipping in cans via trucks. The cost for hazardous waste is clearly the largest cost and represents an opportunity for risk mitigation or for making the choice to ship via rail. Clearly, the cost to manage and dispose of hazardous waste is going to increase. In addition, transportation costs via truck are likely

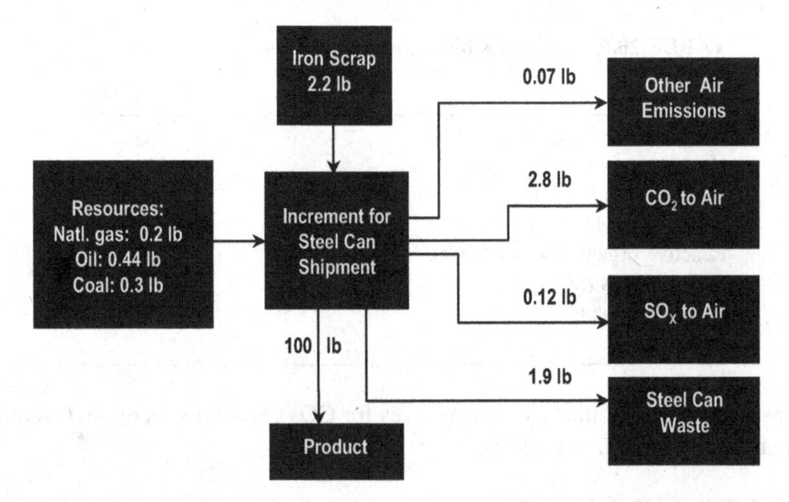

FIGURE 20.6 Additional inputs/outputs for can packaging and shipping by truck.

TABLE 20.4 Financial Inventory[a]

	Rail Shipment	Can Shipment
	Input	
Sand (lb)	31.9	31.9
Coal (lb)	105.4	105.7
Limestone (lb)	3.9	3.9
Oil (lb)	84.8	85.2
Ore rock (lb)	405	405
Natural gas (lb)	15.1	15.3
Iron scrap (lb)		2.2
Water (gal)	195.4	195.4
	Output	
Solid waste (lb)	228	232
Water waste (lb)	2	2
Airborne waste (lb)	391	394
Product (lb)	100	100

[a] Basis: 100 lb of product.

TABLE 20.5 Additional Inputs and Outputs from Steel Can Use

Resources	Wastes
138,600 lb coal	397,100 lb steel can waste
29,300 lb natural gas	402,000 lb CO_2 emissions
43,100 lb oil	15,000 lb SO_x emissions
294,500 lb iron scrap	63,100 lb solid waste

TABLE 20.6 Estimates for Externalities Costs

Pollutant	Cost ($/Ton)
Nonhazardous solid waste	$35
Hazardous waste	$250
Carbon dioxide (CO_2)	$86
Nitrogen oxides (NO_x)	$2,500
Reactive organic gases (VOCs)	$1,750
Sulfur oxides (SO_x)	$135
Particulate matter	$530
Societal costs	?

to increase as fuel prices increase and as taxes for CO_2 emissions increase. Overall, your recommendation would be to ship via rail.

Additional Point to Ponder Who is paying the additional costs when the decision is made to ship by truck?

TABLE 20.7 Incremental Cost for Cans

Cost Category	Cost/Ton	Cost
Nonhazardous solid waste	$35	$1,100
Hazardous waste	$250	$500,000
CO_2	$86	$17,000
NO_x	$2,500	$10,000
SO_2	$135	$1,100
Societal cost		?
Total		$530,000 +

20.6 TOTAL COST ASSESSMENT IN A GREEN CHEMISTRY CONTEXT

Let's now touch on how a life cycle and systems-based perspective very readily reveals what costs are not being borne by the users of a given chemical. If you once again look at Table 20.1 you will readily see that type III, IV, and V costs are not really embedded in the standard cost of a chemical nor generally considered in the production of a chemical. Although it is true that some costing mechanisms are being implemented for emissions, these schemes are in their early days and are not broadly accepted or standardized across economies, regions, states, or continents. As these new costs start to be generally applied, they will become a type II cost.

Another thing worth considering is that for many high-value chemicals or useful materials in commercial use, there are increasingly limited supplies. Supplying these chemicals over time will be associated with increasingly significant costs both monetarily and from the perspective of major disruptions to the environment as sources of rare materials become even scarcer. A good example of this would be in the supply of some transition group metals and platinum group metals that are of great interest to chemists for their use in novel catalyst systems. In addition to the environmental life cycle impacts that occur during mining, extraction, refining, and use of these materials in different catalyst systems, our use of these materials generally results in some portions of these metals being widely dispersed into the environment since complete recovery from process streams is not always possible. In some cases, catalysts are homogeneous and difficult to extract and recover from the reaction mixtures. In those cases where catalyst systems are heterogenized, there can be a 5 to 10% loss of the catalyst with the filter or filter aid. It would be very difficult and probably very costly indeed to collect the homogeneous catalyst from spent mother liquors and equally difficult to collect all the heterogeneous catalyst from a reaction mixture. It is also difficult and expensive to design, synthesize, and employ alternative catalyst systems that perform the same catalytic function as an existing catalyst and achieve identical product quality and cost.

So if we return to what we learned in Chapters 4 and 5 to consider atom economy once again, we may readily see that low atom economical reactions and mass inefficient processes will affect not just the materials cost of synthesizing a new chemical but the total cost as well. As is probably known by the reader, but worth reiterating, this is because:

1. Not all portions of the reactant molecules are incorporated into the molecule we are attempting to synthesize [i.e., our starting materials (and the energy we may consume) are not typically used very efficiently].

2. Our synthetic strategies will affect the length and complexity of the route.
   a. Portions of the molecule may be in the wrong oxidation state.
   b. Protection/deprotection strategies may be required.
   c. We may have multiple chiral centers to set, and once set, we may require chiral resolutions if we are not able to set these assymetrically.
   d. We may have some particularly difficult functional groups to incorporate.
   e. There may be some particularly unstable bonds to deal with.
3. Multiple purification and separations steps may be required to remove by-products, reactants, reagents, solvents, and so on.
4. There are environmental, safety, and health costs associated with the management of materials and the treatment of waste products.

Additional Point to Ponder Imagine that you were trying to make a decision about which synthetic chemical process you should use. Both processes are costed using traditional cost models but then you open the boundary a bit more and include type III, IV, and V costs. Which type of cost do you think will become the biggest differentiator between the two processes?

20.6.1 Relationship Between Traditional Materials Cost and Atom Economy

To illustrate the important relationship between traditional costing methodologies for raw materials and atom economy, let's look at seven different economic models to evaluate costs for materials used in the synthesis of four different drugs. The cost models we use are as follows:

1. *Minimum cost for minimum process stoichiometry + standard yield, reactant stoichiometry, and solvent.* This is the cost when process chemicals are not used in stoichiometric excess, i.e., no more than 1 mol is used.
2. *Minimum cost at 100% atom economy + standard yield, solvent, and process stoichiometry.* Reactant costs may be used to assign a cost to the proportion of each material that is incorporated into the product. From this it is possible to calculate the cost if the atom economy is 100% and standard amounts of solvent and process chemicals are used.
3. *Minimum cost at 100% yield + standard solvent and process stoichiometry.* This is the cost for using standard quantities of reactants, process chemicals, and solvent, but the yield is 100%.
4. *Minimum cost at 100% solvent recovery and standard yield and process stoichiometry.* This is the cost if 100% of all solvents are recovered and reused (assume zero recovery cost).
5. *Minimum cost at 100% atom economy, process stoichiometry, and solvent recovery.* Reactant costs may be used to assign a cost to the proportion of each material that is incorporated into the product, and from this it is possible to calculate the cost if the atom economy is 100%. No stoichiometric excess of process chemicals is used and solvent recovery is assumed.
6. *Minimum cost at 100% yield, solvent recovery, and standard process stoichiometry.* A theoretical minimum cost may be derived assuming a 100% overall yield, 100% solvent recovery, and standard process stoichiometric excess.

TABLE 20.8 Comparison of Cost Scenarios for Drug Synthesis

	% Std. Cost for:			
Cost Model	Drug 1	Drug 2	Drug 3	Drug 4
1, Minimum cost for minimum process stoichiometry + standard yield, reactant stoichiometry, and solvent	86.25	98.61	92.4	97.0
2. Minimum cost at 100% atom economy + standard yield, solvent, and process stoichiometry	87.10	39.58	83.7	69.3
3, Minimum cost at 100% yield + standard solvent and process stoichiometry	70.50	32.15	64.2	56.8
4. Minimum cost at 100% solvent recovery and standard yield and process stoichiometry	63.14	83.79	56.1	54.6
5. Minimum cost at 100% atom economy, process stoichiometry, and solvent recovery	36.49	21.97	39.6	20.9
6. Minimum cost at 100% yield, solvent recovery, and standard process stoichiometry	33.64	15.94	20.3	11.4
7. Minimum cost at 100% yield, solvent recovery, and reactant and process stoichiometry	19.89	14.54	12.1	8.3

7. *Minimum cost at 100% yield, solvent recovery, and reactant and process stoichiometry.* A theoretical minimum cost may be derived assuming no stoichiometric excess, 100% solvent recovery, and a 100% overall yield.

Table 20.8 contains a summary of the costs on a percentage basis, and Figure 20.7 shows these results in graphical form.

As you can see from Figure 20.7, pursuing atom economy from a perspective of current costing considerations may be less desirable than some might think. This costing analysis also suggests that we may best be served by pursuing higher yield reactions, a reduction in

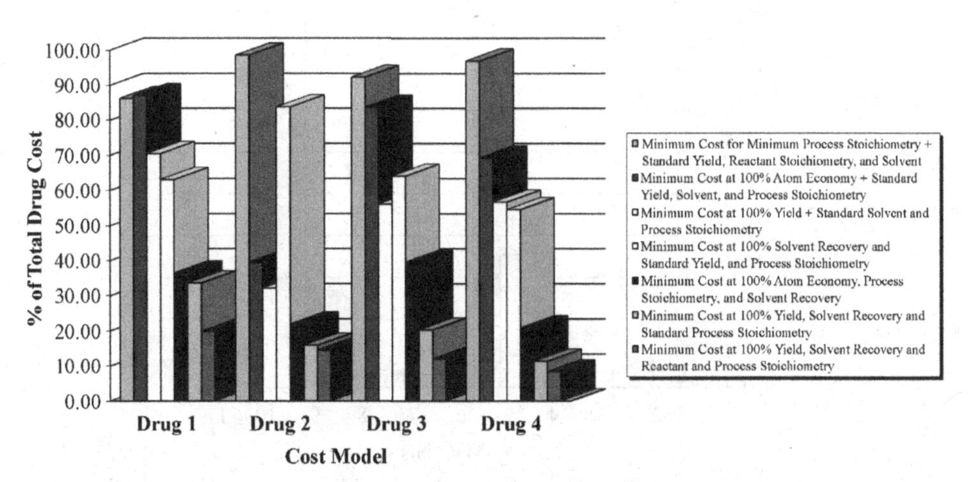

FIGURE 20.7 Comparison of cost models for four drugs. (From Constable et. al, ref. 29. Reproduced by permission of The Royal Society of Chemistry. Copyright © 2002, The Royal Society of Chemistry.)

TABLE 20.9 Comparison of Costs for Drug Substance 3

Reactants	Molar Equivalents Used	Percentage of Molecule in Final Drug[a]	Percentage Contribution to Overall Cost of Drug 3 (%)	Percentage of Total Cost for Nonincorporated Reactants (%)
Intermediate 1	2	43	16.4	12
Reducing agent	4.6	5	38.9	49
Resolving agent	2.2	0	20.4	26
Intermediate 2	2	27	5.8	6
Intermediate 3	1	0	0.8	1
Intermediate 4	1	0	0.8	1
Material 1	3	0	1.5	2
Material 2	1	0	0.1	
Material 3	1	100	13.3	
Material 4	6	0	0.6	1
Material 5	1.2	0	0.6	1
Material 6	1	100	0	
Material 7	10	14.5	0.3	
Material 8	2	0	0.4	
Solvents	—	—	28	
All other materials	—	—	0.4	

[a] This is the wasted cost for each material, due to inefficient incorporation into the product.

the stoichiometric excess of the reactants we use, and elimination of, or at the very least, complete solvent recycle and reuse. Unless and until we can truly use all of our reactants and transform them completely to the products we wish, we will need to pay greater attention to all the materials we use in our syntheses.

A more detailed materials cost analysis for the four drugs considered above is contained in Table 20.9 for drug 3 and in Figures 20.8 and 20.9 for drugs 1 and 2, respectively. This

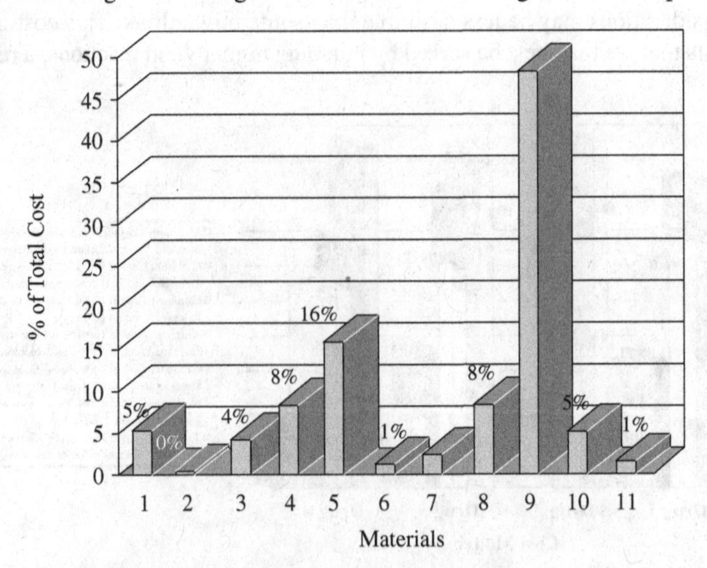

FIGURE 20.8 Materials costs as a percentage of the standard cost in making drug 1. (From Constable et al., ref. 29. Reproduced by permission of The Royal Society of Chemistry. Copyright © 2002, The Royal Society of Chemistry.)

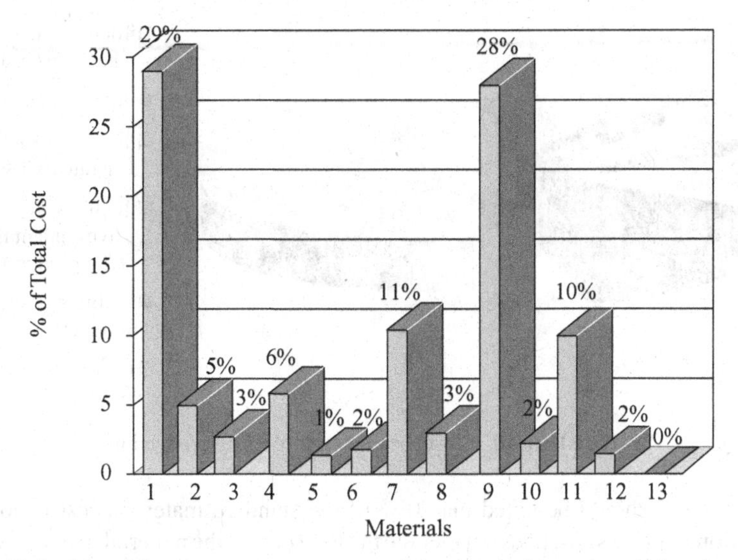

FIGURE 20.9 Materials costs as a percentage of the standard cost in making drug 2. (From Constable et al., ref. 29. Reproduced by permission of The Royal Society of Chemistry. Copyright © 2002, The Royal Society of Chemistry.)

analysis reveals that more than 75% of the material cost for those portions of reactants that do not remain in the final product (column 5, Table 20.9) is usually attributable to a maximum of four materials. It should hopefully come as no surprise at this point that the data in Table 20.9 also provide confirmation that replacing a chiral resolution with a chiral synthesis is a more beneficial economic and atom economical strategy.

As shown in Table 20.10, an interesting aspect of this comparison is that when the total cost of solvents is compared to the cost of poor atom economy, the cost of the solvent is a greater proportion of the materials costs in three of the four syntheses studied. An additional learning is that yield and stoichiometry are the biggest cost drivers and exert significantly more influence on cost than does a poor atom economy.

Another interesting outcome of this comparison came from the results for drug 4, where it was found that neither atom economy nor solvent were significant cost drivers and did not represent opportunities for cost reductions. In the case of this drug synthesis a catalyst was used and it was comparatively costly relative to the other materials used in the synthesis. It was also found in this case that approximately 10% of the catalyst was lost to effluent as part of the process, despite its being a heterogeneous catalyst. A 10% loss of this catalyst to effluent represented 16% of the standard materials costs paid by the company for this drug

TABLE 20.10 Comparison of Solvent and Poor Atom Economy Costs for Drug Substances

Drug	Solvent Cost as % of Total	Cost for Nonincorporated Reactants as % of Total
1	45	32
2	36	21
3	22	61
4	14	10

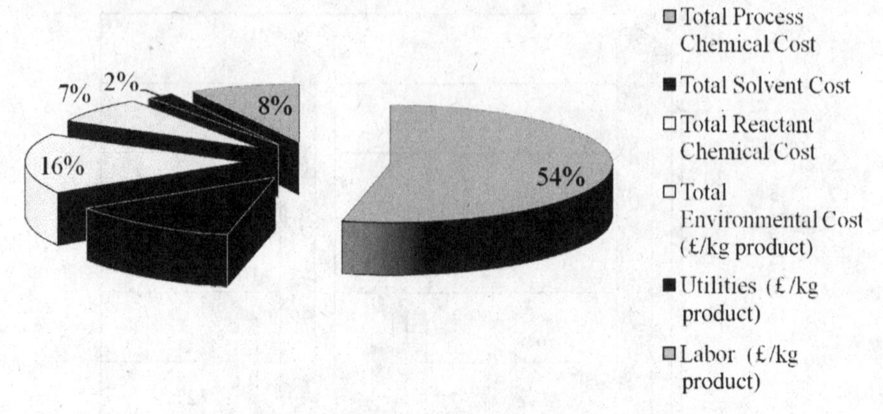

FIGURE 20.10 Costs percentages for drug production.

(Figure 20.10). It should be noted that 16% of the standard materials cost is not a small number! Moreover, this cost does not include the total cost of the material; the life cycle costs were not included (i.e., the cost of raw materials extraction, catalyst production, use, recovery, regeneration, etc.), and the loss of the catalyst to effluent with all the associated issues of metals in effluent were not considered. Although this is not an argument against using catalysts, it is certainly true that the type of catalyst, its potential for reuse, and its recoverability are important features of good process design and environmental and economic performance.

It is recognized that the evaluation above considers only a few industrial processes that represent the current state of affairs for drug manufacture and does not consider costs beyond the simple material costs of drug synthesis. Although it has been shown that the EHS costs in a total cost assessment for many industries can be quite significant, our studies have shown that the EHS costs for high-value-added materials such as pharmaceutical intermediates or pharmaceutical substances are generally less than traditional material costs unless total life cycle costs are included. Until standardized, accepted economic models for life cycle costs are built and agreed upon, it will remain difficult to assess these costs and make acceptable business decisions based on these costs. In addition, unless society forces markets to focus greater attention on these types of costs, they will continue to ignore these costs.

This analysis also ignores the potential benefits from alternative more atom economical routes, where it may be possible to have only two reactants producing a single easily isolated product in a completely recyclable reaction medium at room temperature and pressure. A second alternative would be a synthesis without solvent, but this would undoubtedly increase the energy requirements. Thus, it must be understood that the point of striving for more atom economic reactions in the future is the hope that they use fewer resources (materials and energy) and have higher overall process efficiencies.

PROBLEMS

20.1 There is an old saying that "Hindsight is 20 : 20." Select a compound from the following list (DDT, Alar, PFOA, CFCs) and:

(a) Describe how total cost assessment may have prevented your chosen compounds from becoming items of commerce in the first place.

(b) Which cost type would have been likely to drive the greatest cost?
(c) Which cost type would have been the most difficult to determine?

20.2 Return to Table 16.4.

(a) For each material in the table, which cost type would be represented?
(b) Which cost type would contribute most to the total cost?
(c) Which area—raw material, energy, air emissions, or water emissions—would have the greatest costs not usually accounted for by most industries?

20.3 Problem 5.7 shows the Craig synthesis of nicotine. Find a second synthesis of nicotine and perform a TCA on these to determine which route would be preferred.

20.4 Compare a traditional chiral synthesis employing a resolving agent with an asymmetric synthesis using TCA. Which synthesis is better from a total cost perspective, and why? What are the largest total cost drivers in each synthesis?

20.5 From the situations in Table P20.5, describe how total cost assessment might be applied most effectively and give a reason why, as was done in Example 20.1.

TABLE P20.5

Decision	Rationale
Pollution prevention alternatives	
Materials/supplier selection	
Facility location/layout	
Outbound logistics	
Market-based environmental options	
Public relations/lobbying	
Training	

20.6 Table P20.6 contains an assessment of costs for three different processes for several high-volume production materials used in plastics manufacture. Assume sales of \$1 billion/yr for each chemical.

(a) What general conclusions can you draw regarding each of these processes?
(b) Describe the key vulnerabilities for each process based on the data provided.

TABLE P20.6 Cost Assessment

Metric	Acrylonitrile from Propylene	Adipic Acid from Cyclohexane	Phenol from Cumene
Material intensity	4.35 lb/\$	0.766 lb/\$	1.90 lb/\$
Energy consumption	32.9 kbtu/\$	21.9 kbtu/\$	27.3 kbtu/\$
Water consumption	1.43 gal/\$	114 gal/\$	0.26 gal/\$
Toxic dispersion	0.24 lb/\$	0.003 lb/\$	0.008 lb/\$
Pollutant dispersion	6.37 lb/\$	1.24 lb/\$	0.94 lb/\$

20.7 A drug company is about to produce a drug and is confronted with site costs as shown in Figure 20.10.

(a) From a traditional cost perspective, where should the company attempt to put its resources to reduce costs, and why?
(b) It turns out that the highest-cost material is a catalyst: How might TCA be used to change the companies' practices to address both the use of this particular catalyst and the subsequent loss of some of this catalyst to effluent?

20.8 The cost of a pharmaceutical compound is computed at various stages of development and in the first year of production as shown in Table P20.8.

TABLE P20.8 Cost at a Variety of Stages

	R&D Cost Estimate (1 £/kg product)	R&D Cost Estimate (2 £/kg product)	Site Cost (£/kg product)	Cost of Chemicals in Site Waste Streams (£/kg product)	Site Chemicals in Waste Streams [Cost (£) at 12,000 kg/yr peak production]	Site nonchemical [Cost (£) at 12,000 kg/yr peak production]
Total process chemical cost	0.1	8.7	114.3	105.4	1,265,262.9	
Total solvent cost	30.5	13.1	25.2	25.0	300,081.9	
Total reactant chemical cost	381.8	21.4	30.4	32.5	390,270.9	
Total, site environmental cost (£/kg product)	—	—	14.2			170,861.2
Utilities (£/kg product)	—	—	4.0			474,04.8
Labor (£/kg product)	—	—	15.9			190,491.6
Packaging materials (£/kg product)	—	—	0.6			7,727.9
Filters (£/kg product)	—	—	2.3			27,674.1
Other (£/kg product)	—	—	2.7			32,169.2
Grand total	—	—	203.9	163.0	1,955,615.7	408,757.6

(a) What general conclusions about cost estimation can you derive from this table?
(b) What are the biggest cost drivers once a product is in production?
(c) How might green chemistry and engineering be used to affect the largest cost drivers?
(d) If you were to look at this from a total cost assessment perspective, which items would be the largest cost drivers, and why?

20.9 The costs for different materials used in producing a compound are contained in Table P20.9.

(a) If the process was not changed, which materials would represent the greatest challenge to recovery?
(b) As is typically the case, there are large quantities of low-value materials in the process, such as HCl. What opportunities might there be to address this as a waste? How might TCA help to decide what to do?

TABLE P20.9 Material Costs

Material	Standard Cost/kg of Product (£)	Total Annual Recoverable Material (Potential) (kg)	Potential Recoverable Material Cost (annual) (£)	Average Waste Stream Concentration (w/w)	Stream Composition
A	14.5	219,000	3,175,500	0.25	Aqueous
B	4.36	328,500	1,432,260	0.18	Spent broth
C	33.2	30,849	1,024,189	1.88	IPA M/L
D	13.9	39,943	555,210	1.49	IPE M/L
E	16.8	29,838	501,277	2.61	MeOH M/L
HCl	0.32	2,387,502	328,320	14.17	Aqueous acid
KH_2PO_4	0.07	505,430	317,310	0.35	Spent broth
Mg	0.49	79,404	249,410	1.17	Aqueous acid

(c) Which materials would present the best opportunities for recovery, and why?

(d) What green chemistry approaches might be taken if you wanted to increase the potential for cost reductions?

REFERENCES

1. Bailey, P. E. Full cost accounting for life cycle costs: a guide for engineers and financial analysts. *Environ. Finance*, Spring 1991, pp. 13–29.
2. Surma, J. A survey of how corporate america is accounting for environmental costs. *Understand. Environ. Account. Disclos. Today*, 1992, pp. 85–94.
3. White, A. L., Becker, M., Savage, D. E. Environmentally smart accounting: Using total cost assessment to advance pollution prevention. *Pollut. Prev. Rev.*, 1993, 3(3), 23–35.
4. Kreuze, J., Newell, G. ABC and life cycle costing for environmental expenditures. *Manag. Account.*, Feb. 1994, pp. 38–42.
5. Kennedy, M. L. Getting to the bottom line: how TCA shows the real cost of solvent substitution. *Pollut. Prev. Rev.*, 1994, 4(2), 17–27.
6. White, A. L., Savage, D. E. Budgeting for environmental projects: a survey., *Manag. Account.*, 1995, pp. 48–54.
7. Savage, D. E., White, A. L. New applications of total cost assessment. *Pollut. Prev. Rev.*, 1995, 4, 7–15.
8. Savage, D. E., White, A. L. New applications of total cost assessment. *Pollut. Prev. Rev.*, 1995, 4, 7–15.
9. Epstein, M. J. Improving environmental management with full environmental cost accounting. *Environ. Qual. Manage.*, 1996, 6, 11–22.
10. Epstein, M. J. *Measuring Corporate Environmental Performance: Best Practices for Costing and Managing an Effective Environmental Strategy*. IMA Foundation for Applied Research, Montvale, NJ, 1996.
11. U.S. Environmental Protection Agency, Office of Pollution Prevention and Toxics. *An Introduction to Environmental Accounting as a Business Management Tool: Key Concepts and Terms.* EPA 742-R-95-001. U.S. EPA, Washington, DC, June 1995.
12. American Institute of Chemical Engineers' Center for Waste Reduction Technologies. *Total Cost Assessment Methodology.*, A.D. Little, New York, NY, 1999.

13. Greer, L., Van Loben Sels, C. When pollution prevention meets the bottom line. *Environ. Sci. Technol.*, 1997, 31(9), 418–422.
14. Schaltegger S., Müller, K. Environmental management accounting: current practice and future trends. Geographic focus: global. Calculating the true profitability of pollution prevention. *Greener Manage. Int.*, Spring 1997. 17.
15. Kennedy, M. L. Integrating total cost assessment with new management practices and mandates. *Environ. Qual. Manage.*, 1998, 7(4), 89–97.
16. Reiskin, E., Miller, D., Shapiro, K., Dierks, A., Zinkl, D., White, A., Savage, D. *Strengthening Corporate Commitment to Pollution Prevention in Illinois: Concepts and Case Studies of Total Cost Assessment*. Prepared for Illinois Waste Management and Research Center. Tellus Institute, Boston, 1998.
17. Schaltegger S., Hahn T., Burrit R. Environmental management accounting: overview and main approaches, In *Environmental Management Accounting and the Role of Information Systems*, Seifert, E., Kreeb M. Eds., Kluwer Academic, Dordrecht, The Netherlands, 2000.
18. Schaltegger, S., Burrit, R. *Corporate Environmental Accounting: Issues, Concepts and Practices*. Greenleaf Publishing, Lebanon, TN, 2000.
19. United Nations Division for Sustainable, Development. *Environmental Management Accounting Procedures and Principles*. United Nations, New York, 2001.
20. Gale, R. J. P., Stokoe, P. K. Environmental cost accounting and business strategy. In *Handbook of Environmentally Conscious Manufacturing*, Madu, C., Ed. Kluwer Academic, New York, 2001.
21. Koch, D., Dow Chemical pilot of total "business" cost assessment methodology: a tool to translate EH&S "...right things to do" into economic terms (dollars). *Environ. Prog.*, 2002, 21(1), 20–28.
22. Bengt, S., Environmental costs and benefits in life cycle costing. *Manage. Environ. Qual.*, 2005, 16(2), 107–199.
23. Curkovic, S., Sroufe, R. Total quality environmental management and total cost assessment: an exploratory study, *Int. J. Prod. Econo.*, 2007, 105, 560–579.
24. Global Environmental Management, Initiate. *Finding Cost-Effective Pollution Prevention Initiatives*. GEMT, Washington, DC, 1994.
25. Repetto, R., Austin, D. *Pure Profit: The Financial Implications of Environmental Performance*. World Resources Institute, Washington, DC, 2000.
26. Epstein, M. *Measuring Corporate Environmental Performance*. Irwin Professional Publishing, Chicago, 1996.
27. Badgett, L., Hawke, B., Humphrey, K. *Analysis of Pollution Prevention and Waste Minimization Opportunities Using Total Cost Assessment: A Case Study in the Electronics Industry*. Pacific Northwest Pollution Prevention Research Center, Seattle, WA, 1995.
28. Causing, M., Jensen, S., Haynes, S., Marquardt, W. *Analysis of Pollution Prevention Investments Using Total Cost Assessment: A Case Study in the Metal Finishing Industry*. Pacific Northwest Pollution Prevention Research Center, Seattle, WA, 1996.
29. Constable, D. J. C., Curzons, A. D., Cunningham, V. L., Metrics to green chemistry—which are the best? *Green Chem.*, 2002, 4, 521–527.

PART V

WHAT LIES AHEAD

21

EMERGING MATERIALS

What This Chapter Is About In this chapter we explore some of the advantages and disadvantages of emerging materials that at the time this chapter was written were seen as having the potential to advance different areas of green chemistry and green engineering. We explore a few of the potential advantages and disadvantages of some of the most promising emerging materials.

Learning Objectives At the end of this chapter, the student will be able to:

- Identify a few emerging materials that might advance the state of the art of green engineering and green chemistry.
- Identify some of the advantages and disadvantages of these emerging materials from a green chemistry and engineering perspective.
- Understand how emerging materials may be used to solve current problems.

21.1 EMERGING MATERIALS DEVELOPMENT

In previous chapters we covered metrics for green chemistry and green engineering, many of the basic concepts of environment, health, safety, sustainability, and some of the cost-related aspects of green chemistry and green engineering. In this chapter we review briefly a few of the emerging materials that are being considered by some as parts of the sustainability toolkit of the future.

Green Chemistry and Engineering: A Practical Design Approach, By Concepción Jiménez-González and David J. C. Constable

As is generally the case for many new materials and technologies, not all the desired EHS information to make an informed judgment about the greenness of a product or process is available at first. For example, a comprehensive set of toxicological and life cycle information could in many cases take years to be generated. Despite this lack of information, it is still worth discussing a few of the many emerging materials under active development. In this chapter we illustrate green chemistry and engineering principles through a focus on nanomaterials and some bioplastics. You should realize, however, that this list is very limited in scope, and the authors are quite certain that by the time this book goes to press, there will be more new materials emerging from our universities and research laboratories that some would claim to be green, or at least greener than materials developed previously. These new materials will undoubtedly suffer from a lack of comprehensive data about their EHS and sustainability impacts and benefits and it will take some time to substantiate their claims.

21.2 NANOMATERIALS

Much of the focus of current nanotechnology research, and the area that triggers most concern for human, environmental, and process safety is the development of nanoparticles. Nanoparticles have generally been defined as particles with a mean aerodynamic equivalent diameter of less than 100 nm, as measured by a defined particle sizing method. However, as is true with many new areas of research, agreed definitions tend to be ambiguous, and this is true in the area of nanotechnology as well. The total world market for nanoparticles was expected to exceed $900 million in 2005[1] with electronic, magnetic, and optoelectronic applications having the biggest market share, followed by biomedical, pharmaceutical, and cosmetic applications.[2] For example, some of the current biomedical applications of nanomaterials are in the area of diagnostics, quantitative analysis, and drug delivery. For diagnostic agents, nano-based materials are used to improve the accuracy of detection and the limits of detection of the diseased state. Others have focused on better delivery mechanisms for existing anticancer drugs that reduce their general toxicity because the delivery is more targeted, and in some cases, less drug substance is required to achieve the same benefit, both of which would benefit patients.[3]

Apart from these medical and biomedical applications, the most commercially important nanomaterials today are mainly inorganic oxides such as silica (SiO_2), titania (TiO_2), alumina (Al_2O_3), zinc oxide (ZnO), ceria (CeO_2), zirconia (ZrO_2), and iron oxides (Fe_2O_3, Fe_3O_4). Mixed oxides such as indium–tin or antimonium–tin oxides and some titanates and silicates are also of increasing importance. Currently, there are several other types of nanoparticles under development, such as complex oxides, semiconductors, and nonoxide ceramics.[4] Some of the production methods currently available for the production of nanomaterials are described briefly in Table 21.1.

It depends on the material, but in general, at the nanoscale, the physical properties of materials are frequently different from those of the same material at larger particle sizes. The question at hand then becomes how different are the environmental, health, and safety properties of nanomaterials from their macroscale counterparts? For example, one of the challenges of metal nanoparticles in some applications is the increased pyrophoricity due to the high surface area of the particles, with the obvious process safety challenges. In addition, as particles become smaller, they tend to be more reactive, so one needs to question how this increased reactivity might affect potential process safety hazards in production settings. One

TABLE 21.1 Nanomaterial Production Methods

Method	Description	Disadvantages	Advantages
Reduction (attrition)	Coarse micrometer-sized particles are produced by the application of direct energy, such as in nanomilling. The particle-size range desired is obtained through careful control of shear forces, chemical, and/or thermal environments.	A certain amount of art is involved in achieving desired particle range. Increased potential for: Impurity contamination. Occupational exposure; additional engineering controls might be needed Explosivity/explosibility and/or pyrophoricity; additional safety controls might be needed.	Availability, ease of implementation.
Vapor deposition	Material is vaporized at high temperatures (1500 to 2300 K) and deposited on a surface to form the nanoparticles.	Energy intensity is high.	Potential for fewer impurities and better size control.
Precipitation	Nanoparticles are formed by precipitation from a liquid.	Chemicals used can sinter the particles or lead to agglomeration. In addition, the particles are typically precipitated from an organic solvent, which creates environmental, health, and safety issues that need to be managed.	Generally low cost and can handle high volumes.
Novel	Experimenting with nanoparticle manufacturing using supercritical fluids, microwaves, ultrasound, biomimetics, bioprocesses and electrodeposition.	Long development, lack of reproducibility, scale-up, etc.	Each method has different advantages.

also needs to ask if toxicological and environmental hazards are increased with a reduction in particle size. The latter issue is a particularly thorny question confronting the development of nanoparticle technology.

Given the relatively poor understanding of the potential changes in the hazard profile associated with nanomaterial production and use, many have called for the application of the precautionary principle until we obtain a better understanding of nanomaterial hazards and risks. Although some believe that the application of the precautionary principle would bring the development of nanotechnology to a standstill, it should be possible to continue developing nanotechnology in parallel with research to better understand the health, safety, and environmental hazards and risks of nanomaterials. Uncertainty is inherent to any type of new material and technology research, and without investing in additional research into the EHS hazards and risks of nanomaterials, this uncertainty will continue.[5]

In general, those businesses developing nanotechnology-based products recognize that research is required to better understand the hazards and risks of these materials. To date, there is a flurry of active research at universities and in industries and there are several consortia around the world creating codes of practice for developing and handling nanomaterials. These codes are typically also supported or sponsored by governments as a means of driving policy and regulations.[6]

For laboratory work, the use of conservative standards when working with nanomaterials[7] is normally recommended and warranted. These practices include such things as:

- Communication of any known nanomaterial hazards
- Prevention of:
  Inhalation exposure
  Dermal exposure
  Workspace contamination
  Spills and of exposure during spills
- Minimization and appropriate disposal of nanomaterial waste

Example 21.1 Provide examples of potential advantages and disadvantages of nanomaterials from a sustainability, green chemistry, and green engineering perspective.

Solution Depending on the application, there are several potential advantages in using nanomaterials, such as[8]:

- Reduced energy consumption and the corresponding energy-related emissions
- Cleaner industrial processes (e.g., nanoscale cerium oxide has been developed to decrease diesel engine emissions)
- To help degrade contaminants in soil and groundwater[9] (e.g., iron nanoparticles)
- Curing, managing, and preventing disease by improving diagnostics, drug delivery, and sensing (as in oncology applications[10])
- Providing new materials for safety improvements

However, the same characteristics that make nanoparticles highly desirable for some applications can have some concerns from a green chemistry and green engineering perspective. There is currently uncertainty as to the potential increased hazards of nanomaterials, and some of the issues of concern include:

- The high reactivity of nanoparticles could pose increased process safety hazards.
- Some nanoparticles might be more toxic than micrometer-sized particles of the same materials due to increased surface area and reactivity.
- Small size of nanoparticles might increase dust explosion hazards and might make it easier to penetrate blood–cell barriers causing health hazards.
- Some concern has been expressed as to whether some nanomaterials, such as carbon nanotubes, may have toxicity characteristics similar to asbestos.
- Waste-containing nanomaterials may require special handling and disposal.

Additional Points to Ponder What is the best way to handle the EHS uncertainties of nanomaterials? Are there other disadvantages and advantages of nanomaterials from a green engineering/green chemistry prospective?

21.3 BIOPLASTICS AND BIOPOLYMERS

Plastics are ubiquitous in our economy, especially as packaging materials, and packaging accounts for the largest market share of plastic resin.[11] Compared to materials such as steel or glass, polymers are relatively new materials that have been used in substantial amounts over a short period of time (for about six to seven decades). In some countries, more polymers (on a mass basis) are used than aluminum and glass and in general have had a particularly high growth rate in recent years, representing the fastest-growing rate of bulk materials.[12]

Biopolymers, polymers derived from agricultural feedstocks, are a new category of emerging plastics in the marketplace. These plastics are not the same as the emerging plastics of the 1960s and 1970s. For example, one of the environmental drawbacks of traditional plastics is that they do not biodegrade and they are derived from petroleum sources instead of from renewable resources (we discuss aspects related to materials from renewable sources in Chapter 22). Biopolymers, on the other hand, are derived from renewable sources, have been designed to be biodegradable to a large extent in anaerobic conditions, and tend to have properties similar to those of traditional plastics. Several biopolymers of current interest include thermoplastic starch (TPS), polylactide or poly(lactic acid) (PLA), poly(hydroxybutyric acid) (PHB), and poly(hydroxyalkanoate) (PHA).

In general, there are four main ways to produce bio-based polymers: by modifying natural polymers through chemical, thermal, or mechanical means (e.g., starch polymers); by fermenting bio-based monomers, which are then polymerized [e.g., poly(lactic acid), poly (hydroxyalkanoate)]; from genetically modified crops [e.g., poly(hydroxyalkanoate)]; and from C1 after gasification of biomass, although the latter application is still very limited. Figure 21.1 shows a general overview of the most important emerging biopolymers.

Biopolymers such as PBT, PBS, PURs, and nylons in Figure 21.1 are either not produced commercially, produced in small volumes, or serve niche markets as of this time (e.g., concept automobile parts).[13] Also, the biopolymers shown in Figure 21.1 have different degrees of "bio-based" carbon. For example, PLA and starch polymers have a bio-based carbon backbone, but fossil fuels are used in their processing and conversion to final product.

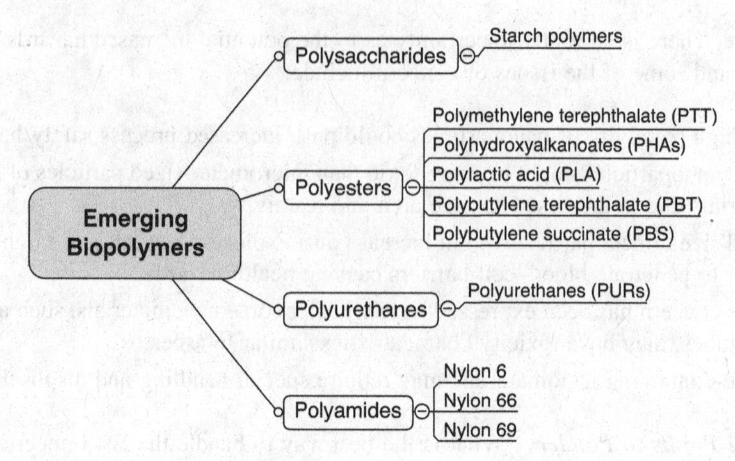

FIGURE 21.1 Overview of the most important emerging biopolymers.

Biopolymers such as PTT have only one of the two monomers derived from renewable sources (PTT is not biodegradable). PHA can be produced either by fermentation or from genetically modified crops (e.g., potatoes). In the following paragraphs we describe the polymers shown in Figure 21.1 in general terms.

1. *Starch polymers.* Starch polymers are biodegradable thermoplastic materials derived from starch that is processed by chemical, thermal, or mechanical means. As you may recall from your study of organic chemistry, starch is the major source of carbohydrates available in nature and is composed mainly of two polymers: poly(saccharide amilose) (linear) and poly(saccharide amylopectin) (highly branched). As of this time, the main agricultural source of starch polymers is corn, although there are other sources, such as potatoes and rice. To date, these polymers are used predominantly in packaging applications, although there are other novel applications, such as using the polymer to encapsulate other chemicals as a means of delivering the chemical or controlling its release.

2. *PLA.* Cargill Inc. launched a project in the late 1980s to develop a new product from natural sources.[14] Processes were developed to produce poly(lactic acid) (PLA) from lactic acid into lactide as well as for purification and polymerization/devolatilization of the lactide. The two major routes to produce polylactic acid are the direct condensation and polymerization of the lactic acid monomer and the ring-opening polymerization through the lactide intermediate. In this case, lactic acid is produced by fermenting glucose obtained enzymatically from cornstarch. In 1994 the company built a 5000-metric ton per year PLA facility in Minnesota, mainly to catalyze the development of a commercial market for PLA. Cargill partnered with Dow Chemical to form Cargill Dow; in 2004, Dow decided to leave the joint venture and the company name was changed to NatureWorks. PLA is a compostable polymer derived from corn, rice, or sugar beet starches. Future expectations are that PLA can be produced via the fermentation of lignocellulosic materials (e.g., bagasse, grass, straws, corn stover) through advances in enzymatic processes.

3. *PTT.* Poly(trimethylene terephthalate) is a polyester produced from 1,3-propanediol (PDO, trimethylene glycol). PDO is condensed and polymerized with either purified terephthalic acid or dimethyl terephthalate. The raw materials are conventionally produced from petrochemical sources, but several companies (e.g., DuPont, Genecor) have developed

processes to produce PDO by fermentation from glucose. Applications for PTT are being developed mainly for the fibers and films sectors.

4. *PHAs.* Poly(hydroxyalkanoates) are aliphatic polyesters produced by microbial fermentation of carbon substrates such as sucrose, oils, and fatty acids. PHAs accumulate inside the microbe and are used as an energy reserve. One particular grouping of PHAs, poly (β-hydroxybutyric acid) (PHB), is produced at ambient temperatures with long residence times, using carbon sources that include glucose, beet, or cane molasses, although starch and whey can also be used to produce PHB.[15]

As mentioned above, PBT, PBS, PURs, and polyamides (PA, nylons) are currently not produced by bioprocesses or produced in very small amounts for niche markets. Poly (butylene succinate) (PBS) can be obtained by using bio-derived succinic acid produced by fermentation and petrochemical terephthalic acid. Polyurethanes can be obtained by using bio-derived polyol produced by fermentation and petrochemically derived isocyanate. Polyamides can be produced from bio-derived caprolactam or adipic acid obtained through fermentation processes. Because petrochemical processes to produce polyamides and polyurethanes tend to have a large number of processing steps, new, less energy-intensive, simpler, and cheaper bioprocesses could gain a clear competitive advantage.

21.3.1 Bio-based Plastics and Sustainability

To fully understand whether or not biopolymers are more sustainable than plastics derived from petroleum, a life cycle assessment approach is needed. There have been several recent studies comparing bio-based materials with materials derived from petrochemical sources. Given the recent worldwide preoccupation with climate change, most of these studies to date have focused on life cycle energy and global warming potential.[16] To obtain a full assessment of the sustainability impacts, as many categories (eutrophication, human toxicity, etc.) as possible need to be analyzed. However, as we have observed with many emerging materials and technologies, data for some of these categories are not currently available publicly. One example of applying LCA more broadly is a study that NatureWorks commissioned to compare PLA with petroleum-derived plastics for clamshell packaging. The comparison included polystyrene (PS), polypropylene (PP), and poly(styrene terephthalate) (PET). Table 21.2 shows the comparative results of the study.

TABLE 21.2 Relative LCA Results for PLA, PS, PP, and PET[a]

LCA impact	PLA	PS	PP	PET
Fossil resource	1	2	3	4
Global warming	1	3	2	4
Summer smog	1	2	3	4
Acidification	3	2	1	4
Eutrophication	3	2	1	4
Human toxicity (carcinogen risk)	1	3	2	4
Human toxicity (PM10)	3	2	1	4
Nonrenewable energy	1	2	3	4

[a]1 represents the smallest and 5 the largest impact compared to the other polymers.

As illustrated in the data in the table, drawing conclusions or making generalizations based on only one or two impact categories is sometimes dangerous, but for this scenario, these results seem to indicate that PLA is the best option from a green perspective, despite the fact that there are some trade-offs in terms of acididfication, eutrophication, and human toxicity.

To truly evaluate the merits of the PLA, a more holistic and comprehensive view is needed. For example, while the use of renewable resources is a desirable goal from a sustainability viewpoint, there is greater complexity in evaluating renewability and the impacts of biomaterials and biopolymers. For example, it is sometimes the case that the material resources and the energy required to produce renewable materials may or may not be renewable. As you know, chemical processing makes use of organic solvents for isolations, the use of nonrenewable feedstocks as part of a material supply chain, and the use of energy that is generated from nonrenewable sources. Yet another factor to consider is how land is used and the competing impacts associated with potential land-use change, which at this time is an arena of ongoing debate.[17,18]

Example 21.2 Vink et al. performed a LCA comparing the first-generation PLA production with the estimated life cycle environmental impacts associated with production of PLA from biomass using wind power as a source of energy.[14] The assessment also compared them with some selected metrics, such as energy, global warming, and water use. What conclusions can you draw from the data presented in Figure 21.2 showing a comparison of LCA energy for several polymers?

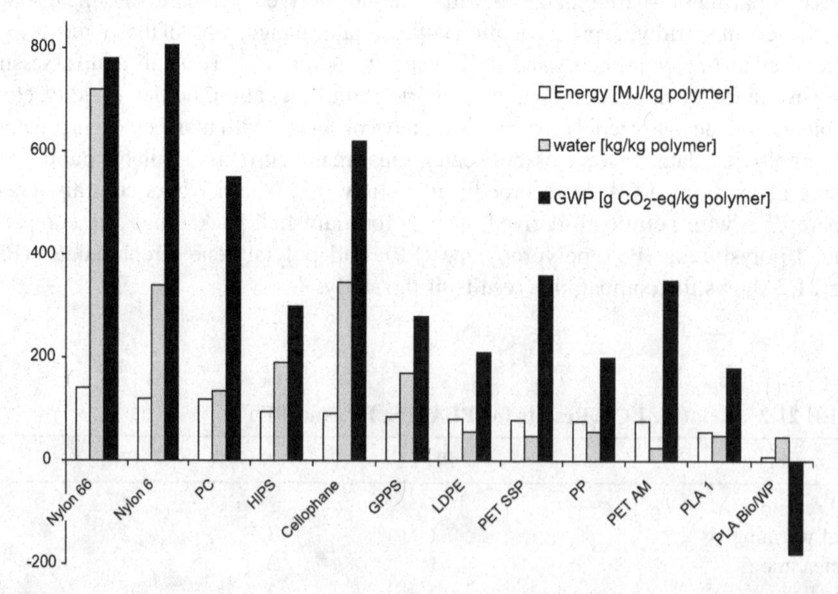

FIGURE 21.2 LCA comparison of selected metrics for bio-PLA and other polymers. PC, Polycarbonate; HIPS, high-impact polystyrene; GPPS, general-purpose polystyrene; LDPE, low-density polyethylene; PET SSP, polyethylene terepthalate, solid-state polymerization (bottle grade); PP, polypropylene; PET AM, polyethylene terepthalate, amorphous (fibers and film grade); PLA1, polylactide (first generation); PLA Bio/WP, polylactide, biomass/wind power scenario.

Solution In general, the first generation of PLA (PLA1) uses 25 to 55% less fossil energy than the polymers derived petrochemically. With the process improvements, the use of fossil energy could be reduced by more than 90%. In terms of global warming potential, the use of corn residue (lignin fraction) and wind energy to supply process heat closes the loop for energy generation and provides a credit for greenhouse gas impact from PLA production. The life cycle water used for the PLA is comparable to that used for the other polymers, even when accounting for irrigation water used in corn crop production.

Additional Points to Ponder What happens to the other life cycle impacts? Would we expect some trade-offs if we had included other impacts in the analysis? What impacts would you expect from land-use changes and trade-offs of crop use for food vs. polymers?

21.4 ABOUT NEW GREEN MATERIALS

These emerging materials represent distinct opportunities for more sustainable product and process designs. As we have seen, some promising materials might be promoted as green or environmentally friendly, but in many instances it is not entirely clear if these materials represent a more sustainable alternative to the traditional materials. For a complete sustainability assessment of new materials, it is necessary to obtain a comprehensive data set for hazard identification, impact assessment, and mitigation. It is necessary for these assessments to be conducted from a life cycle perspective so that the broader implications of new materials are included. Innovation is one of the main drivers for sustainability and cannot be discouraged. At the same time, this innovation needs to be framed by the tenants of sustainability, green chemistry, and green technology, so that the development of new materials accounts for their potential impacts to humans, other species, and the environment.

In Chapter 22 we explore in greater detail the advantages, impacts, needs, and uncertainties of a big sector related to some of the materials we have explored in this chapter: renewable materials.

PROBLEMS

21.1 One potential application for nanoparticles in the electronic manufacturing industry is their use in specialty polishing operations, such as chemical–mechanical planarization (CMP), which is critical to semiconductor and chip manufacture. CMP relies on a slurry of oxide nanoparticles for both chemical reaction and mechanical abrasion between the slurry and the film to obtain smooth and flat silicon wafers. CMP is also used to polish magnetic hard disks. What are the disadvantages and advantages of CMP from a green engineering/green chemistry perspective?

21.2 Some microorganisms have been found to produce nanoparticles. For example, bacterial proteins have been used to produce magnetite in laboratories, yeast cells can create cadmium sulfide nanoparticles, and gold and silver nanoparticles have been obtained from fungus and viral proteins, respectively. What are the advantages and disadvantages of these potential future methods of producing nanoparticles from a green chemistry/green engineering perspective?

21.3 Carbon nanotubes (CNTs) have potential uses in a wide range of applications (e.g., composite materials, batteries, memory devices, electronic displays, transparent conductors, sensors, medical imaging). However, some scientists have expressed concerns regarding the effects of some CNTs. What are those concerns?

21.4 In Section 21.3 it was noted that applications for PTT are being developed mainly in the fibers and films sectors. What specific commercial applications would this mean for the public?

21.5 Bohlmann performed a streamlined LCA comparison of PLA vs. PP yogurt containers using energy as a surrogate for environmental impact. The results he obtained are given in Table P21.5A.

TABLE P21.5A Bohlmann's Results

Energy Category	PLA	PP
Fuel production (MJ/kg)	3.9	6.6
Fuel use (MJ/kg)	47.5	30.8
Transportation (MJ/kg)	4.7	4.8
Feedstock (MJ/kg)	0.6	51.5
Total (MJ/kg)	56.7	93.7
Total (MJ/ton yogurt)	2225	3261

(a) What conclusions can you draw from the results?

(b) Bohlmann also prepared a sensitivity assessment on the energy production for lactic acid, accounting for the base case (using triple-effect evaporators in the calculations) and the worst case (double-effect evaporators) (Table P21.5B).[19] What additional conclusions can you draw from the sensitivity assessment?

TABLE P21.5B Bohlmann's Best and Worst Case Analysis

Gross Energy	Base Case	Worst Case
Energy for lactic acid production (MJ/kg)	30.4	44.1
Total energy for PLA (MJ/kg)	56.7	75.3

21.6 A cradle-to-grave LCA of a few poly(3-hydroxybutyrate) (PHB)-based composites was performed for comparison with petrochemically derived polymers used in cathode-ray tube (CRT) monitor housings [produced conventionally from high-

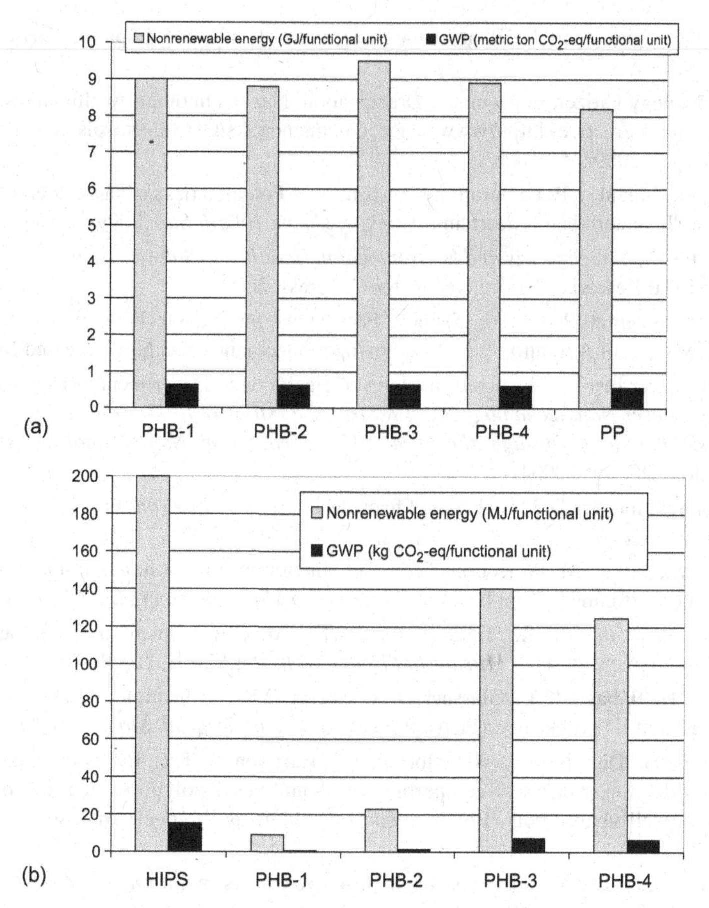

FIGURE P21.6 Comparative LCA for (a) automotive parts and (b) CRT housing. HIPS, high-impact polystyrene; PP, polypropylene; PHB-1 to PHB-4, poly(3-hydroxybutyrate) composites.

impact polystyrene (HIPS)] and internal automotive panels [normally produced from glass-fiber-filled polypropylene (PP-GF)] (Figure P21.6). What conclusions can you draw from the data?

REFERENCES

1. Harper, T., et al. *Nanotechnology Opportunity Report.* Cientifica Ltd., London, Mar. 2002.
2. Ritter, M. N. *GB-201 Opportunities in Nanostructured Materials.* Part A: Electronic, Optoelectronic and Magnetic Applications; Part B: Biomedical, Pharmaceutical and Cosmetic Applications; and Part C: Energy, Catalytic and Structural Applications. Business Communications Co., Norwalk CT, 2001.
3. Cordis. European Technology Platform. NanoMedicine.http://cordis.europa.eu/technology-platforms/pdf/nanomedicine.pdf.
4. Rittner, M. N. Nanoparticles—What's now? What's next? *Chem. Eng. Prog.* 2003, 39S–42S.

5. Lewinski, N. Nanomaterials: What are the environmental and health Impacts? *Chem. Eng. Prog.*, 2008, 37–40.
6. Nanotechnology Citizen Engagement Organization. Nanotechnology Health and Safety Protocols and Good Practices.http://www.nanoceo.net/nanorisks/OHS-Protocols-Best-Practices, accessed May 3, 2009.
7. Hallock, M., Greenley, P., Di Berardinis, L., Kallin, D. Potential risks of nanomaterials and how to safely handle materials of uncertain toxicity. *J. Chem. Health Saf.*, 2009, 1, 16–23.
8. Sargent, J. F. *Nanotechnology and Environmental, Health, and Safety: Issues for Consideration.* Congressional Research Service Report for Congress, 2008.
9. U.S. Environmental Protection Agency. Fact Sheet for Nanotechnology Under the Toxic Substances Control Act. http://www.epa.gov/oppt/nano/nano-facts.htm, accessed May 3, 2009.
10. National Cancer Institute, National Institutes of Health, U.S. Department of Health and Human Services, *Cancer Nanotechnology Plan: A Strategic Initiative to Transform Clinical Oncology and Basic Research Through the Directed Application of Nanotechnology.* U.S. DHHS, Washington, DC, July 2004.
11. Bohlman, G. Biodegradable packaging life cycle assessment. *Environ. Prog.*, 2004, 23(4), 342–346.
12. Patel, M. K., Crank, M. Projections for the production of bulk volume bio-based polymers in europe and environmental implications. *J. Biobased Mater. Bioenergy*, 2007, 1(3), 437–453.
13. Terasawa, I., Tsuneoka, K., Tamura, A., Tanase, M. Development of plant-based plastics technology, "Green Plastic". *Mitsubishi Motors Tech. Rev.*, 2008, 20, 91–96.
14. Vink, E. T. H., Rábago, K. L., Glassner, D. A., Gruber, P. R. Applications of life cycle assessment to NatureWorks™ polylactide (PLA) production. *Polym. Degrad. Stabil.*, 2003, 80, 403–419.
15. Harding, K. G., Dennis, J. S., von Blottnitz, H., Harrison, S. T. L. Environmental analysis of plastic production processes: comparing petroleum-based polypropylene and polyethylene with biologically-based poly-β-hydroxybutyric acid using life cycle analysis. *J. Biotechnol.*, 2007, 130, 57–66.
16. Bohlmann, G. M. Biodegradable packaging life-cycle assessment. *Environ. Prog.*, 2004, 23(4), 342–346.
17. Kløverpris, J. H., Wenzel, H., Nielsen, P.H. Life cycle inventory modelling of land use induced by crop consumption: 1. Conceptual analysis and methodological proposal. *Int. J. Life Cycle Assess.*, 2008, 13(1), 13–21.
18. Fargione, J., Hill, J., Tilman, D., Polasky, S., Hawthorne, P. Land clearing and the biofuel carbon debt. *Science*, 2008, 319, 1235–1238.
19. Bohlmann, G. *Biodegradable Polymer Life Cycle Assessment.* Process Economics Program Report 115D. SRI Consulting, Menlo Park, CA, 2001.

22

RENEWABLE RESOURCES

What This Chapter Is About In this chapter we discuss renewable materials and energy as means of moving toward more sustainable products and production processes. We explore the advantages, impacts, needs, and uncertainties of a large sector of the economy that in most instances is still undergoing considerable development.

Learning Objectives At the end of this chapter, the student will be able to:

- Understand the role that renewable materials and renewable energy play in moving us farther down the path toward more sustainable practices.
- Identify the main renewable materials and their current and potential applications.
- Identify the main sources of renewable energy and their current and potential applications.
- Understand what barriers there are that prevent the use of renewable resources and some of the potential strategies to overcome them.

22.1 WHY WE NEED RENEWABLE RESOURCES

Human beings continue to deplete raw materials at an alarming rate, and we are expending more time and energy to obtain many of the key materials and energy that we require as building blocks in the chemical processes that lead to products. The acquisition of these key dwindling resources also results in an increasing amount of environmental degradation. A large amount of the energy and materials that we use today come from nonrenewable

Green Chemistry and Engineering: A Practical Design Approach, By Concepción Jiménez-González and David J. C. Constable

sources, which by definition are finite and will be depleted at some point. For example, there is a limited and finite amount of mineral ores on Earth, and they normally require large amounts of energy to be extracted, refined, and converted to the desired form for chemical processing. Once all the copper on Earth has been extracted, one must find ways to keep recycling it or find alternatives to its use.

Renewable resources, on the other hand, are derived from plants, animals, or ecosystems and can be regenerated through sustainable management. These resources, both energy and materials, have the potential to be replenished indefinitely if managed properly. You may recall that in Chapter 4 we explored renewability as one of the metrics for green chemistry and green engineering. We discussed how in the current chemical landscape relatively few chemicals available for routine commercial operations are derived from renewable sources. Or, if they are from renewable sources, they are often associated with considerable life cycle impacts or there are significant numbers of trade-offs associated with their use.[1] One example that we discussed was the current debate over the sustainability of producing and using corn-based ethanol as fuel. We emphasized that the key messages from that debate are that assessments of renewability have to be carried out from a life cycle perspective. We also noted that it is challenging to compare highly developed chemical or petrochemical processes that use nonrenewable feedstocks with processes that use potentially renewable feedstocks that are obtained using processing approaches that are comparatively immature or not as fully developed. Remember that the petrochemical industry has been in existence for over 150 years and has been driven toward ever greater processing efficiencies.

Therefore, if this is the current situation, why do we even bother to look into renewable materials and energy if the answer is not that simple and direct? At least part of the answer is that renewable materials and energy do offer one distinct possibility of breaking the linear production cycle that constrains our innovation and drives us away from achieving greater sustainability. Renewable resources offer the opportunity to operate production systems closer to the way that nature works, in a cyclical manner, where materials and energy can be utilized while minimizing waste (see Figure 22.1).

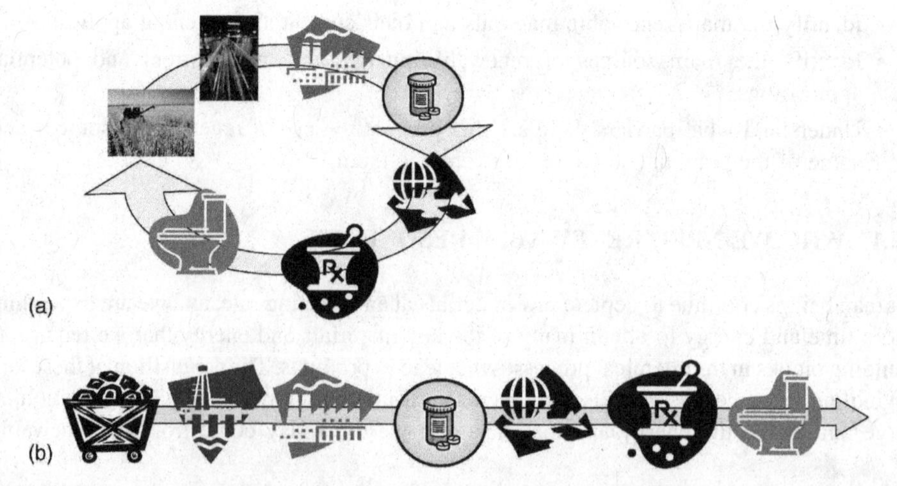

FIGURE 22.1 Bioprocesses (a) might present an opportunity to close the cycle in the way that goods are produced, in some cases utilizing the waste as feedstock. This is in contrast to the linear production systems (b), where waste invariably has to be treated and disposed of without utilization.

Let's recall our discussion in Chapter 8, where we focused on bioprocesses. We noted that there are many additional advantages in using renewable resources beyond conserving materials and energy, such as the minimization of waste through the use of closed-loop systems and biodegradable by-products that are easier to treat. In an ideal system, we will be able to design products that are derived from renewable sources, and those products, at the end of their useful life, will be reassimilated into nature or into closed-loop industrial systems that utilize product waste as inputs, and when waste is generated, as indeed it will be, it is easily biodegradable. Because the use of renewable resources could enable this transformation in the way we make and use the items we need and want, we must strive to utilize them as much as possible. To underscore their importance, we note that the U.S. National Research Council issued a report in 2000 on bio-based industrial products[2] citing the following potential benefits:

- Use of unexploited agricultural productivity
- Use of some agriculture, forestry, or municipal waste
- Potential for more environmentally sustainable products and processes
- Development of better performing or currently unavailable products
- Use of molecular biology to enhance raw materials for easier processing and to reduce the economic and environmental costs of processing

Given these potential benefits and the pivotal role of renewable resources in achieving greater sustainability, the United States and the European Union have set targets to increase the use of renewable resources for products and energy, as shown in Table 22.1. However, to realize the full potential of renewable resources, one must strive to think holistically. For example, the use of renewable resources needs to be managed in a way that ensures long-term availability; that is, they need to be used within their rate of replenishment (e.g., trees grow only so fast) and within the limits of the planet to regenerate itself from the impacts while maintaining the complex equilibrium of living systems. In addition, as we strive to use more renewable materials, we must be aware of and make provisions to counteract the impacts that arise through activities such as agriculture, which is generally material and energy intensive; has significant impacts on the carbon, nitrogen, and phosphorus cycles; and creates competition between arable land for food and for other items of commerce.

Some people might have the misperception that when utilizing renewable materials or energy, one is therefore free to use as much as possible because it is renewable. This is a

TABLE 22.1 U.S. and European Union Goals for the Use of Renewable Resources in Energy Generation, Transportation Fuels, and Products (%)

	United States				European Union			
Renewable Resource	2001	2010	2020	2030	2001	2005	2010	2020–2050
Bioenergy: share of electricity and heat demands	2.8	4	5	5	7.5	—	12.5	26 (2030)
Biofuels: share of demand for transportation	0.5	4	10	20	1.4	2.8	5.8	20 (2020)
Bioproducts: share of biobased chemicals	5	12	18	25	8–10			

Source: ref. 3.

flawed perception on two accounts. First, renewable resources are renewable only if we use them in a sustainable manner, that is, such that the rate of use is less than the rate at which they can be replenished without creating additional stressors on the environment.

22.2 RENEWABLE MATERIALS

Figure 22.2 shows a few materials that can be derived from renewable sources, either through existing or potential technology or biotechnology.[4–6] One can classify renewable materials into three general categories:

1. Commodity chemicals
2. Specialty chemicals
3. General materials

Good examples of renewable commodity chemicals include the biopolymers that we covered in Chapter 21 [poly(lactic acid), propanediol, hydroxybutyrate, etc.] and biofuels. For example, in the last few years, bioethanol has been at the forefront of the popular media, and at this time, bioethanol is still the largest- volume renewable chemical (about 1.6 billion gallons per year in the United States). Other bio-based commodity chemicals that are escalating in use include lactic acid, alcohols (e.g., biobutanol), aldehydes, and organic acids. These oxygen-containing bio-based commodity chemicals have a general advantage over their petrochemically derived counterparts, as it is chemically difficult to oxygenate

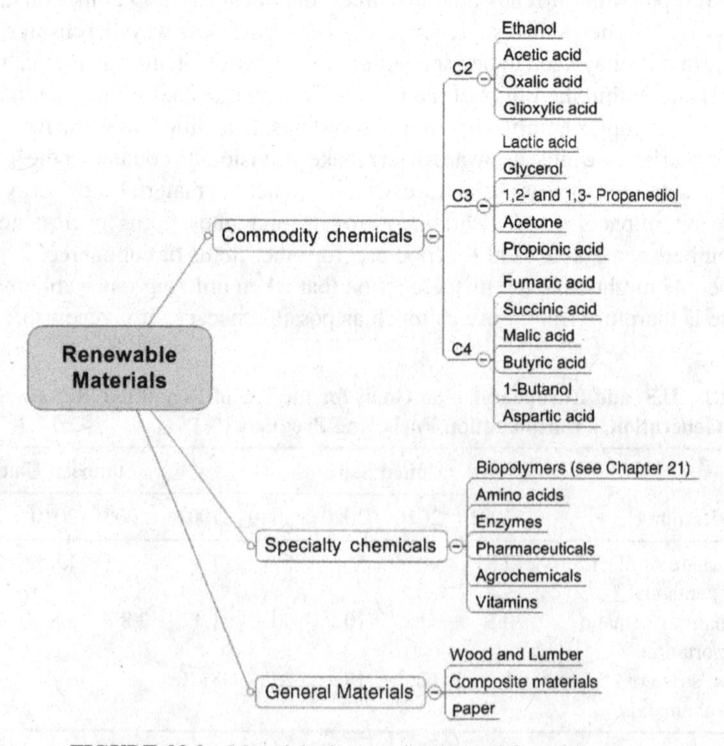

FIGURE 22.2 Materials that can be derived from biomasss.

most petroleum constituents. Because most bio-based chemicals already contain one or more oxygen atoms in their structure, these oxygenated bio-derived chemicals may represent the greatest near-term opportunity for competing successfully with and ultimately replacing commodity chemicals sourced from petroleum.

Another factor to consider when thinking of commodity chemicals such as bioethanol is that this chemical and others have been produced primarily using feedstocks that are also used for food (e.g., sugarcane, corn). However, the greatest opportunity for sustainable commodity chemicals would be to derive these chemicals from lignocellulosic materials (e.g., bagasse, wood waste). Another type of commodity fuel derived from renewable sources are biodiesels. Unlike bioethanol, biodiesels are derived from plant (single cells such as algae and multicellular sources) and vegetable oils. Biodiesel production is unlikely to be sufficient to offset diesel demands, but these fuels may be able to cover the demand for specific markets.

Specialty chemicals derived through large-scale (100 to 200 m^3) industrial fermentation processes cover a wide range of commercially valuable materials such as enzymes, pharmaceuticals, vitamins, amino acids, flavors, fragrances, agrochemicals (e.g., biopesticides, plant growth promoters), thickening agents, and nutraceuticals, among others. The development of these types of biospecialty chemicals are usually driven by the need to introduce specific types of functionality (e.g., chirality) into molecules that are not achievable using traditional synthetic chemistry approaches with petrochemically based feedstocks. They might also be driven by environmental and sustainability goals, such as in the development of enzymes that might be used to produce biofuels or to replace phosphates in detergents.

Example 22.1 The two routes below show chemical synthetic and biotechnological routes for adipic acid:

benzene → ($Ni–Al_2O_3$, 370–800 psi) → cyclohexane → ($Co–O_2$, 120–140 psi) → cyclohexanone + cyclohexanol → (HNO_3) → HO_2C–(CH₂)₄–CO_2H adipic acid

D-glucose → (*E. coli*) → 3-dehydroshikimate → (*E. coli*) → *cis,cis*-muconic acid → (Pt, H_2, 50 psi) → adipic acid

The typical feed solution for the biotech route consists of 6 g of Na_2HPO_4, 0.12 g of $MgSO_4$, 10 g of Bacto tryptone, 3 g of KH_2PO_4, 1 mg of thiamine, 5 g of Bacto yeast, 1 g of NH_4Cl, 10.5 g of NaCl, and 10 g of glucose (62 mmol) per liter of water. From the green chemistry and green engineering standpoint, what advantages and disadvantages does the biotechnology route have in comparison with the chemical route?

TABLE 22.2 Advantages and Disadvantages of the Biotech Route

Advantages of Biotech Route	Disadvantages of Biotech Route
Use of renewable feedstock (glucose)	Uses precious metal catalysts (Pt)
Avoids need for high operating pressures (800 vs. 50 psi)	Potential hydrogenation safety issues
Avoids use of Ni, Co, and Al	
Avoids use of highly hazardous materials (e.g., benzene, nitric acid)	

Solution See Table 22.2.

Additional Point to Ponder What life cycle impacts might be associated with the biotechnology route?

General materials derived from renewable resources include a wide range of products such as chemicals used for or to replace traditional wood and paper products, or composite materials that use sugarcane bagasse and hemp fibers. Another common classification scheme for renewable materials is based on the feedstocks from which products are derived,[7] shown in Table 22.3.

In addition to using bio-derived materials directly, these are and can be used as building blocks for other compounds, which can, in turn, be converted into materials for consumption or can be used to chemically or enzymatically synthesize other materials and chemicals.[8] Table 22.4 presents a few of these derivatives and applications for low-molecular-weight chemicals derived from renewable sources. For more in-depth descriptions of potential bio-based chemicals and current commercially available chemicals and products, consult the BREW[5] report.

The production of materials from renewable sources at this point is at different levels of development, depending on the material and the technological platform. Some processes and their associated products, such as ethanol from sugarcane, are highly developed, whereas others, such as ethanol from lignocellulosic material, are in early phases of industrialization. One needs to take this into account when assessing the future potential of these materials. The concept of a future filled with biorefineries, similar to what we now encounter with modern petrochemical complexes, sits at the core of our vision for the future of renewable materials and provides a road map for the technological step changes that will be required to transition successfully to broader use of renewable materials.

TABLE 22.3 Common Renewably Derived Products and Their Feedstocks

Products	Feedstock
Sugars and starches	Primary feedstocks include sugarcane, sugar beets, corn, wheat, rice, potatoes, barley, sorghum grain, and wood.
Lipids	Obtained from a variety of vegetable oils derived from soybeans, rapeseed, or other oilseeds.
Gum and wood	Include tall oil, alkyd resins, rosins, pitch, fatty acids, and turpentine.
Cellulose derivatives	Bleached wood pulp and cotton linters.
Industrial enzymes	Fermentation processes using a variety of carbon sources, such as rapeseed oil, sugarcane, sugar beets, etc.

TABLE 22.4 Derivative and Applications of Several Bio-based Chemicals

Bio-based Chemical	Bioprocess	Derivative(s)	Production Route(s)	Application(s)
Ethanol	Fermentation of sugars from sugarcane, sugar beets, grains (e.g., corn, rice, wheat, rye) and potatoes to a limited extent	Ethylene Ethyl esters (e.g., ethyl acetate, ethyl lactate) Ethyl *tert*-butyl ether	Dehydration of ethanol from biomass Esterification of ethanol with organic acids Chemical synthesis of ethanol and isobutene	Chemical intermediate Solvents, chemical intermediates Fuel additive
Acetic acid	Fermentation, mainly from ethanol and sucrose	Acetic anhydride Vinyl acetate Ethyl acetate Peracetic acid	Ketene or Halcon (carbonylation of methyl acetate) processes Catalytic reaction of acetic acid, ethylene, and oxygen Esterification of ethanol with acetic acid Reaction with hydrogen peroxide	Chemical intermediate (acetylating and dehydrating agent) Production of polymers and copolymers Solvents, chemical intermediates Epoxide production
Lactic acid	Fermentation, usually from glucose, molasses, sucrose, or starch hydrolysates	Polylactic acid Lactate esters such as ethyl lactate Acrylic acid 1,2-Propanediol Acetaldehyde Propionic acid Oxalic acid	Polymerization or esterification of lactide with alcohols (see Chapter 21) Esterification of lactic acid with alcohols Decarboxylation Dehydration Esterification of lactic acid followed by dehydro-genation; fermentation of sugars Reduction of lactic acid Oxidation, decarboxylation	Packaging Solvents, chemical intermediates Chemical intermediate Plasticizer Plastics, polymers, antifreeze, chemical intermediate Chemical intermediate Chemical intermediate

(continued)

TABLE 22.4 (*Continued*)

Bio-based Chemical	Bioprocess	Derivative(s)	Production Route(s)	Application(s)
Glycerol	By-product of the conversion of fats and oils to fatty acids or fatty acid methyl esters (e.g., rapeseed, methyl ester)	Esters (e.g., nitroglycerin; acetins; mono-, di-, and triglycerides) Acetals or ketals Acrolein 1,3-Propanediol Epichlorhydrin	Esterification with organic and inorganic acids Reactions with aldehydes or ketones Catalytic hydrolysis Fermentation or chemical conversion Reaction with HCl plus dechlorination	Pharmaceuticals, explosives, chemical intermediates Chemical intermediates Chemical intermediate Polymers, solvents, chemical intermediate Epoxy resins, specialty chemicals
Succinic acid	Fermentation from sugars	1,4-Butanediol, tetrahydrofuran, γ-butyrolactone Pyrrolidones (e.g., NMP) Succinic salts	Catalytic routes for these derivatives are technically feasible Selective reductive amination Fermentation by-products	Succinic acid is used as a sweetener or feedstock; derivatives used as solvents in polymer production and chemical intermediates Solvents Coolants, deicers
Furfural	Extraction of pentosan from waste biomass (e.g., corncobs, almond husks, oat hulls, birch wood, bagasse, sunflower husks), followed by hydrolysis and dehydration	Furfuryl alcohol Tetrahydro furfuryl alcohol Tetrahydrofuran Furfuryl amine Resins	Hydrogenation of furfural Catalyzed hydrogenation of furfuryl alcohol Hydrogenation Reductive amination of furfural Acid-catalyzed condensation of furfuryl alcohol or furfural with acetone, formaldehyde, phenol, or urea	Solvent, chemical intermediate, resin production Solvent, chemical intermediate Solvent Chemical intermediate Binders

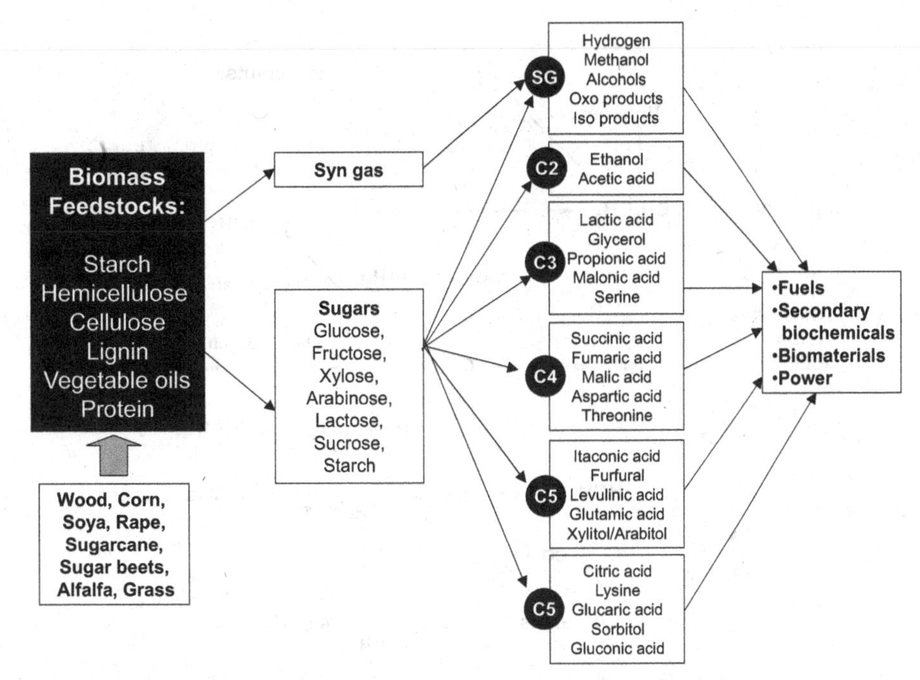

FIGURE 22.3 Potential products of a biorefinery from several biomass sources.

22.3 THE BIOREFINERY

If they were put into widespread use, biorefineries could revolutionize the use of renewable materials and accelerate the transition toward a more sustainable society.[9] In its simplest form, a biorefinery is a facility that produces a series of products from a variety of bio-based feedstocks, including the building blocks for the chemicals of interest and the energy to operate the biorefinery. Figure 22.3 shows a few examples of materials that can be produced from renewable materials as part of the biorefinery

In theory, these biorefineries would use different types of biomass to produce a variety of products. By-products of the biorefinery could be used for internal power generation or cogeneration. In addition, the products derived from a biorefinery will include not only the basic building block chemicals as are obtained from a traditional petrochemical complex, but also other type of building block chemicals that are not easily derived from petroleum feedstocks. The technology required to produce these building blocks will depend on the type of biomass available as feedstocks for the biorefinery, so there will be a need for a suite of technological platforms.[10]

Example 22.2 Figure 22.3 shows methanol as one of the products that can be derived from the syngas fraction in a biorefinery. What would be the commercial significance of this biorefinery product?

Solution Methanol is a building block for many commercial materials; thus, one could produce a variety of commercial chemicals and other materials from renewable sources. Some of these routes to several secondary chemicals produced as part of a biomethanol economy are shown in the product map below:

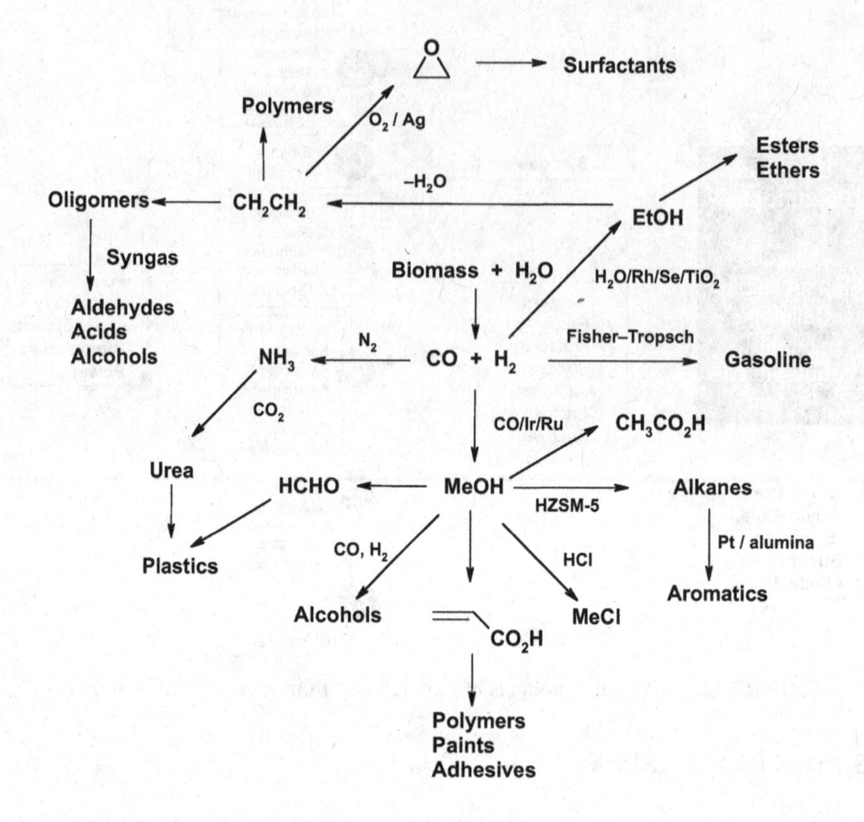

Additional Points to Ponder. What possible disadvantages might there be with a biomethanol-based economy? Describe some of the life cycle impacts related to biomethanol compared to fossil methanol.

Some modern industrial agricultural companies can already lay claim to having proven the practicality of the biorefinery concept, as they produce not only food, feed, and fiber but also, as in the case of corn wet mills, chemicals such as enzymes, lactic acid, citric acids, amino acids, and ethanol. Similar claims are made by wheat millers and soybean processors. Wheat, soybean, and corn facilities already convert about 95% of the mass entering their gates into valuable materials. One might also argue that the pulp and paper industry should be included as a partially integrated biorefinery in which wood is converted into paper, and by-products are used to produce chemicals and fibers.

The biorefinery concept is still evolving, but there have been recent attempts to describe the potential types of biorefineries systematically. For example, three types of biorefineries have been defined as phase I, II, and III biorefineries. A *phase I biorefinery* has specific process capabilities and uses biomass to produce a fixed amount of products with almost no flexibility in processing. One example of this type of facility is a dry mill ethanol plant. A *phase II biorefinery* also uses grain feedstocks but has the additional capability of varying production processes to make a range of products, depending on product demand. Facilities of this type would include corn wet millers that produce ethanol and a few other chemicals. Finally, a *phase III biorefinery*, also known as a fully integrated biorefinery, is a fully developed and integrated chemical manufacturing complex that utilizes a mix of bio-based feedstocks to produce a wide variety of chemicals and materials through a combination of technologies and processing methods. Such a facility has the capability to produce a mix of higher-value chemicals while co-producing biofuels and other lower-value chemicals.[11]

Currently, three different types of phase III biorefinery systems are under active development: the *whole crop biorefinery*, which uses raw materials such as cereals or maize; the *green biorefinery*, which uses "wet" natural biomass such as green grass, lucerne, clover, or immature cereal; and finally, a *lignocellulose biorefinery*, which uses "dry" natural raw materials, such as cellulose-containing biomass and wastes.[12, 13] Another way to classify biorefineries is with a simple classification that distinguishes between facilities that utilize biomass to produce chemicals and facilities that utilize waste materials as feedstocks.[14]

However, the development of a fully integrated biorefinery is still in its infancy, especially as one thinks of the role that these facilities will play in realizing the potential of renewable materials and energy. A fully integrated biorefinery would lower the cost of products produced from renewable materials by producing several bio-based materials at different scales. This means that it would have the capability of producing large volumes of biofuels at the same time that it produces smaller-scale specialty chemicals in a fashion similar to a petroleum refinery. However, since biorefineries will be able to produce food in addition to fuels and a variety of chemicals, the biorefinery should conceivably have an edge over the oil refinery.

From the above it can be concluded that a fully integrated biorefinery will need to be flexible by definition to fully realize its potential. This flexibility is both in terms of being able to operate consistently despite variations in feedstock composition and quality, but also being able to vary the type and quantity of products, depending on market demands, just as current petroleum refineries do.

In addition, a successful integrated biorefinery should be able to utilize lignocellulosic materials as a feedstock. Currently, most of the carbohydrates (sugars) used as carbon sources in fermentation processes come from starches (e.g., corn, soy), and their use as biorefinery feedstocks can compete with their use as food sources and could not only increase the price of the feedstocks but the price of food as well. Lignocellulosic materials, on the other hand, represent a much needed new, less expensive source of carbon as feedstock, but before it can be utilized it will be necessary to develop technologies to fractionate and separate the fermentable sugars at a competitive cost. These technologies include effective enzymes to break down the cellulose and hemicellulose into fermentable carbohydrates, the development of more efficient unit operations and separation trains to extract and purify the desired materials (such as extractive methods, gasification and liquefaction of biomass), and the development of materials that can be produced from some other components of the lignocellulosic biomass, such as lignin and proteins.

Example 22.3 If you think about it, humans have been using renewable materials in the form of food for millennia. Fermentations using yeast, for example, have been employed to produce bread, wine, and beer. Hops are renewable materials used in the production of beer. Can hops be used to produce anything other than beer? The typical composition of hops is shown in Table 22.5.

Solution Hops have, in fact, been used to produce a wide range of products in addition to being used for beer production.[15] From the typical composition of hops given above, resins

TABLE 22.5 Composition of Hops

Component	%
Cellulose and lignins	40
Resins (α and β acids)	15
Protein	15
Water	10
Ash	8
Tannins	4
Fats and waxes	3
Pectins	2
Monosaccharides	2
Essential oils	1

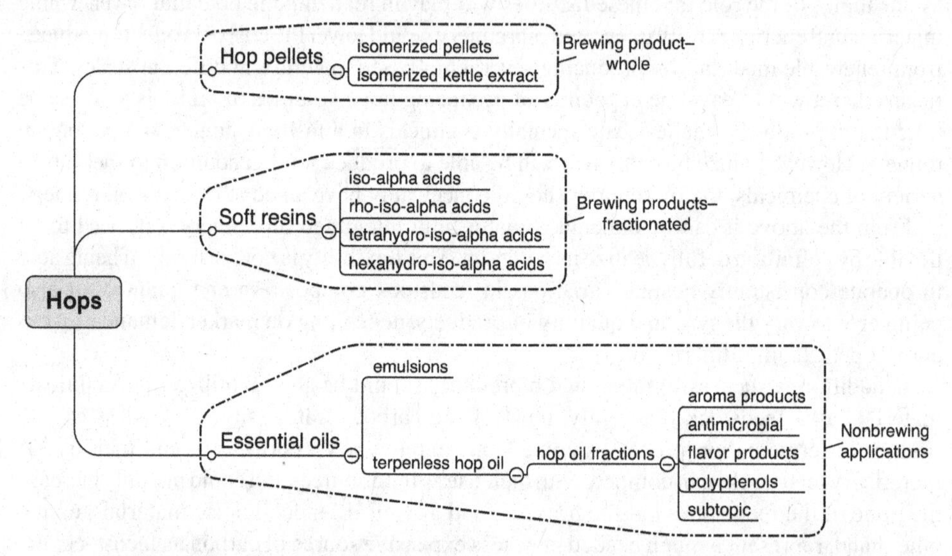

FIGURE 22.4 Some hops derivatives for brewing and nonbrewing applications.

and essential oils have been extracted using supercritical carbon dioxide to make a variety of commercial products currently offered by one company. The product range includes essential oils, herbal medicinal extracts, active pharmaceuticals and specialty chemicals, foam stabilizers and antifoam products for brewing, aroma products, and antimicrobial products. Figure 22.4 shows some of the products derived from hops.

Additional Points to Ponder. What other materials used for food and drink production can be utilized for other purposes? Are there any sustainability impacts that arise from their use? What would be the expected mass efficiencies for these types of products?

To start addressing the challenges noted above, and to identify them in a more strategic manner, the U.S. National Renewable Energy Laboratory (NREL) and the Pacific Northwest National Laboratory (PNNL) commissioned a study on biomass-generated building blocks. This study was commissioned to promote biorefineries that produce multiple products, including higher-value chemicals, fuels, and power. The study identified as many potential bio-based chemicals from all sources of biomass, then reduced the potential candidates to the 30 highest potential candidates. These 30 candidates were further narrowed to a list of the top 12 building blocks that could be obtained from biomass:[16]

1. 1,4-Succinic, 1,4-fumaric, and 1,4-malic acids
2. 2,5-Furan dicarboxylic acid
3. 3-Hydroxypropionic acid
4. Aspartic acid
5. Glucaric acid
6. Glutamic acid
7. Itaconic acid
8. Levulinic acid
9. 3-Hydroxybutyrolactone
10. Glycerol
11. Sorbitol
12. Xylitol/arabinitol

After the top building blocks were identified, the study analyzed the synthesis for each of them and their derivatives in two stages. The first stage was the transformation of the sugars into the building blocks, and the second stage was conversion of the building blocks to secondary chemicals. It was found that biological transformations account for a majority of the transformations from biomass to building blocks, but chemical transformations are the primary synthesis in the conversion of building blocks to secondary chemicals. The study also examined the common technical barriers and the R&D needs that could help improve the economics of producing these building blocks and derivatives.

22.4 RENEWABLE ENERGY

In the first part of this chapter we explored different possibilities for utilizing renewable materials. Another area of intense development for renewable materials is in the area of renewable energy. As world population continues to expand, there will be a considerable corresponding growth in energy demand.[17, 18]

As we have seen earlier, energy use relates not only to global warming, but also to other environmental impacts, such as smog formation, acidification, forest destruction, and others. These impacts need to be considered at the same time if we are to minimize environmental impacts. The first step to minimizing these environmental impacts is, of course, to design systems and processes that are more efficient in terms of materials and energy use, as we have seen earlier. The other step is to maximize the utilization of renewable sources of energy as much as possible.

The Brundtland Commission's Report stated four key elements of sustainable energy:

1. Sufficient growth of energy supplies to meet human needs
2. Energy efficiency and conservation measures
3. Addressing public health and safety issues from energy resources
4. Protection of the biosphere and prevention of more localized forms of pollution

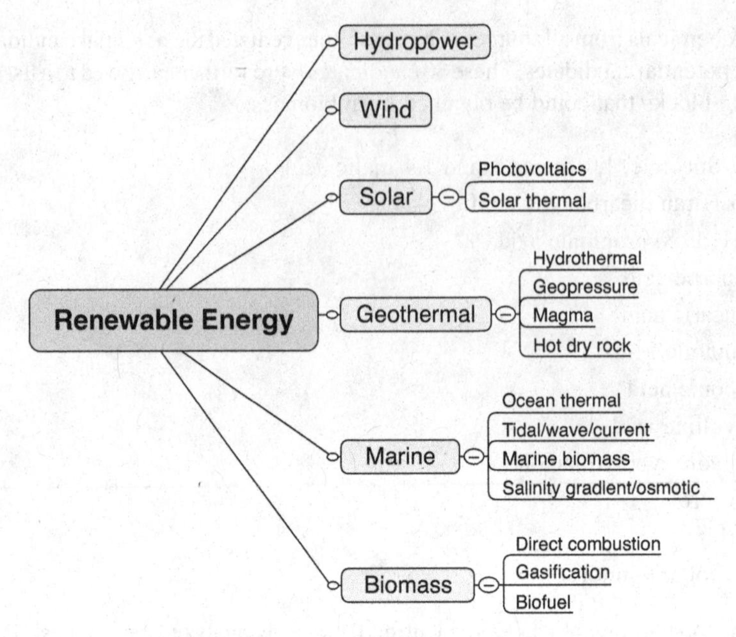

FIGURE 22.5 Renewable energy sources.

Since the energy crisis of the 1970s, there has been a focus on further developing technologies for renewable energy production. This research has built on traditional renewable sources such as hydropower and biomass and has expanded into new renewable technologies such as solar (Figure 22.5).[19] The use of renewable energy still does not account for a large proportion of the primary energy supply in most industrialized countries, and there has been very little improvement since 1990. Figure 22.6 shows the share of renewable energy used in electricity generation within European countries in 1990 and 2003.[20] In the following paragraphs we discuss some examples in the current mix of renewable energy sources that are being explored, both new and old.

22.4.1 Hydropower

Hydropower is still the most developed source of renewable energy, but the current market potential is difficult to establish. Some factors that limit further development of this alternative energy source include the remote locations of hydropower sites and the relatively high front-end capital expenditure. However, hydropower plants do have lower operational costs than thermal and nuclear options. In addition, there are important impacts to and changes in biodiversity, sedimentation, water quality standards, human health deterioration (e.g., malaria, schistosomiasis, Japanese B encephalitis), and downstream impacts. Technologies are being developed to include mitigation strategies for dam construction and design to minimize these impacts.

22.4.2 Wind Energy

Wind energy is broadly available, but at this time accounts for only about 0.1% of total global electricity. Denmark is the main exporter of wind turbine technology and aims to provide

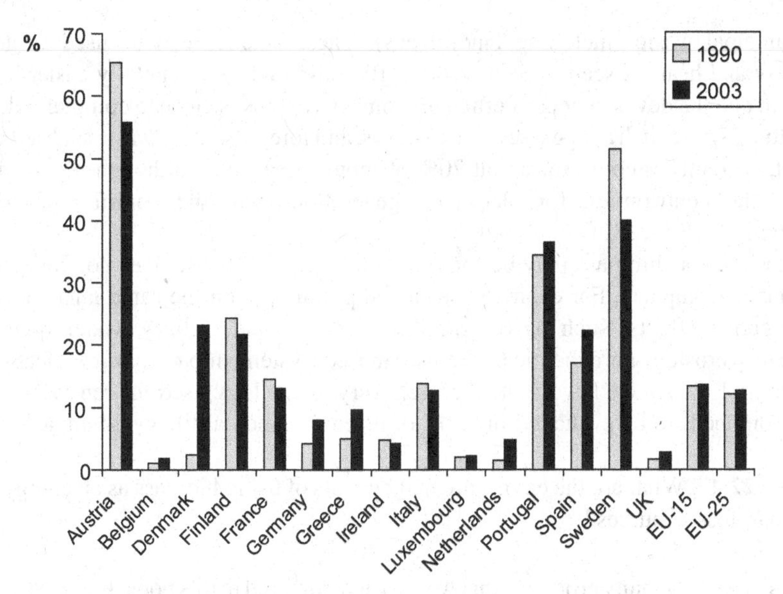

FIGURE 22.6 Percentages of renewable energy used in electricity generation in the European Union during 1990 and 2003. (From ref. 20.)

40 to 50% of the national electricity generation from wind power by 2030. In high-wind areas, wind power is competitive with other forms of electricity, and the global average price is expected to drop to 2.7 to 3 cents per kilowatthour by around 2020, due to economies of scale from mass production and improved turbine designs.[21] Some negative impacts associated with wind technologies include noise, the visual impact on the landscape, impacts on avian populations, and electromagnetic interference.

22.4.3 Geothermal Energy

The most utilized form of geothermal electricity generation involves the use of naturally occurring steam to power turbines. Geothermal energy has been used commercially for about 70 years, both for electricity generation and direct use, with increased use over the past three decades. Naturally occurring geothermal steam resources have been identified in more than 80 countries. The easiest type of steam to use is the "dry steam" found in the form of geysers. Increasing advances in our understanding of geothermal science and improved technology have increased the potential use of geothermal energy. Some of the environmental impacts associated with geothermal fluids include gases such as hydrogen sulfide, ammonia, radon, and mercury, although they are concentrated and disposed of as part of the wastewater. In general, these gaseous emissions represent only a minor fraction of the emissions from fossil or fuel electricity production.

22.4.4 Biomass Resources

Biomass is a rather generic term describing organic materials derived from plants of many different varieties. Biomass resources for energy conversion can be grouped into three main categories: biomass grown for energy content (e.g., biofuels), wood and agricultural waste, and municipal waste (including landfill gas). These sources can be used to generate electricity and heat, as seen in Section 22.4. Biomass fuels are generally easier to gasify than coal, and the development of efficient biomass systems is close to commercialization. Agricultural waste such as bagasse and rice husks, and forest products waste such as bark and sawdust, currently account for about 70% of biomass capacity, although it is difficult to separate the contribution for electricity generation from the contribution for heat production.[22]

As desirable as biomass may be for energy production, its use does not come without environmental impacts. For example, sustained production on the same land surface can have negative effects, such as the diminishment of land fertility, water quality and availability, erosion, agrochemical overuse and ecosystem impacts, eutrophication, and biodiversity loss. As we have stated before, every technology used for renewable energy production needs to be evaluated in terms of its environmental life cycle impacts.

Example 22.4 What are the environmental benefits of using biomass as an energy source rather than fossil sources?

Solution As with many problems that we have encountered in this book, the short answer is "it depends". It depends, of course, on the type of biomass used, if there is a mixture of different fuels, and the type of energy that is being generated. Table 22.6 shows, for example, the savings in global warming potential that are obtained from using wood sources for generating electric power.

Additional Points to Ponder. Why are the savings reduced when using 100% biomass? How do you think the rest of the environmental life cycle impacts will be affected?

22.4.5 Solar Energy

Photovoltaic solar cells convert sunlight directly into electricity. This can be done by flat panel and concentrator systems. Although solar energy is generally available everywhere on the surface of the Earth, there are variations in intensity associated with geographical location, the time of day, and season that will necessitate the use of energy storage. In addition, solar energy is diffuse by nature, so it requires a significant amount of equipment and land to concentrate the solar energy. Solar photovoltaics applications are often employed at the point of use (e.g., for buildings, off-the-grid rural areas, developing countries) to offset the costs and distribution losses. From an environmental life cycle assessment standpoint,

TABLE 22.6 Savings from the Use of Wood

Wood Source	Savings in GWP (g of CO_2-eq./g biomass)
5% co-fired with coal	0.97
15% co-fired with coal	1.02
100%	0.52

solar photovoltaics do not have emissions during the use phase, but there are significant emissions associated with the manufacture and decommissioning of systems. In addition, one of the most controversial issues is whether the amount of energy needed to build systems is smaller or larger than the energy produced during their entire useful lives. There are also concerns about the availability of the materials, as well as health and safety concerns in terms of use of the cadmium in the photovoltaic modules.

Solar energy can also be concentrated into a high-temperature heat source to produce electricity. Low-cloud areas such as desserts are probably most suitable for the use of solar concentrator collectors. It is estimated that if about 1% of the world's desert area were used for solar thermal power plants, there would be sufficient electricity to meet today's worldwide demand. Some examples of solar thermal electric conversion technologies include parabolic trough systems, power towers, dish/engine power plants, and other systems under development. All concentrating solar power technologies have four key elements: the collector/concentrator, a receiver, a mechanism for transport and/or storage, and power conversion. The main environmental concern of solar thermal energy is the requirement for sufficient land to accommodate the facilities.

The easiest application of solar energy is the direct conversion of sunlight for low-temperature heating (up to 100°C). This is achieved using either passive or active solar energy. Active conversion requires a solar converter, and the heat is transported through an appropriate medium, whereas passive conversion heats the process directly and no active components are used.

22.4.6 Renewable Energy Perspective

Table 22.7 is a summary of the barriers and impacts of some renewable energy sources and some potential strategies to overcome them. Use of renewable energy sources has the potential to meet current world energy demand many times over[23, 24] while contributing to sustainability and reducing local and global atmospheric emissions. They could also contribute to increasing the number of energy supply options for the developing world and rural areas. However, we have seen many times that every technology or system needs to be evaluated in terms of the sustainability impacts across the entire life cycle. Renewable energy technologies and options do present several impacts that need to be minimized and managed. However, in many cases these environmental impacts are not of the same magnitude as nonrenewable energy sources. In addition, a number of hurdles will have to be overcome to mainstream these technologies. For example, one common denominator is the need to further develop renewable technologies and increase production capabilities to leverage economies of scale, and this will require continuing research, development, and demonstration. Few renewable energy technologies currently can compete with conventional fossil technologies on a cost basis, so substantial cost reductions are needed.[25, 26]

The fact that technological development is needed to overcome both technical and cost barriers highlights the important role that innovation plays in the development of renewable energy in particular and of more sustainable processes in general. Innovation is, in general, the key that will unleash the potential of developing technologies and that will in turn drive more sustainable process. In Chapter 23 we discuss how to evaluate technologies and processes in a comprehensive manner to determine whether a given technological option would provide a more sustainable alternative.

TABLE 22.7 Some Renewable Energy Sources, Their Potential Barriers, and Potential Strategies to Overcome Those Barriers

Renewable Energy Source	Potential Environmental Impacts	Potential Barriers	Potential Strategies
Wind	Noise emission Visual impact on the landscape Avian population impacts Electromagnetic interference	Fluctuating demand for wind turbines Uncertainties in financing Project preparation time	Favorable tax systems and other economic incentives to stimulate wind energy Further development of the turbine technology
Biomass	Land fertility Water use Erosion Eutrophication Agrochemicals used Biodiversity loss	Uncompetitive costs Characteristics of specific biomass Need for development energy conversion technologies Public acceptability Competition for land use	Internalizing external costs Development of key conversion technologies Improvement of biomass production Development of biorefineries Development of policies to drive R&D
Solar photovoltaics (PVs)	Emissions during manufacturing and possibly on decommissioning Material scarcity (silicon, metals) Concerns with Cd-containing PV modules	Photovoltaic system costs PV electricity costs Health concerns with Cd-containing PVs	Development of a number of key technologies Lower the cost of access to space Solving health and environmental issues related to materials Improving designs for space-based solar systems
Solar thermal	Land use	Technology maturity	Development of technologies
Hydropower	Biodiversity perturbation Sedimentation Human health issues Downstream impacts Water quality standards	High capital investment Downstream social impacts	No flooding of large areas Inclusion of passage for fauna Inclusion of off-takes at various levels Removing leftover biomass in flooded areas and reservoir

PROBLEMS

22.1 A lignocellulosic biorefinery converts cellulosic material into a series of bioproducts. Investigate the types of chemical reactions that would be involved in this process.

22.2 What are some issues associated with the cultivation of renewable feedstock from a green chemistry/green engineering perspective? How can those issues be mitigated?

22.3 What are some issues associated with the cultivation of renewable materials process development from a green chemistry/green engineering perspective? How can those issues be mitigated?

22.4 From a green engineering/green chemistry perspective, what are the advantages and disadvantages of:

(a) A traditional petrochemical refinery
(b) A phase I biorefinery
(c) A lignocellulosic biorefinery
(d) A green biorefinery
(e) A whole crop biorefinery

22.5 Provide three examples of direct solar thermal energy use. Provide advantages and disadvantages of each example from a green chemistry/green engineering perspective.

22.6 The petrochemical and biorefinery processes for syngas are shown in Table P22.6. What can be inferred from the descriptions in the table in terms of green chemistry and green engineering?

TABLE P22.6 Syngas Production Processes

Petrochemical Process	Biorefinery Process
	Pyrolysis:
$CH_4 + H_2O \leftrightarrow CO + 3H_2$	$C_6H_{10}O_5 \rightarrow 5CO + 5H_2 + C$
Nickel oxide catalyst, 300°C, 30 atm	Partial oxdiation:
$CO + 2H_2 \leftrightarrow CH_3OH$	$C_6H_{10}O_5 + O_2 \rightarrow 5CO + CO_2 + 5H_2$
$CO_2 + 3H_2 \leftrightarrow CH_3OH + H_2O$	Steam reforming:
Cu and Zn catalyst, 300°C, 100 atm	$C_6H_{10}O_5 + H_2O \rightarrow 6CO + 6H_2$

22.7 What is the commercial significance of 3-hydroxypropionic acid as one of the top 12 building blocks from biomass? Draw a product map for this biomaterial and highlight chemicals used currently in its commercial production.

22.8 (a) Draw a product map of the vegetable oil biorefinery platform. From the product map, develop a product submap for 9-decenoic acid.

(b) What are the advantages of deriving these products from renewable sources compared to deriving them from petroleum in the green chemistry/green engineering context?

22.9 (a) Could energy from biomass alone satisfy current energy demands? Justify your answer.

(b) How about other sources of renewable energy?

22.10 Methane can be produced from biomass by either thermal gasification or anaerobic digestion. Anaerobic digestion is typically carried out at low temperatures and can convert wet or dry feeds. The products are mainly methane and carbon dioxide.[27] Methane from biomass can, in turn, be used either as a feedstock or as an energy source. Investigate the energy potential of methane derived from biomass and biological waste.

22.11 Cyanobacteria have been proposed as a potential source of renewable energy.[28]

(a) What mechanism has been proposed for using cyanobacteria to generate renewable energy?

(b) Would this type of process qualify as renewable energy? Support your answer.

22.23 How do renewable energy sources compare to fossil energy sources from a life cycle perspective?

(a) Provide your answer in terms of energy, resources, global warming potential, acidification potential, and eutrophication.

(b) What conclusions can be drawn from a green chemistry/engineering perspective?

REFERENCES

1. Elsayed, M. A., Matthews, R., Mortimer, N. D. *Carbon and Energy Balances for a Range of Biofuels Options.* Crown Copyright, Project B/B6/00784/REP. URN 03/386. 2003.
2. National Research Council. *Biobased Industrial Products: Priorities for Research and Commercialization*. National Academies Press,Washington, DC, 2000.
3. Kamm, B., Kamm, M. Principles of biorefineries. *Appl. Microbiol. Biotechnol.*, 2004, 64, 137–145.
4. Dale, B. "Greening" the chemical industry: research and development priorities for biobased industrial products. *J. Chem. Technol. Biotechnol.*, 2003, 78, 1093–1103.
5. Patel, M., et al. *Medium and Long-Term Opportunities and Risks of the Biotechnological Production of Bulk Chemicals from Renewable Resources: The Potential of White Biotechnology.* The BREW Project. Prepared under the European Commission's GROWTH Programme (DG Research), Utrecht, The Netherlands, June 2006.
6. Gavrilescu, M., Chisti, Y. Biotechnology: a sustainable alternative for chemical industry. *Biotechnol. Adv.*, 2005, 23, 471–499.
7. Energetics, Inc. *Industrial Bioproducts: Today and Tomorrow*. Prepared for the U.S. Department of Energy, Office of Energy Efficiency and Renewable Energy, Office of the Biomass Program Washington, DC, July 2003.
8. Rass-Hansen, R., Falsig, H., Jørgensen, B., Christensen, C. H. Bioethanol: fuel or feedstock? *J. Chem. Technol. Biotechnol.*, 2007, 82, 329–333.
9. Clark, J. H. Perspective. Green chemistry for the second generation biorefinery: sustainable chemical manufacturing based on biomass. *J. Chem. Technol. Biotechnol.*, 2007, 82, 603–609.
10. Pollard, G. Catalysis in renewable feedstocks. In *A Technology Roadmap.* CR5676. BHR Solutions Project 180 2421. Prepared on behalf of the Department of Trade and Industry, BHR Group Ltd., Cranfield, Bedfordshire, UK, 2005.
11. Fernando, S., Adhikari, S., Chandrapal, C., Murali, N. Biorefineries: current status, challenges, and future direction. *Energy Fuels*, 2006, 20, 1727–1737.
12. Kamm, B., Kamm, M. Biorefinery: systems. *Chem. Biochem. Eng. Q.*, 2004, 18(1), 1–6.
13. Kamm, B., Kamm, M. Biorefineries: multi product processes. *Adv. Biochem. Eng./Biotechnol.*, 2007, 105, 175–204.
14. Ohara, H. Biorefinery. *Appl. Microbiol. Biotechnol.*, 2003, 62, 474–477.
15. Botanix, Ltd.http://www.botanix.co.uk/index.html, *accessed Sept. 7*, 2009.

16. Werpy, T., Petersen, G. , Eds. DOE/GO-102004-1992. Pacific Northwest National Laboratory (PNNL). National Renewable Energy Laboratory (NREL), Office of Biomass Program (EERE), Springfield, VA, 2004.
17. World Energy Council.Global Energy Perspectives to 2050 and Beyond. Technical report. WEC, London, 1995.
18. World Commission on Environment and Development. *Our Common Future.* Oxford University Press,Oxford, UK, 1987.
19. Dincer, I. Renewable energy and sustainable development: a crucial review. *Renewable Sustainable Energy Rev.* 2000, 4, 157–175.
20. Jefferson, M. Sustainable energy development: performance and prospects. *Renewable Energy*, 2006, 31, 571–582.
21. Sims, R. E. H., Rognerb, H. H., Gregory K. Carbon emission and mitigation cost comparisons between fossil fuel, nuclear and renewable energy resources for electricity generation. *Energy Policy*, 2003, 31, 1315–1326.
22. McVeigh, J., Burtraw, D., Darmstadter, J., Palmer, K. Winner, loser, or innocent victim? Has renewable energy performed as expected? *Solar Energy*, 2000, 68, 237–255.
23. Turkenburg, W. C. , et al.Renewable energy technologies. In *World Energy Assessment: Energy and the Challenge of Sustainability*,Goldemburg, J.,et al., Eds. United Nations Development Project,New York, 2000, Chap. 7, pp.219–272.
24. Turner, J. A. A realizable renewable energy future. *Science*, 1999, 285, 687–689.
25. Haas, R. , et al.How to promote renewable energy systems successfully and effectively. *Energy Policy*, 2004, 32, 833–839.
26. Sims, R. E. H., Rognerb, H. H., Gregory, K. Carbon emission and mitigation cost comparisons between fossil fuel, nuclear and renewable energy resources for electricity generation. *Energy Policy*, 2003, 31, 1315–1326.
27. Chynoweth, D. P., Owens, J. M., Legrand, R. Renewable methane from anaerobic digestion of biomass. *Renewable Energy*, 2001, 22, 1–8.
28. Hansel, A., Lindblad, P. Towards optimization of cyanobacteria as biotechnologically relevant producers of molecular hydrogen, a clean and renewable energy source. *Appl. Microbiol. Biotechnol.*, 1998, 50, 153–160.

23

EVALUATING TECHNOLOGIES

What This Chapter Is About As new and emerging technologies and processes move along the typical development curve, there is a need to assess if the new technologies will perform better than the traditional technologies from a green engineering standpoint. In this chapter we explore an approach that may be used to compare different processes and technologies from a green engineering perspective, and we provide some examples of comparisons between emerging and traditional technologies.

Learning Objectives At the end of this chapter, the student will be able to:

- Understand different factors that may play a role in a given technology's performance from a green engineering standpoint.
- Understand how different metrics can be used to compare technologies and processes.
- Identify trade-offs that might be associated with different metrics and propose alternatives to manage potential disadvantages of the technologies being evaluated.
- Apply a methodology to compare technologies from a green engineering standpoint.

23.1 WHY WE NEED TO EVALUATE TECHNOLOGIES AND PROCESSES COMPREHENSIVELY

As we have seen in previous chapters, if chemists and engineers want to design sustainable technologies and processes, there is a need to consider two interrelated components: green

Green Chemistry and Engineering: A Practical Design Approach, By Concepción Jiménez-González and David J. C. Constable

chemistry and green engineering. It is essential that the two components be developed in parallel to achieve a more sustainable process.

As we have seen before, most chemists tend to focus on reactions rather than the technology around a reaction. That is, if a reaction does not work, chemists are more inclined to change the reaction rather than to investigate different ways in which to perform the reaction. In general, mass and energy (heat/cool) transfer, mixing, phase transfer, and general reactor design, are not pursued as rigorously by the synthetic organic chemist as by the engineer. However, if these process parameters are not considered thoroughly, they may result in poor quality, reduced yields, and rather large inefficiencies during process development and perhaps in the final process. Given increased and ever-increasing pressures to reduce time to market, the diversity of potential products in development, and the demand for greater diversity in products introduced to market, chemists simply must work with engineers in a collaborative manner to increase mass and energy efficiency.

Having said that, chemists are currently faced with the very difficult task of having to select and evaluate the most appropriate technology for each reaction and separation in a synthetic route, and in many cases must select between two synthetic routes that require different reaction and separation technologies. As we saw in Chapter 2, from the green chemistry and green engineering perspective there are many aspects to be considered when selecting different processes and technologies, including product quality, operability, efficiency, environment and safety, and of course, economics. Given all the metrics we have studied in this book, it can be easily seen that one of the challenges that chemists and engineers face in integrating sustainability into process design is how to evaluate and compare a technology or process from a holistic and green point of view so that the best alternatives are chosen.

23.2 COMPARING TECHNOLOGIES AND PROCESSES

Several methodologies have been developed to compare technologies and processes from a green engineering viewpoint.[1–7] Any methodology to compare technologies or processes, either traditional or emerging, needs to have at least the following characteristics:

- A strong scientific and technological basis
- A simple-to-understand approach that is user friendly (simple but not simplistic)
- Employment of an easy-to-use tool that is readily accessible to users
- A flexible design that allows one to compare traditional with existing technologies
- A transparent evaluation process used to develop any guidance

To these general characteristics, we add some thoughts and recommendations for comparing technologies or processes, especially when talking about new and emerging technologies.

1. *Target-oriented comparison.* When comparing different processes or technologies, it is advisable to focus on the desired outcome, objective, or goal of the operation. This is comparable to setting the functional unit in a life cycle assessment. Special care needs to be taken when defining the desired outcome, objective, or goal, as sometimes it is easy to become too focused on comparing only two alternatives closely at hand, A and B, and miss a different opportunity, C, not initially contemplated.

2. *Metrics*. Many sustainability and green chemistry and engineering metrics have been discussed in this book and elsewhere,[8,9] but the main point to highlight is to select metrics that answer the right questions. It will be no surprise that we think the metrics should include environmental, health, safety, and operational aspects within a framework of life cycle thinking. It is also important to have metrics at several levels: high-level metrics to communicate to decision makers, and more detailed metrics to substantiate the high-level metrics and provide additional detail for scientists and engineers to modify and improve their processes.

3. *Case scenario vs. response surfaces*. Different processes or technologies tend to have different applications. One of the big challenges we face in developing and using comparative methodology is the decision as to whether we use a case scenario or generic assessment approach for technologies whose performance may be process-specific. Generic assessments typically require significantly more work, as more assessments will be needed to develop an appropriate data set. One option that can be used to develop such a data set is to choose a case scenario approach to gather a sufficient number of data points. If a case scenario approach is undertaken, one must carefully choose processes that are as relevant and representative as possible. In any case, the information and the assessments are fully documented.

4. *Scores*. Once the metrics we want to use to evaluate various technologies are selected, it is recommended that one use relative scales to produce comparative scores. This is useful to help highlight significant differences and identify trade-offs among the aspects to be evaluated.

23.3 ONE WAY TO COMPARE TECHNOLOGIES

In the paragraphs below we describe one methodology that has been used to compare technologies, the details of which have been published.[10] It has been utilized by a major international pharmaceutical company to build a green technology guide for pharmaceutical unit operations.[11] There are, of course, other methodologies that can be used to undertake these sorts of comparisons, and this comparative assessment methodology is presented here to illustrate one such approach.

23.3.1 Definition of the Target and Alternative Technologies

This methodology utilizes the case scenario approach but applies life cycle assessment metrics. As noted above, the first step of any case scenario is rigorous definition of the particular objective or goal that is desired. For example, the objective may be the removal of solvent on the basis of kilograms of product crystallized or recovered. The goal or objective that defines each case scenario may comprise an entire reaction system (e.g., a separative reactor), a unit operation to achieve separation (e.g., purification, recovery, waste treatment or recycle, or cleanup), or an entire process. The objective should be defined quantitatively and include any important qualitative or quantitative constraints. For example, objectives might include such things as the purification of 1 kg of crystalline product, the removal of 90% of the solvent from a reaction mixture, or the recovery of substance X present in an inlet stream at a concentration of 100 g/L. The added constraints may be that the product is temperature sensitive or that the recovered solvent purity must be 95% or greater.

After the objective is clearly defined, suitable technologies to accomplish the objective are identified. Such technologies could be either traditional or emerging, as long as there is evidence that the case scenario objectives may be achieved reproducibly. In other words, special care needs to be taken to make sure that the alternatives are applicable and can actually be implemented. This step, even though it may appear to be obvious, is crucial for selecting technologies to be evaluated. There also needs to be sufficient information to perform the assessment. Additional technologies not covered in the study that could be employed to accomplish the same objective should be mentioned within the definition of the scenario.

In addition, when dealing with an emerging technology, it is important to highlight its degree of development. It is common for emerging technologies to be downplayed because they are compared with well-developed processes or technologies that have been optimized over a period of years. It is therefore important to level the playing field to assure that the emerging technologies will be compared with their potential performance at a given stage of development. Conversely, in early stages of development the output of the comparison can be used as a benchmark for setting improvement targets.

23.3.2 Metric and Indicator Definition

A group of core and complementary metrics were proposed that provide the best differentiation between the technologies selected. For this methodology, four principal categories are scored: environment, energy, efficiency, and safety. The score for each of these categories is generated based on measurable metrics or indicators, as shown in Fig. 23.1. A life cycle

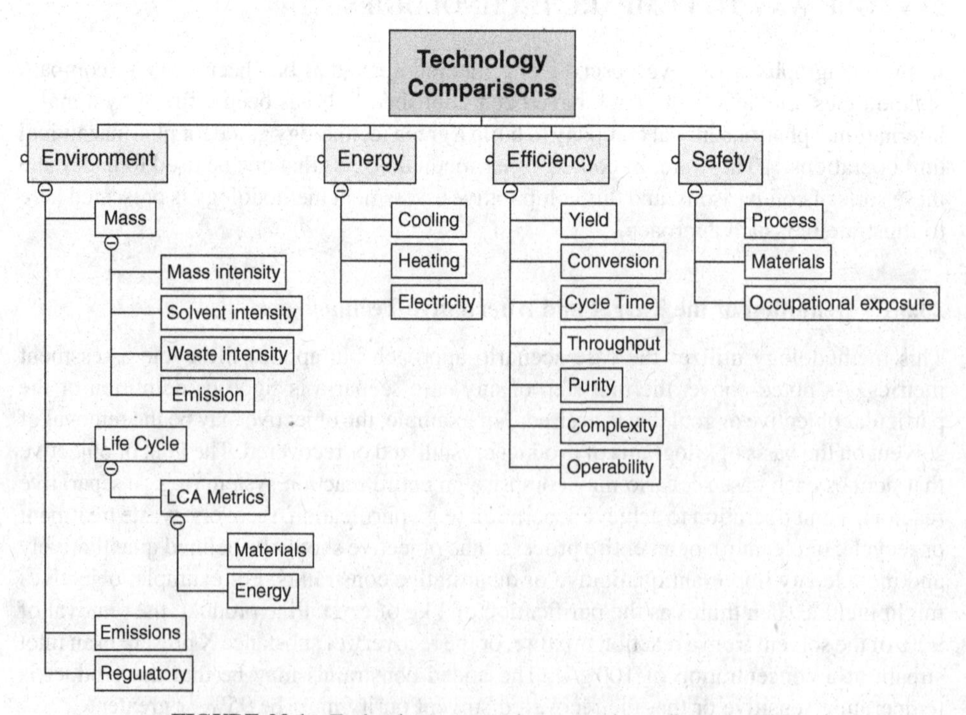

FIGURE 23.1 Technology comparison: metrics and categories.

assessment approach ensures a sufficiently broad view and facilitates a sustainability viewpoint. Where possible and appropriate, these indicators include cradle-to-gate environmental life cycle inventories that take into consideration emissions for the extraction and transformation of raw materials, energy, and even for waste treatment. As we saw in previous chapters, this is especially important when considering energy. Different utilities, such as steam and electricity, have different associated emission profiles. Only a qualitative view for safety and operational considerations was defined, but it is possible to include quantitative measurements for these two categories.

23.3.3 Data Gathering

Once the technologies have been selected, we need to collect data to estimate the metrics. Industrial-scale data or process validation data is extremely valuable and is preferred, but such information is not always available. Full- or pilot-scale data are usually easier to obtain with in-house and traditional technologies and rather difficult to obtain for emerging technologies. In the case of emerging technologies, bench-scale data are normally the only information available. When only bench-scale data are found, it should be noted in the description of the technology and there should be an assessment of the data quality for scaling-up to pilot or full scale. Data required for the assessment are identical to those required for a mass and energy balance for the system, such as principles of operation, process flow diagrams, mass flows, efficiencies, temperatures, pressures, concentrations, toxicological data, and hazard characteristics.

23.3.4 Estimation of Metrics

This step is for calculating the metrics defined above using the data gathered. Table 23.1 shows some examples of a few common metrics, although more (or fewer) metrics may be needed to properly define and compare the technologies. The mass metrics include both environmental impacts and raw material utilization (e.g., emissions, mass intensity). The energy indicators evaluate energy consumption for each alternative.

To derive the mass and energy metrics, the mass and energy balance principles described in Chapters 9 and 12 are used. The basis for the mass and energy balance calculations should be the same as the objective of the scenario (e.g., 1 kg of product, 1 kg of waste treated) and for each technology alternative evaluated. The process emissions metrics should be derived from such things as the chemical reaction conversion efficiency, process separation efficiency, physical properties of the materials, and product and by-product formation. Once the mass and energy balances are completed for each technology alternative, the mass indicators are calculated and tabulated for comparison. All data and any assumptions made should be transparent.

For the life cycle metrics, the principles discussed in Chapters 16, 18, and 19 were applied. The emissions for energy and the production of raw materials are added to the unit process emissions to obtain an estimate of the total emissions generated throughout the life cycle of the unit operation. Life cycle information was extracted from databases and literature to estimate the life cycle burdens associated with raw material and energy production.[12–22]

Significant quantitative and qualitative operational differences that affect process efficiency (e.g., yield, conversion, separation efficiency) and quality (e.g., percent

TABLE 23.1 Examples of Mass and Energy Metris

	Definition
	Mass Metric
Mass intensity (MI)	$\mathrm{MI} = \frac{\text{total mass input to the process, excluding water}}{\text{basis of the mass balance calculations}}$
Waste intensity (WI)	$\mathrm{WI} = \frac{\text{total waste produced}}{\text{basis of the mass balance calculations}}$
Emissions of specific compounds released (E_i)	$E_i = \frac{\text{amount of compound } i \text{ released as an emission}}{\text{basis of the mass balance calculations}}$ (a separate indicator is calculated for each compound released)
	Energy Metric

$$\text{total heating requirements} = \sum_n \frac{mC_p(T_2-T_1)}{\eta}\bigg|_{T_2>T_1} + \sum_n \frac{UA(T_w-T)t}{\eta}\bigg|_{T_w>T} + \sum_n \Delta H_R^T\bigg|_{\text{endothermic}} - \text{heat recovered}$$

$$\text{total cooling requirements} = \sum_n \frac{mC_p(T_2-T_1)}{\eta}\bigg|_{T_1>T_2} + \sum_n \frac{UA(T_w-T)t}{\eta}\bigg|_{T>T_w} + \sum_n \Delta H_R^T\bigg|_{\text{exothermic}} - \text{heat recovered}$$

Sensible heat	$Q = \frac{mC_p(T_2-T_1)}{\eta}$	
Heating or cooling a vessel at a constant temperature	$Q = \frac{UA(T_w-T)t}{\eta}$	
Heat of reaction	$\Delta H_R^T = \frac{m}{\eta}\left(\Delta H_R^{298} + \int_{298}^{T} \nu\, C_p\, dT\right)$	
Refrigeration	$Q = 1.2\,Q'$ $E = \frac{Q'(T_1-T_2)t}{\eta T_1}$	

Variables:
- ν = change in stoichiometric coefficients
- η = efficiency
- ΔH_R^T = heat of reaction at temperature T
- C_p = heat capacity at constant pressure
- E = electricity requirements
- m = mass
- Q = heating or cooling requirements
- Q' = refrigeration requirements
- T = temperature of the system
- t = time
- T_1 = initial temperature
- T_2 = final temperature
- T_w = temperature of heating/cooling media

product, product/impurities ratios) of the outcome or objective must be identified and assessed for each alternative and compared. Other considerations include the operational ranges and limits, typical processing time or reactor or unit operation residence times, process control issues, selectivity issues, the type of operation (e.g., continuous vs. batch), potential operational problems (e.g., fouling, blockages), and ease of scaling up the technology.

Potential material and process safety issues are qualitatively identified. Since the analysis is typically based on calculations and an estimation of the operating conditions, safety issues related to the placement of the process or unit operation in facilities are more difficult to evaluate but need to be included as more information becomes available. Materials with known and reported hazards, such as the potential for flammability, adverse reactions, explosivity, corrosivity, and toxicity should be identified. Material hazard information may be found in a variety of places, including material safety data sheets, SAXs,[23] other published compendia, or a variety of Internet sites. Process conditions that lead to extremes in pressure or temperature should be identified through calorimetry studies and modeling and noted along with any process condition that would tend to make it easier for a process release, uncontrolled reaction, or similar incident to occur.

23.3.5 Comparative Ranking

Once all the technology options are evaluated, a relative rank is assigned to each category (i.e., environmental, energy, safety, efficiency). The ranking is performed by technical experts and is based on the results of the technology analysis and the values for all the metrics. Technical experts are used because technology comparisons need to be made in their appropriate context, and there is a blend of quantitative and qualitative results. In some instances, absolute numbers or ranges (e.g., mass or energy use) may be compared; in other cases, semiquantitative or qualitative relative comparisons must be employed.

The applicable indicators in each category are assigned a numerical value of 0, 5, or 10. This range of 0 to 10 is merely used as a relative scale that is assumed to be easy to apply and understand. A value of zero is given if the technology is perceived as having a disadvantage, 10 if it is perceived as having an advantage, and 5 if the indicator is not perceived as a significant advantage or disadvantage. Once the indicators are assigned a relative numerical ranking, the arithmetic average of the indicators comprising each category is calculated and a color code is assigned. For an average ranking lower than 2.5, the category color is red. If the average ranking is equal to or higher than 2.5 but lower than 7.5, the category color is yellow. Finally, if the average ranking is 7.5 or higher, the category color is green. A visually simple presentation of the comparison is employed using the color coding presented in Table 23.2.

The main purpose of the comparative ranking is to identify adverse issues, favorable characteristics, and possible trade-offs among the alternatives that facilitate an informed and perhaps more sustainable business decision. The ranking is limited by the comparison of

TABLE 23.2 Color Code for Comparative Ranking

Color	If...	Description
Green	Score ≥ 7.5	Technologies considered to have significant advantages
Yellow	$2.5 \leq$ score < 7.5	Technologies with no significant advantages or disadvantages.
Red	Score < 2.5	Technologies considered to have significant disadvantages

TABLE 23.3 Laboratory Conditions

Parameter	Qualification Batch	Laboratory Vortex Mixer
Intermediate A required	125 kg/batch (393.7 mol)	2.5 kg/h (7.9 mol/h)
Weight % of intermediate A in THF	30.2% (w/w)	27.3% (w/w)
Chloroacetonitrile use	1.12 equiv.	1.2 equiv.
PTC use	4.4 kg/batch	90 g/h
Concentration of base	30% (w/w) KOH	30% (w/w) KOH
Intermediate B produced	114.7 kg/batch (325.8 mol)	2.4 kg/h (7.1 mol/h)[2]
Reaction temperature	0–10°C	Room temperature

each case scenario and by the added subjectivity of the ranking. General conclusions about the technologies in all applications and conditions should not be drawn from the rankings in a given case scenario. At this stage a degree of subjectivity is involved, especially for nonnumerical metrics (e.g., operability, material safety).

Example 23.1 Look back at Example 15.1, where we discussed vortex mixers. In that example we estimated the energy requirements of a process using vortex mixers and a batch process. Laboratory conditions for the vortex mixer and batch conditions are given in Table 23.3: How would the two processes compare using the methodology described above? You can refer to Example 15.1 for the process descriptions.

Solution

Mass metrics. The production of 1 kg of intermediate B was used as the basis for calculating the mass and energy balances. The nonisolation and vortex mixer processes have better results for mass, waste, and solvent intensity. The improved performance is due primarily to lower solvent use, but in the case of the vortex mixers, it is also due to an improvement in the yield. The results for the mass intensity, solvent intensity, and waste intensity for each option are provided in Table 23.4.

Energy metrics. In Example 15.1 the energy requirements were estimated, so there is no need to recalculate. Our results then were as shown in Table 23.5:

TABLE 23.4 Mass Metric Results

Mass Metrics	Vortex Mixer	Batch Reactor
Mass intensity (kg/kg intermediate B)	10.2	15.9
Added solvent intensity (kg/kg intermediate B)	0.9	8.9
Waste intensity (kg/kg intermediate B)	9.2	14.9

TABLE 23.5 Energy Requirement Results

Energy Requirements	Vortex Mixer	Batch Reactor
Heating (MJ/kg intermediate B)	0.51	3.17
Cooling with cooling water (MJ/kg intermediate B)	−0.51	−3.23
Refrigeration (MJ/kg intermediate B)	0	−0.5
Electricity (MJ/kg intermediate B)	0.36	0.53

TABLE 23.6 LCI for Selected Solvent and Energy

	Emission (g/kg B)	
	Vortex	Current
Air emissions		
CO_2	1.16×10^4	6.43×10^4
CO	3.59×10^2	1.96×10^3
CH_4	1.78×10	1.04×10^2
NMVOC	4.22×10	2.50×10^2
NO_X	4.84×10	2.73×10^2
SO_X	5.29×10	2.94×10^2
Water emissions		
BOD	6.77×10^{-1}	4.09E
COD	0.189714	1.11×10^2
TDS	3.68×10	2.04×10^2
Solid waste	1.10×10^4	5.96×10^4

Simplified life cycle approach. A simplified life cycle inventory for energy and solvent production was included in the comparison of the three processing options. Life cycle inventory work has demonstrated that solvent production and use account for the greatest proportion of life cycle impacts in a pharmaceutical manufacturing context, so this approach is not without merit.[24] Using this simplified approach, the use of a vortex mixer results in considerably lower life cycle emissions than for the batch and nonisolation processes. For example, as can be seen in Table 23.6, CO_2 emissions associated with vortex mixer use are about 20% of the CO_2 emissions associated with use of the current batch reactor and 30% of the CO_2 emissions associated with use of the nonisolation batch reactor process.

Safety indicators. In general, all processing options described above can be operated safely, and there are only relatively minor safety differences for this case scenario. Both batch processes operate at low temperatures (around 0 °C), whereas the vortex mixer system operates at ambient temperature. It should be noted, however, that vortex mixers represent a potentially safer operating environment, due to the small volumes in the mixing chambers and short residence times of the substances in the system at any given time. This reduces the overall risks of accidental fire or explosion. Furthermore, reduced solvent use in the vortex mixer and nonisolation processes have the potential to reduce volatile organic compound emissions during solvent handling and processing, thereby reducing the potential for occupational exposure to solvent vapors.

Efficiency and operational indicators. The vortex mixer process has the following operational advantages over the current batch reactor process:

- Continuous operation of traditional batch processes
- Improved mass and heat transfer
- Improved control of reactant and reagent residence times and temperature profile
- Reduced solvent use
- The potential for fewer work-ups if continuous separation operations are used

TABLE 23.7 Color-Coded Ranking

Alternative	Environment	Safety	Efficiency	Energy
Batch reactor	Red	Yellow	Yellow	Red
Vortex mixer	Green	Yellow	Green	Green

- Generally higher selectivity, yield, and quality, especially for exothermic and fast reactions
- Linear scale-up (larger-scale vortex mixers will yield the same results given that the residence time and flow velocities used at laboratory scale are matched) or the ability to easily number-up
- Faster development of new processes or substances

Based on the results from an analysis of the three processing options described above, the reaction systems were comparatively ranked. A color-coded summary of the comparative ranking is presented in Table 23.7.

These results suggest that the vortex mixer is a better option for this particular process.

Additional Points to Ponder How would a batch process with no isolations fare in this assessment? How would vortex mixers fare in comparison with static mixers?

23.4 TRADE-OFFS

In Example 23.1, it was very clear that for the case scenario evaluated, the vortex mixer technology performed better than the batch process. However, this is not always the case, and very often there are trade-offs among technologies. How do we deal with trade-offs? It will all depend on the reason for comparing the technologies:

- When evaluating an emerging technology against a well-developed technology, trade-offs can be used to refine the design of the new technology until the trade-off is no longer present.
- When selecting between two technologies, it may become apparent that there is a need to search for additional alternatives that would accomplish the same function. For example, have all the options been considered? Can any of the technologies be modified further?
- Ultimately, most technology alternatives are not perfect, and perhaps the most important part of a standardized approach to comparing alternatives is that issues are highlighted and can be mitigated and/or managed.

Example 23.2 Two technologies, A and B, have been compared using the methodology described above. The high-level comparison is shown in Table 23.8. Which technology would you choose?

Solution As we can see, technologies A and B have obvious trade-offs regarding energy and safety. It is difficult to provide a simple answer given the high-level review and lack of

TABLE 23.8 Technology Comparison

Technology	Environment	Safety	Efficiency	Energy
A	Yellow	Red	Yellow	Yellow
B	Yellow	Yellow	Yellow	Red

detail in the example; however, selection between the technologies ultimately relies on which of the issues might be easier to mitigate or manage. Are there ways to reduce the energy consumption of technology B? Is it relatively straightforward to enhance safety controls for technology A? As the other categories appear to be comparable, several options can be recommended:

- Perform an economic comparison of the investment needed to make the scores for both technologies yellow or green (e.g., how much would it take to make them comparable in all aspects), and select the one that requires the least investment.
- Consider additional technologies to those analyzed here (i.e., technologies C, D, etc.)
- At the end of the day, it might be that we could choose either of the technologies but will do so fully aware of the shortcomings and the need to manage the issues.

Additional Point to Ponder If we considered additional technologies and there were more trade-offs, how would you handle them?

23.5 ADVANTAGES AND LIMITATIONS OF COMPARING TECHNOLOGIES

The main purpose of comparing technologies using a standardized methodology is to provide comparisons that highlight potentially adverse issues, favorable characteristics, and possible trade-offs among unit operations or manufacturing process alternatives. Ultimately, standardized comparisons should at the very least facilitate more informed engineering and business decisions.

The methodology and the ranking process covered in this chapter are limited by the unit operations or manufacturing processing options that are being compared in each case scenario and to a lesser extent, by the subjectivity of the experts ranking the options. These limitations also mean that general conclusions about the technologies should not be drawn from the specific rankings developed for a given case scenario if the technologies are applied using very different conditions or scenarios.

Despite these limitations, this methodology represents a standardized, documented, and systematic approach to performing technology comparisons. In addition, the methodology integrates environmental, health, and safety considerations with efficiency and operational aspects to assist with choosing technologies and equipment.

As has been demonstrated in the examples in this chapter, clear differentiation among technology options is possible, and the results of the analysis may be presented and communicated in a very concise, clear, and easy-to-understand manner. At the same time, the detailed analysis and calculations are transparent and easy to share if desired. As chemists and engineers expand their understanding of what constitutes green or clean technologies, we should be able to move at a faster pace toward more sustainable business practices.

PROBLEMS

23.1 You are evaluating technology A, a well-understood, well developed technology, against technology B, an emerging technology. How would you account for the different degree of development of the new and traditional technologies?

23.2 We noted earlier that other methodologies are available for a systematic comparison of technologies from a green engineering or sustainability perspective.

(a) Investigate two of these technologies and explain at a high level the basis for the comparisons.

(b) How do they differ from the methodology described in this chapter? Highlight the advantages and disadvantages.

23.3 The metrics in Table P23.3 were calculated for a fermentation process and a chemical process for the production of an active pharmaceutical ingredient (API). Both processes use organic solvents in the workup, and there is the possibility of recovering the solvent at different rates. You are the head of technology development and have to select one of the processes. Which would you recommend?

TABLE P23.3 Metrics for API Production

Parameter	Chemical Process	Fermentation Process
Purity (%)	98	98
Yield (mol%)	19	39
Cycle time (months)	18	3
Mass intensity excluding water (kg/kg API)	350	468
Mass productivity	0.3	0.2
Solvent intensity, excluding water (kg/kg API)	332	452
% Solvent	95	97
Water mass (kg)	118	259
Solvent recovery mass	114	405
E-factor kg waste/kg API, including wastewater	467	726
kg waste/kg API, including wastewater, with solvent recovery	353	321
Materials of concern	DCM, hexane	Pentane
Occupational exposure limits	Dichloromethane, 50 ppm	Acetic acid, 10 ppm TWA 8 h
	Diisopropylamine, 5 ppm	Glycerol, 10 μg/m^3 TWA 8 h
	HCl (as gas), 2 ppm ceiling	Peracetic acid, 1 ppm TWA 8 h
	Hexane, 50 ppm	Sodium hydroxide, 2 μg/m^3 ceiling
	MTBE, 50 ppm	Sulfuric acid, 0.2 μg/m^3 TWA 8 h
	Sodium hydroxide, 2 μg/m^3 Ceiling	

TABLE P23.3 *(Continued)*

Parameter	Chemical Process	Fermentation Process
	Sulfuric acid, 0.2 μg/m^3 TWA, 8 h THF, 50 ppm	
Solvents	DCM, ethyl acetate, heptane, Hexane, methanol, methyl acetate, THF	Acetic acid; acetone, heptane; pentane
Process energy (MJ/kg API)	693	1,829
Waste treatment energy (MJ/kg API)	1,382	2,121
Life cycle energy (MJ/kg API)	7,281	15,211

23.4 Use the technology comparison methodology to assess the chemical and biocatalytic routes to 7-ACA, as presented in Example 8.1. Are the conclusions the same as in Example 8.1? The descriptions follow:

(a) Chemical route description. A four-step process is used to convert the potassium salt of cephalosporin C to 7-ACA, as shown in Figure 8.2(a). In the first step, a common protection strategy is used convert the acid to an anhydride and the amine to an amide using chloracetyl chloride in the presence of the base, dimethyl aniline. Next, phosphorous pentachloride is added to the mixed anhydride, which is held at −37 °C to form the imodyl chloride, which is followed sequentially by the addition of methanol to form the transient imodyl ether and then water to form 7-ACA. 7-ACA is precipitated by using ammonia to change the pH to the isoelectric point, and the 7-ACA is recovered methanol wet and then dried under vacuum.

(b) Biocatalytic route. A three-step process is used to convert the potassium salt of cephalosporin C to 7-ACA, as shown in Figure 8.2(b). A solution of cephalosporin C is stirred with the immobilized biocatalyst D-amino acid oxidase (DAO) while air is bubbled through the solution to supply the required oxygen. The by-product of the bioconversion, hydrogen peroxide, reacts spontaneously with the keto intermediate to give glutaryl 7-ACA. The reaction is carried out at a constant temperature (18 °C) and elevated pressure (5 bar) under controlled pH (starting at pH 7.3 and rising to 7.7 at completion) to ensure the desired conversion. Additional hydrogen peroxide may be added to promote greater conversion to glutaryl 7-ACA if desired. Upon completion, the solution containing glutaryl 7-ACA is separated from DAO, and immobilized glutaryl 7-ACA acyclase (GAC) is added at pH 8.4 and temperature 14°C to obtain the desired 7-ACA. Dilution may be required to control the concentration, but upon completion of the reaction, the 7-ACA is separated from GAC and isolated. In both cases, the enzymes may be recovered and reused.

23.5 In Example 15.2 we compared microreactors with full-scale production batch reactors for the reaction between a carbonyl compound and an organometallic agent to produce a fine chemical.[25] How would the technologies compare based on the metrics calculated in Example 15.2? The reaction proceeds as

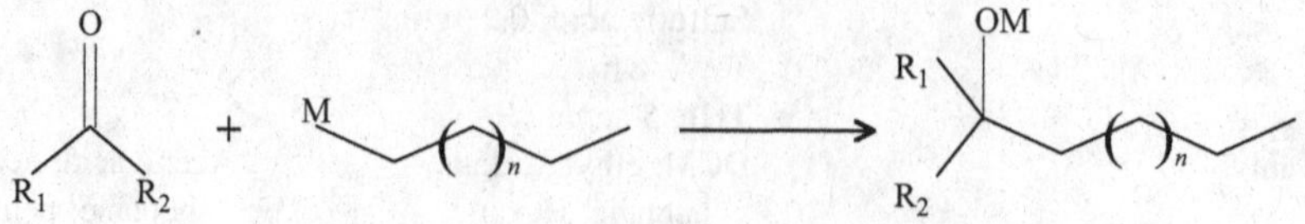

in the liquid phase and is exothermic (standard heat of reaction ca. –300 kJ/mol). The main reaction and most side reactions are fast (<10 s), some parallel and consecutive reactions can occur, and the compounds are sensitive to temperature. Experiments were carried out in microreactors and were compared with laboratory- and full-scale operations. The characteristics of the reactor systems evaluated are given in Table P23.5.

TABLE P23.5 Reactor System Characteristics

Reactor Type	T (°C)	Residence Time	Yield (%)	Volume/Area (m^2/m^3)	Dimensions
Flask	–40	0.5 h	88	80	0.5 L
Stirred vessel (production)	–20	5 h	72	4	6000 L
Microreactor	–10	<10 s	95	10,000	2 × 16 channels of $w \times h = 40 \times 220\,\mu m$

23.6 In Example 15.4 we studied centrifugal separators compared with gravity separators. How would the two technologies compare based on the metrics calculated in Example 15.2? The centrifugal separators were used in the extraction of a pharmaceutical intermediate. The phase separations are used in the mother liquors containing methylene chloride (MDC, specific gravity of 1.33) and water washes. For the batch process, the extraction can be described in a simplified manner as an MDC–water extraction, two 2000-L water washes, and one MDC final wash. The standard batch output is the extraction of 179 kg of intermediate per batch. The data for this comparison are given in Table P23.6.

TABLE P23.6 Data for Equipment Comparison

	Centrifugal Separators	Gravity Separators
Volume of MDC (L/batch)	4,165	4,900
Volume of water washes (L/batch)	10,000	4,000
Yield (% difference from gravity separators due to losses in spent stream)	≈ 10	0
Power usage	7.5 hp at 30 gpm	2 h of agitation
Process time (h)	≈ 12	≈ 22

REFERENCES

1. Saling, P., Kicherer, A., Dittrich-Krämer, B., Wittlinger, R., Zombik, W., Schmidt, I., Schrott, W., Schmidt, S. Eco-efficiency analysis by BASF: the method. *Int. J. Life Cycle Assess.*, 2002, 4, 203–218.
2. Cabezas, H., Bare, J., Mallick, S. Pollution prevention with chemical process simulators: the generalized waste reduction (WAR) algorithm—full version. *Comput. Chem. Eng.*, 1999, 23 (4–5), 623–634.
3. U.S., Environmental Protection, Agency. *Tool for the Reduction and Assessment of Chemical and Other Environmental Impacts* (TRACI): *User's Guide and System Documentation.* EPA 600/R-02/052. National Risk Management Research Laboratory, Cincinnati, OH, 2003.
4. Jiménez-González, C., Curzons, A. D., Constable, D. J. C., Overcash, M. R., Cunningham, V. L. How do you select the "greenest" technology? Development of guidance for the pharmaceutical industry. *Clean Products Process.*, 2001, 3, 35–41.
5. Gani, R., Jørgensen, S. B., Jensen, N. Design of sustainable processes: systematic generation and evaluation of alternatives. Presented at the 7th World Congress of Chemical Engineering, 2005.
6. Carvalho, A., Gani, R., Matos, H. , et al. (2008)Design of sustainable chemical processes: Systematic retrofit analysis generation and evaluation of alternatives. *Process Saf. Environ. Prot.*, doi:10.1016/j.psep.2007.11.003.
7. Carvalho, A., Gani, R., Matos, H. Design of sustainable processes: systematic generation and evaluation of alternatives. In*16th European Symposium on Computer Aided Process Engineering and 9th International Symposium on Process Systems Engineering*, Vol. 2, Marquardt, W., Pantelides, C. (Eds.) Elsevier, New York, 2006, pp. 817–822.
8. Curzons, A. D., Constable, D. J. C., Mortimer, D. N., Cunningham, V. L. So you think your process is green, how do you know? Using principles of sustainability to determine what is green: a corporate perspective. *Green Chem.*, 2001, *3*.1–6.
9. Lapkin, A., Constable, D.J.C.,Eds. *Green Chemistry Metrics.* Blackwell Publishers, Hoboken, NJ, 2008.
10. Jiménez-González, C., Constable, D. J. C., Curzons, A. D., Cunningham, V. L. Developing GSK's Green technology guidance: methodology for case-scenario comparison of technologies. *Clean Techno. Environ. Policy*, 2002, 4, 44–53.
11. Jiménez-González, C., Constable, D. J. C., Henderson, R., De Leeuwe, R., Cardo, L. Embedding sustainability into process development: GlaxoSmithKline's experience. In *Proceedings of the 7th International Conference on Foundations of Computer-Aided Process Design* (FOCAPD): *Design for Energy and the Environment*. Breckenridge, CO, June 12–17, 2009.
12. Swiss Federal Laboratories for Material Testing and, Research. *ECOPRO: Life Cycle Analysis Software.* EMPA, St. Gallen, Switzerland, 1996.
13. *PEMS 4: Life Cycle Assessment Software.* PIRA International. Leatherhead, UK, 1998.
14. Dumas, R. D. *Energy Usage and Emissions Associated with Electric Energy Consumption as Part of a Solid Waste Management Life Cycle Inventory Model.* Department of Civil Engineering, North Carolina State University, Raleigh NC, 1997.
15. Jiménez-González, C., Kim, S., Overcash, M. R. Methodology for developing gate-to-gate life cycle inventory information. *Int. J. Life Cycle Assess.*, 2000, 5, 153–159.
16. Jiménez-González, C., Overcash, M. R. Energy sub-modules applied in life-cycle inventory of processes. *Clean Products Process.*, 2000, 2, 57–66.
17. Ecoinvent: Swiss Centre for Life Cycle Inventories. 2008.http://www.ecoinvent.org, accessed Nov. 10, 2008.
18. Bundesamt für Umwelt, Wald und Landschaft. Ökobilanzen von Packstoffen., Schriftenreihe Umwelt 132. BUWAL, Bern, Swetzerland, 1991.

19. SimaPro. PRé Consultants. http://www.pre.nl/simapro/default.htm.
20. DEAM Database. EcoBilan. http://www.ecobalance.com/uk_deam.php.
21. UMBERTO software. Institute for Environmental Informatics, Hamburg, Germany.http://www.umberto.de/en/.
22. GaBi Software. PE International.http://www.gabi-software.com/.
23. Lewis, R. J. *Sax's Dangerous Properties of Industrial Materials*, 10th ed. Wiley, New York, 2000.
24. Jiménez-González, C., Curzons, A. D., Constable, D. J. C., Cunningham, V. L. *Int. J. Life Cycle Assess.*, 2004, 9, 114.
25. Jenssen, K. F. Micromechanical systems: status, challenges and opportunities. *AIChE J.*, 1999, 45, 2051–2054.

24

INDUSTRIAL ECOLOGY

What This Chapter Is About As a chemist or engineer, it is often difficult to poke your head above the normal fray of academia, government, or business to think about the big picture—beyond what we do on a daily basis within the boundaries of our facilities. It is also difficult for many to think about things from a systems perspective and to see how what we might be doing in a given project or program affects the world. In this chapter we cover industrial ecology, a relatively new way of thinking about industry that studies mass and energy flows through economies, and systems-wide holistic views of the interconnections and interdependencies among industry, the environment, and society. This discipline draws on analogies to the natural world, where the only substantial planetary input is energy, and where materials and energy are moved through and around the environment to support a significant degree of growth and diversity of life. A careful reflection of some of the principles of industrial ecology may be helpful in leading green chemists and green engineers to make better decisions in their work toward sustaining people, places, and the products on which we depend.

Learning Objectives At the end of this chapter, the student will be able to:

- Understand what is meant by industrial ecology.
- Integrate industrial ecology principles into project planning (synthesis, technology, etc.).
- Understand how green chemistry and engineering can be used to facilitate movement toward industrial ecology.

Green Chemistry and Engineering: A Practical Design Approach, By Concepción Jiménez-González and David J. C. Constable

24.1 INDUSTRIAL ECOLOGY BACKGROUND

As with most things in life, it is important that we take some time to provide background or context and to define terms used in industrial ecology. After all, industrial ecology did not arise in a vacuum but grew out of an idea rooted in the environmental movement that began in the 1960s and by the late 1980s and early 1990s carries through to serious reflections on how society moves sustainable development from theory to practice. Clearly, human beings are having an impact on the environment that is not sustainable, and the question remains regarding how we can live such that our impacts are balanced by restoration and regeneration.

If you consider material and energy flows through any given economy, you might come up with a flow similar to that shown in Figure 24.1. Raw materials are extracted from the Earth, processed, converted to a salable good, used, and sold—just as we discussed in Chapter 16. If one looks at the environment or natural ecosystems outside human activity or impacts, you might think about the three types of systems depicted in Figure 24.2. In a *type I system* you have unlimited resources and unlimited waste removal, as is the case for a single cell in an appropriate environment. This is pretty much how many viewed, and to a great extent still do view, the industrial world. In a *type II system* there is a limited input of resources and a limited amount of waste that must be taken out of the system in order for the ecosystem to survive over time. In a *type IIII system*, the only external input is energy from the sun. Nutrients and energy are cycled through the system in a closed loop with no material from outside the system required to enter or leave to sustain the system over long periods of time. As a result of thinking about closed-loop cyclical type III systems as are found in nature, some began to think about whether or not humans might be able to live in a manner that replicates, to the greatest extent possible, a type III system.

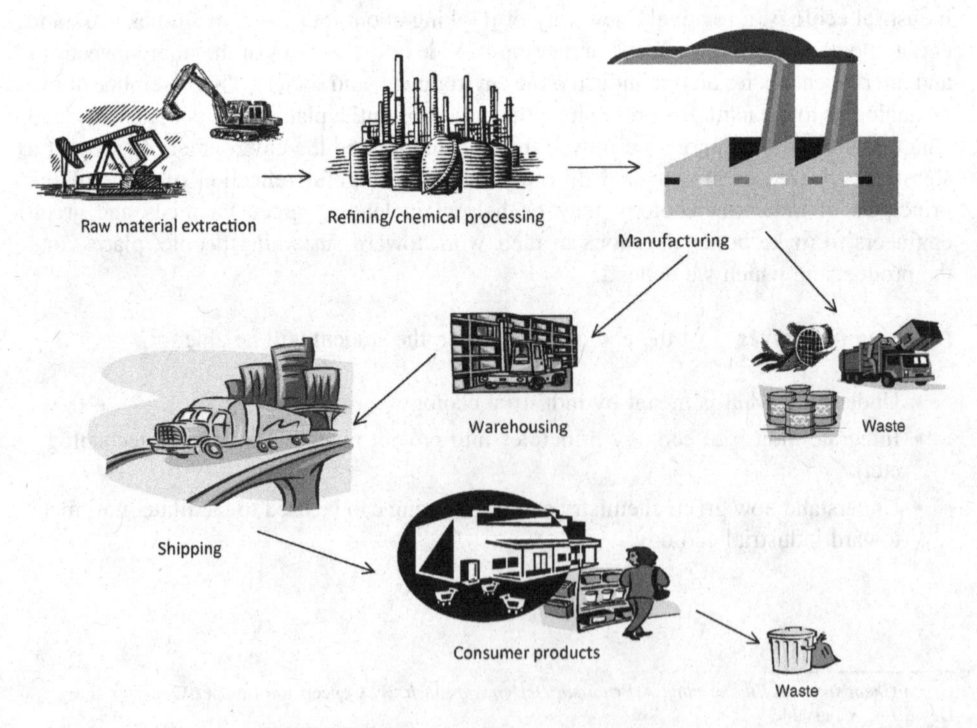

FIGURE 24.1 Linear production and waste.

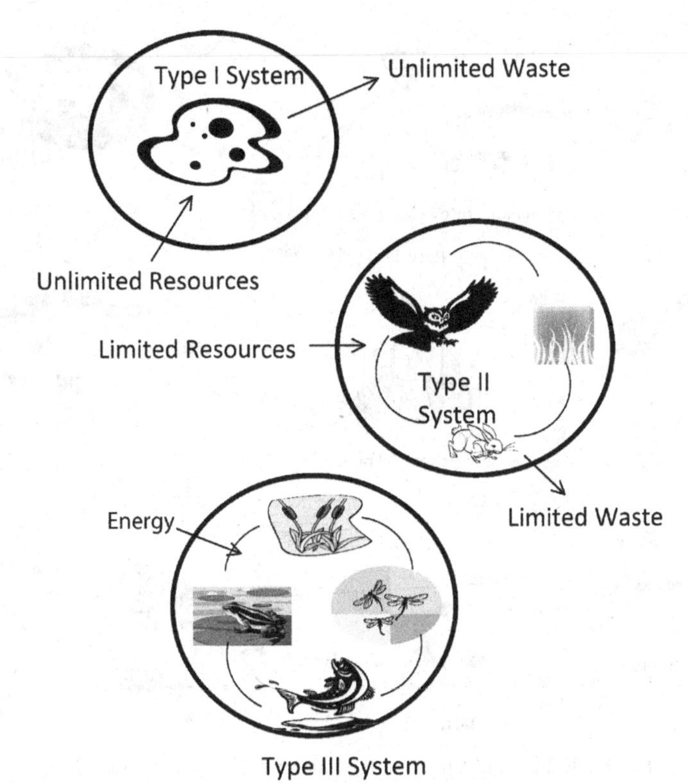

FIGURE 24.2 Type I, II, and III, systems.

Among the early thinkers about the potential to use ecosystems as models for industrial activity were Frosch and Gallopoulos.[1] They looked at the current ways in which modern manufacturing was carried out and noted that, with a few exceptions, it is generally not carried out in a fashion that employs cyclical flows of material and energy. That is, different industries buy raw materials and energy (or the raw materials to produce energy); produce products, by-products, and waste; and these by-products and waste flow linearly through an economy before they are almost completely dispersed into the environment. They therefore argued that rather than a linear one pass-through system, industries should organize themselves such that mass and energy considered to be waste for one industry become the mass and energy for another, as depicted in Figure 24.3.

If one pauses for a moment to reflect on material and energy flows through our current economies, the facts are rather staggering. Von Weizacker et al.[2] observe that "actually we are more than ten times better at wasting resources than at using them," citing a National Academy of Engineering study that found 93% of the materials that are purchased by industry for conversion to salable products do not end up in the product at all. You may recall that we covered mass and energy efficiency in Chapter 2, and Table 2.1 shows the material efficiencies of a variety of industries.

To add to the basic inefficiency of production, what is bought is often discarded at a breathtaking rate. Paul Hawken estimated that within six weeks of sale, over 90% of the materials that make up a product or that are used to make a product are discarded. Our performance in energy efficiency is not much better. The standard incandescent light bulb, now being outlawed in the European Union but an American standby, converts only 3% of the

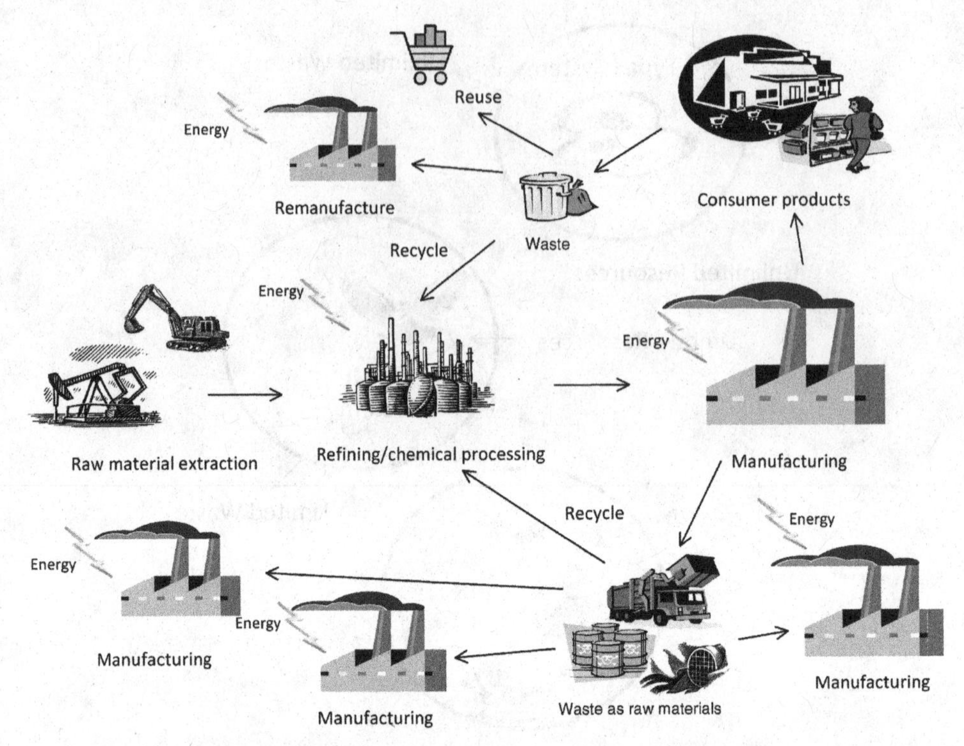

FIGURE 24.3 Idealized industrial ecology production model.

energy originally generated at a fossil fuel or nuclear plant into usable light. Indeed, 70% of the energy originally obtained is lost in conversion and transmission before it ever reaches a house. In a standard automobile, 80 to 85% of the energy in gasoline is lost in the drive train before it gets to the wheels.[2] Further energy is wasted for energy inefficient and poorly insulated buildings in the United States, where 1 Btu in 12 of the total world energy production is used for heating and cooling.[3]

Recognition of these facts led to the development of ideas about how industry might be changed, and these ideas eventually coalesced into a field of study that became known as *industrial ecology*. As you might imagine would be the case for a new field, a number of different definitions for industrial ecology have been proposed over the past 20 years. Graedel and Allenby proposed the following definition in the first textbook written for in the field:[4]

> Industrial Ecology is the means by which humanity can deliberately and rationally approach and maintain a desirable carrying capacity, given continued economic, cultural, and technological evolution. The concept requires that an industrial system be viewed not in isolation from its surrounding systems, but in concert with them. It is a systems view in which one seeks to optimize the total materials cycle from virgin material, to finished material, to component, to product, to obsolete product, and to ultimate disposal. Factors to be optimized include resources, energy, and capital.

Of note in this definition is the emphasis on deliberate and rational actions that humans may take from a holistic, systems-wide perspective or view of the world. Implicit in much of what Allenby has written[4, 5] is that we as humans are engaged in earth systems engineering whether or not we are doing so deliberately. It is therefore arguably better for us to understand exactly what we are doing and control our Earth systems engineering rationally rather than being at the mercy of a multitude of systems we have no idea how to control and no idea how to mitigate our impacts.

The Institute of Electrical and Electronic Engineers (IEEE) offered the following definition of *industrial ecology* in their "White Paper on Sustainable Development and Industrial Ecology"[6]:

> Industrial ecology is the objective, multidisciplinary study of industrial and economic systems and their linkages with fundamental natural systems. It incorporates, among other things, research involving energy supply and use, new materials, new technologies and technological systems, basic sciences, economics, law, management, and social sciences. Although still in the development stage, it provides the theoretical scientific basis upon which understanding, and reasoned improvement, of current practices can be based. Oversimplifying somewhat, it can be thought of as the "the science of sustainability." It is important to emphasize that industrial ecology is an objective field of study based on existing scientific and technological disciplines, not a form of industrial policy or planning system.

Of note in this definition is the explicit emphasis on viewing industrial ecology as an objective field of study, not a normative one. This means that analysis or study from an industrial ecology perspective would pull in economic and social issues and attempt to treat them as objective factors in the overall system. They did not view industrial ecology as a means for developing or promoting policy and planning.

The UN Environment Program proposed the following more succinct definition for industrial ecology[7]:

> Systems oriented study of the physical, chemical, and biological interactions and interrelationships both within industrial systems and between industrial and natural ecological systems.

This definition, apart from being the simplest, emphasizes the interactions and interrelationships that should characterize industrial complexes based on the model of biological ecosystems, where interdependencies are essential to maintaining healthy and vigorous systems. In the industrial ecosystem, a group of enterprises would be organized so that there would be no waste in the materials (mass) and energy used; waste from one industrial entity would be used as a raw material for another industrial entity.[8] It is important to remind ourselves at this point that there are limits to the extent to which we are able to use waste material or energy before waste must be removed from a system. Despite our potential desire to circumvent them, clearly the laws of thermodynamics cannot be violated.[9–11]

24.2 PRINCIPLES AND CONCEPTS OF INDUSTRIAL ECOLOGY AND DESIGN

As we saw in Chapter 2, principles of green chemistry and engineering are helpful in organizing our thoughts about what actions we should take to implement green chemistry and green engineering in our work. In general, there has not been quite the focus on industrial ecology principles as there has been for green chemistry and green engineering, but Allenby has offered several principles of industrial ecology, shown in the accompanying box.

PRINCIPLES OF INDUSTRIAL ECOLOGY

1. Products, processes, services, and operation can produce residuals, but not waste.
2. Every process, product, facility, constructed infrastructure, and technological system should be planned to the extent possible to be easily adapted to foreseeable, environmentally preferable innovations.
3. Every molecule that enters a specific manufacturing process should leave that process as part of a salable product.
4. Every erg of energy used in manufacture should produce a desired material transformation.
5. Industries should make minimal use of materials and energy in products, processes, services, and operations.
6. Materials used should be the least toxic for the purpose, all else being equal.
7. Industries should get most of their needed materials through recycling streams (theirs or those of others) rather than through raw material extraction, even in the case of common materials.
8. Every process and product should be designed to preserve the embedded utility of the materials used. This might involve designs that extend the life of the product, or facilitate recycling of subassemblies or components, rather than just materials.
9. Every product should be designed so that it can be used to create other useful products at the end of its current life.
10. Every industrial landholding, facility, or infrastructure system or component should be developed, constructed, or modified with attention given to maintaining or improving local habitats and species diversity and to minimizing impacts on local or regional resources.
11. Close interactions should be developed with material suppliers, customers, and representatives of other industries, with the aim of developing cooperative ways of minimizing packaging and of recycling and reusing materials.

Source: Adapted from ref.5.

As you can see, there are elements of these principles that are very similar to those we saw for green chemistry and green engineering; indeed, these should not be seen as competing principles but as complementary ones. Garner and Keoleian[12] suggest that are certain key concepts that help to distinguish industrial ecology from other fields of study. These include:

- Systems analysis
- Material and energy flows and transformations
- A multidisciplinary approach
- Analogies to natural systems
- Linear (open-loop) vs. cyclical (closed-loop) systems

It is worth expanding these ideas a bit. *Systems analysis* or a systems view of things encompasses a broad view of the interrelationships or interplay between the range of human activities and the environment. There are multiple systems that one may choose to include in an analysis, from global geopolitical and economic systems to organizations and structures on a more local level. Similarly, there are multiple systems and levels of systems that regulate industry, from high-level institutional hierarchies to systems that regulate product development, manufacturing, marketing, and sales. Whether one looks at broad societal systems or more constrained systems within a given company, there is an equally broad array of environmental and ecological systems that will be affected, depending on the level that one chooses to study. Systems analysis enables one to think about sustainability and sustainable products on both a global and a local level through a judicious choice of systems affected. The trick is to choose the scope in such a way that the results are meaningful and useful.

Material and energy flows and transformations are an integral part of industrial ecology. It is generally true that most individuals, societies, and organizations are not at all aware of how material and energy flows through economies, regions, companies, or households. We are also unaware of all the various material and energy transformations that we undertake when making products or delivering a particular service. Nor do we understand how these flows and transformations create waste. A central theme of industrial ecology is to undertake a sufficiently detailed study of material and energy flows such that one can begin to see where the various products, by-products, and wastes may be used, reused, or turned into useful products or services.

A multidisciplinary approach is as central to industrial ecology as it is to most other green chemistry, green engineering, or sustainability initiatives. As with these other disciplines, there is a certain level of complexity that one must manage and master to be truly successful in thinking about and undertaking industrial ecology. If we intend to understand the interplay of human systems, we need to be at least familiar in passing with economic, political, legal, human health, engineering, and other fields of study. To understand environmental and ecological systems, there should be some familiarity with fundamental sciences such as chemistry and biology, rounded out by an understanding of environmental systems, ecological principles, and natural resource management. Knowing who to involve at which point is key to arriving at an optimal solution.

As mentioned earlier, industrial ecology requires one to look at natural systems and draw analogies to industrial systems so that human and natural resources, and material and energy flows, are optimized in such a way that there is a move toward increasingly closed-loop systems. The chemical and biological systems at work in naturally occurring ecosystems are used to build, degrade, and recycle all the nutrients and building blocks within a system; nothing is wasted and nothing leaves the system. It is this notion of a closed system, where waste becomes food for another process, that acts as a conceptual framework for grouping industrial systems in such a way that materials and wastes are passed from one industrial entity to the next so as to ensure the continual survival of all members of the system. There is a dynamic equilibrium between these industrial entities that requires a high degree of interconnectedness and integration between industries that is often difficult to achieve in practice.

24.3 INDUSTRIAL ECOLOGY AND DESIGN

Given the discussion above regarding principles and concepts of industrial ecology, it is hopefully relatively simple to see that design is a key feature in the practical implementation of industrial ecology. But what do we know about product design; that is, what are the general

TABLE 24.1 Traditional Product Design Goals

Acronym	Meaning	Why Do We Do This?
DfM	Design for manufacturability	Product can be made easily and at reasonable cost
DfL	Design for logistics	Production and supply activities are well orchestrated
DfT	Design for testability	Product quality may be checked conveniently
DfP	Design for pricing	Ensures that product will sell
DfSL	Design for safety and liability	Product is safe to use and company's liability risk is reduced
DfR	Design for reliability	Product works as intended over an extended period
DfS	Design for serviceability	Maintenance services can be offered at a reasonable cost to the customer and company

guiding design principles that we might find in industry today? Table 24.1 contains a traditional view of what product designers normally think about when conceiving a product.

Additional Points to Ponder From what you know about green chemistry and green engineering, is there anything that you would change about these traditional design goals? How might these meet the principles of industrial ecology that we just described?

Given the emphasis on integration and interdependency in an industrial ecosystem that we noted above, one would be hard-pressed to achieve the goals of industrial ecology without significant thought given to the overall design of a product or service we might be thinking about developing. General design considerations for industrial ecology have been adapted from Cohen-Rosenthal[12] and are shown in the accompanying box. Although design for extended use is generally a good thing in many applications for durable goods, it should be understood that there are some areas where design for extended use can create significant issues, as has been the case for chlorofluorocarbons that were found to be ozone depleting, for plastics that have been found to be effectively nonbiodegradable, for pharmaceuticals that are designed to have stable shelf lives, and for early generations of herbicides and pesticides that were very nonspecific and were designed to remain effective for long periods of time when applied.

GENERAL DESIGN CONSIDERATIONS FOR INDUSTRIAL ECOLOGY

Design for extended use

Development of "smart materials"

Design for reuse, repair, and remanufacturing

Design for disassembly

Demanufacturing

Disassembly

Recycling—material loops

Design for environment or design from first principles
- Dematerialization
- Molecules
- Chemical reactions
- Nanochemistry

Design for waste
- Landfill mining
- Infill
- Energy conversion
- Pollution control residuals

Source: ref.12.

What is in view for extended use are materials such as clothing, other textiles, and material for furniture or building coatings, that are by design long-lived through the application of "smart," adaptive, and/or self-repairing mechanisms. These materials would have properties that would change with environmental conditions so as to be more durable or adaptable to changes in temperature, pH, humidity, and so on. As you might imagine, there are not many examples of these sorts of materials in existence, but there is certainly much talk of their development, especially since the advent of recent advances in nanotechnology and biomimicry. One could imagine the development of materials that change color to absorb heat or reflect it, materials that change porosity depending on humidity or temperature, and materials that release certain chemicals that repulse a potential pest.

The next tier of the hierarchy would be to design materials for disassembly, demanufacturing, or recycling. McDonough and Baumgarten[13] speak of creating *technical nutrients* for some materials in common use so that they may be used over and over again, being returned to their original state through low-impact processes. A great example of this ethos in action may be found with DuPont's Petretec process.[14] DuPont has found a way to take any polyester and unzip the polymer to release the virgin monomer, which may then be reused to make "new" polyester products of many kinds. The process can be used to reclaim monomer from mixed-material streams containing polyester and is readily integrated into existing polyester manufacturing facilities. The overall result is that polyester is diverted from landfills or from waste to energy applications, thereby achieving a lower life cycle environmental impact.

Another example of this attention to recycle and reuse may be found in how modern Xerox machines are developed and marketed. A graphic showing the general outline of the Xerox program for recycle and reuse is shown in Figure 24.4. Xerox uses and reuses electronics, optics, and other components over and over in new products or products that are changed, upgraded, and expanded over time. Materials that cannot be reused are either recycled into other product streams or become raw materials for new parts.

A third area where disassembly, demanufacturing, and recycling are used extensively is automobile, manufacture, where about 95% of the average vehicle is recycled in one form or another, thereby avoiding huge quantities of waste finding their way to landfills. From a chemist's perspective, Table 24.2 contains recycling options for chemicals, solvents, and materials that would be encountered more frequently.

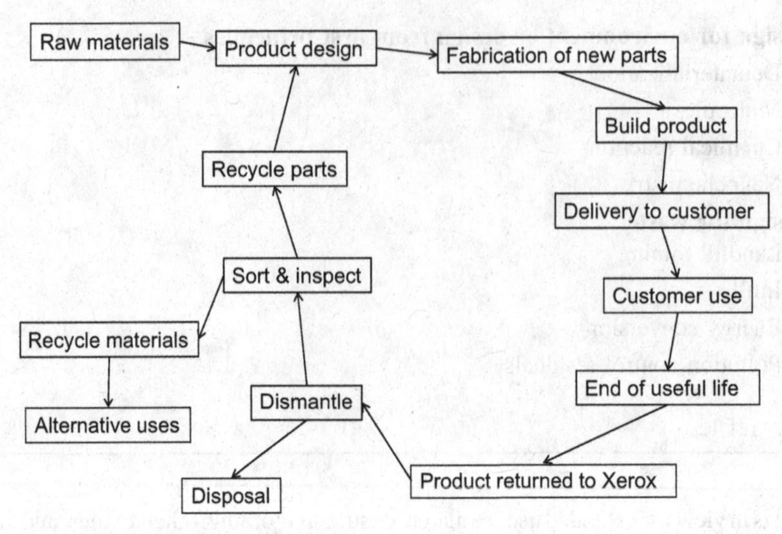

FIGURE 24.4 XEROX equipment recovery and parts reuse/recycle process.

Design from first principles is where green chemistry and green engineering hold the greatest promise for achieving the objectives of industrial ecology. In the case of dematerialization, one attempts to use as little material and energy as possible to achieve the desired function in a product or service. Through judicious redesign of materials of construction, or through the development of new materials containing the desired properties for existing or new applications, green chemistry holds great promise for removing toxic materials and larger volumes of materials from current product designs. As has been noted throughout this book, target molecules and the synthetic processes employed to produce those molecules can and should be changed to achieve more sustainable outcomes. We can see that the simple

TABLE 24.2 Recycling Options for Various Materials

Class of Nonrenewable Material	Recycling Technically Feasible?	Recycling Economically Feasible?	Examples
I	Yes	Yes	Catalysts, some solvents, most industrial metals
II	Yes	No	Refrigerants, some solvents, packaging materials
III	No	No	Coatings, pigments, fuels, lubricants, pesticides, herbicides, fertilizers, reagents, explosives, detergents

Source: ref.15.

product design goals in Table 24.1 might need to be expanded a bit to accommodate our need to design from first principles and realize the goals of industrial ecology. Table 24.3 contains additional design goals, normally referred to as design for the environment goals, and these may help to get us a bit closer to the objectives of industrial ecology. The table is not exhaustive but is illustrative of the types of things that we should be thinking about when we think about design for the environment.

TABLE 24.3 Goals in Design for the Environment

Acronym	Meaning	Why Do We Do This?
DfM	Design for manufacturability	To enable pollution prevention/source reduction during manufacturing through use of: Less material Fewer materials Safer materials and processes
DfEE	Design for energy efficiency	To achieve reductions in energy consumption during the use of products or services: Promote flexible energy use Design that promotes renewable energy Design for reduced life cycle emissions Design for carbon neutrality
DfZT	Design for zero toxics	To remove toxic materials from the supply chain and in products: Reduce incidence of acute and chronic human and environmental risks Reduce management costs for high-hazard materials Reduce potential for product liability
DfD	Design for dematerialization	To reduce embodied material and energy intensity of a product or service: Less material per unit produced
DfP	Design for packaging	To minimize amount of packaging required so there is less: Material used to ensure robust sales Nonrecylcable packaging Impactful (toxic, nonbiodegradable or reusable, etc.) packaging Life cycle impact
DfL	Design for logistics	Arrange supply chain to: Use locally manufactured materials Require less or lower-impact (e.g., rail) transportation of components and/or products
DfL	Design for longevity	Design for modularity to ease: Upgrading and delay ultimate need for complete replacement Serviceability and, later, disassembly Design for serviceability to ease: Repairs that lead to longer life Recapture of used and/or broken parts
DfD	Design for disassembly	To promote reuse of components:

(continued)

TABLE 24.3 (*Continued*)

Acronym	Meaning	Why Do We Do This?
		Quicker and cheaper disassembly More complete disassembly Dismantling by simple tools
DfR	Design for recycling	To promote greater materials recovery: Increase content of recyclable materials Increase content of no- or low-toxicity materials Reduce variety of different materials Use materials that can be locally recycled Provide easier materials identification Ensure easy separability of nonrecyclables so that they may be disposed of safely
DfC	Design for compostability	In appropriate applications, increase content of materials that are: Biodegradable over short periods of time Contain no or few toxic materials
DfER	Design for energy recovery	To promote greater energy recovery, increase content of materials that: May be incinerated safely Produce low- or no-toxicity residues Allow for composting of residues or alternative productive uses
DfC	Design for compliance	So that materials do not: Fall under stringent regulatory requirements Require special handling, storage, and management So that you are ahead of future regulatory constraints

If you have spent any time in industry or have read technical or industrial journals, or perhaps have seen items in the popular press, you have no doubt heard of the idea of achieving *zero waste*. As the rhetoric goes, you need a goal, a goal should challenge us, and a goal to produce zero waste may not be achievable, but if we don't strive for perfection, we are admitting defeat. Fortunately, as scientists and engineers we know a little something about the laws of thermodynamics,[16] and if we just think about it for a few minutes, we realize that zero waste is not an achievable goal. In the general design goals we saw from Cohen-Rosenthal above, he included a design for waste goal. This is arguably a good thing to think about because waste is something we still produce in abundance!

Although it is not discussed routinely, landfills do, in fact, hold large quantities of high-quality plastics, metals, and other materials whose recovery and reentry into our raw material supplies may at some point become economically attractive, especially as we are confronted by the fact that minerals are both dwindling and increasingly difficult to obtain. Moreover, even though we continue to develop our capability to recycle, downcycle (i.e., reuse a material not as virgin material but to convert it to another use that is of inherently lower value), and reuse materials, we are invariably left with something that is truly waste. In some instances these residuals may be feedstocks for waste-to-energy facilities, and that may provide some benefit to society. Unfortunately, these residuals, together with pollution prevention residuals, usually also contain toxic or unusable materials. In some instances these residuals may be sequestered in a material such as concrete and reused, or they may be stabilized in some fashion and used as infill for roads or other construction projects. Therefore, it is imperative that during the design phase we should be thinking about our potential for creating waste so as to avoid it to the greatest extent possible. When all other options for recycle and reuse fail, we must find ways of minimizing the societal and environmental impacts of our waste.

Additional Points to Ponder With all these design considerations to keep in mind, how would you determine which design consideration is likely to be the most important? Can you imagine a situation where pursuit of one design consideration actually interferes with the achievement of another?

24.4 INDUSTRIAL ECOLOGY IN PRACTICE

You may be inclined to think that the discussion above is all well and good, perhaps a bit theoretical and in the "nice to do" category, but not very practical to implement. Happily, there are a number of notable examples and case studies where the principles of industrial ecology have been applied and eco-industrial parks (EIPs) have been created. The most famous example is in Kalundborg, Denmark, an industrial complex or park that has been growing steadily in size while integrating various industrial partner's mass and energy inputs and outputs over many years. What exactly is an eco-industrial park? Glavic and Lukman[17] provide the following definition:

> It is a community of businesses, manufacturing and services, located together in a common property, seeking enhanced environmental, economic, and societal performance through collaboration in managing environmental and resource issues including information, energy, water, materials, infrastructure, and natural habitats.[6, 7] The benefit of such a community is greater than the sum of individual benefits.[18] This community benefits from the relationships and interconnectedness between environs and their environment.

Korhonen and Snäkin[19] expand on this definition and tell us that:

> IEs are complex systems, and they are self-organizational,[20, 21] i.e., very difficult to intentionally plan, design or manage.[22, 23] IEs are impossible to create "from scratch".[24] . . . But one must note that all industrial ecosystems are different with their own economic, social, cultural and ecological characteristics, which makes comparison and learning difficult.[25]

Although Kalundborg is perhaps the most famous and well-studied EIP, others have been created throughout the world and on most continents in the past 10 to 15 years. Christensen[26] provides an excellent overview of Kalundborg that illustrates the evolutionary nature of an EIP and documents its growth in size and integration since its beginning in 1961. As of 2006, the Kalundborg EIP consisted of more than 20 commercial agreements between six industries and some of the utilities of the local municipality. Korhonen[25] also provides an excellent overview of the evolution of an EIP, the Uimaharju forest industry park in eastern Finland, a small park containing six industrial actors owned by three different companies. The park began in 1951 with a sawmill and associated landfill, and as of 2003 had grown to encompass a pulp mill, a combined heat and power plant, a gas plant, and waste plants to treat the water and waste ash. The Uimaharju EIP has a high degree of diversity and integration, employing energy cascades and extensive raw material recycling and recovery.

So what are some of the areas in which industries can partner to achieve a greater level of integration and symbiosis? A few come readily to mind.

1. It should be easy to see that sharing utilities (steam, energy, cooling, water treatment or conditioning, etc.) used in primary production may offer economies of scale, greater overall efficiency, and opportunities for evolutionary technical improvements.
2. By-product or waste exchanges, energy exchanges, and so on, are all of potential mutual benefit to industrial partners.
3. There is the possibility of building waste treatment facilities that may, as with utilities sharing, offer economies of scale, greater efficiencies, or may even generate salable products.

What Kalundborg, Uimaharju, and other examples of EIPs clearly demonstrate is that it is extremely difficult to create an EIP from scratch. There are a number of reasons for this.

1. Industries that are not part of a larger parent organization generally are not very good at collaborating, and there aren't always obvious economic benefits seen at a high level or through traditional accounting systems.
2. It is difficult for any industry to become reliant on a partner for a long-term relationship because businesses can and do fail, production changes in volume or ceases, and new products having different inputs and outputs may be introduced that create major technical problems or potentially affect the quality or efficacy of a product.
3. There may be major regulatory barriers, as in the United States, where laws such as the Resource Conservation and Recovery Act (RCRA) prevent the storage or transfer of certain types of wastes. There are a number of laws that make transfer of materials across company boundaries difficult or impossible.
4. There may be technical difficulties; for examples, the outputs from one industrial partner just may not fit with another business's needs, or a given waste stream may have a constituent that creates a problem in the downstream processing of the receiving partner.
5. EIPs require a far greater degree of communication between partners than is generally practiced between unrelated businesses, and these communication issues are sometimes greater than the technical issues. Changes in production, product lines, and so on, need to be communicated well in advance, and potential implications extend beyond the fence line of the business.

Despite these challenges, the potential opportunity that exists in creating eco-industrial parks is enormous. Clearly, there are a great many benefits that many companies achieve through mass and energy integration within their own boundaries, so it is not too much of a stretch to see the benefit it could have beyond the fence line. In addition, our need to become more sustainable will drive a higher degree of integration and collaboration, or we will never achieve our objectives.

PROBLEMS

24.1 Find an example of an eco-industrial Park or industrial or agricultural symbiosis other than Kalundborg or Uimaharju in the literature or on the Web.

(a) Describe the industries in the park and how they are collaborating.

(b) What environmental benefits are achieved through this park?

(c) What are the major barriers to successful collaboration in the park?

(d) Describe any areas where you think new integration and collaboration might be possible.

24.2 Choose a production facility that you know something about or one about which you can find good mass and energy balance information. You can choose an example from another chapter in this book.

(a) Describe how you might create an eco-industrial park around this facility.

(b) What by-product streams may be used as is?

(c) Where would it be difficult to find a consumer for the waste streams?

(d) What technical difficulties might you encounter?

24.3 Find a commercially available product that has been designed for extended use.

(a) What principles of industrial ecology are evident in this product?

(b) What changes in chemistry or process were required to extend the life of this product?

(c) What parts of this product could be recycled or reused?

24.4 Find a commercially available product that might be in its entirety or in parts considered to be a "smart" material or product.

(a) What are the unique features of this product?

(b) How might the smart features be a problem in the future?

24.5 How would you design a molecule for disassembly?

(a) What would be the synthetic design features of such a molecule?

(b) What types of chemical processes would be required to disassemble the molecule?

(c) How does the design for disassembly create future issues or benefits?

24.6 Consider a solvent such as acetone or methylene chloride.

(a) How would you change the molecule so that it has fewer environmental safety or health concerns?

(b) How would you change the molecule to improve its recyclability?

(c) Considering the principles of industrial ecology, what are the benefits and disadvantages of the solvent?

24.7 Consider an automobile.

(a) How is it designed for endurance, disassembly, reuse, and/or remanufacture?

(b) Where does chemistry enable these operations?

(c) How could an automobile be improved from an industrial ecology perspective?

24.8 Plastics are ubiquitous, high-volume commodity items.

(a) Which plastics are recyclable?

(b) Which plastics are convertible back to virgin monomers and reusable?

(c) Find one of the newer polymers on the market that are made from renewable sources. Are these designed for recycle and/or reuse?

24.9 There is always a bit of tension in the green chemistry field between those who design molecules and those who design "stuff" (e.g., consumer products such as vacuums, appliances, and cars). Using the table of design principles, develop a table of actions that you would take for designing molecules compared to stuff. What does this tell you about the principles?

REFERENCES

1. Frosch, R. A., Gallopoulos, N. Strategies for manufacturing. *Sci. Am.*, 1989, 261 (3), 144–152.
2. Von Weizacker, E., Lovins, A., Lovins, L. H. *Factor Four: Doubling Wealth, Halving Resource Use*. Earthscan Publications, London, 1998.
3. Wellesley-Miller, S. Towards symbiotic architecture. In: *Earth's Answer: Explorations of Planetary Culture at the Lindisfarne Conferences*. Katz, M., Marsh, W. P., Thompson, G. G. Eds. Harper & Row, New York, 1977, p.9.
4. Graedel, T. E., Allenby, B. R. *Industrial Ecology*. Prentice Hall, Upper Saddle River, NJ, 1995.
5. Allenby, B. R. *Industrial Ecology: Policy Framework and Implementation*. Prentice Hall, Upper Saddle River, NJ, 1995.
6. United Nations Environment Programme, Division of Technology, Industry, and Economics. Cleaner production (CP) Activities.http://www.uneptie.org, accessed June 2, 2004.
7. Manahan, S. E. *Environmental Chemistry*, 8th ed., CRC Press, Boca Raton, FL, 2004.
8. Ehrenfeld, J., Gertler, N. The evolution of interdependence at Kalundborg.*Ind. Ecol.*, 1997, 1, 67–80.
9. Ayres, R. U. On the life cycle metaphor: where ecology and economics diverge.*Ecol. Econ.*, 2004, 48, 425–438.
10. Daly, H. *Beyond Growth: The Economics of Sustainable Development*. Beacon Press, Boston, 1996.
11. Garner, A., Keoleian, G. A. *Industrial Ecology: An Introduction*. National Pollution Prevention Center for Higher Education, University of Michigan, Ann Arbor, MI, Nov., 1995.
12. Cohen-Rosenthal, E. Making sense out of industrial ecology: a framework for analysis and action. *J. Cleaner Prod.*, 2004, 12, 1111–1123.
13. McDonough, W., Baungarten, M. *Cradle to Cradle*. North Point Press, New York, 2002.
14. http://dupont.t2h.yet2.com/t2h/page/techpak?keyword=Petretec&id=35545&qid=1507320&sid=10&args=3%25091%25091507320%2509%2509Petretec%2509%2509%2509-1.
15. Ayres, R. U., Simonis, Udo. E. Eds., *Industrial Metabolism: Restructuring for Sustainable Development*. UN University Press, Tokyo, 1994.
16. Huesemann, M. H. The limits of technological solutions to sustainable development.*Clean Technol. Environ. Policy*, 2003, 5, 21–34.

17. Glavic, P., Lukman, R. Review of sustainability terms and their definitions.*J. Cleaner Prod.*, 2007, 15, 1875–1885.
18. Mirata, M. Experiences from early stages of a national industrial symbiosis programme in the UK: determinants and coordination challenges.*J. Cleaner Prod.*, 2004, 12, 967.
19. Korhonen, J., Snäkin, J.-P. Analyzing the evolution of industrial ecosystems: concepts and application.*Ecol. Econ.*, 2005, 52, 169–186.
20. Allenby, B., Cooper, W. E. Understanding industrial ecology from a biological systems perspective.*Total Qual. Environ. Manage.*, 1994,343–354.
21. Desrochers, P. Industrial symbiosis: the case for market coordination.*J. Cleaner Prod.*, 2004, 12 (8–10), 1099–1110.
22. Ehrenfeld, J., Gertler, N. The evolution of interdependence at Kalundborg.*J. Ind. Ecol.*, 1997, 1, 67–80.
23. Johansson, A., Ehrenfeld J.andGertler, N. Response to: "Industrial ecology in practice: the evolution of interdependence in Kalundborg." Online letters to the editor.*J. Ind. Ecol.*, 1997, 1, 1.
24. Chertow, M. Eco-industrial park model reconsidered.*J. Ind. Ecol.*, 1998, 2, 8–10.
25. Korhonen, J. Some suggestions for regional industrial ecosystems.*Eco-Manage. Audit.*, 2001, 8, 57–69.
26. http://continuing-education.epfl.ch/webdav/site/continuing-education/shared/Industrial%20Ecology/Presentations/11%20Christensen.pdf.

25

TYING IT ALL TOGETHER: IS SUSTAINABILITY POSSIBLE?

What This Chapter Is About We have already covered most of the green chemistry and green engineering concepts that will allow us to design more sustainable manufacturing processes. In this chapter we explore how the concepts of innovation, science, green chemistry, and green engineering enable the adoption of sustainability principles. In addition, we explore the role that culture, behaviors, and the sociopolitical environment play in the quest for sustainability and to influence more sustainable practices. Finally, we explore some specific actions that can be taken to influence organizations and society to embed sustainability principles in their practices.

Learning Objectives At the end of this chapter, the student will be able to:

- Understand the role that innovation, green chemistry, and green engineering play in the development of a sustainable society.
- Understand the role that culture, mindset, and policy play in fulfilling sustainability aspirations.
- Identify crucial behaviors and strategies to influence the adoption of sustainable practices.
- Identify actions that can be taken in the business context to move toward more sustainable practices.

Green Chemistry and Engineering: A Practical Design Approach, By Concepción Jiménez-González and David J. C. Constable

25.1 CAN GREEN CHEMISTRY AND GREEN ENGINEERING ENABLE SUSTAINABILITY?

As we have seen in the previous chapters, if chemists and engineers want to design more sustainable processes, green chemistry and green engineering are two interrelated components that are essential to deliver innovative, more sustainable processes. We also have seen that it is essential that green chemistry and green engineering be developed in parallel to achieve more sustainable processes.

Sustainability is a global concept, and the strategies to achieve balance between environmental, economic, and social realms have to be implemented at the local level without losing sight of the global implications and the interrelationships between Earth systems and human society. Thus, the very concept of sustainability implies that systems thinking and systems analysis are the cornerstone of sustainability and sustainable thinking. This is particularly important when we are focusing on technological options and the innovations that will be required of green chemistry and green engineering as we work toward a more sustainable future. But is sustainability achievable?

Sustainability has to be seen very much as a work in progress.[1] As we saw in Chapters 1 and 2, sustainability is a balancing act. One of the typical analogies to explain sustainability calls for a three-legged table, where each one leg represents one aspect of sustainability (see Figure 25.1). If one of the legs is missing, the sustainability table falls down; if one of the legs is shorter, the sustainability table is uneven.

In order to design and deliver more sustainable chemical and manufacturing processes, one needs to integrate sustainability principles into all the disciplines that are involved in the design and implementation of processes, products, and systems. In other words, to design and build more sustainable chemical processes we need to embed sustainability principles into mainstream chemistry and engineering until we arrive at the day when chemistry is by default green chemistry and engineering is by default green engineering. However, we will not achieve a more sustainable society if we do not strive constantly for greater innovation in chemistry and engineering. To be successful in green chemistry and green engineering, one must overcome greater and more difficult challenges for innovation than does much of

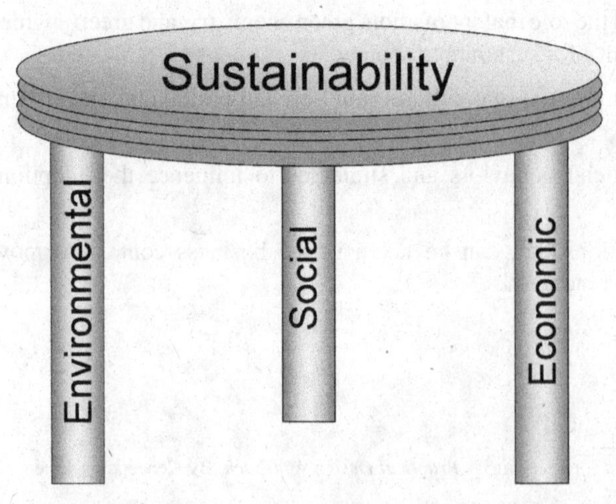

FIGURE 25.1 Balancing the sustainability table.

traditional synthetic organic chemistry or engineering on its own. It can be argued that green chemistry and green engineering are indeed "smart" chemistry and engineering, insofar as through these disciplines, chemists and engineers attempt to design safer, more mass- and energy-efficient process, and avoid hazardous processes by design, thereby avoiding problems that have plagued chemical processes for decades. We are not suggesting that synthetic chemists and chemical engineers have historically set out to design less than optimal processes; rather, they just have not been aware of nor accountable for the consequences of their chemistry outside an exceedingly narrow range of interests and accountabilities.

Green chemistry and engineering therefore require the best and the brightest to rethink and challenge existing paradigms and push the limits of traditional synthetic chemistry and engineering knowledge. New and better ways of designing and building molecules are required; the toolset of the past 150 or more years is no longer acceptable and is very far from sustainable. We require people who understand and embrace different academic disciplines, something that is traditionally anathema to a synthetic chemist, to design systems, processes, and products that are more sustainable by design.

25.2 SUSTAINABILITY: CULTURE AND POLICY

With the foregoing thoughts in mind, you might see how the technological advances of green chemistry and green engineering cannot be developed in isolation of society and societal needs, culture, and government (policy). Governments, and society in general, can and do play a huge role in implementing policies, strategies, and practices that enable sustainability practices, and influence movement toward more sustainable behaviors. For example, a traditional policy approach has been to regulate and, in recent legislation, to eliminate, the use and disposal of hazardous materials and waste. By the same token, there is movement away from the concept of "waste management" toward the concept of *materials management*. This means that we not only include how we manage materials brought within the boundaries of a plant, but also materials that are part of the wider boundary that includes the entire supply chain for a cradle-to-gate, more holistic view.[2] As you are no doubt aware, this is normally easier said than done. Policies move slowly through established consultation periods, they are analyzed and reanalyzed to assess their potential economic implications and technical feasibility, and of course, they need to survive the legislative debate. It is quite clear then that as global society needs to move toward more sustainable practices, government and intergovernmental policies and strategies need to support technological innovations that will deliver sustainability.

As important as government and society are, another important arena where policies are established and implemented is in industry and academia. Businesses must strive toward more sustainable practices from every level of the organization, starting at the top. To drive sustainability within operations, organizations need to embrace sustainability widely, as part of company values that drive business operations. Sustainability principles need to become embedded in an organization as a normal way of doing business. Forward-looking businesses have already seen this not only as a requirement, a need, or a demand, but most important, they have seen it as a means to obtain competitive advantage and as a driver to ensure ethical profitability.[3] Another part of the equation is that once businesses set policies these cannot be driven in isolation. There is some evidence that government and business sustainability strategies and policies are beginning to converge.[4]

Finally, academia plays a fundamental role in developing future scientists, engineers, and societal leaders who are able to drive sustainability into every part of the economy. It is imperative that the concepts discussed in this book be implemented not only as part of a separate course on green chemistry or green engineering, but as part of the normal curricula in schools and universities. Some universities have started programs in green chemistry and green engineering, which is a good start and precisely the right thing to do. It is especially important that academia spur innovation, especially in terms of graduate education and Ph. D.-level research. However, a more sustainable approach is to include these topics systematically in the curricula of students, from kindergarten through college, with specialization done at the upperclass and graduate levels.[5]

25.3 INFLUENCING SUSTAINABILITY

Coupling technological innovation with policies and strategies is not sufficient to bring about the kind of change that is needed to lead society toward more sustainable practices. To influence people to make sustained changes in the face of pervasive, persistent, and prolonged resistance, we need to use all the spheres of influence at our disposal. Influencing is an essential part of driving sustainability; as we have seen, the achievement of greater sustainability won't happen through people working in isolation and requires the interaction and collaboration of many areas.

Patterson et al. have studied so-called "masters of influence," people who can exert tremendous influence by changing a few key behaviors at crucial moments.[6] These people are not necessarily corporate leaders, although they could be, and they come from many walks of life. There were two main discoveries that resulted from studying these masters of influence that were seen as being critical to their success in influencing others:

Success in influencing depends on a focus on a few key critical behaviors that would make a difference to the outcome desired.

Influence masters utilize all spheres of influence at their disposal to change those critical behaviors.

These two findings need to be utilized when one attempts to advance sustainability in organizations. First, let's focus on critical behaviors. Although there may be many behaviors that might help to move an organization toward more sustainable practices, there are only a few critical behaviors that are going to drive the desired change at the crucial time. To leverage these few critical behaviors, one must become a social scientist in addition to being a chemist or an engineer. One must observe the culture of the organization and determine at what time decisions are made that can mean the difference between a green, sustainable process and a process that is not as efficient.

To influence the application of green chemistry, green engineering, and sustainability, one must pursue actively such behaviors as:

- Reviewing EHS aspects of reagents and selecting the greenest ones
- Identifying by-products and waste during process development
- Using life cycle thinking to select materials

- Reducing mass and energy requirements during process design
- Increasing the use of renewable resources in raw materials

In addition to identifying crucial behaviors, one needs to observe and understand how people make decisions, especially what enables or hinders decisions about pursuing the right behaviors (i.e., more sustainable ones), so one can act upon across several spheres of influence.

Regarding these spheres of influence, the authors of the study propose six areas that are going to help drive the critical behaviors. People generally do, or don't do, things for two main reasons:

- Because they can (their ability to do it)
- Because they want to (their motivation to do it)

So, to influence someone to adopt the desired behaviors, one must utilize spheres of influence that touch both ability and motivation. Motivation and ability can be enabled or hindered by the individual (i.e., self), other people (e.g., society, colleagues, managers, the community), or things (e.g., policies, structures, incentives). A representation of the spheres of influence is shown in Figure 25.2.

As mentioned earlier, to influence some of the sustainability behaviors mentioned before and perhaps others, one needs to effectively employ tools and strategies that will cover all six spheres of influence. Figure 25.3 shows some examples of tools that can be employed to influence sustainability principles within corporations.

It is important to stress that all tools and strategies need to be applied in a coordinated manner in each area of influence. Each area of influence by itself would not render the synergy that would be needed to advance green chemistry, green engineering, and sustain-

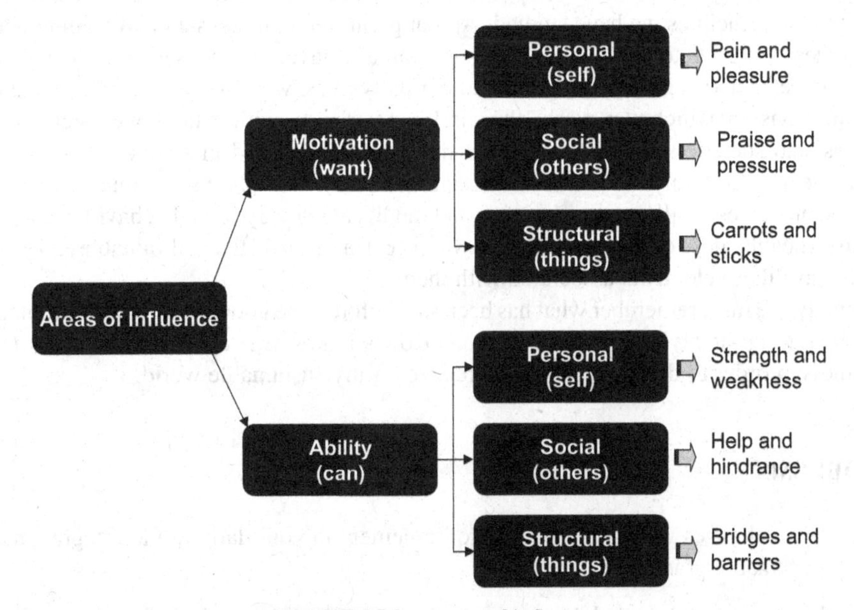

FIGURE 25.2 Areas of influence.

	Motivation (want)	Ability (can)
Personal (individual)	Values Believes Experiences	Sustainability Education and Training Practice (e.g., secondments)
Social (group)	CEO & Upper Management Direction Sustainability Benchmarking Sustainability Networks External Stakeholder Demands Customer Demands Recognition	Sustainable Manufacturing Teams Sustainability Council Sustainability Champions Centers of Excellence Sustainability Networks Industrial Collaborations Culture
Structural (things)	Cost Reduction Increased Efficiency Competitive Advantage Sustainability Policy and Standards Sustainability Strategy Sustainability Principles Sustainability Awards	Sustainability Metrics Sustainability Tools Science and Technology EHS Data Resources Facilities

FIGURE 25.3 Examples of tools to influence sustainability.

ability in corporations. It is important that a sustainability implementation plan encompass strategies and tactics in each area of influence.

25.4 MOVING TO ACTION

After you close this book and start with your new coursework, your research, or with your life as a professional, you will have the opportunity to utilize the concepts you have learned and to apply the principles we have studied. At that point, the real test starts. We can begin to apply these concepts to our own lives, as some already have.[7] Then, as we start to embrace these concepts and begin applying them to our daily lives, we can think of influencing the organizations in which we work, the circles of friends with whom we interact, the professional societies in which we spend time and energy, and in general, the world in which we live. Certainly our journey will require the use of all spheres of influence (perhaps even on ourselves), will require changing and modifying deeply rooted behaviors, and will require a healthy dose of patience and perseverance. Having said that, all valuable endeavors in life have these elements associated with them.

Finally, we must remember what has been said before: The good solve today's problems, the great design today's world to avoid tomorrow's issues. It is up to us to live up to our greatness in order to dream, design, and realize a truly sustainable world.

PROBLEMS

25.1 What changes would you be able to implement in your daily life to integrate more sustainable practices?

25.2 You have been elected to serve as a representative in your local legislature. What modifications would you suggest to make your constituency more sustainable?

25.3 You are a very active member of your professional organization (e.g., ACS, AIChE) and have just been elected president of the organization. As part of the change agents of technological innovation, this association believes in sustainability as a core value. How would you ingrain the concepts of sustainability into the culture of the organization even more?

25.4 You are the newly appointed dean of the College of Engineering at your university. What high-level changes would you propose to embed green chemistry, green engineering, and sustainability aspects into the curriculum you have already covered in your area of study?

25.5 You have been named the sustainability chief executive officer of your corporation. Your company owns and operates chemical plants that produce and market a wide range of products: from organic chemicals to fine chemicals to consumer goods.

(a) Develop a new sustainability plan to propose to the company's board of directors.

(b) Who are the key stakeholders that would need to be involved, consulted, and informed according to your plan?

REFERENCES

1. Sikdar, S. Achieving technical sustainability. *Chem. Eng. Prog.*, 2009, 1, 34–45.
2. Hecht, A. Government perspective on sustainability. *Chem. Eng. Prog.*, 2009, 1, 41–46.
3. Holliday, C. O., Schmidheiny, S., Sir Philip Watts KCMG. *Walking the Talk: The Business Case for Sustainable Development*. Greenleaf Publishing, Bath, UK, Aug. 2002.
4. Hecht, A. D. The next level of environmental protection: business strategies and government policies converging on sustainability. *Sustainable Dev. Law Policy*, 2007, VIII (1), 19–26.
5. Allen, D. T., Murphy, C. F., Allenby, B. R., Davidson, C.I. Incorporating Sustainability into Chemical Engineering Education. *Chem. Eng. Prog.*, 2009, 1, 47–53.
6. Patterson K., Grenny, J., Maxfield, D., McMillan, R., Switzler, A. *Influencer: The Power to Change Anything*. McGraw-Hill, New York, 2007.
7. Sylvester, R. W. A. Personal perspective on sustainability through energy efficiency. *Chem. Engi. Prog.*, 2009, 1, 54–59.

INDEX

Green Chemistry and Engineering: A Practical Design Approach, By Concepción Jiménez-González and David J. C. Constable

Printed in the United States
By Bookmasters